Engineering Electromagnetics Essentials

B N Basu

Emeritus Professor, Sir J. C. Bose School of Engineering
Mankundu, West Bengal, India

Formerly, Professor and Head, Department of Electronics Engineering
Banaras Hindu University, Varanasi, India

Universities Press

UNIVERSITIES PRESS (INDIA) PRIVATE LIMITED

Registered Office
3-6-747/1/A & 3-6-754/1, Himayatnagar
Hyderabad 500 029, Telangana, India
email: info@universitiespress.com, www.universitiespress.com

Distributed by
Orient Blackswan Private Limited

Registered Office
3-6-752 Himayatnagar, Hyderabad 500 029, Telangana, India

Other Offices
Bengaluru / Bhopal / Chennai / Ernakulam / Guwahati Hyderabad /
Jaipur / Kolkata / Lucknow / Mumbai / New Delhi / Noida / Patna

ISBN 978 81 7371 956 1

Typeset in Aldine 401 at 10.5pt *by*
OSDATA, Hyderabad 500 029

Printed in India by
Graphica Printers, Hyderabad

Published by
Universities Press (India) Private Limited
3-6-747/1/A & 3-6-754/1, Himayatnagar
Hyderabad 500 029, Telangana, India

To

Professor N B Chakrabarti

The great philosopher and visionary who mentored my research work

and

Bharati Basu Dey

My Didimoni

Ex-Scientist (Library) CSIR-NML, Jamshedpur

Contents

Foreword

ELectromagnetics has traditionally been an unpopular subject among students as well as teachers. Students find it difficult to understand because of the preponderance of mathematics, vector analysis in particular, and the three-dimensional concepts dominating field configurations. Teachers find it difficult to create an interest in students for the subject. The thought that the subject—which is one of the foundations of the discipline of electrical engineering—can be and should be made simpler and interesting is relatively recent, and a few textbooks have been written that attempt to make an innovative presentation of the material. The present book is an attempt in this direction, and I believe a successful one too. Throughout the preparation of the book, I have had the privilege of interacting with the author; in fact, I read some of the sample chapters and gave my frank comments on what I liked, and more so on what I did not, putting myself in the role of a student. The end result, as I see in the completed manuscript, is much more than what I expected. The author places heavy emphasis on basic concepts and makes all efforts to present them in as simple a manner as possible without sacrificing the rigour needed to grasp and apply them in the applications of the concepts. The author then gradually guides students in their journey to engineering applications with ample and clear illustrations, and innovative and thought-provoking examples. It is my hope that the book will serve the valuable purpose of motivating students to pursue the subject further in their future studies and professional career.

S C Dutta Roy
Formerly, Professor of Electrical Engineering
Indian Institute of Technology Delhi, India

Preface

It is rather a happenstance that Professor S C Dutta Roy asked me to write a book on electromagnetics while I was delivering a series of lectures on the same subject to the MTech students on the initiative of Dr Chandra Shekhar and Dr S N Joshi at Central Electronics Engineering Research Institute (CEERI), Pilani (CSIR), under its Academy of Scientific and Innovative Research programme. Professor Dutta Roy, popularly known as the 'Digital Signal Processing Man', insisted that I should bring out the essentials of the subject concisely in simple and interactive language, catering to the requirements of undergraduate and postgraduate students. I had a thorough discussion with him before zeroing in on the objectives of writing this book as mentioned in the introductory chapter of the book. I believe in the following quote from *Swahili's Collected Works*, vol. IV, p. 183: "The only true teacher is he who can immediately come down to the level of the student, and transfer his soul to the student's soul and see through the student's eyes and hear through his ears and understand through his mind. Such a teacher can really teach and none else."

School-level knowledge of static electricity and magnetism is prerequisite enough for comprehending this book. The derivation of vector calculus expressions in the beginning of the book makes the book self-contained. Vector calculus makes a concise understanding of the subject simple. Recall how Oliver Heaviside used this handy toolkit to re-express Maxwell's twenty equations as a set of only four equations, which form the backbone of electromagnetics. Professor Prasad Khastgir, in a Conference Souvenir at Varanasi, wrote: "Several disciplines hang as gems on one priceless necklace which it was Maxwell's privilege and honour to recognise as capricious Nature s enduring ornament." I am sure when students have the tool of vector calculus with them, it becomes greatly manageable for them to appreciate these equations owing to the two geniuses and the several concepts derived from their work, and to subsequently apply them to practical engineering problems.

Simple concepts are built up in this book around (i) the basic current density relation involving volume charge density and velocity of charge flow, (ii) the current continuity equation, (iii) the displacement current conceived in one of Maxwell's equations, (iv) the wave equation obtainable by decoupling the electric and magnetic field quantities in

Maxwell's equations, (v) electromagnetic boundary conditions, (vi) the solution of the wave equation subject to electromagnetic boundary conditions with reference to a given problem, and (vii) the flow, loss, and radiation of electromagnetic power predictable by Poynting theorem. A base knowledge of the subject can be developed by using these concepts to study the propagation of electromagnetic waves through an unbounded medium, for instance, through a free-space medium and a semi-infinite conducting medium, the latter giving the concept of ac or RF resistance. These concepts can be further exemplified by studying the propagation of electromagnetic waves through a bounded medium such as a conducting hollow-pipe waveguide, as well as the storage of electromagnetic energy in a cavity resonator, the latter in the form of a short-ended or open-ended waveguide. Thus, such selected concepts are brought out within the purview of this book with the conviction that students, after studying the book, will become prepared to apply such concepts to understand and design electromagnetics-based devices and systems for engineering applications.

I have taken care to restrict the length of the book to a plausible limit that is accessible by a student interested in appreciating the essence of the subject within a reasonable amount of time. Illustrative examples provided in the text of chapters will acquaint the students with interesting problems and their solutions, thereby strengthening their concepts of the subject. Some stimulating and thought-provoking review questions along with answers/solutions/hints have also been added at the end of chapters. It is hoped that by practising these questions students will get an opportunity to summarise and revise the knowledge gained, and be inspired to undertake further challenging problems pertaining to even more of practical relevance. I would lastly like to address that I have provided the timeline of progress in electromagnetics, with a special reference as accolade to Sir J C Bose to motivate young practitioners of electromagnetics to strive forward.

"*Was there ever a more horrible blasphemy than the statement that all the knowledge of God is confined to this or that book? How dare men call God infinite, and yet try to compress Him within the covers of a little book*!"

Swami Vivekananda, *Raja-Yoga*

The aforementioned philosophy of Swami Vivekananda is equally applicable to the knowledge of other disciplines as well including the one elucidated in the present book.

Acknowledgements

PROFESSOR S C Dutta Roy inspired me to write this book. He waited patiently for the manuscript which he meticulously reviewed spending his precious time, and graciously recommended several changes to make the book more clearly presentable. He kindly wrote the foreword of the book. No words will suffice to express my gratitude to him.

I thank Vishant Dwivedi for motivating me to write the book while I was teaching the subject at CSIR-CEERI, Pilani. He has immensely helped me in selecting problems and solving them. I sincerely thank Dr Mukul Agrawal of Stanford University, who was my student at Banaras Hindu University, for his helpful criticisms on some portions of the manuscript.

Dr Vishal Kesari, DRDO scientist with whom I carried out research at Banaras Hindu University, formatted the entire manuscript, developed the pilot design of the front and back covers, generously prepared the index of the book; and corrected the proof of the manuscript. Above all, he took up the added responsibility of keeping updated progress report of the preparation of the manuscript. Sreelatha Menon, on behalf of Universities Press, joins me in acknowledging the support and good will from Dr Kesari.

To Dr Somnath Chatterjee, Amit Varshney and Sumit Varshney—who spared their valuable time to draw the figures of the book, without which the manuscript could not have been completed—go my acknowledgement.

Dr S K Datta, DRDO scientist, who was my doctoral student and who has been helping me at every stage of the preparation of the manuscript, is beyond giving thanks. I have no words to thank Professor P K Saha, Institute of Radiophysics and Electronics, Calcutta University, with whom I freely discussed several conceptual issues often at odd hours. I express my gratitude to Professor Manfred Thumm of Karlsruhe Institute of Technology for his helpful suggestions and Professor Edl Schamiloglu of New Mexico University and Professor Richard Temkin of Massachusetts Institute of Technology for their inspiration.

I keep in record my gratitude to Dr S S S Aagarwala, Advocate Subhas Chand, Dr R M Dubey, B G Mallick, Dr R U Khan, Dr Chandra Shekhar, Professor B N Biswas, Professor S K Srivastava, Professsor R K Jha, Professor S K Kak, Professor P K Jain,

Professor P Chakrabarti, Professor S Jit, Dr G Singh, Dr S N Joshi, Dr L M Joshi, Dr R S Raju, Dr A K Sinha, Dr R K Sharma, Dr U N Pal, Dr P Dalela, Dr Chhaya Dalela, Dr Sarit Pal, Dr A Bera and Dr R Barik, for their kind moral support during the preparation of the manuscript.

I am also grateful to Professor G S Park, Dr L Kumar, Dr K S Bhat, Dr S Kamath, Professor R Mahanty, Dr M I H Ansari, Professor Ashoke Ghosh, Dr S K Ghosh, Subhradeep Chakraborty, Madan Lal, S Guhathakurta, Debnath Mukherjee, Swapna Goswami and Swapan K Ghosh for their keen interest in my work.

I sincerely acknowledge Smriti Srivastava and Thomas Mathew Rajesh, both of Universities Press, for coordinating our cooperative efforts and helping me reorganise problems and solutions, and add review questions. I express my acknowledgments to Sreelatha Menon of Universities Press for encouraging me right from the word go and for her scholarly and meticulous editing of the manuscript. I was immensely inspired by her comparison of the important role that Thomas assumed with that of Peter Guzzardi, one of the editors of Stephen Hawking's *My Brief History*, who made Professor Hawking modify his write-up over and over again to make it easily comprehensible; at times Professor Hawking thought the process would never end, but he admits that Peter Guzzardi was right about the book having a much better composition as a result.

I sincerely thank my son Dr Indraneel Basu, his wife Sudeshna Basu and their daughters Shreya and Niya, for their interest, persuasion, and adjustments on the domestic front. I also took the privilege of taking assistance from my son Dr Priyoneel Basu from time to time in getting the peripherals of the book reviewed. I thank him wholeheartedly and also I acknowledge his wife Dr Sweta Arora Basu and their son Krishanu for allowing me to exempt myself from household duties while writing this book.

Above all, I profoundly thank my dear wife Manikuntala for encouraging me during difficult moments and for her love, kindness, forbearance and understanding.

B N Basu
Varanasi
16 May 2015

Timeline of Progress in Electromagnetic Theory and Related Areas

Ancient Period

In the *Vedas*, one of the oldest scriptures of Hinduism composed in the Vedic period—roughly estimated between 1700 BCE and 1100 BCE (terminus ante quem for all Vedic Sanskrit literature being 150 BCE)—there is a reference to 'Ayas' (meaning iron) and 'Akarshan' (meaning attraction) signifying iron-magnet relationship.

Subsequently, we get the reference with credits to Thales of Miletus—a pre-Socratic Greek philosopher and one of the Seven Sages of Greece, regarded by many including Aristotle as the first philosopher in the Greek tradition—for discovering in the **6th Century BC** the attraction/spark between fur/hair and a substance like amber when rubbed against each other (the Greek word for amber being 'elektron'). Around the same time in India, renowned physician Sushruta, believed to be world's first specialist in plastic surgery, was known to have performed surgical operations such as rhinoplasty, removal of bladder stones, hydrocele and removal of metallic splinters with the help of magnets. In **5th Century BC**, Kalidasa, in his Sanskrit epic poem *Kumarasambavam*, wrote '*ayaskantena lohavat akarshtum*' referring to the magnet-like attraction of iron to describe the disrupted state of mind of Lord Shiva that was fixated on intense penance. The earliest literary reference in ancient China to magnetism associated with the loadstone attracting iron lies in the **4th Century BC** book *The Master of Demon Valley* authored by Wang Li, a Taoist recluse. The **3rd Century BC** witnessed the development of the Baghdad Battery or Parthian Battery resembling the galvanic cell believed to have been used for electroplating. A story in *The Natural History: Book 1*, authored by Pliny the Elder (Gaius Plinius Secundus, a Roman author, naturalist, and natural philosopher as well as naval and army commander) in the **1st Century BC** records the incidence of a shepherd named Magnes discovering that 'the nails of his shoes and the iron ferrule of his staff adhered to the ground', referring to the magnetic properties of some iron stones.

Later Medieval Period

In the **11th Century AD**, Shen Kuo, a Chinese scientist and statesman, mentioned the role of a magnetic needle compass for improving the accuracy of navigation by employing the astronomical concept of true north in terms of magnetic declination towards the north pole.

By **12th Century AD**, the Chinese had already begun using lodestone compass for navigation. In 1187, Alexander Neckam, an English schoolman, scientist and the author of *De Utensilibus*, meaning 'On Instruments', was the first in Europe to describe the magnetic compass as an aid to navigation. Archbishop Eustathius of Thessalonica of Greece recorded that *Woliver*, king of the Goths, was able to draw sparks from his body. In **13th Century AD**, Petrus Peregrinus de Maricourt, a French scientist, conducted experiments on magnetism and invented an improved nautical compass. In 1269, he wrote the treatise named *Epistola de Magnete* in which he expounded the properties of a magnet. Furthermore, Flavio Gioja, an Italian scientist as well as Al-Ashraf, a Yemeni sultan, physicist, astronomer and geographer, invented and discussed the dry compass. In the work of the latter, the reference to the wet compass in the form of a magnetic needle placed in a bowl of water can also be found. The reference to iron fish-like compass in the Persian book of tales of this period appears to have been derived from a particular Chinese design. The iron fish-like compass in the wet format (in which the metallic fish is placed in a cup of oil with an aim to achieve stability against vibration) came into being in the name of Matsya Yantra ('Matsya' meaning fish and 'Yantra' meaning instrument) in India for navigational purpose in the **4th** and **5th** Century BC. Ibn Sim'ūn (an Egyptian astronomer and muezzin) describes in an Arabic exposition the instrument known as 'Qibla (Kabba) indicator' to find the direction of the holy city of Mecca. In **14th Century AD**, Ibn al-Shatir, a Syrian astronomer and timekeeper, invented a timekeeping device incorporating both a universal sundial and a magnetic compass for the purpose of finding the times of Salah prayers.

Modern Period

16th Century AD

1550: Girolamo Cardano, an Italian mathematician, physician, and astrologer who formulated the fundamental rules of probability (which is known to have aided him in his habit of gambling), distinguished between electrical and magnetic forces as recorded in his book *De Subtilitate*.

1600: William Gilbert, an English physician, physicist and natural philosopher who is regarded by some as the father of electrical engineering or electricity and magnetism, describes in his book, *De Magnete, Magneticisque Corporibus, et de Magno Magnete Tellure* (On the Magnet and Magnetic Bodies, and on the Great Magnet the Earth) many experiments with his model of Earth called the 'terrella' to conclude that Earth itself is a magnet, while at the same time explaining why the needle of the magnetic compass points to the north. The unit of magnetomotive force or magnetic potential has been named Gilbert after him.

17th Century AD

1646: Sir Thomas Browne, an English scientist, used the word 'electricity' in the context of the field of medicine in his work *Pseudodoxia Epidemica.*

1663: Otto von Guericke, a German scientist who invented the air pump and worked on the prediction of weather using barometer, is acclaimed for inventing a prototypical form of an electrostatic generator using a continuously rotating sulphur globe that could be rubbed with hand or a piece of cloth.

1675: Robert Boyle, an Irish, English or Anglo-Irish scientist best known for Boyle's law for the absolute pressure and the volume of a given mass of confined gas within a closed system of a constant temperature, stated that electric attraction and repulsion can act across vacuum.

18th Century AD

1705: Francis Hauksbee, an English scientist widely recognised for his work on electricity and electrostatic repulsion, improved von Guericke's electrostatic generator using a glass globe.

1729: Stephen Gray, an English dyer and amateur astronomer, and Granville Wheler, an an English clergyman and scientist, carried out an experiment (which may be described as the beginning of electrical communication) in which they could transmit electrical 'virtue' produced by grazing a glass tube through thin iron wire to deflect brass leaves at a distance of about 900 feet.

1734: Charles François de Cisternay Du Fay, a French scientist, discovered 'vitreous' and 'resinous', now known as positive and negative charges respectively, the former produced by rubbing glass, rock crystal or precious stones with hair or wool, and the latter by rubbing amber, copal or gum-lac with silk or paper. He also distinguished between 'electrics' and 'non-electrics', which were later came to be termed conductors and insulators respectively by William Watson and Benjamin Franklin.

1745: Pieter van Musschenbroek of Leiden or Leyden, a Dutch scientist, invented the so-called Leyden jar, a primitive capacitor or condenser for storing electrical energy. Also, Ewald Georg von Kleist, a German physicist, independently invented the capacitor known as the Kleistian jar.

Subsequently, Daniel Gralath, a scientist and Mayor of Danzig in Poland, further improved the design of such a condenser and combined several jars to invent the 'battery'.

1752: Benjamin Franklin, a political theorist and statesman, considered as one of the founding fathers of the USA (who used to conduct the dangerous experiment of flying a kite through storm-laden sky) established the link between lightning and electricity and also gave the concept of charge neutrality due to the presence of equal amounts of positive and negative charges. He is partially credited with the development of the concept of electric potential.

1753: One "C. M", claimed to be a Scottish scientist, either Charles Morrison of Greenock or Charles Marshall of Aberdeen mentioned in the 17th February edition of Scots Magazine, recorded possibly the first suggestion of an electric telegraph system. It consisted of 26 insulated wires, each corresponding to a letter of the English alphabet and each connected to electrostatic machines in such manner that the far end of the charged wire attracted a disc of paper marked with the corresponding letter. A message could be sent through such wires using the 26 alphabets, charged one at a time. This system was also developed in the form of the first ever acoustic telegraph, in which a set of bells, each of a different tone, are produced by the spark of a corresponding charged wire.

1774: Georges-Louis Le Sage, a Swiss physicist known for his theory of gravitation and who anticipated the kinetic theory of gases, built a static electricity telegraph using the principle of Gray and Wheler (as mentioned before).

1785: Charles Coulomb, a French physicist, introduced the inverse-square law of electrostatics, and in his honour the SI unit of electric charge is called coulomb.

1791: Luigi Galvani, an Italian physician, physicist, medical practitioner and philosopher, discovered galvanic electricity. He is referred to as the father of bioelectricity, a subject on which he carried out extensive experiments, regarded by Volta as the 'most beautiful and important discoveries', following his accidental observation that 'touching exposed muscles in frogs' legs with a scalpel that had been close to a static electrical machine caused them to jump', which he called 'animal electricity'.

1799: Alessandro Volta, an Italian physicist, created a cell called the voltaic cell after him, which produced an electrical energy by the chemical action of several pairs of alternating copper (or silver) and zinc discs piled up and separated by cloth or cardboard soaked in brine (salt water) or acid for increased conductivity. He also demonstrated the glowing of visible light from a wire carrying electric current, and subsequently constructed a battery by combining a number of voltaic cells. The unit of electric potential is called volt in his honour.

19th Century AD

1802: Gian Domenico Romagnosi, an Italian legal scholar, published a report in an Italian newspaper that a magnetic needle can be deflected by a voltaic pile. His claim, however, was overlooked by the scientific community.

1803: Thomas Young, an English scientist, performed his famous double-slit experiment and concluded that light is a wave, thereby challenging the belief of Isaac Newton (an English natural philosopher) that light, a manifestation of electromagnetism, is a particle.

1819: Hans Christian Oersted, a professor at the University of Copenhagen, Denmark, developed an understanding of the relationship between electricity and magnetism based on his accidental discovery of the movement of a magnetic compass in the vicinity of a nearby electric current. The unit of magnetic field is called oersted in his honour.

1820: André-Marie Ampère, a professor of mathematics at the Ecole Polytechnique in France, discovered that the magnetic field circulating in a closed-path was related to the current flowing through the perimeter of the path (Ampere's circuital law). He also described the force between current-carrying conductors and thus developed the concept that like currents attract and unlike currents repel. He was one of the first to develop electrical measurement techniques such as the galvanometer using a current coil and a magnetic needle.

Jean-Baptiste Biot, a physicist, astronomer and mathematician, who studied the polarisation of light and Félix Savart, a physicist who, albeit a student of medicine, worked with Biot in the area of the theory of electric current and medicine, both from France, worked together to come up with the Biot-Savart's law for the generation of magnetic field caused by an electric current. Johann Salomo Christoph Schweigger, a German chemist, physicist, and professor, built a sensitive galvanometer for detecting small amounts of current using a coil wrapped around a graduated compass and named it after Luigi Galvani.

1824: François Arago, a French physicist, developed the concept of the rotating magnetic field.

1825: William Sturgeon (founder of the first English electric journal, *Annals of Electricity*) developed a relatively powerful electromagnet by putting an iron core inside a helical coil.

1826: Georg Simon Ohm, a German physicist and mathematician, expounded Ohm's law of electrical resistance. The unit of electrical resistance/impedance is called ohm in his honour.

1827: Ányos Jedlik, a Hungarian physicist, started experimenting with electromagnetic coils.

1829: Francesco Zantedeschi, an Italian priest and physicist, predicted the production of electric currents in closed circuits by an approaching or receding magnet, thereby anticipating the law of electromagnetic induction attributed to Michael Faraday.

1831: Michael Faraday, an English scientist, made astounding discoveries in the fields of electromagnetism and electrochemistry encompassing electromagnetic, diamagnetism and electrolysis. He is credited with the discovery of the law of electromagnetic induction, possibly influenced by the work carried out in 1829 by Francesco Zantedeschi.

1832: Baron Pavel L'vovitch Schilling, also known as Paul Schilling, a Russian diplomat and scientist of Baltic German origin, built the electromagnetic telegraph (employing a single-needle system in which a code was used to indicate the characters) and is acclaimed for putting into practice a binary system for signal transmission. (Carl Friedrich Gauss and Wilhelm Weber, both of Germany, also constructed a telegraph around the same time, and it was successfully operating two years before Schilling could put his into practice).

William Sturgeon, a British scientist, invented the first commutator-type direct current electric motor capable of turning machinery.

Hippolyte Pixii, a French engineer, built an archetypal form of alternating current electrical generator (the first alternator using a rotating magnet).

1833: Heinrich Lenz, a Russian physicist of Baltic German origin, predicted the law of relevance to electromagnetic induction (referring to the opposition to the cause to which such induction is due) named after him, and is accredited by the Russians for independently discovering what is known as Joule's law of electric power loss which is referred to in Russia as 'Joule-Lenz law'.

1835: Joseph Henry, an American scientist and the first secretary of the Smithsonian Institution, invented the electric relay and incepted the practical design of electric telegraph. He also construed the phenomena of self inductance and mutual inductance, the latter devised independently of Michael Faraday's work on the same. Henry, the SI unit of inductance, is named in his honour.

1836: Dr David Alter, an American inventor and scientist with German and Swiss ancestry, demonstrated to witnesses the first American electric telegraph in Elderton, Pennsylvania. Later he invented an electric buggy (predecessor of the automobile), electric bell and short-range telephone (forerunner to Alexander Graham Bell's telephone).

1837: Johann Friedrich Carl Gauss and Wilhelm Eduard Weber, both German physicists, jointly invented the first electromagnetic telegraph using a reflecting galvanometer. The units Gauss and Weber, the CGS and SI units of magnetic flux density, are named respectively in their honour.

Samuel Finley Breese Morse, an American inventor and accomplished painter, designed an improved single-wire telegraph system and co-invented, with Alfred Vail, the Morse code or signalling alphabet. ("What hath God wrought?" was the first ever message sent by Morse from Baltimore to Washington, D.C. on 24th May 1844.)

Thomas Davenport, an American inventor, patented a commutator-type direct-current electric motor based on the work of William Sturgeon (refer to his discovery in 1832 mentioned earlier).

1839: Michael Faraday (refer to his discovery in 1831 mentioned earlier) concluded that the division between the various kinds of electricity are illusory and that there is underlying relationship between electricity and magnetism.

1840: James Prescott Joule, an English physicist and brewer, formulated Joule's Law (referred to as the Joule-Lenz law in Russia) for resistive power loss. He worked on the correlation between mechanical and thermal energy and his work led to the development of the first law of thermodynamics. He also contributed to the development of the absolute scale of temperature with Lord Kelvin (William Thomson, eminent English mathematical physicist and engineer). The SI unit of energy is named joule in his honour.

1845: Michael Faraday (refer to his discovery in 1831 mentioned earlier) discovered the influence of an external magnetic field on the propagation of light. To be more precise, it is the rotation of the plane of polarisation of a plane-polarised wave passing through a gyro-magnetic material in a steady magnetic field, a phenomenon known as Faraday rotation.

1849: Hippolyte Fizeau and Jean-Bernard Foucault, French physicists, measured the speed of light to be about 298,000 km/s.

1850: Gustav Robert Kirchhoff, a German physicist, contributed to the fundamental understanding of electrical circuits. Interestingly, two independent concepts, one in circuit theory and the other in thermal radiation, are named as Kirchhoff's laws.

1855: Guillaume Duchenne, a French neurologist, used alternating current in electrotherapeutic triggering of muscle contractions.

1864: James Clerk Maxwell, a Scottish physicist, unified electricity, magnetism, and optics into the field of electromagnetism which resulted in the formulation of 20 equations in 20 variables. Later around 1885, the theory was reformulated by Oliver Heaviside, a self-taught English electrical engineer. (See the developments in 1885.) It is also claimed that German physicist Heinrich Hertz also independently established these equations at Karlsruhe, Germany

Thus, Maxwell unified electromagnetics in the same way Isaac Newton did mechanics. He authored his famous work, *A Treatise on Electricity and Magnetism* (Clarendon, Oxford, 1872).

1866: Giovanni Caselli, an Italian professor at the University of Florence, invented the pantelegraphe (a term suggesting the amalgamation of two words, 'pantograph', a tool that copies drawing and 'telegraph', a machine) that sends messages over a wire and thus established the first Telefax (commonly used today to verify signatures in banking transactions) between Lyons and Paris.

1873: James Clerk Maxwell (refer to his discovery in 1864 mentioned earlier) revealed that light is an electromagnetic phenomenon.

1875: G.R. Carey, an American scientist, invented the phototube.

1876: Alexander Graham Bell, a Scottish scientist and engineer, and Elisha Gray, an American inventor, independently invented the telephone. Johann Philipp Reis, a self-educated German scientist and inventor, had already constructed a working telephone structure in 1861 in Germany employing the so-called knitting-needle receiver. He however, failed to file for a patent.

Karl Ferdinand Braun in Germany demonstrated the rectification effect of metal-semiconductor contacts (whiskers) PbS.

1878: Thomas Alva Edison, an American inventor and businessman, invented improved, long lasting and commercial incandescent light bulbs.

The Ganz Vállalatok or 'Ganz Companies', a Hungarian company based in Budapest, started working with single-phase ac power systems.

1879: William Crookes (an English physicist and chemist) invented the cathode-ray tube.

Walter Baily (an American professor) devised a copper disc to rotate using alternating current (an early stage of ac motor and the first prototypical induction motor).

1880: Pierre Curie, a French scientist, invented piezoelectricity.

Edison (refer to his inventions in 1878 mentioned earlier) switched on the world's first electrical power distribution system providing 110 volts dc power connection to 59 customers.

1884: Lucien Gaulard, an Italian scientist, developed the power transmission system from Lanzo to Turino.

1885: Oliver Heaviside, English self-taught electrical engineer, re-theorised 20 of Maxwell's equations in 20 variables and condensed them into 4 equations using his concepts of vector calculus; the same was achieved independently by Heinrich Hertz at Karlsruhe in Germany, already mentioned under the developments in 1864.

1886: American engineers George Westinghouse, William Stanley and Oliver B. Shallenberger. developed the first ac power system in USA using step-up and step-down transformers.

1887: Heinrich Rudolf Hertz, a German physicist, was the first to observe photoelectric effect while experimenting with a spark-gap generator. He was also the foremost to discover electromagnetic waves in an experiment at the University of Karlsruhe, Germany. The unit of frequency is termed as hertz (Hz) in his honour.

1888: Galileo Ferraris, Italian physicist and engineer, hailed as 'the father of three-phase current, developed the efficient and commutator-free ac poly-phase motor, competitive with its dc counterpart.

Heinrich Hertz (refer to his discovery in 1887 mentioned earlier) generated and detected electromagnetic waves, typically at 30 and 450 MHz, and demonstrated known optical phenomena of reflection, refraction, polarisation and interference. He also measured the velocity of electromagnetic waves. (He used a spark gap connected to an induction coil as the source and a circular loop of thick wire with ends separated by a very tiny gap for a small observable spark as the detector. He interpreted the distance between consecutive nulls in the standing-wave pattern in a room, in which one of the walls was covered with a large zinc sheet to obtain the velocity of electromagnetic waves). Further, his experiment in which sparks were produced at the centre of an end-loaded dipole to radiate electromagnetic waves and a resonant square-loop to receive them may be considered as reception by an antenna.

1891: Mikhail Dolivo-Dobrovsky, a German engineer, built the first full three-phase ac generation and long-distance power transmission system, from Lauffen to Frankfurt, covering a distance of 109 miles.

Nikola Tesla, a Serbian-American inventor, transmitted electric energy without wires.

1895: J. C. Bose, Indian physicist and botanist, established the first ever wireless signalling by giving public demonstration of transmission of electromagnetic waves that involved the ringing of a bell and the explosion of some gunpowder planted at a distance.

Alexander Stepanovich Popov, a Russian physicist who was conducting an experiment of a similar nature to that of Bose's, possibly unaware of the work of the latter, wrote that he was still entertaining the hope of remote-signalling radio waves that time.

Russian physicist Wilhelm Conrad Röntgen discovered X-ray radiation.

1896: French physicist Henri Becquerel discovered radioactivity.

1897: Guglielmo Marconi, an Italian inventor, performed his wireless signalling experiment on Salisbury Plain in England, apparently unaware of the work of Bose and that of Popov (refer to their respective observations in 1895 mentioned earlier).

Joseph John (J. J.) Thomson, an English physicist, is credited with the discovery of the electron (besides that of the isotope and the mass spectrometer).

Karl Ferdinand Braun, a German scientist, (refer to his discovery in 1897 mentioned earlier) invented the cathode-ray tube.

1898: J. C. Bose (refer to his discovery in 1895 mentioned earlier) established the rotation of plane of polarisation of electric waves by twisted structures—bundles of jute fibres twisted to the right (positive) and to the left (left)—thereby showing their behaviours as electro-optic analogues of two varieties of sugar solutions, typically, dextrose and levulose, in an experiment that is believed to be the first ever millimetre-wave experiment in artificial materials that exhibit the chiral characteristics for electromagnetic applications.

1899: Jonathan Adolf Wilhelm Zenneck, a German physicist and electrical engineer, developed the oscilloscope following the work of Braun (refer to his discovery in 1897 mentioned earlier). He also worked on surface slow waves that are presently known as plasmons.

20th Century AD

1900: Max Karl Ernst Ludwig Planck, a German scientist, instituted the theoretical foundation of quantum mechanics.

1904: John Ambrose Fleming, an English electrical engineer and physicist, invented the vacuum tube diode.

1905: Albert Einstein, a German-born-American theoretical physicist, demonstrated that Maxwell's equations are not required to explain electromagnetic radiation if special relativity is taken into account.

Karl Ferdinand Braun in Germany invented the phased-array antenna (employing 3 wires).

1906: Lee De Forest, an American scientist widely honoured as the 'father of radio' and the 'grandfather of television', invented the vacuum tube triode.

1909: Robert Andrews Millikan, an American scientist, measured the charge of an electron.

Karl Ferdinand Braun invented coupled resonance circuits for transmitting and receiving in Germany. For this he, along with Marconi, was awarded the prestigious Nobel Prize in Physics in 1909.

1911: Heike Kamerlingh Onnes, a Dutch scientist, discovered superconductivity.

1913: William D. Coolidge, an American physicist, developed the first practical X-ray tube.

1914: K. F. Lindman studied chiral media made by embedding many randomly oriented small wire helices in a host medium.

1920: German physicists Heinrich Georg Barkhausen and K. Kurz invented the first 'transit-time' microwave tube (Barkhausen-Kurz oscillator).

1921: Albert W. Hull, an American physicist, invented the 'smooth-wall' magnetron.

1922: Philo T. Farnsworth, an American engineer, developed the tube scanning system for television.

1923: Vladimir Kosmich Zworykin, a Russian scientist, developed the iconoscope or cathode-ray tube.

1929: Robert J. Van de Graaff , an American physicist, invented the Van de Graaff generator (an electrostatic generator which uses a moving belt to accumulate very high voltages, as much as ~5 megavolts).

1930: Karl Guthe Jansky, an American physicist and radio engineer, instituted a new area of electromagnetic fields, namely, radio astronomy. He is often deem d to be the 'father of radio astronomy.'

1935: Rozhansky, Agnessa Heil, and Oskar Heil, scientists of Russian heritage, propounded the theory of velocity and density modulation and that of electron bunching in microwave tubes.

1936: W. Hahn and G. Metcalf , American physicists, invented the floating drift klystron.

1937: Russel Varian and Siguard Varian, American inventors known by the name of Varian brothers, founded the Varian Associates in 1948, now known as Communications & Power Industries. Independent of Hahn and Metcalf (refer to their discoveries in 1936 mentioned above), they invented the klystron based on the microwave resonator called the Rumbatron based on earlier works by W. W. Hansen, a U.S. physicist, immigrant from Denmark.

1939: John Randall and Harry Boot, English physicists, invented the 'cavity' magnetron.

1940: N. E. Lindenblad, an American engineer, invented the travelling-wave tube.

1942: Rudolf Kompfner, an Austrian-born American engineer and physicist, unaware of the invention of Lindenblad (refer to his discovery in 1940 mentioned above) invented the travelling-wave tube.

1946: J. D. Krauss, motivated by listening to a lecture in which the role of the helix in travelling-wave tube was discussed, invented the helix antenna.

Herrald T. Friis, a Danish-born scientist working at Bell Telephone Laboratory in USA, brought forth his famous 'transmission formula' for the power received by a receiving antenna for a given power from a transmitting antenna and for a given distance between the two antennas at a given wavelength.

1948: W. E. Cock arranged conducting spheres, discs and strips periodically and effectively tailored the effective refractive index of this artificial media to make lightweight microwave lenses.

And thus electromagnetic theory and pertinent areas made a progress in the rest of the period up to the present time in the above timeline. This glimpse of timeline of progress (which has been based on the various internet sources listed separately in the bibliography of this book) should inspire all of you who are interested in studying electromagnetics and using it for practical applications.

In this context, a special mention may be made about the ground-breaking work of Hertz (refer to his discovery in 1888 in the timeline chronicled above) concerning the generation and detection of electromagnetic waves, typically at 30 and 450 MHz. J. C. Bose of India (refer to his discovery in 1895 in the timeline chronicled above) on the other hand, demonstrated these phenomena in the millimetre-wave frequency range.

A Few More Words about Sir J. C. Bose

All Indians should be inspired by the momentous life and work of Sir J. C. Bose, aforementioned chronologically in the year 1895 in the historical timeline. He, a true pioneer and man of great stature, shaped the journey of India's scientific progress and uplifted our country on the global platform with his stellar achievements. He is described as the first ever scientist in the world to demonstrate the wireless transmission of electromagnetic waves.

Sir J. C. Bose (1858-1937)

- Born on 30th November 1858 at Bikrampur in Mymensingh near Dhaka in erstwhile East Bengal, India, presently in Munshiganj District, Bangladesh.
- Recipient of the awards/distinction: Companion of the order of the Indian Empire (1903, Companion of the order of the Star of India (1912), Knighthood (1917), Fellow of Royal Society (FRS) (1920) (the first Indian to become an FRS) and Member of the Vienna Academy of Sciences (1928).
- Graduated from Calcutta University (1879).
- Graduated from Cambridge University in Natural Sciences with Tripose (1884) after having abandoned his studies in medicine in London University due to ill health. He also had a B.A. degree from Cambridge along with a B.Sc. degree from London University.
- Profoundly influenced by Lord Rayleigh, one of his educators.
- Influenced his students, for instance, Professor S. N. Bose, who is well known for Bose-Einstein statistics.
- Professor of Physical Science at Presidency College, Calcutta (1885-1915).
- Laid the foundation of experimental science in India with meagre facilities, carrying out experiments in an adjoining lavatory in Presidency College, Calcutta (for instance, in his experiments on the rotation of plane of polarisation, he used the bundle of jute fibres as the medium and the pages of a book (Railway Timetable) as analyser/polariser).
- Founded Bose Research Institute, Calcutta (November 30, 1917), owing to the great support extended to him by Sara Bull, Sister Nivedita and Swami Vivekananda, who asserted about the said foundation: 'This is not a laboratory but a temple.'
- Studied the properties of very short electromagnetic (radio) waves, detecting them using improved version of Oliver Lodge's coherer and showing them to be similar to those of visible light waves.
- Put forth a general theory of the properties of contact-sensitive materials.
- Developed spiral-spring receiver for millimetre-wave radiation.
- Speculated on the existence of electromagnetic radiation from sun.
- Father of artificial chiral material (using twisted jute fibres: Jagadish Chunder Bose, MA, DSc, Professor of Physical Science, Presidency College, Calcutta, "On the rotation of plane of polarisation of electric waves by a twisted structure," *Proc. R. Soc. Lond* **a 63** (1898), 146-152; communicated by Lord Raleigh, FRS).

On the above, R. W. Ziolkwoski and N. Enghta remarked: "To the best of our knowledge, the first attempt to explore the concept of 'artificial' materials appears to trace back to the late part of the nineteenth century when in 1898 Jagadish Chunder Bose conducted the first microwave experiment on twisted structures—geometries that were essentially artificial chiral elements by today's technology."

- First to anticipate, according to Sir Neville Mott, Noble Laureate, the existence of p-type and n-type semiconductors.
- US-patented for the development of a semi-conducting galena crystal as a detector of radio waves, owing to the initiative taken by Sister Nivedita (1904).
- Investigated into plant physiology and the similarities between the behavioural response of plant and animal tissue.
- Demonstrated the minute movements of plants to external stimuli and measured their rate of growth using a few highly sensitive instruments such as 'Crescograph'.
- Presented in 1895 the first ever public demonstration of electromagnetic waves in India to ring a bell remotely (nearly a mile away passing through walls and body of the erstwhile Lieutenant Governor of Bengal) to detonate some gunpowder. This was much before Popov in Russia harboured the hope of remote signalling with radio waves and also much before Marconi carried out his wireless signalling experiment in Salisbury Plain in England in 1897.
- Recognised by IEEE as one of the fathers of radio science for his pioneering experimental work in millimetre-band radio (September 14, 2012).
- Regarded as the father of science fiction writers in Bengali literature.
- Quoted by him: "The true laboratory is the mind where behind illusions we uncover the laws of truth."
- Described thus by other great men:

> J. C. Bose was at least 60 years ahead of his time.
>
> — Sir Neville Mott

> Across the oceans, on the western shore,
> Reigns the temple of Goddess
> Of wealth and science,
> There you have journeyed, my friend,
> And returned richly crowned.
>
> —Kavi Guru Rabindranath Tagore

1 Introduction

Electromagnetic theory or electromagnetics, in short, is an interesting subject based on the fundamental principles of electricity and magnetism. It has applications in electrical, electronics, and communication engineering and related disciplines. Electromagnetics, therefore, is included as one of the compulsory subjects in these engineering disciplines.

After reading this book, you will, hopefully, develop a clear understanding of the subject in a reasonable period of time. If mathematical treatments are essential for this purpose at any stage, it need not scare you—I have taken care to derive and explain all the necessary steps and made all efforts to see that you enjoy the mathematics.

1.1 What is the Objective of the Book?

Many of my colleagues in the teaching profession find the subject of electromagnetics heavily mathematical and rather difficult to understand. Some of them advised me to write an easily understandable treatise on the subject. This motivated me to write this book on electromagnetics, based on my familiarity with the subject as well as the experience of writing another specialised book on the subject a few years ago (B. N. Basu, *Electromagnetic Theory and Applications in Beam-Wave Electronics*, Singapore: World Scientific, 1996). Hence, I set the objective of writing a book on electromagnetics that would

- make the subject easy to understand;
- give students an overall essence of the subject;
- uncover in a reasonable period of time the elementary concepts of the subject as is required to appreciate the engineering application of the subject; and
- make the study of the subject enjoyable and entertaining.

1.2 Why is the Title of the Book So Chosen?

With the objective of writing this book made clear, I debated on what should be the title of the book. Could it be 'Engineering Electromagnetics Made Simple'? Here, I apprehended that the phrase 'Made Simple' in the title would perhaps strengthen the popular impression that electromagnetics is a difficult subject to learn and that I am making an effort to make it simple. Please do not believe in the first place that electromagnetics is at all a difficult subject to learn. I assure you, and will convince you through this book, that electromagnetics is an interesting subject that is simple to learn. In fact, it is the approach to teaching the subject—by encompassing too many topics at a time and illustrating them with lengthy exercises—that often makes the learning of the subject rather tiresome and tedious. This is perhaps the reason for branding the subject as difficult. In fact, this approach prevents many a beginner from getting an overall feel of the subject and from developing a love for it. If it is argued that electromagnetics is difficult because it is mathematical, then I must counter-argue that *electromagnetics is simple because it is mathematical, and it is mathematics that makes the subject simple*. However, the question is whether the subject of electromagnetics is at all mathematical. If you take mathematics as a language to concisely express some physical facts and explanations, then certainly the subject of electromagnetics, and for that matter many other subjects of engineering, too, are mathematical. Even then, I assure you that electromagnetics involves the so-called mathematics to a much lesser extent than many other subjects of electrical, electronics, or communication engineering.

Therefore, discarding the title *Engineering Electromagnetics Made Simple*, and keeping in view the already set objective of writing this book, I thought it suitable to choose one out of the following E^3 titles for this book: Entertaining Engineering Electromagnetics, Elementary Engineering Electromagnetics, Essence of Engineering Electromagnetics, and Engineering Electromagnetics Essentials. Finally, I zeroed down to the last one: *Engineering Electromagnetics Essentials*, while taking care to see that the treatise fulfils all the aspects of the set objective.

1.3 What Mathematical Background is Required?

There is no prerequisite mathematics prescribed for the present book; it is good enough for the reader to know high-school-level algebra, trigonometry, and calculus. Since a quantity to be dealt with may be a scalar (for instance, electric potential) or a vector (for instance, electric field), the concepts of vector algebra (including the dot and cross products of two vectors) and those of vector calculus (more specifically the gradient of a scalar, divergence of a vector and curl of a vector) are required to be understood. You will find the vector calculus expressions deduced in this book from first principles (Chapter 2) very useful in understanding several concepts in the book.

1.4 What Are You Going to Get in This Book?

The book is a self-contained treatise on electromagnetics. The selected topics include most of the fundamentals of electromagnetics of engineering relevance. Care is taken not to bury the subject in too many details on each topic so that most of the relevant topics are covered in a reasonable amount of time with a complete feel and understanding of the subject. Thus, when you reach the end of this book, you will hopefully have reached a level of understanding to enthuse you to explore more advanced material related to the subject.

From this point until the end of the present chapter, I will describe the scope of the book. Some of the terminologies used for describing this scope might sound new to some readers. Do not feel lost, however; the meanings of these terminologies will become clear as we go ahead. The book contains eleven chapters in addition to the present introductory chapter. The grammar of the language of electromagnetics is based on the system of coordinates as appropriate to the problem under study and the fundamental concepts of vector calculus, namely the gradient of a scalar quantity (like electric potential) and the divergence as well as the curl of a vector quantity (such as electric field). It becomes necessary to use these vector calculus expressions in rectangular, cylindrical or spherical polar coordinate systems depending on whether the problem enjoys rectangular, cylindrical or spherical symmetry. It is somewhat easier to derive and remember these expressions in the rectangular coordinate system than in the cylindrical or spherical polar coordinate systems. Interestingly, it is much simpler to arrive at these expressions in the generalised curvilinear coordinate system than in cylindrical or spherical polar systems. Moreover, we will see that the generalised curvilinear coordinate system enjoys symmetry in that if we write the first term of the expression for the gradient, divergence or curl, we can easily permute it to write the remaining terms of the expression. This harmony of the generalised curvilinear coordinate system enables us to memorise a vector calculus expression more easily in the generalised curvilinear system than in the cylindrical or spherical polar systems. Once you are able to write the expression for the gradient, divergence or curl in the generalised curvilinear system, you will be able to easily translate it, in a straightforward manner, to the corresponding expression in the desired rectangular, cylindrical or spherical coordinate systems (Chapter 2).

We have only to recapitulate the basic laws of electricity and magnetism associated with the names of Coulomb, Gauss, Laplace, Poisson, Biot-Savart, Ampere, Lorentz, Faraday, Ohm, and Joule, already learnt in school. We will thus revisit some of the basic concepts of static electric field (Chapter 3) and static magnetic field (Chapter 4). The electric field and the electric potential will be related using a vector calculus expression (Chapter 3). Basic vector calculus (Chapter 2) has also been used to derive Poisson's and Laplace's equations, the former in the presence of electric charge and the latter in its absence (Chapter 3). We will see these two equations expressed in terms of either the divergence of the electric field or the Laplacian of the electric potential (Chapter 3). We will also be recapitulating the concept of the steady magnetic field due to a dc current through a conductor (Biot-Savart's law) as well as the concept of force on a current-carrying conductor placed in a steady magnetic field. The latter, in turn, is also developed from the concept of the Lorentz force experienced by a charge moving in a steady magnetic field (Chapter 4).

The understanding of time-varying magnetic field leading to Faraday's law and that of time-varying electric field leading to the concept of displacement current are dealt with in Chapter 5. However, the more generalised equations which follow from the study of time-varying electric and magnetic fields and which are extensively used in electromagnetic problems are the four Maxwell's equations (Chapter 5). These four equations can be expressed in the integral form in terms of the line and surface integrals of the electric and magnetic fields. Alternatively, these equations can be expressed in the differential form in terms of the divergence and curl of the electric and magnetic fields. You will definitely feel confident if you are able to write Maxwell's equations in the differential form in the generalised curvilinear system and are then able to translate them for the desired coordinate system (Chapter 5). Though Maxwell's equations are simple, yet they can describe electromagnetic laws concisely and help in formulating and solving a wide range of electromagnetic problems. Appreciating the significant role played by Maxwell's equations, J. R. Whinnery rightly describes these equations as 'Simple enough to imprint on a T-shirt, and yet rich enough to provide new insights throughout a lifetime of study'[1] (Chapter 5).

The flow of electrically charged particles like electrons obeys the continuity equation, which can be developed from the concept of the conservation of electric charge. Further, the continuity equation can be used to develop the concept of relaxation time and hence understand that while a conductor can be charged only at its surface, a dielectric can be charged throughout its bulk (Chapter 5). We will see that there is a need to introduce the concept of the vector potential along with that of the scalar potential in order to relate the electric field to the electric potential for treating time-varying field problems. The concept of scalar and vector retarded potentials is also introduced in Chapter 5; the application of this concept to conduction current antennas, however, is uncovered only in Chapter 8.

Maxwell's equations can be combined to form a wave equation in either electric or magnetic field (Chapter 6). The solution of the wave equation for an unbounded medium is helpful in understanding the transverse electromagnetic wave (TEM) mode of propagation in the unbounded free-space. We will see that for such a mode of propagation, the directions of the electric field, the magnetic field, and the propagation of electromagnetic power are mutually perpendicular to each other. We will also see that the analysis of the problem leads us to the well-known value of the speed of light, c. Similarly, the solution of the wave equation for an unbounded conducting medium helps us in developing the concept of surface resistance of a conducting medium, and hence also the concept of the ac resistance or RF resistance of a conductor (Chapter 6). You will be able to appreciate how the RF resistance of a conductor depends on the conductivity, or, alternatively, on the resistivity of the conducting material as well as on the wave frequency, and is thus not the same as

[1] J. R. Whinnery, The Teaching of Electromagnetics, *IEEE Transactions on Education*, vol. 33, pp. 3-7 (1990).

its dc resistance (Chapter 6). The simple analysis will show us, from the standpoint of the attenuation of waves, how to select the operating frequency for communication through a medium like sea-water, whose conductivity and permittivity are known (Chapter 6).

The wave equation in electric or magnetic field can be solved for a bounded medium subject to the electromagnetic boundary conditions of the problem. There are four generalised electromagnetic boundary conditions at a point on the interface between two media, two of which are for the tangential components of the electric and magnetic fields respectively, and two for the corresponding normal components. These boundary conditions can be derived, with the help of the four Maxwell's equations in the integral form, in terms of the electric and magnetic field vectors and the unit vector normal to the interface at the point where the boundary conditions are sought. We can subsequently apply these generalised electromagnetic boundary conditions to a point at the interface between a dielectric and another dielectric (or a free-space) medium as well as at the interface between a conductor and a dielectric (or a free-space) medium. The phenomenon of reflection of electromagnetic waves at these interfaces can also be understood with the help of these boundary conditions (Chapter 7).

Further, in the chapter to follow, Chapter 8, we will develop from first principles the concept of energy stored in electric and magnetic fields. We will use these concepts to derive an energy balance theorem called the Poynting theorem as well as its complex form that can be used to study the time-averaged power flow through a bounded or an unbounded medium and the associated power loss due to the presence of a lossy conducting medium. The concept is further exemplified in finding the power loss per unit area at a conducting boundary; this concept has been used later in Chapter 9 for finding the attenuation constant of a waveguide. The application of the concept of retarded vector potential to conduction current antennas is also studied in Chapter 8.

Next, in Chapter 9, you will be introduced to a basic problem of electromagnetic engineering (to be more specific, microwave engineering), namely electromagnetic wave propagation through a hollow-pipe metal waveguide. (During World War II, microwave engineers in USA used to be called, rather jokingly, as 'plumbers' for working with hollow pipes.) Further, in Chapter 9, the concepts of waveguide parameters, namely guide wavelength, phase propagation constant, phase velocity, group velocity and wave impedance are developed. The characteristic equation or the dispersion relation of the waveguide that relates either the phase velocity with wave frequency or the wave frequency with phase propagation constant is obtained. You will find it interesting to note that the waveguide behaves as a high-pass filter, allowing wave propagation through the waveguide only above a cutoff frequency. Below the cutoff frequency the waveguide supports the evanescent mode corresponding to an attenuated wave. Two kinds of excitation of modes in a waveguide, namely the TE mode and the TM mode, are brought within the purview of study. The significance of mode numbers vis-à-vis the field pattern is discussed. The analysis of the waveguide developed mostly considers a rectangular waveguide, though the salient features of the analysis of a cylindrical waveguide are also provided. The analysis includes power flow through the waveguide as well as an estimate of the power-handling capability of the waveguide. The propagation of a wave through the waveguide is also associated with a loss of power due to the finite resistivity of the conducting material of the waveguide. The analysis developed estimates the loss in a rectangular waveguide in terms of the attenuation constant of the waveguide (Chapter 9).

Similarly, with reference to a cavity resonator, the concepts of resonance frequency and cavity length are developed in Chapter 10. We will see how the transmission line theory can be used to find the length of a waveguide with its both ends either short-circuited or open-ended such that the waveguide behaves as a cavity resonator. This analysis has given an occasion to elaborate the theory of transmission line and the application of Smith chart to transmission line problems (Chapter 10). Further, we will see that the concept of the energy stored in a cavity resonator can be combined with the effect of the loss contributed by the finite resistivity of the conducting material of the resonator for predicting the quality factor of the cavity (Chapter 10).

Finally, the book is concluded with a summary chapter and a discussion on the extent to which the objective set at the outset of this chapter is fulfilled. We have selectively chosen topics that are necessary to project the essentials of electromagnetic theory for its engineering application. For instance, in microwave engineering, the applications encompass antenna and wave propagation, microwave circuits, and vacuum as well as solid-state electron devices (Chapter 11).

A good number of appendixes are included at the end of chapters where relevant. The concepts developed in the text are illustrated by worked-out examples. In addition, students get the opportunity to summarise their learnings by answering review questions of the narrative, numerical and objective (multiple-choice) types, for which answers/solutions/hints are provided at the end of the chapters.

2

Vector Calculus Expressions for Gradient, Divergence and Curl

ELectromagnetic problems usually enjoy rectangular, cylindrical, or spherical symmetry, and accordingly, these problems are treated in *rectangular, cylindrical, or spherical-polar coordinate systems*. It is therefore necessary to define at the very outset a point in the desired coordinate system, where, for instance, one would like to find an electromagnetic quantity like electric field. Similarly, in the statement and formulation of a problem, we may have to define *area* and *volume elements*, and *element of a distance vector* directed from one point to another in the desired coordinate system. These definitions, together with the definitions of the *line integral* and the *surface integral of a vector quantity*, would also in turn be necessary to define and deduce the *vector calculus* expressions of relevance to electromagnetic problems. Often, the natural course of deduction of electromagnetic concepts leads to results in terms of these vector calculus expressions. The vector calculus expressions that are often encountered in electromagnetic problems are the *gradient of a scalar quantity*, the *divergence of a vector quantity*, the *curl of a vector quantity*, and the *Laplacian of a scalar quantity* (which is the divergence of the gradient of a scalar quantity). If these terminologies, namely gradient, divergence, and curl sound new to some of you at this stage, you need not worry—I assure you that they will become clear as you proceed to read the rest of this chapter. These expressions, which refer to both scalar quantities[1], such as the electric potential, and vector quantities, such as the electric field[1], will help in formulating the electromagnetic problems found in the rest of this treatise.

The primary objective of the present chapter is to first deduce the vector calculus expressions in the *generalised curvilinear coordinate system* and then interpret them for

[1]A scalar quantity is defined by its magnitude while a vector quantity by both its magnitude and direction.

reading the corresponding expressions in the rectangular, cylindrical, or spherical-polar coordinate systems.

Why curvilinear coordinate system?

Though we need not actually use the generalised curvilinear coordinate system to formulate a practical electromagnetic problem, yet vector calculus expressions in the generalised system are very easy to deduce and remember. Also, they can be translated, in a simple and straightforward manner, to the corresponding expression in the desired system, namely rectangular, cylindrical or spherical system. This is due to the symmetry in the arrangement of terms that each of these vector calculus expressions enjoy in the curvilinear coordinate system. To be more specific, in the curvilinear coordinate system, each of the vector calculus expressions comprise three groups of terms, and by permuting the terms of the first group we can easily write the terms of the remaining two groups. In fact, while treating an electromagnetic problem, you will feel really confident when you can write these expressions easily without referring to any external source. Furthermore, those who have already come across the deductions of these expressions in the cylindrical or spherical coordinate systems are going to find that the corresponding deduction in the generalised curvilinear system is much more simple. So, let us break the ice and hit the right spot by deducing the vector calculus expressions in the generalised curvilinear coordinate system.

2.1 Representation of a Point in Different Coordinate Systems

Irrespective of whether the coordinate system considered is rectangular, cylindrical, or spherical, let us define three mutually perpendicular axes —X, Y, and Z —which meet at a point called the *origin* of the coordinates (Figs. 2.1*a*-*c*). We can then define the coordinates of the origin as (0,0,0) and those of the position of a point P as (x, y, z) in rectangular system, (r, θ, z) in cylindrical system, and (r, θ, ϕ) in spherical system.

In rectangular system (Fig. 2.1*a*),

- the coordinate x of the point P represents the perpendicular distance of the point from the YZ-plane ($x = 0$-plane);
- the coordinate y of the point P represents the perpendicular distance of the point from the ZX-plane ($y = 0$-plane);
- the coordinate z of the point P represents the perpendicular distance of the point from the XY-plane ($z = 0$-plane).

In this system, the YZ-plane is defined as the plane passing through the Y and Z axes. Similarly, the ZX- and XY-planes are defined. Over the YZ-plane, the values of the coordinates y and z vary from point to point while the value of x remains constant at $x = 0$.

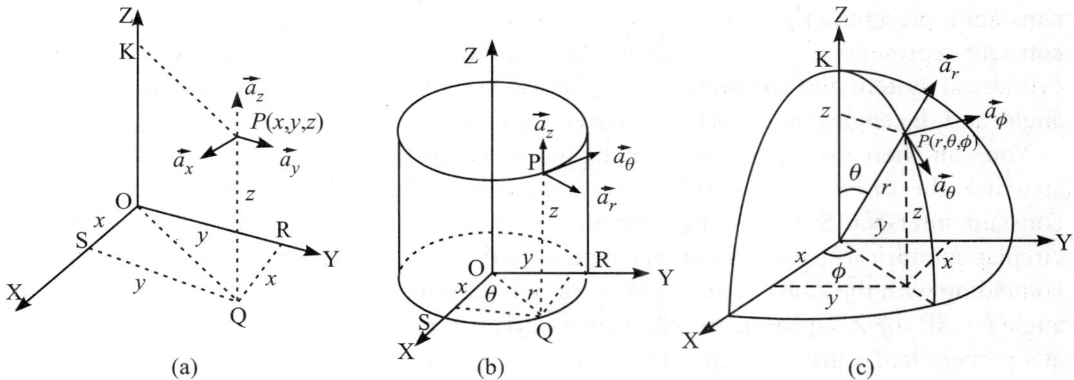

Fig. 2.1 *Representation of a point P in (a) rectangular, (b) cylindrical, and (c) spherical coordinate systems.*

Therefore, the YZ-plane is also referred to as the $x = 0$-plane. Similarly, the ZX-plane is referred to as the $y = 0$-plane, and the XY-plane as the $z = 0$-plane. The plane $x =$ constant (with a value other than 0) represents a plane parallel to and at a perpendicular distance x from the YZ-plane ($x = 0$-plane). Similarly, the plane $y =$ constant represents a plane parallel to and at a perpendicular distance y from the ZX-plane ($y = 0$-plane), and the plane $z =$ constant represents a plane parallel to and at a perpendicular distance z from the XY-plane ($z = 0$-plane).

In the cylindrical system (Fig. 2.1*b*),

- the coordinate r represents the perpendicular distance of the point P from the Z-axis;
- the coordinate θ represents the angle between (i) the plane passing through the point P and the Z-axis and (ii) the ZX-plane;
- the coordinate z represents the perpendicular distance of the point P from the XY-plane.

With a little thought, you will easily appreciate that in the cylindrical system, $r =$ constant represents a cylindrical surface of radius r with its axis coincident with the Z-axis; $\theta =$ constant represents a plane that subtends an angle θ with the ZX-plane; and $z =$ constant represents a plane parallel to and at a distance z from the XY-plane (Fig. 2.1*b*).

In spherical system (Fig. 2.1*c*),

- the coordinate r represents the distance of the point P from the origin O;
- the coordinate θ represents the angle between (i) the line joining the origin O with the point P and (ii) the Z-axis;
- the coordinate ϕ represents the angle between (i) the plane passing through the point P and the Z-axis and (ii) the ZX-plane.

Again, with a little thought, you will see that in spherical system $r =$ constant represents a sphere of radius r with its centre at the origin O $(0, 0, 0)$ of the X, Y, and Z axes; $\theta =$

constant represents a right-circular cone of angle θ with its vertex at the origin O; and $\phi =$ constant represents a plane that subtends an angle ϕ with the ZX-plane. The angle θ in cylindrical system and the angle ϕ in spherical system are each known as the azimuthal angle; and the angle θ in spherical system is known as the zenith angle.

You can then easily visualise that in the rectangular system, P (x, y, z) is the point at which the three planar surfaces represented by $x=$ constant, $y=$ constant, and $z =$ constant intersect. Similarly, in cylindrical system, P (r, θ, z) is the point at which the circular cylindrical surface of radius r represented by $r =$ constant (the axis of the cylinder coinciding with the Z-axis), the planar surface represented by $\theta =$ constant (subtending an angle θ with the ZX-plane), and the planar surface represented by $z =$ constant (which is at a perpendicular distance z from the XY-plane) intersect. Similarly, in spherical system, P (r, θ, ϕ) is the point at which the spherical surface of radius r represented by $r =$ constant (which has its centre at the origin O), the conical surface represented by $\theta =$ constant (with cone angle θ), and the planar surface represented by $\phi =$ constant (subtending an angle ϕ with the ZX-plane) intersect.

Furthermore, you can easily relate cylindrical coordinates (r, θ, z) to rectangular coordinates (x, y, z) as follows (refer to Fig. 2.1*b*; also see Appendix A2.1):

$$\left.\begin{aligned} x &= r\cos\theta \\ y &= r\sin\theta \\ z &= z \end{aligned}\right\} . \text{ (Cylindrical vis-à-vis rectangular coordinates)} \tag{2.1}$$

Similarly, refer to Fig. 2.1c to obtain the following relation between spherical coordinates (r, θ, ϕ) and rectangular coordinates (x, y, z) (see Appendix A2.1):

$$\left.\begin{aligned} x &= r\sin\theta\cos\phi \\ y &= r\sin\theta\sin\phi \\ z &= r\cos\theta \end{aligned}\right\} . \text{ (Spherical vis-à-vis rectangular coordinate)} \tag{2.2}$$

Thus, X, Y, and Z axes of the rectangular coordinate system continue to be important as the reference axes even while representing a point in the cylindrical and spherical coordinate systems (Fig. 2.1). However, these three coordinate systems have their own coordinate axes in three mutually perpendicular directions according to relations that involve unit vectors and their cross products[2] as follows (Fig. 2.1):

$$\vec{a}_x \times \vec{a}_y = \vec{a}_z \text{ (Rectangular system)} \tag{2.3}$$

$$\vec{a}_r \times \vec{a}_\theta = \vec{a}_z \text{ (Cylindrical system)} \tag{2.4}$$

$$\vec{a}_r \times \vec{a}_\theta = \vec{a}_\phi \text{ (Spherical system)} \tag{2.5}$$

[2]The cross product of the two vectors $\vec{A} \times \vec{B}$ is a vector quantity, the magnitude of which is $AB\sin\theta$ and the direction of which is normal to the plane containing the vectors $\vec{A}$ and $\vec{B}$ such that the sense of its direction is that of the axial motion of a screw moved from the direction of $\vec{A}$ to that of $\vec{B}$ through the angle θ, which is the angle between vectors $\vec{A}$ and $\vec{B}$.

where

- $\vec{a}_x$, $\vec{a}_y$, and $\vec{a}_z$ are unit vectors in the directions of increasing x, y, and z respectively in rectangular system (Fig. 2.1*a*);
- $\vec{a}_r$, $\vec{a}_\theta$, $\vec{a}_z$ are unit vectors in the directions of increasing r, θ, and z respectively in cylindrical system (Fig. 2.1*b*); and
- $\vec{a}_r$, $\vec{a}_\theta$, $\vec{a}_\phi$ are unit vectors in the directions of increasing r, θ, and ϕ respectively in spherical system (Fig. 2.1*c*).

2.2 Volume Elements in Different Coordinate Systems

Now that we know how to define a point by a system of three coordinates, namely (x, y, z) in rectangular, (r, θ, z) in cylindrical, and (iii) (r, θ, ϕ) in spherical systems of coordinates, let us define the coordinates of the points that are infinitesimally close to this point by bringing about elemental or incremental changes in the coordinates of the point. For instance, by making such incremental changes in the rectangular system, we can define, close to the point A (x, y, z), seven neighbouring points: B $(x, y + dy, z)$, C $(x, y, z + dz)$, D $(x + dx, y, z)$, E $(x + dx, y + dy, z)$, F $(x, y + dy, z + dz)$, G $(x + dx, y, z + dz)$, and H $(x + dx, y + dy, z + dz)$, where dx, dy, and dz are differential increments (infinitesimal changes) in x, y, and z respectively (see Fig. 2.2*a* and Table 2.1). Similarly, we can generate all such points (B, C, D, E, F, G, and H) in the neighbourhood of A in cylindrical (Fig. 2.2*b*) as well as spherical-polar (Fig. 2.2*c*) coordinate systems and tabulate their coordinates (Table 2.1). These eight points, namely A, B, C, D, E, F, G, and H, constitute the eight corners of an element of a differential parallelepiped volume in rectangular, cylindrical or spherical coordinate systems.

Table 2.1 *Coordinates of the reference point (A) and those of seven neighbouring points (B, C, D, E, F, G, and H), in rectangular, cylindrical, and spherical coordinates, the points representing the corners of a differential parallelepiped volume.*

Corner point	Rectangular (Fig. 2.2*a*)	Cylindrical (Fig. 2.2*b*)	Spherical-polar (Fig. 2.2*c*)
A	(x, y, z)	(r, θ, z)	(r, θ, ϕ)
B	$(x, y + dy, z)$	$(r, \theta + d\theta, z)$	$(r, \theta + d\theta, \phi)$
C	$(x, y, z + dz)$	$(r, \theta, z + dz)$	$(r, \theta, \phi + d\phi)$
D	$(x + dx, y, z)$	$(r + dr, \theta, z)$	$(r + dr, \theta, \phi)$
E	$(x + dx, y + dy, z)$	$(r + dr, \theta + d\theta, z)$	$(r + dr, \theta + d\theta, \phi)$
F	$(x, y + dy, z + dz)$	$(r, \theta + d\theta, z + dz)$	$(r, \theta + d\theta, \phi + d\phi)$
G	$(x + dx, y, z + dz)$	$(r + dr, \theta, z + dz)$	$(r + dr, \theta, \phi + d\phi)$
H	$(x + dx, y + dy, z + dz)$	$(r + dr, \theta + d\theta, z + dz)$	$(r + dr, \theta + d\theta, \phi + d\phi)$

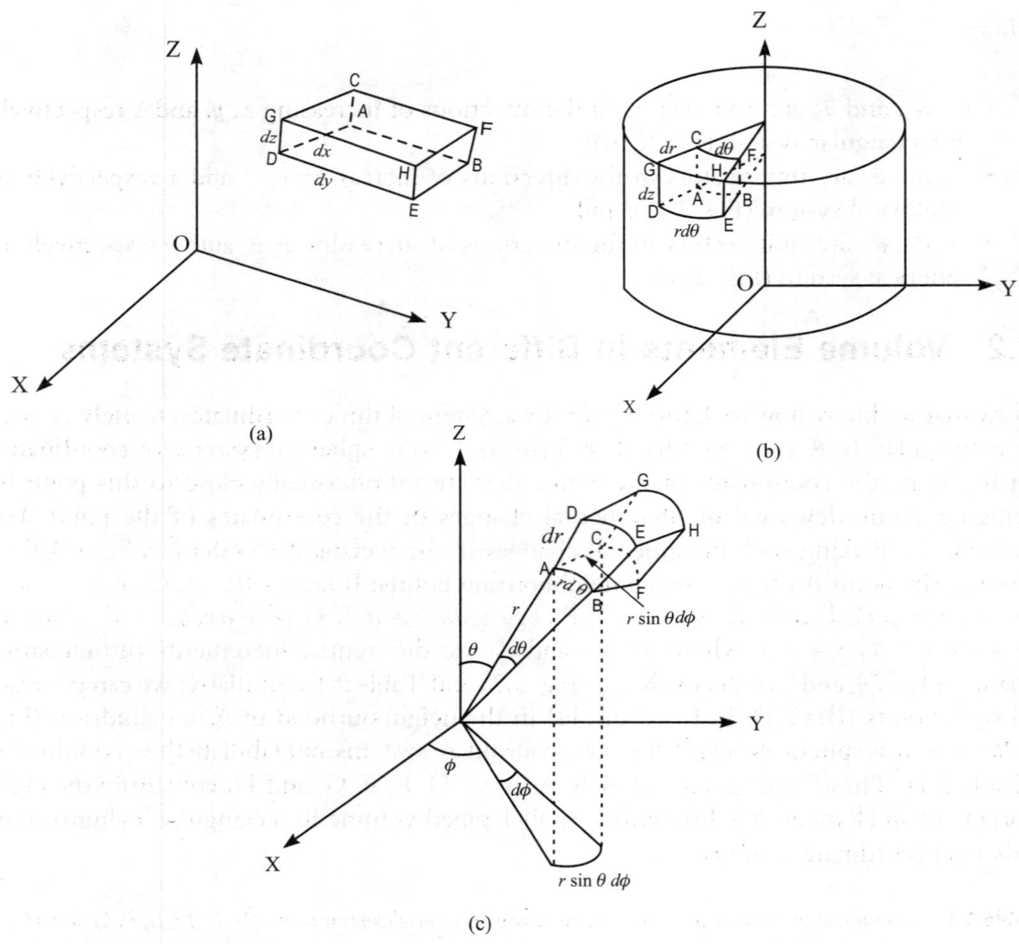

Fig. 2.2 *Element of volume in (a) rectangular, (b) cylindrical, and (c) spherical coordinate systems.*

If we choose to vary one of the three coordinates of a point while keeping the remaining two constant, then we will generate either a straight or a curved line of infinitesimal length, depending on which of the coordinates we vary and in which coordinate system we do so. This is how we generate the length elements of a differential volume . Thus, in the rectangular coordinate system (Fig. 2.2*a*), we can generate a straight line of infinitesimal length dx by changing x by a differential amount dx while keeping y and z constant. The length dx lies on the one hand on the plane $y =$ constant, while on the other hand it lies on the plane $z =$ constant. In the differential volume, this length dx can be identified as the side AD (taking A as the point of reference (x, y, z) (Fig. 2.2*a*). Similarly, we can generate a straight line of infinitesimal length dy by changing y by a differential amount dy while keeping $x$and z constant, and another straight line of infinitesimal length dz while keeping x and y constant. The infinitesimal length elements, dy and dz, can be identified as AB and AC respectively of the differential volume. Let us then put these length elements in the rectangular system for future reference as follows:

$$\left.\begin{array}{l} \mathrm{AD} = dx \\ \mathrm{AB} = dy \\ \mathrm{AC} = dz \end{array}\right\}. \quad \text{(Rectangular system)} \tag{2.6}$$

In cylindrical coordinate system (Fig. 2.2*b*), we can generate a straight line of infinitesimal length dr by changing r by a differential amount dr while keeping θ and z constant. On the one hand, the infinitesimal length dr lies on the plane θ = constant, while on the other hand it lies on the plane z = constant. However, a curved line of infinitesimal arc length $rd\theta$ is generated when we bring about a differential change $d\theta$ in θ while keeping r and z constant, r being the radius of the arc. This infinitesimal length $rd\theta$ lies on the circular-cylindrical surface r = constant on the one hand, while on the other hand it lies on the plane z = constant. Similarly, we can generate a straight line of infinitesimal length dz when we change z by a differential amount dz while keeping r and θ constant. This infinitesimal length dz lies on the one hand on the circular-cylindrical surface r = constant, while on the other hand it lies on the plane θ = constant. Thus, we can put together the length elements of the differential volume in cylindrical system as follows (Fig. 2.2*b*):

$$\left.\begin{array}{l} \mathrm{AD} = dr \\ \mathrm{AB} = rd\theta \\ \mathrm{AC} = dz \end{array}\right\}. \quad \text{(Cylindrical system)} \tag{2.7}$$

In spherical-polar coordinate system (Fig. 2.2*c*), similarly, a straight line of infinitesimal length dr is generated by changing r by a differential amount dr while keeping θ and ϕ constant. The infinitesimal length dr lies on the one hand on the cone θ = constant, while on the other hand it lies on the plane ϕ = constant. We can further generate a curved line of infinitesimal arc length $rd\theta$ by bringing about a differential change $d\theta$ in θ while keeping r and ϕ constant, the radius of the arc being taken as r. The infinitesimal length $rd\theta$ lies on the one hand on the spherical surface r = constant, while on the other hand it lies on the plane ϕ = constant. We can generate the infinitesimal arc length $r \sin\theta \, d\phi$ by changing ϕ by a differential amount $d\phi$ while keeping r and θ constant, the radius of the arc being $r \sin\theta$, which is the perpendicular distance of the point (r, θ, ϕ) from the Z-axis. Clearly, the infinitesimal length $r \sin\theta \, d\phi$ lies on the plane z = constant, which is parallel to, and at a distance z from, the XY-plane. Thus, we can put together the length elements of the differential volume in the spherical system as (Fig. 2.2*c*):

$$\left.\begin{array}{l} \mathrm{AD} = dr \\ \mathrm{AB} = rd\theta \\ \mathrm{AC} = r\sin\theta\, d\phi \end{array}\right\}. \quad \text{(Spherical system)} \tag{2.8}$$

The basic length elements AD, AB, and AC of the differential volume found as in (2.6)–(2.8) for the three systems of coordinates, each originating from the reference point A, constitute the *basic length elements* of a *differential volume*. Similarly, the lengths of the remaining nine sides of the differential volume can also be obtained (Appendix A2.2). We can also find from these basic length elements the area elements of all the six faces of the differential volume (Appendix A2.2). Thus, we can find the volume element $d\tau$ by multiplying the basic length elements AD, AB, and AC (found as above and tabulated in Appendix A2.2) as follows:

$$d\tau = dxdydz \qquad \text{(Rectangular system)} \tag{2.9}$$

$$d\tau = dr\, rd\theta\, dz \qquad \text{(Cylindrical system)} \tag{2.10}$$

$$d\tau = dr\, rd\theta\, r\sin\theta\, d\phi. \qquad \text{(Spherical-polar system)} \tag{2.11}$$

Let us now introduce the generalised curvilinear coordinates u_1, u_2, u_3 (Fig. 2.3), similar to the coordinates (x, y, z) in rectangular, (r, θ, z) in cylindrical and (r, θ, ϕ) in spherical coordinate systems. Correspondingly, we can define three mutually perpendicular unit vectors, $\vec{a}_1$, $\vec{a}_2$, and $\vec{a}_3$ in the increasing directions of u_1, u_2, u_3 respectively (similar to the unit vectors $\vec{a}_x$, $\vec{a}_y$, and $\vec{a}_z$ in rectangular; $\vec{a}_r$, $\vec{a}_\theta$, and $\vec{a}_z$ in cylindrical; and $\vec{a}_r$, $\vec{a}_\theta$, and $\vec{a}_\phi$ in spherical coordinate systems (Table 2.2). As in rectangular, cylindrical, or spherical coordinate system, we can represent in the curvilinear system a point A (u_1, u_2, u_3), as well as its seven neighbouring points: B $(u_1, u_2 + du_2, u_3)$, C $(u_1, u_2, u_3 + du_3)$, D (u_1+du_1, u_2, u_3), E $(u_1+du_1, u_2+du_2, u_3)$, F$(u_1, u_2+du_2, u_3+du_3)$, G$(u_1+du_1, u_2, u_3+du_3)$, and H$(u_1 + du_1, u_2 + du_2, u_3 + du_3)$, all constituting the corners of a differential volume (Fig. 2.3).

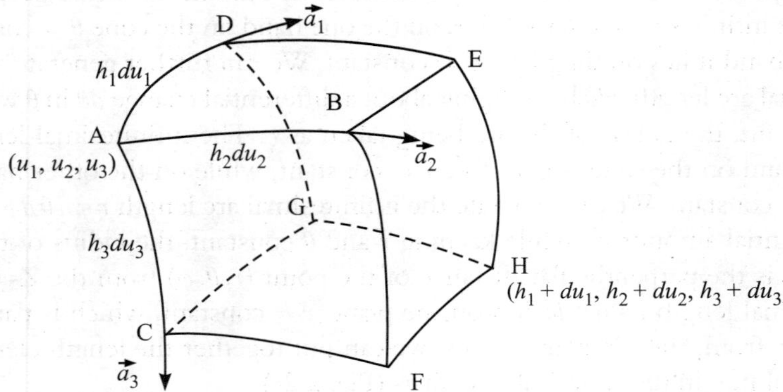

Fig. 2.3 *An element of volume with its corners at A, B, C, D, E, F, G and H in the generalised curvilinear coordinate system, enclosing the point where the divergence of the vector quantity is to be found.*

If we define the basic length elements in the generalised curvilinear system as AD $= h_1 du_1$, AB $= h_2 du_2$, and AC $= h_3 du_3$, we can find by multiplying the basic length elements AD, AB, and AC the following expression for the volume element $d\tau$:

$$d\tau = h_1 du_1 h_2 du_2 h_3 du_3. \tag{2.12}$$

We can use the generalised expression (2.12) for rectangular, cylindrical, and spherical systems as well if we interpret h_1, h_2, h_3 and u_1, u_2, u_3, suitably. For this purpose, compare (2.12) with (2.9)–(2.11). Thus, for rectangular system, we have to take $h_1 = 1$, $h_2 = 1$, $h_3 = 1$; $u_1 = x$, $u_2 = y$, $u_3 = z$ to make (2.12) identical to (2.9). Similarly, we can compare (2.12) with (2.10) and (2.11) in order to interpret (2.12) for cylindrical and spherical systems respectively (see Table 2.2).

Table 2.2 *Quantities pertaining to the differential volume element in different systems of coordinates with respect to those of curvilinear system*

Quantity	Rectangular	Cylindrical	Spherical-polar
u_1	x	r	r
u_2	y	θ	θ
u_3	z	z	ϕ
h_1	1	1	1
h_2	1	r	r
h_3	1	1	$r \sin\theta$
$\vec{a}_1$	$\vec{a}_x$	$\vec{a}_r$	$\vec{a}_r$
$\vec{a}_2$	$\vec{a}_y$	$\vec{a}_\theta$	$\vec{a}_\theta$
$\vec{a}_3$	$\vec{a}_z$	$\vec{a}_z$	$\vec{a}_\phi$
AD$= h_1 du_1$	dx	dr	dr
AB$= h_2 du_2$	dy	$rd\theta$	$rd\theta$
AC$= h_3 du_3$	dz	dz	$r \sin\theta\, d\phi$
$d\tau = h_1 du_1 h_2 du_2 h_3 du_3$ (2.12)	$dxdydz$ (2.11)	$dr\, rd\theta\, dz$ (2.10)	$dr\, rd\theta\; r\sin\theta\, d\phi$ (2.9)

2.3 Element of Distance Vector Directed from One Point to Another

Let us consider two points which are separated by an infinitesimal distance. Let us choose two points, typically, A and H, situated at two corners of a differential volume, which have already been defined in terms of their respective coordinates in rectangular, cylindrical, and spherical systems as well as in generalised curvilinear coordinate systems. To recapitulate, the coordinates of thevpoint A are (x, y, z), (r, θ, z), and (r, θ, ϕ) in rectangular, cylindrical, and spherical coordinate systems respectively, and the corresponding coordinates of the point H are $(x+dx, y+dy, z+dz)$, $(r+dr, \theta+d\theta, z+dz)$, and $(r+dr, \theta+d\theta, \phi+d\phi)$ respectively (Table 2.3). Similarly, in generalised curvilinear system, the coordinates of the point A are (u_1, u_2, u_3), and those of the point H are $(u_1 + du_1, u_2 + du_2, u_3 + du_3)$ (Table 2.3).

We can now define an elemental distance vector $d\vec{R}$ such that its magnitude is dR, equal to the length of the line AH, with its direction from A to H (refer to Table 2.3). Taking rectangular coordinate system (Fig. 2.2*a*), we can reach H from A by first moving through the element of distance dx in the direction of $\vec{a}_x$ from A to D; then through dy in the direction of $\vec{a}_y$ from D to E; and, finally, through dz in the direction of $\vec{a}_z$ from E to H. In cylindrical system, the corresponding elemental distances to move along would be dr from A to D in the direction of $\vec{a}_r$; $rd\theta$ from D to E in the direction of $\vec{a}_\theta$; and dz from E to H in the direction of $\vec{a}_z$ (Fig. 2.2*b*). Similarly, in spherical system, the corresponding elemental distances to move along would be dr from A to D; $rd\theta$ from D to E; and $r\sin\theta d\phi$ from E to H, in the directions of $\vec{a}_r$, $\vec{a}_\theta$, and $\vec{a}_\phi$ respectively (Fig. 2.2*c*). Similarly, in the curvilinear system the corresponding elemental distances to move from A to reach H would be through $h_1 du_1$ from A to D; $h_2 du_2$ from D to E; and $h_3 du_3$ from E to H, in the directions of $\vec{a}_1$, $\vec{a}_2$, and $\vec{a}_3$ respectively (Fig. 2.3). Hence, we can now define $d\vec{R}$, in terms of the unit vectors in the different systems of coordinates, as follows:

$$\left.\begin{aligned} d\vec{R} &= dx\bar{a}_x + dy\vec{a}_y + dz\bar{a}_z && \text{(Rectangular system)} \quad \text{(a)} \\ d\vec{R} &= dr\bar{a}_r + rd\theta\,\vec{a}_\theta + dz\bar{a}_z && \text{(Cylindrical system)} \quad \text{(b)} \\ d\vec{R} &= dr\bar{a}_r + rd\theta\,\vec{a}_\theta + r\sin\theta\bar{a}_\phi && \text{(Spherical system)} \quad \text{(c)} \\ d\vec{R} &= h_1du_1\bar{a}_1 + h_2du_2\vec{a}_2 + h_3du_3\bar{a}_3 && \text{(Curvilinear system).} \quad \text{(d)} \end{aligned}\right\}. \quad (2.13)$$

Table 2.3 *Coordinates of the point A and H, located at an infinitesimal distance from each other, chosen typically as the two corners points of a differential volume.*

Points	Rectangular	Cylindrical	Spherical	Curvilinear
A	x y z	r θ z	r θ ϕ	u_1 u_2 u_3
H	$x + dx$ $y + dy$ $z + dz$	$r + dr$ $\theta + d\theta$ $z + dz$	$r + dr$ $\theta + d\theta$ $\phi + d\phi$	$u_1 + du_1$ $u_2 + du_2$ $u_3 + du_3$

2.4 Line Integral and Surface Integral or Flux of a Vector Quantity

A physical vector quantity, say, the electric field $\vec{E}$, in general, can vary from point to point over a line of finite length, be it a straight line or a curved line. However, an infinitesimal length of the line can be regarded as a straight line even though it is a part of a curved line, and we can regard, over this infinitesimal length, the quantity $\vec{E}$ as constant with respect to both its magnitude and direction. Further, in general, the direction of $\vec{E}$ need not be the same as that of the alignment of the infinitesimal length of the line. With reference to this infinitesimal length of line, we can now define an *element of line integral* dI_E as the dot product of the vectors $\vec{E}$ and $d\vec{l}$, and put it as (Fig. 2.4):

$$dI_E = \vec{E} \cdot d\vec{l} \tag{2.14}$$

where $d\vec{l}$ is an infinitesimal length element vector the magnitude dl aligned along the direction in which we wish to take the integral (Fig. 2.14). If $\vec{E}$ subtends an angle θ with $d\vec{l}$, then we can express (2.14) as

$$dI_E = Edl \cos\theta. \tag{2.15}$$

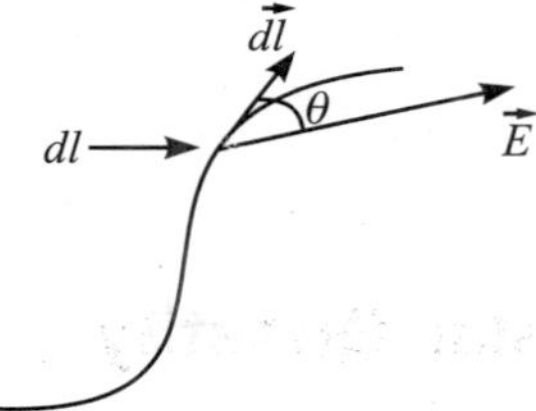

Fig. 2.4 *A vector, typically $\vec{E}$, and the length element vector $d\vec{l}$ considered on a line. The angle between the vectors is θ.*

The concept of the line integral becomes further relevant when we take it over a finite length extending from, say A to B. Then we can define the finite line integral as

$$I_{E(\mathrm{A}\to\mathrm{B})} = \int_A^B dI_{E(\mathrm{A}\to\mathrm{B})} = = \int_A^B \vec{E} \cdot d\vec{l} \tag{2.16}$$

which is obtained by integrating (2.14) over the finite length between A and B (Fig. 2.4). The direction of the length vector follows the movement from A to B of the point on which the length element is considered.

Similarly, the vector $\vec{E}$, even though it varies over a finite area of a surface, may be regarded as constant with respect to both its magnitude and direction over an infinitesimal area of the finite surface, the latter considered plane even though the finite surface may be curved. Thus we can define an element of surface integral of $\vec{E}$ over an infinitesimal area of the surface (also referred to as the element of flux dF_E of $\vec{E}$ through the infinitesimal surface area as follows:

$$dF_E = \vec{E}.d\vec{S} \tag{2.17}$$

which may be alternatively expressed as

$$dF_E = \vec{E}.\vec{a}_n dS \tag{2.18}$$

where $d\vec{S}$ is the infinitesimal area vector, the magnitude of which is dS, and the direction of which is that of the normal to the infinitesimal area, and $\vec{a}_n$ is the unit vector in the direction of the normal.

We can then find the flux through a finite surface area S by integrating (2.17) or (2.18) over the area, that is, taking the surface integral, as

$$F_E = \int_S \vec{E}.d\vec{S} = \int_S \vec{E}.\vec{a}_n dS. \tag{2.19}$$

If the finite surface area encloses a volume, the flux F_E is called the outward or the inward flux, depending on whether $\vec{a}_n$ is directed outward or inward to the closed volume. For such an enclosure, the surface integral representing the flux F_E given by (2.19) becomes the 'closed' surface integral, which may be expressed as

$$F_E = \oint_S \vec{E}.d\vec{S} = \oint_S \vec{E}.\vec{a}_n dS. \tag{2.20}$$

2.5 Gradient of a Scalar Quantity

The gradient of a scalar quantity, such as electric potential V at a point, is a vector quantity. In general, the scalar quantity varies in space. It does vary with distance, with different rates in different directions. The magnitude of the vector representing the gradient of the scalar V is the maximum rate of variation of V, and it is directed along where the maximum variation of V takes place.

In this section, let us proceed to deduce from first principles an expression for the gradient of V representing a scalar quantity in the generalised curvilinear coordinate system. We will subsequently interpret this generalised expression for the corresponding expressions in the rectangular, cylindrical, and spherical-polar coordinate systems.

Although the scalar V varies in space, one can find a surface on which V remains constant at all the points. Such a surface is defined as equipotential V. Similarly, we can define equipotential $V + dV$ on which the potential remains constant at $V + dV$. Clearly, these two equipotentials, V and $V + dV$, will be separated in space but by an infinitesimal extent (Fig. 2.5).

Let us take two points A(u_1, u_2, u_3) and H($u_1 + du_1, u_2 + du_2, u_3 + du_3$) on the equipotentials V and $V + dV$ respectively (Fig. 2.5). These two points, A and H, in general, differ with respect to each of their coordinates (see Section 2.3). The minimum distance between the point A(u_1, u_2, u_3) situated on the equipotential V and the equipotential $V + dV$ is the perpendicular distance AN $= dn$, say, where N is the foot of the perpendicular (normal) drawn from A to the equipotential $V + dV$. Since the difference between the potentials of any two points on the equipotentials V and $V + dV$ is the same, being equal to dV, the maximum rate of variation of V with distance will take place in the direction of the normal AN. The magnitude of this maximum rate of variation of V is obtained by dividing

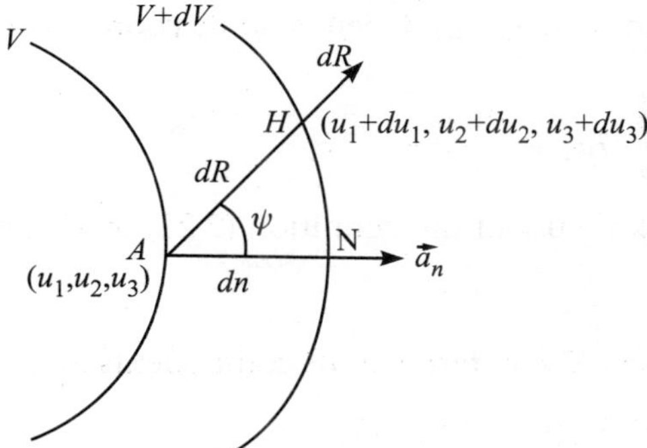

Fig. 2.5 *Elemental distance vector $d\vec{R}$ separating two nearby points situated on two equipotentials, V and $V + dV$ respectively.*

dV by AN $= dn$, the latter being the minimum distance between the equipotentials at point A(u_1, u_2, u_3). Therefore, we can now define a quantity called the gradient of V, abbreviated as grad V (which is a vector quantity as mentioned in the beginning of this section) and denoted by the symbol ∇V, and express it as follows:

$$\text{grad}\, V = \nabla V = \frac{\partial V}{\partial n}\vec{a}_n \tag{2.21}$$

where $\partial V/\partial n$ is the magnitude of the gradient, and $\vec{a}_n$ is the unit vector (of magnitude unity) in the direction of the normal AN, which is also the direction of the gradient.

At this stage, let us recall the expression (2.13*d*) in generalised curvilinear coordinate system for the vector $d\vec{R}$ of magnitude dR (=AH) that is directed from A to H (Section 2.3), and rewrite it in the following form:

$$d\vec{R} = dR\vec{a}_R = h_1 du_1 \vec{a}_1 + h_2 du_2 \vec{a}_2 + h_3 du_3 \vec{a}_3 \qquad [(2.13d) \text{ rewritten}] \tag{2.22}$$

where $\vec{a}_R$ is the unit vector (magnitude unity) which has the same direction as that of $d\vec{R}$.

We can express AN$= dn$ [the minimum distance between the equipotentials V and $V + dV$ at the point A (u_1, u_2, u_3)] in terms of AH $= dR$ as

$$\text{AN} = \text{AH} \cos\psi,$$

that is,

$$dn = dR\cos\psi\ ,$$

where ψ is the angle between AN and AH. Alternatively, this can be expressed as

$$dn = \vec{a}_n \cdot d\vec{R}. \tag{2.23}$$

With the help of (2.23), we can now express the incremental change dV in the potential as

$$dV = \frac{\partial V}{\partial n}dn = \frac{\partial V}{\partial n}\vec{a}_n \cdot d\vec{R}. \tag{2.24}$$

Further, since V at a point (u_1, u_2, u_3) depends on its coordinates, we can also express dV as

$$dV = \frac{\partial V}{\partial u_1}du_1 + \frac{\partial V}{\partial u_2}du_2 + \frac{\partial V}{\partial u_3}du_3. \tag{2.25}$$

Also, independently, making use of the definition (2.21) of ∇V, we can express (2.24) as

$$dV = \nabla V \cdot d\vec{R}. \tag{2.26}$$

Let us, at this stage, express ∇V in terms of its components as

$$\nabla V = (\nabla V)_1\vec{a}_1 + (\nabla V)_2\vec{a}_2 + (\nabla V)_3\vec{a}_3. \tag{2.27}$$

We can then substitute (2.27) for ∇V and (2.22) for $d\vec{R}$ into (2.26) to obtain

$$dV = [(\nabla V)_1\vec{a}_1 + (\nabla V)_2\vec{a}_2 + (\nabla V)_3\vec{a}_3] \cdot [h_1 du_1\vec{a}_1 + h_2 du_2\vec{a}_2 + h_3 du_3\vec{a}_3],$$

which, after evaluating the dot products ($\vec{a}_1 \cdot \vec{a}_1 = 1,\ \vec{a}_2 \cdot \vec{a}_2 = 1,\ \vec{a}_3 \cdot \vec{a}_3 = 1;\ \vec{a}_1 \cdot \vec{a}_2 = 0,\ \vec{a}_1 \cdot \vec{a}_3 = 0,\ \vec{a}_2 \cdot \vec{a}_3 = 0$, etc.) becomes

$$dV = (\nabla V)_1 h_1 du_1 + (\nabla V)_2 h_2 du_2 + (\nabla V)_3 h_3 du_3. \tag{2.28}$$

Now, if we compare (2.28) with (2.25), or more precisely equate the coefficients of du_1, du_2 and du_3, we obtain

$$\left.\begin{aligned} \frac{\partial V}{\partial u_1} &= (\nabla V)_1 h_1 \\ \frac{\partial V}{\partial u_2} &= (\nabla V)_2 h_2 \\ \frac{\partial V}{\partial u_3} &= (\nabla V)_3 h_3 \end{aligned}\right\},$$

whence we can write

$$\left.\begin{aligned} (\nabla V)_1 &= \frac{1}{h_1}\frac{\partial V}{\partial u_1} \\ (\nabla V)_2 &= \frac{1}{h_2}\frac{\partial V}{\partial u_2} \\ (\nabla V)_3 &= \frac{1}{h_3}\frac{\partial V}{\partial u_3} \end{aligned}\right\}. \tag{2.29}$$

The expression for the gradient of the scalar V in generalised curvilinear coordinate system is finally obtained by substituting the components (2.29) into (2.27) as follows:

$$\nabla V = \frac{1}{h_1}\frac{\partial V}{\partial u_1}\vec{a}_1 + \frac{1}{h_2}\frac{\partial V}{\partial u_2}\vec{a}_2 + \frac{1}{h_3}\frac{\partial V}{\partial u_3}\vec{a}_3. \tag{2.30}$$

We can interpret the generalised expression (2.30) in the curvilinear system to write the corresponding expressions in rectangular, cylindrical and spherical-polar coordinate systems with the help of Table 2.2 as follows.

$$\nabla V = \frac{\partial V}{\partial x}\vec{a}_x + \frac{\partial V}{\partial y}\vec{a}_y + \frac{\partial V}{\partial z}\vec{a}_z \quad \text{(Rectangular)} \tag{2.31}$$

$$\nabla V = \frac{\partial V}{\partial r}\vec{a}_r + \frac{1}{r}\frac{\partial V}{\partial \theta}\vec{a}_\theta + \frac{\partial V}{\partial z}\vec{a}_z \quad \text{(Cylindrical)} \tag{2.32}$$

$$\nabla V = \frac{\partial V}{\partial r}\vec{a}_r + \frac{1}{r}\frac{\partial V}{\partial \theta}\vec{a}_\theta + \frac{1}{r\sin\theta}\frac{\partial V}{\partial \phi}\vec{a}_\phi. \quad \text{(Spherical-polar)} \tag{2.33}$$

2.6 Divergence of a Vector Quantity

The divergence of a vector quantity, such as electric field $\vec{E}$ at a point, is a scalar quantity. It is defined as the outward flux F_E of $\vec{E}$ through the surface of an elemental or differential volume enclosing the point divided by the volume element $\Delta\tau$, the volume element shrinking to a point in the limit of $\Delta\tau$ tending to zero. (Refer to Section 2.4 and equation (2.20) for closed surface integral of $\vec{E}$ over the surface of an enclosure.) Mathematically, this definition of divergence of $\vec{E}$, abbreviated as div $\vec{E}$ and represented by the symbol $\vec{\nabla}\cdot\vec{E}$, can be put as

$$\text{div}\vec{E} = \nabla\cdot\vec{E} = \lim_{\Delta\tau\to 0}\frac{\oint_S \vec{E}\cdot\vec{a}_n\,dS}{\Delta\tau}. \tag{2.34}$$

Let us now proceed to find an expression for $\nabla\cdot\vec{E}$ in curvilinear coordinate system with the help of (2.34). For this purpose, let us take a differential volume $\Delta\tau$ of a parallelepiped in generalised curvilinear coordinate system (Fig. 2.3). We can identify eight corner points (Table 2.4) as well as three pairs of opposite faces (Table A2.1) of the differential volume $\Delta\tau$ in the generalised curvilinear system (Table 2.5).

In order to find the numerator of the right-hand of (2.34), let us find the contributions of fluxes of $\vec{E}$ through all the faces of the differential volume and add them together (Fig. 2.3). We already know how to find the flux of a vector quantity through an infinitesimal area, and so, we can now find the contributions to the flux of $\vec{E}$ through all the faces of the infinitesimal volume $\Delta\tau$ (Section 2.4). Note that the unit vector $\vec{a}_1$ is directed inwards towards the differential volume $\Delta\tau$; it is in the direction of increasing u_1. Therefore, we can write the 'inward' flux F_E of $\vec{E} = (E_1\vec{a}_1 + E_2\vec{a}_2 + E_3\vec{a}_3)$ through the front face ABFC (u_1 = constant), in the direction of $\vec{a}_1$', as (Fig. 2.3):

$$F_E = (E_1\vec{a}_1 + E_2\vec{a}_2 + E_3\vec{a}_3).(dS_1)\vec{a}_1 \tag{2.35}$$

('Inward' flux through the front face ABFC in the direction of $\vec{a}_1$)

where dS_1 is the area of the front face ABFC. (Recall how the unit vectors $\vec{a}_1$, $\vec{a}_2$ and $\vec{a}_3$ in the generalised curvilinear system have been compared with those of the rectangular cylindrical and spherical systems; see Table 2.2). We have considered here the flux through the infinitesimal element of area dS_1 over which $\vec{E} = (E_1\vec{a}_1 + E_2\vec{a}_2 + E_3\vec{a}_3)$ can be regarded as approximately constant.

Substituting $dS_1 = (\text{AB})(\text{AC}) = (h_2du_2)(h_3du_3)$ in (2.35) (see Table 2.2), we can express it as

$$F_E = (E_1\vec{a}_1 + E_2\vec{a}_2 + E_3\vec{a}_3).(h_2du_2h_3du_3)\vec{a}_1, \tag{2.36}$$

Table 2.4 *Corner points of a differential volume of parallelepiped in generalised curvilinear coordinate system (Fig. 2.3)*

Corner point	Curvilinear system
A	(u_1, u_2, u_3)
B	$(u_1, u_2 + du_2, u_3)$
C	$(u_1, u_2, u_3 + du_3)$
D	$(u_1 + du_1, u_2, u_3)$
E	$(u_1 + du_1, u_2 + du_2, u_3)$
F	$(u_1, u_2 + du_2, u_3 + du_3)$
G	$(u_1 + du_1, u_2, u_3 + du_3)$
H	$(u_1 + du_1, u_2 + du_2, u_3 + du_3)$

Table 2.5 *Pairs of opposite faces of a differential volume of parallelepiped in generalised curvilinear coordinate system*

Pairs of opposite faces	Curvilinear system
ABFC	$u_1 = \text{constant}$
DEHG	$u_1 + du_1 = \text{constant}$
ADGC	$u_2 = \text{constant}$
BEHF	$u_2 + du_2 = \text{constant}$
ADEB	$u_3 = \text{constant}$
CGHF	$u_3 + du_3 = \text{constant}$

which on taking the dot product simplifies to

$$F_E = E_1h_2du_2h_3du_3. \tag{2.37}$$

('Inward' flux through the front face ABFC in the direction of $\vec{a}_1$)

We can see that the 'outward' flux $\vec{E}$ through the back face DEHG ($u_1 + du_1 = \text{constant}$) is 'in the direction of $\vec{a}_1$' (Fig. 2.3), and that it is different by an infinitesimal extent $(\partial F_E/\partial u_1)du_1$ from F_E, the 'inward' flux through the front face ABFC, which is also 'in the direction of $\vec{a}_1$'. Therefore, we can write the outward flux 'in the direction of $\vec{a}_1$'through the back face DEHG of the differential volume as

$$= F_E + \frac{\partial F_E}{\partial u_1}du_1. \tag{2.38}$$

('Outward' flux through the back face DEHG in the direction of $\vec{a}_1$)

However, we have yet to find the 'outward' flux through the front face ABFC

(u_1 = constant), which is 'in the direction of $-\vec{a}_1$' (and not in the direction of $\vec{a}_1$). Following the steps similar to those used to obtain (2.36), we get the outward flux as

$$= (E_1\vec{a}_1 + E_2\vec{a}_2 + E_3\vec{a}_3).[-(h_2du_2h_3du_3)\vec{a}_1] = -E_1h_2du_2h_3du_3 = -F_E. \tag{2.39}$$

('Outward' flux through the front face ABFC in the direction of $-\vec{a}_1$)

Therefore, by adding (2.39) and (2.38), we obtain the 'outward' flux through the front and the back faces of the differential elemental volume $\Delta\tau$ as

$$= (-F_E) + (F_E + \frac{\partial F_E}{\partial u_1}du_1) = \frac{\partial F_E}{\partial u_1}du_1. \tag{2.40}$$

Substituting F_E from (2.37) in (2.40), we can then write

Contributions of the front and the back faces (each normal to $\vec{a}_1$) to the outward flux

$$= (du_1du_2du_3)\frac{\partial}{\partial u_1}(E_1h_2h_3). \tag{2.41}$$

We can now easily permute (2.41) to write the contributions from the other faces of the elemental volume to the outward flux as:

Contributions of the two side faces (each normal to $\vec{a}_2$) to the outward flux

$$= (du_1du_2du_3)\frac{\partial}{\partial u_2}(E_2h_3h_1) \tag{2.42}$$

and contributions of the top and the bottom faces (each normal to $\vec{a}_3$) to the outward flux

$$= (du_1du_2du_3)\frac{\partial}{\partial u_3}(E_3h_1h_2). \tag{2.43}$$

Now, adding (2.41)-(2.43), we obtain

Outward flux through all the faces of the differential volume

$$= (du_1du_2du_3)\left[\frac{\partial}{\partial u_1}(E_1h_2h_3) + \frac{\partial}{\partial u_2}(E_2h_3h_1) + \frac{\partial}{\partial u_3}(E_3h_1h_2)\right]. \tag{2.44}$$

We can obtain the expression for div$\vec{E} = \nabla \cdot \vec{E}$ from (2.34) by plugging (2.44) and (2.12) in the numerator and the denominator of (2.34) respectively, as follows:

$$\nabla \cdot \vec{E} = \lim_{(du_1,du_2,du_3)\to 0} \frac{(du_1du_2du_3)\left[\frac{\partial}{\partial u_1}(E_1h_2h_3) + \frac{\partial}{\partial u_2}(E_2h_3h_1) + \frac{\partial}{\partial u_3}(E_3h_1h_2)\right]}{h_1du_1h_2du_2h_3du_3}. \tag{2.45}$$

Taking the limit and after a little rearrangement of terms, (2.45) simplifies to

$$\nabla \cdot \vec{E} = \frac{1}{h_1 h_2 h_3}\left[\frac{\partial(h_2 h_3 E_1)}{\partial u_1} + \frac{\partial(h_3 h_1 E_2)}{\partial u_2} + \frac{\partial(h_1 h_2 E_3)}{\partial u_3}\right]. \tag{2.46}$$

We can interpret the generalised divergence expression (2.46) in curvilinear system as the corresponding expressions for the divergence in the rectangular, cylindrical, and spherical-polar coordinate systems, with the help of Table 2.2 as follows:

$$\nabla \cdot \vec{E} = \frac{\partial E_x}{\partial x} + \frac{\partial E_y}{\partial y} + \frac{\partial E_z}{\partial z}. \quad \text{(Rectangular)} \tag{2.47}$$

$$\nabla \cdot \vec{E} = \frac{1}{r}\left[\frac{\partial(rE_r)}{\partial r} + \frac{\partial E_\theta}{\partial \theta} + \frac{\partial(rE_z)}{\partial z}\right]$$

$$= \frac{1}{r}\frac{\partial(rE_r)}{\partial r} + \frac{1}{r}\frac{\partial E_\theta}{\partial \theta} + \frac{\partial E_z}{\partial z}. \quad \text{(Cylindrical)} \tag{2.48}$$

$$\nabla \cdot \vec{E} = \frac{1}{r^2 \sin\theta}\left[\frac{\partial(r^2 \sin\theta\, E_r)}{\partial r} + \frac{\partial(r \sin\theta\, E_\theta)}{\partial \theta} + \frac{\partial(rE_\phi)}{\partial \phi}\right]$$

$$= \frac{1}{r^2}\frac{\partial(r^2 E_r)}{\partial r} + \frac{1}{r\sin\theta}\frac{\partial(\sin\theta\, E_\theta)}{\partial \theta} + \frac{1}{r\sin\theta}\frac{\partial(E_\phi)}{\partial \phi}. \quad \text{(Spherical-polar)} \tag{2.49}$$

2.7 Curl of a Vector Quantity

The curl of a vector quantity, such as electric field $\vec{E}$ at a point, is essentially a vector quantity. It is represented as curl $\vec{E}$ or $\nabla \times \vec{E}$. It may be written, in terms of its three components, $(\nabla \times \vec{E})_1$, $(\nabla \times \vec{E})_2$ and $(\nabla \times \vec{E})_3$, as

$$\nabla \times \vec{E} = (\nabla \times \vec{E})_1 \vec{a}_1 + (\nabla \times \vec{E})_2 \vec{a}_2 + (\nabla \times \vec{E})_3 \vec{a}_3, \tag{2.50}$$

where the directions of $(\nabla \times \vec{E})_1$, $(\nabla \times \vec{E})_2$ and $(\nabla \times \vec{E})_3$ are the same as those of $\vec{a}_1$, $\vec{a}_2$ and $\vec{a}_3$ respectively.

Mathematically, then, $(\nabla \times \vec{E})_3$, the component of $\nabla \times \vec{E}$ taken typically in the direction of $\vec{a}_3$ (Fig. 2.6) may be put as

$$(\nabla \times \vec{E})_3 = \lim_{\Delta S_3 \to 0} \frac{\oint_l \vec{E} \cdot d\vec{l}}{\Delta S_3} \tag{2.51}$$

where ΔS_3 is the area of the differential area element that is normal to $\vec{a}_3$ and enclosing the point where the component of the curl is to be found. The closed line integral occurring in the numerator of (2.51) may be found by the method discussed earlier (Section 2.4). Here, we have to take care to see that the sense of the path of the line integral is such that if a screw were to be rotated in this sense, it would have a translational motion in the direction of the component of $\nabla \times \vec{E}$ to be found. We have to then divide the closed line integral thus found by the area element and take the limit of the area element as tending to zero, whereby the area element shrinks to the point where the component of $\nabla \times \vec{E}$ is being sought.

Hence, in order to evaluate the numerator of (2.51), let us take the closed line integral around the periphery of the area element ADEB (u_3 = constant) perpendicular to $\vec{a}_3$—from A to D, D to E, E to B, and finally from B to A—such that a screw, if rotated in the sense of this path, would have a translational motion along the direction of $\vec{a}_3$ (Fig. 2.6). Thus, we can use (2.51) to write:

$$(\nabla \times \vec{E})_3 = \lim_{\Delta S_3 \to 0} \frac{dI_{E(\mathrm{A}\to\mathrm{D})} + dI_{E(\mathrm{D}\to\mathrm{E})} + dI_{E(\mathrm{E}\to\mathrm{B})} + dI_{E(\mathrm{B}\to\mathrm{A})}}{\Delta S_3} \tag{2.52}$$

where ΔS_3 is the area of the differential area element ADEB perpendicular to $\vec{a}_3$. The area element occurring in the denominator of (2.51) is given by

$$\Delta S_3 = h_1 du_1 h_2 du_2 \tag{2.53}$$

which is the product of the infinitesimal lengths of the sides AD $(= h_1 du_1)$ and AB $(= h_2 du_2)$ of the area element ADEB (Table 2.4; Figs. 2.3 and 2.6).[3]

Over the infinitesimal lengths of the element of area, $\vec{E}$ can be regarded as constant. Hence, in order to find the closed line integral, we proceed as follows. Since along the length AD (from A to D) the direction is the same as that of $\vec{a}_1$ (Fig. 2.4), we can write

$dI_{E(\mathrm{A}\to\mathrm{D})}$ =Line integral of $\vec{E}$ from A to D along AD$(= h_1 du_1)$ $(u_2 = \text{constant})$ is

$$\vec{E} \cdot (h_1 du_1 \vec{a}_1) = (E_1\vec{a}_1 + E_2\vec{a}_2 + E_3\vec{a}_3).(h_1 du_1 \vec{a}_1) = E_1 h_1 du_1 = L, \text{ say.} \tag{2.54}$$

Further, from (2.54), we can write, in view of the variation of L with u_2:

$$\text{Line integral of } \vec{E} \text{ from B to E along BE } (u_2 + du_2 = \text{constant}) = L + \frac{\partial L}{\partial u_2} du_2\,. \tag{2.55}$$

Since the direction of EB is that of $-\vec{a}_1$, being opposite to that of BE in the direction of $\vec{a}_1$, we can write from (2.55)

[3]The length of the side AD of the area element is $h_1 du_1$ and that of the side AB of the area element is $h_2 du_2$ (Table 2.4; Fig. 2.6). The length of the side BE, opposite to AD $(= h_1 du_1)$, is also taken as $h_1 du_1$. Similarly, the length of the side DE, opposite to AB $= h_2 du_2$, is also taken approximately as $h_2 du_2$. (We have to take h_1, h_2, u_1, and u_2 for the coordinate system concerned from Table 2.5. For instance, $h_1 = 1$, $h_2 = r$, $u_1 = r$, and $u_2 = \theta$ in the cylindrical system. In cylindrical system, the lengths of the two opposite sides AD$= h_1 du_1$ and BE are to be taken as the same, being equal to dr. Similarly, the length of the side AB is equal to $rd\theta$ and that of the side DE is $(r + dr)d\theta \approx rd\theta$, so that again the lengths of the two opposite sides AB and DE are to be taken as the same).

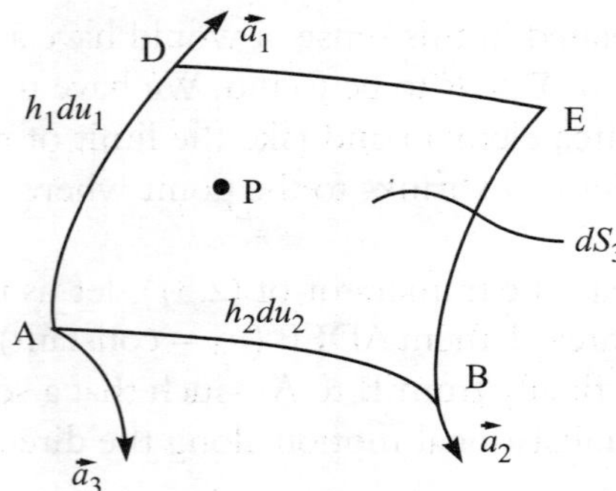

Fig. 2.6 *Area element dS_3, with its corners at A, D, E, and B in the generalised curvilinear coordinate system, enclosing the point where the component of the curl of a vector quantity is to be found. Here, the component is taken typically in the direction of $\vec{a}_3$, which is normal to the area element.*

$dI_{E(E\to B)}$ =Line integral of $\vec{E}$ from E to B along EB ($u_2 + du_2 =$ constant)

$$= -(L + \frac{\partial L}{\partial u_2} du_2). \tag{2.56}$$

Line integral of $\vec{E}$ from A to B along AB= h_2du_2 (u_1 =constant) is $\vec{E} \cdot (h_2du_2\vec{a}_2)$

$$= (E_1\vec{a}_1 + E_2\vec{a}_2 + E_3\vec{a}_3).(h_2du_2\vec{a}_2) = E_2h_2du_2 = L', \text{ say.} \tag{2.57}$$

Since the direction of BA is that of $-\vec{a}_2$, being opposite to that of AB in the direction of $\vec{a}_2$, we can write from (2.57):

$$dI_{E(B\to A)} = \text{Line integral of } \vec{E} \text{ from B to A along BA} = -L'. \tag{2.58}$$

Further, from (2.57), we can write in view of the variation of L' with u_1:

$dI_{E(D\to E)}$ =Line integral of $\vec{E}$ from D to E along DE ($u_1 + du_1$ =constant)

$$= L' + \frac{\partial L'}{\partial u_1} du_1. \tag{2.59}$$

Therefore, we can complete the evaluation of the closed line integral by adding the line integrals $dI_{E(A\to D)}$, $dI_{E(D\to E)}$, $dI_{E(E\to B)}$ and $dI_{E(B\to A)}$, that is by adding (2.54),(2.59), (2.56) and (2.58) to obtain

$$dI_{E(A\to D)} + dI_{E(D\to E)} + dI_{E(E\to B)} + dI_{E(B\to A)}$$

$$= L + L' + \frac{\partial L'}{\partial u_1} du_1 - \left(L + \frac{\partial L}{\partial u_2} du_2\right) - L' = \frac{\partial L'}{\partial u_1} du_1 - \frac{\partial L}{\partial u_2} du_2,$$

which, on substituting L and L' from (2.54) and (2.57) respectively becomes

$$dI_{E(A\to D)} + dI_{E(D\to E)} + dI_{E(E\to B)} + dI_{E(B\to A)} = \frac{\partial(h_2E_2)}{\partial u_1}du_1du_2 - \frac{\partial(h_1E_1)}{\partial u_2}du_1du_2. \tag{2.60}$$

We can now substitute (2.53) and (2.60) into (2.52) to obtain the following expression for $(\nabla \times \vec{E})_3$, that is, the component of $\nabla \times \vec{E}$ in the direction of $\vec{a}_3$:

$$(\nabla \times \vec{E})_3 = \frac{1}{h_1h_2}\left(\frac{\partial(h_2E_2)}{\partial u_1} - \frac{\partial(h_1E_1)}{\partial u_2}\right). \tag{2.61}$$

Following the same procedure as followed in deducing (2.61), we can obtain the components $(\nabla \times \vec{E})_1$ and $(\nabla \times E)_2$ of $(\nabla \times \vec{E})$ in the directions of $\vec{a}_1$ and $\vec{a}_2$ as well, taking the area elements perpendicular to these directions respectively. Thus, we obtain the following expressions permuting (2.61):

$$(\nabla \times \vec{E})_1 = \frac{1}{h_2h_3}\left(\frac{\partial(h_3E_3)}{\partial u_2} - \frac{\partial(h_2E_2)}{\partial u_3}\right) \tag{2.62}$$

and

$$(\nabla \times \vec{E})_2 = \frac{1}{h_3h_1}\left(\frac{\partial(h_1E_1)}{\partial u_3} - \frac{\partial(h_3E_3)}{\partial u_1}\right). \tag{2.63}$$

Therefore, putting (2.61)-(2.63) in (2.50), we can write the following expression for $\nabla \times \vec{E}$:

$$\nabla \times \vec{E} = \frac{1}{h_2h_3}\left(\frac{\partial(h_3E_3)}{\partial u_2} - \frac{\partial(h_2E_2)}{\partial u_3}\right)\vec{a}_1 + \frac{1}{h_3h_1}\left(\frac{\partial(h_1E_1)}{\partial u_3} - \frac{\partial(h_3E_3)}{\partial u_1}\right)\vec{a}_2 + \frac{1}{h_1h_2}\left(\frac{\partial(h_2E_2)}{\partial u_1} - \frac{\partial(h_1E_1)}{\partial u_2}\right)\vec{a}_3, \tag{2.64}$$

which may alternatively be expressed as the following determinant:

$$\nabla \times \vec{E} = \frac{1}{h_1h_2h_3}\begin{vmatrix} h_1\vec{a}_1 & h_2\vec{a}_2 & h_3\vec{a}_3 \\ \dfrac{\partial}{\partial u_1} & \dfrac{\partial}{\partial u_2} & \dfrac{\partial}{\partial u_3} \\ h_1E_1 & h_2E_2 & h_3E_3 \end{vmatrix}. \tag{2.65}$$

We can easily interpret the expression (2.65) for curl in generalised curvilinear coordinate system for the corresponding expressions in rectangular, cylindrical, and spherical-polar coordinate systems with the help of Table 2.2 as follows:

$$\nabla \times \vec{E} = \left(\frac{\partial E_z}{\partial y} - \frac{\partial E_y}{\partial z}\right)\vec{a}_x + \left(\frac{\partial E_x}{\partial z} - \frac{\partial E_z}{\partial x}\right)\vec{a}_y + \left(\frac{\partial E_y}{\partial x} - \frac{\partial E_x}{\partial y}\right)\vec{a}_z. \quad \text{(Rectangular)} \tag{2.66}$$

$$\nabla \times \vec{E} = \frac{1}{r}\left(\frac{\partial E_z}{\partial \theta} - \frac{\partial (rE_\theta)}{\partial z}\right)\vec{a}_r + \left(\frac{\partial E_r}{\partial z} - \frac{\partial E_z}{\partial r}\right)\vec{a}_\theta + \frac{1}{r}\left(\frac{\partial (rE_\theta)}{\partial r} - \frac{\partial E_r}{\partial \theta}\right)\vec{a}_z$$

$$= \left(\frac{1}{r}\frac{\partial E_z}{\partial \theta} - \frac{\partial E_\theta}{\partial z}\right)\vec{a}_r + \left(\frac{\partial E_r}{\partial z} - \frac{\partial E_z}{\partial r}\right)\vec{a}_\theta + \frac{1}{r}\left(\frac{\partial (rE_\theta)}{\partial r} - \frac{\partial E_r}{\partial \theta}\right)\vec{a}_z. \quad \text{(Cylindrical)} \tag{2.67}$$

$$\nabla \times \vec{E} = \frac{1}{r\sin\theta}\left(\frac{\partial(\sin\theta\, E_\phi)}{\partial \theta} - \frac{\partial E_\theta}{\partial \phi}\right)\vec{a}_r + \frac{1}{r}\left(\frac{1}{\sin\theta}\frac{\partial E_r}{\partial \phi} - \frac{\partial (rE_\phi)}{\partial r}\right)\vec{a}_\theta$$

$$+\frac{1}{r}\left(\frac{\partial (rE_\theta)}{\partial r} - \frac{\partial E_r}{\partial \theta}\right)\vec{a}_\phi. \quad \text{(Spherical-polar)} \tag{2.68}$$

However, we can choose to continue retaining the determinant form of the expression for curl (2.65) with the help of Table 2.2, as follows in rectangular, cylindrical, and spherical-polar coordinate systems:

$$\nabla \times \vec{E} = \begin{vmatrix} \vec{a}_x & \vec{a}_y & \vec{a}_z \\ \dfrac{\partial}{\partial x} & \dfrac{\partial}{\partial y} & \dfrac{\partial}{\partial z} \\ E_x & E_y & E_z \end{vmatrix}. \quad \text{(Rectangular)} \tag{2.69}$$

$$\nabla \times \vec{E} = \frac{1}{r}\begin{vmatrix} \vec{a}_r & r\vec{a}_\theta & \vec{a}_z \\ \dfrac{\partial}{\partial r} & \dfrac{\partial}{\partial \theta} & \dfrac{\partial}{\partial z} \\ E_r & rE_\theta & E_z \end{vmatrix}. \quad \text{(Cylindrical)} \tag{2.70}$$

$$\nabla \times \vec{E} = \frac{1}{r^2\sin\theta}\begin{vmatrix} \vec{a}_r & r\vec{a}_\theta & r\sin\theta\vec{a}_\phi \\ \dfrac{\partial}{\partial r} & \dfrac{\partial}{\partial \theta} & \dfrac{\partial}{\partial \phi} \\ E_r & rE_\theta & r\sin\theta\, E_\phi \end{vmatrix}. \quad \text{(Spherical-polar)} \tag{2.71}$$

2.8 Laplacian of Scalar and Vector Quantities

The gradient of a scalar quantity is a vector quantity (Section 2.5), and the divergence of a vector quantity is a scalar quantity (Section 2.6). Therefore, if we take the divergence of the gradient of a scalar quantity, say V, we get a scalar quantity called the Laplacian of V, for which we use the symbol $\nabla^2 V$. Thus, we may write

$$\nabla \cdot \nabla V = \nabla^2 V. \tag{2.72}$$

The expression for the gradient of a scalar quantity, say V (grad$V = \nabla V$) and that for the divergence of a vector, say $\vec{E}$ (div$\vec{E} = \nabla.\vec{E}$) in the curvilinear coordinate system are given by (2.30) and (2.46) respectively.

Note: Note that, in the discussion preceding (2.72), we have taken $\vec{E}$ to represent any vector quantity and V any scalar quantity. They are not to be specifically taken as the electric field and the electric potential quantities, respectively. We have shown later in Chapter 4 that $\vec{E}$ is related V as $E = -\nabla V$.

Let us find an expression for the Laplacian in the generalised curvilinear coordinate system and subsequently read it for rectangular, cylindrical, and spherical-polar systems. For this purpose, first recall expression (2.46) for $\nabla \cdot \vec{E}$ in terms of the components E_1, E_2, and E_3 of the vector $\vec{E}$. The expression (2.46) is valid for any vector and hence, for the vector ∇V as well. Therefore, we can write the expression for $\nabla \cdot \nabla V$ by replacing in (2.46) the vector $\vec{E}$ by the vector ∇V and also the vector components here, E_1, E_2 and E_3 of $\vec{E}$ by the components $(\nabla V)_1$, $(\nabla V)_2$ and $(\nabla V)_3$ of the vector ∇V respectively as follows:

$$\nabla \cdot \nabla V = \frac{1}{h_1 h_2 h_3}\left(\frac{\partial(h_2 h_3 (\nabla V)_1)}{\partial u_1} + \frac{\partial(h_3 h_1 (\nabla V)_2)}{\partial u_2} + \frac{\partial(h_1 h_2 (\nabla V)_3)}{\partial u_3}\right). \tag{2.73}$$

Substituting (2.29) in (2.73) and using (2.72), we obtain the following expression for the Laplacian of V in the curvilinear coordinate system:

$$\nabla^2 V = \nabla \cdot \nabla V = \frac{1}{h_1 h_2 h_3}\left[\frac{\partial}{\partial u_1}\left(\frac{h_2 h_3}{h_1}\frac{\partial V}{\partial u_1}\right) + \frac{\partial}{\partial u_2}\left(\frac{h_3 h_1}{h_2}\frac{\partial V}{\partial u_2}\right) + \frac{\partial}{\partial u_3}\left(\frac{h_1 h_2}{h_3}\frac{\partial V}{\partial u_3}\right)\right]. \tag{2.74}$$

The corresponding expressions for the rectangular, cylindrical and spherical-polar coordinate systems can be interpreted from (2.74), with the help of Table 2.2 as:

$$\nabla^2 V = \nabla \cdot \nabla V = \frac{\partial}{\partial x}\left(\frac{\partial V}{\partial x}\right) + \frac{\partial}{\partial y}\left(\frac{\partial V}{\partial y}\right) + \frac{\partial}{\partial z}\left(\frac{\partial V}{\partial z}\right)$$

$$= \frac{\partial^2 V}{\partial x^2} + \frac{\partial^2 V}{\partial y^2} + \frac{\partial^2 V}{\partial z^2}. \quad \text{(Rectangular)} \tag{2.75}$$

$$\nabla^2 V = \nabla \cdot \nabla V = \frac{1}{r}\left[\frac{\partial}{\partial r}\left(r\frac{\partial V}{\partial r}\right) + \frac{\partial}{\partial \theta}\left(\frac{1}{r}\frac{\partial V}{\partial \theta}\right) + \frac{\partial}{\partial z}\left(r\frac{\partial V}{\partial z}\right)\right]$$

$$= \frac{1}{r}\frac{\partial}{\partial r}\left(r\frac{\partial V}{\partial r}\right) + \frac{1}{r^2}\frac{\partial^2 V}{\partial \theta^2} + \frac{\partial^2 V}{\partial z^2}. \quad \text{(Cylindrical)} \tag{2.76}$$

$$\nabla^2 V = \nabla \cdot \nabla V = \frac{1}{r^2 \sin\theta}\left[\frac{\partial}{\partial r}\left(r^2 \sin\theta \frac{\partial V}{\partial r}\right) + \frac{\partial}{\partial \theta}\left(\sin\theta \frac{\partial V}{\partial \theta}\right) + \frac{\partial}{\partial \phi}\left(\frac{1}{\sin\theta}\frac{\partial V}{\partial \phi}\right)\right]$$

$$= \frac{1}{r^2}\frac{\partial}{\partial r}\left(r^2 \frac{\partial V}{\partial r}\right) + \frac{1}{r^2 \sin\theta}\frac{\partial}{\partial \theta}\left(\sin\theta \frac{\partial V}{\partial \theta}\right) + \frac{1}{r^2 \sin^2\theta}\frac{\partial^2 V}{\partial \phi^2} \quad . \qquad \text{(Spherical-polar)} \qquad (2.77)$$

The Laplacian refers to a scalar quantity such as the electric potential V. However, one can also define the Laplacian with reference to a component of a vector quantity such as the electric field $\vec{E}$ as $\nabla^2 E_1$, $\nabla^2 E_2$, or $\nabla^2 E_3$. Consequently, then we can also define the Laplacian of the vector, here $\vec{E}$, as the vector sum of the Laplacian of the vector components as:

$$\nabla^2 \vec{E} = (\nabla^2 E_1)\vec{a}_1 + (\nabla^2 E_2)\vec{a}_2 + (\nabla^2 E_3)\vec{a}_3, \qquad (2.78)$$

where $(\nabla^2 \vec{E})_1$, $(\nabla^2 \vec{E})_2$ and $(\nabla^2 \vec{E})_3$ are obtainable simply by replacing V by E_1, E_2 and E_3 respectively in (2.74)–(2.77) for different coordinate systems.

Appendix A2.1

Relation of cylindrical and spherical coordinates with rectangular coordinates

With reference to rectangular, cylindrical and spherical coordinate systems (Fig. 2.1), the three perpendicular distances are

QR $= x$

QS $= y$

PQ $= z$ (Figs. 2.1*a*–*c*),

where Q is the foot of the perpendicular from the point P on the XY-plane; R is the foot of the perpendicular from the point Q to the Y-axis; and S is the foot of the perpendicular from the point Q to the X-axis. Further, if K is the foot of perpendicular from the point P to the Z-axis, then

KO (= PQ)= z

RQ = OS = x

SQ = OR = y

PK = QO = $[(\text{OS})^2 + (\text{SQ})^2]^{1/2} = (x^2 + y^2)^{1/2}$.

In addition, for cylindrical coordinate system, from the simple geometry (Fig. 2.1*b*), the following relations hold good:

$$r = \text{KP} = \text{OQ};$$

and

$$\left.\begin{aligned} x &= \text{OS} = \text{RQ} = (\text{OQ}\cos\theta = r\cos\theta \\ y &= \text{OR} = \text{SQ} = (\text{OQ}\sin\theta = r\sin\theta \\ z &= z(= \text{PQ}) \end{aligned}\right\}. \text{(Cylindrical vis-à-vis rectangular)}$$

Similarly, from the geometry of spherical coordinate system(Fig. 2.1*c*), the following relations are obtained:

$$\text{OP} = r, \quad \text{KP} = \text{OQ} = \text{OP}\sin\theta = r\sin\theta;$$

and

$$\left.\begin{aligned} x &= \text{OS} = \text{RQ} = \text{OQ}\cos\phi = r\sin\theta\cos\phi \\ y &= \text{OR} = \text{SQ} = \text{OQ}\sin\phi = r\sin\theta\sin\phi \\ \text{z} &= \text{KO} = \text{PQ} = \text{OP}\cos\theta = r\cos\theta \end{aligned}\right\}.\text{(Cylindrical vis-à-vis rectangular)}$$

Appendix A2.2

Description of the volume element

The differential volume has twelve sides (Fig. 2.2). The three pairs of opposite faces may be identified as (i) ABFC and DEHG; (ii) ADGC and BEHF; and (iii) ADEB and CGHF (Table A2.1) (Fig. 2.2). Each of these faces is either a planar or a curved surface defined by a constant value of one of the coordinates (Table A2.1).

Table A2.1 *Opposite faces of the elemental volume in rectangular, cylindrical and spherical-polar coordinate systems*

Pairs of opposite faces	Rectangular	Cylindrical	Spherical-polar
ABFC	$x = \text{constant}$	$r = \text{constant}$	$r = \text{constant}$
DEHG	$x + dx = \text{constant}$	$r + dr = \text{constant}$	$r + dr = \text{constant}$
ADGC	$y = \text{constant}$	$\theta = \text{constant}$	$\theta = \text{constant}$
BEHF	$y + dy = \text{constant}$	$\theta + d\theta = \text{constant}$	$\theta + d\theta = \text{constant}$
ADEB	$z = \text{constant}$	$z = \text{constant}$	$\phi = \text{constant}$
CGHF	$z + dz = \text{constant}$	$z + dz = \text{constant}$	$\phi + d\phi = \text{constant}$

The pair of faces ABFC and DEHG in the rectangular coordinate system, are planar surfaces over which the values of the x coordinate remain constant at x and $x + dx$ respectively. Hence, they are also called x = constant and $x + dx$ = constant surfaces respectively. Similarly, we can define other pairs of surfaces of the differential volume and define them in rectangular, cylindrical, and spherical-polar coordinate systems (see Table A2.1). Each pair of opposite faces has in turn a pair of opposite sides (see Table A2.2).

Table A2.2 *Opposite sides of three pairs of opposite faces of the differential volume labelled as (i), (ii) and (iii)*

Opposite faces											
(i) ABFC		(i) DEHG		(ii) ADGC		(ii) BEHF		(iii) ADEB		(iii) CGHF	
AB	CF	DE	GH	AD	CG	BE	FH	AD	BE	CG	FH
AC	BF	DG	EH	AC	DG	BF	EH	AB	DE	CF	GH

Further, each side of these faces is common with two adjacent faces. While tabulating the opposite sides of these faces, each side is mentioned twice as it is common with two adjacent faces (Table A2.2). However, the opposite sides of the differential volume, irrespective of the coordinate system (rectangular, cylindrical, or spherical-polar) considered, do not appreciably differ in lengths, if they differ at all, from the 'basic infinitesimal length elements' AD, AB, and AC as mentioned before, which respectively lead the three groups of sides tabulated (see Table A2.3). Although the opposite sides of a face of the differential volume have the same infinitesimal length in rectangular system, they may or may not be exactly the same in cylindrical and spherical-polar coordinate systems (see Table A2.3).

Table A2.3 *Infinitesimal lengths of the opposite sides of the differential volume, in three groups parallel to three mutually perpendicular directions—rectangular, cylindrical and spherical-polar coordinate systems*

Group	Opposite sides	Rectangular	Cylindrical	Spherical-polar
1	AD	dx	dr	dr
	CG	dx	dr	dr
	BE	dx	dr	dr
	FH	dx	dr	dr
2	AB	dy	$rd\theta$	$rd\theta$
	CF	dy	$rd\theta$	$rd\theta$
	DE	dy	$(r + dr)d\theta$	$(r + dr)d\theta$
	GH	dy	$(r + dr)d\theta$	$(r + dr)d\theta$
3	AC	dz	dz	$r \sin\theta \, d\phi$
	BF	dz	dz	$r \sin\theta \, d\phi$
	DG	dz	dz	$(r + dr) \sin\theta \, d\phi$
	EH	dz	dz	$(r + dr) \sin\theta \, d\phi$

Thus, in rectangular coordinate system, the differential volume is a rectangular parallelepiped. However, in cylindrical system and spherical-polar coordinate systems, unlike in rectangular system, the elemental volume formed by making incremental changes in the coordinates of the point are not any more a rectangular parallelepiped (Fig. 2.2). The sides of any of the faces of the elemental volume in this case may or may not be straight lines. The opposite faces of the element may or may not have the same area, and the opposite sides of each face of the element may or may not have the same lengths (Fig. 2.2).

Examples

In this chapter, we have derived vector calculus expressions. Although examples have not been provided separately in the text of this chapter, as has been done in the chapters to follow, these expressions have been amply exemplified by their application in the rest of the book.

Summarising Notes

- ✓ Electromagnetics, a subject that helps in developing the understanding of many concepts in electrical, electronics and communication engineering, is based on the fundamental principles of electricity and magnetism.
- ✓ Vector calculus expressions can concisely describe the concepts of the subject. You can hit the right spot by very easily deriving, in the curvilinear system of coordinates, the expressions for the gradient of a scalar quantity, divergence of a vector quantity, curl of a vector quantity, and Laplacian of a scalar or a vector quantity.
- ✓ Vector symmetry of the terms of the vector calculus expressions must have fascinated you in that from one of the terms of an expression you can permute to write the remaining terms of the expression.
- ✓ Vector calculus expressions in the curvilinear system of coordinates can easily be translated to write the corresponding expressions in the rectangular, cylindrical and spherical systems of coordinates.
- ✓ Vector calculus expressions immensely help us develop concisely the concepts of electromagnetics.

Review Questions

The review questions in this chapter are meant to ensure that you have learnt the basic concepts of vector calculus expressions that are going to be extensively applied in the rest of the book. You should satisfy yourself by answering the following:

2.1 Derive from first principles the expressions for gradient, divergence and curl in generalised curvilinear coordinate system.

2.2 There is harmony and symmetry in the vector calculus expressions for gradient, divergence and curl when expressed in the generalised curvilinear coordinate system. Validate this by writing from memory one of the terms of the concerned expression, and writing down the remaining by permutation.

2.3 Reproduce the expressions for the gradient, divergence and curl in the generalised curvilinear coordinate system without consulting any external source.

2.4 State the physical interpretation of gradient, divergence and curl.

3

Basic Concepts of Static Electric Fields

Some of the basic concepts of static electric fields or electrostatics are recapitulated in this chapter with a view to using them in developing electromagnetic theory dealt with in the rest of this book. Two basic laws of electrostatics are Coulomb's law and Gauss's law (Section 3.1). Coulomb's law tells us about the force between electric charges (Section 3.1.1) as well as about the electric field associated with electric charges (Section 3.1.2). Gauss's law is applicable to the problems that enjoy geometrical symmetry—rectangular, cylindrical or spherical. We can derive Gauss's law from Coulomb's law. Alternatively, if we take Gauss's law as granted, then we can derive Coulomb's law from Gauss's law (Section 3.1.3).

Further, the concept of electric potential is developed in this chapter from the work involved in moving a charge in the presence of an electric field (Section 3.2). In addition, making use of the basic concepts of vector calculus covered in Chapter 2, two more concepts of electrostatics, expressed in the form of Poisson's and Laplace's equations, are also developed (Section 3.3).

3.1 Coulomb's Law and Gauss's Law of Electrostatics

3.1.1 Coulomb's law of electrostatics

We can express Coulomb's law of electrostatics for the force between two stationary point charges in an unbounded homogeneous insulating medium (that is, an insulating medium of uniform property extending to infinity in all directions) in the following vectorial form (Fig. 3.1):

$$\vec{F} = \frac{q_1 q_2}{4\pi\varepsilon r^2}\vec{a}_r \tag{3.1}$$

where $\vec{F}$ is the force in newton (N) on a point charge q_2 coulomb (C) due to a point charge q_1 coulomb. Also, r is the distance in metres (m) between the point charges, $\vec{a}_r$ the unit vector (of magnitude unity) along the line joining the two point charges directed from q_1 to q_2 (Fig. 3.1), and ε represents a physical property of the medium called permittivity. The electrostatic force between the two charges is known as the Coulomb force.

The ratio of the permittivity of a medium ε to that of free-space or empty medium ε_0 is called the relative permittivity ε_r $(= \varepsilon/\varepsilon_0)$ or the dielectric constant of the medium. The presence of the rationalisation factor 4π in the Coulomb force expression (3.1) excludes its appearance from many extensively used expressions derived from (3.1), for instance, from the expression stating Gauss's law that will be discussed later in this chapter. Equation (3.1) is expressed in the rationalised MKS (metre, kilogram, second) system. Thus, the non-appearance of the factor 4π in the expressions derived from the basic expression (3.1) makes life simple when calculating many of the electrical quantities we come across in practice.

The Coulomb force on the point charge q_2 due to the point charge q_1, given by (3.1), is essentially a repulsive force if $\vec{F}$ has the same direction as that of $\vec{a}_r$. Thus, $\vec{F}$ is repulsive if q_1 and q_2 are like charges, that is, if both charges are either each positive or each negative, making the product q_1q_2 occurring in (3.1) a positive quantity. Similarly, the force is attractive when $\vec{F}$ is oppositely directed to $\vec{a}_r$, that is, if q_1 and q_2 are unlike charges, making the product q_1q_2 occurring in (3.1) a negative quantity. In other words, this means that like charges repel and unlike charges attract, a concept that you are already familiar with from your school days.

Let the rectangular coördinates of the point A (Fig. 3.1) where the point charge q_1 is located, be x_1, y_1, and z_1 and those of the point B where the point charge q_2 is located be x_2, y_2, and z_2. Then we can express the unit vector $\vec{a}_r$ directed from A to B in terms of a distance vector $\vec{r}$ and its magnitude r (being the shortest distance between the points A and B) as:

$$\vec{a}_r = \frac{\vec{r}}{r} \tag{3.2}$$

where

$$r = [(x_2 - x_1)^2 + (y_2 - y_1)^2 + (z_2 - z_1)^2]^{1/2} \tag{3.3}$$

$$\vec{r} = (x_2 - x_1)\vec{a}_x + (y_2 - y_1)\vec{a}_y + (z_2 - z_1)\vec{a}_z. \tag{3.4}$$

The free-space medium has the lowest permittivity, $\varepsilon = \varepsilon_0$, of all media, where $\varepsilon_0 = 1/(36\pi) \times 10^{-9} = 8.854 \times 10^{-12}$ C^2/N-m^2. Consequently, of all media, the free-space medium provides the largest Coulomb force between two point charges, since the quantity ε occurs in the denominator of the force equation (3.1). You can easily see from (3.1) that the unit of permittivity ε is C^2/N-m^2, which we have already used above while writing the value of ε_0. However, the practical unit of ε is F/m (Farad per metre). This you can appreciate from the expression for the capacitance C (Farad or F) of a parallel-plate capacitor:

$$C = \frac{\varepsilon A}{d} \tag{3.5}$$

Fig. 3.1 *Two point charges q_1 and q_2 located at points A and B respectively, separated by a distance r. The unit vector $\vec{a}_r$ is shown directed from A to B.*

where ε is the permittivity of the medium filling the space between the two identical conducting parallel plates, each of area A (m^2), separated by a distance d (m), it being assumed that any of the dimensions of the plate is much larger than d. Many of you must already be familiar with the expression (3.5). It is easy to derive (3.5). However, let us keep the proof of (3.5) in abeyance for the time being until we take up Gauss's law later in this chapter. We will then rewrite this expression as (3.100) after deriving it from first principles.

At this stage, let us proceed to exemplify the 'vectorial' form (3.1) of Coulomb's law by taking a very simple problem. The problem is to find the force on a point charge q placed at the corner of a square situated on a plane due to three identical point charges, each q, placed at the remaining three corners of the square. We will take the rectangular coordinate system, and let one of the corners, say A $(0,0)$, be the origin of the coordinates. Let B $(a, 0)$, C (a, a), and D $(0, a)$ be the other three corners of the square (Fig. 3.2).

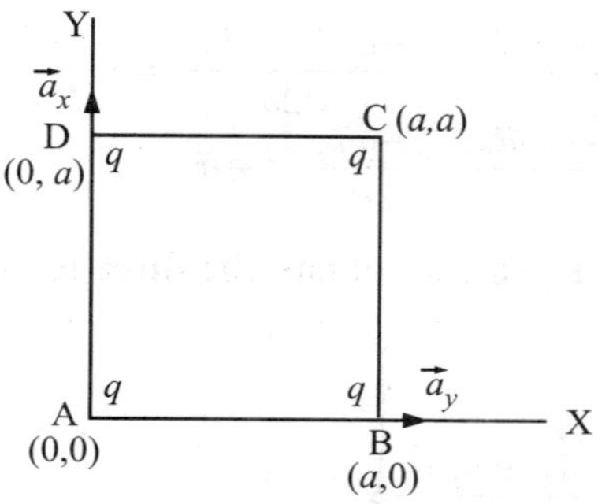

Fig. 3.2 *Identical point charges at the corners of a square.*

The charges are placed in free space ($\varepsilon = \varepsilon_0$). We can easily find the forces $\vec{F}_{\mathrm{BA}}$, $\vec{F}_{\mathrm{CA}}$, and

$\vec{F}_{\rm DA}$ on the point charge at A due to the point charges at B, C, and D respectively with the help of (3.1), (3.2) and the coordinates of the points provided as follows:

$$\left.\begin{aligned} \vec{F}_{\rm BA} &= \frac{(q)(q)}{4\pi\varepsilon_0\,({\rm BA})^2}\vec{a}_{\rm BA} \\ \vec{F}_{\rm CA} &= \frac{(q)(q)}{4\pi\varepsilon_0\,({\rm CA})^2}\vec{a}_{\rm CA} \\ \vec{F}_{\rm DA} &= \frac{(q)(q)}{4\pi\varepsilon_0\,({\rm DA})^2}\vec{a}_{\rm DA} \end{aligned}\right\}. \tag{3.6}$$

Here, BA, CA, and DA are the distances between B and A, C and A, and D and A respectively, while $\vec{F}_{\rm BA}$, $\vec{F}_{\rm CA}$ and $\vec{F}_{\rm DA}$ represent the forces directed from B to A, C to A, and D to A respectively. $\vec{a}_{\rm BA}$, $\vec{a}_{\rm CA}$, and $\vec{a}_{\rm DA}$ are the unit vectors directed from B to A, C to A, and D to A respectively. The x and the y coordinates of A, B, C, and D being A$(0,0)$, B$(a,0)$, C(a,a), and D$(0,a)$, we can easily obtain with the help of (3.3) the following distances:

$$\left.\begin{aligned} {\rm BA} &= [(0-a)^2+(0-0)^2]^{1/2} = a \\ {\rm CA} &= [(0-a)^2+(0-a)^2]^{1/2} = \sqrt{2}a \\ {\rm DA} &= [(0-0)^2+(0-a)^2]^{1/2} = a \end{aligned}\right\}. \tag{3.7}$$

You will be able to appreciate (3.7) directly from the geometry of the problem from Fig. 3.2. Further, note that while finding $\vec{F}_{\rm BA}$ through $\vec{a}_{\rm BA}$ with the help of (3.6), we have to interpret the coordinates of A as x_2, y_2, z_2 and those of B as x_1, y_1, z_1. We have to make similar interpretations while finding $\vec{F}_{\rm CA}$ through $\vec{a}_{\rm CA}$ and $\vec{F}_{\rm DA}$ through $\vec{a}_{\rm DA}$ [see (3.6)]. Thus, for this purpose, we can use (3.2)-(3.4) to obtain

$$\left.\begin{aligned} \vec{a}_{\rm BA} &= \frac{(0-a)\vec{a}_x+(0-0)\vec{a}_y}{\rm BA} = \frac{-a\vec{a}_x}{a} = -\vec{a}_x & \text{(a)} \\ \vec{a}_{\rm CA} &= \frac{(0-a)\vec{a}_x+(0-a)\vec{a}_y}{\rm CA} = \frac{-a(\vec{a}_x+\vec{a}_y)}{\sqrt{2}a} = \frac{-(\vec{a}_x+\vec{a}_y)}{\sqrt{2}} & \text{(b)} \\ \vec{a}_{\rm DA} &= \frac{(0-0)\vec{a}_x+(0-a)\vec{a}_y}{\rm DA} = \frac{-a\vec{a}_y}{a} = -\vec{a}_y & \text{(c)} \end{aligned}\right\}. \tag{3.8}$$

Substituting (3.2) and (3.8) in (3.6), we obtain the three force expressions as

$$\left.\begin{aligned} \vec{F}_{\rm BA} &= -\frac{(q)(q)}{4\pi\varepsilon_0\,(a)^2}\vec{a}_x \\ \vec{F}_{\rm CA} &= -\frac{(q)(q)}{4\pi\varepsilon_0\,(\sqrt{2}a)^2}\frac{(\vec{a}_x+\vec{a}_y)}{\sqrt{2}} \\ \vec{F}_{\rm DA} &= -\frac{(q)(q)}{4\pi\varepsilon_0\,(a)^2}\vec{a}_y \end{aligned}\right\},$$

which when added together yields

$$\vec{F}_{\text{BA}} + \vec{F}_{\text{CA}} + \vec{F}_{\text{DA}} = -\frac{q^2}{4\pi\varepsilon_0\, a^2}(1 + \frac{1}{(2\sqrt{2})})(\vec{a}_x + \vec{a}_y). \tag{3.9}$$

The sum total of the forces given by (3.9) is the force on the point charge q at A, one of the corner points of the square of side a, due to the point charges q at the remaining three corners of the square. Putting this sum total of forces on the point charge q at A as $\vec{F}_{\text{A}} = \vec{F}_{\text{BA}} + \vec{F}_{\text{CA}} + \vec{F}_{\text{DA}}$, we can express (3.9) as

$$\vec{F}_{\text{A}} = \vec{F}_{\text{BA}} + \vec{F}_{\text{CA}} + \vec{F}_{\text{DA}} = \frac{q^2}{4\pi\varepsilon_0\, a^2}\left(\sqrt{2} + \frac{1}{2}\right)\left[\frac{-(\vec{a}_x + \vec{a}_y)}{\sqrt{2}}\right].$$

With the help of (3.8*b*), this can be expressed as:

$$\vec{F}_{\text{A}} = \underbrace{\left[\frac{q^2}{4\pi\varepsilon_0\, a^2}\left(\sqrt{2} + \frac{1}{2}\right)\right]}_{\text{(magnitude)}} \quad \underbrace{[\vec{a}_{\text{CA}}]}_{\text{(unit vector)}}. \tag{3.10}$$

We can infer from (3.10) that the magnitude of the force $\vec{F}_{\text{A}}$ on the point charge q at one of the corners of the square of side a due to the point charges q at the remaining three corners is $q^2/4\pi\varepsilon\, a^2(\sqrt{2} + 1/2)$, and that this force is directed along $\vec{a}_{\text{CA}}$, the diagonal of the square (Fig. 3.2). Through this simple problem, you will be able to easily appreciate how to use the vectorial form of Coulomb's law (3.1).

It is easier to work out the above two-dimensional problem which led to (3.10) rather than a three-dimensional one where the point charges, each q, are situated at the corners of a cube instead of a square (Fig. 3.3). In this three-dimensional problem, too, as in the two-dimensional problem above, the sum total of the forces on a point charge q placed at any of the eight corners of the cube due to the point charges, each q, placed at the remaining seven corner points is given by

$$\vec{F} = \underbrace{\left[\frac{q^2}{4\pi\varepsilon_0 a^2}\left(\sqrt{3} + \sqrt{\frac{3}{2}} + \frac{1}{3}\right)\right]}_{\text{(magnitude)}} \quad \underbrace{\left[-\frac{\vec{a}_x + \vec{a}_y + \vec{a}_z}{\sqrt{3}}\right]}_{\text{(unit vector)}}. \tag{3.11}$$

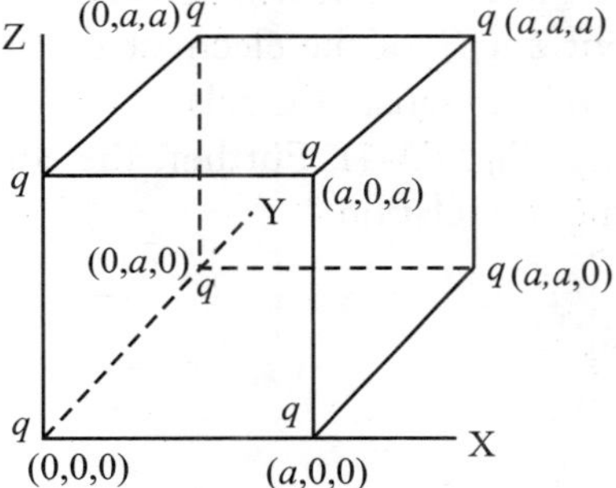

Fig. 3.3 *Identical point charges at the corners of a cube.*

The direction of the force given by (3.11) is the same as that of the unit vector indicated in (3.11), which can be identified as the unit vector directed from the charge located at the body diagonal corner point (a, a, a) of the cube to the charge at the corner point $(0, 0, 0)$ where the force is to be found. The problem is left to you to work out at leisure.

EXAMPLE E3-1

It is observed in an experiment that two small identical metallic charged balls, each of charge q and mass m, hung by a long thread of length l from a common point (hook), get separated from each other by a distance d due to electrostatic repulsion and is set to an equilibrium on the vertical plane. Find a formula for the value of the charge q in terms of m, l and d, taking $l >> d$. Numerically, appreciate the formula by finding the value of the charge if the values of m, l and d are 1 gm, 1 m and 2 cm respectively.

SOLUTION

The Coulomb force F between the charged balls is given by

$$F = \frac{q^2}{4\pi\varepsilon_0 d^2}. \tag{E3-1.1}$$

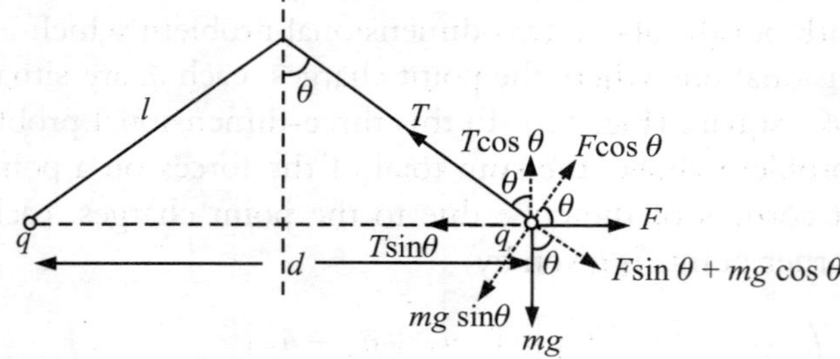

Fig. E3-1

At equilibrium, the tension T on the thread will be balanced by the force acting opposite to it caused by the component $F \sin\theta$ of the electrostatic force F and the component of gravitational force, $mg\cos\theta$, so as to satisfy the relation $T = F\sin\theta + mg\cos\theta$, where g is the acceleration due to gravity (Fig. E3-1). Further, the component $F\cos\theta$ will balance the component $mg\sin\theta$, giving the relation

$$F\cos\theta = mg\sin\theta,$$

whence we get

$$\tan\theta = \frac{F}{mg}. \tag{E3-1.2}$$

Substituting (E3-1.1) in (E3-1.2), we obtain the expression for $\tan\theta$ as

$$\tan\theta = \frac{q^2}{4\pi\varepsilon_0 d^2 mg}. \tag{E3-1.3}$$

We can also get an alternative expression for $\tan\theta$ from the geometry of the problem as (Fig. E3-1):

$$\tan\theta = \frac{d/2}{\sqrt{(d/2)^2 + l^2}}. \tag{E3-1.4}$$

Equating the right-hand sides of (E3-1.3) and (E3-1.4), we obtain:

$$q = \left(\frac{2\pi\varepsilon_0 d^3 mg}{\sqrt{d^2/4 + l^2}}\right)^{1/2},$$

which, for $l >> d$ approximately becomes

$$q = \left(\frac{2\pi\varepsilon_0 d^3 mg}{\sqrt{d^2/4 + l^2}}\right)^{1/2}. \tag{E3-1.5}$$

Putting $\varepsilon_0 = 8.854 \times 10^{-12}$F/m, $m = 0.001$ kg, $l = 1$ m, $d = 0.02$ m and $g = 9.8$ m/s^2 in (E3-1.5), we get $q = 2.088 \times 10^{-9}$ C= 2.088 nC.

3.1.2 Electric field and electric displacement due to a point charge

Let us now express the basic Coulomb force equation (3.1) in the following form:

$$\vec{F} = q_2\vec{E}, \tag{3.12}$$

where the quantity $\vec{E}$ is given by

$$\vec{E} = \frac{q_1}{4\pi\varepsilon r^2}\vec{a_r} \tag{3.13}$$

and which may be defined as the electric field intensity or simply, the electric field due to point charge q_1 at a distance r at the location of the point charge q_2. We can generalise the Coulomb force expression (3.12) by dropping the subscript from q_2 as

$$\vec{F} = q\vec{E} \tag{3.14}$$

to read it as the expression for the force on any point charge q placed in a region of electric field $\vec{E}$, though while formulating (3.12) we specifically considered the electric field caused by a point charge q_1.

EXAMPLE E3-2

Show that the trajectory of an electron of charge $-|e|$ is parabolic if it is shot with a constant velocity v in a region of uniform electric field of magnitude E at right angle to the field.

SOLUTION

Let the electron (of negative charge $-|e|$) be shot with velocity v along the x direction from the origin $x = 0$, $y = 0$ of the rectangular coordinate system to the region of electric field directed along y (Fig. E3-2). We can write the force equation of electronic motion with the help of (3.14) as

$$m\frac{d^2y}{dt^2} = -|e|E \tag{E3-2.1}$$

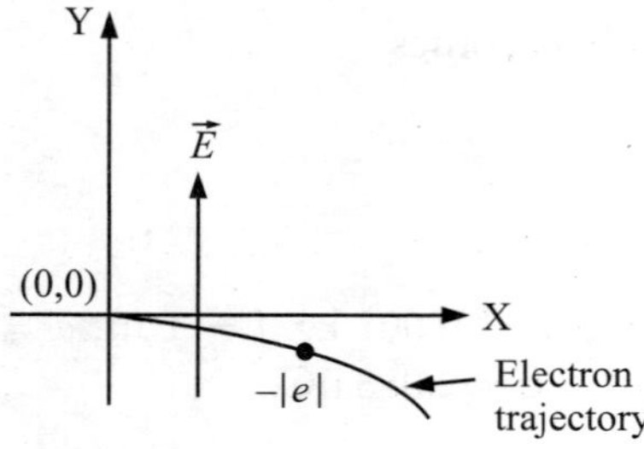

Fig. E3-2

whence we obtain the acceleration of the electron along y by rearrangement of terms as

$$\frac{d^2y}{dt^2} = \frac{-|e|}{m}E \tag{E3-2.2}$$

for the electric field directed along y acting on the negative charge $-|e|$ of the electron. The force due to the field is acting in the negative y direction.

Integrating (E3-2.2), we get the velocity of the electron along y as

$$\frac{dy}{dt} = \frac{-|e|}{m}Et + C_1 \tag{E3-2.3}$$

where C_1 is an integration constant. At $t = 0$, the electron is at the origin $x = 0$, $y = 0$, where it has a velocity only along the x direction and none along y, that is, $dy/dt = 0$. So, putting $t = 0$ and $dy/dt = 0$, we get $C_1 = 0$. We can then write (E3-2.3) as

$$\frac{dy}{dt} = \frac{-|e|}{m}Et. \tag{E3-2.4}$$

Integrating (E3-2.4), we obtain

$$y = \frac{-|e|}{m} E \frac{t^2}{2} + C_2 \tag{E3-2.5}$$

where C_2 is an integration constant. At $t = 0$, the electron is at the origin $x = 0$, $y = 0$. Putting $t = 0$ and $y = 0$ in (E3-2.5), we get $C_2 = 0$, in view of which (E3-2.5) becomes

$$y = \frac{-|e|}{m} E \frac{t^2}{2}. \tag{E3-2.6}$$

Since the electric field is directed along y, the displacement x of the electron along the X-axis is not affected and there is no acceleration along x. We can therefore write

$$x = vt. \tag{E3-2.7}$$

Substituting $t = x/v$ from (E3-2.7) in (E3-2.6), we obtain

$$y = \frac{-|e|\, E}{2mv^2} x^2. \tag{E3-2.8}$$

The expression (E3-2.8) represents a parabolic trajectory of the electron.

You will have observed from (3.14) that the unit of the electric field $\vec{E}$ is N/C and that numerically it is equal to the force on a unit charge (corresponding to $q = 1$). The expression (3.14) is known as the Coulomb force on a point charge, and it clearly indicates that the Coulomb force $\vec{F}$ takes on the same direction as that of the electric field $\vec{E}$.

We can also generalise (3.13) by dropping the subscript from q_1 to define the expression for the electric field $\vec{E}$ due to any point charge q at a distance r from it as:

$$\vec{E} = \frac{q}{4\pi\varepsilon\, r^2} \vec{a}_r \tag{3.15}$$

where $\vec{a}_r$ is the unit vector directed from the point charge q to the point where the electric field is so defined. From (3.15), you can see that while the magnitude of this electric field $\vec{E}$ is

$$E = \frac{q}{4\pi\varepsilon\, r^2} \tag{3.15a}$$

and its direction is that of $\vec{a}_r$ if the sign of the charge q is taken as positive.

At this stage we can define another quantity called the electric displacement $\vec{D}$ involving electric field $\vec{E}$ and permittivity of the medium ε as (see Appendix A3.1 to know more about the physical significance of $\vec{D}$)

$$\vec{D} = \varepsilon\, \vec{E}. \tag{3.16}$$

Here, we assume the medium to be 'isotropic', which means that the electric displacement component along a coordinate axis is equal to the permittivity ε times the electric field component along the same coordinate, being independent of the field components along the other axes. However, the relation (3.16) needs to be modified for unconventional media such as anisotropic and chiral media (see Appendix A4.3).

With the help of (3.15) and (3.16), we can write the following expression for the electric displacement $\vec{D}$ due to the point charge q at a distance r as:

$$\vec{D} = \frac{q}{4\pi r^2}\vec{a}_r. \tag{3.17}$$

Let us now see how to use the expressions (3.13) or (3.17) for the electric field (or displacement) due to a point charge in the simple problem of finding the electric field near a long uniform line-charge distribution (a charge uniformly sprayed over a long line). The medium is taken as a free-space ($\varepsilon = \varepsilon_0$) (Fig. 3.4).

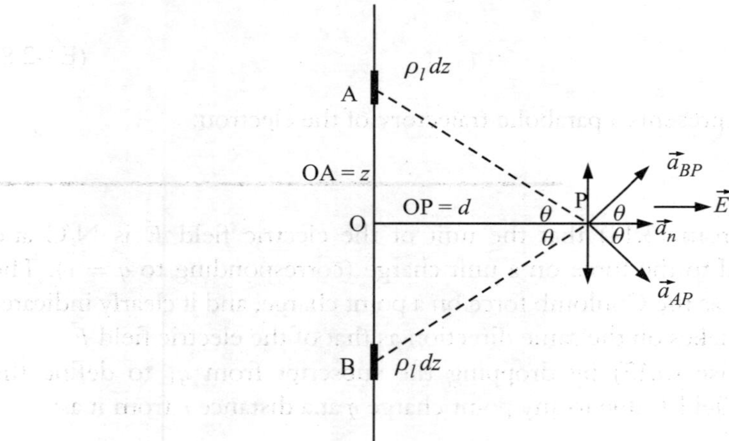

Fig. 3.4 *A long line charge distribution and two symmetrically placed charge elements placed on it considered for the purpose of finding the electric field near the line.*

Let OP $= d$ be the perpendicular distance of the point P where we are required to find the electric field from the line charge. Point O is taken as the origin of the Z-axis, being the foot of perpendicular drawn from P to the line charge (Fig. 3.4). We can take l, the length of the line charge, to be very large compared to d. Let ρ_l be the line charge density, that is, the charge per unit length (C/m). Therefore, for every charge element $\rho_l dz$ spread over an element dz on the line located at a point A at a distance z from O in the 'positive' z direction, we can consider another symmetrically placed identical charge element $\rho_l dz$ located at the point B at a distance z from O in the 'negative' z direction such that OA=OB. Let us thus find the contributions to the electric field at P due to such pairs of charge elements located over the entire line charge length, and subsequently sum up such contributions to find the electric field at P due to the entire line charge distribution (Fig. 3.4). This approach, as we are going to see in the analysis to follow, will simplify the problem, although the electric field components at P contributed by such charge elements will, in general, be different with respect to their magnitudes and/or directions.

The element of electric field $d\vec{E}_{\text{AP}}$ due to the charge element $\rho_l dz$ considered at A, located at a distance z from O in the 'positive' z direction, can be written with the help of (3.15) as:

$$d\vec{E}_{\text{AP}} = \frac{\rho_l dz}{4\pi\varepsilon_0(\text{AP})^2}\vec{a}_{\text{AP}}, \tag{3.18}$$

where $\vec{a}_{\text{AP}}$ is the unit vector directed from A to P. We can take the two components of $d\vec{E}_{\text{AP}}$, one along the direction of 'negative' z, $(dE_{\text{AP}})_{(-z)}$ and the other perpendicular to the Z-axis, $(dE_{\text{AP}})_{\perp}$ (Fig. 3.4). Similarly, we can write the element of electric field $d\vec{E}_{\text{BP}}$ due to the charge element $\rho_l dz$ considered at B, at a distance z from O in the 'negative' z direction with the help of (3.15) as

$$d\vec{E}_{\text{BP}} = \frac{\rho_l dz}{4\pi\varepsilon_0(\text{BP})^2}\vec{a}_{\text{BP}} \tag{3.19}$$

where $\vec{a}_{\text{BP}}$ is the unit vector directed from B to P (Fig. 3.4). We can take the two components of $d\vec{E}_{\text{BP}}$, one along 'positive' z direction, $(dE_{\text{BP}})_{(+z)}$ and the other perpendicular to the Z-axis, $(dE_{\text{BP}})_{\perp}$.

The magnitude of the element of electric field $d\vec{E}_{\text{AP}}$ given by (3.18) is

$$dE_{\text{AP}} = \frac{\rho_l dz}{4\pi\varepsilon_0(\text{AP})^2},$$

and, therefore, its component $(dE_{\text{AP}})_{\perp}$ in the direction normal to the z direction is

$$(dE_{\text{AP}})_{\perp} = dE_{\text{AP}}\cos\theta = \frac{\rho_l dz}{4\pi\varepsilon_0(\text{AP})^2}\cos\theta \tag{3.20}$$

where θ is the angle between $\vec{a}_{\text{AP}}$ and $\vec{a}_n$, the unit vector at point P in a direction normal to the direction of the Z-axis (Fig. 3.4). Similarly, we can find

$$(dE_{\text{BP}})_{\perp} = dE_{\text{BP}}\cos\theta = \frac{\rho_l dz}{4\pi\varepsilon_0(\text{BP})^2}\cos\theta. \tag{3.21}$$

In view of the symmetry of the problem, we have AP = BP, which in turn makes the right-hand sides of (3.20) and (3.21) equal, giving

$$(dE_{\text{AP}})_{\perp} = (dE_{\text{BP}})_{\perp}. \tag{3.22}$$

Similarly, we may see that

$$(dE_{\rm BP})_{(+z)} = (dE_{\rm AP})_{(-z)}. \tag{3.23}$$

We can note from the geometry of the problem (Fig. 3.4) that the components $(dE_{\rm AP})_{(-z)}$ and $(dE_{\rm BP})_{(+z)}$ are oppositely directed and they being equal in magnitude, will nullify each other. On the other hand, the components $(dE_{\rm AP})_{\perp}$ and $(dE_{\rm BP})_{\perp}$ are in the same direction and will add up to the value of the electric field at P.

Therefore, we can integrate $(dE_{\rm AP})_{\perp}$ given by (3.20) to find the contribution to the magnitude of the electric field at P from all the charge elements considered between $z = 0$ and $z = \infty$ (noting that the length of the line charge is considered very large compared to the distance d). The contribution $(E)_{(+)\,\rm half}$ to the electric field at P from the positive half of the line charge is obtained by the integration :

$$(E)_{(+)\rm half} = \int (dE_{\rm AP})_{\perp} == \int_{z=0}^{z=\infty} \frac{\rho_l dz}{4\pi\varepsilon_0 ({\rm AP})^2} \cos\theta. \tag{3.24}$$

Further, the geometry of the problem gives the following relations that will help to evaluate the integral of (3.24) (by reducing the two variables z and θ to one variable, namely θ):

$$\left.\begin{aligned} & z = d\tan\theta \\ & dz = d\sec^2\theta\, d\theta \\ & ({\rm AP})^2 = d^2 + z^2 = d^2 + (d\tan\theta)^2 \\ & \quad = d^2 + d^2\tan^2\theta = d^2(1+\tan^2\theta) = d^2\sec^2\theta \end{aligned}\right\}. \tag{3.25}$$

Substituting (3.25) in (3.24), we obtain

$$(E)_{(+)\,\rm half} = \int_{z=0}^{z=\infty} \frac{\rho_l d\sec^2\theta\, d\theta}{4\pi\varepsilon_0 d^2\sec^2\theta} \cos\theta = \frac{\rho_l}{4\pi\varepsilon_0 d} \int_{z=0}^{z=\infty} \cos\theta\, d\theta. \tag{3.26}$$

We may also note from the geometry of the problem that $z = 0$ corresponds to $\theta = 0$ and $z = \infty$ corresponds to $\theta = \pi/2$. Therefore, changing accordingly the limits of integration in (3.26), we can write

$$(E)_{(+)\,\rm half} = \frac{\rho_l}{4\pi\varepsilon_0 d} \int_{\theta=0}^{\theta=\pi/2} \cos\theta\, d\theta = \frac{\rho_l}{4\pi\varepsilon_0 d}[\sin\theta]_0^{\pi/2} = \frac{\rho_l}{4\pi\varepsilon_0 d}[1-0] = \frac{\rho_l}{4\pi\varepsilon_0 d}. \tag{3.27}$$

Similarly, in view of the symmetry of the problem, you will easily appreciate that the contribution from the negative half of the line charge $(E)_{(-)\,\rm half}$ will be the same as that contributed by the positive half of the line charge $(E)_{+\rm half}$ given by (3.27), and thus we can write:

$$(E)_{(-)\text{ half}} = \frac{\rho_l}{4\pi\varepsilon_0 d}. \tag{3.28}$$

Therefore, adding (3.27) and (3.28) we obtain the magnitude E of the electric field $\vec{E}$ due to the entire line charge distribution as

$$E = (E)_{(+)\text{ half}} + (E)_{(-)\text{ half}} = \frac{\rho_l}{2\pi\varepsilon_0 d}. \tag{3.29}$$

Since $\vec{E}$ is contributed only by the component of the electric field in the direction normal to the line charge, we may express the electric field due to the line charge distribution as

$$\vec{E} = \frac{\rho_l}{2\pi\varepsilon_0 d}\vec{a}_n \tag{3.30}$$

where $\vec{a}_n$ is a unit vector in the direction normal to the line charge distribution.

The problem of the long line charge distribution treated here enjoys cylindrical symmetry (Fig. 3.4). We may therefore assign the point P to the radial coordinate r. At every point on a circle of radius r passing through point P on a plane at right angle to the Z-axis, the magnitude of the electric field would be given by (3.30), with the interpretation of $d = r$, and, further, the direction of the electric field $\vec{E}$ would be the same as that of the unit vector $\vec{a}_r$ in the radially outward direction. Thus, we can alternatively write (3.30) to obtain the following expression for the electric field due to a long line charge:

$$\vec{E} = \frac{\rho_l}{2\pi\varepsilon_0 r}\vec{a}_r. \tag{3.31}$$

EXAMPLE E3-3

Obtain an expression for the electric field due to a charge of surface charge density ρ_s uniformly spread over a plane bounded by a circle of radius a at a point situated at a perpendicular distance z from the centre. How can you interpret the result to find an expression for the electric field close to a planar charge distribution of infinite cross section?

SOLUTION

Consider the charge to be uniformly distributed over the XY-plane within a circular boundary of radius a. The point where the electric field is to be found is located at P $(0, 0, z)$, at a perpendicular distance z from the centre O of the circular charge distribution taken as the origin of the coordinates. We will proceed to solve the problem in two steps. In the first step, let us find the element of the electric field due to an element of annular-ring charge distribution of radius r and thickness dr (Fig. E3-3). In the second step, let us integrate the elemental field obtained in the first step between the limits $r = 0$ and $r = a$ to obtain the electric field due to the entire circular charge distribution. For this purpose, let us proceed as follows.

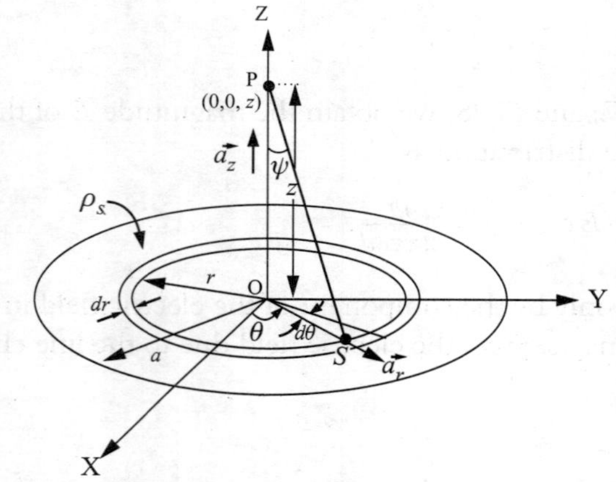

Fig. E3-3

Step (i): Consider a sub-element of arc of length $rd\theta$ and thickness dr at an arbitrarily considered point at S, say, on the element of annular charge ring of radius r. The area of the ring sub-element is then $rd\theta dr$, and the charge on it is $\rho_s rd\theta dr$, ρ_s being the surface charge density, supposedly uniform over the area over which the charge is distributed.

Step (ii): Write, with the help of (3.15), the sub-element of electric field $d\vec{E}'$ due to the charge distributed on the arc sub-element carrying charge $\rho_s rd\theta dr$ as described in (i) as

$$d\vec{E}' = \frac{\rho_s rd\theta dr}{4\pi\varepsilon_0 s^2}\vec{a}_{SP} \tag{E3-3.1}$$

where $\vec{a}_{SP}$ is the unit vector directed from the point S to the point P, the latter situated at a perpendicular distance z from the centre of the ring. $s =$ SP is the distance between S and P. We can then substitute (refer to Fig. E3-3)

$$\vec{a}_{SP} = \frac{\overrightarrow{\text{SP}}}{\text{SP}} = \frac{\overrightarrow{\text{SO}} + \overrightarrow{\text{OP}}}{\text{SP}} = \frac{-r\vec{a}_r + z\vec{a}_z}{s}$$

in (E3-3.1) to write

$$d\vec{E}' = \left(\frac{\rho_s rd\theta dr}{4\pi\varepsilon_0 s^2}\right)\left(\frac{-r\vec{a}_r + z\vec{a}_z}{s}\right),$$

which may be expressed as

$$d\vec{E}' = \left(\frac{\rho_s r^2 d\theta dr}{4\pi\varepsilon_0 s^3}\right)\vec{a}_r + \left(\frac{\rho_s rzd\theta dr}{4\pi\varepsilon_0 s^3}\right)\vec{a}_z. \tag{E3-3.2}$$

Step (iii): Find the element of electric field $d\vec{E}$ due to the entire annular charge ring of radius r and thickness dr by integrating the sub-element of electric field given by (E3-3.2) as follows:

$$d\vec{E} = \int d\vec{E}' = \int \left(\frac{\rho_s r^2 d\theta dr}{4\pi\varepsilon_0 s^3}\right)\vec{a}_r + \left(\frac{\rho_s rzd\theta dr}{4\pi\varepsilon_0 s^3}\right)\vec{a}_z.$$

This may be evaluated as

$$d\vec{E} = \left(\frac{\rho_s r^2 dr}{4\pi\varepsilon_0 s^3}\right)\int_0^{2\pi} d\theta\vec{a}_r + \left(\frac{\rho_s rzdr}{4\pi\varepsilon_0 s^3}\right)\vec{a}_z\int_0^{2\pi} d\theta. \tag{E3-3.3}$$

Here, we will notice that $\vec{a}_r$ occurring in the first term of the right-hand side of (E3-3.3) changes from sub-element to sub-element on the charge ring due to its changing direction, and therefore, $\vec{a}_r$ is brought within the integral of the first term. However, correspondingly, $\vec{a}_z$ does not change in the second term of the right-hand side of (E3-3.3), and therefore, $\vec{a}_z$ is brought outside the integral of the second term. Moreover, the contribution to $d\vec{E}$ by the first term of the right-hand side of (E3-3.3) due to a sub-element charge of the annular ring will be balanced by the equal and opposite contribution due to the diametrically opposite sub-element charge of the annular ring. This is because the unit vector $\vec{a}_r$ at a sub-element charge, for instance at S, will be equal in magnitude (unity) but oppositely directed to the unit vector $\vec{a}_r$ at a sub-element charge at the point diametrically opposite to S. Therefore, the first term of the right-hand side of (E3-3.3) becomes null and that enables us to write (E3-3.3) as

$$d\vec{E} = \left(\frac{\rho_s rzdr}{4\pi\varepsilon_0 s^3}\right)\vec{a}_z\int_0^{2\pi} d\theta = \left(\frac{\rho_s rzdr}{4\pi\varepsilon_0 s^3}\right)\vec{a}_z \times 2\pi,$$

which may be simplified and rearranged as

$$d\vec{E} = \left(\frac{\rho_s z}{2\varepsilon_0}\right)\left(\frac{rdr}{s^3}\right)\vec{a}_z. \tag{E3-3.4}$$

Let us, at this stage, define a variable $\psi = \angle\text{OPS}$, for which we can write, putting OS $= r$, OP $= z$ and SP $= s$:

$$\tan\psi = \frac{\text{OS}}{\text{OP}} = \frac{r}{z} \tag{E3-3.5}$$

and

$$\cos\psi = \frac{\text{OP}}{\text{SP}} = \frac{z}{s}. \tag{E3-3.6}$$

From (E3-3.6), we then obtain

$$s = z\sec\psi \tag{E3-3.7}$$

and from (E3-3.5), we obtain $r = z\tan\psi$ which on differentiation gives

$$dr = z\sec^2\psi d\psi. \tag{E3-3.8}$$

Substituting (E3-3.7) and (E3-3.8) in (E3-3.4), we can express the latter as

$$d\vec{E} = \left(\frac{\rho_s z}{2\varepsilon_0}\right)\left(\frac{rz\sec^2\psi}{(z\sec\psi)^3}\right) d\psi\vec{a}_z,$$

which simplifies to

$$d\vec{E} = \left(\frac{\rho_s}{2\varepsilon_0}\right)\left(\frac{r}{z\sec\psi}\right) d\psi\vec{a}_z. \tag{E3-3.9}$$

Further, substituting (E3-3.5) in (E3-3.9) we can express the latter as

$$d\vec{E} = \left(\frac{\rho_s}{2\varepsilon_0}\right)\left(\frac{\tan\psi}{\sec\psi}\right) d\psi\vec{a}_z,$$

which simplifies to

$$d\vec{E} = \left(\frac{\rho_s}{2\varepsilon_0}\right)\sin\psi\, d\psi\,\vec{a}_z. \tag{E3-3.10}$$

Step (iv): The contributions due to the annular charge rings considered between $r = 0$ and $r = a$ will give us the expression for $\vec{E}$ due to the charge sprayed over an area bounded by a circle of radius a. Thus, integrating the electric field due to the annular ring of charge distribution obtained in (iii) and given by (E3-3.10), we obtain

$$\vec{E} = \int d\vec{E} = \left(\frac{\rho_s}{2\varepsilon_0}\right)\int_0^{\tan^{-1}a/z} \sin\psi\, d\psi\,\vec{a}_z \tag{E3-3.11}$$

where the upper and lower limits are written with the help of (E3-3.5) putting therein $r = a$ and $r = 0$ respectively. Further, we can evaluate (E3-3.11) as

$$\vec{E} = \left(\frac{\rho_s}{2\varepsilon_0}\right)[-\cos\psi]_0^{\tan^{-1}a/z}\vec{a}_z = -\left(\frac{\rho_s}{2\varepsilon_0}\right)\left[\cos\left(\tan^{-1}\frac{a}{z}\right) - \cos 0\right]\vec{a}_z,$$

which then becomes

$$\vec{E} = -\left(\frac{\rho_s}{2\varepsilon_0}\right)\left(\frac{z}{\sqrt{a^2+z^2}} - 1\right)\vec{a}_z = \left(\frac{\rho_s}{2\varepsilon_0}\right)\left(1 - \frac{z}{\sqrt{a^2+z^2}}\right)\vec{a}_z. \tag{E3-3.12}$$

In (E3-3.12), we obtain the desired expression for the electric field due to the charge distributed over a plane bounded by a circle. As a special case of $z << a$, (E3-3.12) gives the following expression for the electric field close to a planar charge distribution of infinite cross section:

$$\vec{E} = \frac{\rho_s}{2\varepsilon_0}\vec{a}_z,$$

an expression that becomes identical to (3.47) if we interpret $\vec{a}_z = \vec{a}_n$, the unit normal vector perpendicular to the plane of the charge distribution.

3.1.3 Gauss's law of electrostatics

Closely related to Coulomb's law is Gauss's law, which can be stated in terms of the flux of electric field $\vec{E}$ or that of electric displacement $\vec{D}$. (Recall how the flux of a vector quantity is defined in Section 2.4). In general, $\vec{E}$ or $\vec{D}$ is not uniform in a region, and it can vary with space coordinates. However, such a quantity may be considered to be constant over an infinitesimal area dS. Hence, we can write the element of flux $d\phi_D$ of the electric displacement $\vec{D}$ through the element of area dS as (refer to (2.17)and (2.18) for the definition of flux):

$$d\phi_D = \vec{D}\cdot d\vec{S} = \vec{D}\cdot\vec{a}_n dS \tag{3.32}$$

where $d\vec{S}$ is the area element vector which has the magnitude dS and a direction that is normal to the area element, being the same as the direction in which the flux is sought. $\vec{a}_n$ is a unit vector normal to the area element taken in the direction of $d\vec{S}$ (Fig. 3.5). Though $\vec{D}$ remains constant over the infinitesimal area dS, it varies, in general, over a finite area S. Therefore, in order to find the flux ϕ_D of the electric displacement $\vec{D}$ over a finite area S, we have to integrate (3.32) over the surface area S (Section 3.4) as follows:

$$\phi_D = \int_S d\phi_D = \int_S \vec{D}\cdot d\vec{S} = \int_S \vec{D}\cdot\vec{a}_n dS. \tag{3.33}$$

Similarly, you can find the flux ϕ_E of the electric field $\vec{E}$ as

$$\phi_E = \int_S d\phi_E = \int_S \vec{E}\cdot d\vec{S} = \int_S \vec{E}\cdot\vec{a}_n dS. \tag{3.34}$$

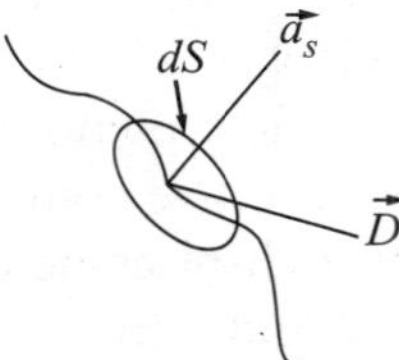

Fig. 3.5 *The element of area dS showing the electric displacement $\vec{D}$.*

The integrals in (3.33) and (3.34) are the surface integrals, as indicated by the subscript S

(Section 2.4). If the surface were a closed surface over the area S, the surface integral would change from 'surface integral' ($\int_S$) to 'closed surface integral' ($\oint_S$) [see Section 2.4; equation (2.20)]. In terms of such closed surface integral of either the electric displacement or the electric field, it becomes convenient to state another law of electrostatics, namely, Gauss's law.

Gauss's law states that the outward flux of electric displacement $\vec{D}$ through the surface of a closed volume enclosing a charge Q—estimated by a closed surface integral of $\vec{D}$ over the area of the volume enclosure—is equal to the charge enclosed, Q. Mathematically, with the help of (3.33), we can state Gauss's law as

$$\phi_D = \oint_S \vec{D} \cdot d\vec{S} = \oint_S \vec{D} \cdot \vec{a}_n dS = Q \tag{3.35}$$

where we have to take the direction of $d\vec{S}$ or that of $\vec{a}_n$ as the direction of the 'outward' normal to the surface. Substituting (3.16) in (3.35), we can alternatively express Gauss's law mathematically in terms of the outward flux of electric field $\vec{E}$ or closed surface integral of $\vec{E}$ [see (3.10)] as

$$\phi_E = \oint_S \vec{E} \cdot d\vec{S} = \oint_S \vec{E} \cdot \vec{a}_n dS = \frac{Q}{\varepsilon} \tag{3.36}$$

which in words states that the outward flux of electric field $\vec{E}$ through the surface of a closed volume enclosing a charge Q—estimated by a closed surface integral of $\vec{E}$ over the area of the volume enclosure—is equal to the charge enclosed Q divided by the permittivity of the medium, ε. Later in this section, we will take up the proof of Gauss's law stated by (3.35) or (3.36) (see Section 3.1.3.A).

The flux of electric displacement or electric field, given by (3.35) or (3.36), going out of a closed volume enclosing a charge is positive if the charge enclosed is positive. In other words, the flux of electric displacement or electric field 'goes out of' a volume enclosing a 'positive charge'. On the contrary, if the charge enclosed were negative, then the outward flux going out of a closed volume of electric displacement or electric field would be negative, that is, the inward flux would be positive, which, in other words, means that the flux of electric displacement or electric field 'goes into' a volume that encloses a 'negative charge' (Fig. 3.6). As a corollary, if a volume does not enclose any charge, the flux of electric displacement or electric field going out of or into the volume enclosure would be zero. Therefore, if you have a positive point charge separated by a negative point charge, the electric 'flux lines' of electric displacement or electric field will emanate from the positive charge (point A in Fig. 3.6) and terminate on the negative charge (point B in Fig. 3.6). If you consider in the vicinity of these two charges a point where no charges exist (typically, point C in Fig. 3.6), and enclose this point by a volume, then as many flux lines will enter the volume enclosure as they leave. Consequently, this makes the net flux of electric displacement or field going out of or into the volume enclosing a point

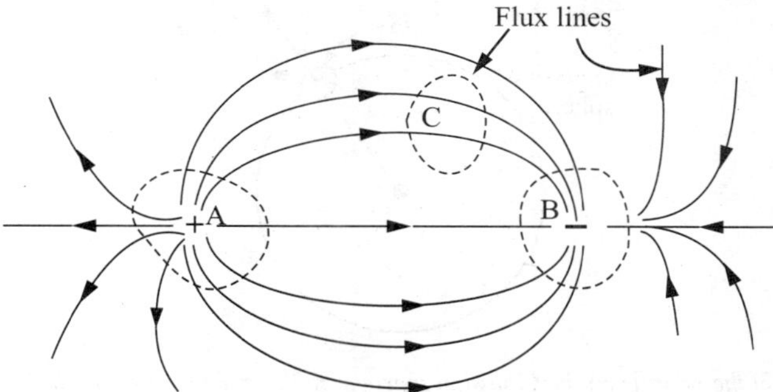

Fig. 3.6 *Flux lines in the vicinity of typical three points: A and B locating a positive charge and a negative charge respectively, and C representing a charge-free point.*

unoccupied by any charge zero. Further, recall the concept of electric 'lines of force' learnt in school. The density of these lines at a point is a measure of the intensity of the electric displacement or field at the point, and the tangent to a line of force passing through the point represents the direction of the electric displacement or field at the point. Further, if the magnitudes of these two equal and opposite charges, separated by a distance, are equal, they are said to constitute an electric dipole (see Examples E3-5 and E3-6).

Gauss's law makes it easier to find the magnitude of the electric field vector in problems that enjoy rectangular, cylindrical or spherical symmetry. It follows that once the magnitude of the electric field in such a problem has been found by Gauss's law, the direction of the field could also be found as well by a physical reasoning. Some illustrative examples are given below.

Illustrative Examples of Applications of Gauss's Law

(a) Electric field due to a point charge from Gauss's law

Let us, for instance, proceed to find with the help of Gauss's law (3.36) the magnitude E of the electric field $\vec{E}$ due to a point charge q, obtained earlier from Coulomb's law (3.15a). The problem enjoys spherical symmetry in that the magnitude of electric field all over the spherical surface of radius r due to a point charge q placed at the centre of the sphere will be the same. Such a surface is called a Gaussian surface, a *Gaussian sphere* in the present problem (Fig. 3.7). It is worth mentioning here that the Gaussian surface is a fictitious mathematical proposition that is brought into the formulation of problems that enjoy geometrical symmetry. It turns out that while applying Gauss' law (3.36) to a problem of symmetry, the evaluation of the surface integral (3.36) becomes rather simple because the magnitude of the electric field in the integrand of the surface integral can be regarded as constant, enabling it to be taken outside the integrand. While the Gaussian surface in the present problem is a sphere enjoying spherical symmetry, in the immediately following problem—that of finding the electric field due to a uniform line charge distribution—it is

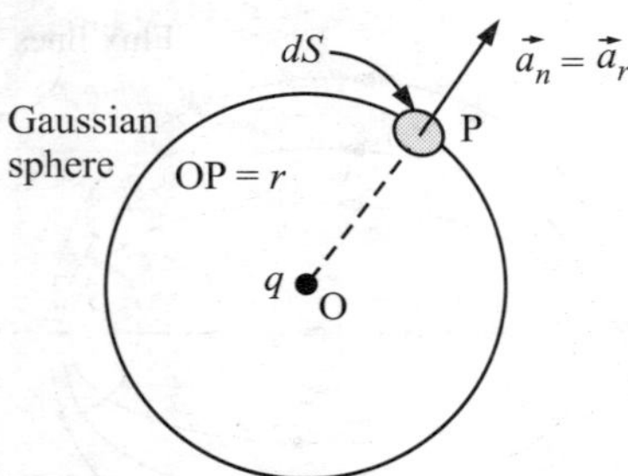

Fig. 3.7 *The location of the point P on the Gaussian sphere of radius r where the electric field due to a point charge q located at the centre O of the sphere is to be found.*

a circular-cylindrical surface that enjoys azimuthal symmetry. In yet another problem of finding the electric field due to a planar sheet charge distribution, the Gaussian surface is taken as a rectangular parallelepiped.

In the present problem, we can interpret q to be positive, in which case the electric field $\vec{E}$ will be in the direction of the point (P) on the Gaussian sphere where the electric field is to evaluated, that is, in the radially outward direction, normal to such a surface. Therefore, we may take $\vec{a}_r = \vec{a}_n$ for this case, where $\vec{a}_n$ is the unit vector normal to the Gaussian sphere at point P, and $\vec{E} = E\vec{a}_r = E\vec{a}_n$, which we may now apply to Gauss's law (3.36), putting $Q = q$, to write

$$\oint_S \vec{E} \cdot d\vec{S} = \oint_S E\vec{a}_n \cdot \vec{a}_n dS = \oint_S EdS = \frac{q}{\varepsilon}, \tag{3.37}$$

where dS is an elemental area on the Gaussian surface.

Since E is constant at every point on the surface of the Gaussian sphere, we may take it outside the surface integral of (3.37) to write

$$E \oint_S dS = \frac{q}{\varepsilon}. \tag{3.38}$$

Further, putting $\oint_S dS = 4\pi r^2$, where r is the radius of the Gaussian sphere, in (3.38), we get

$$E 4\pi r^2 = \frac{q}{\varepsilon},$$

whence we obtain

$$E = \frac{q}{4\pi\varepsilon r^2} \tag{3.39}$$

which is identical to the expression for the magnitude of the electric field due to a point charge (3.15a). The direction of the electric field is, however, not predictable by Gauss's law; it has to be found by physical reasoning. As has been discussed preceding (3.37), the electric field is given by $\vec{E} = E\vec{a}_r$, which may be written with the help of (3.39) as

$$\vec{E} = \frac{q}{4\pi\varepsilon r^2}\vec{a}_r. \tag{3.40}$$

We thus see that (3.40) obtained by Gauss's law is identical to (3.15) obtained starting from Coulomb's law (3.1).

(b) Electric field due to a long line charge from Gauss's law

Next, let us take the example of a long line-charge distribution for which we obtained (3.31) as the expression of electric field near the long line charge starting from the expression for the electric field due to a point charge (3.15), which in turn was obtained from Coulomb's law (3.1) (Fig. 3.8). We take the medium as the free-space region ($\varepsilon = \varepsilon_0$). The problem enjoys azimuthal symmetry. In order to treat the problem of symmetry, let us consider a closed Gaussian cylinder of length l and radius r passing through the point (P) where we have to find the electric field due to the line charge. From the physical reasoning discussed preceding (3.31) with reference to the present problem, the direction of the electric field at any point on the cylindrical surface is normal to the line charge, that is, in the radially outward direction. Therefore, we may take $\vec{a}_r = \vec{a}_n$, where $\vec{a}_n$ is the unit vector normal to the Gaussian cylindrical surface at the point P, and $\vec{E} = E\vec{a}_r = E\vec{a}_n$ (Fig. 3.8). Putting this in Gauss's law (3.36) and taking $Q = \rho_l l$ for the charge enclosed by the Gaussian cylinder, where ρ_l is the charge per unit length of the line, we can write Gaussian cylinder

$$\oint_S \vec{E}.d\vec{S} = \oint_S E\vec{a}_n.\vec{a}_n dS = \int_S EdS = \frac{\rho_l l}{\varepsilon_0}. \tag{3.41}$$

Note that in the present problem, since the electric field is only in the radial direction, the top and the bottom faces of the Gaussian cylinder do not contribute to the surface integral of (3.41).

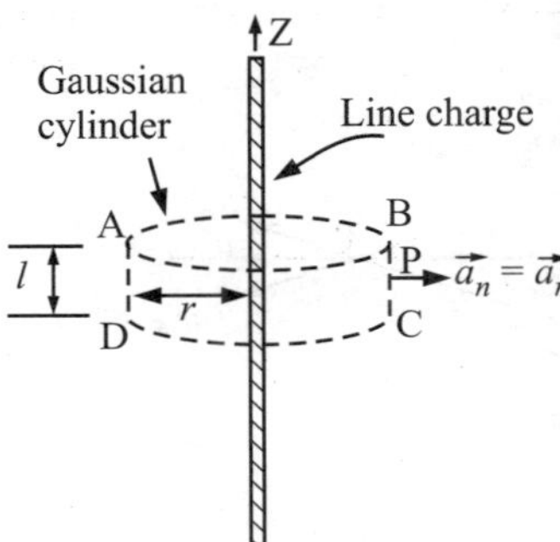

Fig. 3.8 *Circular Gaussian cylinder ABCD of radius r passing through the point P where the electric field due to a long line charge distribution is to be found.*

Since the electric field E is constant over the circularly curved cylindrical surface of radius r in view of the cylindrical symmetry of the present problem, we may take E outside the surface integral of (3.41) to obtain

$$E\int_S dS = \frac{\rho_l l}{\varepsilon_0}. \tag{3.42}$$

Further, putting the area of the cylindrical surface of radius r and length l in (3.42), we obtain

$$E2\pi rl = \frac{\rho_l l}{\varepsilon_0},$$

which gives the magnitude of the electric field at P due to the line charge as

$$E = \frac{\rho_L}{2\pi\varepsilon_0 r}. \tag{3.43}$$

Since the electric field due to the long line charge distribution is radial, , we may write $\vec{E} = E\,\vec{a}_r$, which may be further expressed, with the help of (3.43), as

$$\vec{E} = \frac{\rho_L}{2\pi\varepsilon_0 r}\vec{a}_r. \tag{3.44}$$

You will see that expression (3.44) for the electric field due to a long line charge distribution obtained by Gauss's law is identical to (3.31), the latter obtained earlier by a rather involved method starting from Coulomb's law.

(c) Electric field due to a large sheet of charge from Gauss's law

Closely related to the problem of a long line charge distribution is the interesting problem of electric charge uniformly distributed over a plane—a planar sheet of charges. This problem enjoys rectangular symmetry (Fig. 3.9).

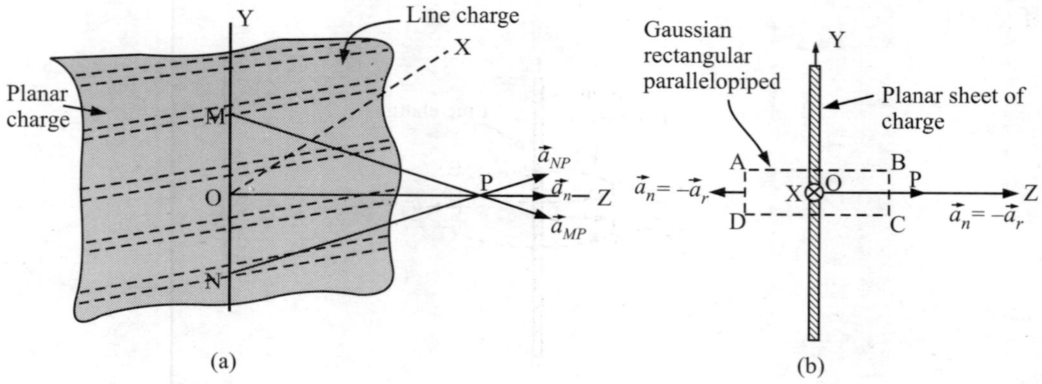

Fig. 3.9 *An array of parallel lines of charge representing a planar sheet of charge on the XY plane. Part (a) shows the electric field components contributed by any two lines of charge symmetrically placed with respect to a point P considered on the Z-axis, while part (b) shows the YZ cross section of the planar sheet of charge with the point P where the electric field is to be found, close to the sheet on the Z-axis. The Gaussian rectangular parallelopiped is symmetrical with respect to the sheet and passes through P.*

Let us assume the plane of the charge sheet to be in the XY-plane (Fig. 3.9*a*). The perpendicular distance OP of the point P from the plane of the charge sheet is taken to be much smaller than the dimensions of the planar charge distribution. Also, the point O is taken as the origin of the rectangular coordinate system considered. Let us now consider that the planar charge distribution comprises an array of long lines of charge, each parallel to the X-axis. Hence, we can see that, similar to the problem of a single long line charge distribution (Fig. 3.7), the electric field at point P due to any one of such line charges will be normal to the length of the line charge and directed towards point P. Let us consider two parallel line charges, symmetrically placed with respect to the X-axis, crossing the Y-axis at points M and N such that OM = ON (Fig. 3.9*a*). Then the electric field at P due to the line charge passing through M will have the direction of the unit vector $\vec{a}_{\mathrm{MP}}$ directed from M to P; likewise, the electric field at P due to the line charge passing through N will have the direction of the unit vector $\vec{a}_{\mathrm{NP}}$ directed from N to P. However, due to symmetry, the magnitudes of these fields are the same. Each of these fields at P may be resolved into two perpendicular components, one normal to the planar charge distribution, from O to P in the direction of the unit vector $\vec{a}_n$, and the other parallel to the planar charge distribution. You will be able to easily appreciate that the parallel component of the field contributed by the line charge passing through M is in the positive y direction and that through N is in the negative y direction. Further, as they have the same magnitude due to symmetry, they will cancel out. The perpendicular components, however, is in the direction of $\vec{a}_n$ and will add up (Fig. 3.9*a*). Since the entire planar charge distribution can be considered to be comprised of such symmetrically placed line charge distributions, we can say that the electric field due the planar charge distribution at a nearby point will be along $\vec{a}_n$, in the direction normal to the plane of charge distribution passing through point P where the field is to be found. We may then use this understanding to find the electric field at a point close to such a sheet of charges with the help of Gauss's law.

For this purpose, let us consider a closed rectangular parallelopiped Gaussian volume (Fig. 3.9*b*) such that its left face AD and its right side face BC, each of area S, say (each shown as a line in the cross-sectional view), are equidistant from the planar charge distribution. Further, one of the faces, here BC, passes through the point P where the electric field is to be found. Note that while the normal unit vector with reference to the right face BC is $\vec{a}_n = \vec{a}_z$, with reference to the left face AD it is $\vec{a}_n = -\vec{a}_z$. Accordingly, the electric field on the right face BC is along $\vec{a}_n = \vec{a}_z$ and that on the left face AD is along $\vec{a}_n = -\vec{a}_z$. Both fields, however, have the same magnitude due to the symmetry of the problem. Let the plane carry a uniform charge density (charge per unit of area) ρ_s. Applying Gauss's law (3.36) to the closed rectangular Gaussian volume (Fig. 3.9*b*) by taking $Q = \rho_s S$ for the charge enclosed by the Gaussian volume, we can write

$$\int_{S\,(\text{right face})} E\vec{a}_n \cdot \vec{a}_n dS + \int_{S\,(\text{left face})} (-E\vec{a}_n) \cdot (-\vec{a}_n dS) = \frac{\rho_s S}{\varepsilon_0}$$

which then becomes

$$\int_S 2E dS = \frac{\rho_s S}{\varepsilon_0}. \tag{3.45}$$

We have taken the medium to be free space and put $\varepsilon \Rightarrow \varepsilon_0$. Further, since the electric field is normal to the surface charge distribution, the top and the bottom faces of the Gaussian rectangular volume have not contributed to the surface integral of (3.45). Now, the magnitude of the electric field E being constant on each of the left and the right faces of the Gaussian volume, we may take E outside the integral of (3.45) to express it as:

$$2E \int_S dS = 2ES = \frac{\rho_s S}{\varepsilon_0},$$

which gives the magnitude of the electric field at P as

$$E = \frac{\rho_s}{2\varepsilon_0}. \tag{3.46}$$

Since the electric field at P is directed normal to the surface charge distribution, we may express the former in the vectorial form, with the help of (3.46), as

$$\vec{E} = \frac{\rho_s}{2\varepsilon_0} \vec{a}_n. \tag{3.47}$$

(d) Electric field due to a charged plane conductor from Gauss's law

The electric field due to a uniformly charged plane conductor can be found using Gauss's law in view of the rectangular symmetry enjoyed by the problem, as discussed in the preceding problem. We will, therefore, consider a Gaussian rectangular parallelopiped enclosure such that the left face of it now passes through the conductor and the right face through the point P where the electric field is to be found (Fig. 3.10).

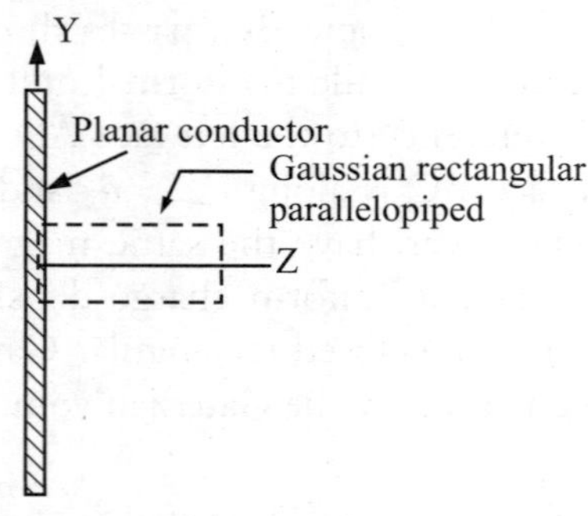

Fig. 3.10 *Cross section of the Gaussian rectangular parallelopiped on the YZ plane. The left face of the parallelopiped is inside the planar conductor while the right face passes through point P close to the conductor where the electric field is to be found.*

Further, as in the problem of the sheet of charge, the top and the bottom faces of the Gaussian enclosure does not contribute to the evaluation of the surface integral of (3.45), Moreover, the left face will also not contribute to the surface integral, taking the electrical field inside the conductor to be zero[1]. Therefore, considering the contribution only of the right face of the Gaussian enclosure (Fig. 3.10) to the surface integral of (3.45) and taking the charge enclosed by the enclosure to be $Q = \rho_s S$, we can write, for the present problem the following expression:

$$\int\limits_{S\,(\text{right face})} E\vec{a}_n \cdot \vec{a}_n dS = \int\limits_{S} E dS = \frac{\rho_s S}{\varepsilon_0}. \tag{3.48}$$

In view of the symmetry of the problem, the magnitude of the electric Field E is constant over the side faces of the Gaussian volume (rectangular parallelopiped), and hence, it can be taken outside the surface integral of (3.48) to write

$$E \int\limits_S dS = ES = \frac{\rho_s S}{\varepsilon_0},$$

whence we obtain the magnitude of the electric field at P as

$$E = \frac{\rho_s}{\varepsilon_0}. \tag{3.49}$$

Noting that the electric field at P is directed normal to the planar conductor, we may express the electric field at P in the vectorial form with the help of (3.49) as

$$\vec{E} = \frac{\rho_s}{\varepsilon_0}\vec{a}_n. \tag{3.50}$$

(e) Electric field inside a long beam of electrons from Gauss's law

In problems related to the formation and confinement of a beam of electrons or to the interaction of such an electron beam with electromagnetic waves, you will need to know

[1] The electric field inside a good conductor is zero. This can be appreciated with the help of Gauss's law after we introduce the concept of relaxation time of a physical medium in Section 5.1. It will be shown that there are no free charges inside the bulk of a good conductor, and that they decay very fast to appear as the surface charge on the boundary of the conductor. If we construct an element of Gaussian sphere in the bulk of a good conductor enclosing a point at its centre, then the charge enclosed by the element of spherical volume enclosure would be zero. The electric field if it were present in the spherical Gaussian volume would have the same magnitude at all the points on its surface and have the outward normal direction at any point. Therefore, the outward flux of electric field through the closed surface of the element of Gaussian sphere becomes

$$\phi_E = \oint\limits_S \vec{E} \cdot \vec{a}_n dS = \oint\limits_S E\vec{a}_n \cdot \vec{a}_n dS = E \oint\limits_S dS = E\Delta S,$$

where ΔS is the surface area of the element of Gaussian sphere. The flux ϕ_E thus found as above, according to Gauss's law (3.36) becomes zero since the charge enclosed is zero ($Q = 0$ in (3.36), giving $\phi_E = E\Delta S = 0$ and hence $E = 0$ since $\Delta S \neq 0$). (See also Appendix A5.1 in Chapter 5).

the electric field due to electrons inside the electron beam (also known as the space-charge field) or at the edge of the electron beam. Let us apply Gauss's law (3.36) to a closed Gaussian cylinder of circular cross section inside the electron beam, passing through the point P where the electric field is to be found (Fig. 3.11). The Gaussian cylinder is aligned along the Z-axis, and is coaxial with the circular cross section of the beam of electrons. If ρ is the volume charge density of the electron beam (which can be considered as a charged fluid of electrons, all moving with the same velocity), r is the radius of the Gaussian cylinder and l its length, its volume is given by $\pi r^2 l$. Noting that the charge enclosed by the closed Gaussian volume is $Q = (\pi r^2 l)(\rho)$, we have by Gauss's law,

$$\oint_S \vec{E} \cdot d\vec{S} = \int_S E\vec{a}_n \cdot \vec{a}_n dS = \int_S EdS = \frac{(\pi r^2 l)(\rho)}{\varepsilon_0}. \tag{3.51}$$

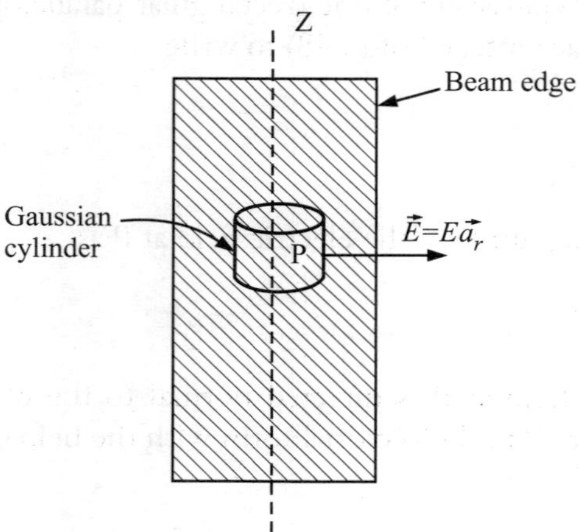

Fig. 3.11 *Gaussian cylinder of circular cross section inside an electron beam of circular cross section, coaxial with it, passing through point P where the space-charge electric field of the beam is to be found.*

It may be noted that as in the problem of a long line charge density, the electric field at any point on the Gaussian curved surface in this problem too is radial, allowing us to take $\vec{a}_r = \vec{a}_n$ and $\vec{E} = E\,\vec{a}_r = E\,\vec{a}_n$ in the left-hand side of (3.51). Also, since the electric field is radial, the top and the bottom faces of the Gaussian cylinder does not contribute to the surface integral of (3.51). Further, the electric field E is constant over the Gaussian curved cylindrical surface, which allows us to take E outside the surface integral of (3.51) to write:

$$E \int_S dS = \frac{(\pi r^2 l)(\rho)}{\varepsilon_0},$$

which, on putting $\int_S dS = 2\pi rl$, becomes

$$E 2\pi rl = \frac{\pi r^2 l \rho}{\varepsilon_0}.$$

After rearranging terms, the amplitude of the electric field E can be written as

$$E = \frac{r\rho}{2\varepsilon_0}. \tag{3.52}$$

Since the space-charge electric field of the electron beam is radial and expressible as $\vec{E} = E\vec{a}_r$, we may write the vector form of this field with the help of (3.52) as

$$\vec{E} = \frac{r\rho}{2\varepsilon_0}\vec{a}_r. \tag{3.53}$$

EXAMPLE E3-4

Consider alternate layers of negatively charged electrons and positively charged ions in a region of charge neutrality, called plasma (see Fig. E3-4). If the electrons, which are relatively freer to move than ions, are displaced from their mean position with respect to the ions due to some extraneous disturbing electric field, they would tend to restore their mean position due to the attraction of ions. In the process, they would overshoot the mean position and again tend to restore their mean position and further overshoot the mean position. This would result in the electrons oscillating about their mean position. Show that the angular frequency of oscillation of electrons, called the angular plasma frequency ω_p, is given by

$$\omega_p = \left(\frac{|\eta|\,|\rho_0|}{\varepsilon_0}\right)^{1/2},$$

where $|\rho_0|$ and $|\eta|$ are the magnitude of the charge volume density and the specific charge (charge-to-mass ratio) of an electron respectively.

SOLUTION

Let the layers of the positive and the negative charges (composed of electrons and ions) be distributed over planes parallel to the XY plane (Fig. E3-4). If the electrons are displaced from their mean position to the extent ξ within the ensemble of charges (called plasma) over a length d, say, along z due to a disturbing external electric field, then there would be, at the ends of the length, a layer of positive charges (composed of ions) in the form of a sheet of positive charge parallel to the XY plane. Similarly, there would be another layer of negative charge (composed of electrons) in the form of another sheet of negative charge, also parallel to the XY plane, each formed at the boundary of the collection of charges (Fig. E3-4). You may find the electric field due to each of these sheets of charge with the help of (3.46) in terms of the magnitude of the surface charge density $|\rho_s|$, and add them up to find the electric field $\vec{E}$ to which an electron is subjected.

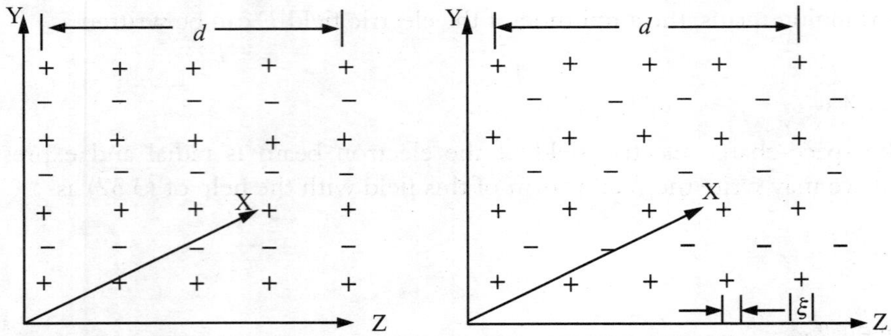

Fig. E3-4

Thus, we have

$$E\left(=\frac{|\rho_s|}{2\varepsilon_0}+\frac{|\rho_s|}{2\varepsilon_0}\right)=\frac{|\rho_s|}{\varepsilon_0}. \tag{E3-4.1}$$

We can also to relate $|\rho_s|$ to the magnitude of the volume charge density $|\rho_0|$ as

$$|\rho_s|=|\rho_0|\,\xi \tag{E3-4.2}$$

obtained by equating the charge $|\rho_0|\,\alpha\xi$ within the volume $\alpha\xi$ between two identical cross-sectional planes of the electron flow, each parallel to the XY-plane and each of area α, to the charge $|\rho_s|\,\alpha$, where ξ is the displacement of the electrons along z. However, an electron that gets displaced through ξ will now be subjected to the restoring field E. We can then easily set up the force equation for the electron as

$$m\frac{d^2\xi}{dt^2}=-\,|e|\,E,$$

which may be put as

$$\frac{d^2\xi}{dt^2}=-\,|\eta|\,E, \tag{E3-4.3}$$

where e, m and η are the charge, mass, and charge-to-mass ratio of an electron respectively. Further, with the help of (E3-4.1) and (E3-4.2), we can express (E3-4.3) as

$$\frac{d^2\xi}{dt^2}=-\,|\eta|\,\frac{|\rho_s|}{\varepsilon_0}=-\frac{|\eta|\,|\rho_0|\,\xi}{\varepsilon_0}=-\omega_p^2\xi \tag{E3-4.4}$$

where we have put

$$\omega_p^2=\frac{|\eta|\,|\rho_0|}{\varepsilon_0},$$

which gives

$$\omega_p = \left(\frac{|\eta||\rho_0|}{\varepsilon_0}\right)^{1/2}. \tag{E3-4.5}$$

Equation (E3-4.4) can be identified as that of the simple harmonic oscillatory motion of an electron with angular frequency ω_p [called the angular plasma frequency of the electron, given by (E3-4.5)].

3.1.3.A Gauss's law from Coulomb's law

We had obtained the expression (3.17) for the electric displacement $\vec{D}$ due to a point charge starting from Coulomb's law. Let us now use this expression to find the outward flux of electric displacement through a closed surface of area S, of arbitrary shape and volume, which encloses a point charge q (Fig. 3.12). For this purpose, let us first find the element of flux $d\phi_D$ through an element of area dS considered on the area S. Though the electric displacement $\vec{D}$, in general, varies over the surface S, we can regard it as being a constant over the infinitesimal area dS considered. Hence, substituting the expression (3.17) for the electric displacement in the expression (3.32) for the element of the flux of electric displacement, we obtain

$$d\phi_D = \vec{D}.d\vec{S} = \vec{D}.\vec{a}_n dS = \frac{q}{4\pi r^2}\vec{a}_r.\vec{a}_n dS \tag{3.54}$$

where $\vec{a}_r$ is the unit vector directed from the point charge q to the element of area dS and $\vec{a}_n$ the unit vector directed normally outward from the element of area dS. In general, the direction of the element of area from the point charge is different from the direction of the outward normal from the area element dS. If the angle between $\vec{a}_r$ and $\vec{a}_n$ is θ, we have $\vec{a}_r \cdot \vec{a}_n = (1)(1)\cos\theta = \cos\theta$ (Fig. 3.12), which may be then substituted in (3.54) to write

$$d\phi_D = \frac{q}{4\pi\, r^2}\cos\theta\, dS = \frac{q}{4\pi}\frac{dS\cos\theta}{r^2}. \tag{3.55}$$

Further, with the help of the following expression for the element of solid angle $d\Omega$ subtended by the element of area dS at the point charge:

$$d\Omega = \frac{dS\cos\theta}{r^2},$$

we may express (3.55) as

$$d\phi_D = \frac{q}{4\pi}d\Omega. \tag{3.56}$$

Taking the closed surface integral of (3.56) over the surface enclosing the point of charge q, we then obtain

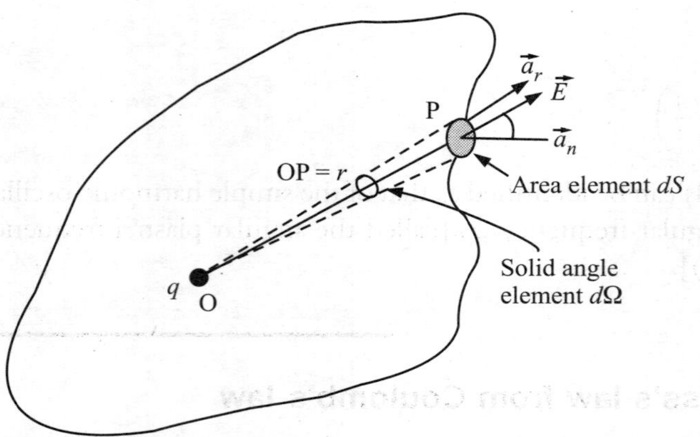

Fig. 3.12 *Showing a Gaussian volume of arbitrary shape enclosing a point charge q at O, a point P on the surface of the volume enclosure, the electric field $\vec{E}$ at P, an element of area dS over P, and the element of solid angle $d\Omega$ subtended by dS at O, the location of the point charge.*

$$\phi_D = \oint_S d\phi_D = \oint_S \frac{q}{4\pi} d\Omega = \frac{q}{4\pi} \oint_S d\Omega \,. \tag{3.57}$$

Since any closed surface makes a solid angle of 4π at any point inside the enclosure, we have

$$\oint_S d\Omega = 4\pi. \tag{3.58}$$

Therefore, substituting (3.58) in (3.57), we obtain

$$\phi_D = \frac{q}{4\pi} \oint_S d\Omega = \left(\frac{q}{4\pi}\right)(4\pi) = q. \tag{3.59}$$

It is easy to extend (3.57) to the case in which the closed surface encloses a distribution of point charges instead of a single point charge. Let the surface enclose, instead of a single point charge q, a distribution of n point charges, $q_1, q_2, q_3, \ldots, q_n$, for which we can, with the help of (3.57), write

$$\phi_D = \frac{q_1}{4\pi} \oint_S d\Omega_1 + \frac{q_2}{4\pi} \oint_S d\Omega_2 + \frac{q_3}{4\pi} \oint_S d\Omega_3 + \ldots\ldots\frac{q_n}{4\pi} \oint_S d\Omega_n \tag{3.60}$$

where $d\Omega_1, d\Omega_2, d\Omega_3, \ldots\ldots d\Omega_n$ are the elements of solid angle subtended by the element of area dS at the point charges $q_1, q_2, q_3, \ldots\ldots q_n$ respectively. Further, since, as in (3.57),

$$\oint_S d\Omega_1 = \oint_S d\Omega_2 = \oint_S d\Omega_3 = \ldots\ldots = \oint_S d\Omega_n = 4\pi \,,$$

we may replace each of the closed surface integrals in (3.60) by 4π and express (3.60) as

$$\phi_D = \frac{q_1}{4\pi}4\pi + \frac{q_2}{4\pi}4\pi + \frac{q_3}{4\pi}4\pi + \ldots\ldots\ldots \frac{q_n}{4\pi}4\pi = q_1 + q_2 + q_3 + \ldots\ldots\ldots + q_n. \tag{3.61}$$

Putting $Q = q_1 + q_2 + q_3 + \ldots\ldots + q_n$ in (3.61), we thus obtain

$$\phi_D = Q, \tag{3.62}$$

which is identical with Gauss's law (3.35). Similarly, starting from the expression (3.15) for the electric field, instead of the expression (3.17) for the electric displacement, we can easily obtain

$$\phi_E = \frac{Q}{\varepsilon} \tag{3.63}$$

which is identical with (3.36), another form of Gauss's law. Thus, in (3.62) or (3.63) we have deduced Gauss's law starting from Coulomb's law.

3.1.3.B Gauss's law in terms of a volume charge density

We can generalise the charge distribution $Q\,(= q_1 + q_2 + q_3 + \ldots\ldots\ldots + q_n)$ within a closed volume τ by a volume charge density ρ, which in general varies within the volume but can be regarded as being constant within an element of volume $d\tau$ within the volume τ. Thus, we may replace Q occurring in Gauss's law by the volume integral of the volume charge density ρ:

$$Q = \int_\tau \rho d\tau.$$

This enables us to write Gauss's law given by (3.35) and (3.36) in terms of the volume charge density with respect to the electric displacement and electric field respectively as

$$\oint_S \vec{D} \cdot d\vec{S} = \oint_S \vec{D} \cdot \vec{a}_n dS = \int_\tau \rho d\tau \tag{3.64}$$

$$\oint_S \vec{E} \cdot d\vec{S} = \oint_S \vec{E} \cdot \vec{a}_n dS = \frac{1}{\varepsilon}\int_\tau \rho d\tau. \tag{3.65}$$

3.1.3.C Gauss's law for predicting Coulomb's law

If we take Gauss's law to be a more fundamental law than Coulomb's law, it is possible to derive Coulomb's law from Gauss's law as well. For this purpose, let a spherical surface of area S and radius r enclose a point charge q_1 at its centre (Fig. 3.13). The volume enclosure so obtained is a Gaussian sphere. The problem enjoys a spherical symmetry since it is implied that the magnitude E of the electric field $\vec{E}$ will be the same at every point on the Gaussian sphere. However, the direction of $\vec{E}$ will not be the same everywhere on the sphere and it will be the direction of the point on the sphere from the centre of the sphere. Obviously then, the direction of the electric field would be radial at the point P considered on the sphere. We can then write the electric field $\vec{E}$ at the point P as (Fig. 3.13):

$$\vec{E} = E\,\vec{a}_r. \tag{3.66}$$

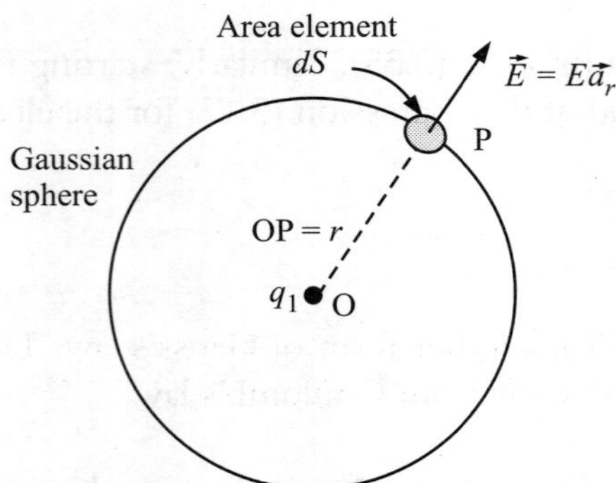

Fig. 3.13 *A Gaussian sphere of radius r enclosing the point charge q_1 at its centre, showing an element of area dS on its surface at point P.*

$$\oint_S E\vec{a}_r \cdot \vec{a}_n dS = \frac{q_1}{\varepsilon}. \tag{3.67}$$

Further, here $\vec{a}_n = \vec{a}_r$ (Fig. 3.13), in view of which we can write (3.67) as

$$\oint_S E\vec{a}_r \cdot \vec{a}_r dS = \frac{q_1}{\varepsilon}.$$

Since $\vec{a}_r \cdot \vec{a}_r = 1$, this simplifies to

$$\oint_S E dS = \frac{q_1}{\varepsilon}. \tag{3.68}$$

Since the magnitude of the electric field E is constant everywhere on the Gaussian sphere, we can take E outside the surface integral of the left-hand side of (3.68) to write

$$E \oint_S dS = \frac{q_1}{\varepsilon}. \tag{3.69}$$

Now, substituting

$$\oint_S dS = 4\pi r^2,$$

the area of the Gaussian sphere, we obtain

$$E 4\pi r^2 = \frac{q_1}{\varepsilon},$$

whence we may write the magnitude of the electric field at any point on the sphere and hence also at P as

$$E = \frac{q_1}{4\pi\varepsilon r^2}. \tag{3.70}$$

Putting (3.70) in (3.66), we may then write the electric field at the point P due to the point charge q_1 located at the centre of the sphere as

$$\vec{E} = \frac{q_1}{4\pi\varepsilon r^2}\vec{a}_r. \tag{3.71}$$

Now, if we place a point charge q_2 at the point P, then the Coulomb force on the charge due to the electric field may be written with the help of (3.14) as

$$\vec{F} = q_2 \frac{q_1}{4\pi\varepsilon r^2}\vec{a}_r,$$

which, after the rearrangement of terms, takes the form

$$\vec{F} = \frac{q_1 q_2}{4\pi\varepsilon r^2}\vec{a}_r. \tag{3.72}$$

This is identical to (3.1). Thus, you can see from (3.72) that Coulomb's law (3.1) has followed from Gauss's law.

3.2 Electric Field and Potential

In order to move a charged particle against an electric field, we have to do some work or spend an amount of energy, which is stored in the form of electric potential energy of the charged particle. This is analogous to the gravitational potential energy stored in a particle when it is lifted to a height from the ground (surface of the earth) against gravitational force. In the case of gravitational potential energy, the potential of the ground is taken as the zero reference, while in the case of the charged particle being moved against an electric field to be brought to a particular point, the zero reference of potential is taken at an infinite distance from that point. The potential V at a point is related to the potential energy W of a charged particle, treated as a point charge, carrying charge Q placed at the point as follows:

$$V = \frac{W}{Q} \tag{3.73}$$

where the potential energy W is interpreted as the work done in bringing the charge against the electric field up to the point. We can see from (3.73) that the unit of the potential V

is J/C, the unit of W being joule (J) and that of charge Q coulomb (C). The unit, J/C, of potential is widely known by its practical unit, the volt (V). We can also appreciate from (3.73) that the potential at a point is numerically equal to the potential energy of a unit point charge placed at the point.

3.2.1 Electric potential due to a point charge

Let us take a simple example in which we will find the electric potential due to a point charge q at a distance from the point charge (Fig. 3.14). The expression for the electric field $\vec{E}$ at a point P due to the point charge q at a distance r from the point has already been derived in the form of (3.15), which may be rewritten here as

$$\vec{E} = \frac{q}{4\pi\varepsilon_0 r^2}\vec{a}_r. \qquad \text{(Equation (3.15) rewritten)} \qquad (3.74)$$

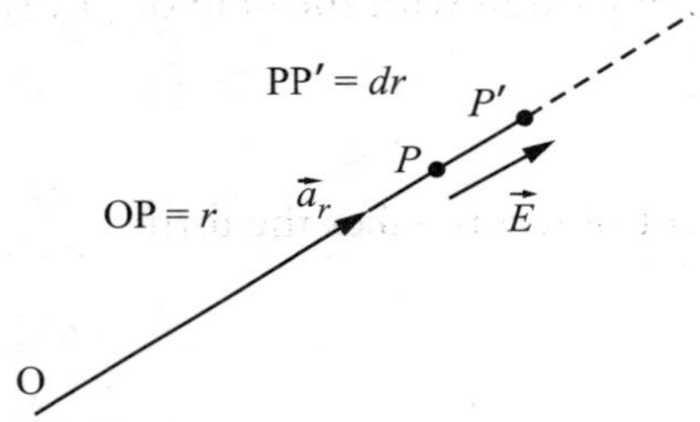

PP′ = dr

P'

P

$\vec{a}_r$

OP = r

$\vec{E}$

O

Fig. 3.14 *Showing a point charge q at point O and a point P at a distance r from it where the electric potential is to be found. P′ is a nearby point at a distance $r + dr$.*

The direction of $\vec{E}$ is that of the unit vector $\vec{a}_r$ directed from the point charge q to the point P where the electric field is so defined. We will consider the medium to be free-space ($\varepsilon = \varepsilon_0$). Let us consider another point charge Q at point P, at a distance r from the point charge q. The electrostatic force $\vec{F}$ on another point charge Q at this point, at a distance r from the point charge q, due to the electric field given by (3.74) can be written with the help of (3.14) interpreting $q = Q$ and taking $\vec{E}$ from (3.74), as follows:

$$\vec{F} = Q\vec{E} = Q\frac{q}{4\pi\varepsilon_0 r^2}\vec{a}_r = F\,\vec{a}_r, \qquad (3.75)$$

where F is the magnitude of the force given by

$$F = Q\frac{q}{4\pi\varepsilon_0 r^2}. \qquad (3.76)$$

In fact, we could have written expressions (3.75) and (3.76) with the help of Coulomb's law (3.1). The force $\vec{F}$, as can be seen from (3.75), takes on the direction of $\vec{a}_r$. In other words, this force is directed along the line joining the point charges q and Q, and it is so from the former to the latter.

Let us now find the amount of work W that needs to be done in bringing a point charge Q from infinity ($r = \infty$) [considered to be the zero reference potential, as discussed following preceding (3.73)], to the to the point P situated at a distance r from the point charge q where the potential V is sought in the presence of the electric field in the region due to the point charge q. Though the magnitude of F varies with r according to (3.76), we may take it as a constant over an infinitesimal distance dr. Therefore, the element of work done dW in moving the point charge Q through an infinitesimal distance dr from point P$'$ located at a distance $r + dr$ from the point charge q to point P at a distance r against the electrostatic force F is (Fig. 3.14)

$$dW = F dr \tag{3.77}$$

Here, it is implied that the electric field is directed from P to P$'$ for the charge interpreted as positive (Fig. 3.14). Substituting (3.76) in (3.77), we obtain

$$dW = Q\frac{q}{4\pi\varepsilon_0 r^2}dr. \tag{3.78}$$

By integrating (3.78), we obtain the work done W in bringing the point charge Q from infinity, corresponding to zero reference potential, up to the point P where the potential V is sought, against the electrostatic force, as

$$W = \int_r^\infty Q\frac{q}{4\pi\varepsilon_0 r^2}dr = \frac{Qq}{4\pi\varepsilon_0}\int_r^\infty \frac{1}{r^2}dr = \frac{Qq}{4\pi\varepsilon_0}\left[-\frac{1}{r}\right]_r^\infty = \frac{Qq}{4\pi\varepsilon_0}\frac{1}{r}. \tag{3.79}$$

Substitution of (3.79) in (3.73) then yields the following expression for the potential V due to the point charge q at a distance r from it:

$$V = \frac{W}{Q} = \frac{\frac{Qq}{4\pi\varepsilon_0}\frac{1}{r}}{Q} = \frac{q}{4\pi\varepsilon_0 r}. \tag{3.80}$$

3.2.2 Electric field in terms of potential gradient

The potential, in general, varies with space coordinates. However, we may have points that have the same potential, and the surface on which the locus of such points lie is known as an equipotential. Let us consider two nearby points, A and B. Let their potentials be $V_A = V$ and $V_B = V + dV$ respectively, differing from each other by an infinitesimal extent dV. Let point A lie on the equipotential V and the point B on the equipotential $V + dV$ (Fig. 3.15).

If we draw a normal to the equipotential V at point A, then, in general, point B will not lie in the direction of this normal. Further, the direction of the electrical field $\vec{E}$ in the region, in general, will be different from the direction of the line joining the points A and B (Fig. 3.15). Let us see what this direction will be with respect to equipotential V. Let us write the element of work done dW in moving a point charge Q against the electrical field $\vec{E}$ as

$$dW = \vec{F} \cdot d\vec{R}, \tag{3.81}$$

where $d\vec{R}$ is the distance vector, the magnitude of which is dR =AB (the length of the line joining A and B) and the direction of which is from the point A to the point B. $\vec{F}$ is the force on the point charge Q, which, with the help of (3.14) may be written as

$$\vec{F} = Q\vec{E}. \tag{3.82}$$

The element of work dW then may be expressed by substituting (3.82) in (3.81) as

$$dW = Q\vec{E} \cdot d\vec{R}. \tag{3.83}$$

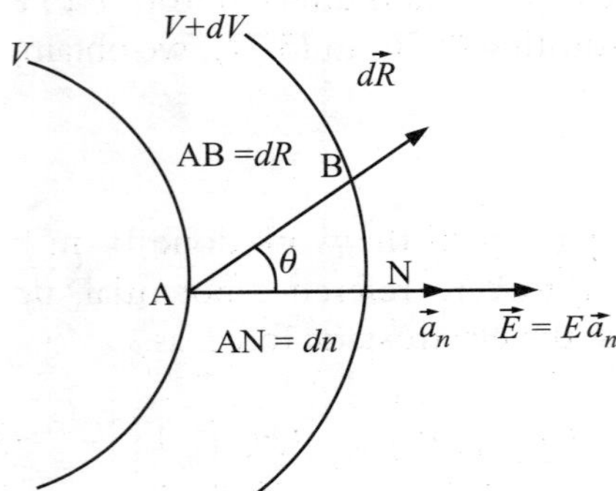

Fig. 3.15 *Depicting two nearby equipotentials, V and $V + dV$, passing through two nearby points A and B respectively in the region of an electric field for the purpose of finding a relation between the electric field and the gradient of potential.*

Further, with the help of (3.73) we can see that the work done in moving a point charge Q against the electric field $\vec{E}$ from infinity to point A is $QV_A = QV$. Similarly, the work done in moving the point charge Q to point B is $QV_B = Q(V + dV)$. Thus, we obtain the expression for the element of work done in bringing the point charge from point B to point A as:

$$dW = QV_A - QV_B = QV - Q(V + dV) = -QdV. \tag{3.84}$$

We can note from (3.84) that the element of work done dW against the electric field $\vec{E}$ would be positive for a postive point charge Q and a negative value of dV corresponding to the point B at a lower potential $[V_B(= V_B + dV)]$ than point A at a higher potential $[V_A(= V)]$. Comparing (3.83) and (3.84), we get

$$dV = -\vec{E} \cdot d\vec{R}. \tag{3.85}$$

Now, if point B lying on the equipotential $V_B = V + dV$ is chosen very close to point A lying on the equipotential $V_A = V$ $(B \to A)$, in the limit both A and B lie on the same equipotential V, the distance vector $d\vec{R}$ becomes tangential to the equipotential V at the point A. In addition, in this limit ($B \to A$), we can take $dV \to 0$, so that from (3.85) we have

$$\vec{E} \cdot d\vec{R} = 0_{(\mathrm{B} \to \mathrm{A})}. \tag{3.86}$$

Since, according to (3.86), the dot product of the vector quantities $\vec{E}$ and $d\vec{R}$ is zero for the limiting situation of B $\to$ A, the quantities $\vec{E}$ and $d\vec{R}$ become normal to each other. Therefore, since $d\vec{R}$ is tangential to the equipotential V at the point A , it turns out that $\vec{E}$ is normal to the tangent to the equipotential V at the point A. Consequently, we can express $\vec{E}$ in terms of a unit vector $\vec{a}_n$ in the direction of the normal to the equipotential V at the point A (Fig. 3.15) as follows:

$$\vec{E} = E\,\vec{a}_n. \tag{3.87}$$

Further, for the nearby points A and B lying on the equipotentials V and $V + dV$ respectively, we can write with the help of (3.85) the following:

$$dV = -E\,dR\cos\theta \tag{3.88}$$

where θ is the angle between $d\vec{R}$ and $\vec{E}$. For the nearby points A and B, we can also write $dn = dR\cos\theta$, where AN $= dn$ is the perpendicular distance between the equipotentials V and $V + dV$ at point A (Fig. 3.15). We can therefore alternatively express (3.88) as

$$dV = -E\,dn,$$

which can be further expressed as

$$\frac{\partial V}{\partial n} dn = -E\,dn.$$

This simplifies to

$$E = -\frac{\partial V}{\partial n}. \tag{3.89}$$

Plugging (3.89) in (3.87), we then obtain the following expression for the electric field:

$$\vec{E} = -\frac{\partial V}{\partial n}\vec{a}_n. \tag{3.90}$$

At this stage, let us recall the definition (2.21) of the gradient of a scalar quantity, here V, and express (3.90) as

$$\vec{E} = -\nabla V, \tag{3.91}$$

where the expression for ∇V, the gradient of V, can be obtained from (2.30)-(2.33) for curvilinear, rectangular, cylindrical and spherical polar coordinate systems respectively.

The relation (3.91) which states that the electric field is the negative gradient of electric potential is extensively used in electromagnetics. We will see later (Section 5.5) how for time-varying situations the relation (3.91) would be modified to involve a term representing the time rate of variation of a quantity called the vector potential.

Let us illustrate the relation (3.91) by taking up the simple problem of finding the capacitance of a parallel-plate capacitor, consisting of two identical large parallel conducting plates separated by a distance small compared to the dimensions of the conducting plates. We will consider the region between the plates to be filled up with a dielectric of permittivity ε (Fig. 3.16). The problem enjoys rectangular symmetry, and we can therefore use expression (2.31) for the gradient of V in rectangular coordinates (x, y, z) and hence write (3.91) as

$$\vec{E} = -\nabla V = -\left(\frac{\partial V}{\partial x}\vec{a}_x + \frac{\partial V}{\partial y}\vec{a}_y + \frac{\partial V}{\partial z}\vec{a}_z\right). \tag{3.92}$$

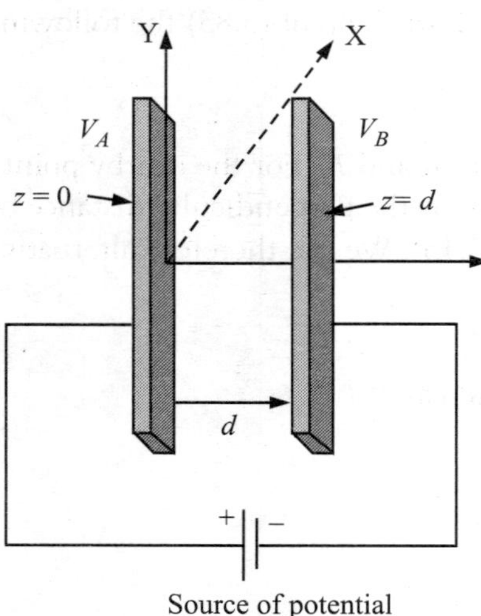

Fig. 3.16 *Two identical large parallel conducting plates separated by a small distance, forming a parallel-plate capacitor supplied by a source of potential.*

Let one of the plates of the capacitor be placed at $z = 0$ and the other at $z = d$, making the distance between the plates the plates equal to d. In this problem, since the x and y dimensions of the plates are large compared to d, we can, in general, take the varying quantity, here the potential V, to vary along z and not along x and y. In other words, the problem now becomes 'one-dimensional', and we can take $\partial/\partial z \neq 0$ and $\partial/\partial x = \partial/\partial z = 0$, which, in turn makes (3.92) as:

$$\vec{E} = -\frac{\partial V}{\partial z}\vec{a}_z. \tag{3.93}$$

As can be seen from (3.93), the electric field in the region between the plates is directed along z, and it has the magnitude E given by

$$E = -\frac{\partial V}{\partial z}. \tag{3.94}$$

If we integrate (3.94), we can find the difference between the potential V_A at the conducting plate A at $z = 0$, and the potential V_B at the conducting plate B at $z = d$ as follows :

$$\int_{V_B}^{V_A} dV = -\int_{z=d}^{z=0} E dz,$$

which, after evaluating its left side, may be expressed as

$$\int_{V_B}^{V_A} dV = [V]_{V_B}^{V_A} = V_A - V_B = -\int_{z=d}^{z=0} E dz. \tag{3.95}$$

Replacing ε_0 by ε, the permittivity of the region between the conducting plates of the capacitor, in the expression for the electric field (3.49) and then substituting this expression in (3.95), we obtain

$$V_A - V_B = -\int_{z=d}^{z=0} \frac{\rho_s}{\varepsilon} dz \tag{3.96}$$

where ρ_s is the surface charge density developed on the conducting plate A due to a source of potential that charges the conducting plates A and B of the capacitor to equal and opposite extents. Since ρ_s/ε_0 is constant, it can be taken outside the integral of (3.96) to obtain

$$V_A - V_B = -\frac{\rho_s}{\varepsilon}\int_{z=d}^{z=0} dz = -\frac{\rho_s}{\varepsilon}[z]_d^0 = -\frac{\rho_s}{\varepsilon}(0 - d) = \frac{\rho_s}{\varepsilon} d. \tag{3.97}$$

If Q is the charge on the conducting plate A and A is the area of the conducting plate, then we have $\rho_s = Q/A$, which we can substitute in (3.97) to write

$$V_A - V_B = \frac{\rho_s}{\varepsilon} d = \frac{Q}{\varepsilon A} d. \tag{3.98}$$

Therefore, the expression for the capacitance C of the parallel-plate capacitor, which is defined as

$$C = \frac{Q}{V_A - V_B} \tag{3.99}$$

can be then obtained by substituting (3.98) in (3.99) as

$$C\left(=\frac{Q}{V_A - V_B}\right) = \frac{\varepsilon A}{d} \tag{3.100}$$

which is identical to (3.5) introduced earlier in this chapter to appreciate that the practical unit of permittivity ε of a medium is F/m.

EXAMPLE E3-5

Obtain an expression for the electric field due to a short dipole in terms of the dipole moment at a point off from the axis of the dipole.

SOLUTION

The electric dipole consists of two equal and opposite point charges, q and $-q$, separated by a distance l, called the dipole length. For the short dipole, we consider l to be far less than the distance r of the point $P(r, \theta, \phi)$, where the electric field intensity is sought, from the middle of the dipole ($l << r$). In view of the condition $l << r$, we can treat the dipole as a point which enjoys spherical symmetry and hence, we can formulate the problem in spherical coordinates (r, θ, ϕ). (Recall that in the beginning of Chapter 2 we have stated that the electromagnetic problems which enjoy rectangular, cylindrical, or spherical symmetry can be treated in rectangular, cylindrical, or spherical-polar coordinate systems, respectively. Thus, a point charge can be treated in spherical-polar coordinate system, a line charge in cylindrical coordinate system and a sheet charge in rectangular coordinate system.)

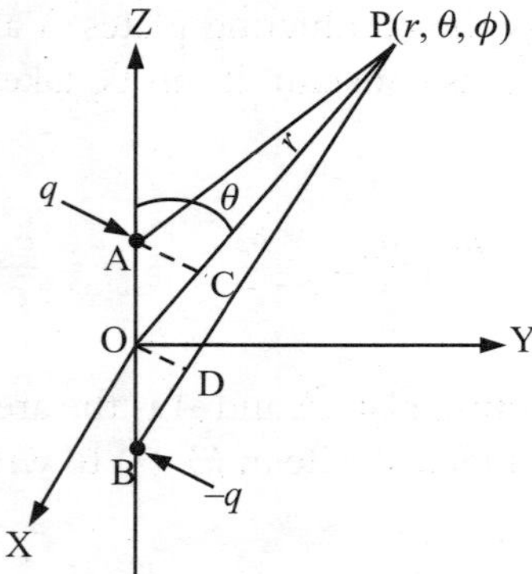

Fig. E3-5

With reference to the present problem, let us now consider the dipole to be aligned along the Z-axis. The point charges q and $-q$ are located at $z = l/2$ and $z = -l/2$, at points A and B respectively. The distance of P from the dipole where the electric field is sought is

OP $= r$, which is measured from the middle of the dipole located at the origin of O the coordinates (Fig. E3-5).

We will first find the potential V due to the dipole and then obtain the electric field $\vec{E}$ from V. The sum of the potentials due to the point charges q and $-q$ can be written with the help of (3.80) as

$$V = \frac{q}{4\pi\varepsilon_0}\left(\frac{1}{\mathrm{AP}} - \frac{1}{\mathrm{BP}}\right) \cong \frac{q}{4\pi\varepsilon_0}\left(\frac{1}{\mathrm{CP}} - \frac{1}{\mathrm{BP}}\right) \tag{E3-5.1}$$

$$= \frac{q}{4\pi\varepsilon_0}\left(\frac{1}{\mathrm{OP}-\mathrm{OC}} - \frac{1}{\mathrm{BD}+\mathrm{DP}}\right),$$

where C and D are the feet of perpendiculars drawn from A and O on OP and BP respectively. While obtaining (E3-5.1), we have taken AP $\approx$ CP, a valid approximation for a short dipole as long as $r >> l$ (Fig. E3-5). For such an approximation one can also take DP $\approx$ OP. Further, noting OP $= r$; OC $=$ OA $\cos\theta = \frac{l}{2}\cos\theta$; BD $= \frac{l}{2}\cos\theta$; and DP $\approx$ OP $= r$; we may express (E3-5.1) as

$$V = \frac{q}{4\pi\varepsilon_0}\left(\frac{1}{r-(l/2)\cos\theta} - \frac{1}{(l/2)\cos\theta + r}\right),$$

which can be simplified as

$$V = \frac{q}{4\pi\varepsilon_0}\left(\frac{(l/2)\cos\theta + r - [r-(l/2)\cos\theta]}{[r-(l/2)\cos\theta][(l/2)\cos\theta + r]}\right)$$

$$= \frac{q}{4\pi\varepsilon_0}\left(\frac{l\cos\theta}{r^2-(l^2/4)\cos^2\theta}\right),$$

which, for $r >> l$, further simplifies to

$$V = \frac{ql\cos\theta}{4\pi\varepsilon_0 r^2} = \frac{p\cos\theta}{4\pi\varepsilon_0 r^2} \tag{E3-5.2}$$

where $p = ql$ is called the dipole moment.

From the expression (E3-5.2) for V thus obtained in terms of the dipole moment p and the distance r of the point from the middle of the dipole, we can find the electric field with the help of (3.91) and (2.33) as follows:

$$\vec{E} = -\nabla V = -\left(\frac{\partial V}{\partial r}\vec{a}_r + \frac{1}{r}\frac{\partial V}{\partial\theta}\vec{a}_\theta + \frac{1}{r\sin\theta}\frac{\partial V}{\partial\phi}\vec{a}_\phi\right). \tag{E3-5.3}$$

Since the problem enjoys azimuthal symmetry, $\partial/\partial\phi = 0$ (the quantity V varying with r and θ but not with ϕ), the above equation can be expressed as

$$\vec{E} = -\nabla V = -\left(\frac{\partial V}{\partial r}\vec{a}_r + \frac{1}{r}\frac{\partial V}{\partial \theta}\vec{a}_\theta\right). \tag{E3-5.4}$$

The partial derivatives of (E3-5.2) are

$$\left.\begin{aligned} \frac{\partial V}{\partial r} &= \frac{-2p\ \cos\theta}{4\pi\varepsilon_0 r^3} \\ \frac{\partial V}{\partial \theta} &= \frac{-p\ \sin\theta}{4\pi\varepsilon_0 r^2} \end{aligned}\right\}. \tag{E3-5.5}$$

Substitution of (E3-5.5) in (E3-5.4) gives

$$\vec{E} = -\left(\frac{\partial V}{\partial r}\vec{a}_r + \frac{1}{r}\frac{\partial V}{\partial \theta}\vec{a}_\theta\right) = -\left(\frac{-2p\ \cos\theta}{4\pi\varepsilon_0 r^3}\vec{a}_r + \frac{1}{r}\frac{-p\ \sin\theta}{4\pi\varepsilon_0 r^2}\vec{a}_\theta\right). \tag{E3-5.6}$$

We can simplify (E3-5.6) to obtain the the following expression for the electric field due to the dipole:

$$\vec{E} = \frac{1}{4\pi\varepsilon_0 r^3}\, p\,(2\ \cos\theta\,\vec{a}_r + \sin\theta\ \vec{a}_\theta). \tag{E3-5.7}$$

EXAMPLE E3-6

Obtain an expression for the electric field due to a short dipole at a point situated on the axis of the dipole in terms of the dipole moment.

SOLUTION

We obtained the generalised expression (E3-5.7) for the electric field $\vec{E}$ due to a short dipole at a point away from the dipole at an angle θ from the axis of the dipole (see Example E3-5). For the point on the axis of the dipole, $\theta = 0$ (see Fig. E-3.5). So, we can put $\cos\theta = 1$, $\sin\theta = 0$ and $\vec{a}_r = \vec{a}_z$ in the generalised field expression (E3-.5.6) to obtain the desired expression as

$$\vec{E} = \frac{1}{4\pi\varepsilon_0}\frac{1}{r^3}(p(2\times 1)\vec{a}_r + 0\ \vec{a}_\theta) = \frac{p}{2\pi\varepsilon_0 r^3}\vec{a}_r = \frac{p}{2\pi\varepsilon_0 r^3}\vec{a}_z.$$

EXAMPLE E3-7

A charge of uniform volume charge density ρ is uniformly distributed inside a sphere radius R. Show that the potential V at a point r inside the sphere ($0 \le r \le R$) is given by $V = (3R^2 - r^2)\rho/6\varepsilon_0$.

SOLUTION

Using Gauss's law , we can find the expression for the magnitude of electric field E at a point r inside the spherical charge distribution ($0 \le r \le R$). The charge enclosed by the sphere of radius r is equal to the volume of the sphere times the volume charge density ρ. The charge so obtained is $(4\pi/3)r^3\rho$. According to Gauss's law, you can equate this charge to the flux of electric field through the area $4\pi r^2$ as follows [refer to (3.39)]:

$$\varepsilon_0 4\pi r^2 E = \frac{4\pi}{3} r^3 \rho,$$

whence you obtain

$$E = \frac{\rho r}{3\varepsilon_0}. \quad (0 \le r \le R) \tag{E3-7.1}$$

Now, we can equate E given by (E3-7.1) to the negative potential gradient (Section 3.2) to write

$$E = -\frac{dV}{dr} = \frac{\rho r}{3\varepsilon_0}. \quad (0 \le r \le R). \tag{E3-7.2}$$

Integrating (E3-7.2), we get

$$V = \int dV = -\int \frac{\rho r}{3\varepsilon_0} dr,$$

which can be evaluated as

$$V = -\frac{\rho}{3\varepsilon_0}\frac{r^2}{2} + C \qquad (0 \le r \le R) \tag{E3-7.3}$$

where C is the integration constant. The potential $(V)_{r=R}$ at the periphery ($r = R$) of the spherical charge distribution is obtained by putting $r = R$ in (E3-7.3) as follows:

$$(V)_{r=R} = -\frac{\rho}{3\varepsilon_0}\frac{R^2}{2} + C. \tag{E3-7.4}$$

We can obtain the integration constant C if we know $(V)_{r=a}$.

In order to find $(V)_{r=R}$, let us find the potential V outside the charge distribution at $R \le r \le \infty$ and then interpret V so found at say, $r = R$. With the help of (3.80), we can find V $(R \le r \le \infty)$ by considering the charge within the sphere of uniform volume charge density ρ as being concentrated at its centre at $r = 0$. Therefore, putting $q = (4\pi/3)R^3\rho$ (charge within the sphere of radius R) in (3.80), we obtain

$$V = \frac{\left(\frac{4\pi}{3}\right) R^3 \rho}{4\pi\varepsilon_0 r} = \frac{R^3\rho}{3\varepsilon_0 r}. \qquad (R \le r \le \infty) \tag{E3-7.5}$$

Putting $r = R$ in (E3-7.5), we obtain

$$(V)_{r=R} = \frac{R^3\rho}{3\varepsilon_0 R} = \frac{R^2\rho}{3\varepsilon_0}. \tag{E3-7.6}$$

Now, we can find the integration constant C by substituting (E3-7.6) in (E3-7.4) as follows:

$$\frac{R^2\rho}{3\varepsilon_0} = -\frac{\rho}{3\varepsilon_0}\frac{R^2}{2} + C,$$

whence we get

$$C = \frac{R^2\rho}{3\varepsilon_0} + \frac{R^2\rho}{6\varepsilon_0} = \frac{R^2\rho}{2\varepsilon_0}. \tag{E3-7.7}$$

Substituting (E3-7.7) in (E3-7.3), then we obtain

$$V = -\frac{r^2\rho}{6\varepsilon_0} + \frac{R^2\rho}{2\varepsilon_0},$$

which simplifies to

$$V = \frac{(3R^2 - r^2)\rho}{6\varepsilon_0}. \ (0 \le r \le R)$$

EXAMPLE E3-8

A parallel-plate capacitor is filled with a dielectric medium, the relative permittivity of which varies linearly from one plate of the capacitor to the other, separated by a distance d, from ε_{r1} to ε_{r2}. Obtain an expression for the capacitance of the capacitor C in terms of ε_{r1}, ε_{r2}, d, and A, the plate area, the dimensions of the plate considered large compared to d.

SOLUTION

Since the relative permittivity ε_r of the dielectric of the parallel-plate capacitor in the present problem varies linearly from $\varepsilon_r = \varepsilon_{r1}$ at one of its plates at $z = 0$ to $\varepsilon_r = \varepsilon_{r2}$ at its other plate at $z = d$, we can put the linear rate of variation of the relative permittivity of the dielectric as $(\varepsilon_{r2} - \varepsilon_{r1})/d$. Accordingly, we can express the permittivity $\varepsilon = \varepsilon_0\varepsilon_r$ of the dielectric at a distance z $(0 \le z \le d)$ as follows:

$$\varepsilon = \varepsilon_0\left(\varepsilon_{r1} + \frac{\varepsilon_{r2} - \varepsilon_{r1}}{d}z\right) \tag{E3-8.1}$$

where the direction z is perpendicular to the plates. With the help of (3.49), we can write the expression for the magnitude of the z-directed electric field E at a point inside the parallel-plate capacitor in terms of the permittivity given by (E3-8.1) and σ_s, the surface charge density on the plate of the capacitor as follows:

$$E = \frac{\sigma_s}{\varepsilon_0\left(\varepsilon_{r1} + \frac{\varepsilon_{r2} - \varepsilon_{r1}}{d}z\right)}. \tag{E3-8.2}$$

Further, with the help of (3.94) you can also express E as the negative potential gradient as

$$E = -\frac{dV}{dz}. \tag{E3-8.3}$$

Equating the right-hand sides of (E3-8.2) and (E3-8.4), we can then write

$$\frac{dV}{dz} = -\frac{\sigma_s}{\varepsilon_0\left(\varepsilon_{r1} + \frac{\varepsilon_{r2}-\varepsilon_{r1}}{d}z\right)}. \tag{E3-8.4}$$

Now, let us integrate (E3-8.4) between the limits $V = V_0$ at $z = 0$ and $V = 0$ at $z = d$, where the limiting values of of z correspond to the locations of the plates of the capacitor as mentioned preceding (E3-8.1). For a positive value of V_0, this would correspond to the plate at $z = 0$ raised to a higher potential with respect to the plate at $z = d$. Thus, the integration of (E3-8.4) so carried out within the limits as above lead to the following:

$$\int_{V=0}^{V=V_0} dV = -\frac{\sigma_s}{\varepsilon_0}\int_{z=d}^{z=0}\frac{dz}{\varepsilon_{r1} + \frac{\varepsilon_{r2}-\varepsilon_{r1}}{d}z}. \tag{E3-8.5}$$

We can evaluate (E3-8.5) as follows:

$$[V]_0^{V_0} = -\frac{\sigma_s}{\varepsilon_0}\left[\frac{\ln(\varepsilon_{r1} + \frac{\varepsilon_{r2}-\varepsilon_{r1}}{d}z)}{\frac{\varepsilon_{r2}-\varepsilon_{r1}}{d}}\right]_{z=d}^{z=0},$$

which, on substitution of the limits will give us the following expression:

$$V_0 = -\frac{\sigma_s}{\varepsilon_0}\frac{d}{\varepsilon_{r2}-\varepsilon_{r1}}[\ln(\varepsilon_{r1} - \ln(\varepsilon_{r1}+\varepsilon_{r2}-\varepsilon_{r1})],$$

which further simplifies first to

$$V_0 = -\frac{\sigma_s}{\varepsilon_0}\frac{d}{\varepsilon_{r2}-\varepsilon_{r1}}[\ln\varepsilon_{r1} - \ln\varepsilon_{r2}]$$

and then to

$$V_0 = \frac{\sigma_s}{\varepsilon_0}\frac{d}{\varepsilon_{r2}-\varepsilon_{r1}}\ln\frac{\varepsilon_{r2}}{\varepsilon_{r1}}. \tag{E3-8.6}$$

By dividing the charge on the plate ($= \sigma_s A$) by V_0 given by (E3-8.6), we can obtain the expression for C as:

$$C\left(=\frac{\sigma_s A}{V_0}\right) = \left(\frac{\varepsilon_0 A}{d}\right)\left(\frac{\varepsilon_{r2}-\varepsilon_{r1}}{\ln\frac{\varepsilon_{r2}}{\varepsilon_{r1}}}\right).$$

EXAMPLE E3-9

Find an expression for the electric field in the region between the inner and outer conductors of a coaxial cable in terms of the charge developed on a conductor, and hence obtain an expression for the capacitance per unit length of the cable.

SOLUTION

Treat the problem as one-dimensional in which the potential varies only in the radial direction. Applying Gauss's law, we can obtain the expression for the electric field at a point $a \le r \le b$ in the region between the inner and outer conductors of the coaxial cable, where a is the radius of the inner conductor and b is the inner radius of the outer conductor. The electric field so obtained is radial and becomes identical to (3.43), which may be rewritten as

$$E = \frac{\rho_L}{2\pi\varepsilon_0 r} \tag{E3-9.1}$$

where ρ_L is the charge per unit length of the inner conductor. Now, we can rewrite (3.91) relating the electric field and potential as

$$\vec{E} = -\nabla V. \tag{E3-9.2}$$

Considering the angular symmetry of the problem, we can read (E3-9.2) with the help of (2.32) as

$$\vec{E} = -\frac{\partial V}{\partial r}\vec{a}_r,$$

whence we can write

$$E = -\frac{\partial V}{\partial r}. \tag{E3-9.3}$$

Putting (E3-9.1) in (E3-9.3), we obtain

$$\frac{\partial V}{\partial r} = -\frac{\rho_L}{2\pi\varepsilon_0 r}. \tag{E3-9.4}$$

Integrating (E3-9.4) between the limits $r = a$, where the potential is considered as $V = V_0$, and $r = b$, where the potential is considered as $V = 0$, we can write

$$\int_{V=0}^{V=V_0} dV = -\frac{\rho_L}{2\pi\varepsilon_0}\int_b^a \frac{1}{r}dr. \tag{E3-9.5}$$

We can then evaluate (E3-9.5) as

$$[V]_0^{V_0} = -\frac{\rho_L}{2\pi\varepsilon_0}[\ln r]_b^a,$$

giving

$$V_0 = -\frac{\rho_L}{2\pi\varepsilon_0}(\ln a - \ln b) = \frac{\rho_L}{2\pi\varepsilon_0}(\ln b - \ln a) = \frac{\rho_L}{2\pi\varepsilon_0}\ln\frac{b}{a}. \tag{E3-9.6}$$

Thus, in (E3-9.6) we obtain a relation between the charge per unit length ρ_L developed on the conductor and the potential difference V_0 between the conductors. We can rearrange to write the expression for ρ_L as

$$\rho_L = \frac{2\pi\varepsilon_0}{\ln\frac{b}{a}} V_0. \tag{E3-9.7}$$

Dividing ρ_L by V_0, we then derive the following expression for the capacitance per unit length C of the coaxial cable:

$$C\left(=\frac{\rho_L}{V_0}\right) = \frac{2\pi\varepsilon_0}{\ln\frac{b}{a}}. \tag{E3-9.8}$$

3.3 Poisson's and Laplace's Equations

Poisson's equation is the differential or point form of Gauss's law obtained earlier in integral form in terms of a volume charge distribution as

$$\oint_S \vec{D}\cdot d\vec{S} = \oint_S \vec{D}\cdot\vec{a}_n dS = \int_\tau \rho d\tau. \quad \text{[Equation (3.64) rewritten]} \tag{3.101}$$

Let us now take an infinitesimal volume element $\Delta\tau$ over which we can take the volume charge density ρ to be fairly constant. If we now apply Gauss's law (3.101) to this volume element, we obtain

$$\oint_S \vec{D}\cdot\vec{a}_n dS \cong \rho\Delta\tau,$$

which may be expressed, by dividing it by $\Delta\tau$, as

$$\frac{\oint_S \vec{D}\cdot\vec{a}_n dS}{\Delta\tau} \cong \rho. \tag{3.102}$$

A sign of approximation is put in (3.102) since the volume charge density ρ, in general, varies in a volume and it has been taken as constant over the infinitesimal volume. We can, however, remove the sign of approximation from (3.102) by shrinking the volume element $\Delta\tau$ to zero to write the exact equation as follows:

$$\lim_{\Delta\tau\to 0} \frac{\oint_S \vec{D}\cdot\vec{a}_n dS}{\Delta\tau} = \rho. \tag{3.103}$$

However, by definition, the left-hand side of (3.103) is $\nabla\cdot\vec{D}$, the divergence of the electric displacement vector $\vec{D}$ (see the definition (2.34) of the divergence of a vector). Therefore, replacing the left-hand side of (3.103) by $\nabla \cdot \vec{D}$, we may write it as

$$\nabla \cdot \vec{D} = \rho \tag{3.104}$$

which may also be expressed, by putting $\vec{D} = \varepsilon \vec{E}$ [see (3.16)], as

$$\nabla \cdot \vec{E} = \frac{\rho}{\varepsilon}. \tag{3.105}$$

Thus, in (3.104) or (3.105) we obtain what is known as Poisson's equation that gives the expression for the volume charge density ρ at a point in terms of the electric displacement $\vec{D}$ or the electric field $\vec{E}$. In the special case of the charge density ρ being zero at a point, the Poisson's equation becomes what is known as Laplace's equation, which may be written by putting $\rho = 0$ in (3.104) and (3.105) respectively, as

$$\nabla \cdot \vec{D} = 0 \tag{3.106}$$

$$\nabla \cdot \vec{E} = 0. \tag{3.107}$$

The expression for the divergence occurring on the left-hand side of (3.104)-(3.107), in the desired system of coordinates, may be taken from (2.46)-(2.49).

EXAMPLE E3-10

Show $\nabla \cdot \vec{D} = 0$ outside a long thin line charge distribution.

SOLUTION

The electric field for the line charge distribution given by (3.31) can be rewritten as

$$\vec{E} = E\vec{a}_r = \frac{\rho_L}{2\pi\varepsilon_0 r}\vec{a}_r,$$

where ρ_l is the line charge density. From this, we can write E_r as

$$E_r = \frac{\rho_L}{2\pi\varepsilon_0 r}. \tag{E3-10.1}$$

Substitute (E3-10.1) in the following expression for divergence in the cylindrical coordinates;

$$\nabla \cdot \vec{E} = \frac{1}{r}\frac{\partial(rE_r)}{\partial r} = \frac{1}{r}\frac{d(rE_r)}{dr},$$

obtained with the help of expression (2.48) for the divergence of a vector in cylindrical coordinate system taken for the present problem, where $E_\theta = E_z = 0$ as well as $\partial/\partial\theta = 0$ (azimuthal symmetry). Thus, we obtain

$$\nabla\cdot\vec{E}=\frac{1}{r}\frac{d\left(r\frac{\rho_L}{2\pi\varepsilon_0 r}\right)}{dr}=\frac{1}{2\pi\varepsilon_0 r}\frac{d\rho_L}{dr}=\frac{1}{2\pi\varepsilon_0 r}\times 0=0. \qquad \text{(E3-10.2)}$$

Multiplying (E3-10.2) by ε_0, we obtain $\nabla\cdot\vec{D}$, the divergence of $\vec{D}=\varepsilon_0\vec{E}$ as

$$\nabla\cdot\vec{D}=\nabla\cdot\varepsilon_0\vec{E}=\varepsilon_0\nabla\cdot\vec{E}=\varepsilon_0\times 0=0.$$

3.3.1 Laplacian form of Poisson's and Laplace's Equations

Poisson's equation may be expressed in terms of electric potential V by substituting $\vec{E}=-\nabla V$ [See (3.91)] in (3.105) as

$$\nabla\cdot(-\nabla V)=\frac{\rho}{\varepsilon},$$

which can be expressed, in terms of the Laplacian operator [see (2.72)], as

$$\nabla^2 V=-\frac{\rho}{\varepsilon}. \qquad (3.108)$$

For the special case of $\rho=0$, we can then write from (3.108) the Laplace's equation in terms of the Laplacian operator as

$$\nabla^2 V=0. \qquad (3.109)$$

Illustrating the applications of the Laplacian form of Poisson's and Laplace's equations in electrostatic problems

Let us illustrate, in the following problems (a) and (b), the Laplacian form of Poisson's and Laplace's equations applicable to regions where the charges are present and absent respectively.

(a) Electric field in a region between the plates of a parallel-plate capacitor

Let us take up Laplace's equation (3.109) valid in a charge-free region and apply it to the case of a parallel-plate capacitor to find an expression for the electric field in the region between the plates of the capacitor (Fig. 3.16). The problem enjoys rectangular symmetry and, as discussed following (3.92), for plates of the capacitor large compared to the distance between the plates, the problem becomes one-dimensional, enabling us to take $\partial/\partial z\neq 0; \partial/\partial x=\partial/\partial x=0$ in the expression for $\nabla^2 V$ given by (2.75). Thus, we can write $\nabla^2 V=\partial^2 V/\partial z^2$ for this problem, which may be then substituted in (3.109) to obtain

$$\frac{\partial^2 V}{\partial z^2}=0.$$

This can be written as total derivative since the potential V depends on z as

$$\frac{d^2V}{dz^2} = 0. \tag{3.110}$$

On integration, (3.110) gives

$$\frac{dV}{dz} = A, \tag{3.111}$$

where A is an integration constant. Integrating (3.111), we obtain

$$\int dV = \int A dz,$$

which may be evaluated as

$$V = Az + B, \tag{3.112}$$

where B is an integration constant. By putting $V = V_A$, the potential at the plate A, and $z = 0$, the coordinate of the plate A in (3.112), we get the integration constant B as

$$B = V_A. \tag{3.113}$$

Similarly, putting $V = V_B$, the potential at the plate B, and $z = d$, the coordinate of the plate B in (3.112), we obtain

$$V_B = Ad + B,$$

which, making use of (3.113) becomes

$$V_B = Ad + V_A.$$

This on algebraic simplification gives the integration constant A as

$$A = -\frac{V_A - V_B}{d}. \tag{3.114}$$

Therefore, substituting (3.113) and (3.114) in (3.112), we obtain the expression for the potential in the region between the plates as

$$V = -\frac{V_A - V_B}{d}z + V_A. \tag{3.115}$$

The electric field $\vec{E}$ in the region between the plates may be obtained with the help of the following expression for the electric field that is obtained from (3.93) by changing the partial derivative to total derivative, since the potential depends only on z:

$$\vec{E} = -\frac{dV}{dz}\vec{a}_z. \tag{3.116}$$

Thus, substituting (3.115) in (3.116) and carrying out the differentiation, we obtain

$$\vec{E} = \frac{V_A - V_B}{d}\vec{a}_z. \tag{3.117}$$

From (3.117), we can easily appreciate that the electric field in the region between the plates of the capacitor is constant and that it is in the direction of $\vec{a}_z$, that is, it is directed along z, perpendicular to the plates. Further, it has the magnitude $(V_A - V_B)/d$, the potential difference between the plates divided by the distance d between the plates.

(b) Child-Langmuir's law for a planar space-charge limited diode

Let us take the space-charge-limited diode to exemplify Poisson's equation (3.108) in Laplacian form in a region where the charges are present. The diode consists of a planar cathode that emits electrons and a planar anode raised to a potential higher than that of the cathode so that the electrons emitted from the cathode are accelerated and collected at the anode. The cathode and anode dimensions are considered very large compared to the distance between the cathode and the anode so that the problem can be treated as one-dimensional and we can take the potential V in the cathode-anode region to vary only with z ($\partial/\partial z \neq 0$; $\partial/\partial x = \partial/\partial y = 0$). Under space-charge-limited condition, with the increase of the cathode temperature, even though the number of electrons emitted can increase, all the electrons emitted by the cathode are not collected by the anode. Consequently, under this condition, the anode current cannot be increased by increasing the cathode temperature; it can be increased only by increasing the anode potential. Let us now find the potential distribution in the diode between the cathode and the anode, and hence also know how the anode current depends on the anode potential with respect to the cathode. Since the cathode–anode region is now not devoid of charges, being constituted by the charge of the electrons in flight between the cathode and the anode, we have to use Poisson's equation (3.108), taking the medium permittivity ε as that of free-space ε_0. Thus, in the present one-dimensional case ($\partial/\partial z \neq 0; \partial/\partial x = \partial/\partial y = 0$), we may write (3.108) with the help of (2.75) as

$$\frac{\partial^2 V}{\partial z^2} = -\frac{\rho}{\varepsilon_0},$$

which may be expressed as follows by changing the partial derivative to the total derivative since the potential V depends only on z:

$$\frac{d^2 V}{dz^2} = -\frac{\rho}{\varepsilon_0}, \tag{3.118}$$

where ρ is the volume charge density of the space-charge constituted by the electrons. Let us next proceed to solve (3.118) to find the potential distribution V as a function of $0 \leq z \leq d$ in the region between the cathode at $z = 0$ and the anode at $z = d$, subject to the following boundary conditions at the cathode ($z = 0$):

$$z = 0, \; V = 0, \tag{3.119}$$

$$z = 0, \frac{dV}{dz} = 0. \tag{3.120}$$

Condition (3.119) implies that the zero reference of potential is taken at the cathode. Condition (3.120) is implicit in the assumption of space-charge limited operation of the diode, for which we take, under equilibrium condition, the rate of the variation of V with z or the slope of the V-versus-z plot as zero at the cathode $z = 0$, that is, $dV/dz = 0$. This also means, in view of the relation $\vec{E} = -dV/dz\vec{a}_z$, given by (3.116), that the electric field is zero at the cathode. This further implies that the negative space-charge field at the cathode due to the electrons in flight between the cathode and the anode is neutralised by the positive electrostatic field due to the applied anode potential. If the slope of potential variation were positive, the electric field, according to (3.118), would be in the negative z direction, that is, directed towards the cathode, and more electrons that carry negative charges would leave the vicinity of cathode, thereby increasing the negative space charge in the region and hence depressing the V-versus-z plot towards the zero slope at the cathode. On the other hand, if the slope were negative, the electric field according to (3.118) would be in the positive z direction, that is, directed away from the cathode. As a result, the emitted electrons carrying negative charges would be forced back to the cathode, which would reduce the negative space charge in the region and consequently lift the V-versus-z plot towards the zero slope at the cathode.

In order to solve (3.118) subject to the boundary conditions (3.119) and (3.120), let us multiply both sides of (3.118) by $2dV/dz$ to obtain the following expression:

$$2\frac{dV}{dz}\frac{d^2V}{dz^2} = -\frac{2\rho}{\varepsilon_0}\frac{dV}{dz}. \tag{3.121}$$

The left side of (3.121) may be identified as $(d/dz)[(dv/dz)^2]$, so that we may express (3.121) as

$$\frac{d}{dz}\left(\frac{dV}{dz}\right)^2 = -\frac{2\rho}{\varepsilon_0}\frac{dV}{dz}. \tag{3.122}$$

At this stage, we have to note that ρ is not independent of V. The two relations that can be used to find the relation between ρ and V are

$$\frac{1}{2}mv^2 = |e|\, V \tag{3.123}$$

and

$$J = \rho v, \tag{3.124}$$

where v is the velocity of electrons and V is the electric potential, and J is the current density, being the current per unit cross-sectional area of the electron flow, all taken at a distance z from the cathode, m the electronic mass and $|e|$ the magnitude of the electronic charge e. It is assumed that all the electrons in flight between the cathode and the anode have the same velocity at the same distance z from the cathode.

Here, (3.123) simply states the balance of the kinetic energy gained by an electron with the potential energy lost by it when it falls through a potential V, it being assumed that the initial velocity of electrons or the velocity of emission at the cathode ($z = 0$) is zero. With a little argument, the relation (3.124) will follow from the calculation of the number of electrons in a volume between two identical cross-sectional planes of the electron flow, each of area α. The time an electron would take to move from one plane to the other downstream the flow is one second, since the distance traversed by an electron is numerically equal to v, being the distance between the two cross-sectional planes considered. Therefore, you will agree that all the electrons lying within the volume $(v)(\alpha)$, between the cross-sectional planes (each of area α and separated by a distance v) will cross the down-stream cross-sectional plane within a time span of one second. (For instance, the electrons midway between the two planes would take half a second to cross the down-stream cross-sectional plane). The number of such electrons crossing the down-stream cross-sectional plane in one second is $(v)(\alpha)(n)$, which is obtained by multiplying the volume $(v)(\alpha)$ by the number per unit volume, that is, the number density of electrons n, say. The current constituted by the flow of this number of electrons is obtained by multiplying this number by the charge e carried by each electron, as $(v)(\alpha)(n)(e)$. Dividing this current by the cross-sectional area α of the electron flow, we then obtain the expression for the current density J as: $J = (v\,\alpha\,n\,e)/\alpha = nev$, which on putting $ne = \rho$ becomes $J = \rho v$, which is identical with (3.123). (The charge per unit volume or the volume charge density ρ is clearly the product of the number of electrons per unit volume n and the charge e carried by each electron).

We can then express ρ with a little algebra from (3.123) and (3.124) as

$$\rho = \frac{J}{v} = \frac{J}{\sqrt{2\frac{|e|}{m}V}} = \frac{J}{\sqrt{2\,|\eta|\,V}}, \tag{3.125}$$

where $|\eta|$ $(= |e|\,/m)$ is the magnitude of the charge-to-mass ratio of an electron. Substituting (3.125) in (3.122), we can then easily obtain

$$\frac{d}{dz}\left(\frac{dV}{dz}\right)^2 = -\frac{2\frac{J}{\sqrt{2|\eta|V}}}{\varepsilon_0}\frac{dV}{dz} = \frac{2\,|J|}{\varepsilon_0\sqrt{2\,|\eta|}}V^{-1/2}\frac{dV}{dz}dz \tag{3.126}$$

where we have put $J = -\,|J|$, keeping in mind that $J\,(= \rho v)$ carries its sign (due to the volume charge density of electrons, ρ, carrying its sign). We may now integrate (3.126) to obtain

$$\int \frac{d}{dz}\left(\frac{dV}{dz}\right)^2 dz = \int \frac{2\,|J|}{\varepsilon_0\sqrt{2\,|\eta|}}V^{-1/2}\frac{dV}{dz}dz,$$

which may be expressed as

$$\int d\left(\frac{dV}{dz}\right)^2 = \int \frac{2\,|J|}{\varepsilon_0\sqrt{2\,|\eta|}} V^{-1/2} dV. \tag{3.127}$$

This can easily be integrated to obtain

$$\left(\frac{dV}{dz}\right)^2 = \frac{2\,|J|}{\varepsilon_0\sqrt{2\,|\eta|}} \frac{V^{1/2}}{1/2} + G \tag{3.128}$$

where G is an integration constant. However, at the cathode $z = 0$, according to the boundary conditions (3.119) and (3.120), we have $V = 0$ and $dV/dz = 0$, respectively, which when substituted in (3.128) makes $G = 0$, which, in turn, makes (3.128) read as

$$\left(\frac{dV}{dz}\right)^2 = \frac{4\,|J|}{\varepsilon_0\sqrt{2\,|\eta|}} V^{1/2}. \tag{3.129}$$

Taking the square root of (3.129), we then get

$$\frac{dV}{dz} = \left(\frac{4\,|J|}{\varepsilon_0\sqrt{2\,|\eta|}}\right)^{1/2} V^{1/4},$$

which, for the purpose of carrying integration may be put as

$$\frac{dV}{V^{1/4}} = \left(\frac{4\,|J|}{\varepsilon_0\sqrt{2\,|\eta|}}\right)^{1/2} dz. \tag{3.130}$$

This may be integrated as

$$\int \frac{dV}{V^{1/4}} = \int \left(\frac{4\,|J|}{\varepsilon_0\sqrt{2\,|\eta|}}\right)^{1/2} dz$$

which, upon carrying out the integration, yields

$$\frac{V^{-1/4+1}}{-1/4+1} = \left(\frac{4\,|J|}{\varepsilon_0\sqrt{2\,|\eta|}}\right)^{1/2} z + H \tag{3.131}$$

where H is an integration constant. However, substituting the boundary condition $V = 0$ at the cathode, $z = 0$, given by (3.119), into (3.131), we get $H = 0$, which when substituted in (3.131) gives the following expression after rearrangement of terms:

$$V^{3/4} = \frac{3}{4}\left(\frac{4\,|J|}{\varepsilon_0\sqrt{2\,|\eta|}}\right)^{1/2} z. \tag{3.132}$$

If we now square (3.132) and rearrange the terms, we obtain the expression for $|J_0|$, the magnitude of current density in the diode, known as the Child-Langmuir's law as follows:

$$|J| = \frac{4}{9}\sqrt{2\,|\eta|}\varepsilon_0 \frac{V^{3/2}}{z^2}. \tag{3.133}$$

Thus, according to (3.133), $|J|$ is directly proportional to $V^{3/2}$ and inversely proportional to z^2. It is because of the index of power 3/2 with V that Child-Langmuir's law (3.133) is also known as 3/2-power law. Further, for identical planar cathode and anode having the same area, say, A, we can find the anode current per unit area, $|J_0|\ (= I_0/A)$ of the diode by putting $z = d$ and $V = V_0$ in (3.133) as follows:

$$|J_0|\,(= I_0/A) = \frac{4}{9}\sqrt{2\,|\eta|}\varepsilon_0 \frac{V_0^{3/2}}{d^2},$$

where I_0 is the anode current and V_0 the anode voltage and d is the distance between the cathode and the anode.

In Example (a) above, we have seen how we can solve the Laplacian form of Laplace's equation to obtain the electric field in the region between the plates of a parallel-plate capacitor where there are no charges by remembering that the Laplace's equation is valid in a charge-free region. Similarly, in Example (b) above, we have seen how to solve the Laplacian form of Poisson's equation, valid for a charge-present region, to derive Child–Langmuir's law for a space-charge-limited diode.

EXAMPLE E3-11

Obtain an expression for the electric field at a point located between a pair of large conducting plates forming a wedge separated by an infinitesimal insulating gap in terms of the potential difference between the plates and the wedge angle.

SOLUTION

This problem enjoys cylindrical symmetry, allowing us to treat it in cylindrical coordinates. Let the two plates be considered as hinged at Z-axis with an infinitesimal insulating separation between them to form a wedge (Fig. E3-11). Let the plate that is at a higher potential $V = V_0$ at $\theta = 0$ lie on the XZ plane, while the other plate at the potential $V = 0$ lie on a plane at $\theta = \theta_0$, where θ_0 is the wedge angle. We also consider the problem as one-dimensional, considering the potential to vary only with θ $(\partial/\partial r = \partial/\partial z = 0; \partial/\partial\theta \neq 0)$.

We can write Laplace's equation for the region between the conducting plates with the help of (3.109) in conjunction with the expression for the Laplacian of potential given by (2.77) in view of the potential varying only with θ $(\partial/\partial r = \partial/\partial z = 0)$ as

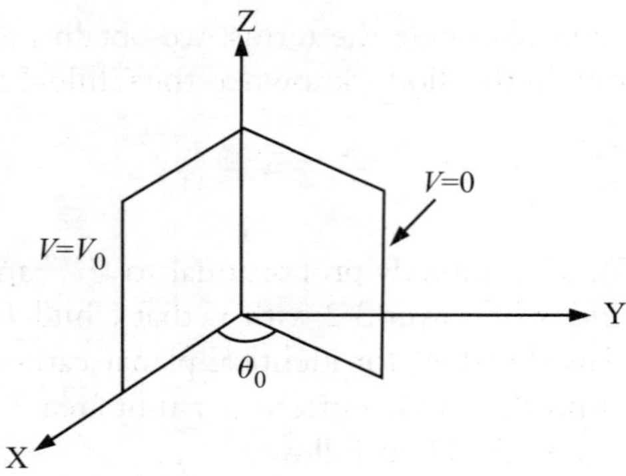

Fig.E3-11

$$\nabla^2 V = \frac{1}{r^2}\frac{\partial^2 V}{\partial \theta^2} = 0,$$

whence we get

$$\frac{d^2 V}{d\theta^2} = 0. \qquad \text{(E3-11.1)}$$

Integrating (E3-11.1), we get

$$\frac{dV}{d\theta} = C_1,$$

which on second integration gives

$$V = C_1\theta + C_2, \qquad \text{(E3-11.2)}$$

where C_1 and C_2 are integration constants that can be determined from the potential values defined. Thus, putting $V = V_0$ and $\theta = 0$ in (E3-11.2), we get

$$C_2 = V_0, \qquad \text{(E3-11.3)}$$

and also putting $V = 0$ and $\theta = \theta_0$ in (E3-11.2), we get

$$0 = C_1\theta_0 + C_2. \qquad \text{(E3-11.4)}$$

Substituting (E3-11.3) in (E3-11.4) we then obtain

$$C_1 = -\frac{V_0}{\theta_0}. \qquad \text{(E3-11.5)}$$

Plugging (E3-11.3) and (E3-11.5) in (E3-11.2) we then obtain

$$V = -\frac{\theta}{\theta_0}V_0 + V_0 = V_0\left(1 - \frac{\theta}{\theta_0}\right). \qquad \text{(E3-11.6)}$$

We can find $\vec{E}$ in the region between the hinged plates with the help of (3.91) in conjunction with (2.32), the latter, in the present context, to be read taking $\partial/\partial r = \partial/\partial z = 0$ as follows:

$$\vec{E} = -\nabla V = -\frac{1}{r}\frac{dV}{d\theta}\vec{a}_\theta. \tag{E3-11.7}$$

Substituting the derivative of (E3-11.6) in (E3-11.7) we then obtain

$$\vec{E} = \frac{1}{r}\frac{V_0}{\theta_0}\vec{a}_\theta.$$

EXAMPLE E3-12

Two long identical coaxial conducting cones are placed tip-to-tip with an infinitesimal gap between them to form a bi-conical geometry. Obtain an expression for the electric field at a point located between the cones in terms of the potential difference between the plates and half-cone angle.

SOLUTION

The problem enjoys spherical symmetry, allowing us to treat it in spherical coordinates (r, θ, ϕ). Let the common axis of the cones coincide with the Z-axis such that the tip of the cones coincides with the origin of the coordinates, where the cones are electrically insulated by an infinitesimal gap (Fig. E3-12). Thus, the two cones, each of half-cone angle θ_0 , are situated respectively in the regions of positive and negative z. Further, let the potential of the cone $(\theta = \theta_0)$ lying in the positive z region be $V = V_0$ and that of the cone $(\theta = \pi - \theta_0)$ lying in the negative z region be $V = 0$. We also consider the problem as one-dimensional, considering the potential to vary only with θ $(\partial/\partial r = \partial/\partial\phi = 0;\ \partial/\partial\theta \neq 0)$.

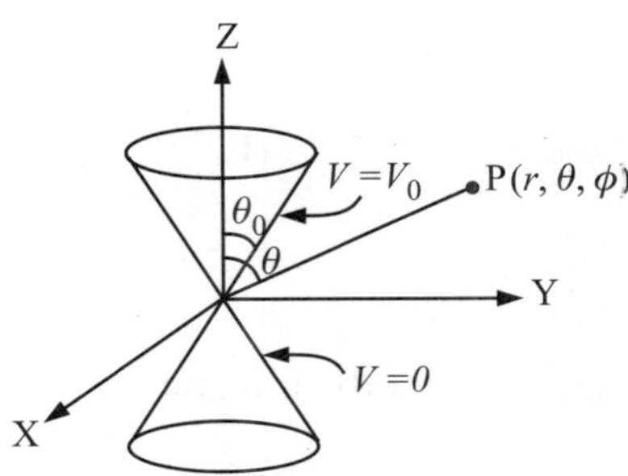

Fig. E3-12

Laplace's equation for the region between the cones may be written with the help of (3.109) in conjunction with the expression for the Laplacian of potential given by (2.77), in view of the potential varying only with θ $(\partial/\partial r = \partial/\partial\phi = 0)$ as

$$\nabla^2 V = \frac{1}{r^2 \sin\theta}\frac{\partial}{\partial\theta}\left(\sin\theta\frac{\partial V}{\partial\theta}\right)$$

$$= \frac{1}{r^2 \sin\theta}\frac{d}{d\theta}\left(\sin\theta\frac{dV}{d\theta}\right) = 0,$$

whence we get

$$\frac{d}{d\theta}(\sin\theta\frac{dV}{d\theta}) = 0. \tag{E3-12.1}$$

Integrating (E3-12.1), we obtain

$$\sin\theta\frac{dV}{d\theta} = C_1,$$

which may be written as

$$\frac{dV}{d\theta} = \frac{C_1}{\sin\theta} = C_1 \csc\theta, \tag{E3-12.2}$$

where C_1 is an integration constant. Further integration of (E3-12.2) results in

$$V = C_1\int \csc\theta d\theta = C_1 \ln\left|\tan\frac{\theta}{2}\right| + C_2 \tag{E3-12.3}$$

where C_2 is another integration constant. We can find the constants from the potentials on the cones: $V = V_0$ at $\theta = \theta_0$ and $V = 0$ at $\theta = \pi - \theta_0$.

Putting $V = V_0$ and $\theta = \theta_0$ in (E3-12.3), we get

$$V_0 = C_1 \ln\left|\tan\frac{\theta_0}{2}\right| + C_2, \tag{E3-12.4}$$

and, similarly, putting $V = 0$ and $\theta = \pi - \theta_0$ in (E3-12.3), we get

$$C_2 = -C_1 \ln\left|\tan\frac{\pi - \theta_0}{2}\right|. \tag{E3-12.5}$$

Solving (E3-12.4) and (E3-12.5), we find the integration constants as

$$\left.\begin{aligned} C_1 &= \frac{V_0}{2}\frac{1}{\ln\left|\tan\frac{\theta_0}{2}\right|} \\ C_2 &= -\frac{V_0}{2}\frac{\ln\left|\cot(\frac{\theta_0}{2})\right|}{\ln\left|\tan\frac{\theta_0}{2}\right|} \end{aligned}\right\}. \tag{E3-12.6}$$

Substituting (E3-12.6) in (E3-12.3), we then obtain

$$V = \frac{V_0}{2}\frac{\ln\left|\tan\frac{\theta}{2}\right|}{\ln\left|\tan\frac{\theta_0}{2}\right|} - \frac{V_0}{2}\frac{\ln\left|\cot(\frac{\theta_0}{2})\right|}{\ln\left|\tan\frac{\theta_0}{2}\right|},$$

which can be simplified to obtain the following expression for the potential V in the region between the conducting cones:

$$V = \frac{V_0}{2}\left[1 + \frac{\ln\left|\tan\frac{\theta}{2}\right|}{\ln\left|\tan\frac{\theta_0}{2}\right|}\right]. \qquad \text{(E3-12.7)}$$

We can find $\vec{E}$ in the region between the conducting cones with the help of (3.91) in conjunction with (2.33), the latter in the present context to be read, taking $\partial/\partial r = \partial/\partial\phi = 0$, as follows:

$$\vec{E} = -\nabla V = -\frac{1}{r}\frac{\partial V}{\partial\theta}\vec{a}_\theta = -\frac{1}{r}\frac{dV}{d\theta}\vec{a}_\theta. \qquad \text{(E3-12.8)}$$

We can now easily find $dV/d\theta$, which is to be put in (E3-12.8) to find $\vec{E}$ as

$$\frac{dV}{d\theta} = \frac{V_0}{2\ln\left|\tan\frac{\theta_0}{2}\right|}\frac{d}{d\theta}\ln\left|\tan\frac{\theta}{2}\right|$$

$$= \frac{V_0}{2\ln\left|\tan\frac{\theta_0}{2}\right|}\frac{1}{\tan\frac{\theta}{2}}\frac{1}{2}\sec^2\frac{\theta}{2}$$

The above, in view of the relation $\sin\theta = 2\sin(\theta/2)\cos(\theta/2)$, can be expressed as

$$\frac{dV}{d\theta} = \frac{1}{2}\frac{V_0}{\ln\left|\tan\frac{\theta_0}{2}\right|}\frac{1}{\sin\theta}. \qquad \text{(E3-12.9)}$$

Substituting (E3-12.9) in (E3-12.8), we then obtain

$$\vec{E} = -\frac{1}{r}\frac{dV}{d\theta}\vec{a}_\theta = -\frac{V_0}{2\ln\left|\tan\frac{\theta_0}{2}\right|}\frac{1}{r\sin\theta}\vec{a}_\theta.$$

EXAMPLE E3-13

Derive the following expression for the depletion width X of a pn-junction diode in terms of (i) the donor and the acceptor number densities N_d and N_a in the n-type and the p-type regions of the diode respectively, and (ii) the barrier potential V_B:

$$X = \left(\frac{2\varepsilon V_B}{|e|}\right)^{1/2}\left[\left(\frac{1}{N_a(1+N_a/N_d)}\right)^{1/2} + \left(\frac{1}{N_d(1+N_d/N_a)}\right)^{1/2}\right],$$

where $|e|$ is the magnitude of the electron charge and ε is the permittivity of the diode region.

SOLUTION (HINT)

For the necessary background, you may go through a book on semiconductor devices (for instance, Denisse Le Croissettee, *Transistors*, Prentice-Hall of India, New Delhi, 1968).

Treat the problem in the rectangular coordinate system. In the one-dimensional case, take the potential to vary only along x, taking the origin $x = 0$ at the junction of the diode. Subject to the boundary condition implied in the problem, solve Poisson's equation in the Laplacian form: $d^2V/dx^2 = -\rho/\varepsilon$ (3.118), where ρ is the volume charge density of the depletion regions, to be interpreted as $\rho = -\,|e|\,N_a$ in the p-type region and $\rho = |e|\,N_d$ in the n-type region. N_a is the number density of the acceptors, equal to the equilibrium number density of holes in the p-type region and N_d is the number density of the donors, equal to the equilibrium number density of electrons in the n-type region. (You have already used one-dimensional Poisson's equation while deducing Child–Langmuir's law in Section 3.3). In this problem, there are the negatively and the positively charged depletion regions in the p-type and the n-type regions respectively, each adjacent to the junction of the two regions. At the end of the depletion region comprising negative charges in the p-type region, take the electric potential as V_1 , which is essentially negative. Similarly, at the end of the depletion region comprising positive charges in the n-type region, take the electric potential as V_2, which is essentially positive. Thus, the barrier potential becomes $V_B = |V_1| + V_2$ (V_1 being negative). As for the boundary conditions of the problem, take the potentials $V = 0$ at $x = 0$, the junction of the two regions, $V = V_1$ at $x = -X_1$, the end of the depletion region in the p-type region, and $V = V_2$ at $x = X_2$, the end of the depletion region in the n-type region. Also, take the potential gradient $dV/dX = 0$ at both the ends of the depletion regions: $x = -X_1$ and $x = X_2$. Hence, you can find X_1 and X_2, and, therefore, you can add them to obtain the desired expression for the depletion width $X(= X_1 + X_2)$.

Appendix A3.1: Electric displacement

A dielectric material is said to be polarised in the presence of an external electric field $\vec{E}$ which exerts a force to slightly separate the bound positive charges in the atomic nuclei from the bound negative charges of electrons of the material by a 'displacement', small compared to atomic dimensions. These opposite charges move in opposite directions thereby creating electric dipoles. With reference to this phenomenon of polarisation of a dielectric, we can define the electric displacement $\vec{D}$ as

$$\vec{D} = \varepsilon_0\vec{E} + \vec{P}, \qquad \text{(A3.1)}$$

where $\vec{P}$ is called the polarisation density of dipole moments. In many substances $\vec{P}$ is proportional to $\vec{E}$, enabling us to express it as

$$\vec{P} = \varepsilon_0 \chi_e \vec{E}, \tag{A3.2}$$

where χ_e is a dimensionless quantity called the electric susceptibility of the dielectric, the factor ε_0 ensuring that the dimension of $\vec{P}$ which occurs as the second term of the right-hand side of (A3.1) becomes the same as that of its first term.

We can rather easily appreciate the definition (A3.1) of electric displacement through (A3.2) with reference to a simple example of a parallel-plate capacitor in which the medium between the plates is filled in by a dielectric material of permittivity ε. However, let us first consider the situation in which there exists a free-space region between the plates. The magnitude of the electric field E in this region directed from the positive plate to the negative plate may be written, recalling (3.49) as:

$$E = \frac{\rho_s}{\varepsilon_0}. \tag{A3.3}$$

Now, if a dielectric replaces the free-space region between the plates in the presence of an external electric field, the phenomenon of polarisation of the dielectric will take place [as explained preceding (A3.1)] and consequently, due to induced electric dipoles, the surface charges with induced surface charge densities $-\rho_i$ and $+\rho_i$, say, on the surfaces in contact with the positive and negative plates respectively of the capacitor will be developed. As a result, the magnitude of the electric field E between the plates will be reduced from that given by (A3.3) to the following:

$$E = \frac{\rho_s - \rho_i}{\varepsilon_0}. \tag{A3.4}$$

Treating the induced surface charge density ρ_i as proportional to the electric field E, and taking it as $\rho_i = \varepsilon_0 \chi_e E$, we can express (A3.4) as

$$E = \frac{\rho_s - \varepsilon_0 \chi_e E}{\varepsilon_0}, \tag{A3.5}$$

where χ_e is the electric susceptibility of the dielectric, a dimensionless quantity. Notice that the factor ε_0 introduced in the second term of the numerator of the right-hand side of (A3.5) makes the second term have the dimension of the electric field, that is, the dimension of the left-hand side.

We can now rearrange the terms of (A3.5) to express it as

$$E = \frac{\rho_s}{\varepsilon_0(1 + \chi_e)}. \tag{A3.6}$$

Further, interpreting

$$\varepsilon = \varepsilon_0(1+\chi_e) = \varepsilon_0\varepsilon_r \tag{A3.7}$$

as the permittivity of the dielectric and

$$\varepsilon_r = 1+\chi_e \tag{A3.8}$$

as the relative permittivity of the dielectric, we can alternatively express (A3.6) as

$$E = \frac{\rho_s}{\varepsilon_0(1+\chi_e)} = \frac{\rho_s}{\varepsilon_0\varepsilon_r} = \frac{\rho_s}{\varepsilon}. \tag{A3.9}$$

At this stage, if we define the electric displacement D as

$$D = \varepsilon E \tag{A3.10}$$

we can then write with the help of (A3.7) and (A3.10)

$$D = \varepsilon_0 E + \varepsilon_0 \chi_e E,$$

which can be identified as

$$D = \varepsilon_0 E + P \tag{A3.11}$$

interpreting

$$P = \varepsilon_0 \chi_e E. \tag{A3.12}$$

Thus, we have seen that in the present example of the parallel-plate capacitor that the expressions for the electric displacement (A3.11) through (A3.12) are identical to (A3.1) through (A3.2) except for the non-appearance of the vector notation, since the example of the parallel-plate capacitor considered is essentially a one-dimensional problem, the electric field being directed from one plate to another in a direction normal to the plates.

SUMMARISING NOTES

✓ Basic concepts of static electric field or electrostatics have been recapitulated in this chapter.

✓ Rationalised MKS system has been introduced while writing the expression for the electrostatic force between two point charges called Coulombs law. The rationalisation factor 4π has been introduced in Coulomb's force expression, thereby removing the appearance of the factor 4π from many extensively used expressions derived from Coulomb force expression, for instance, Gauss's law.

✓ Coulomb's law has been used to find

- electrostatic force on a point charge due to the distribution of other point charges; and

– electric field due to a point charge as well as the electric field due to a distribution of point charges.

✓ Finding the electric field by Coulomb's law has been illustrated in examples of charge distribution such as line and surface charge distributions.

✓ Gauss's law greatly simplifies the finding of electric field in problems that enjoy rectangular, cylindrical or spherical symmetry, which otherwise would become quite involved if Coulomb's law were used instead.

✓ Concept of electric potential has been introduced.

✓ Expression for the electric field in terms of the gradient of electric potential has been derived and illustrated in examples, for instance, in the problem of finding electric field due to a short dipole as well as in the problem of finding the capacitance of a parallel-plate capacitor and the capacitance of a coaxial cable.

✓ Poisson's and Laplace's equations have been derived from first principles and their use illustrated in the problems of finding the electric field in the region between a pair of conductors at different electric potentials by solving Laplace's equation expressed in terms of Laplacian of potential.

✓ Poisson's equation in terms of Laplacian of potential has been solved to deduce Child-Langmuir's law for a space-charge-limited diode, which relates the anode current of the diode to the anode potential as well as anode-to-cathode distance of the diode.

Review Questions

Narrative type

3.1 Write down the units of charge, electric field, electric potential, electric displacement (electric flux density), line charge density, surface charge density, volume charge density, capacitance, and permittivity.

3.2 From the formula of the capacitance of a parallel-plate capacitor, deduce that the unit of permittivity is F/m.

3.3 How can you validate that the unit of electric field is N/C and, alternatively, it is V/m?

Hint: See (3.14) and (3.93).

3.4 Show that the unit of permittivity $C^2/N\text{-}m^2$ obtainable from Coulomb's law is equivalent to its practical unit V/m.

3.5 Verify that the expression (3.30), that is, $\vec{E} = \rho_L/(2\pi\varepsilon_0 d)\vec{a}_\perp$ is dimensionally correct.

3.6 Use Coulomb's force equation (3.1) to explain the phenomenon that the unlike charges attract and like charges repel.

3.7 Express Coulomb's force equation as $\vec{F} = q_1q_2/(4\pi\varepsilon r^3)\vec{r}$ in terms of the distance vector $\vec{r}$.

3.8 Prove Gauss's law from Coulomb's law.

3.9 Prove Coulomb's law from Gauss's law.

3.10 Give examples of the application of Gauss's law in finding the electric field in each of the problems which enjoy the rectangular, cylindrical and spherical symmetries respectively.

3.11 Prove that the electric field $\vec{E}$ is equal to the negative gradient of the electric potential V $(\vec{E} = -\nabla V)$.

3.12 Expand the expression for the electric field in terms of the gradient of potential in generalised curvilinear coordinates system and interpret it in the rectangular, cylindrical and spherical coordinate systems.

3.13 Derive from first principles the expression for the capacitance of a parallel-plate capacitor in terms of the area of the plates, distance between plates, and permittivity of the dielectric filling the region between the plates. Assume large plate dimensions compared to the distance between plates.

3.14 Show that the magnitude of electric field in the region between the plates of a parallel-plate capacitor connected to a constant-voltage source becomes doubled if the distance between the plates is reduced to half.

Hint: The magnitude of the electric field is V/d, where V is the voltage across the capacitor connected to a constant-voltage source and d is the distance between the plates. The value of V remains the same when the distance between the plates is reduced to $d/2$.

3.15 Deduce an expression for the capacitance per unit length of a coaxial cable in terms of the radii of the inner and outer conductors and the permittivity of the dielectric filling the region between the conductors.

3.16 Derive Poisson's equation in terms of the divergence of electric displacement: $\nabla \cdot \vec{D} = \rho$, and express it in generalised curvilinear coordinate system and hence, read the expression in rectangular, cylindrical and spherical coordinate systems.

3.17 Write down Poisson's equation in 'Laplacian' form in generalised curvilinear coordinate system and interpret it for the rectangular, cylindrical and spherical coordinate systems.

3.18 Show that the static electric field $\vec{E}$ due to a point charge is 'irrotational', that is, $\nabla \times \vec{E} = 0$.

Hint: From the expression for the electric field due to a point charge (3.15), you may see that, in this problem the following relations hold good: $E_r \neq 0$, $E_\theta = 0$, $E_\phi = 0$ and $\partial/\partial r \neq 0$, $\partial/\partial\theta = 0$, $\partial/\partial\phi = 0$. Invoking these relations in (2.68), you can easily show that $\nabla \times \vec{E} = 0$, thereby proving that $\vec{E}$ is irrotational.

3.19 Use one-dimensional Poisson's equation $d^2V/dz^2 = -\rho/\varepsilon_0$ to derive Child-Langmuir's law for a space-charge-limited diode.

3.20 Assuming that the angular momentum of an electron of a hydrogen atom moving in a circular orbit around the nucleus of the atom can, according to Bohr's model, take the discrete values $mvr = nh/2\pi$, where n is a positive integer, r is the orbital radius of the electron, m and v are the mass and the velocity of the orbital electron and h $(= 6.62 \times 10^{-34}$ Js) is Planck's constant, show that the permitted values of the radius r_n and the energy W_n of the nth orbital electron are given by

$$r_n = \frac{n^2 h^2 \varepsilon_0}{\pi\, m\, |e|^2} \text{ and } W_n = -\frac{m\,|e|^4}{8n^2 h^2 \varepsilon_0^2}$$

respectively, where $|e|$ is the magnitude of the electron's charge, the subscript n referring to the nth orbit of the electron.

Hint: The nucleus of the hydrogen atom has the charge $|e|$, which attracts the orbital electron of the atom of charge $-|e|$. Balance the electrostatic force of attraction (centripetal force) on the orbital electron of charge $-|e|$ due to the positive charge $|e|$ of the nucleus of the atom with the centrifugal force mv^2/r to find v , which in turn can be put in the given discrete values of the angular momentum of the orbital electron to obtain the desired expression for r_n. Next, find the potential energy of charge $-|e|$ of the orbital electron due to the positive charge $|e|$ of the nucleus of the atom and the kinetic energy $(1/2)mv^2$ of the orbital electron. Then add up these two energies for the total energy, thereby obtaining the desired expression for W_n.

Review questions with solutions

3.21 Verify Gauss's law with reference to the problem of a long line charge of uniform line charge distribution.

Answer: Consider a Gaussian cylindrical surface of length l, say, and of radius r, coaxial with the uniform line charge distribution of line charge density ρ_l. Let the electric displacement at a point distant r from the axis of the Gaussian surface, that is, at a perpendicular distance r, be $\vec{D}$. The left-hand side of Gauss's law (3.35), interpreting therein $\vec{a}_n = \vec{a}_r$ and $\vec{D} = D\vec{a}_r$ for the Gaussian cylindrical surface can be found as

Left-hand side of Gauss's law

$$= \oint_S \vec{D} \cdot \vec{a}_n dS = \oint_S (D\vec{a}_r) \cdot (\vec{a}_r) dS = \oint_S D \cdot dS = DS, \quad (1)$$

where $S = 2\pi r l$ (2)

is the area of the cylindrical surface of radius r and length l of the Gaussian volume enclosure. Substituting (2) in (1) we may then write

$$\text{Left-hand side of Gauss's law} = \oint_S \vec{D} \cdot \vec{a}_n dS = DS = 2\pi r l D. \quad (3)$$

The value of $D(= \varepsilon_0 E)$ that can be found from Coulomb's law may be written with the help of (3.31) as

$$\vec{D} = \varepsilon_0 \vec{E} = \frac{\rho_L}{2\pi r}\vec{a}_r,$$

which gives $D = \dfrac{\rho_L}{2\pi r}$. (4)

Substituting (4) in (3), we obtain

$$\text{Left-hand side of Gauss's law} = \oint_S \vec{D}\cdot\vec{a}_n dS = \frac{\rho_L}{2\pi r}S. \quad (5)$$

Substitution of (2) in (5) results in

$$\text{Left-hand side of Gauss's law} = \oint_S \vec{D}\cdot\vec{a}_n dS = \frac{\rho_L}{2\pi r}2\pi r l = \rho_L l. \quad (6)$$

Express (6) in term of the charge Q within the Gaussian volume enclosure, which is given by

$$Q = \rho_L l. \quad (7)$$

Substituting (7) in (6), we obtain

$$\text{Left-hand side of Gauss's law} = Q. \quad (8)$$

The right-hand side of Gauss's law (3.35), which is the charge Q within the Gaussian volume, may be put as

$$\text{Right-hand side of Gauss's law} = Q. \quad (9)$$

Comparing (8) and (9), we can infer that Gauss's law is verified with reference to the given problem.

3.22 Show that the electric field observed at a point P due to a point charge remains unaffected if it is surrounded by a thin conducting sphere such that the centre of the sphere is located at the point charge and the radius of the sphere is less than the distance between the point charge and the point of observation of the field.

Answer: The point charge q being at the centre of the conducting spherical shell, of thickness much smaller than the radius R of the shell, will induce uniformly distributed charge on the spherical surface. The total induced charge is q , with uniform charge density $q/(4\pi R^2)$. So, although the original charge is shielded, the induced charge will produce the same electric field at P as that due to the original charge as if there were no conducting shell present.

(We can also appreciate the above understanding by a simple analytical approach using the concept of electromagnetic boundary conditions (developed later as well; see Chapter 7) as follows. We consider three regions: 1, i and 2. The regions 1 and 2 refer to the free-space regions inside and outside the sphere, while the region i refers to the material region of the conducting shell. We take the finite thickness of the conducting sphere to be much lesser than the radius of the sphere, so that the inner and the outer radii can be taken as approximately the same.

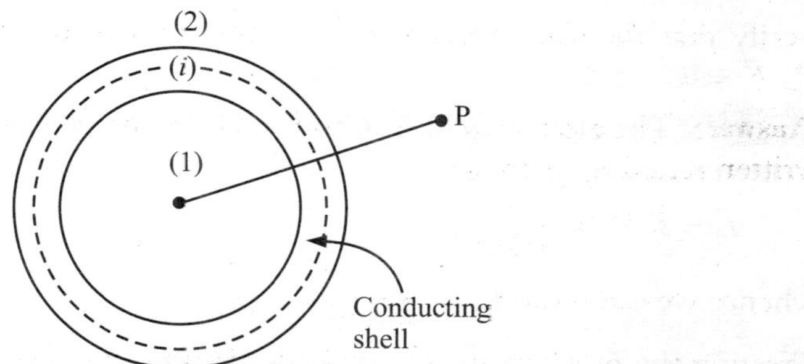

We can use the general electromagnetic boundary condition (also see (7.6) in Chapter 7) to write

$$D_{ri} - D_{r1} = \rho_{s1} \tag{1}$$

and $$D_{r2} - D_{ri} = \rho_{s2}, \tag{2}$$

where subscripts 1 and 2 refer to the free-space regions 1 and 2 respectively, and subscript i to the material region of the conducting shell of finite thickness. ρ_{s1} and ρ_{s2} are the surface charge densities induced at the inner and outer surfaces of the spherical shell respectively.

You can put $D_{ri} = 0$ considering the material shell to be a perfect conductor. This enables us to write (1) and (2) as

$$-D_{ri} = \rho_{s1} \tag{3}$$

$$D_{r2} = \rho_{s2}. \tag{4}$$

Adding (3) and (4), we can write

$$D_{r2} - D_{ri} = \rho_{s2} + \rho_{s1}. \tag{5}$$

If we take the inner and outer radii of the shell to be the same, we can approximately take ρ_{s1} and ρ_{s2} to be equal in magnitude but with opposite signs, enabling us to put$\rho_{s2} + \rho_{s1} = 0$ in (5) and write the latter as

$$D_{r2} - D_{ri} = 0. \tag{6}$$

Putting

$$\left.\begin{aligned} D_{ri} &= \varepsilon_0 E_{ri} \\ D_{r2} &= \varepsilon_0 E_{r2} \end{aligned}\right\}$$

in (6), we then obtain

$$E_{r2} = E_{ri}. \tag{7}$$

From (7) we can infer that the conducting shell is not "seen" by the observation point outside the spherical shell. This analysis corroborates the earlier found conclusion by physical reasoning that although the original charge is 'shielded', the induced charge produces the same electric field at P as if there were no conducting shell present.

3.23 Verify that the static electric field $\vec{E}$ due to a point charge is solenoidal, that is, $\nabla \cdot \vec{E} = 0$.

Answer: The electric field $\vec{E}$ outside a point charge q at a distance r from it may be written recalling (3.15) as

$$\vec{E} = E_r \vec{a}_r = \frac{q}{4\pi\varepsilon_0 r^2}\vec{a}_r,$$

whence we can write $E_r = \dfrac{q}{4\pi\varepsilon_0 r^2}$. (1)

Treating the problem in spherical coordinates, we can now write with the help of (2.49), noting that in the present problem $E_\theta = E_\phi = 0$:

$$\nabla \cdot \vec{E} = \frac{1}{r^2}\frac{\partial(r^2 E_r)}{\partial r} = \frac{1}{r^2}\frac{d(r^2 E_r)}{dr}. \quad (2)$$

Substituting (1) in (2), we then obtain

$$\nabla \cdot \vec{E} = \frac{1}{r^2}\frac{d\left(r^2 \frac{q}{4\pi\varepsilon_0 r^2}\right)}{dr} = \frac{1}{4\pi\varepsilon_0 r^2}\frac{dq}{dr} = \frac{1}{4\pi\varepsilon_0 r^2} \times 0 = 0.$$

Since $\nabla \cdot \vec{E} = 0$, the electric field is solenoidal. (The solenoidal vector, here $\vec{E}$, is a 'divergence-free vector' corresponding to zero net flux through a closed volume. The nomenclature originates from the Greek word sōlēn for 'solenoid', meaning 'pipe', representing a fixed volume in the present context. Thus, $\vec{E}$ is solenoidal if $\nabla \cdot \vec{E} = 0$.)

Numerical type

3.24 Show 1 C = 3×10^9 statC.

Answer: Statcoulomb (statC) or franklin (Fr) is electrostatic unit (esu) of electric charge used in the CGS (centimetre-gram-second) system of unit.

Take the force as 1 dyne = 10^{-5}N between two point charges, each q, separated by a distance of 1 cm = 10^{-2}m. Hence, find q with the help of Coulomb's law (3.1) as

$$q = \frac{1}{3 \times 10^9}\ \text{C},$$

giving

$$1\ \text{statC} = \frac{1}{3 \times 10^9}\ \text{C},$$

whence you obtain

$$1\ \text{C} = 3 \times 10^9\ \text{statC (esu)}.$$

3.25 Two positive point charges separated by a distance of 1 m are repelled by the Coulomb force of 18×10^9N. Calculate the values of the individual charges if the combined charge value is given as 3C.

Answer: Denoting by q_1 and q_2 the amounts of individual point charges, we can use Coulomb's law (3.1) to find the product of the charges as

$$q_1 q_2 = 4\pi\varepsilon_0 r^2 F, \tag{1}$$

where $F$is the magnitude of the Coulomb force between the point charges. In the present problem, $F = 18 \times 10^9$ N and $r = 1$ m, which may be put in (1) to get

$$q_1 q_2 = 4\pi\varepsilon_0 18 \times 10^9 \text{C}^2. \tag{2}$$

Putting $4\pi\varepsilon_0 = 1/(9 \times 10^9)$ F in (1), we can express (2) as

$$q_1 q_2 = \frac{18 \times 10^9}{9 \times 10^9} = 2\ \text{C}^2. \tag{3}$$

Further, in this problem,

$$q_1 + q_2 = 3\ \text{C}. \tag{4}$$

Substituting q_2 from (3) in (4), we obtain

$$q_1^2 - 3q_1 + 2 = 0. \tag{5}$$

The quadratic equation (5) may be solved for q_1 as

$$q_1 = \frac{3 \pm 1}{2}\ \text{C},$$

giving $q_1 =$ either 2 C or 1 C. If we take $q_1 = 2$ C, we get from (4) $q_2 = 1$ C. On the other hand, if we take $q_1 = 1$ C, we get from (4) $q_2 = 2$ C. Therefore, from either of these possible solutions, we obtain that one of the point charges has a charge of 1 C while the other has a charge of 2 C.

3.26 Two point charges each $q = 10^{-19}$ C are located at A $(0, 0, a)$ and B $(0, 0, -a)$. Calculate the electric field at P (a, a, a) taking $a = 10^{-9}$m.

Answer: With the help of (3.1)-(3.3), you can find the electric field $\vec{E}_{\text{PA}}$ at P (a, a, a) due to the point charge q at A $(0, 0, a)$ as

$$\vec{E}_{\text{PA}} = \frac{q}{4\pi\varepsilon_0(\sqrt{2}a)^3} a(\vec{a}_x + \vec{a}_y). \tag{1}$$

Similarly, the electric field $\vec{E}_{\text{PB}}$ at P (a, a, a) due to the point charge q at B $(0, 0, -a)$ is

$$\vec{E}_{\text{PB}} = \frac{q}{4\pi\varepsilon_0(\sqrt{6}a)^3} a(\vec{a}_x + \vec{a}_y + 2\vec{a}_z). \tag{2}$$

Therefore, adding (1) and (2) we obtain the electric field $\vec{E}_{\text{P}} (= \vec{E}_{\text{PA}} + \vec{E}_{\text{PB}})$ at P due to both the point charges put together as

$$\vec{E}_P(=\vec{E}_{PA}+\vec{E}_{PB})=\frac{q}{4\pi\varepsilon_0(\sqrt{2}a)^3}a(\vec{a}_x+\vec{a}_y)+\frac{q^2}{4\pi\varepsilon_0(\sqrt{6}a)^3}a(\vec{a}_x+\vec{a}_y+2\vec{a}_z),$$

which may be simplified as

$$\vec{E}_P=\frac{q}{4\pi\varepsilon_0 a^2}[(\frac{1}{(\sqrt{2})^3}+\frac{1}{(\sqrt{6})^3})(\vec{a}_x+\vec{a}_y)+\frac{2}{(\sqrt{6})^3}\vec{a}_z],$$

and calculated as

$$\vec{E}_P=\frac{q}{4\pi\varepsilon_0 a^2}[(0.354+0.068)(\vec{a}_x+\vec{a}_y)+0.136\vec{a}_z] \quad (3)$$

Putting $q = 10^{-19}$C (given), $a = 10^{-9}$m (given) and $4\pi\varepsilon_0 = 4\pi \times (1/36\pi) \times 10^{-9} = 1/(9 \times 10^9)$F/m in (3), we can calculate

$$\vec{E}_P = 9 \times [(0.422)(\vec{a}_x+\vec{a}_y)+0.136\vec{a}_z] \times 10^8 \text{ V/m},$$

which simplifies to

$$\vec{E}_P = (3.798\vec{a}_x + 3.798\vec{a}_y + 1.224\vec{a}_z) \times 10^8 \text{ V/m}.$$

3.27 Show that a maximum amount of charge equal to 330μ C can be distributed uniformly over a sphere of radius 1 m in air while avoiding an electric breakdown, taking the breakdown field of air as 30 kV/cm.

Hint: The magnitude of electric field immediately outside the sphere may be found with the help of, for instance Gauss's law as $E = q/(4\pi\epsilon_0 r^2)$, in which we have to put $E(= 30$ kV/cm$) = 3 \times 10^6$V/m (breakdown threshold field) and $\varepsilon_0 = (1/36\pi) \times 10^{-9}$ F/m to obtain the value $q = 3.3 \times 10^{-4}$ C $= 330\mu$C.

3.28 An electron moves along x with a velocity 2 cm/s in a region of uniform electric field of 5 kV/cm along z. What is the force $\vec{F}$ on the electron?

Answer: The force $\vec{F}$ on the electron carrying charge $-|e|$ due to the electric field $\vec{E} = |e| E\vec{a}_z$ is $\vec{F} = -|e|\vec{E} = -|e| E\vec{a}_z$. Putting $E = 5$ kV/cm $= 5 \times 10^3 \times 10^2 = 5 \times 10^5$ V/m in the foregoing expression for $\vec{F}$ and remembering $|e| = 1.6 \times 10^{-19}$ C, we get

$$\vec{F} = -5 \times 10^5 \times 1.6 \times 10^{-19}\vec{a}_z = -8 \times 10^{-14}\vec{a}_z \text{ N}.$$

3.29 What is the area of the plates required for constructing a parallel-plate capacitor of capacitance 100 pF using two metal plates with mica spacer ($\varepsilon_r = 5.4$) that separates the plates by a distance of 10^{-2} cm.

Answer: With the help of (3.100) we can write

$$A = \frac{Cd}{\varepsilon} = \frac{Cd}{\varepsilon_0\varepsilon_r}. \quad (1)$$

Substitute the given data are: $d = 10^{-2}$cm$= 10^{-4}$m, $\varepsilon_r = 5.4$ and 100 pF $= 100\times10^{-12}$F in (1) and calculate

$$A = \frac{Cd}{\varepsilon} = \frac{Cd}{\varepsilon_0 \varepsilon_r} = \frac{100 \times 10^{-12} \times 10^{-4}}{8.854 \times 10^{-12} \times 5.4} = 2.1 \times 10^{-4}\ \text{m}^2 = 2.1\ \text{cm}^2.$$

Multiple-choice questions

3.30 A large sheet of charge of surface charge density -4 C/m^2 is uniformly distributed over XY plane ($z = 0$) in free space. The electric displacement $\vec{D}$ at a distance $z = -3$ cm from the sheet of charge is

(a) $\vec{D} = 2\vec{a}_z$ C/m^2 (b) $\vec{D} = -4\vec{a}_z$ C/m^2

(c) $\vec{D} = -1.33\vec{a}_z$ C/m^2 (d) $\vec{D} = 1.33\vec{a}_z$ C/m^2.

Hint: Recall the expression (3.47) to find $\vec{E}$ in which to interpret $\vec{a}_n = -\vec{a}_z$. From $\vec{E}$ you can obtain $\vec{D} = \varepsilon_0 \vec{E}$.

3.31 The electric field in the region between two large plane sheets perpendicular to Z-axis of uniform charge densities ρ_s and $-\rho_s$, respectively, separated by a small distance compared to the sheet dimensions, is

(a) $\vec{E} = 0$ (b) $\vec{E} = \frac{\rho_s}{\varepsilon_0}\vec{a}_z$

(c) $\vec{E} = \frac{\rho_s}{2\varepsilon_0}\vec{a}_z$ (d) $\vec{E} = \frac{2\rho_s}{\varepsilon_0}\vec{a}_z$

3.32 A point charge q is placed in a region of uniform electric field of magnitude E in a direction that makes an angle 45^0 with either of z and y axes on the YZ ($x = 0$) plane. If the point charge now moves along z, the force $\vec{F}$on it is

(a) $\vec{F} = q\dfrac{E}{\sqrt{2}}(\vec{a}_z + \vec{a}_y)$ (b) $\vec{F} = q\dfrac{E}{\sqrt{2}}\vec{a}_z$

(c) $\vec{F} = q\frac{E}{\sqrt{2}}\vec{a}_y$ (d) $\vec{F} = qE\vec{a}_z$

3.33 Consider two point charges: one at A ($z = -d$) and the other at B ($z = d$), and take the following fours situations: (i) the charge at A is q and that at B is $-q$; (ii) the charge at A is q and that at B is q; (iii) the charge at A is $-q$ and that at B is $-q$; and (iv) the charge at A is $-q$ and that at B is q. The electric field at the mid-point ($z = 0$) between A and B is zero for the situations

(a) (i) and (ii) (b) (ii) and (iii)

(c) (iii) and (iv) (d) (iv) and (i)

3.34 The point charge q' at the centre of a system of point charges, each q, situated at the corners of a square will keep the system in equilibrium if

(a) $q' = -4q$ (b) $q' = 0$

(c) $q' = -[(1 + 2\sqrt{2})/4]q$ (d) $q' = -[(1 + 4\sqrt{2})/4]q$

3.35 The force $\vec{F}$ on a point charge 2 C, which moves. with a velocity 3 m/s in the direction of x and placed in a region of uniform electric field 4 V/m along y, is

(a) $\vec{F} = 6\vec{a}_x$ N (b) $\vec{F} = 6\vec{a}_y$ N

(c) $\vec{F} = 8\vec{a}_x$ N (d) $\vec{F} = 8\vec{a}_y$ N

3.36 The current density in a space-charge-limited diode

(a) increases with the potential difference between the anode and cathode and the decrease in the distance between them.

(b) decreases with the potential difference between the anode and cathode and the decrease in the distance between them.

(c) increases with the potential difference between the anode and cathode and the decrease in the distance between them.

(d) increases with the potential difference between the anode and cathode and is independent of the distance between them.

Answers

3-30 (a) 3-31 (b) 3-32 (a) 3-33 (b) 3-34 (c) 3-35 (d) 3-36 (c)

4

Basic Concepts of Static Magnetic Fields

In this chapter we will see that Coulomb's and Gauss's laws of static magnetic fields or magnetostatics follow analogously from the corresponding laws of static electric fields or electrostatics. Two more basic laws involving steady current and steady magnetic field will be dealt with in this chapter, namely Biot-Savart's law and Ampere's circuital law. You will also be recapitulating the concept of Lorentz force law experienced by a moving charge in a magnetic field and extending it to find the force experienced by a current-carrying conductor placed in a magnetic field.

4.1 Laws of Magnetostatics

In this section we will develop two useful concepts of magnetostatics (static magnetic field) that follow directly from the corresponding concepts of electrostatics: Coulomb's law (Section 4.1.1) and Gauss's law (Section 4.1.2).

4.1.1 Coulomb's law of magnetostatics, magnetic field and flux and flux density due to a magnetic charge

We had introduced in Section 3.1.3 the electric dipole comprising two equal and opposite charges separated by a distance; this came within the purview of electrostatics. Analogously, a permanent magnet or a magnetised material is considered a magnetic dipole in magnetostatics. For instance, if we take a bar magnet, its magnetic properties are concentrated near its two ends known as the poles of the magnet. We can assign a pole strength or 'magnetic *point* charge' to each of this pair of poles. The two poles of a bar magnet are called the (geographical) north seeking, or simply the 'north pole', and the (geographical) south seeking, or simply the 'south pole' of the magnet. The north pole of a magnet is associated with a 'positive' magnetic point charge, and similarly, the south pole with a 'negative' magnetic point charge. Interestingly, by breaking a magnet into two pieces, we cannot separate out the north and the south poles; rather, these two pieces would each become an independent magnet with individual north and south poles at its two ends. In other words, although both electric positive and negative charges can freely exist without requiring both to co-exist, magnetic positive and negative charges cannot freely exist—they have to always exist together. Further, two equal and opposite point charges, each of value q_m and separated by a distance, can be referred to as a magnetic dipole, analogous to an electric dipole (see also Examples E3-4 and E3-5 in Chapter 3). The concept of the magnetic dipole will be further illustrated in Section 4.2.

The effects of a magnet using the laws of magnetostatics can be developed in analogy with the laws of electrostatics. Thus, the force between a pole of a magnet and a pole of another magnet can be quantified by a law analogous to Coulomb law (3.1) of electrostatics as follows:

$$\vec{F} = \frac{\mu}{4\pi}\frac{q_{m1}q_{m2}}{r^2}\vec{a}_r, \tag{4.1}$$

where q_{m1} and q_{m2} are the magnetic charges assigned to the two poles of the two magnets respectively, separated by a distance r, and μ is a medium property called the permeability. Thus, the permeability μ of a medium in magnetostatics is analogous to the permittivity ε of the medium in electrostatics. For a free-space as well as for a non-magnetic medium, the permeability is $\mu = \mu_0 = 4\pi \times 10^{-7}$ H/m, which is lower than the permeability of a magnetic medium. That the practical unit of μ is H/m would become clear later when we develop the expression for the inductance of a solenoid, the unit of the inductance being Henry or H [see the discussion following (4.48)]. The ratio of the permeability of a medium to the permeability of free space, $\mu_r (= \mu/\mu_0)$, is called the relative permeability of the medium. While the electrostatic Coulomb force between two electrostatic point charges is the largest in free-space medium owing to ε occurring in the denominator of the electrostatic Coulomb force expression (3.1), the magnetostatic Coulomb force is lower in free-space or a non-magnetic medium than in a magnetic medium as μ appears in the numerator of the magnetostatic Coulomb force expression (4.1). Further, you will be able to easily appreciate with the help of (4.1) and by following the same line of argument of electrostatics [refer to the explanation preceding equation (3.3)] that like magnetic poles or magnetic charges repel, while unlike magnetic poles or magnetic charges attract each other.

The magnetic flux density $\vec{B}$ is defined in terms of magnetic field $\vec{H}$, analogous to the definition of electric displacement $\vec{D}$ in terms of the electric field $\vec{E}$ (3.16) as follows:

$$\vec{B} = \mu\vec{H}. \tag{4.2}$$

In this expression defining the dependence of $\vec{B}$ on $\vec{H}$, we had taken μ to be isotropic for conventional media, just as we had taken the medium permittivity ε to be isotropic in the expression defining the dependence of $\vec{D}$ on $\vec{E}$ given by (3.16). However, these simple relations, (3.16) and (4.2), need to be modified for non-conventional media such as anisotropic, ferrite and chiral media (see Appendix A4.3). (The present book deals with electromagnetic behaviours of conventional media. The electromagnetic behaviours of non-conventional media have been kept outside the scope of the book).

Further, if a magnetic point charge q_m is placed in a region of magnetic flux density $\vec{B}$, then we can write an expression analogous to the electrostatic expression (3.14) for the force experienced by the magnetic charge as follows:

$$\vec{F} = q_m \vec{B}. \tag{4.3}$$

Similarly, we can define the magnetic flux density $\vec{B}$ due to a magnetic charge q_m at a distance r by an expression analogous to the electrostatic expression (3.15) as follows:

$$\vec{B} = \frac{\mu}{4\pi} \frac{q_m}{r^2} \vec{a}_r. \tag{4.4}$$

Further, in view of (4.2), we can write the expression for the magnetic field $\vec{H}$ due to a magnetic charge q_m, analogous to the electrostatic expression (3.17), as:

$$\vec{H} = \frac{q_m}{4\pi r^2} \vec{a}_r. \tag{4.5}$$

If the electric charge q in electrostatics and the magnetic charge q_m in magnetostatics are analogous, then, comparing the force equations (3.14) and (4.3), we find that the electric field $\vec{E}$ and the magnetic flux density $\vec{B}$ are analogous. This is also apparent when we compare (3.15) and (4.4), the expressions for $\vec{E}$ and $\vec{B}$ involving a medium property, namely permittivity ε and permeability μ respectively. The expression (3.17) for the electric displacement $\vec{D}$ and the expression (4.5) for the magnetic field $\vec{H}$, on the other hand, are each devoid of the medium property. Thus, $\vec{D}$ and $\vec{H}$ can be considered to be analogous quantities. It is worth summarising the analogous electrostatic and magnetostatic quantities and expressions (Table 4.1).

4.1.2 Gauss's law and Poisson's equation of magnetostatics

Now, before writing Gauss's law of magnetostatics in a form analogous to (3.35) or (3.36), let us recall that free magnetic charges do not exist. A magnetic charge is always accompanied by an equal and opposite magnetic charge as in a permanent magnet. Consequently, if we take a bar magnet that can be considered a magnetic dipole and surround its north pole (representing its positive magnetic point charge) by a closed volume, then the magnetic flux lines originating from the north pole leave the magnet after emerging out of the closed volume and then renter the closed volume to pass through the south pole (negative magnetic point charge) and, finally, end up at the north pole (positive magnetic point charge) from where they started. Thus, as many flux lines enter a volume enclosing the north pole of a magnet as the numbers that leave it. Similarly, you will appreciate that as many flux lines leave a volume enclosing the south pole (negative

Table 4.1 *Analogous electrostatic and magnetostatic quantities and expressions*

Electrostatic quantities/expressions	Magnetostatic quantities/expressions
ε	μ
$\vec{E}$	$\vec{B}$
$\vec{D}$	$\vec{H}$
$\vec{F} = \frac{q_1 q_2}{4\pi\varepsilon\, r^2}\vec{a}_r$	$\vec{F} = \frac{\mu}{4\pi}\frac{q_{m1}q_{m2}}{r^2}\vec{a}_r$
$\vec{F} = q\vec{E}$	$\vec{F} = q_m\vec{B}$
$\vec{E} = \frac{q}{4\pi\varepsilon\, r^2}\vec{a}_r$	$\vec{B} = \frac{\mu\, q_m}{4\pi\, r^2}\vec{a}_r$
$\vec{D} = \frac{q}{4\pi\, r^2}\vec{a}_r$	$\vec{H} = \frac{q_m}{4\pi\, r^2}\vec{a}_r$

magnetic charge) of a magnet as the numbers that enter it. Thus, magnetic flux lines always form closed loops. In other words, magnetic flux lines are continuous (unlike electric flux lines that are discontinuous, originating from the positive charge of an electric dipole and terminating on its negative charge). Therefore, in view of the continuity of magnetic flux lines, we can put the net magnetic flux going out of a closed volume as zero and write the following Gauss's law of magnetostatics—corresponding to Gauss's law of electrostatics expressed earlier as (3.35) and (3.36)—in terms of the closed surface integral of the magnetic flux density $\vec{B}$ or magnetic field $\vec{H}$, that is, in terms of the flux of $\vec{B}$ and $\vec{H}$ respectively:

$$\oint_S \vec{B}\cdot d\vec{S} = \oint_S \vec{B}\cdot\vec{a}_n dS = 0 \tag{4.6}$$

and

$$\oint_S \vec{H}\cdot d\vec{S} = \oint_S \vec{H}\cdot\vec{a}_n dS = 0. \tag{4.7}$$

Recall that in (4.2) we had introduced magnetic flux density $\vec{B}$ in terms of magnetic field $\vec{H}$, just as we had in (3.16) introduced electric displacement $\vec{D}$ in terms of electric field $\vec{E}$. Similarly, as we can appreciate from (3.35) that the unit of ϕ_D, the electric flux of $\vec{D}$, is C and that of electric displacement or electric flux density $\vec{D}$ is C/m^2 (see Review Question 3.1). We can now refer to (4.6) to put the unit of the magnetic flux as weber or Wb and that of magnetic flux density $\vec{B}$ as Wb/m^2 or tesla or T.

Now, if we begin from (4.6) and proceed to obtain its differential form in the way we did while deriving Poisson's equation of electrostatics (3.104) from Gauss's law of electrostatics (3.101), we obtain the expression

$$\nabla \cdot \vec{B} = 0. \tag{4.8}$$

Similarly, beginning from (4.7), we obtain

$$\nabla \cdot \vec{H} = 0. \tag{4.9}$$

We can interpret (4.8) or (4.9) as the Poisson's equation of magnetostatics.

Thus, you will appreciate that Gauss's law of magnetostatics (equations (4.6) or (4.7) and Poisson's equation of magnetostatics [equations (4.8) or (4.9)] are, in fact, manifestations of the absence of free magnetic charges (poles) or that of the continuity of magnetic flux lines.

4.2 Biot-Savart's Law

The element of magnetic field $d\vec{H}$and the element of magnetic flux density $d\vec{B} = (\mu_0 d\vec{H})$ due to a direct current through an element of length of a filamentary conductor (say, a solid-cylindrical conductor of circular cross-section of radius approaching zero) is given by the following expression known as the Biot-Savart's law:

$$\left.\begin{aligned} d\vec{H} &= \frac{i\,d\vec{l} \times \vec{a}_R}{4\pi R^2} \\ d\vec{B} &= (\mu_0 d\vec{H}) = \mu_0 \frac{i\,d\vec{l} \times \vec{a}_R}{4\pi R^2} \end{aligned}\right\}, \tag{4.10}$$

where i is the current through the conductor and $d\vec{l}$ the element of the length vector of magnitude dl whose direction is the same as that of the current through the conductor. R is the distance of a the point P from the current element where the magnetic field is sought , $\vec{a}_R$ being a unit vector directed from the current element to the point (Fig. 4.1). Thus, $i\,d\vec{l}$ here represents the current element.

One can use (4.10) to find the magnetic field $d\vec{H}$ due to a current element and then multiply it by the permeability μ of a magnetic material to find the magnetic flux density $d\vec{B}$ in the material [see (4.2)], thereby removing the ambiguity about what part of the field comes from an external current and what comes from the material itself.

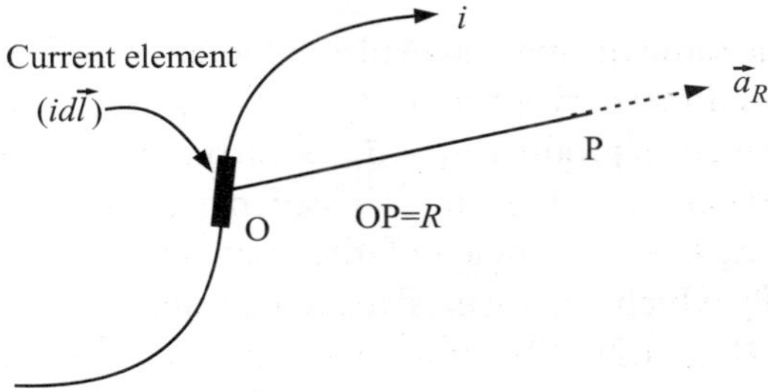

Fig. 4.1 *A current element and the point at a distance where the magnetic field is sought.*

In order to illustrate (4.10), let us take the example of a thin straight conducting wire of infinite length carrying a current and find the magnetic field due to it at a distance from it. The approach we will use is to find the magnetic field at a point due to an infinitesimal length of the wire with the help of Biot-Savart's law (4.8) and integrate it for the contribution from the entire length of the wire (Fig. 4.2).

The problem enjoys cylindrical symmetry since the magnetic field has the same magnitude at any perpendicular distance r from the wire, all around a circle of radius r; however, its direction changes from point to point on the circle, taking the azimuthal direction at the point where the magnetic field is sought.

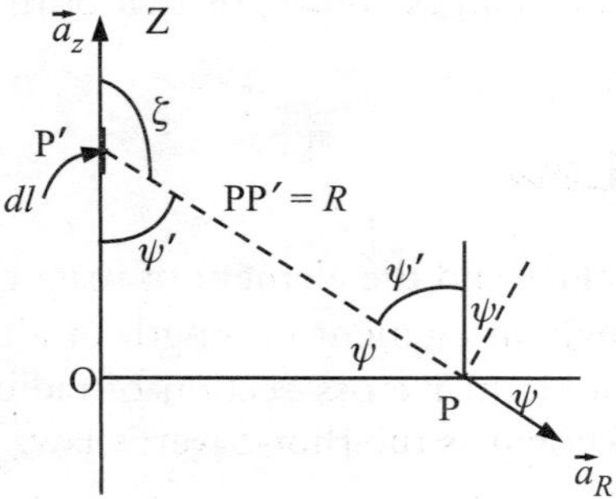

Fig. 4.2 *Showing an infinitely long filamentary current-carrying straight conducting wire carrying a direct current and a nearby point where the magnetic field due to it is to be computed with the help of Biot-Savart's law.*

Let the wire in the present problem be coincident with the Z-axis, extending from $z = -\infty$ to $z = +\infty$; the origin $z = 0$ is at O. Let the point where the magnetic field is to be found be located at point P at a perpendicular distance OP$= r$ from the wire, that is, from the Z-axis (Fig. 4.2). The magnetic field due to a current element extending over length dz located at P′ at a distance z from the origin on the Z-axis can be then written with the help of (4.8), putting therein $d\vec{l} = dz\vec{a}_z$, where $\vec{a}_z$ is a unit vector in the direction of increasing z, as follows (Fig. 4.2):

$$d\vec{H} = \frac{i\,dz\,\vec{a}_z \times \vec{a}_R}{4\pi R^2} \tag{4.11}$$

where $\vec{a}_R$ and R have the same meaning as explained following (4.10). Thus, here, P′P $= R$ and the unit vector $\vec{a}_R$ is directed from P′ to P.

The magnitude of the cross product $\vec{a}_z \times \vec{a}_R$ occurring in the right-hand side of (4.11) is $(1)(1)\sin\varsigma = \sin\varsigma$, where ς is the angle between the vectors $\vec{a}_z$ and $\vec{a}_R$ (Fig. 4.2). Further, angle ς is related to the angles of the triangle ΔOP′P as $\varsigma = \pi - \psi'$, where ψ' is the angle $\angle$OP′P, which, in turn is related to angle ψ as $\psi' = \pi/2 - \psi$, where ψ is the angle $\angle$OPP′ (Fig. 4.2). Therefore, we can write the magnitude of $\vec{a}_z \times \vec{a}_R$ as $\sin\varsigma = \sin(\pi - \psi') = \sin\psi' = \sin(\pi/2 - \psi) = \cos\psi$. Further, the cross product $\vec{a}_z \times \vec{a}_R$ is in the azimuthal direction, in the direction of $\vec{a}_\theta$, where $\vec{a}_\theta$ is the unit vector in the azimuthal direction. That is, it is in the direction of increasing θ, which is perpendicular to and going into the plane of the paper (Fig. 4.2), it being implied that the problem is treated in the cylindrical coordinate system (r, θ, z). Therefore, in view of the above magnitude and direction of the cross product $\vec{a}_z \times \vec{a}_R$, we can express

$$\vec{a}_z \times \vec{a}_R = \cos\psi\, \vec{a}_\theta. \tag{4.12}$$

Substituting (4.12) in (4.11), we can then write the element of magnetic field at point P due to the current element at P′as (Fig. 4.2):

$$d\vec{H} = \frac{i\, dz \cos\psi}{4\pi\, R^2}\, \vec{a}_\theta. \tag{4.13}$$

Further, from the geometry of the problem (Fig. 4.2), we obtain the trigonometric relations:

$$R = r \sec\psi \tag{4.14}$$

and

$$z = r \tan\psi,$$

which upon differentiation gives

$$dz = r \sec^2\psi d\psi. \tag{4.15}$$

Substituting (4.14) and (4.15) in (4.13), the expression for the element of magnetic field due to the current element of a straight wire considered at a distance z from the origin is then obtained as:

$$d\vec{H} = \frac{i\, r \sec^2\psi\, d\psi\, \cos\psi}{4\pi\, r^2 \sec^2\psi}\, \vec{a}_\theta = \frac{i}{4\pi\, r} \cos\psi\, d\psi\, \vec{a}_\theta,$$

which can be straightaway integrated for the entire length of the current-carrying straight wire to obtain the expression for the magnetic field due to it at point P as follows:

$$\vec{H} = \int d\vec{H} = \frac{1}{4\pi} \int \frac{i}{r} \cos\psi\, d\psi\, \vec{a}_\theta. \tag{4.16}$$

The quantities i and the perpendicular distance r of the point P from the straight wire remain the same irrespective of the location of the current element, and therefore, these quantities can be taken outside the integration of (4.16), which can then be expressed as

$$\vec{H} = \int d\vec{H} = \frac{i}{4\pi\, r} \int \cos\psi\, d\psi\, \vec{a}_\theta. \tag{4.17}$$

While carrying out the integration of (4.17), we have to take the appropriate limits of integration for the infinitely long wire, that is, for the entire length of the wire that extends from $z = -\infty$ to $z = +\infty$. It easily follows from the simple geometry of the problem as well as from the relation (4.12) that the limit $z = -\infty$ corresponds to $\psi = -\pi/2$, and the limit $z = +\infty$ corresponds to $\psi = +\pi/2$ (Fig. 4.2). Accordingly, then, we can insert the limits of integration in (4.17) to write the following expression for the magnetic field due to the infinitely long current-carrying wire:

$$\vec{H} = \frac{i}{4\pi\, r} \int_{-\pi/2}^{+\pi/2} \cos\psi \; d\psi \, \vec{a}_\theta. \tag{4.18}$$

The evaluation of (4.18) gives

$$\vec{H} = \frac{i}{4\pi r}[\sin\psi]_{-\pi/2}^{+\pi/2}\vec{a}_\theta = \frac{i}{4\pi r}[\sin(\pi/2) - \sin(-\pi/2)]\vec{a}_\theta = \frac{i}{4\pi r}(\sin\pi/2 + \sin\pi/2)\vec{a}_\theta,$$

which, on substituting $\sin\pi/2 = 1$ becomes

$$\vec{H} = \frac{i}{2\pi\, r}\vec{a}_\theta. \tag{4.19}$$

The expression (4.19) for the magnetic field due to an infinitely long straight conducting wire carrying a current, however, will get modified for a wire of a finite length (see Appendix A4.1).

EXAMPLE E4-1

Find the magnetic field due to a direct current passing through a straight wire of a 'finite' length at a point equidistant from the end points of the wire in terms of the length of the wire.

SOLUTION

We can use the nomenclature of Appendix A4.1 and recall (A4.7) to write

$$\vec{H} = \frac{i}{4\pi\, r}\left(\frac{1}{\sqrt{1 + r^2/\,|l_1|^2}} + \frac{1}{\sqrt{1 + r^2/\,|l_2|^2}}\right)\vec{a}_\theta, \tag{E4-1.1}$$

where $r = \mathrm{OP}$, the perpendicular distance of the point P from the wire where the magnetic field is sought (Fig. E4-1). In the present context, we can take $|l_1| = |l_2| = l/2$, since the point P where the magnetic field is to be found is equidistant from the end points of the wire and $l = |l_1| + |l_2|$, l being the wire length. Equation (E4-1.1) can then be written as follows (Fig. E4-1):

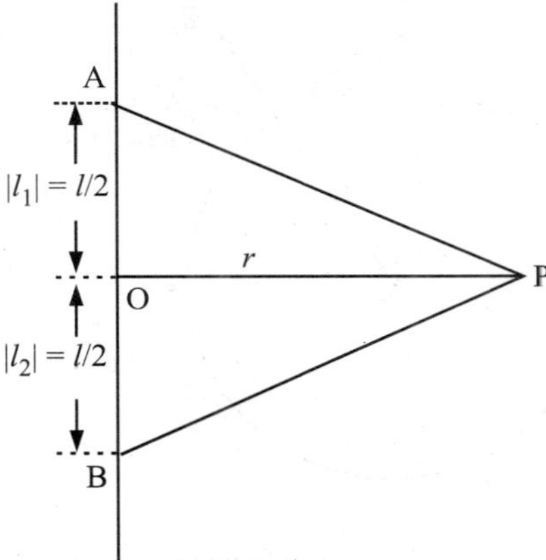

Fig. E4-1

$$\vec{H} = \frac{i}{4\pi r}\left(\frac{2}{\sqrt{1 + r^2/(l/2)^2}}\right)\vec{a}_\theta$$
$$= \frac{i}{2\pi r}\left(\frac{1}{\sqrt{1 + (2r/l)^2}}\right)\vec{a}_\theta, \tag{E4-1.2}$$

which is the required expression for the magnetic field at the perpendicular distance r from the wire in terms of the length l of the wire.

As another very simple problem to illustrate Biot-Savart's law (4.8), let us find, with the help of this law, the magnetic field at the centre of a circular loop of radius a carrying a current i (Fig. 4.3). Using the nomenclature of the cylindrical coordinate system (Section 2.1), we consider the circular current loop to lie on a plane perpendicular to the Z-axis. Further, we take the current to be in the azimuthal direction so that we can interpret $d\vec{l}$ in (4.10) as $d\vec{l} = dl\vec{a}_\theta$. Similarly, since the point where the magnetic field (centre of the loop) is to be found is directed radially inward from the current element, we can interpret $\vec{a}_R$ in (4.10) as $\vec{a}_R = -\vec{a}_r$ (since $\vec{a}_r$ is directed radially outward) (Fig. 4.3). We have also to interpret R in (4.8) as $R = a$, the distance between the current element and the point where the magnetic field (centre of the loop) is to be found. With these interpretations for $d\vec{l}$, $\vec{a}_R$, and R, we can then express (4.10), with reference to the present problem (Fig. 4.3), as

$$d\vec{H} = \frac{i\,d\vec{l} \times \vec{a}_R}{4\pi\,R^2} = \frac{i(dl\,\vec{a}_\theta) \times (-\vec{a}_r)}{4\pi\,a^2} = \frac{-idl\,\vec{a}_\theta \times \vec{a}_r}{4\pi\,a^2} = \frac{idl\,\vec{a}_z}{4\pi\,a^2}, \tag{4.20}$$

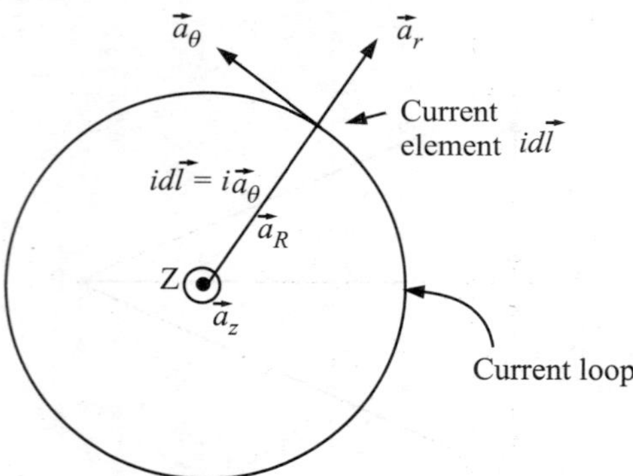

Fig. 4.3 *A circular current loop and a current element on it for finding the magnetic field at the centre of the loop considered on a plane perpendicular to the Z-axis.*

where we have used the relation (2.4) between the unit vectors in the cylindrical coordinate system that gives $\vec{a}_\theta \times \vec{a}_r = -(\vec{a}_r \times \vec{a}_\theta) = -\vec{a}_z$. The magnetic field at the centre of the loop due to the entire current in the loop is obtained by integrating (4.20) as

$$\vec{H} = \int d\vec{H} = \frac{i\vec{a}_z}{4\pi\, a^2} \int dl. \tag{4.21}$$

Now, since the integration in (4.21) over the circular loop simply becomes the circumference of the circle, $\int dl = 2\pi\, a$, we can express (4.21) as

$$\vec{H} = \int d\vec{H} = \frac{i\vec{a}_z}{4\pi a^2} 2\pi a = \frac{i}{2a} \vec{a}_z. \tag{4.22}$$

EXAMPLE E4-2

Find an expression for the magnetic field due to a regular polygonal loop of wire carrying a direct current at the centre of the loop and hence deduce, as its special case, the expression (4.22) for the magnetic field at the centre of a circular loop of wire carrying a direct current.

SOLUTION

The centre of the polygonal loop of the wire is equidistant from the end points of each side of the polygon or from the corners of the polygon. We can rewrite (E-1.2) (see Example E4-1) as follows:

$$\vec{H} = \frac{i}{2\pi\, r} \left(\frac{1}{\sqrt{1 + (2r/l)^2}} \right) \vec{a}_\theta, \tag{E4-2.1}$$

where r is the perpendicular distance of the point where the magnetic field is to be found from the wire of a finite length l. In the present problem, in order to find the contribution to the magnetic field due to a single side of the polygonal loop of current, let us replace

l with Δl, the length of each side of the polygonal. Let us also replace r with a, the perpendicular distance of the centre O of the polygon current loop (where the magnetic field is to be found) from a side of the polygon (Fig. E4-2). Thus, we obtain the following expression for the contribution to the magnetic field due to a single side of the polygonal loop of current:

$$\vec{H} = \frac{i}{2\pi\, r}\left(\frac{1}{\sqrt{1+(2a/\Delta l)^2}}\right)\vec{a}_\theta. \qquad \text{(E4-2.2)}$$

We can then sum up the individual contributions to the magnetic field each given by (E4-2.2) to obtain the following expression for the polygonal loop of current:

$$\vec{H} = \sum_n \frac{i}{2\pi\, a}\left(\frac{1}{\sqrt{1+(2a/\Delta l)^2}}\right)\vec{a}_z, \qquad \text{(E4-2.3)}$$

where the summation extends over all n sides of the polygon current loop.

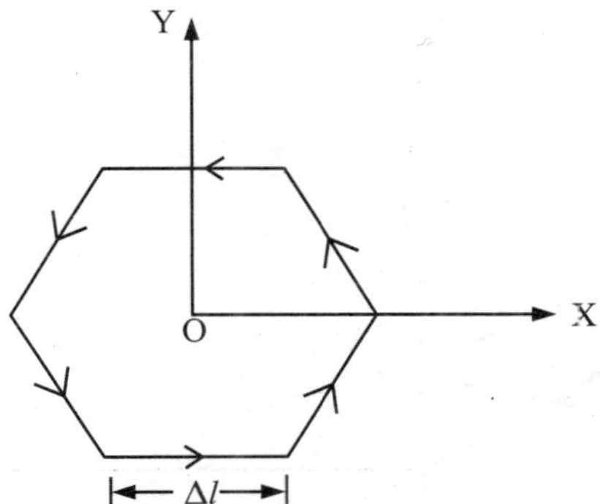

Fig. E4-2

For a large number of sides of the polygon ($n >> 1$), the polygon tends to become a circle, and we can take $a >> \Delta l$ to write (E4-2.1) as follows:

$$\vec{H} = \sum_n \frac{i}{2\pi\, a}\left(\frac{1}{\sqrt{(2a/\Delta l)^2}}\right)\vec{a}_z = \frac{i}{4\pi\, a^2}\sum_n \Delta l\, \vec{a}_z,$$

which, putting $\sum_n \Delta l = 2\pi\, a$ for a circle, gives the required expression

$$\vec{H} = \frac{i}{4\pi\, a^2} 2\pi\, a\, \vec{a}_z = \frac{i}{2a}\vec{a}_z,$$

which is found to be identical to (4.22).

We have applied Biot-Savart's law so far to derive, in the form of (4.22), the expression for magnetic field due to a circular loop of current at the centre of the loop. The same expression also follows as a special case of the problem of finding the magnetic field due to a regular polygonal loop of current (Example E4-2). In Example E4-3 to follow , we will use Biot-Savart's law to find the magnetic field at a point on the axis of the loop, perpendicular to the plane containing the loop and passing through the centre of the loop, though off from the centre.

EXAMPLE E4-3

Apply Biot-Savart's law to find an expression for the magnetic field due to a circular loop of direct current at a point lying on the axis of the loop off from its centre, the axis being perpendicular to the plane of the loop and passing through its centre. Hence also obtain, as a special case, expression (4.22) for the magnetic field at the centre of the loop.

SOLUTION

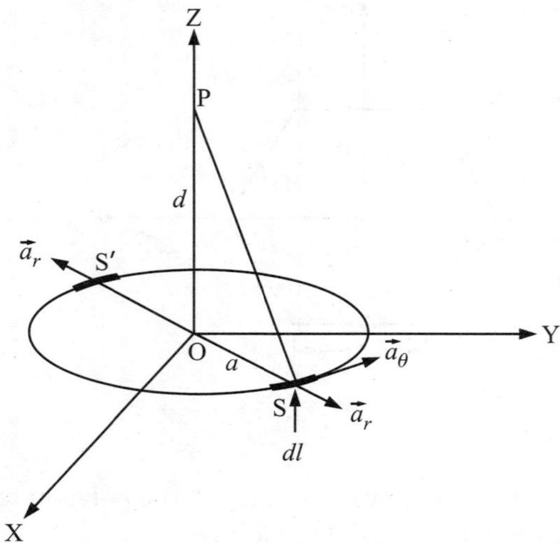

Fig. E4-3

Let us take a circular loop of wire of radius a carrying a direct current, lying on the XY plane with its centre at the origin O. Let P be the point on the Z-axis at a distance d from the origin (OP $= d$) where the magnetic field is to be found (Fig. E4-3). We can find with the help of Biot-Savart's law (4.10) the element of magnetic field $d\vec{H}$ at point P due to the current element $i\,d\vec{l} = idl\vec{a}_\theta$ considered arbitrarily at point S on the loop, i being the current through the loop. However, in the present context, it is more convenient to interpret the unit vector $\vec{a}_R$ in terms of the distance vector $\vec{R} = R\vec{a}_R$, or in other words in terms of $\vec{a}_R = \vec{R}/R$, so that we can read (4.10) as

$$d\vec{H} = \frac{i\,d\vec{l} \times \vec{a}_R}{4\pi\, R^2} = \frac{idl\,\vec{a}_\theta \times \vec{R}}{4\pi\, R^3}. \tag{E4-3.1}$$

Taking $\vec{R} = \overrightarrow{\text{SP}}$ and $R = \text{SP}$, with reference to the present problem, we can put (E4-3.1) as:

$$d\vec{H} = \frac{idl\,\vec{a}_\theta \times \overrightarrow{\text{SP}}}{4\pi(\text{SP})^3}. \tag{E4-3.2}$$

However, as can be seen from the geometry of the problem, $\overrightarrow{\text{SP}} = \overrightarrow{\text{SO}} + \overrightarrow{\text{OP}} = -\text{SO}\vec{a}_r + \text{OP}\vec{a}_z = -a\vec{a}_r + d\vec{a}_z$, and $\text{SP} = R = (a^2 + d^2)^{1/2}$, which can be then substituted in (E4-3.2) to write

$$d\vec{H} = \frac{i\,dl\vec{a}_\theta \times (-a\vec{a}_r + d\vec{a}_z)}{4\pi(a^2 + d^2)^{3/2}},$$

which, in view of the relation $\vec{a}_\theta \times \vec{a}_r = -\vec{a}_z$, $\vec{a}_\theta \times \vec{a}_z = \vec{a}_r$ (cylindrical system) can be read as

$$d\vec{H} = \frac{i\,dl(a\vec{a}_z + d\vec{a}_r)}{4\pi(a^2 + d^2)^{3/2}}. \tag{E4-3.3}$$

The above expression (E4-4.3) is for the element of magnetic field $d\vec{H}$ at point P due to a current element considered at any arbitrary point on the circular loop of wire. In order to find the magnetic field due to the entire circular loop, we have to divide the loop into a number of such current elements. However, you will notice that the direction of $\vec{a}_r$ at current element $i\,d\vec{l}$ considered at the point S is opposite to that considered at the diametrically opposite point S′. Therefore, the contributions of this pair of current elements due to the second term of the above expression for $d\vec{H}$ containing $\vec{a}_r$ will cancel out; however, the contributions due to the first term containing $\vec{a}_z$will be identical. Therefore, the magnetic field at point P due to the entire coil, which can be divided in such pairs of current elements at diametrically opposite points on the circular loop of the wire, can be found integrating only the first term of (E4-4.3) as

$$\vec{H} = \int d\vec{H} = \int \frac{i\,dla\vec{a}_z}{4\pi(a^2 + d^2)^{3/2}} = \frac{i\,a\vec{a}_z}{4\pi(a^2 + d^2)^{3/2}} \int dl. \tag{E4-3.4}$$

The integration of the line elements becomes the length of the circumference of the circular loop, that is, $\int dl = 2\pi a$, which substituted in (E4-3.4) gives

$$\vec{H} = \frac{i\,a}{4\pi(a^2 + d^2)^{3/2}} 2\pi\, a\, \vec{a}_z = \frac{i\,a^2}{2(a^2 + d^2)^{3/2}} \vec{a}_z. \tag{E4-3.5}$$

We can also alternatively express (E4-3.5) as

$$\vec{H} = \frac{i\,a}{4\pi(a^2+d^2)^{3/2}}2\pi a\vec{a}_z = \frac{i\,\pi a^2}{2\pi(a^2+d^2)^{3/2}}\vec{a}_z = \frac{i\,\alpha}{2\pi(a^2+d^2)^{3/2}}\vec{a}_z, \qquad \text{(E4-3.6)}$$

where $\alpha = \pi a^2$ is the area of the circular loop.

Putting $d = 0$ in (E4-3.5), we obtain the following expression for the magnetic field at the centre of the loop:

$$\vec{H} = \frac{i\,a^2}{2(a^2)^{3/2}}\vec{a}_z = \frac{i\,a^2}{2a^3}\vec{a}_z = \frac{i}{2a}\vec{a}_z, \qquad \text{(E4-3.7)}$$

which is identical to (4.22). By multiplying the numerator and denominator of (E4-3.7) by $\pi a^2 (= \alpha)$, we can also express (E4-3.7) as

$$\vec{H} = \frac{i}{2a}\vec{a}_z = \frac{i}{2a}\frac{\pi a^2}{\pi a^2} = \frac{i\,\alpha}{2\pi a^3}\vec{a}_z. \qquad \text{(E4-3.8)}$$

The quantity $i\alpha$ appearing in the numerator of (E4-3.8) will be identified in Example 4-6 as the magnetic dipole moment of the current loop.

Let us take up a couple of more problems to exemplify the application of Biot-Savart's law in finding magnetic field.

EXAMPLE E4-4

Find an expression for the magnetic field at the middle point of a square loop of side equal to a carrying a direct current i.

SOLUTION

For a regular polygon loop of direct current i, we have already obtained in Example E4-2 the following expression for the magnetic field $\vec{H}$ at the centre of the loop:

$$\vec{H} = \sum_n \frac{i}{2\pi\,a}\left(\frac{1}{\sqrt{1+(2a/\Delta l)^2}}\right)\vec{a}_z \qquad \text{(E4-4.1)}$$

where, the centre of the loop is considered at the origin of the coordinates ($x = y = 0$); Δl is the length of each side of the polygonal loop of wire and the summation extends over the n sides of the polygon; and a is the perpendicular distance of each side of the polygon from the centre of the polygon where the magnetic field is sought.

In the present problem of a square loop of current (Fig. E4-4), we have to take $n = 4$ while using (E4-4.1). Also, the summation on the right-hand side is equal to four times the quantity within the summation, however, with the interpretation that $\Delta l = a$ and that the perpendicular distance is $a/2$ instead of a (since in the present context the symbol a stands for the side of the polygon). Thus, we obtain the following expression for the magnetic field at the middle point of the square loop:

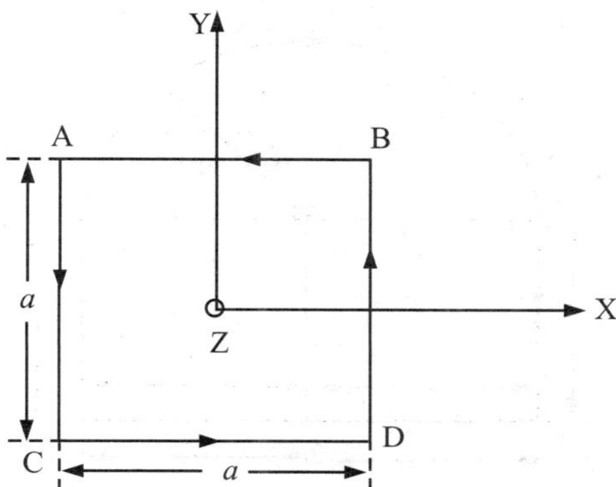

Fig. E4-4

$$\vec{H} = 4 \times \frac{i}{2\pi \frac{a}{2}} \left(\frac{1}{\sqrt{1 + (2(a/2)/a)^2}} \right) \vec{a}_z,$$

which simplifies to

$$\vec{H} = \frac{2\sqrt{2}i}{\pi a} \vec{a}_z.$$

EXAMPLE E4-5

A direct current i passes through a loop that consists of two parts: one half a semicircle (first part) connected to one half of a square (second part), both joined together and lying on the same plane so as to form a single current loop of a composite half-circle and half-square configuration. Find the magnetic field due to such a loop at the centre of the semicircular part of the loop, taking the radius a of the semicircular part as equal to the shorter side of the half-square part.

SOLUTION

The expression for the magnetic field at the centre of a circular loop of radius a carrying direct current i has been obtained earlier (see the expression (4.22) and also Example E4-2) as

$$\vec{H} = \frac{i}{2a} \vec{a}_z. \tag{E4-5.1}$$

Also, the expression for the magnetic field at the middle point of a square loop of side equal to a carrying a direct current i has been obtained earlier (see Example E4-4) as

$$\vec{H} = \frac{2\sqrt{2}i}{\pi a} \vec{a}_z. \tag{E4-5.2}$$

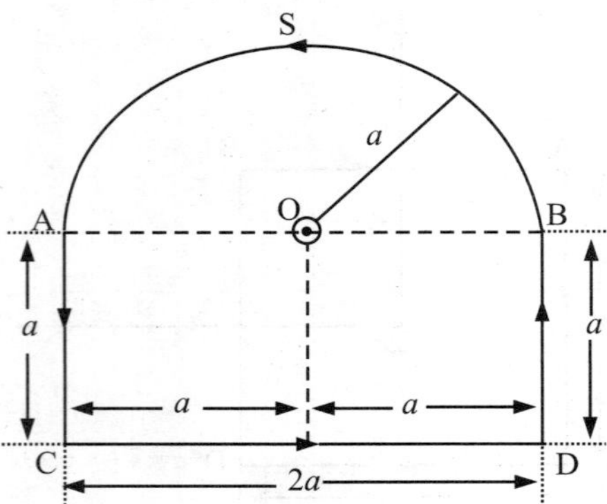

Fig. E4-5

In the present problem (Fig. E4-5), the contribution of the first part (semi-circular part) of the loop (ASB) to the centre (O) of the semicircle is half the value given by (E4-5.1) and can be written as

$$\vec{H} = \left(\frac{1}{2}\right)\frac{i}{2a}\,\vec{a}_z = \frac{i}{4a}\,\vec{a}_z. \qquad \text{(first part)} \tag{E4-5.3}$$

Similarly, the contribution of the second part of the loop (ACDB) is half the value given by (E4-5.2), in which, however, we have to replace a by $2a$, now the side of the square (CD=AB). This can be written as

$$\vec{H} = \left(\frac{1}{2}\right)\frac{2\sqrt{2}i}{\pi(2a)}\,\vec{a}_z = \frac{2\sqrt{2}i}{4\pi a}\,\vec{a}_z. \qquad \text{(second part)} \qquad \text{(E4-5.)}$$

Adding (E4-5.3) and (E4-5.4), we obtain the desired expression

$$\vec{H} = \frac{i}{4a}\,\vec{a}_z + \frac{2\sqrt{2}i}{4\pi\, a}\,\vec{a}_z = \frac{i}{4a}\left(1 + \frac{2\sqrt{2}i}{\pi}\right)\vec{a}_z.$$

Now that we are familiar with the use of Biot-Savart's law in finding the magnetic field due to a circular loop of current with the help of (4.22) (Examples E4-2 – E4-5), we are in a position to illustrate further the concept of magnetic dipole introduced in Section 4.1.1 with the help of the examples E4-6 and E4-7.

EXAMPLE E4-6

Show that a circular loop of wire carrying a direct current can be considered as a magnetic dipole of moment equal to the area times the current of the coil in analogy with an electric dipole.

SOLUTION

The expression for the electric field due to a short electric dipole at a point situated on the axis of the dipole field can be obtained in terms of the dipole moment p as (see Example E3-5 in Chapter 3)

$$\vec{E} = \frac{p}{2\pi\varepsilon_0 r^3}\vec{a}_z,$$

whence we obtain the corresponding expression for the electric displacement $\vec{D}(=\varepsilon_0\vec{E})$ as

$$\vec{D} = \varepsilon_0\vec{E} = \frac{p}{2\pi\, r^3}\vec{a}_z. \qquad \text{(E4-6.1)}$$

The above expression for $\vec{D}$ due to an electric dipole becomes analogous (see Table 4.1) to the following expression for $\vec{H}$ due to a circular loop of current at the centre of the loop carrying current i, given by (4.22), which can be rewritten in the following form (see Example E4-3):

$$\vec{H} = \frac{i}{2a}\vec{a}_z = \frac{i\,\alpha}{2\pi a^3}\vec{a}_z = \frac{p_m}{2\pi a^3}\vec{a}_z, \qquad \text{(E4-6.2)}$$

where we have defined a quantity called 'magnetic' dipole moment $p_m = i\,\alpha$ in terms of the loop current i and the area of the loop $\alpha(=\pi a^2)$ of the loop of radius a. The expression (E4-6.2) is identical to the expression (E4-3.8) deduced in Example E4-3 with the help of Biot-Savart's law.

EXAMPLE E4-7

Obtain an expression for the magnetic field due to a direct current through a circular loop at a point off the axis perpendicular to the plane containing the loop and passing through its centre by treating the loop as a magnetic dipole analogous to an electric dipole.

SOLUTION

The electric field due to a 'short' electric dipole off the axis of the dipole at an angle θ can be obtained in terms of the electric dipole moment p as (see Example E3-5):

$$\vec{E} = \frac{1}{4\pi\varepsilon_0 r^3}\, p(2\,\cos\theta\,\vec{a}_r + \sin\theta\,\vec{a}_\theta),$$

whence we obtain the corresponding expression for the electric displacement $\vec{D}(=\varepsilon_0\vec{E})$ as

$$\vec{D} = \frac{1}{4\pi\, r^3}\, p(2\,\cos\theta\,\vec{a}_r + \sin\theta\,\vec{a}_\theta). \qquad \text{(E4-7.1)}$$

Extending the analogy established in Example E4-6, we can then write the following expression for the analogous magnetic field $\vec{H}$ (Table 4.1) due to circular loop of direct current by replacing p by $p_m(=i\,\alpha)$ and $\vec{D}$ by $\vec{H}$ in (E4-7.1), where $\alpha(=\pi a^2)$ is the area of the loop.

$$\vec{H} = \frac{1}{4\pi\, r^3}\, p_m(2\,\cos\theta\,\vec{a}_r + \sin\theta\;\vec{a}_\theta). \tag{E4-7.2}$$

Remembering the relation $p_m(=i\,\alpha)$ and $\alpha(=\pi a^2)$, we can express (E4-7.2) in terms of the current i and radius a of the loop as

$$\vec{H} = \frac{\pi\, a^2 i}{4\pi\, r^3}(2\,\cos\theta\,\vec{a}_r + \sin\theta\;\vec{a}_\theta). \tag{E4-7.3}$$

Multiplying (E4-7.2) by μ_0(see Table 4.1), we obtain

$$\vec{B} = \frac{\mu_0}{4\pi} p_m \frac{1}{r^3}(2\,\cos\theta\,\vec{a}_r + \sin\theta\;\vec{a}_\theta).$$

EXAMPLE E4-8

Identify the analogous quantities relevant to electric and magnetic dipoles.

SOLUTION

From Examples 4-6 and 4-7, we can identify the following analogous quantities relevant to electric and magnetic dipoles. That $\vec{E}$ is analogous to $\vec{B}$ and $\vec{D}$ to $\vec{H}$ can be recalled from Table 4.1.

Electric dipole	Magnetic dipole
$\vec{E} = \frac{1}{4\pi\varepsilon_0 r^3}\, p(2\,\cos\theta\,\vec{a}_r + \sin\theta\;\vec{a}_\theta)$	$\vec{B} = \frac{\mu_0}{4\pi} p_m \frac{1}{r^3}(2\,\cos\theta\,\vec{a}_r + \sin\theta\;\vec{a}_\theta)$
$1/4\pi\varepsilon_0$	$\mu_0/4\pi$
ε_0	$1/\mu_0$
$\vec{D} = \frac{1}{4\pi\, r^3}\, p(2\,\cos\theta\,\vec{a}_r + \sin\theta\;\vec{a}_\theta)$	$\vec{H} = \frac{1}{4\pi\, r^3}\, p_m(2\,\cos\theta\,\vec{a}_r + \sin\theta\;\vec{a}_\theta)$
$p = ql$	$p_m(= i\,\alpha = i\pi\, a^2)$

4.3 Ampere's Circuital Law

4.3.1 Ampere's circuital law in integral form

According to Ampere's circuital law, the line integral of the magnetic field around a closed path of any arbitrary shape on a plane perpendicular to a current is equal to the current enclosed by the path. Mathematically, the law can be stated in terms of a closed line integral as

$$\oint_l \vec{H}\cdot d\vec{l} = i, \tag{4.23}$$

where i represents the current enclosed. (For the definition of the line integral of a vector quantity, see Section 2.4). $\vec{H}$ is the magnetic field at a point P on the closed path considered, and $d\vec{l}$ is a length element vector at the point such that the magnitude of $d\vec{l}$ is dl and its direction is the sense of rotation of a screw that will cause an axial motion of the screw in the direction of the current i (Fig. 4.4). Further, we have used in (4.23) the symbol of a closed line integral (see Section 2.4), since the path around which the line integral is taken is essentially closed here.

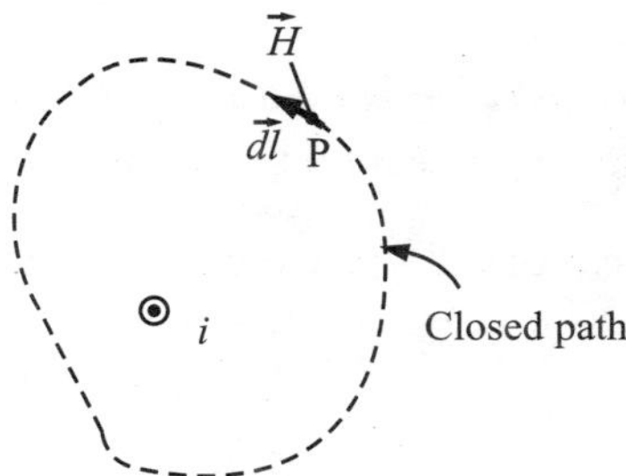

Fig. 4.4 *A closed path of an arbitrary shape enclosing a current i showing an element of length vector and the magnetic field at any point P on the path.*

Like Gauss's law of electrostatics (see Section 3.1.3), we can apply Ampere's circuital law in integral form (4.23) to problems that enjoy geometrical symmetry. For instance, the problem of an infinitely long current-carrying wire (dealt with in Section 4.2 by Biot-Savart's law) enjoys cylindrical symmetry. Hence the magnitude of the magnetic field all around the periphery of a circular path of radius r, considered on a plane perpendicular to the current, will be the same at every point on the path with the direction of the field being azimuthal. Further, the element of the length vector $d\vec{l}$ at any point on the circular path will also be azimuthal in this case (Fig. 4.5). Therefore, we can represent $\vec{H}$ and $d\vec{l}$ at any point P on the circular loop in this problem by

$$\left.\begin{aligned} \vec{H} &= H\vec{a}_\theta \\ d\vec{l} &= dl\vec{a}_\theta \end{aligned}\right\}, \tag{4.24}$$

which, when substituted in (4.23) gives

$$\oint H\vec{a}_\theta \cdot dl\vec{a}_\theta = i,$$

which, in view of the dot product $\vec{a}_\theta \cdot \vec{a}_\theta = 1$ becomes

$$\oint Hdl = i. \tag{4.25}$$

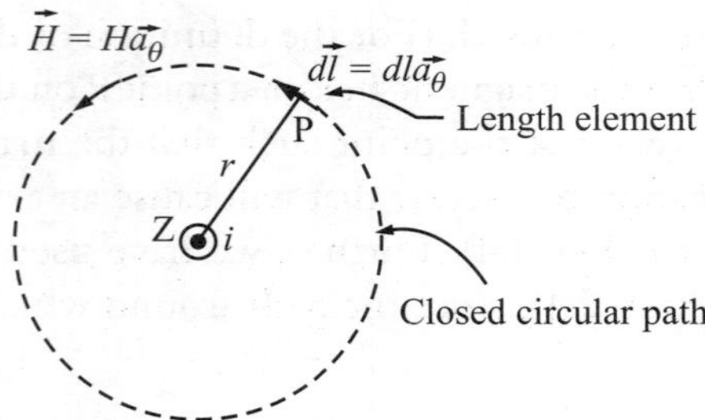

Fig. 4.5 *A long straight wire along z carrying current i and a circular closed path of radius r considered on a plane perpendicular to the Z-axis passing through the point P where the magnetic field is to be found.*

Since H is uniform over the circular loop, we can take it outside the integral of (4.25), which will simplify it to

$$H \oint dl = I,$$

which, on putting $\oint dl = 2\pi r$, the circumference of a circular loop of radius r, gives

$$H = \frac{i}{2\pi r}. \tag{4.26}$$

Now, substituting (4.26) in the expression for $\vec{H}$ given by (4.24), we obtain the following expression for the magnetic field due to an infinitely long current carrying wire:

$$\vec{H} = \frac{i}{2\pi r}\vec{a}_\theta. \tag{4.27}$$

Interestingly, the expression (4.27) for the magnetic field due to an infinitely long current-carrying wire obtained by Ampere's circuital law in integral form becomes identical to the expression (4.19) obtained earlier by Biot-Savart's law that involved the evaluation of a definite integral. Thus, through this simple example we have demonstrated how the use of Ampere's circuital law in integral form can make the analysis of a problem of static magnetic fields rather easy, analogous to Gauss's law for a problem of electrostatics (Section 3.1.3).

Let us now take up another simple example in which we apply Ampere's circuital law in integral form to find the magnetic field due to a long solenoid at a point on the axis of the solenoid (Fig. 4.6). Subsequently, we will use this example to find the inductance of a solenoid.

Let us take the length of the solenoid to be very large compared to the radius of the solenoid as well as to the distance between two consecutive turns of the solenoid. We take the axis of the solenoid to be along z. Since the solenoid is taken as long, we can take the magnetic field due to the solenoid to be independent of z. The direction of the magnetic field due to currents, both out of and into the plane of the paper will be azimuthal, which you will be able to appreciate, for instance, with the help of (4.19). Therefore, the magnetic flux lines are anticlockwise for the current out of the plane of the paper and clockwise for the current into the plane of the paper (Fig. 4.6). Furthermore, one can also appreciate the direction of the magnetic flux lines (clockwise or anticlockwise) by examining the vector cross product term $d\vec{l} \times \vec{a}_R$ occurring in Biot-Savart's law (4.10) as applied to the current through the solenoid. In developing this understanding, refer to the analogy of a moving screw, in which the direction of the angular motion of the screw gives the direction of the magnetic flux lines corresponding to the axial motion of the screw in the direction of the current.

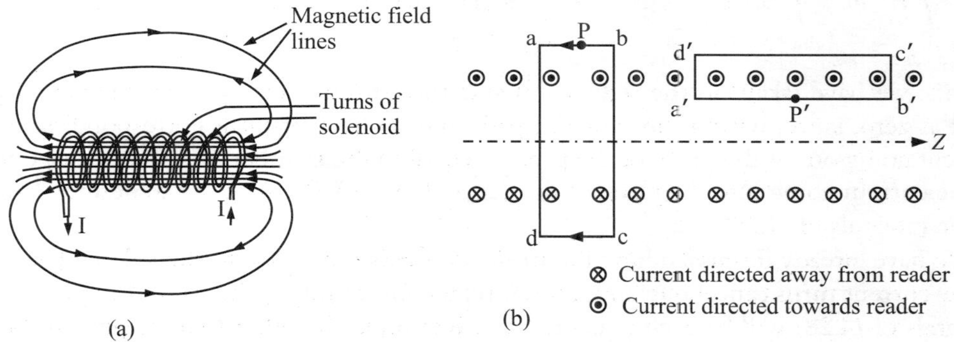

Fig. 4.6 *Schematic of a long solenoid showing (a) the magnetic field lines and (b) its cross-section, the latter showing the currents in and out of the plane of the paper and two closed rectangular paths, abcd and $a'b'c'd'$, around which line integrals of the magnetic field are taken to find the magnetic fields outside and inside the solenoid respectively.*

In a direction transverse to the axis of the solenoid (Z-axis), the anticlockwise magnetic flux line due to one turn will be opposite in direction to the anticlockwise magnetic flux line due to the next turn, and their contributions to the magnetic field will cancel out each other for the currents out of the plane of the paper. Similarly, the contributions to the magnetic field due to the clockwise magnetic flux lines between two consecutive turns caused by the currents directed into the plane of the paper will cancel out in a direction transverse to the Z-axis (Fig. 4.6).

At a point P outside the solenoid, at a radial distance much larger than the solenoid radius, the magnetic field contributed by the anticlockwise flux lines due to the currents out of the plane of the paper will balance the magnetic field contributed by the clockwise flux lines due to currents into the plane of the paper, thereby making the magnetic field outside the solenoid zero (Fig. 4.6). This can be appreciated from the following reasoning. In the cross-sectional view of the solenoid carrying a direct current (Fig. 4.6*b*), the magnetic flux lines outside the solenoid at P due to a current represented by a cross surrounded by a circle will be in the clockwise direction, while those due to a current represented by a dot surrounded by a circle will be in the anticlockwise direction. One can appreciate these clockwise and anticlockwise directions of flux lines with the help of Biot-Savart's law or the screw rule as discussed earlier. Hence, the directions of the flux lines, being clockwise and anticlockwise outside the solenoid at the point P, will tend to cancel one another. It corresponds to a magnetic field at P tending to a null in the limit of the point P being at a large distance from the solenoid compared to the radius of the solenoid.

Using an alternative approach, this can be further appreciated by applying Ampere's circuital law (4.23) to a closed rectangular path abcd, in which the paths da and bc are transverse to the axis of the solenoid and the paths ab and cd are parallel to and considered at unequal radial distances from the axis of the solenoid, as follows (Fig. 4.6):

$$\int_{ab} \vec{H}\cdot d\vec{l} + \int_{bc} \vec{H}\cdot d\vec{l} + \int_{cd} \vec{H}\cdot d\vec{l} + \int_{da} \vec{H}\cdot d\vec{l} = 0. \tag{4.28}$$

In this, we have taken on the right-hand side the current i enclosed by the closed path abcd as zero, since, within the enclosed path, the number of turns corresponding to the current going out of the plane of the paper is equal to the number of turns corresponding to the current going into the plane of the paper (Fig. 4.6). We have yet to define $\vec{H}$ and $d\vec{l}$ in the integrals of (4.28).

We have already discussed how the magnetic fields transverse to the solenoid axis due to the current turns cancel out, and, therefore, the integrands of the second and the fourth integrals of (4.28) will be each equal to zero. Retaining the other two integrals in (4.28), the expression becomes

$$\int_{ab} \vec{H}\cdot d\vec{l} + \int_{cd} \vec{H}\cdot d\vec{l} = 0. \tag{4.29}$$

If, for the sake of argument, we consider that there exists a magnetic field outside the solenoid, say at point P, then the magnetic field will have the direction along the positive z direction for the current out of the plane of the paper, and in the negative z direction for the current into the plane of the paper. Let us, therefore, by suitably interpreting the magnitude of the magnetic field, represent the magnetic field at point P as $\vec{H} = H_{\text{outside}}\vec{a}_z$. This will also be the value of the field all along the path ab. Further, we can take $d\vec{l} = dl\vec{a}_z$ along path ab. Since path cd is at a different radial distance from the solenoid than path ab, the magnetic field along path ab will be different from that along path cd. Let us represent the magnetic field along this path as $\vec{H} = H'_{\text{outside}}\vec{a}_z$ such that

$$H_{\text{outside}} \neq H'_{\text{outside}}, \tag{4.30}$$

and also take $d\vec{l} = -dl\vec{a}_z$ along this path cd, the negative sign being on account of the direction of the path along cd being in the negative z direction (Fig. 4.6). Therefore, substituting these values of $\vec{H}$ and $d\vec{l}$ in (4.29) for paths ab and cd, we can write:

$$\int_{ab} H_{\text{outside}}\vec{a}_z\cdot(dl\vec{a}_z) + \int_{cd} H'_{\text{outside}}\vec{a}_z\cdot(-dl\vec{a}_z) = 0,$$

which simplifies, after taking H_{outside} and H'_{outside} outside their respective integrals, since they are constant along paths ab and cd respectively, to

$$H_{\text{outside}} \int_{\text{ab}} dl - H'_{\text{outside}} \int_{\text{cd}} dl = 0. \tag{4.31}$$

Taking the length of the path ab = cd = l, each of the integrals in (4.31) becomes equal to l and therefore we have, from (4.30):

$$H_{\text{outside}} l - H'_{\text{outside}} l = 0,$$

which, since $l \neq 0$, gives

$$H_{\text{outside}} = H'_{\text{outside}} \, ,$$

a relation that contradicts (4.30) unless we take the magnetic field not to exist outside the solenoid, that is,

$$H_{\text{outside}} = H'_{\text{outside}} = 0. \tag{4.32}$$

Since the path abcd has been taken arbitrarily, we can conclude from (4.32) that no magnetic field exists outside a long solenoid.

Now, let us proceed to find the magnetic field at a point inside the solenoid; the point is arbitrarily taken at P′ (Fig. 4.6). Let us now draw a closed rectangular path a′b′c′d′ such that it passes through P′ and encloses the turns of the solenoid corresponding to the currents going out of the plane of the paper. Applying Ampere's circuital law (4.23) to the closed path, we can write

$$\int_{\text{a}'\text{b}'} \vec{H} \cdot d\vec{l} + \int_{\text{b}'\text{c}'} \vec{H} \cdot d\vec{l} + \int_{\text{c}'\text{d}'} \vec{H} \cdot d\vec{l} + \int_{\text{d}'\text{a}'} \vec{H} \cdot d\vec{l} = nli \, , \tag{4.33}$$

in which we have taken, on the right-hand side, the current enclosed by the closed path a′b′c′d′ as nli, where n is the number of turns per unit length of the solenoid; l is the length of the paths parallel to the axis of the solenoid (Z-axis): a′b′ =d′c′ = l; and i is the solenoid current passing through each turn (Fig. 4.6). As we have already discussed, the magnetic fields transverse to the solenoid axis due to the current turns cancel out. Hence, the integrands of the second and the fourth integrals of (4.33) will be each equal to zero. Hence,

$$\int_{\text{a}'\text{b}'} \vec{H} \cdot d\vec{l} + \int_{\text{c}'\text{d}'} \vec{H} \cdot d\vec{l} = nli. \tag{4.34}$$

However, since the magnetic field outside the solenoid is zero, a result that we have obtained already [see the finding following (4.32)], the integrand of the second integrand

and hence also the second integral of (4.34) become zero, so that we have from (4.34)

$$\int_{a'b'} \vec{H} \cdot d\vec{l} = nli. \tag{4.35}$$

It is easy to appreciate that the magnetic flux lines due to the solenoid turns considered, corresponding here to the current out of the plane of the paper, cause the magnetic field at P′ to be aligned along z. We can express this magnetic field as $\vec{H} = H\vec{a}_z$; this is also the magnetic field along the path a′b′ passing through P′, since for a long solenoid the magnetic field $\vec{H}$ will not vary along z and can be regarded as constant along the path a′b′. Further, the element of the length vector $d\vec{l}$ along the path a′b′ is directed along z, and we can represent it as $d\vec{l} = dl\vec{a}_z$. Therefore, substituting $\vec{H} = H\vec{a}_z$ and $d\vec{l} = dl\vec{a}_z$ in (4.35), we can write

$$\int_{a'b'} H\vec{a}_z \cdot dl\vec{a}_z = nli,$$

whence we get

$$\int_{a'b'} Hdl = nli. \tag{4.36}$$

Since the magnetic field is constant over the path a′b′, we can take H outside the integral of (4.36) to write

$$H\int_{a'b'} dl = nli. \tag{4.37}$$

Since the integral of (4.36) is simply the length of the path a′b′ $= l$, we obtain the following expression from (4.37)

$$Hl = nli,$$

which simplifies to

$$H = ni. \tag{4.38}$$

Making use of the expression $\vec{H} = H\vec{a}_z$ [see the discussion following (4.35)], we can then write from (4.38) the following expression for the magnetic field at a point inside the solenoid:

$$\vec{H} = ni\vec{a}_z. \tag{4.39}$$

We can now easily derive the expression for the inductance of the solenoid with the help of (4.39). For this, let us first write the magnetic flux density, putting $\mu = \mu_0$ in (4.2), as

$$\vec{B} = \mu_0 \vec{H}. \tag{4.40}$$

Substituting (4.39) in (4.40), we then get the following expression for the magnetic flux density inside the solenoid

$$\vec{B} = \mu_0 \vec{H} = \mu_0 n i \vec{a}_z, \tag{4.41}$$

which shows that the magnetic flux density is independent of the radial coordinate, that is, it remains uniform over the cross section of the solenoid. In order to find the flux $\phi_{B,\,\text{singlet turn}}$ of $\vec{B}$ let us, for the sake of convenience, recall (2.19) and rewrite it as

$$F_E = \int_S \vec{E}.d\vec{S} = \int_S \vec{E}.\vec{a}_n dS, \tag{4.42}$$

where the symbols are as explained in Chapter 2. In the present context, interpreting $F_E = \phi_{B,\,\text{singlet turn}}$ and $\vec{E} = \vec{B}$ in (4.42), we can write

$$\phi_{B,\,\text{singlet turn}} = \int_S \vec{B} \cdot \vec{a}_n dS, \tag{4.43}$$

in which the area S is the cross-sectional area α of the inside of the solenoid over which $\vec{B}$ remains constant. $\vec{a}_n$ is the unit vector normal to the cross-sectional area α of the solenoid, that is, in the direction of the axis of the solenoid or in the z direction, so that we can take

$$\vec{a}_n = \vec{a}_z. \tag{4.44}$$

Putting (4.41) and (4.44) in (4.43), we get the flux of $\vec{B}$ per single turn as

$$\phi_{B,\,\text{singlet turn}} = \int_S \mu_0 n i \vec{a}_z \cdot \vec{a}_z dS = \mu_0 n i \int_S dS = \mu_0 n i \alpha \tag{4.45}$$

in which we have put the integration of the element of area over the cross-sectional area of the solenoid as

$$\int_S dS = \alpha.$$

If we take the length of the solenoid as l, there will be nl turns in the solenoid and the flux ϕ_B of $\vec{B}$ linked with all the turns of the solenoid is obtained by multiplying (4.45) by the number of turns nl. Thus, we obtain

$$\phi_B = (\phi_{B,\,\text{singlet turn}})(nl). \tag{4.46}$$

Substituting (4.45) in (4.46), we then get

$$\phi_B = (\mu_0 n i \alpha)(nl) = \mu_0 n^2 \alpha l i. \tag{4.47}$$

The self inductance of the solenoid L is obtained by dividing ϕ_B given by (4.47) by the solenoid current i as

$$L = \frac{\phi_B}{i} = \frac{\mu_0 n^2 \alpha\, l i}{i} = \mu_0 n^2 \alpha\, l. \tag{4.48}$$

The unit of magnetic flux is weber or Wb. Therefore, it follows from (4.48) that the unit of inductance is Wb/A or henry or H. Thus, with the unit of the inductance being H, the units of n (turns per unit length) being m^{-1}, the unit of α (cross-sectional area of the solenoid) being m^2, and the unit of l (length of the solenoid) being m, it can be seen from (4.48) that the unit of the permeability of the medium (here, free-space) is H/m, as mentioned following (4.1).

EXAMPLE E4-9

Show that the magnetic field at a point inside a toroidal coil is jointly proportional to the number of turns of the coil and the current through the coil and is inversely proportional to the radial distance of the point from the centre of the toroid.

SOLUTION

A toroidal coil is made by bending a solenoid to make a coil in the shape of a toroid. We can now apply Ampere's circuital law around a closed circular path of radius r passing through the point inside the toriodal coil to find the magnitude of the magnetic field intensity H at the point as $H = NI/(2\pi r)$ in terms of the number of turns N and the current through the coil I (Fig. E4-9). The expression can then be interpreted to show the relationship required by the question.

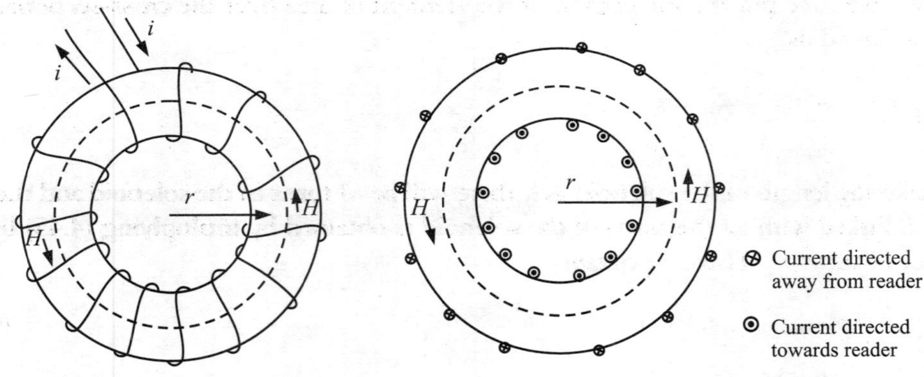

Fig. E4-9

EXAMPLE E4-10

Obtain the following expression for the inductance L per unit length of a coaxial cable in terms of the radius a of its inner conductor and the inner radius b of its outer conductor:

$$L = \frac{\mu_0}{2\pi} \ln \frac{b}{a}.$$

SOLUTION

Let us consider a rectangular element of strip of length l and elemental cross-sectional area ldr considered in a radial cross-sectional plane of the coaxial cable passing through the point at the radial coordinate r, extending over an elemental radial thickness dr lying between r and $r + dr$ in the region between the inner and outer conductors (Fig. E4-10). With the help of Ampere's circuital law (4.23) we can find the azimuthal magnetic field H_θ and hence the azimuthal magnetic flux density $B_\theta (= \mu_0 H_\theta)$ at r as

$$B_\theta (= \mu_0 H_\theta) = \frac{\mu_0 I}{2\pi r}. \tag{E4-10.1}$$

The element of magnetic flux $d\phi_B$ of B_θ through the rectangular element of strip of length l and elemental cross-sectional area ldr is

$$d\phi_B = B_\theta l dr,$$

in which we can substitute (E4-10.1) to write

$$d\phi_B = \frac{\mu_0 Il}{2\pi} \frac{dr}{r}. \tag{E4-10.2}$$

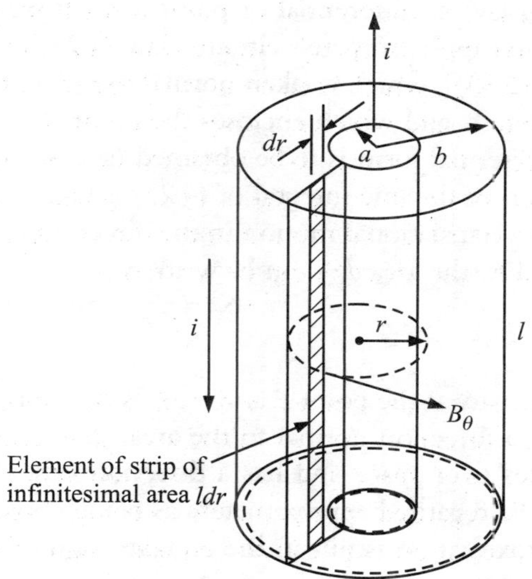

Fig. E4-10

Integrating (E4-10.2) between the limits $r = a$, the radius a of the inner conductor, and $r = b$, the inner radius b of the outer conductor, we obtain the magnetic flux ϕ_B through the radial cross-sectional area between the conductors as

$$\phi_B = \int d\phi_B = \frac{\mu_0 I}{2\pi} l \int_a^b \frac{dr}{r} = \frac{\mu_0 I}{2\pi} l [\ln r]_a^b \tag{E4-10.3}$$

$$= \frac{\mu_0 I}{2\pi} l (\ln b - \ln a) = \frac{\mu_0 I}{2\pi} l \ln \frac{b}{a}.$$

Dividing (E4-10.3) by I we obtain the following expression for the inductance of length l of the coaxial cable:

$$= \frac{\phi_B}{I} = \frac{\frac{\mu_0 I}{2\pi} l \ln \frac{b}{a}}{I} = \frac{\mu_0}{2\pi} l \ln \frac{b}{a},$$

which, when divided by l, gives the expression for the inductance L per unit length of the coaxial cable as:

$$L = \frac{\mu_0 I}{2\pi} \ln \frac{b}{a}.$$

4.3.2 Ampere's circuital law in differential form

In electrostatics, we obtained Poisson's equation (3.104) in point form or differential form from Gauss's law (3.35) or in integral form (3.64). Similarly, in magnetostatics, we can obtain Ampere's circuital law in differential or point form from its integral form (4.23). For this purpose, let us first apply Ampere's circuital law (4.23) to a closed element of area $dS_3 (= h_1 du_1 h_2 du_2)$ [see (2.53)], which is taken normal to the unit vector $\vec{a}_3$ in generalised curvilinear coordinate system, and which encloses the point (P) where Ampere's circuital law in point form or differential form is to be obtained (see Section 2.7 and Figs. 2.4 and 4.7). The sense of rotation of the line integral of (4.23) is taken such that a screw rotated in this sense would have a translational motion in the direction of $\vec{a}_3$ (Fig. 4.7).

The current i enclosed by the area dS_3 can be written as

$$i \cong \vec{J}.\vec{a}_3 dS_3, \tag{4.49}$$

where $\vec{J}$ is the current density at the point P and $\vec{J}.\vec{a}_3$ is the component J_3 of the current density $\vec{J}$ at the point in a direction normal to the area, that is, in the direction of $\vec{a}_3$. In general, although $\vec{J}$ varies over space and has a direction other than the normal to the area element dS_3, it can be regarded approximately as being constant over the element of area dS_3. Hence an approximation is put in the equality sign of (4.49). Putting (4.49) in the right-hand side of (4.23), we can then write Ampere's circuital law to the closed area element as (Fig. 4.7):

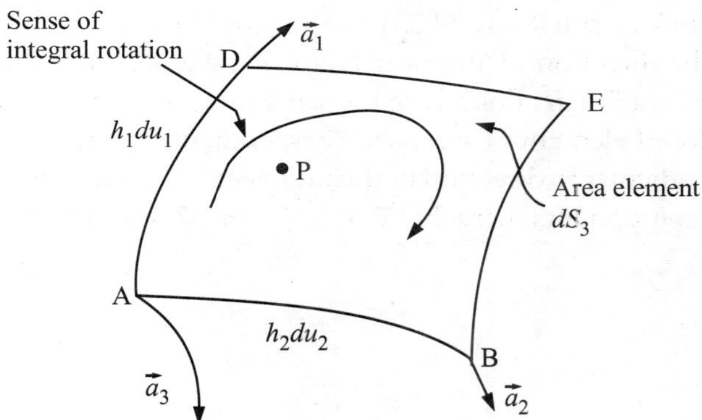

Fig. 4.7 *Element of area dS_3 perpendicular to the unit vector $\vec{a}_3$ showing the sense of rotation ADEBA. The line integral is taken around periphery of the element dS_3 and the current density vector $\vec{J}$ has a component J_3 in the direction of $\vec{a}_3$.*

$$\oint_l \vec{H} \cdot d\vec{l} \cong \vec{J}.\vec{a}_3 dS_3. \tag{4.50}$$

Dividing both sides of (4.50) by dS_3, we can then write

$$\frac{\oint_l \vec{H} \cdot d\vec{l}}{dS_3} \cong \vec{J}.\vec{a}_3. \tag{4.51}$$

We can remove the sign of approximation from (4.51) by taking the limit $dS_3 \to 0$ as follows:

$$\lim_{dS_3 \to 0} \frac{\oint_l \vec{H} \cdot d\vec{l}}{dS_3} = \vec{J}.\vec{a}_3. \tag{4.52}$$

Noting that $J_3 = \vec{J}.\vec{a}_3$, the component J_3 of $\vec{J}_3$ in the direction of $\vec{a}_3$, we can express (4.52) as

$$\lim_{dS_3 \to 0} \frac{\oint_l \vec{H} \cdot d\vec{l}}{dS_3} = J_3. \tag{4.53}$$

By definition, the left-hand side of (4.52) or (4.53) is $(\nabla \times \vec{H})_3$, the component of $\nabla \times \vec{H}$ in the direction of $\vec{a}_3$ [see the definition (2.51) for the component of the curl of a vector]. Therefore, replacing the left-hand side of (4.53) by $(\nabla \times \vec{H})_3$, we can write

$$(\nabla \times \vec{H})_3 = J_3. \tag{4.54}$$

The expansion of $(\nabla \times \vec{H})_3$ is obtainable from (2.61), in which, however, we have to replace $\vec{E}$ by $\vec{H}$ and $\vec{E}_1, \vec{E}_2$ by $\vec{H}_1, \vec{H}_2$.

Applying Ampere's circuital law (4.23) to the closed element of area dS_3, considered perpendicular to the direction of the unit vector $\vec{a}_3$, we obtained the expression for the component $(\nabla \times \vec{H})_3$ of $\nabla \times \vec{H}$. Following the same method, if we apply Ampere's circuital law (4.23) to the closed element of areas dS_2 (perpendicular to direction of the unit vector $\vec{a}_2$), and dS_1 (perpendicular to direction of the unit vector $\vec{a}_1$), we will obtain the expression for the remaining components, namely $(\nabla \times \vec{H})_2$ and $(\nabla \times \vec{H})_1$ of $\nabla \times \vec{H}$ corresponding to (4.54) as:

$$(\nabla \times \vec{H})_2 = J_2 \tag{4.55}$$

and

$$(\nabla \times \vec{H})_1 = J_1. \tag{4.56}$$

Combining the components (4.54)-(4.56), we then obtain

$$(\nabla \times \vec{H})_1\vec{a}_1 + (\nabla \times \vec{H})_2\vec{a}_2 + (\nabla \times \vec{H})_3\vec{a}_3 = J_1\vec{a}_1 + J_2\vec{a}_2 + J_3\vec{a}_3,$$

which, in view of the relations

$$\nabla \times \vec{H} = (\nabla \times \vec{H})_1\vec{a}_1 + (\nabla \times \vec{H})_2\vec{a}_2 + (\nabla \times \vec{H})_3\vec{a}_3 \qquad \text{[see (2.5)]}$$

and

$$\vec{J} = J_1\vec{a}_1 + J_2\vec{a}_2 + J_3\vec{a}_3$$

takes the following differential form:

$$\nabla \times \vec{H} = \vec{J}. \tag{4.57}$$

The expression (4.57), derived from the integral form of Ampere's circuital law (4.23), can be interpreted as the point form of Ampere's circuital law. The expressions for $\nabla \times \vec{H}$ occurring in (4.57) in the different coordinate systems can be read from (2.65) for the generalised curvilinear coordinate system, and from (2.69)-(2.71) for the rectangular, cylindrical, and spherical-polar coordinate systems respectively, interpreting therein $\vec{E}$ as $\vec{H}$ and E_1, E_2, E_3 as H_1, H_2, H_3.

4.4 Lorentz Force on a Moving Charge and Force on a Current-carrying Conductor in a Magnetic Field

4.4.1 Lorentz force

We have seen how a static charge when placed in an electric field experiences a force (Section 3.1). However, a charge placed in a magnetic field does not experience a force unless it moves in a direction different from that of the magnetic field. The force $\vec{F}$ that

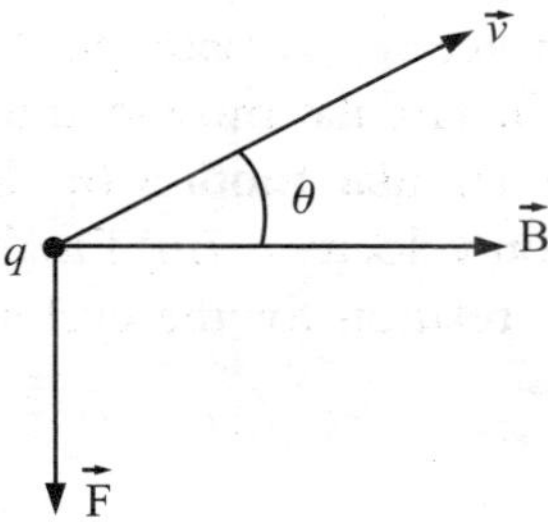

Fig. 4.8 *A point charge moving with a velocity $\vec{v}$ in a region of magnetic flux density $\vec{B}$ experiencing Lorentz force $\vec{F}$.*

a point charge q experiences while moving with a velocity $\vec{v}$ in a region of magnetic field flux density $\vec{B}$ is given by the following Lorentz force expression:

$$\vec{F} = q\vec{v} \times \vec{B},$$

where $\vec{F}$ has the magnitude $F = qvB\sin\theta$ (θ is the angle between $\vec{v}$ and $\vec{B}$) and the direction perpendicular to the plane containing $\vec{v}$ and $\vec{B}$ (Fig. 4.8).

There are practical electron devices and systems in which we need to form a flow of charge, say, a beam of electrons. We can also need to confine such a charge flow. In some of these devices, the magnetic field takes a dominant role in the interaction between an electron beam and electromagnetic waves. In the analysis of such devices and systems in the presence of a magnetic field, the Lorentz force expression (4.25) proves to be useful.

EXAMPLE E4-11

Show that an electron, when it is shot with a dc velocity perpendicular to a uniform dc magnetic field of flux density of magnitude B, executes a circular motion with an angular frequency ω_c (called the electron cyclotron frequency) given by $\omega_c = |\eta|\, B$, where $|\eta|$ is the magnitude of the charge-to-mass ratio of an electron.

SOLUTION

We can consider the point charge and the velocity vector to lie on the plane of the paper (page of the book) and the magnetic field vector to be perpendicular to this plane and directed away from the reader. The Lorentz force given by (4.58) will help us to understand that the force, mutually perpendicular to the magnetic field, and the velocity of the point charge would cause the point charge to execute an anticlockwise circular motion in the plane of the paper if the point charge is positive. For a negative point charge representing an electron, this circular motion would be clockwise, which would however be changed to anticlockwise motion if the direction of the magnetic field is reversed from the direction away from the reader to towards the reader.

(i) We can balance the Lorentz force $|e|\, vB$ with the centrifugal force $mv^2 r$ to obtain a relation between the radius r of the circular motion and the electron velocity v. (ii) Next, obtain the time period T of this circular motion by dividing the circumference of the circular motion $2\pi r$ by the electron velocity v. (iii) Finally, dividing 2π by the time period T so obtained, we can derive the relation for the cyclotron angular frequency ω_c. These steps are as follows:

(i) $$|e|\, vB = \frac{mv^2}{r}$$

giving $$v = \frac{|e|\, Br}{m} = |\eta|\, Br.$$

(ii) $$T = \frac{2\pi r}{v} = \frac{2\pi r}{|\eta|\, Br} = \frac{2\pi}{|\eta|\, B}.$$

(iii) $$\omega_c = \frac{2\pi}{T} = \frac{2\pi}{\frac{2\pi}{|\eta| B}} = |\eta|\, B.$$

4.4.2 Force on a current-carrying conductor

Let us now consider a very useful manifestation of the application of Lorentz force given by (4.58)—the force on a current-carrying conductor that is placed in a magnetic field. Let us take a thin conducting wire carrying a current in a magnetic field. Let us consider a steady flow of current comprised of charged particles regarded as point charges, each experiencing Lorentz force given by (4.58).

We have already gathered from the discussion following (3.124) that the current, if constituted by electrons, is given by $i = (v)(\alpha)(n)(e)$, which, in terms of $|e|$, the magnitude of electronic charge $[e = -\,|e|]$, can be expressed as

$$i = -n\ |e|\ v\,\alpha. \tag{4.59}$$

Here, v is the velocity of electrons, taken as the same for all the electrons; n is the number of electrons per unit volume; and α is the cross-sectional area of the charge flow, taken here as the cross-sectional area of the wire.

In general, the magnetic field varies over the wire. However, we can regard the magnetic field as constant over an infinitesimal length dl of the wire that occupies a volume $dl\alpha$ and contains $n\, dl\, \alpha$ number of electrons, each of which is subjected to Lorentz force $-\,|e|\ \vec{v} \times \vec{B}$ obtained by putting $q = e = -\,|e|$ in (4.58). Therefore, multiplying this force by the number of electrons $n\, dl\, \alpha$ contained in the length element dl, we obtain the expression for the force $d\vec{F}$ on the current element occupying the length element dl as follows:

$$d\vec{F} = [n\, dl\, \alpha][-\,|e|\ \vec{v} \times \vec{B}] = -n\ |e|\ \ dl\, \alpha\, \vec{v} \times \vec{B}. \tag{4.60}$$

Since the electrons move with velocity $\vec{v}$ in the length element dl in the direction of the alignment of the length element, we can define an element of length vector as

$$d\vec{l} = dl\vec{a}_v \tag{4.61}$$

where $\vec{a}_v$ is a unit vector in the direction of the velocity of the electrons $\vec{v}$ through the element of length dl.

$\vec{v}$ is thus expressible as

$$\vec{v} = v\vec{a}_v \tag{4.62}.$$

With the help of (4.62), we can express (4.60) as

$$d\vec{F} = -n\ |e|\ dl\, \alpha\, v\vec{a}_v \times \vec{B} = -n\ |e|\ \ \alpha\, v\, dl\vec{a}_v \times \vec{B},$$

which, we can be further expressed, with the help of (4.61) as

$$d\vec{F} = -n\ |e|\ \ v\, \alpha\, d\vec{l} \times \vec{B}.$$

Substituting (4-59) into (4.63), we then obtain the following expression for the element of force $d\vec{F}$ experienced by a thin wire occupying a length dl and carrying a current i in a region of magnetic flux density $\vec{B}$ (Fig. 4.9):

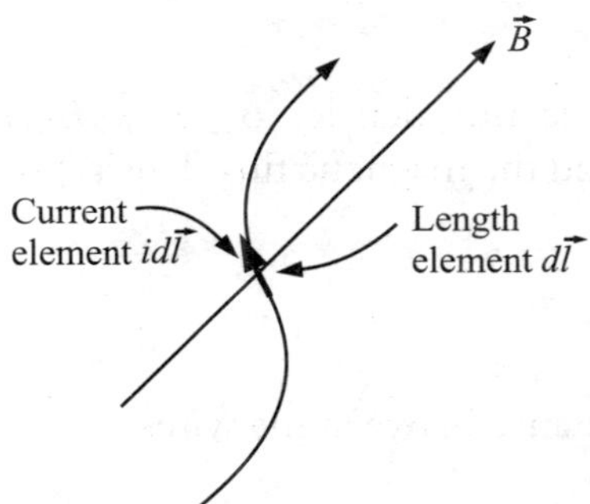

Fig. 4.9 *A thin conducting wire carrying current i placed in a magnetic flux density $\vec{B}$; the current element $id\vec{l}$ spread over the length element $d\vec{l}$ experiences a force $\vec{F}$.*

$$d\vec{F} = i\, d\vec{l} \times \vec{B}. \tag{4.64}$$

In general, the magnetic field can vary from point to point on the current-carrying wire of finite length. In order to find the force $\vec{F}$ on the complete length of the wire, we have to integrate (4.64) over the length of the wire and evaluate it as

$$\vec{F} = \int d\vec{F} = \int i\, d\vec{l} \times \vec{B}. \tag{4.65}$$

Let us now take the simple example of a pair of current-carrying wires with a view to appreciating both (i) Biot-Savart's law (Section 4.2) and Ampere's circuital law (Section 4.3.2) used for finding the magnetic field due to a current-carrying conductor, and (ii) the force experienced by a current-carrying conductor placed in a magnetic field discussed in the present subsection (Section 4.4.2), which, in fact, is the manifestation of Lorentz force experienced by a charge moving in a magnetic field. The example is also aimed at appreciating that like currents attract and unlike currents repel. For this purpose, let us take two long parallel straight wires, the first wire carrying current i_1 and the second i_2 (Fig. 4.10). In the rectangular coordinate system, let the first wire be located parallel to the Z-axis and the second wire, parallel to the first wire (and hence to Z-axis) be located at a perpendicular distance d from the first wire. Let us consider a length element dl_2 on the second wire. The element of force $d\vec{F}_{21}$ on the length element dl_2 on the second wire carrying current i_2 due to the entire first wire can be found with the help of (4.64) as

$$d\vec{F}_{21} = i_2\, d\vec{l_2} \times \vec{B}_{21}, \tag{4.66}$$

where length vector $d\vec{l_2}$ has a magnitude of dl_2 and is directed along positive z; and $\vec{B}_{21}$ is the magnetic flux density at the second wire due the current i_1 of the first wire, which is valid at all points of the second wire including the location of the length element dl_2. Accordingly, we can express $d\vec{l_2}$ as

$$d\vec{l_2} = dl_2\vec{a}_z \tag{4.67}$$

and also express the magnetic flux density $\vec{B}_{21}(= \mu_0\vec{H}_{21})$ (see (4.40) for the relation between the magnetic field and the magnetic flux density) with the help of (4.19) or (4.27) as:

$$\vec{B}_{21} = \mu_0\frac{i_1}{2\pi\, d}\vec{a}_\theta, \tag{4.68}$$

d being the perpendicular distance between the wires.

Substituting (4.67) and (4.68) into (4.66), we get

$$d\vec{F}_{21} = i_2(dl_2\vec{a}_z) \times (\mu_0\frac{i_1}{2\pi d}\vec{a}_\theta) = \frac{\mu_0 i_1 i_2 dl_2}{2\pi d}(\vec{a}_z \times \vec{a}_\theta). \tag{4.69}$$

However, at the location of the element dl_2 (Fig. 4.10), we can easily interpret $\vec{a}_\theta = -\vec{a}_x$ in view of $\vec{B}_{21}$—which has been substituted in (4.66)—involving $\vec{a}_\theta$, the unit vector in the direction of increasing θ in the cylindrical system, which at the second wire is in the direction of x decreasing, that is, in the direction away from the reader, while noting that $\vec{a}_x$ is directed towards the reader. Therefore, substituting $\vec{a}_\theta = -\vec{a}_x$ so interpreted in (4.69) we obtain

$$d\vec{F}_{21} = -\frac{\mu_0 i_1 i_2 dl_2}{2\pi d}(\vec{a}_z \times \vec{a}_x),$$

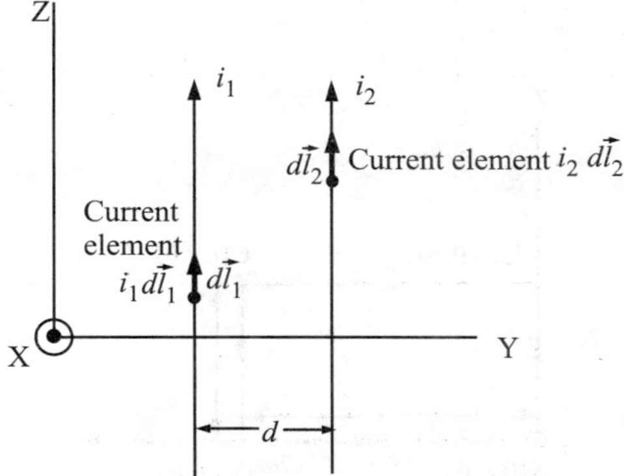

Fig. 4.10 *A pair of parallel straight current-carrying wires experiencing force due to each other.*

which, in view of the relation $\vec{a}_z \times \vec{a}_x = \vec{a}_y$, becomes

$$d\vec{F}_{21} = -\frac{\mu_0 i_1 i_2 dl_2}{2\pi d}\vec{a}_y. \tag{4.70}$$

While deriving (4.70), we have considered both the currents i_1 and i_2 to be in the z direction or, in other words, considered to be "like currents". Consequently, in (4.70) we obtained the direction of the force $d\vec{F}_{21}$ on the length element dl_2 on the second wire as that of $-\vec{a}_y$, that is, towards the first wire (Fig. 4.10). Thus, the force turns out to be that of attraction for the like currents. On the other hand, if the direction of current in the second wire is reversed, that is, taken in the negative z direction so that the currents i_1 and i_2 become "unlike currents", then the length element $d\vec{l}_2$ becomes $d\vec{l}_2 = -dl_2\vec{a}_z$ instead of $d\vec{l}_2 = dl_2\vec{a}_z$ given by (4.67). Consequently, we obtain the expression for $d\vec{F}_{21}$ as one identical to (4.70) but with a reversal of sign on its right-hand side. This reversal of sign on the right-hand side of (4.70) for unlike currents corresponds to the direction of $d\vec{F}_{21}$, being away from the first wire (same as that of $\vec{a}_y$), corresponding to a force of repulsion (Fig. 4.10).

EXAMPLE E4-12

A rectangular loop of direct current i of length l, aligned along X-axis, and breadth b, aligned along Z-axis, is placed in a region of uniform magnetic field of flux density $\vec{B}$ applied parallel to the plane of the loop in the direction of x. If freely suspended, the current loop will rotate due to the force exerted by the magnetic field. Typically, the magnetic field can be generated by curved pole pieces such that even if the loop rotates to take a new position, the magnetic field continues to be parallel to the plane of the loop.

Hence, for the situation, verify the relation $\vec{\tau} = i\vec{\alpha} \times \vec{B}$ in terms of the torque $\vec{\tau}$ developed on the current loop and the area vector $\vec{\alpha}$, whose magnitude is the loop area $\alpha = lb$ and which has the direction same as that of the linear motion of a screw rotated in the sense of the current in the loop.

SOLUTION

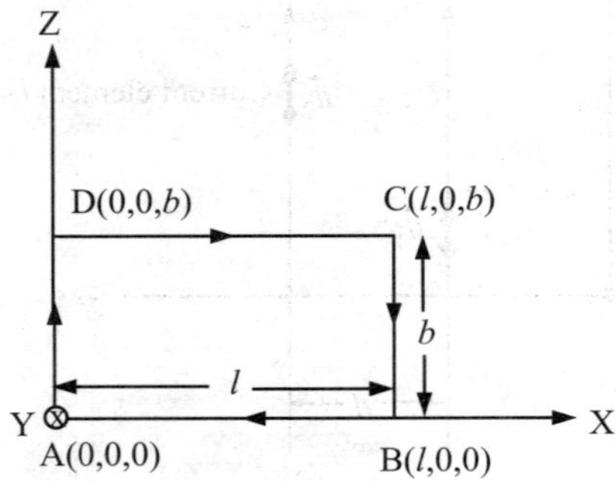

Fig. E4-12

Let us define the rectangular current loop ABCD such that it lies in the XZ plane ($y = 0$) (Fig. E-4.12). It has four corners A$(0, 0, 0)$ (at the origin), B$(l, 0, 0)$, C$(l, 0, b)$ and D$(0, 0, b)$. The length AB coincides with the X-axis and the breadth AD coincides with the Z-axis. The direction of current is from B to A; from A to D; from D to C; and from C to B. The flux density $\vec{B}$ of the given magnetic field in which the loop is placed can be represented as

$$\vec{B} = B\vec{a}_x. \tag{E4-12.1}$$

With the help of (4.65), we can find the forces $\vec{F}_{\mathrm{AB}}$ on the arm AB of the loop as

$$\left.\begin{aligned}\vec{F}_{\mathrm{AB}} &= i(\vec{\mathrm{BA}}) \times B\vec{a}_x = i(\mathrm{AB})(-\vec{a}_x) \times (B\vec{a}_x) = -ilB\,(\vec{a}_x \times \vec{a}_x) = 0\\ \vec{F}_{\mathrm{BC}} &= i(\vec{\mathrm{CB}}) \times B\vec{a}_x = i(\mathrm{CB})(-\vec{a}_z) \times (B\vec{a}_x) = -ibB\,\vec{a}_y\\ \vec{F}_{\mathrm{CD}} &= i(\vec{\mathrm{DC}}) \times B\vec{a}_x = i(\mathrm{DC})(\vec{a}_x) \times (B\vec{a}_x) = ilB\,(\vec{a}_x \times \vec{a}_x) = 0\\ \vec{F}_{\mathrm{DA}} &= i(\overrightarrow{\mathrm{DA}}) \times B\vec{a}_x = i(\mathrm{AD})(\vec{a}_z) \times (B\vec{a}_x) = ibB\,\vec{a}_y\end{aligned}\right\}. \tag{E4-12.2}$$

From (E4-12.2) we notice that the forces $\vec{F}_{\mathrm{AB}}$ and $\vec{F}_{\mathrm{CD}}$ are each null, producing no influence on the current loop. However, the forces $\vec{F}_{\mathrm{BC}}$ and $\vec{F}_{\mathrm{DA}}$ are equal in magnitude but opposite in directions, causing a rotational effect on the loop. We can represent these unbalanced forces, given by (E4-12.2) as follows:

$$\left.\begin{aligned}\vec{F}_{\mathrm{BC}} &= -ibB\,\vec{a}_y = -F\vec{a}_y\\ \vec{F}_{\mathrm{DA}} &= ibB\,\vec{a}_y = F\vec{a}_y\end{aligned}\right\}, \tag{E4-12.3}$$

where

$$F = ibB. \tag{E4-12.4}$$

The loop, as a result of the unbalanced forces given by (E4-12.3), will rotate about the negative Z-axis. The consequent torque $\vec{\tau}$ developed on the current loop then has the direction of negative z, and hence it can be represented as

$$\vec{\tau} = -\tau\vec{a}_z, \tag{E4-12.5}$$

where τ is given by the product of the force F and the distance l between the sides AD and BC of the loop given as follows:

$$\tau = Fl,$$

which, in view of (E4-12.4) becomes

$$\tau = ilbB. \tag{E4-12.6}$$

Substituting (E4-12.6) in (E4-12.5), we can write

$$\vec{\tau} = -ilbB\vec{a}_z. \text{ (E4-12.7)}$$

At this stage, it is convenient to define an area vector

$$\vec{\alpha} = \alpha\vec{a}_y, \tag{E4-12.8}$$

where $\alpha = lb$ (E4-12.9)

is the magnitude of the area vector, being the area of the loop, and the direction of the area vector is along y, being the direction in which a screw would linearly move if it were rotated in the sense of the direction of the current through the loop. Further, with the help of (E4-12.1) and (E4-12.8), we can write

$$\vec{\alpha} \times \vec{B} = \alpha\vec{a}_y \times B\vec{a}_x = \alpha B\vec{a}_y \times \vec{a}_x = -\alpha B\vec{a}_z. \tag{E4-12.10}$$

Also, with the help of (E4-12.9) we can express $\vec{\tau}$ given by (E4-12.7), as

$$\vec{\tau} = -i\alpha B\vec{a}_z.$$

which, in view of the relation $\vec{a}_x \times \vec{a}_y = \vec{a}_z$ can be put as

$$\vec{\tau} = -i\alpha B\vec{a}_x \times \vec{a}_y = -i(B\vec{a}_x) \times (\alpha\vec{a}_y). \tag{E4-12.11}$$

We can then finally express (E4-12.11), with the help of (E4-12.1) and (E4-12.8), as

$$\vec{\tau} = -i(B\vec{a}_x) \times (\alpha\vec{a}_y) = -i\vec{B} \times \vec{\alpha} = i\vec{\alpha} \times \vec{B}.$$

4.5 Magnetic Vector Potential due to a Steady Current

We have seen in Section 4.2 how Biot-Savart's law given by (4.10) can be used to find the magnetic field due to a steady current through a conductor. We will now develop the concept of the magnetic vector potential that can also be used for the same purpose; the approach is similar to that of finding the electric field from the scalar electric potential in electrostatics (see Section 3.2.2). (The generalised concept of the magnetic vector potential for time-varying fields, from which the present concept also follow as a special case, has been developed later in Section 5.5).

Let us for the sake of convenience rewrite the expression (4.10) for the element of magnetic flux density $d\vec{B}$ due to the current element $i\,d\vec{l}$ as follows:

$$d\vec{B} = \mu_0 d\vec{H} = \frac{\mu_0}{4\pi}\frac{i\,d\vec{l} \times \vec{a}_R}{R^2}. \tag{4.71}$$

Putting $\vec{a}_R = \vec{R}/R$, we can express (4.71) as

$$d\vec{B} = \mu_0 d\vec{H} = \frac{\mu_0}{4\pi}\frac{i\,d\vec{l} \times \vec{R}}{R^3} = \frac{\mu_0}{4\pi} i\,d\vec{l} \times \frac{\vec{R}}{R^3}, \tag{4.72}$$

where $\vec{R}(= R\vec{a}_R)$ is the distance vector, the magnitude of which is R—the distance between the current element (source point) and the point where the magnetic flux density is sought (field observation point)—and the direction of which is from the source point to the field observation point. Further, we can substitute the relation $\nabla\left(\frac{1}{R}\right) = -\frac{1}{R^3}\vec{R}$ (see Appendix A4.2 for the deduction of this relation) in (4.72) to obtain

$$d\vec{B} = -\frac{\mu_0}{4\pi} i\,d\vec{l} \times \nabla\left(\frac{1}{R}\right). \tag{4.73}$$

Now, rearranging the vector identity

$$\nabla \times (\psi\vec{G}) = \psi\nabla \times \vec{G} - \vec{G} \times \nabla\psi$$

as

$$\vec{G} \times \nabla\psi = \psi\nabla \times \vec{G} - \nabla \times (\psi\vec{G})$$

and putting therein $\vec{G} = i\,d\vec{l}$ and $\psi = 1/R$, we can write the following relation:

$$i\,d\vec{l} \times \nabla\left(\frac{1}{R}\right) = \frac{1}{R}\nabla \times i\,d\vec{l} - \nabla \times \left(\frac{1}{R} i\,d\vec{l}\right). \tag{4.74}$$

Substituting (4.74) in (4.73) we can then write

$$d\vec{B} = \frac{\mu_0}{4\pi}\left[\nabla \times \left(\frac{1}{R} i\,d\vec{l} - \frac{1}{R}\nabla \times i\,d\vec{l}\right)\right]. \tag{4.75}$$

However, note that the second term of the right-hand side of (4.75) becomes zero since in that term the curl operation is taken with respect to the field-point variables (x, y, z) while the current element $i\,d\vec{l}$ term on which this operation is taken is represented by the source point variables $(x', y', z'$, say) (see Appendix A4.2). Therefore, putting the second term of (4.75) equal to zero, we can write:

$$d\vec{B} = \nabla \times \left(\frac{\mu_0}{4\pi} \frac{i\,d\vec{l}}{R} \right) = \nabla \times d\vec{A}, \tag{4.76}$$

where $d\vec{A}$ is the element of vector defined as

$$d\vec{A} = \frac{\mu_0}{4\pi} \frac{i\,d\vec{l}}{R}. \tag{4.77}$$

By adding the contributions from individual current elements, we can then obtain the magnetic flux density $\vec{B}$ due to an entire current distribution by integrating (4.76) as

$$\vec{B} = \int d\vec{B} = \int \nabla \times d\vec{A} = \nabla \times \vec{A}, \tag{4.78}$$

where $\vec{A}$, in turn, is obtained as the integration of (4.77) as

$$\vec{A} = \int d\vec{A} = \int \frac{\mu_0}{4\pi} \frac{i\,d\vec{l}}{R}, \tag{4.79}$$

called the magnetic vector potential.

It becomes often convenient to find the magnetic flux density due to a current passing through a conductor by first finding the magnetic vector potential $\vec{A}$ due to the current with the help of (4.79) and then substituting it in (4.78). Let us illustrate this by taking the task of finding the magnetic field near an infinitely long straight current-carrying filamentary wire, for which we had earlier deduced (Section 4.2) the expression (4.19) using Biot-Savart's law.

For the purpose of the above illustration, let us consider the wire carrying a direct current i to be aligned along the Z-axis (Fig. 4.11). Consider a current element idz' of infinitesimal length dz' on the Z-axis at M at a distance z' from the origin O $(0, 0, 0)$ such that OM $= z'$. The element of magnetic vector potential $d\vec{A}$ at a point P (r, θ, z) due to a current element idz' can be written, with the help of (4.77), interpreting $d\vec{l} = dz'\vec{a}_z$ as

$$d\vec{A} = \frac{\mu_0}{4\pi} \frac{i\,dz'\,\vec{a}_z}{R}, \tag{4.80}$$

where $R(=$ PM) is the distance of the point P (r, θ, z) from the current element at M. Then from the right-angled triangle MNP, we can write

$$R = \text{PM} = [(\text{NM})^2 + (\text{NP})^2]^{1/2}, \tag{4.81}$$

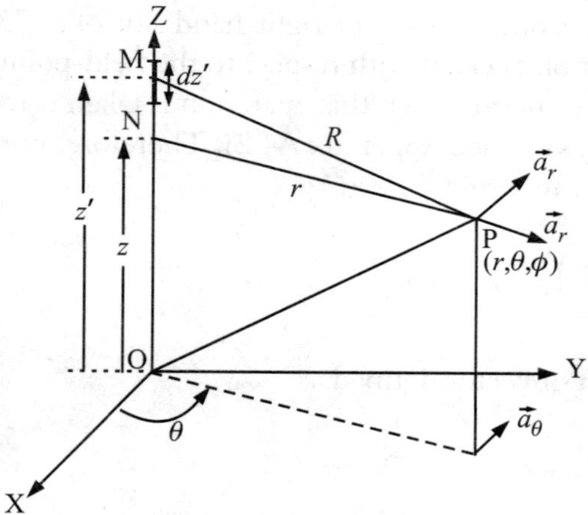

Fig. 4.11 *An infinitely long straight filamentary conducting wire carrying a direct current and a point where the magnetic field due to it is to be found using the concept of vector potential.*

where the point N is the foot of perpendicular dropped from P on the Z-axis that makes NP $= r$, and NM = OM – ON, which, since here OM $= z'$ and ON $= z$, becomes NM $= z' - z$ (Fig. 4.11). Therefore, substituting NP $= r$ and NM $= z' - z$ in (4.81), we can write

$$R = \text{PM} = [(z' - z)^2 + r^2]^{1/2}. \tag{4.82}$$

The element of magnetic vector potential $d\vec{A}$ at point P can then be expressed by substituting (4.82) in (4.80) as

$$d\vec{A} = \frac{\mu_0}{4\pi} \frac{i\, dz'\, \vec{a}_z}{[(z' - z)^2 + r^2]^{1/2}}. \tag{4.83}$$

The element of magnetic flux density $\vec{B}$ at point P can be then written by putting (4.83) in (4.78) as

$$\vec{B} = \int \nabla \times d\vec{A} = \frac{\mu_0 i}{4\pi} \int \nabla \times \frac{dz' \vec{a}_z}{[(z' - z)^2 + r^2]^{1/2}}. \tag{4.84}$$

The integrand in the right-hand side of (4.84) can be written with the help of (2.67), noting that the problem enjoys the azimuthal symmetry ($\partial/\partial\theta = 0$), as

$$\nabla \times \frac{dz' \vec{a}_z}{[(z' - z)^2 + r^2]^{1/2}} = -\frac{\partial}{\partial r} \frac{dz'}{[(z' - z)^2 + r^2]^{1/2}} \vec{a}_\theta. \tag{4.85}$$

We can carry out the differentiation on the right-hand side of (4.85) to obtain

$$\nabla \times \frac{dz'\vec{a}_z}{[(z'-z)^2+r^2]^{1/2}} = \frac{rdz'}{[(z'-z)^2+r^2]^{3/2}}\vec{a}_\theta,$$

which can be written, with the help of (4.82) as

$$\nabla \times \frac{dz'\vec{a}_z}{[(z'-z)^2+r^2]^{1/2}} = \frac{r}{R^3}dz'\vec{a}_\theta. \tag{4.86}$$

Plugging (4.86) in (4.84), we obtain

$$\vec{B} = \frac{\mu_0 i}{4\pi}\int \nabla \times \frac{dz'\vec{a}_z}{[(z'-z)^2+r^2]^{1/2}} = \frac{\mu_0 i}{4\pi}\int \frac{r}{R^3}dz'\vec{a}_\theta. \tag{4.87}$$

In the present context, the integration in (4.87) has to be carried out for the infinitely long wire by taking the integration from $z' = -\infty$ to $z' = \infty$. Therefore, we have to take (4.87) as

$$\vec{B} = \frac{\mu_0 i}{4\pi}\int_{z'=-\infty}^{z'=\infty} \frac{r}{R^3}dz'\,\vec{a}_\theta. \tag{4.88}$$

We can evaluate the integral of (4.88) by introducing the variable $\psi = \angle\text{MPN}$, which would satisfy the relation (Fig. 4.11):

$$\tan\psi = \frac{\text{NM}}{\text{NP}} = \frac{z'-z}{r},$$

which can be rearranged as

$$z' - z = r\tan\psi. \tag{4.89}$$

We then take the differential of (4.89) to obtain

$$dz' = r\sec^2\psi d\psi. \tag{4.90}$$

Further, from the geometry of the problem (Fig. 4.11), we have the relation

$$R = r\sec\psi. \tag{4.91}$$

Therefore, substituting (4.90) and (4.91) in (4.88) and the changing the integration limits $z' = -\infty$ and $z' = \infty$ to $\psi = \pi/2$ and $\psi = -\pi/2$ respectively, we then obtain

$$\vec{B} = \frac{\mu_0 i}{4\pi}\int_{\psi=-\pi/2}^{\psi=\pi/2} \frac{r}{r^3\sec^3\psi} r\sec^2\psi\, d\psi\,\vec{a}_\theta = \frac{\mu_0 i}{4\pi r}\int_{\psi=-\pi/2}^{\psi=\pi/2} \cos\psi\, d\psi\,\vec{a}_\theta. \tag{4.92}$$

We can evaluate the integral in (4.92) giving

$$\vec{B} = \frac{\mu_0 i}{4\pi r}[\sin\psi]_{-\pi/2}^{\pi/2}\vec{a}_\theta = \frac{\mu_0 i}{4\pi r}[\sin(\pi/2) - \sin(-\pi/2)]\vec{a}_\theta,$$

which, on substituting the values $\sin(\pi/2) = 1$ and $\sin(-\pi/2) = -1$, results in the following expression for the magnetic flux density due to an infinitely long straight current-carrying filamentary wire

$$\vec{B} = \frac{\mu_0 i}{2\pi r}\vec{a}_\theta. \tag{4.93}$$

The corresponding expression for $\vec{H}(= \vec{B}/\mu_0)$ then follows by dividing (4.93) by μ_0 as

$$\vec{H} = \frac{i}{2\pi r}\vec{a}_\theta. \tag{4.94}$$

We thus see that the expression (4.94) obtained here by the vector potential approach becomes one and the same as the expression (4.19) obtained earlier applying Biot-Savart's law. As another example of finding the magnetic field from the concept of vector potential, we will find the magnetic flux density due to a circular loop of current at a point off its axis in Example E4-13 next. Also, we have later used in this book (Chapter 8; Section 8.4) the concept of vector potential to predict the radiation characteristics of a conduction current antenna.

EXAMPLE E4-13

Obtain an expression for the magnetic field due to a circular loop carrying a direct current at a point off the axis of the loop using the concept of magnetic vector potential. Verify that it agrees with the expression for the same obtained by treating the loop as a magnetic dipole, obtained while drawing an analogy between the electric and magnetic dipoles (see Example E4-7).

SOLUTION

The present problem is more generalised than Example E4-3 in that the point where the magnetic field is sought is now off the axis of the circular loop. Moreover, for the present purpose, we will use the concept of the vector potential rather than Biot-Savart's law, unlike Example E4-3.

We consider the current-carrying circular loop to be placed on the XY plane with its centre at the origin of coordinates ($x = y = z = 0$) (Fig. E4-13.1). Let us take now two identical current elements on the loop, each $Id\bar{l}$, one at A and the other at B, symmetrically placed at an angle ϕ' with respect to the X-axis. We can find the contribution to the magnetic vector potential due to these current elements at a point P (r, θ, ϕ) off the axis of the loop using the following approach (Fig. E4-13.1).

With the help of (4.77), we can then find the element of vector potential due to the current elements at A and B. The contribution to the element of vector potential

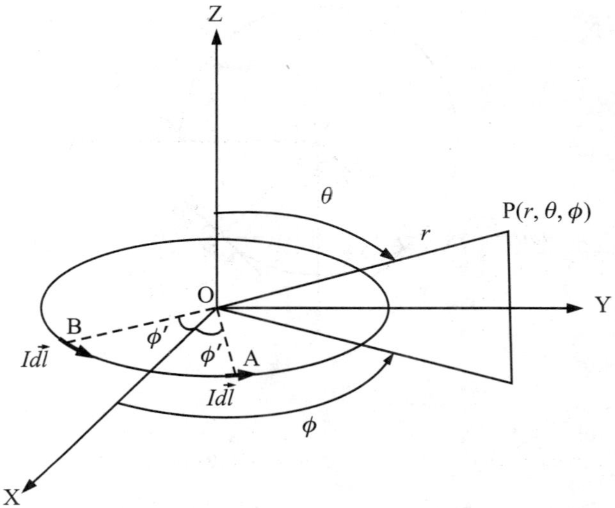

Fig. E4-13.1

$d\vec{A}$ from the current element at A has two components, one along positive x and the other along positive y, and, similarly, the contribution from the current element at B has two components, one along negative x and the other along positive y. Therefore, the contributions from the current elements A and B will cancel out in x direction and add up in positive y direction. We can find the contribution from B along positive y direction as $dA_y = dA\cos\phi'$, where ϕ' is the angle of the direction of B from the X-axis. No generality is lost if we choose the X-axis such that for the point P (r, θ, ϕ), ϕ is taken as zero so that the projection of the point P on the XY plane would lie on the X-axis itself. The direction of $\vec{a}_\phi$ then becomes the same as that of $\vec{a}_y$. Therefore, we can take the element of vector potential due to a current element on the circular loop as $dA_\phi (= dA_y) = dA\cos\phi'$.

Further, since the rectangular coordinates of the point P are [see (2.2)] $x = r\sin\theta\cos\phi = r\sin\theta$, $y = r\sin\theta\sin\phi = 0$, $z = r\cos\theta$, and those of the current element (B) on the circular loop of radius a are (Fig. E4-13.2):

$$x = a\cos\phi',\ y = a\sin\phi',\ z = 0,$$

we can find the distance R between the current element and the point P as

$$R = [(r\sin\theta - a\cos\phi')^2 + (0 - a\sin\phi')^2 + (r\cos\theta - 0)^2]^{1/2} \qquad \text{(E4-13.1)}$$

$$= (r^2 + a^2 - 2ra\sin\theta\cos\phi')^{1/2}.$$

The expression (E4-13.1) for $a << r$, applicable to a small loop of current, becomes

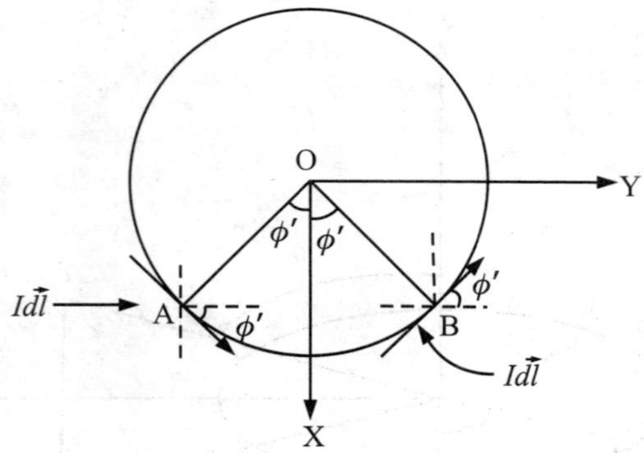

Fig. E4-13.2

$$R = (r^2 - 2ra\sin\theta\cos\phi')^{1/2} \cong r(1 - 2(a/r)\sin\theta\cos\phi')^{1/2}. \tag{E4-13.2}$$

Using this expression (E4-13.2) for R, we can then find dA with the help of (4.77) and hence dA_ϕ as

$$\begin{aligned} dA_\phi &= dA\cos\phi' = \frac{\mu_0 idl}{4\pi R}\cos\phi' \qquad \text{(E4-13.3)} \\ &= \frac{\mu_0 idl}{4\pi r(1 - 2(a/r)\sin\theta\cos\phi')^{1/2}}\cos\phi' \\ &= \frac{\mu_0 i}{4\pi r}(1 - 2(a/r)\sin\theta\cos\phi')^{-1/2} dl\cos\phi. \end{aligned}$$

On binomial expansion and ignoring higher powers of a/r (in view of $a << r$), the expression (E4-13.3) becomes

$$dA_\phi = \frac{\mu_0 i}{4\pi r}(1 + (a/r)\sin\theta\cos\phi')dl\cos\phi'. \tag{E4-13.4}$$

Putting $dl = ad\phi'$, we can express (E4-13.4) as

$$\begin{aligned} dA_\phi &= \frac{\mu_0 ia}{4\pi r}\left(1 + \frac{a}{r}\sin\theta\cos\phi'\right)\cos\phi'\,d\phi' \qquad \text{(E4-13.5)} \\ &= \frac{\mu_0 ia}{4\pi r}\left(\cos\phi' d\phi' + \frac{a}{r}\sin\theta\cos^2\phi' d\phi'\right). \end{aligned}$$

The contribution to the vector potential at P from all the current elements of the circular loop of current is then obtained by integrating the above expression (E4-13.5) over ϕ' extending from 0 to 2π .

$$A_\phi = \frac{\mu_0 ia}{4\pi r}\left(\int_0^{2\pi} \cos\phi' d\phi' + \frac{a}{r}\sin\theta \int_0^{2\pi} \cos^2\phi' d\phi'\right). \tag{E4-13.6}$$

We can evaluate (E4-13.6) as

$$A_\phi = \frac{\mu_0 ia}{4\pi r}\left(0 + \frac{a}{r}\sin\theta \frac{1}{2} 2\pi\right) = \frac{\mu_0 i\pi a^2}{4\pi r^2}\sin\theta. \tag{E4-13.7}$$

Further, with the help of the expression (2.68) for the curl of a vector in spherical polar coordinates, interpreted here for the vector potential $\vec{A}$, and the expression (4.78) for $\vec{B} = \nabla \times \vec{A}$

$$\vec{B} = \nabla \times \vec{A} = \frac{1}{r\sin\theta}\left(\frac{\partial(\sin\theta A_\phi)}{\partial\theta}\right)\vec{a}_r - \frac{1}{r}\left(\frac{\partial(rA_\phi)}{\partial r}\right)\vec{a}_\theta. \tag{E4-13.8}$$

can be evaluated by substituting (E4-13.7) in (E4-13.8):

$$\vec{B} = \frac{1}{r\sin\theta}\frac{\partial}{\partial\theta}\left(\sin\theta \frac{\mu_0 i\pi a^2}{4\pi r^2}\sin\theta\right)\vec{a}_r - \frac{1}{r}\frac{\partial}{\partial r}\left(r\frac{\mu_0 i\pi a^2}{4\pi r^2}\sin\theta\right)\vec{a}_\theta,$$

which can be easily evaluated as

$$\vec{B} = \frac{\mu_0 i\pi a^2}{4\pi r^3}(2\cos\theta\,\vec{a}_r + \sin\theta\,\vec{a}_\theta). \tag{E4-13.9}$$

Note that the expression (E4-13.9), by putting $p_m = i\pi a^2$ becomes identical to the expression obtained earlier by treating the loop as a magnetic dipole of moment p_m while drawing an analogy between the electric and magnetic dipoles (see Example E4-7).

Appendix A4.1

Magnetic field due to a direct current passing through a straight wire of a finite length

Let us take a current-carrying straight wire aligned along the Z-axis (Fig. A4.1 and Fig. 4.2). However, instead of a wire of infinite length considered in the illustration given following Biot-Savart's law (4.10) in the text, we take now a wire of a finite length l. Let the coordinates of the end points of the wire be $z = z_1$ and $z = z_2$ respectively, the origin O ($z = 0$) being chosen at an arbitrary point on the length of the wire. The wire is geometrically segmented into two portions of lengths $|l_1|$ and $|l_2|$ respectively at O such that $z_1 = |l_1|$ and $z_2 = -|l_2|$, making the wire length $l = |l_1| + |l_2|$ (Fig. A4.1). We can now express the magnetic field at a perpendicular distance r from the wire, with the help of (4.17) as (Figs. 4.2 and A4.1):

$$\vec{H} = \int d\vec{H} = \frac{i}{4\pi r}\int_{\psi_2}^{\psi_1} \cos\psi \, d\psi \, \vec{a}_\theta, \tag{A4.1}$$

where the angle ψ and the unit vector $\vec{a}_\theta$ have been defined preceding (4.12), in which ψ, for the sake of convenience, is rewritten here as

$$\tan\psi = \frac{z}{r}. \tag{A4.2}$$

We can evaluate the definite integral of (A4.1) to obtain

$$\vec{H} = \frac{i}{4\pi r}(\sin\psi_1 - \sin\psi_2)\,\vec{a}_\theta, \tag{A4.3}$$

where $\psi = \psi_1$ and $\psi = \psi_2$ are the angles subtended by the end points of the finite-length wire at the point where the magnetic field (Fig. A4.1) is to be found. These are, with the help of (A4.2),

$$\left.\begin{aligned} \tan\psi_1 &= \frac{z_1}{r} \\ \tan\psi_2 &= \frac{z_2}{r} \end{aligned}\right\}. \tag{A4.4}$$

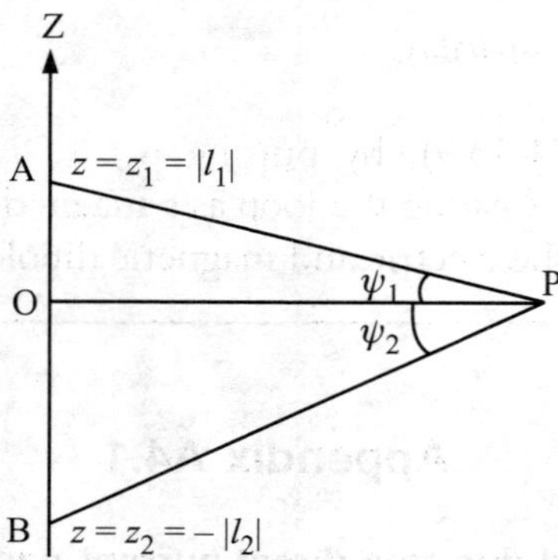

Fig. A4.1 *Finite-length current-carrying straight wire aligned along the Z-axis and the point P where the magnetic field due to the wire is to be found.*

However, putting $z_1 = |l_1|$ and $z_2 = -|l_2|$ [as stated preceding (A4.1)], we can write (A4.4) as

$$\left.\begin{aligned} \tan\psi_1 &= \frac{|l_1|}{r} \\ \tan\psi_2 &= \frac{-|l_2|}{r} \end{aligned}\right\},$$

whence we can also write

$$\left.\begin{aligned} \sin\psi_1 &= \frac{|l_1|}{\sqrt{|l_1|^2 + r^2}} \\ \sin\psi_2 &= \frac{-|l_2|}{\sqrt{|l_2|^2 + r^2}} \end{aligned}\right\}. \tag{A4.5}$$

Substituting (A4.5) in (A4.3) we then obtain

$$\vec{H} = \frac{i}{4\pi r}\left(\frac{|l_1|}{\sqrt{|l_1|^2 + r^2}} - \frac{-|l_2|}{\sqrt{|l_2|^2 + r^2}}\right)\vec{a}_\theta. \tag{A4.6}$$

With a little algebra, we can also express (A4.6) as

$$\vec{H} = \frac{i}{4\pi r}\left(\frac{1}{\sqrt{1 + r^2/|l_1|^2}} + \frac{1}{\sqrt{1 + r^2/|l_2|^2}}\right)\vec{a}_\theta. \tag{A4.7}$$

For an infinitely long wire, we can put $|l_1| = |l_2| = \infty$ to obtain

$$\vec{H} = \frac{i}{4\pi r}\left(\frac{1}{\sqrt{1+0}} + \frac{1}{\sqrt{1+0}}\right)\vec{a}_\theta = \frac{i}{2\pi r}\vec{a}_\theta. \tag{A4.8}$$

We can see, as expected, that (A4.8) becomes identical with (4.19).

Appendix A4.2

Deduction of the relation: $\nabla\left(\frac{1}{R}\right) = -\frac{1}{R^3}\vec{R}$

Let us take the gradient operation on the scalar $1/R$ with respect to the field point variables: x, y, z, the current element being located at the source point variables: x', y', z'. To make life simple, let us take the rectangular coordinate system and with the help of these variables represent the distance vector $\vec{R}$ (defined in Section 2.3) as

$$\vec{R} = (x - x')\vec{a}_x + (y - y')\vec{a}_y + (z - z')\vec{a}_z \tag{A4.9}$$

Its magnitude is given by

$$R = [(x - x')^2 + (y - y')^2 + (z - z')^2]^{1/2}. \tag{A4.10}$$

Now, with the help of (2.31), we can write

$$\nabla\left(\frac{1}{R}\right)=\frac{\partial}{\partial x}\left(\frac{1}{R}\right)\vec{a}_x+\frac{\partial}{\partial y}\left(\frac{1}{R}\right)\vec{a}_y+\frac{\partial}{\partial z}\left(\frac{1}{R}\right)\vec{a}_z=-\frac{1}{R^2}\left(\frac{\partial R}{\partial x}\vec{a}_x+\frac{\partial R}{\partial y}\vec{a}_y+\frac{\partial R}{\partial z}\right)\vec{a}_z. \tag{A4.11}$$

We can find $\partial R/\partial x$occurring in (A4.11) by taking the partial derivative of R given by (A4.10) as follows:

$$\frac{\partial R}{\partial x}=\frac{1}{2}[(x-x')^2+(y-y')^2+(z-z')^2]^{-1/2}\times 2(x-x')$$

$$=\frac{x-x'}{[(x-x')^2+(y-y')^2+(z-z')^2]^{1/2}}=\frac{x-x'}{R},$$

which can be read using (A4.10) as

$$\frac{\partial R}{\partial x}=\frac{x-x'}{R}. \tag{A4.12}$$

Similarly, we get

$$\frac{\partial R}{\partial y}=\frac{y-y'}{R} \tag{A4.13}$$

and $$\frac{\partial R}{\partial z}=\frac{z-z'}{R}. \tag{A4.14}$$

Substituting (A4.12)-(A4.14) in (A4.11), we obtain

$$\nabla\left(\frac{1}{R}\right)=-\frac{1}{R^2}\left(\frac{x-x'}{R}\vec{a}_x+\frac{y-y'}{R}\vec{a}_y+\frac{z-z'}{R}\right)\vec{a}_z,$$

which simplifies to

$$\nabla\left(\frac{1}{R}\right)=-\frac{1}{R^3}[(x-x')\vec{a}_x+(y-y')\vec{a}_y+(z-z')\vec{a}_z]. \tag{A4.15}$$

In view of (A4.9), we thus obtain from (A4.13)

$$\nabla\left(\frac{1}{R}\right)=-\frac{1}{R^3}\vec{R}. \tag{A4.16}$$

Note: We have used (A4.16) in the text preceding (4.73).

Appendix A4.3

Behaviours of some non-conventional media: Anisotropic, ferrite/gyromagnetic and bi-isotropic/chiral

Anisotropic dielectric material: For an isotropic dielectric material of permittivity ε, the electric displacement components (D_x, D_y, D_z) in terms of the electric field components (E_x, E_y, E_z) in rectangular coordinates is represented as

$$\left.\begin{aligned} D_x &= \varepsilon E_x \\ D_y &= \varepsilon E_y \\ D_x &= \varepsilon E_x \end{aligned}\right\}. \qquad \text{(A4.17)}$$

Thus, for an isotropic dielectric, the electric displacement component in a direction depends only on the electric field component in that direction and not on the electric field components in the other directions.

However, for an anisotropic dielectric material, the dependence of displacement components on electric field components as represented by (A4.17) gets modified as

$$\left.\begin{aligned} D_x &= \varepsilon_{xx} E_x + \varepsilon_{xy} E_y + \varepsilon_{xz} E_z \\ D_y &= \varepsilon_{yx} E_x + \varepsilon_{yy} E_y + \varepsilon_{yz} E_z \\ D_x &= \varepsilon_{zx} E_x + \varepsilon_{zy} E_y + \varepsilon_{zz} E_z \end{aligned}\right\}. \quad \text{(Anisotropic dielectric)} \qquad \text{(A4.18)}$$

For such a material medium, the electric displacement componentsD_x, D_y and D_z each depend on all the electric field components E_x, E_y and E_z. In a matrix form, we can express (A4.18) as

$$\begin{bmatrix} D_x \\ D_y \\ D_z \end{bmatrix} = \begin{bmatrix} \varepsilon_{xx} & \varepsilon_{xy} & \varepsilon_{xz} \\ \varepsilon_{yx} & \varepsilon_{yy} & \varepsilon_{yz} \\ \varepsilon_{zx} & \varepsilon_{zy} & \varepsilon_{zz} \end{bmatrix} \begin{bmatrix} E_x \\ E_y \\ E_z \end{bmatrix} \quad \text{(Anisotropic dielectric)} \qquad \text{(A4.19)}$$

where all the elements of the permittivity matrix are different. The behaviour of the 'biaxial' dielectric can also be represented by (A4.19) by taking all the elements except the diagonal elements of the permittivity matrix as zero as follows:

$$\begin{bmatrix} D_x \\ D_y \\ D_z \end{bmatrix} = \begin{bmatrix} \varepsilon_{xx} & 0 & 0 \\ 0 & \varepsilon_{yy} & 0 \\ 0 & 0 & \varepsilon_{zz} \end{bmatrix} \begin{bmatrix} E_x \\ E_y \\ E_z \end{bmatrix}. \quad \text{(Biaxial dielectric)} \qquad \text{(A4.20)}$$

The biaxial dielectric for which two of the diagonal elements of the permittivity matrix of (A4.20) representing its behaviour are the same while one is different is called the uniaxial dielectric. For instance, putting, typically, $\varepsilon_{xx} = \varepsilon^{\text{ord}}$ and $\varepsilon_{yy} = \varepsilon_{yy} = \varepsilon^{\text{extraord}}$, the behaviour of the uniaxial dielectric can be represented by

$$\begin{bmatrix} D_x \\ D_y \\ D_z \end{bmatrix} = \begin{bmatrix} \varepsilon^{\text{ord}} & 0 & 0 \\ 0 & \varepsilon^{\text{extraord}} & 0 \\ 0 & 0 & \varepsilon^{\text{extraord}} \end{bmatrix} \begin{bmatrix} E_x \\ E_y \\ E_z \end{bmatrix} \quad \text{(Uniaxial dielectric)} \qquad \text{(A4.21)}$$

where the superscripts 'ord' and 'extraord' stand for the ordinary and extraordinary axes of the dielectric such that

(i) if the electric field is only along x, that is, $E_x \neq 0, E_y = E_z = 0$, we get $D_x = \varepsilon^{\text{ord}} E_x$;

(ii) if the electric field is only along y, that is, $E_y \neq 0, E_z = E_x = 0$, we get $D_y = \varepsilon^{\text{extraord}} E_y$, and

(iii) if the electric field is only along z, that is, $E_z \neq 0, E_y = E_z = 0$, we get $D_z = \varepsilon^{\text{extraord}} E_z$.

Ferrite/gyromagnetic medium: In a gyromagnetic medium, the left-circularly and right-circularly polarised waves (refer to Appendix A6.1 and Example 6-2 for the definition of left-circular and right-circular types) have different propagation constants, which, in turn, depend on the steady or static magnetic field applied to the medium. One such medium is the ferrite, which is a ceramic-like compound having a very low conductivity ~100 mho/m (typical) as compared to ~10^7 mho/m (S/m) of iron. The ferrite is made by sintering a mixture of metal oxides having the general composition $MO \cdot Fe_2O_3$, where M stands for a divalent metal or a mixture of such metals such as Fe, Zn, Ni, Mn, Mg, and Cd. The ferrite is a gyromagnetic medium, in which the right-circularly and left-circularly polarised waves have different phase propagation constants as stated earlier. The plane of polarisation of a linearly polarised plane wave, which can be treated as the sum of the right-circularly and left-circularly polarised waves, rotates when it propagates through a ferrite medium immersed in a static magnetic field. Such a rotation is referred to as Faraday rotation. The phenomenon has practical application, for instance, in making a non-reciprocal microwave component called the isolator.

For a ferrite/gyromagnetic medium, the relation between the magnetic flux density and magnetic field components, similar to the relation (A4.19) between the electric displacement and electric field components for an anisotropic dielectric, can be written as

$$\begin{bmatrix} B_x \\ B_y \\ B_z \end{bmatrix} = \begin{bmatrix} 1+\chi_m & j\varsigma & 0 \\ -j\varsigma & 1+\chi_m & 0 \\ 0 & 0 & 1 \end{bmatrix} \begin{bmatrix} H_x \\ H_y \\ H_z \end{bmatrix} \quad \text{(Magnetised ferrite)} \qquad \text{(A4.22)}$$

where

$$\left.\begin{aligned} \chi_m &= \frac{\omega_0 \omega_m}{\omega_0^2 - \omega^2} \\ \varsigma &= \frac{\omega \omega_m}{\omega_0^2 - \omega^2} \end{aligned}\right\} \tag{A4.23}$$

with

$$\left.\begin{aligned} \omega_0 &= \gamma B_0 = \frac{m}{P} B_0 \\ \omega_m &= \mu_0 \gamma M_0 = \mu_0 \frac{m}{P} M_0 \end{aligned}\right\} \tag{A4.24}$$

where $m = 9.27 \times 10^{-24}$ Am^2 and $P = 0.527 \times 10^{-34}$Js are the magnetic moment and the angular momentum respectively of an electron. M_0 is the static dipole moment per unit volume of the material contributed by the dipole moments of spinning electrons.

Bi-isotropic/chiral: A chiral medium is characterised by different phase velocities for the right- and left-circularly polarised waves. It is a particular type of medium, called the bi-isotropic medium, in which $\vec{D}$ and $\vec{B}$ each are related to both $\vec{E}$ and $\vec{H}$ as follows:

$$\left.\begin{aligned} \vec{D} &= \varepsilon \vec{E} + (\chi - j\kappa)\sqrt{\mu_0 \varepsilon_0}\vec{H} \\ \vec{B} &= \mu \vec{H} + (\chi + j\kappa)\sqrt{\mu_0 \varepsilon_0}\vec{E} \end{aligned}\right\} \quad \text{(Bi-isotropic medium)} \tag{A4.25}$$

where χ and κ are dimensional magneto-electric parameters. For the special case in which the parameters $\chi = 0$ and $\kappa = 0$, the general expression (A4.24) becomes (3.16) and (4.2), that is, $\vec{D} = \varepsilon \vec{E}$ and $\vec{B} = \mu \vec{H}$ respectively, which are the relations valid for a general isotropic medium. The special case of $\kappa \neq 0$ refers to a chiral medium. We can put the various cases of (A4.24) corresponding to zero and non-zero values of χ and κ for referring to different media as follows:

1. $\chi = 0, \kappa = 0$: simple isotropic medium;
2. $\chi = 0, \kappa \neq 0$: Pasteur chiral medium;
3. $\chi \neq 0, \kappa \neq 0$: general bi-isotropic medium;
4. $\chi \neq 0, \kappa = 0$: Tellegen medium.

Note: Sir J. C. Bose carried out the first ever microwave experiment in artificial materials (on twisted structures) for electromagnetic applications which exhibit the chiral characteristics. Bose in his experiment used bundles of jute fibres twisted to the right and to the left to represent electro-optic analogues of two varieties of sugar solutions, typically, dextrose and levulose.

(Jagadish Chunder Bose, On the rotation of plane of polarisation of electric waves by a twisted structure, *Proc. R. Soc. Lond,* vol. a 63, pp. 146–152, 1898).

Summarising Notes

✓ Basic concepts of static magnetic fields or magnetostatics have been developed.

✓ Magnetostatic quantities analogous to the corresponding electrostatic quantities have identified.

✓ Coulomb's and Gauss's laws as well as Poisson's and Laplace's equations of magnetostatics have been developed on the same line of electrostatics.

✓ Biot-Savart's law predicts magnetic field due to a current element as does Coulomb's law predict electric field due to a point charge.

✓ Use of Biot-Savart's law has been demonstrated in illustrative examples, for instance, in the problem of finding the magnetic fields due to a long filamentary steady current, a filamentary current of finite length, a circular loop of current and a polygonal loop of current.

✓ Ampere's law has made the problem of finding the magnetic field due to a steady current simpler in problems that enjoy geometrical symmetry, as has Gauss's law done in electrostatic problems.

✓ Force experienced by a moving point charge placed in a magnetic field is given by Lorentz force equation.

✓ Concept of Lorentz force equation has been extended to find the force on a current carrying conductor placed in a magnetic field with illustrative examples.

✓ Absence of free magnetic charges (poles) makes it possible to write Gauss' law and Poisson's equation of magnetostatics analogous to those of electrostatics.

✓ Gauss's law of magnetostatics and Poisson's equation of magnetostatics have been appreciated as the manifestation of the absence of free magnetic charges (poles) or that of the continuity of magnetic flux lines.

Review Questions

Narrative type

4.1 Write down the units of magnetic field, magnetic flux, magnetic flux density, magnetic vector potential, current density, surface current density, inductance, and permeability.

4.2 While recognising that the unit of magnetic field is A/m, also explain how it is alternatively Wb/A Wb/(Hm).

Hint: Use (4.2) and (4.48) and note that the unit of permeability is H/m.

4.3 Examining the formula of the inductance of a solenoid, validate that the unit of permeability is H/m or, alternatively, W/Am.

Hint: See (4.48) for the formula.

4.4 Verify the dimensional correctness of the expression (4.84):

$$\vec{B} = \frac{\mu_0 i}{4\pi} \int \nabla \times \frac{dz'\vec{a}_z}{[(z'-z)^2 + r^2]^{1/2}} .$$

4.5 Identify the magnetostatic quantities analogous to the following electrostatic quantities: electric field $\vec{E}$ and electric flux density or displacement $\vec{D}$ and medium property ε.

4.6 Explain the statement that the magnetic flux lines are continuous.

4.7 Derive an expression for the magnetic field due to an infinitely long straight wire carrying a direct current using Biot-Savart's law.

4.8 Derive an expression for the magnetic field due to a straight wire of a finite length carrying a direct current using Biot-Savart's law.

4.9 Using Biot-Savart's law, deduce an expression for the magnetic field at the centre of a circular loop carrying a direct current.

4.10 Obtain an expression for the magnetic field due to a regular polygonal loop of wire carrying a direct current at the centre of the loop. Hence, from this expression, as a special case, deduce the expression for the magnetic field at the centre of a circular loop of wire carrying a direct current.

4.11 Derive an expression for the magnetic field at the middle point of a square loop carrying a direct current.

Hint: Use the expression obtained in question 4-10 to get the answer as a special case (see Example E4-4). However, you should also attempt to derive the expression using the expression for magnetic field due to a straight wire of a 'finite length' carrying a direct current as obtained in Question 4.8.

4.12 With the help of Biot-Savart's law, find an expression for the magnetic field due to a current-carrying circular loop at a point on the axis of the loop perpendicular to the plane containing the loop but off from its centre. Hence, from this expression, deduce as a special case the expression for the magnetic field at the centre of the loop.

4.13 How can you verify Ampere's circuital law with reference to a long current-carrying straight wire for which you know the magnetic field near it from Biot-Savart's law?

4.14 Apply Ampere's circuital law to find an expression for the magnetic field at the axis of a long solenoid.

4.15 Explain why the magnetic field is zero outside a long solenoid.

4.16 State Gauss' law and Poisson's equation of magnetostatics and derive the latter from the former.

4.17 Write down Poisson's equation of magnetostatics in generalised curvilinear coordinate system and hence interpret it in rectangular, cylindrical and spherical systems.

4.18 Show that the magnetic field at a point inside a toroidal coil of direct current is jointly proportional to the number of turns and the current through the coil and is inversely proportional to the radial distance of the point from the centre of the toroid.

Hint: A toroidal coil is made by bending a solenoid to make a coil in the shape of a toroid. Apply Ampere's circuital law around a closed circular path of radius r passing through a point inside the toriodal coil to find the magnitude of the magnetic field intensity at the point as $H = NI/(2\pi r)$, the expression that can be interpreted to find the answer.

4.19 Prove Ampere's circuital law in differential form from its integral form.

4.20 Express Ampere's circuital law in differential form in generalised curvilinear coordinate system . Hence, also express it in rectangular, cylindrical, and spherical-polar coordinate systems.

4.21 Deduce from first principles an expression for the element of magnetic vector potential starting from Biot-Savart's law.

4.22 Find the magnetic dipole moment of a circular loop of wire carrying a direct current by treating it as analogous to an electric dipole.

Hint: See Example E4-6.

4.23 Obtain an expression for the magnetic field due to a circular loop carrying a direct current for two situations: (a) when the point is on the axis, the latter being perpendicular to the plane containing the loop and passing through the centre of the loop, and (b) when the point is off the said axis, by treating the problem as analogous to that of an electric dipole and also, alternatively, treating the problem by the approach of magnetic vector potential.

4.24 Appreciate that the expression for the magnetic field due to a circular current loop at a point off the axis of the loop obtained by treating the loop as a magnetic dipole, analogous to an electric dipole, becomes identical with that obtained using the concept of magnetic vector potential.

Hint: Compare the findings of Example E4-7 with those of Example E4-13.

4.25 Elaborate, with relevant theory, the observation that like currents attract and unlike currents repel.

Hint: See the discussion following (4.65).

Review questions with solutions

4.26 Apply Ampere's circuital law to find the magnetic field $\vec{H}$ in the following regions of a coaxial cable: (i) $0 \leq r \leq a$, (ii) $a \leq r \leq b$ and (iii) $b \leq r \leq c$, where a is the radius of the inner conductor, b is the inner radius of the outer conductor, and c is the outer radius of the outer conductor.

Answer: (i) $\vec{H} = \frac{Ir}{2\pi a^2}\vec{a}_\theta (0 \leq r \leq a)$; (ii) $\vec{H} = \frac{I}{2\pi r}\vec{a}_\theta (a \leq r \leq b)$; and (iii) $\vec{H} = \frac{I}{2\pi r}\frac{c^2-r^2}{c^2-b^2}\vec{a}_\theta (b \leq r \leq c)$.

4.27 Prove that the magnetic flux density $\vec{B}$ due to a long straight current-carrying conducting wire is solenoidal, that is, $\nabla.\vec{B} = 0$.

Answer: The problem enjoys cylindrical symmetry. So, you can formulate the problem in cylindrical coordinate system (r, θ, z). With reference to this problem, take help of (4.19) to write the following expression for the magnetic flux density, $\vec{B} = \mu_0 \vec{H}$, due to a long straight conducting wire carrying a current i at a distance r from the wire:

$$\vec{B} = \frac{\mu_0 i}{2\pi\, r}\vec{a}_\theta. \tag{1}$$

From (1) we get

$$\left.\begin{aligned} &B_r = B_z = 0 \\ &B_\theta = \mu_0 i/(2\pi\, r) \\ &\partial/\partial\theta = \partial/\partial z = 0 \\ &\partial/\partial r \neq 0 \end{aligned}\right\}. \tag{2}$$

Now, with the help of (2.48), we can write

$$\nabla \cdot \vec{B} == \frac{1}{r}\frac{\partial (rB_r)}{\partial r} + \frac{1}{r}\frac{\partial B_\theta}{\partial \theta} + \frac{\partial B_z}{\partial z}. \tag{3}$$

Putting (2) in (3), we then obtain

$$\nabla \cdot \vec{B} = \frac{1}{r}\frac{\partial B_\theta}{\partial \theta}. \tag{4}$$

Substituting B_θ from (2) in (4), we obtain

$$\nabla \cdot \vec{B} = \frac{1}{r}\frac{\partial B_\theta}{\partial \theta} = \frac{1}{r}\frac{\partial}{\partial \theta}(\frac{\mu_0 i}{2\pi\, r}) = \frac{1}{r} \times 0 = 0. \tag{5}$$

Thus, in this problem, as can be seen from (5),$\nabla \cdot \vec{B} = 0$ and, therefore, $\vec{B}$ is solenoidal.

4.28 Deduce an expression for the voltage, called Hall voltage V, developed across a rectangular block of an n-type semiconductor immersed in a uniform dc magnetic field of flux density B applied perpendicular to its length when a current I passes in the direction of its length. Express it in terms of the number density of electrons n and the separation h between those opposite faces of the block across which the voltage is developed.

Answer: Let the length of the rectangular semiconductor block be aligned along y in the direction of current I . In an n-type semiconductor, electrons constitute the charge carriers. Let the direction of magnetic field of flux density $\vec{B}$ be positive z. Therefore, we can express the Lorentz force $\vec{F_B}$ on an electron due to the magnetic field as

$$\vec{F_B} = (-|e|(-v\vec{a}_y) \times B\vec{a}_z = |e|\, vB\vec{a}_x. \tag{1}$$

Under the influence of the Lorentz force given by (1), the electrons will deflect along positive x , which will set up an electric field $\vec{E} = E\vec{a}_x$ which, in turn, exerts electrostatic force

$$\vec{F_E} = |e|\, E\vec{a}_x. \tag{2}$$

In equilibrium, $\vec{F_E}$ and $\vec{F_B}$ will balance each other , giving $\vec{F_E} = \vec{F_B}$, which can be written with the help of (1) and (2) as follows:

$$vB = E. \tag{3}$$

Further, we can express the magnitude of electric field E in terms of the voltage V developed across the faces of the semiconductor perpendicular to the X-axis, caused by the deflection of electrons as

$$E = \frac{V}{b}, \tag{4}$$

where b is the separation between the faces representing the breadth of the semiconductor block. Substituting (4.99) in (4.98), we then obtain

$$v = \frac{E}{B} = \frac{V}{bB}. \tag{5}$$

Now, multiplying the current density $J = |\rho|\, v = n\,|e|\, v$ (see (3.124) by the cross-sectional area bh of the block through which current I passes, we can express $I(= Jbh)$ as

$$I = Jbh = n\,|e|\, vbh, \tag{6}$$

where h is the height of the block, and ρ and n are respectively the volume charge density and the number density of electrons in the block.

Substituting (1) in (6), we get

$$I = \frac{n\,|e|\, hV}{B},$$

which gives the Hall voltage V as

$$V = \frac{IB}{n\,|e|\,h}.$$

Numerical type

4.29 What value of current has to be passed through a long straight wire that will give a magnetic field of magnitude 1 A/m at a perpendicular distance of 1 m?

Answer: 6.28 A.

4.30 Find the magnetic field due to a 6 m straight length of a filamentary conductor carrying a direct current of 5 A at a perpendicular distance of 3 m from the middle of the conductor.

Answer: We can use the following expression for the magnetic field derived in Example 4.1:

$$\vec{H} = \frac{i}{2\pi r}\left(\frac{1}{\sqrt{1+(2r/l)^2}}\right)\vec{a}_\theta,$$

in which substitute $i = 5$ A, $r = 3$ m and $l = 6$ m and calculate $\vec{H} = 0.188\,\vec{a}_\theta$ A/m.

4.31 Show that a magnetic field of magnitude 1 A/m can be obtained at the centre of a circular current loop of radius 50 cm that carries a direct current of 1 A.

Hint: Recall (4.22).

4.32 Show that a current per side length value of 1.1 A/m of a square loop carrying a direct current will give a magnetic field magnitude of 1 A/m at the middle point of the loop.

Hint: Use the expression for the magnitude of magnetic field: $H = 2\sqrt{2}i/(\pi a)$ obtainable from Example E4-4.

4.33 A particle carrying a charge of 3 C is shot with a velocity of 4 m/s along x in a region where it is simultaneously subjected to an electric field of $\vec{E} = 2\vec{a}_x + 3\vec{a}_y + 4\vec{a}_z$ V/m and a magnetic field of $\vec{B} = 2\vec{a}_x + 3\vec{a}_y$ T. Calculate the force $\vec{F}$ on the particle due to these fields.

Answer: $\vec{F} = 6\vec{a}_x + 9\vec{a}_y + 48\vec{a}_z$ N

4.34 At what velocity in the x direction should an electron be shot in a region where a static electric field of 10 V/m and a static magnetic field of 2 T simultaneously exist in the y and z directions respectively so that the electron moves without deviation?

Answer: 5 m/s.

Hint: The force $\vec{F}$on an electron carrying charge $-\,|e|$in the simultaneous presence of the electric field $\vec{E}$ and magnetic field of flux density $\vec{B}$ is

$$\vec{F} = -\,|e|\,(\vec{E} + \vec{v}\times\vec{B}) = -\,|e|\,(E\vec{a}_y + v\vec{a}_x \times B\vec{a}_z) = -\,|e|\,v(E\vec{a}_y + B(\vec{a}_x\times\vec{a}_z)$$
$$= -\,|e|\,v(E - vB)\vec{a}_y.$$

Putting $\vec{F} = 0$, we can then obtain the condition for the electron to move without deviation as $E - vB = 0$, whence we can calculate $v(= E/B) = 10/2 = 5$ m/s.

4.35 A straight filamentary conductor of length 1 m lying on the XY plane ($z = 0$) and oriented at an angle of 45^0 from either of the X and Y axes is immersed in a magnetic field of flux density 1 T along z. Show that the force on the conductor exerted by the magnetic field is $(\vec{a}_x - \vec{a}_y)/\sqrt{2}$ N if the conductor carries a direct current of 1 A. Hence, appreciate that the 'magnitude' of this force remains the same if the conductor is aligned along either the X- or Y-axis.

4.36 Calculate the required flux density of the magnetic field in a region where an electron will have a rotation with cyclotron frequency of 42 GHz .

Answer: 1.5 T. (Use the expression for the angular cyclotron frequency $\omega_c = |\eta|\, B$, where $|\eta| = 1.76 \times 10^{11}$ C/kg is the magnitude of the charge-to-mass ratio of an electron; see Example E4-11).

4.37 A circular loop of radius 2 m lies on the XY plane ($z = 0$). The loop carries a direct current 5 A round the loop from the direction of the positive x to positive y. What will be the magnetic force on the loop due to a long straight wire aligned with the Z-axis carrying a direct current of 10 A in the positive z direction?

Answer: Magnetic force = 0. (You will appreciate that the element of length $d\vec{l}$ on the circular loop and the magnetic flux density $\vec{B}$ at the elemental length are in the same direction. Consequently, this would make the magnetic force $id\vec{l} \times \vec{B} = 0$ on a current element of the loop, irrespective of the data given in the problem).

4.38 With reference to the magnetic field vector $\vec{H} = 5\sin\theta\vec{a}_\phi$ over the surface defined by $r = 6,\ 0 \le \theta \le \pi/2,\ 0 \le \phi \le \pi/2$ (quarter of a spherical surface), verify Stokes' theorem:

$$\int_S (\nabla \times \vec{H}) \cdot d\vec{S} = \oint_l \vec{H} \cdot d\vec{l}.$$

Answer: From the given magnetic field $\vec{H} = 5\sin\theta\vec{a}_\phi$, we can write

$$\left.\begin{aligned} H_r &= H_\theta = 0 \\ H_\phi &= 5\sin\theta \\ \partial/\partial r &= \partial/\partial\phi = 0 \end{aligned}\right\} . \tag{1}$$

In view of (1), we can expand $\nabla \times \vec{H}$occurring on the left-hand side of the Stokes' theorem with the help of (2.68):

$$\nabla \times \vec{H} = \frac{1}{r\sin\theta}\left(\frac{\partial(\sin\theta\, H_\phi)}{\partial\theta}\right)\vec{a}_r = \frac{5}{r\sin\theta}\left(\frac{\partial(\sin^2\theta)}{\partial\theta}\right)\vec{a}_r$$

$$= \frac{5}{r\sin\theta}(2\sin\theta\cos\theta)\vec{a}_r$$

$$= \frac{5 \times 2\cos\theta}{r}\vec{a}_r. \qquad (2)$$

Substituting (2) on the left-hand side of the expression representing Stokes' theorem, we obtain

$$\int_S (\nabla \times \vec{H}) \cdot d\vec{S} = \int_S \frac{5 \times 2\cos\theta}{r}\vec{a}_r \cdot d\vec{S}. \qquad (3)$$

In view of the expression for the elementary area $d\vec{S} = r^2 \sin\theta d\theta d\phi \vec{a}_r$, expression (4.98) becomes

$$\int_S (\nabla \times \vec{H}) \cdot d\vec{S} = \int_S \left(\frac{5 \times 2\cos\theta}{r}\vec{a}_r\right) \cdot (r^2 \sin\theta d\theta d\phi \vec{a}_r) = \int_S (5r \times 2\sin\theta\cos\theta d\theta d\phi)$$

$$= 5r \int_S (\sin 2\theta d\theta d\phi) = 5r \int_{\theta=0}^{\pi/2} \sin 2\theta d\theta \int_{\theta=0}^{\pi/2} d\phi = 5r \left[\frac{-\cos 2\theta}{2}\right]_0^{\pi/2} \left[\phi\right]_0^{\pi/2},$$

which, on substitution of the limits $r = 6$ (given) becomes

$$\int_S (\nabla \times \vec{H}) \cdot d\vec{S} = 5r\tfrac{\pi}{2} = 5 \times 6 \times \tfrac{\pi}{2} = 5 \times 6 \times \tfrac{3.142}{2} = 47.13\,\text{A}. \qquad (4)$$

On the other hand, substituting $\vec{H} = 5\sin\theta\vec{a}_\phi = 5\vec{a}_\phi$ and $d\vec{l} = dr\vec{a}_r + rd\theta\vec{a}_\theta + r\sin\theta d\phi\vec{a}_\phi = dr\vec{a}_r + rd\theta\vec{a}_\theta + rd\phi\vec{a}_\phi$ on the right-hand side of the expression representing Stokes' theorem, with due care to take $\theta = \pi/2$ so that $\sin\theta = 1$ over the semicircle contour (of the quarter of the spherical surface bounded by $r = 6,\ 0 \le \theta \le \pi/2,\ 0 \le \phi \le \pi/2$) along which the line integral is to be taken, we get

$$\oint_l \vec{H} \cdot d\vec{l} = \oint_l (5\vec{a}_\phi) \cdot (dr\vec{a}_r + rd\theta\vec{a}_\theta + rd\phi\vec{a}_\phi) = \oint_l (5\vec{a}_\phi) \cdot (rd\phi\vec{a}_\phi) = 5r \int_{\phi=0}^{\pi/2} d\phi = 5r\frac{\pi}{2},$$

which, by putting $r = 6$ (given) becomes

$$\oint_l \vec{H} \cdot d\vec{l} = 5 \times 6 \times \frac{3.142}{2} = 47.13. \qquad (5)$$

Comparing (4) and (5), we can verify

$\int_S (\nabla \times \vec{H}) \cdot d\vec{S} = \oint_l \vec{H} \cdot d\vec{l}$ (Stokes' theorem)

Multiple-choice questions

4.39 The unit of magnetic vector potential is

(a) Wb/m (b) Wb/m^2 (c) T/m (d) T/m^2.

4.40 The unit of $\int_S (\nabla \times \vec{H}) \cdot d\vec{S}$ is

(a) Am^2 (b) A (c) T (d) Wb.

4.41 If ϕ_B is the magnetic flux through a volume enclosing positive magnetic point charge q_m representing the north pole of a bar magnet which, in turn, represents a magnetic dipole, and ϕ_E is the electric flux through a volume enclosing positive electric point charge q of an electric dipole, then

(a) $\phi_E = 0$ and $\phi_B = 0$ (b) $\phi_E = q$ and $\phi_B = q_m$
(c) $\phi_E = q$ and $\phi_B = -q_m$ (d) $\phi_E = q$ and $\phi_B = 0$.

4.42 A charged particle moving with a velocity $\vec{v}$ in a region of uniform magnetic field of flux density $\vec{B}$ experiences Lorentz force $\vec{F} \neq 0$if

(a) $\vec{v} = 0$ and $\vec{B} \neq 0$ (b) $\vec{v} \neq 0$ and $\vec{B} = 0$
(c) $\vec{v} \neq 0$ and $\vec{B} \neq 0$, and $\vec{v}$ makes an angle of $\pi/4$ with $\vec{B}$ (d) $\vec{v} \neq 0$ and $\vec{B} \neq 0$, and $\vec{v}$ and $\vec{B}$ have the same direction.

4.43 A charged particle moves along positive x in a region of static electric and magnetic fields applied in the y and x directions respectively. The force exerted on the particle due to the electric and magnetic fields simultaneously present is

(a) in the direction of positive x
(b) in the direction of positive y
(c) in the direction of positive z
(d) in a direction that would depend on the magnitudes of the electric field and the speed of the particle.

4.44 A filamentary current element is placed with its length aligned along x in a region of uniform magnetic field in the direction of z. The magnetic force on the current element will be along

(a) positive y (b) negative y (c) positive x (d) negative x.

Answers

4-39 (a) 4-40 (b) 4-41 (d) 4-42 (c) 4-43 (b) 4-44 (c)

5

Basic Concepts of Time-varying Electric and Magnetic Fields

Time-dependent electromagnetic phenomena can be studied using the basic concepts involving time-varying electrical and magnetic fields that constitute the subject of study of this chapter. We begin the chapter with the description of a phenomenon that is based on the principle of the conservation of charge. In static electric field phenomena, with reference to the flow of electric charges, it is implied that the amount of flow of charge into a given closed volume is the same as the flow out of the volume. This amounts to nil variation in electrical charge density with time at a point inside the volume. However, in case the charge density varies with time at the point, the inflow and the outflow of charge would not be equal for the closed volume enclosing the point. This phenomenon leads to the derivation of continuity equation and the concept of relaxation time of a medium (Section 5.1). Though a dielectric does not conduct direct current, there would be a current called the displacement current through a dielectric for a time-varying electric field established through the dielectric. This concept will be developed in Section 5.2. We will formulate Faraday's law for a time-varying magnetic field and subsequently express it in point form or differential form (Section 5.3). Furthermore, for time-varying phenomena, we will see that electric and magnetic fields are coupled, that is, there will be a magnetic field associated with a time-varying electric field, and vice versa. For coupled electric and magnetic fields, we can put together Poisson's equation for electric fields, Poisson's equation for magnetic fields, Ampere's circuital law in the differential form, and Faraday's law in the differential form to write a set of four equations known as Maxwell's equations in point form or differential form. These equations can be considered as the four valves of the heart of electromagnetic theory. The basic integral form from which the point or the differential forms of Maxwell's equations are derived is known as Maxwell's equations in the integral form (Section 5.4).

5.1 Continuity Equation and Relaxation Time in Physical Media

5.1.1 Continuity equation

The principle of conservation of electrical charge implies that all the charges, either at rest or in motion, must be accounted for at all times. The transport of a conserved quantity, here the electric charge, can be described by a differential equation called the continuity equation. In this section, we will derive the continuity equation involving current density and volume charge density from first principles. Current, as we know, is the time rate of flow of charge. The current density at a point is a vector quantity whose magnitude is the current per unit area and whose direction is the direction of the velocity of the charge flow at the point. The relation between the magnitudes of current density J, the volume density ρ and velocity of charge flow v, has already been introduced as $J = \rho v$ (3.124). In vector form, the current density vector $\vec{J}$ can be expressed in terms of the velocity vector $\vec{v}$ as $\vec{J} = \rho \vec{v}$.

According to the principle of conservation of charge, if a charge (interpreted as the positive charge) flows out of a volume, the quantity of the charge within the volume will decrease. Conversely, if the charge flows into the volume, the quantity of the charge within the volume will increase. Keeping this principle in mind, we can proceed to derive the continuity equation for a charge flow. For this purpose, consider a flow of electrical charge through a region (constituting an electric current) and assign a charge density ρ and a current density $\vec{J}$ to a point in the region. Within this region, let us consider an arbitrary closed volume bounded by surface S enclosing the point. The flux of $\vec{J}$ through the closed surface S can be found with the help of (2.20), replacing therein $\vec{E}$ by $\vec{J}$ as follows:

$$\oint_S \vec{J}.\, d\vec{S} = \oint_S \vec{J}.\, \vec{a}_n \, dS. \tag{5.1}$$

For a dc current, which does not vary with time, according to the principle of conservation of charge, the charge flowing into the volume enclosure must be equal to the charge flowing out, thereby making the flux of current density $\vec{J}$ through the closed surface S equal to zero. This enables us to write the following expression for a dc current:

$$\oint_S \vec{J}.\, \vec{a}_n \, dS = 0. \tag{5.2}$$

However, for a time-varying current, again, according to the conservation of charge, the amount of charge in the volume enclosure must decrease with time if the amount of outflow of charge exceeds the inflow, and conversely, the amount of charge in the volume enclosure must increase with time if the amount of charge inflow exceeds the charge outflow. Thus, for a time-varying current, we can equate the flux of $\vec{J}$ through the volume enclosure (interpreted as out of the volume) to the negative time rate of change of charge Q in the volume, which we can mathematically express as

$$\oint_S \vec{J}.\vec{a}_n\,dS = -\frac{dQ}{dt}. \tag{5.3}$$

Obviously, for a positive value of the flux of $\vec{J}$, the time rate of change of charge dQ/dt within the volume has to be negative so as to make the right-hand side of (5.3) positive. Now, putting $Q = \int_\tau \rho\,d\tau$ (see Section 3.1.3.B), we can express (5.3) as

$$\oint_S \vec{J}.\vec{a}_n dS = -\frac{d}{dt}\int_\tau \rho d\tau \tag{5.4}$$

where τ represents the volume enclosure. Choosing the surface area of the volume enclosing the charge to be constant, and since the integral of (5.4) is convergent, we can replace the total derivative on the right-hand side of (5.4) with partial derivative and place it under the definite integral. Thus, we can express (5.4) as

$$\int_S \vec{J}.\vec{a}_n dS = \int_\tau -\frac{\partial\rho}{\partial t}\,d\tau. \tag{5.5}$$

The volume integral on the right-hand side of (5.5) can be evaluated if the volume charge distribution and hence the volume charge density ρ as a function of space coordinates is known a priori. The problem is simple if the volume enclosure τ essentially shrinks to an element of volume enclosure $\Delta\tau$ bounded by an element of area ΔS, such that the enclosure continues to enclose the point where the charge density ρ and current density $\vec{J}$ are defined. Within this enclosure element $\Delta\tau$, we can regard the charge density ρ to be approximately constant over space coordinates. This makes it possible to take $-\partial\rho/\partial t$ outside the integral of the right-hand side of (5.5), to enable us to write the following approximate expression:

$$\int_{\Delta S} \vec{J}.\vec{a}_n\,dS \cong -\frac{\partial\rho}{\partial t}\int_{\Delta\tau} d\tau. \tag{5.6}$$

A straightforward integration of the right-hand side of (5.6) gives

$$\int_{\Delta S} \vec{J}.\vec{a}_n\,dS \cong -\frac{\partial\rho}{\partial t}\Delta\tau,$$

which when divided by $\Delta\tau$ becomes

$$\frac{\int_{\Delta S} \vec{J}.\vec{a}_n\,dS}{\Delta\tau} \cong -\frac{\partial\rho}{\partial t}. \tag{5.7}$$

The sign of approximation can be removed from (5.7) and replaced by the sign of equality if we take the limit of the volume element as tending to zero. That is, $\Delta\tau \to 0$.

$$\lim_{\Delta\tau\to 0} \frac{\int_{\Delta S} \vec{J}.\vec{a}_n\, dS}{\Delta\tau} = -\frac{\partial\rho}{\partial t}. \tag{5.8}$$

However, by definition, the left-hand side of (5.8) becomes $\nabla.\vec{J}$, the divergence of $\vec{J}$, which can be easily appreciated by recalling (2.34). Therefore, replacing the left-hand side of (5.8) by $\nabla.\vec{J}$, we obtain the following expression known as the continuity equation:

$$\nabla.\vec{J} = -\frac{\partial\rho}{\partial t}. \tag{5.9}$$

By rearrangement of terms, we can also express the continuity equation as

$$\nabla.\vec{J} + \frac{\partial\rho}{\partial t} = 0. \tag{5.10}$$

Using an alternative approach, we can easily deduce the continuity equation (5.10) starting from Maxwell's equations. However, we keep such deduction in abeyance until we take up Maxwell's equations later in this chapter (Section 5.4).

5.1.2 Relaxation time in physical media

Relaxation time is a measure of how fast or slow a medium of uniform conductivity and permittivity approaches electrostatic equilibrium. Starting from the continuity equation and taking the help of Poisson's equation (Section 3.3) and Ohm's law, we are going to derive here an expression for relaxation time, which, we will see, depends upon the conductivity and the permittivity of the medium. The conduction current density $\vec{J}$ at a point in the medium due to electric field $\vec{E}$ at the point in terms of σ, the conductivity of the medium, is given by Ohm's law[1]:

$$\vec{J} = \sigma\vec{E}. \tag{5.11}$$

If we substitute (5.11) in (5.10), we get

$$\nabla.(\sigma\vec{E}) + \frac{\partial\rho}{\partial t} = 0 \tag{5.12}$$

ρ being the volume charge density at the point. The divergence of a vector quantity (here $\sigma\vec{E}$) depends on space coordinates, as can be seen from the expression for divergence for different coordinate systems given by equations (2.46) to (2.49). However, since we are taking the conductivity σ to be uniform in the medium, we can take it outside the expression for the divergence in (5.12) and express the same as

$$\sigma\nabla.\vec{E} + \frac{\partial\rho}{\partial t} = 0. \tag{5.13}$$

[1]See the footnote following (6.100) for the circuit theory expression for the law in terms of the voltage, current and resistance.

Now, if we substitute Poisson's equation (3.105) in (5.13), we obtain

$$\sigma\frac{\rho}{\varepsilon} + \frac{\partial \rho}{\partial t} = 0,$$

which can be rearranged as

$$\frac{\partial \rho}{\partial t} = -\frac{\sigma}{\varepsilon}\rho. \tag{5.14}$$

By further rearrangement of the terms of (5.14) and then integrating it, we get

$$\int \frac{d\rho}{\rho} = -\frac{\sigma}{\varepsilon}\int dt. \tag{5.15}$$

The evaluation of the integration of (5.15) is straightforward and we obtain, after the integration,

$$\ln \rho = -\frac{\sigma}{\varepsilon}t + \text{constant}. \tag{5.16}$$

If we choose to put the constant of integration on the right-hand side of (5.16) as constant $= \ln \rho_0$, where ρ_0 itself is another constant, then we can express (5.16) as

$$\ln \rho = -\frac{\sigma}{\varepsilon}t + \ln \rho_0\,,$$

which can be rearranged as

$$\ln \rho - \ln \rho_0 = -\frac{\sigma}{\varepsilon}t\,,$$

and which can be further simplified as

$$\ln \frac{\rho}{\rho_0} = -\frac{\sigma}{\varepsilon}t. \tag{5.17}$$

Furthermore, we can put (5.17) in exponential form as

$$\frac{\rho}{\rho_0} = \exp\left(-\frac{\sigma}{\varepsilon}t\right),$$

which can be put as

$$\rho = \rho_0 \exp\left(-\frac{t}{T}\right), \tag{5.18}$$

where

$$T = \frac{\varepsilon}{\sigma} = \frac{\varepsilon_r \varepsilon_0}{\sigma} \tag{5.19}$$

is called the relaxation time of the medium, ε_r being the relative permittivity (or dielectric constant) of the medium and $\varepsilon_0 = 8.54 \times 10^{-18}$ F/m being the permittivity of free-space [see the discussion following (3.4) and (3.5)]. Thus, the relaxation time T of a medium is the time constant of the decay of the volume charge density ρ in the medium. Further, it can be easily appreciated from (5.18) that the constant ρ_0 is the equilibrium volume charge density, being the value of ρ at $t = 0$.

By examining (5.18) and (5.19), we find that for smaller values of relaxation time T of the medium, the volume charge density $\rho \to 0$ rather fast. This suggests that within a negligible time the excess charge injected into the bulk of a good conductor will disappear ($\rho \to 0$), only to reappear on the outer surface of the conductor in compliance with the conservation of charge. This is what happens within a good conducting medium like copper that has a small value of the relaxation time T (Table 5.1). Therefore, a good conductor like copper ($T = 1.5 \times 10^{-19}$) cannot be charged throughout its bulk; it can be charged only at its surface, while materials with large values of the relaxation time like dielectric mica ($T \sim 1/2$ a day) or fused quartz ($T \sim 50$ days) can be charged throughout its bulk.

Table 5.1 *Conductivity, permittivity, and relaxation time of typical medium materials*

Medium material	σ (mho/m)	ε_r	$T = \dfrac{\varepsilon_r \varepsilon_0}{\sigma}$
Copper	5.8×10^7	1	1.5×10^{-19} s
Sea-water	4	81	2×10^{-10} s
Corn oil	5×10^{-4}	3.1	0.55 s
Mica	10^{-15}	5.8	$\sim 1/2$ a day
Quartz (fused)	10^{-17}	5	~ 50 days

The concept of the relaxation time developed in this section will be very useful in understanding the electromagnetic boundary conditions at the interface between two dielectrics (Section 7.2) as well as those at the interface between a conductor and a dielectric (Section 7.3). You will see that the corollaries to a very small value of the relaxation time of a good conductor are the presence of a surface charge density on the surface of the conductor and the non-existence of the electric field inside the conductor (also refer to the note preceding (3.48) and Appendix A5.1).

5.2 Time-Varying Magnetic Field and Faraday's Law

According to Faraday's law of electromagnetic induction, a time-varying magnetic field generates an electric field. The law shows how an electromotive force induced in a closed circuit, abbreviated as EMF[2] (which in turn is associated with the electric field), is

[2]EMF is a measure of the capability of a source of energy to drive an electric charge around a circuit, that is, generate a circuit current; it is estimated as the energy per unit charge that is imparted by the energy source. The unit of EMF is,

proportional to the time rate of change of magnetic flux density that is linked with the circuit.

According to Faraday's law, the circuit current i is given by:

$$i = \frac{\mathcal{E} + \mathcal{E}_{\text{induced}}}{R} \tag{5.20}$$

where R is the circuit resistance, $\mathcal{E}$ is the EMF of the energy source, and $\mathcal{E}_{\text{induced}}$ is the induced EMF .

Further, we can also obtain the following expression for the circuit current i in terms of the time rate of increase of flux of magnetic flux density ϕ_B, henceforth referred to as magnetic flux, as follows:

$$i = \frac{\mathcal{E} - \frac{d\phi_B}{dt}}{R} \tag{5.21}$$

where ϕ_B is the magnetic flux (that is, the flux of magnetic flux density $\vec{B}$ through the closed area bounding the circuit) linked up with the circuit.

The derivation of expression (5.21) can be appreciated with reference to an electrical circuit considered in Appendix A5.2. The circuit consists of a straight conducting rod that can freely slide on a pair of parallel conducting rails, the motion of the conducting rod being caused by the force applied on it by a uniform magnetic field in which conducting rod and the pair of rails are immersed. The circuit is completed by an external source of the EMF sending a current through the circuit (see Fig. A5.1a). It is implied in the analysis that the energy supplied by the source of EMF is balanced by the energy that is spent in doing work to move the conducting rod plus the energy that is lost in the circuit resistance. With the help of this circuit example, we can easily derive (5.21) in a few steps.

Now, by comparing (5.20) and (5.21), we can easily appreciate that the following relation holds good for representing one of the forms of Faraday's law:

$$\mathcal{E}_{\text{induced}} = -\frac{d\phi_B}{dt}. \tag{5.22}$$

According to this formulation (also with reference to the example of Appendix A5.2), the quantity $d\phi_B/dt$ is positive, representing an increase in the magnetic flux with time so as to make $\mathcal{E}_{\text{induced}}$ given by (5.22) a negative quantity. We can thus see that the induced EMF, $\mathcal{E}_{\text{induced}}$, sends a current in a direction opposite to the one caused by the EMF $\mathcal{E}$ of the source in the circuit. Thus, the current in the circuit due to the applied EMF is responsible for the force on the conducting rod, which, consequently, makes the conducting rod to move and thereby bring about an increase in magnetic flux through the circuit with time. This, in turn, causes an electromagnetic induction or induced EMF to take place (see Appendix A5.2), which results in an oppositely directed induced current causing an opposing force on the conducting rod. The latter opposes any increase in the

---

therefore, that of potential, that is, volt or V (see (3.73) and the discussion following it).

magnetic flux with time through the circuit, thereby opposing the cause responsible for the electromagnetic induction (see Appendix A5.2). This is what is known as Lenz's law, according to which electromagnetic induction takes place such that it opposes the cause to which it is due.

In the present formulation, the change of magnetic flux with time has been designed as the increase in magnetic flux with time, and hence, accordingly, we can express (5.21) and (5.22) as

$$\left.\begin{aligned} i &= \frac{\mathcal{E} - \left|\dfrac{d\phi_B}{dt}\right|}{R} \\ \mathcal{E}_{\text{induced}} &= -\left|\frac{d\phi_B}{dt}\right| \end{aligned}\right\},$$

numbered in Appendix A5.2 as equation (A5.26), where $|d\phi_B/dt|$ represents the magnitude of the rate of change of magnetic flux with time. Interestingly, these expressions for i and $\mathcal{E}_{\text{induced}}$ remain unchanged even if the formulation were changed such that the change of magnetic flux with time represents a decrease rather than an increase in magnetic flux with time (See Appendix A5.2 for detailed explanation). Thus, the statement of Lenz's provided at the end of the preceding paragraph continues to hold, irrespective of whether the change in magnetic flux with time responsible for electromagnetic induction is due to an increase or a decrease in the magnetic flux with time.

EXAMPLE E5-1

A straight conducting rod of length l moves parallel to its length with a velocity v on a plane perpendicular to a uniform magnetic field region of flux density B. Obtain an expression for the amount of induced voltage developed across the length of the rod.

SOLUTION

The area generated by the movement of the rod through an element of distance Δz is ϕ. The element of flux ($\Delta\phi_B$) of magnetic flux density B through the element of area $l\Delta z$ then is obtained by multiplying it with B, as $\Delta\phi_B = Bl\Delta z$. The required expression for the amount of voltage induced across the rod, in view of (5.21) and (5.22) [see also (A5.26)], is then

$$\text{voltage induced} = \frac{d\phi_B}{dt} = \lim_{\Delta t\to 0}\frac{\Delta\phi_B}{\Delta t} = \lim_{\Delta t\to 0}\frac{Bl\Delta z}{\Delta t} = Bl\left(\lim_{\Delta t\to 0}\frac{\Delta z}{\Delta t}\right) = Blv.$$

EXAMPLE E5-2

A circular conducting disc of radius a, immersed in a dc magnetic field of flux density B perpendicular to it, is rotated with an angular velocity ω to make the so-called Faraday disc generator. Obtain an expression for the amount of voltage developed at the rim with respect to the axis or axle of the disc. Discuss the polarity of the voltage.

SOLUTION

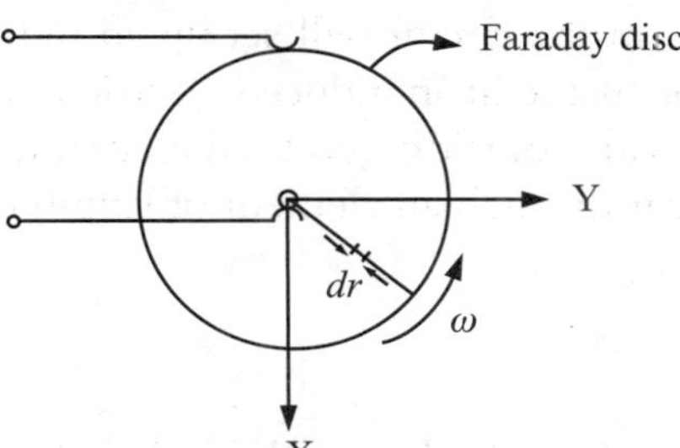

Fig. E5-2

Let the conducting disc lie on the XY plane, taken as the plane of the paper. Let the magnetic field be directed along the positive z, that is, towards the reader (Fig. E5-2). We can smear out the conducting disc into conducting radial rods. The element of voltage developed along an element of radial rod length dr considered at a radial distance r from the axis of the disc will be $Bdrv$, where v is the velocity of the element of the radial rod length (see Example E5-1). Putting $v = \omega r$, the element of voltage then becomes $Bdr\omega r = B\omega rdr$. Therefore, integrating the foregoing expression over the length of the rod, which happens to be the radius of the disc a, we can obtain the required expression for the amount of voltage developed at the rim with respect to the axis or axle of the generator as follows:

$$\text{Voltage developed} = \int_0^a B\omega rdr = B\omega \int_0^a rdr = B\omega \left[r^2/2\right]_0^a = B\omega a^2/2.$$

In a practical Faraday disc generator, two brushes make contact with the disc at the axle and the rim of the disc, which are connected to the external voltage terminals of the generator. We can comment on the polarity of the voltage developed (given by the deduced expression as above) by noting the Lorentz force on a free moving electron of charge $-|e|$ (see Sections 4.4.1 and 4.4.2). The Lorentz force $\vec{F}$ on the free charge moving in the azimuthal direction of $\vec{a}_\theta$ due to rotation of the disc in 'anticlockwise' direction is given by

$$\vec{F} = -|e|\, v\vec{a}_\theta \times B\vec{a}_z = -|e|\, vB\vec{a}_\theta \times \vec{a}_z = -|e|\, vB\vec{a}_r,$$

which in view of the negative sign in the expression indicates that the Lorentz force on the free charge will be radially inward. This will cause the accumulation of the negative charge at the axis or axle of the disc, leaving the positive charge at the rim of the disc. Thus, in this situation a positive voltage will be developed at the rim with respect to the axis or axle of the disc. The polarity will be reversed if either the rotation of the disc is changed from the anticlockwise to the clockwise direction or if the direction of the magnetic field is reversed from the positive z direction (towards the reader) to the negative z direction (away from the reader).

5.2.1 Integral form of Faraday's law

The induced EMF, $\mathcal{E}_{\text{induced}}$, in the circuit will set up an induced current i_{induced}, which can be attributed to a flow of charge in an induced electric field $\vec{E}$. If we move a test point charge q round the closed path coinciding with the circuit, the element of work done dW on the charge in moving it through an element of length dl of the path can be written as

$$dW = \vec{F} \cdot d\vec{l} \tag{5.23}$$

where $\vec{F}$ is the force on the charge q due to the induced electric field $\vec{E}$ and $d\bar{l}$ is the element of length vector whose magnitude is dl and whose direction is the direction in which the charge is moved. Substituting the expression for the force as [see (3.14)]

$$\vec{F} = q\vec{E} \tag{5.24}$$

in (5.23), we can write

$$dW = q\vec{E} \cdot d\vec{l}. \tag{5.25}$$

The work W done in moving the charge q round the circuit is then obtained by taking the closed line integral of (5.25) as (see Section 2.4):

$$W = \oint_l dW = \oint_l q\vec{E} \cdot d\vec{l} = q \oint_l \vec{E} \cdot d\vec{l}. \tag{5.26}$$

Alternatively, we can equate W to the energy imparted by the induced EMF to the charge q as

$$W = q\,\mathcal{E}_{\text{induced}}. \tag{5.27}$$

Comparing (5.26) with (5.27), we obtain

$$q \oint_l \vec{E} \cdot d\vec{l} = q\mathcal{E}_{\text{induced}},$$

which simplifies to

$$\mathcal{E}_{\text{induced}} = \oint_l \vec{E} \cdot d\vec{l}. \tag{5.28}$$

The expression for the electric field generated by a time-varying magnetic flux is then obtained by substituting (5.28) in (5.22) as follows:

$$\oint_l \vec{E} \cdot d\vec{l} = -\frac{d\phi_B}{dt} \tag{5.29}$$

which is one of the ways of expressing Faraday's law of electromagnetic induction.

Further, the magnetic flux ϕ_B through the closed area S bounding the circuit can be written, in terms of a surface integral of magnetic flux density $\vec{B}$, as [see Section 2.4]:

$$\phi_B = \int_S \vec{B} \cdot \vec{a}_n \, dS. \tag{5.30}$$

Substituting (5.30) in (5.29), we can alternatively express Faraday's law (5.29) as

$$\oint_l \vec{E} \cdot d\vec{l} = -\frac{d}{dt} \int_S \vec{B} \cdot \vec{a}_n \, dS. \tag{5.31}$$

The magnetic flux ϕ_B linked up with the circuit given by (5.30) can be made to vary with time by varying either the area of the path S or the magnetic field $\vec{H}$, and hence also magnetic flux density $\vec{B}$ with time so as to generate an electric field $\vec{E}$ according to (5.29). However, in situations where the area S remains constant and the magnetic flux density $\vec{B}$ varies with time, we can replace the total time-derivative on the right-hand side of (5.32) by its partial derivative and put it under a convergent, definite integral. In view of this, therefore, we can write another expression for Faraday's law as

$$\oint_l \vec{E} \cdot d\vec{l} = -\int_S \frac{\partial \vec{B}}{\partial t} \cdot \vec{a}_n \, dS. \tag{5.32}$$

Thus, in (5. 29), (2.31), and (5.32), we obtain three alternative ways of stating Faraday's law of electromagnetic induction, all in the integral form. It would be worth putting together all the different ways of expressing Faraday's law by recalling (5.22), (5.29), (2.31), and (5.32) as follows:

$$\left.\begin{aligned} \mathcal{E}_{\text{induced}} &= -\frac{d\phi_B}{dt} \\ \oint_l \vec{E} \cdot d\vec{l} &= -\frac{d\phi_B}{dt} \\ \oint_l \vec{E} \cdot d\vec{l} &= -\frac{d}{dt} \int_S \vec{B} \cdot \vec{a}_n dS \\ \oint_l \vec{E} \cdot d\vec{l} &= -\int_S \frac{\partial \vec{B}}{\partial t} \cdot \vec{a}_n \, dS \end{aligned}\right\}. \tag{5.33}$$

Let us illustrate the application of Faraday's law (5.33) with the following few examples.

EXAMPLE E5-3

Show that the voltage induced on a circular loop of radius a lying on the z plane immersed in magnetic flux density $\vec{B} = (B_{ox}\vec{a}_x + B_{oy}\vec{a}_y)\cos\omega t$ is $B_{oz}\pi a^2\omega\sin\omega t$.

SOLUTION

The given magnetic flux density as a function of time t is

$$\vec{B} = (B_{ox}\vec{a}_x + B_{oy}\vec{a}_y)\cos\omega t. \tag{E5-3.1}$$

With the help of (5.33), we can find the voltage induced $V(t)$ as

$$V(t) = -\frac{d\phi_B}{dt} = -\frac{d}{dt}\vec{B}\cdot(\pi a^2\vec{a}_z). \tag{E5-3.2}$$

Substituting (E5-3.1) in (E5-3.2) and remembering the relations $\vec{a}_x \cdot \vec{a}_z = 0$ and $\vec{a}_z \cdot \vec{a}_z = 1$ we obtain

$$V(t) = -\frac{d}{dt}(B_{oz}\cos\omega t)(\pi a^2). \tag{E5-3.3}$$

Carrying out the differentiation in (E5-3.3), we thus get

$$V(t) = B_{oz}\pi a^2\omega\sin\omega t.$$

EXAMPLE E5-4

A rectangular $a \times b$ conducting loop of dimension $(a \times b)$ carrying current i is immersed in a non-uniform magnetic field of flux density $\vec{B}$. One of the arms of the loop contains a resistor of resistance R. Find the current i through the resistor if $a = 1.2$ m, $b = 0.2$ m; $R = 10\ \Omega$, and $\vec{B} = 6\cos(5\pi \times 10^7 t - \frac{2}{3}\pi x)\ \mu$T.

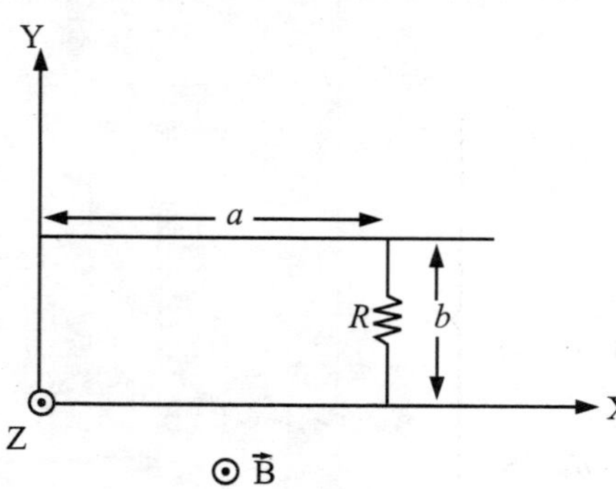

Fig. E5-4

SOLUTION

Let the loop lie on the XY plane (Fig. E5-4). The given magnetic flux density can be put as

$$\vec{B} = 10^{-6} \times 6\cos\left(5\pi \times 10^7 t - \frac{2}{3}\pi x\right)\vec{a}_z\ \text{T}. \tag{E5-4.1}$$

We can find the magnetic flux through the loop as

$$\phi_B = \iint \vec{B} \cdot (dxdy\vec{a}_z). \tag{E5-4.2}$$

Substituting (E5-4.1) in (E5-4.2), we can write

$$\phi_B = \iint 10^{-6} \times 6\cos(5\pi \times 10^7 t - \frac{2}{3}\pi x)dxdy \text{ Wb} \tag{E5-4.3}$$

$$= 10^{-6} \times 6 \int_0^a \cos(5\pi \times 10^7 t - \frac{2}{3}\pi x)dx \int_0^b dy \text{ Wb}.$$

$$\phi_B = 10^{-6} \times \frac{6}{-\frac{2}{3}\pi} \left[\sin(5\pi \times 10^7 t - \frac{2}{3}\pi x)\right]_0^a [y]_a^b \text{ Wb} \tag{E5-4.4}$$

$$= -10^{-6} \times \frac{9}{\pi} \left[\sin(5\pi \times 10^7 t - \frac{2}{3}\pi x)\right]_0^a [y]_a^b \text{ Wb}.$$

On substitution of the limits in (E5-4.4), we obtain:

$$\phi_B = -10^{-6} \times \frac{9}{\pi} b \left[\sin(5\pi \times 10^7 t - \frac{2}{3}\pi a) - \sin(5\pi \times 10^7 t)\right] \text{ Wb}. \tag{E5-4.5}$$

Putting $a = 1.2$ m, $b = 0.2$ m in (E5-4.5), we obtain

$$\phi_B = -10^{-6} \times \frac{1.8}{\pi} [\sin(5\pi \times 10^7 t - 0.8\pi) - \sin(5\pi \times 10^7 t)] \text{ Wb}. \tag{E5-4.6}$$

The time differentiation of (E5-4.6) next yields

$$\frac{d\phi_B}{dt} = -10^{-6} \times \frac{1.8}{\pi} \times 5\pi \times 10^7 [\cos(5\pi \times 10^7 t - 0.8\pi) - \cos(5\pi \times 10^7 t)] \text{ Wb/s},$$

which gives

$$\left|\frac{d\phi_B}{dt}\right| = 10^{-6} \times 9.0 \times 10^7 [\cos(5\pi \times 10^7 t - 0.8\pi) - \cos(5\pi \times 10^7 t)] \text{ Wb/s}. \tag{E5-4.7}$$

As explained following (5.22), the emf induced $\mathcal{E}_{\text{induced}}$ on the loop can then be written, with the help of (E5-4.7), as

$$\mathcal{E}_{\text{induced}} = -\left|\frac{d\phi_B}{dt}\right| = -10^{-6} \times 9.0 \times 10^7 [\cos(5\pi \times 10^7 t - 0.8\pi) - \cos(5\pi \times 10^7 t)] \text{ V} \tag{E5-4.8}$$

noting that Wb/s is synonymous with V (volt).

Since the given resistance in the circuit is $R = 10\ \Omega$, we obtain the induced loop current

i by dividing (E5-4.8) by $R = 10$ and dropping the negative sign that conforms to Lenz's law [see the discussion following (5.22)] as follows:

$$i = \frac{10^{-6} \times 9.0 \times 10^7[\cos(5\pi \times 10^7 t - 0.8\pi) - \cos(5\pi \times 10^7 t)]}{10} \text{ A},$$

which simplifies to

$$i = 9[\cos(5\pi \times 10^7 t - 0.8\pi) - \cos(5\pi \times 10^7 t)] \text{ A}.$$

EXAMPLE E5-5

A square loop of side a is located near a long straight wire carrying a current $I = I_0 \sin \omega t$ such that the wire and the loop lie on the same plane and the wire is parallel to one of the sides of the square loop. Find the amount of voltage induced in the loop, taking d as the distance of separation between the wire and the nearer parallel side of the loop.

SOLUTION

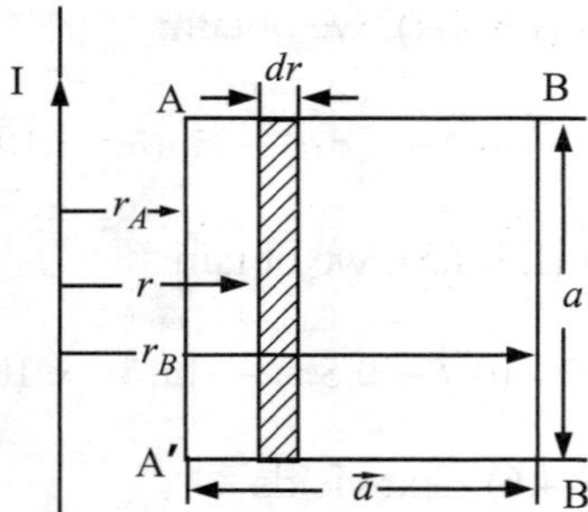

Fig. E5-5

Let the square current loop (ABB′A′) lie on the plane of the paper at a distance $r_A = d$. Let us consider an element of strip on the loop of infinitesimal area adr at a distance r from the wire (Fig. E5-5). With the help of either (4.19) or (4.27), we can write the following expression for magnetic field at the position of the element of strip due to the long straight wire carrying the given current $I = I_0 \sin \omega t$:

$$\vec{H} = \frac{I_0 \sin \omega t}{2\pi r} \vec{a}_\theta. \tag{E5-5.1}$$

For the loop lying on the plane of the paper, the instantaneous direction of $\vec{H}$, that is, the instantaneous direction of $\vec{a}_\theta$, is perpendicular to the plane of paper and directed away from the reader, taking the direction of the instantaneous current as vertically upward. Now, from (E5-5.1) we can write the magnitude H of $\vec{H}$ at the element of strip as

$$H = \frac{I_0 \sin \omega t}{2\pi r}. \tag{E5-5.2}$$

Multiplying (E5-5.2) by μ_0, we obtain the expression for the magnitude of the magnetic flux density B at the at the element of strip as

$$B = \frac{\mu_0 I_0 \sin \omega t}{2\pi r}. \tag{E5-5.3}$$

The element of magnetic flux $d\phi_B$ of B through the element of strip on the loop can then be written by multiplying (E5-5.3) by the infinitesimal area adr of the loop as

$$d\phi_B = Badr = \frac{\mu_0 I_0 \sin \omega t a}{2\pi} \frac{dr}{r}. \tag{E5-5.4}$$

The magnetic flux ϕ_B through the entire loop is then obtained by integrating (E5-5.4) as follows:

$$\phi_B = \int d\phi_B = \frac{\mu_0 I_0 \sin \omega t a}{2\pi} \int_{r_A}^{r_B} \frac{1}{r} dr = \frac{\mu_0 I_0 \sin \omega t a}{2\pi} (\ln r_B - \ln r_A) \tag{E5-5.5}$$

$$= \frac{\mu_0 I_0 \sin \omega t a}{2\pi} \ln \frac{r_B}{r_A},$$

where $r = r_A$ and $r = r_B$ are the distances of the two sides AA′ and BB′ respectively of the loop parallel to the straight wire. Putting $r_A = d$ (given) and $r_B = d + a$ (the side a of the loop given) in (E5-5.5), we obtain

$$\phi_B = \frac{\mu_0 I_0 \sin \omega t a}{2\pi} \ln \left(\frac{d+a}{d} \right) = \frac{\mu_0 I_0 \sin \omega t a}{2\pi} \ln \left(1 + \frac{a}{d} \right). \tag{E5-5.6}$$

With the help of (5.22), now we can find the amount of voltage induced in the loop from the time differentiation of (E5-5.6) as

$$\text{Induced voltage} = \frac{\mu_0 a \omega}{2\pi} \ln \left(1 + \frac{a}{d} \right) I_0 \cos \omega t.$$

5.2.2 Differential form of Faraday's law

In Section 4.3.2, we had derived Ampere's circuital law in the differential form (4.57) starting from its integral form (4.23). Following the same approach now, we can obtain Faraday's law in the differential form from its integral form.

For this purpose, let us first express the right-hand side of Ampere's circuital law (4.23) as

$$i = \int_S \vec{J} \cdot \vec{a}_n \, dS. \tag{5.34}$$

In view of (5.34), we can then express the integral form of Ampere's circuital law (4.23) in the following form:

$$\oint_l \vec{H} \cdot d\vec{l} = \int_S \vec{J} \cdot \vec{a}_n \, dS. \tag{5.35}$$

Recall that, starting from its integral form (4.23), interpreted here as (5.35), we had obtained the differential form of Ampere's circuital law (4.57), rewritten here as

$$\nabla \times \vec{H} = \vec{J}. \tag{5.36}$$

Further, for the sake of convenience of the present formulation, let us rewrite Faraday's law (5.32) by shifting the minus sign from the outside of the definite integral of (5.32) to its inside, as follows:

$$\oint_l \vec{E} \cdot d\vec{l} = \int_S -\frac{\partial \vec{B}}{\partial t} \cdot \vec{a}_n \, dS. \tag{5.37}$$

At this stage, it is possible to see an analogy between Ampere's circuital law (5.35) and Faraday's law (5.37) by comparing these two expressions. Thus, you will notice that $\vec{E}$ of Faraday's law (5.37) is analogous to $\vec{H}$ of Ampere's circuital law (5.35), and, similarly, $-\frac{\partial \vec{B}}{\partial t}$ of Faraday's law (5.37) is analogous to $\vec{J}$ of Ampere's circuital law (5.35). Therefore, we can straightaway write the expression for Faraday's law in the differential form analogous to the differential form (5.36) of Ampere's circuital law, by replacing $\vec{H}$ therein by $\vec{E}$ and $\vec{J}$ by $-\frac{\partial \vec{B}}{\partial t}$, as follows:

$$\nabla \times \vec{E} = -\frac{\partial \vec{B}}{\partial t}. \tag{5.38}$$

Alternatively, following the same procedure as followed in Section 4.3.2, we could have first deduced the differential form of Faraday's law and then obtained Ampere's circuital law in the differential form using the analogy between $\vec{H}$ and $\vec{E}$ as well as between $\vec{J}$ and $-\frac{\partial \vec{B}}{\partial t}$ as has been done above.

5.3 Time-varying Electric Field and Displacement Current

We can see both in the integral and the differential or point forms of Faraday's law (5.32) and (5.38) respectively, how an electric field, in electromagnetic induction, is associated with a time-varying magnetic field and thus how electric and magnetic fields are coupled in the time-varying phenomena (Section 5.2). However, such an association of a magnetic field with a time-varying electric field is not explicit in the integral and the point forms of Ampere's circuital law, (4.23) and (4.57) respectively, which involves the current i and the current density $\vec{J}$ respectively. Both of these do not explicitly show the variation of electric field with time. Now, let us see how Ampere's circuital law gets modified for time-varying situations. We will develop the required modification here as a requirement of the consistency of Ampere's circuital law with the continuity equation in time-varying situations.

For this purpose and for the sake of the convenience of the present analysis, let us rewrite here (5.10), the continuity equation in time-varying situations, as:

$$\nabla . \vec{J} + \frac{\partial \rho}{\partial t} = 0. \tag{5.39}$$

Substituting Poisson's equation (3.104) in (5.39), we get

$$\nabla . \vec{J} + \frac{\partial}{\partial t} \nabla \cdot \vec{D} = 0. \tag{5.40}$$

Since the divergence of a vector quantity, here $\vec{D}$, involves the partial derivatives of space coordinates, as can be seen from (2.46)-(2.49), we can interchange the differential operators in the second term of (5.40) to write it as

$$\nabla . \vec{J} + \nabla \cdot \frac{\partial \vec{D}}{\partial t} = 0,$$

which, by rearranging the terms, yields the following expression for the continuity equation:

$$\nabla . \left(\vec{J} + \frac{\partial \vec{D}}{\partial t} \right) = 0. \tag{5.41}$$

Now, recall Ampere's circuital law in the differential or the point form (4.57) and rewrite it as follows:

$$\nabla \times \vec{H} = \vec{J}. \tag{5.42}$$

Further, recall also the well known vector identity that the divergence of the curl of a vector quantity is zero, which can be written here applying it to the magnetic field vector $\vec{H}$ as follows:

$$\nabla . (\nabla \times \vec{H}) = 0. \tag{5.43}$$

Substituting (5.42) in (5.43), we get

$$\nabla . \nabla \times \vec{H} = \nabla . \vec{J} = 0. \tag{5.44}$$

However, note that (5.44) conflicts with the continuity equation (5.41) in the time-varying situations. In order to make the continuity equation (5.44) consistent with (5.41), we must then understandably modify (5.44) by adding the term $\partial \vec{D}/\partial t$ to $\vec{J}$ in (5.44), so as to write it as follows:

$$\nabla . \nabla \times \vec{H} = \nabla . \left(\vec{J} + \frac{\partial \vec{D}}{\partial t} \right) = 0,$$

which implies that we have to equate the quantities under the divergence [under the operator $\nabla \cdot$ (del dot)]. Hence,

$$\nabla \times \vec{H} = \vec{J} + \frac{\partial \vec{D}}{\partial t}. \tag{5.45}$$

Thus, in (5.45) we get the differential or point from of Ampere's circuital law that is consistent with the continuity equation in time-varying situations.

The first term $\vec{J}$ of the right-hand side of (5.45) is the current density constituted by the flow of charges through a medium. If the medium has a finite conductivity, the current density represented by $\vec{J}$ is the conduction current density, which is given by Ohm's law (5.11), which can be rewritten here as

$$\vec{J} = \sigma \vec{E}. \tag{5.46}$$

If the medium is a free-space medium, the current density represented by $\vec{J}$ is referred to as the convection current density, which can be expressed, with the help of (3.124) in the following vectorial form:

$$\vec{J} = \rho \vec{v} \tag{5.47}$$

where $\vec{v}$ is the velocity of charge-flow, supposed to be the same for all the charged particles, and ρ is the volume charge density of the charge-flow.

The second term, $\partial \vec{D}/\partial t$ of the right-hand side of (5.45), is called the displacement current density. Thus, we see that the displacement current density is not contributed to by any flow of charges, and that it exists in a medium when there is a variation of electric displacement $\vec{D}$ or electric field $\vec{E}$ with time t [remember the relation $\vec{D} = \varepsilon \vec{E}$; see (3.16)]. Further, you can see in the expression (5.45) how a magnetic field is associated with a time-varying electric field, just as in the association of an electric field with a time-varying magnetic field in the point or the differential form of Faraday's law (5.38).

We have thus modified the differential or the point form of Ampere's circuital law (5.42) into (5.45), in which the displacement current density $\partial \vec{D}/\partial t$ has been added to its right-hand side in order to make the law consistent with the continuity equation. This immediately prompts us to modify the integral form of Ampere's circuital law (5.35) too by adding to its right-hand side the displacement current (interpreted as the surface integral of the displacement current density $\partial \vec{D}/\partial t$) as follows:

$$\oint_l \vec{H} \cdot d\vec{l} = \int_S \vec{J} \cdot \vec{a}_n \, dS + \int_S \frac{\partial \vec{D}}{\partial t} \cdot \vec{a}_n \, dS = \int_S \left(\vec{J} + \frac{\partial \vec{D}}{\partial t} \right) \cdot \vec{a}_n \, dS. \tag{5.48}$$

For a medium made of a perfect dielectric, we can take the conductivity $\sigma = 0$, which makes the conduction current density $\vec{J}$ given by (5.46) zero, so that there will exist only a displacement current density in the dielectric medium for a time-varying electric field $\vec{E}$ or a time-varying electric displacement $\vec{D}$ $(= \varepsilon \vec{E})$. This is evident in a capacitor, known to block a dc current that does not vary with time and allows an ac current that varies with time. If we take a parallel-plate capacitor, for example, the current $i_{\text{capacitor}}$ through the capacitor in a circuit can be written as

$$i_{\text{capacitor}} = \frac{dQ}{dt} \tag{5.49}$$

where Q is the charge on a plate of the capacitor. If C represents the capacitance of the capacitor, we can express it as [see (3.99)] as

$$Q = CV \tag{5.50}$$

where V is the potential difference between the plates of the capacitor. We can differentiate (5.50) with respect to time to obtain

$$i_{\text{capacitor}} = \frac{dQ}{dt} = C\frac{dV}{dt}. \tag{5.51}$$

Taking the dimensions of the plates of the capacitor large compared to the distance d between the plates, we can treat the problem as one-dimensional and take the magnitude of the electric field E in the region between the plates to be constant and directed perpendicular to the plates (Section 3.2.2). Therefore, we can express E with the help of (3.117) as

$$E = \frac{V}{d}. \tag{5.52}$$

Now, substituting V from (5.52) into (5.51), we can write

$$i_{\text{capacitor}} = Cd\frac{dE}{dt}. \tag{5.53}$$

If we recall the expression (3.100) for the capacitance, $C = \varepsilon A/d$, where ε is the permittivity of the dielectric filling the region between the plates of area A each, and substitute it into (5.53), we get

$$i_{\text{capacitor}} = \varepsilon A\frac{dE}{dt}. \tag{5.54}$$

The current density $J_{\text{capacitor}}$ through the capacitor can be obtained by dividing (5.54) by the area A as:

$$J_{\text{capacitor}} = \varepsilon\frac{dE}{dt}. \tag{5.55}$$

Substituting the relation $D = \varepsilon E$ [see (3.16)] in (5.55), we obtain

$$J_{\text{capacitor}} = \frac{dD}{dt}. \tag{5.56}$$

Further, the displacement current density $J_{\text{displacement}}$ through the dielectric filling the region between the capacitor plates, which is the second term of the right-hand side of (5.45), can be written as

$$J_{\text{displacement}} = \frac{\partial D}{\partial t}. \tag{5.57}$$

The electric field E and hence the electric displacement D are uniform in the region between the plates, that is, D is independent of the space coordinates. E and hence D vary only with the time coordinate, it being assumed that a time-varying potential difference is applied between the plates of the capacitor by an external source in the circuit. Therefore, we can replace the partial derivative of D with time by its total derivative in (5.57) to express it as

$$J_{\text{displacement}} = \frac{dD}{dt}. \tag{5.58}$$

Further, multiplying (5.58) by the area A, we obtain the displacement current $i_{\text{displacement}}$ through the dielectric medium of the capacitor as

$$i_{\text{displacement}} = A\frac{dD}{dt}. \tag{5.59}$$

Substituting the relation $D = \varepsilon E$ in (5.59), we can write

$$i_{\text{displacement}} = \varepsilon A\frac{dE}{dt}. \tag{5.60}$$

Finally, comparing (5.60) with (5.54), we get

$$i_{\text{capacitor}} = i_{\text{displacement}} \tag{5.61}$$

and, similarly, comparing (5.56) with (5.58), we get

$$J_{\text{capacitor}} = J_{\text{displacement}}. \tag{5.62}$$

Thus, deductions (5.61) and (5.62) demonstrate that, for a time-varying situation, the current as well as the current density through a capacitor is identical to the displacement current as well as the displacement current density through the medium filling the space between the plates of the capacitor.

Further, unlike the conduction current through a conducting medium or the convection current through free-space, the displacement current through the region between the plates of the capacitor does not arise due to the actual flow of charges; it is caused by the time variation of the electric displacement or electric field in the region. However, when the capacitor is connected to an external closed circuit, the displacement current in the region between the plates of the capacitor manifests itself as the conduction current through the connecting wire of the circuit.

EXAMPLE E5-6

Find the displacement current density in free space where the magnetic flux density is given by $\vec{B} = B_0 \cos 2y \sin(\omega t - \pi z)\vec{a}_x$.

SOLUTION

We can write with the help of (5.45), putting therein $\vec{J} = 0$ and using the expression for curl given by (2.66), while remembering that in the given problem $\vec{B}$ has only the non-zero x-component ($B_x \neq 0, B_y = 0, B_z = 0$):

$$\frac{\partial \vec{D}}{\partial t} = \nabla \times \vec{H} = \frac{1}{\mu_0}\nabla \times \vec{B} = \frac{1}{\mu_0}\left(\frac{\partial B_x}{\partial z}\vec{a}_y - \frac{\partial B_x}{\partial y}\vec{a}_z\right). \tag{E5-6.1}$$

Putting the given magnetic flux density expression $B_x = B_0 \cos 2y \sin(\omega t - \pi z)$ and remembering $\mu_0 = 4\pi \times 10^{-7}$ H/m, we can carry out the differentiation in the right-hand side of the above equation to write

$$\frac{\partial \vec{D}}{\partial t} = B_0[-2.5 \times 10^6 \cos 2y \cos(\omega t - \pi z)\vec{a}_y + \frac{20 \times 10^6}{4 \times \pi}\,\mathrm{ccs}\, 2y \sin(\omega t - \pi z)\vec{a}_z]\ \mathrm{A/m^2},$$

which, putting the value of π, simplifies to the following expression for the displacement current density $\partial \vec{D}/\partial t$:

$$\frac{\partial \vec{D}}{\partial t} = B_0[-2.5 \times 10^6 \cos 2y \cos(\omega t - \pi z)\vec{a}_y + 1.6 \times 10^6\,\mathrm{ccs}\, 2y \sin(\omega t - \pi z)\vec{a}_z]\ \mathrm{A/m^2}.$$

Let us next consider a dielectric which is not perfect. In other words, let the dielectric have some finite conductivity so that it can allow both conduction and displacement currents to pass through it for a time-varying situation. In this case, we can write the current density through the dielectric as

$$\bar{J} = \sigma \vec{E} + \frac{\partial \vec{D}}{\partial t} \tag{5.63}$$

where the first term is contributed by the conduction current density given by (5.46) and the second term by the displacement current density given by (5.57). In view of the relation $\vec{D} = \varepsilon \vec{E}$, we can express (5.63) as

$$\vec{J} = \sigma \vec{E} + \varepsilon \frac{\partial \vec{E}}{\partial t}. \tag{5.64}$$

Considering $\vec{E}$ to be time-periodic and taking the time dependence of $\vec{E}$ as $\exp(j\omega t)$, using the phasor description of a time periodic function, we can take $\partial/\partial t \equiv j\omega$ (see Appendix A5.3) and express (5.64) as

$$\vec{J} = \sigma \vec{E} + j\omega\varepsilon \vec{E}. \tag{5.65}$$

Further, we can express (5.65) as

$$\vec{J} = \sigma \vec{E} + j\omega\varepsilon \vec{E} = j\omega \left(\varepsilon + \frac{\sigma}{j\omega} \right) \vec{E} = j\omega \left(\varepsilon - j\frac{\sigma}{\omega} \right) \vec{E} \tag{5.66}$$

an expression in a form that helps one to express the current density $\vec{J}$ in terms of a quantity called the complex permittivity ε^* of an imperfect dielectric, characterised by $\sigma \neq 0$, as follows:

$$\vec{J} = j\omega\varepsilon^* \vec{E} \tag{5.67}$$

where

$$\varepsilon^* = \varepsilon' - j\varepsilon'' \tag{5.68}$$

in which we have put the real and imaginary parts as $\varepsilon' = \varepsilon$ and $\varepsilon'' = \sigma/\omega$ respectively, choosing to express them in terms of single-primed or double-primed quantities (for an imperfect dielectric ($\sigma \neq 0$) to distinguish them from the permittivity ε of a perfect dielectric, which is essentially a real quantity).

Had the dielectric been perfect, we would have put $\sigma = 0$ in (5.65) and have read this relation as

$$\vec{J} = j\omega\varepsilon \vec{E}. \tag{5.69}$$

Comparing (5.67) with (5.69), we thus see that for an imperfect dielectric ($\sigma \neq 0$), the permittivity has changed from the conventional real permittivity ε to the complex permittivity $\varepsilon^*(= \varepsilon' - j\varepsilon'')$ given by (5.68). The ratio of the imaginary to the real parts of the complex permittivity is called the loss tangent ($\tan\delta$) of an imperfect dielectric, given by:

$$\tan\delta = \frac{\varepsilon''}{\varepsilon'} = \frac{\sigma/\omega}{\varepsilon} = \frac{\sigma}{\omega\varepsilon} \tag{5.70}$$

where δ is the angle by which the current density in an imperfect dielectric ($\sigma \neq 0$) falls short of leading the electric field by the phase angle of $\pi/2$, the value of the phase lead in a perfect dielectric ($\sigma = 0$). (See Appendix A5.4 for an elaboration with the help of the phasor diagram depicting the current density and the electric field in an imperfect dielectric.) Furthermore, we can find an alternative expression for the complex dielectric constant ε^* with the help of (5.68) and (5.70) as follows:

$$\varepsilon^*(= \varepsilon' - j\varepsilon'') = \varepsilon' \left(1 - j\frac{\varepsilon''}{\varepsilon'} \right) = \varepsilon'(1 - j\tan\delta) = \varepsilon(1 - j\tan\delta). \tag{5.71}$$

5.4 Maxwell's Equations

Let us recall the following four integral equations that we have already derived and, for the sake of convenience, rewrite them as follows:

$$\oint_S \vec{D} \cdot d\vec{S} = \oint_S \vec{D} \cdot \vec{a}_n \, dS = \int_\tau \rho \, d\tau \qquad \text{(3.64) rewritten} \qquad (5.72)$$

$$\oint_S \vec{B} \cdot d\vec{S} = \oint_S \vec{B} \cdot \vec{a}_n \, dS = 0 \qquad \text{(4.6) rewritten} \qquad (5.73)$$

$$\oint_l \vec{E} \cdot d\vec{l} = -\int_S \frac{\partial \vec{B}}{\partial t} \cdot \vec{a}_n \, dS \qquad \text{(5.32) rewritten} \qquad (5.74)$$

$$\oint_l \vec{H} \cdot d\vec{l} = \int_S \left(\vec{J} + \frac{\partial \vec{D}}{\partial t} \right) \cdot \vec{a}_n \, dS. \qquad \text{(5.48) rewritten} \qquad (5.75)$$

Similarly, we may recall and rewrite the following differential equations derived earlier, corresponding to (5.72)-(5.75) respectively:

$$\nabla \cdot \vec{D} = \rho \qquad \text{(3.104) rewritten} \qquad (5.76)$$

$$\nabla \cdot \vec{B} = 0 \qquad \text{(4.8) rewritten} \qquad (5.77)$$

$$\nabla \times \vec{E} = -\frac{\partial \vec{B}}{\partial t} \qquad \text{(5.38) rewritten} \qquad (5.78)$$

$$\nabla \times \vec{H} = \vec{J} + \frac{\partial \vec{D}}{\partial t} \qquad \text{(5.45) rewritten.} \qquad (5.79)$$

The integral equations (5.72)-(5.75) and the differential equations (5.76)-(5.79), which the electric and the magnetic vector quantities $\vec{E}$, $\vec{D}$, $\vec{H}$, and $\vec{B}$ satisfy everywhere, and which are valid for both static and time-varying situations are called the integral and the differential forms of Maxwell's equations respectively. In these equations, ρ represents the volume charge density (Section 3.3). $\vec{J}$ is the conduction current density (Section 5.1) or the convection current density (Section 5.3), and $\partial \vec{D}/\partial t$ is the displacement current density (Section 5.3). Remember that we had deduced the integral forms (5.72) and (5.73) as well as the differential forms (5.76) and (5.77) for static fields. However, they are valid for time-varying situations as well. Further, the integral form (5.74) and the differential form (5.78) were obtained for time-varying situations; however, they are valid for static fields as well, interpreting the time derivative of $\vec{D}$ as zero, which amounts to the displacement current density being absent in the medium thereby making the right-hand sides of (5.74) and (5.78) each equal to zero. Further, remember that the introduction of the displacement current density in (5.75) and (5.79) was a necessary condition for these two equations to be consistent with the continuity equation for time-varying fields (Section 5.3).[3]

In the rest of this book, Maxwell's four equations will be used frequently, as and when necessary, in order to develop several concepts of electromagnetic theory and their applications. However, we choose two simple examples to exemplify one of Maxwell's equations expressed in both the integral and the differential formats: (5.74) and (5.78) respectively.

We take up first the example of a parallel-plate capacitor consisting of a pair of identical circular plates (Fig. 5.1). The dimensions of the plates are considered large compared to the distance between the plates (Section 3.2). Let us apply an electric field across the plates of the capacitor that varies with time; it, however, remains uniform everywhere in the region between the plates at a given instant of time. Let us proceed to find the magnetic field generated due to this time-varying electric field inside the capacitor at a radial distance from the axis of the circular plates, first with help of the integral Maxwell's equation (5.75), and then with the help of the differential Maxwell's equation (5.79).

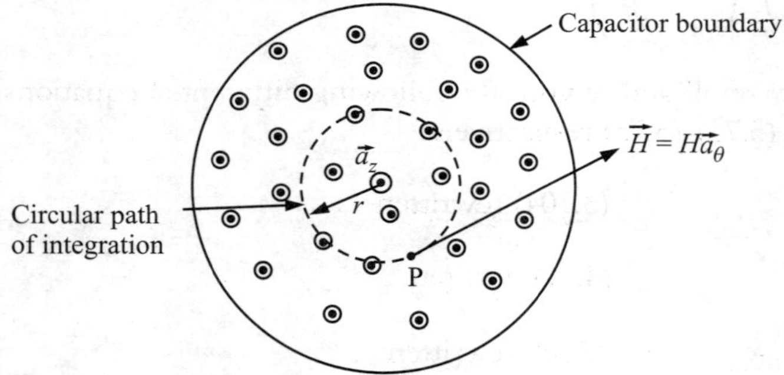

Fig. 5.1 *Cross section perpendicular to the Z-axis of a parallel-plate capacitor consisting of circular plates showing the instantaneous time varying electric field represented by dots with circles (pointing towards the reader) and the corresponding azimuthal magnetic field at a point on a coaxial circular path.*

We will use the integral equation (5.75) and put the conduction current density $\vec{J} = 0$ for the dielectric medium filling the region between the capacitor plates. Taking into account the contribution of only the displacement current density, the equation can be written as follows:

$$\oint_l \vec{H} \cdot d\vec{l} = \int_S \frac{\partial \vec{D}}{\partial t} \cdot \vec{a}_n \, dS. \tag{5.80}$$

Let us then apply (5.80) over a circular path of radius r passing through the point where the magnetic field in the region between the capacitor plates is to be found, at a radial distance

[3]James Clerk Maxwell originally gave as many as twenty equations in twenty variables and it was Oliver Heaviside, who is one of the founders of vector calculus (Chapter 2), who reduced these equations to four equations.

r from the axis of the capacitor with circular plates (Fig. 5.1). Earlier, in the problem of a current-carrying straight conducting wire, while exemplifying Ampere's circuital law, we had taken the magnetic field to be azimuthal (Section 4.3.1). In the present problem, analogously, we will take the magnetic field to be azimuthal and write $\vec{H} = H_\theta \vec{a}_\theta$ and $d\vec{l} = dl\,\vec{a}_\theta$ in (5.80). The electric field is applied along the axis of the capacitor (with circular plates), and we can put $\vec{D}\,(= \varepsilon\,\vec{E}) = D\,\vec{a}_z$ in (5.80). The direction of the normal to the area enclosing the circular path is along the positive Z-axis, noting that a screw moves axially in this direction when it is rotated, here, in anticlockwise direction (Fig. 5.1). Therefore, we can take $\vec{a}_n = \vec{a}_z$ in (5.80). Hence, we can apply (5.80) over the circular path of radius r to write

$$\oint_l H_\theta\,\vec{a}_\theta \cdot dl\,\vec{a}_\theta = \int_S \frac{\partial D\,\vec{a}_z}{\partial t} \cdot \vec{a}_z\,dS,$$

which simplifies to

$$H_\theta \oint_l dl = \frac{dD}{dt} \int_S dS \tag{5.81}$$

in which we have taken both H_θ and $\partial D/\partial t$ outside the integration since they are both uniform over the region of integration. Also, we have replaced the partial derivative $\partial D/\partial t$ by the total derivative dD/dt. Further, since the circumference of the circular path (passing through the point where we intend to find the magnetic field) is $2\pi\,r$ and the area of the circular region enclosed by the circular path is $\pi\,r^2$, we can evaluate the closed integrals in (5.81) as

$$\left.\begin{aligned} \oint_l dl &= 2\pi\,r \\ \int_S dS &= \pi\,r^2 \end{aligned}\right\}. \tag{5.82}$$

Substituting (5.82) in (5.81), we then obtain

$$H_\theta 2\pi\,r = \frac{dD}{dt}\pi\,r^2,$$

whence we get

$$H_\theta = \frac{dD}{dt}\frac{r}{2},$$

which in vectorial notation can be expressed as

$$\vec{H} = H_\theta\,\vec{a}_\theta = \frac{dD}{dt}\frac{r}{2}\vec{a}_\theta\,. \tag{5.83}$$

Putting $D = \varepsilon\,E$, the above equation can be written as

$$\vec{H} = \varepsilon\frac{dE}{dt}\frac{r}{2}\vec{a}_\theta\,. \tag{5.84}$$

Thus, starting from the integral Maxwell's equation (5.75), we obtained the expression for the magnetic field inside a circular parallel-plate capacitor as (5.83) or (5.84). Now, let us apply the differential Maxwell's equation (5.79) to the same problem, which, on putting the conduction current density as $\vec{J} = 0$ for the dielectric medium filling the region between the capacitor plates, can be written as

$$\nabla \times \vec{H} = \frac{\partial \vec{D}}{\partial t}, \tag{5.85}$$

whose z-component can be expressed as

$$(\nabla \times \vec{H})_z = \frac{\partial D}{\partial t}. \tag{5.86}$$

With the help of (2.67), we can write the z-component of (5.85) in the cylindrical coordinates (r, θ, z) as

$$\frac{1}{r}\left(\frac{\partial (rH_\theta)}{\partial r} - \frac{\partial H_r}{\partial \theta}\right) = \frac{\partial D}{\partial t}. \tag{5.87}$$

However, in this problem, the magnetic field is in the azimuthal direction such that $H_\theta \neq 0, H_r = 0$ and, moreover, the problem enjoys azimuthal symmetry and hence $\partial/\partial\theta = 0$. In view of this, we can simplify (5.87) as

$$\frac{1}{r}\frac{\partial (rH_\theta)}{\partial r} = \frac{\partial D}{\partial t}. \tag{5.88}$$

Since, in this problem H_θ depends only on r and D depends only on t, we can replace the partial derivatives by the total derivatives in (5.88) to write the latter as

$$\frac{1}{r}\frac{d(rH_\theta)}{dr} = \frac{dD}{dt},$$

which, upon integration becomes

$$\int d(rH_\theta) = \int \frac{dD}{dt} r dr. \tag{5.89}$$

Since D and hence also dD/dt in this problem are uniform in the region between the plates and thus independent of r, we can take dD/dt outside the integration of (5.89) and express it as

$$\int d(rH_\theta) = \frac{dD}{dt} \int r dr,$$

which, upon integration becomes

$$rH_\theta = \frac{dD}{dt}\frac{r^2}{2} + C \tag{5.90}$$

where C is an integration constant. Dividing (5.90) by r, we then obtain

$$H_\theta = \frac{dD}{dt}\frac{r}{2} + \frac{C}{r}. \tag{5.91}$$

If the integration constant C has a non-zero value, then from (5.91), $H_\theta \to \infty$ as $r \to 0$, meaning that the magnetic field blows up to infinity at the axis of the capacitor, a finding which is not physically possible, Therefore, we must put $C = 0$ in (5.91) to write it as

$$H_\theta = \frac{dD}{dt}\frac{r}{2} \tag{5.92}$$

which, in vectorial form becomes

$$\vec{H} = \frac{dD}{dt}\frac{r}{2}\vec{a}_\theta\,. \tag{5.93}$$

Putting $D = \varepsilon E$ in (5.93), we can also express it as

$$\vec{H} = \varepsilon\frac{dE}{dt}\frac{r}{2}\vec{a}_\theta\,. \tag{5.94}$$

The expression for the magnetic field at a radial distance r from the axis of a circular parallel-plate capacitor, which we obtain here as (5.93) or (5.94) starting from the differential form of Maxwell's equation (5.75), is found to be identical to the corresponding expressions (5.83) or (5.84) obtained earlier from the integral form of Maxwell's equation (5.79).

As a second example, let us take an analogous problem of a long solenoid in which the magnetic field remains uniform over its cross section (Section 4.3.1) at a given instant of time but varies with time at a known time rate of variation. A circular or azimuthal electric field will be generated at a point inside the solenoid due to the variation of the magnetic field with time, just as a circular or azimuthal magnetic field is generated due to the variation of electric field with time. Hence, let us find the value of the electric field inside the solenoid at a given radial distance from the axis of the solenoid in terms of the time rate of variation of the magnetic flux density inside the solenoid. The approach to the problem of the solenoid is exactly the same as that was followed in the problem of the circular parallel-plate capacitor.

With reference to a parallel-plate capacitor with circular plates, we had obtained the expression

$$\vec{H} = \frac{dD}{dt}\frac{r}{2}\vec{a}_\theta \qquad [(5.83) \text{ or } (5.93) \text{ rewritten}], \tag{5.95}$$

starting either from the following integral form of Maxwell's equation

$$\oint_l \vec{H}\cdot d\vec{l} = \int_S \frac{\partial\vec{D}}{\partial t}\cdot\vec{a}_n dS \qquad [(5.80) \text{ rewritten}] \tag{5.96}$$

or from the following differential form of Maxwell's equation

$$\nabla \times \vec{H} = \frac{\partial \vec{D}}{\partial t}. \qquad \text{(5.85) rewritten.} \tag{5.97}$$

Now, with reference to the problem of a solenoid, analogously, we can start either from the following integral form of Maxwell's equation

$$\oint_l \vec{E} \cdot d\vec{l} = -\int_S \frac{\partial \vec{B}}{\partial t} \cdot \vec{a}_n \, dS \qquad \text{(5.74) rewritten} \tag{5.98}$$

or from the differential form of Maxwell's equation

$$\nabla \times \vec{E} = -\frac{\partial \vec{B}}{\partial t} \qquad \text{(5.78) rewritten.} \tag{5.99}$$

Therefore, if we compare either (5.98) with (5.96) or (5.99) with (5.97), we can immediately note that the quantities $\vec{E}$ and $-\partial\vec{B}/\partial t$ for the problem of the solenoid are analogous to the quantities $\vec{H}$ and $\partial\vec{D}/\partial t$ respectively for the problem of the circular parallel-plate capacitor. Therefore, we can write the expression for the electric field inside the solenoid at a radial distance r from the axis of the solenoid by replacing $\vec{H}$ by $\vec{E}$ as well as $d\vec{D}/dt$ by $-d\vec{B}/dt$ in (5.95) as follows:

$$\vec{E} = -\frac{dB}{dt}\frac{r}{2}\vec{a}_\theta\,. \tag{5.100}$$

In the three examples to follow consecutively, you will see the use of Maxwell's equations treated in the three coordinate systems: rectangular, cylindrical and spherical respectively.

EXAMPLE E5-7

Derive, with the help of the relevant Maxwell's equation, the expression for the magnetic field $\vec{H}$ for a uniform plane wave propagating along z ($E_z = H_z = 0$; $\partial/\partial x = \partial/\partial y = 0$) in free space if the electric field is given by

$$\vec{E} = (20\vec{a}_x + 30\vec{a}_y)\cos(3\pi \times 10^{10}t - 100\pi\, z).$$

SOLUTION

From the given expression for the electric field:

$$\vec{E} = (20\vec{a}_x + 30\vec{a}_y)\cos(3\pi \times 10^{10}t - 100\pi\, z) \text{ V/m},$$

we identify

$$\left.\begin{aligned} E_x &= 20\cos(3\pi \times 10^{10}t - 100\pi\, z) \text{ V/m} \\ E_y &= 30\cos(3\pi \times 10^{10}t - 100\pi\, z) \text{ V/m} \end{aligned}\right\}. \tag{E5-7.1}$$

Let us recall Maxwell's equation, namely $\nabla \times \vec{E} = -\partial \vec{B}/\partial t = -\mu_0 \partial \vec{H}/\partial t$ (5.38) and write it with the help of (2.66) while noting that $E_z = H_z = 0$; $\partial/\partial x = \partial/\partial y = 0$ (given):

$$\left.\begin{aligned} \frac{\partial E_y}{\partial z} &= -\mu_0 \frac{\partial H_x}{\partial t} \\ \frac{\partial E_x}{\partial z} &= -\mu_0 \frac{\partial H_y}{\partial t} \end{aligned}\right\}. \tag{E5-7.2}$$

Differentiating E_y given by (E5-7.1) with respect to z, we obtain

$$\frac{\partial E_y}{\partial z} = 30 \times 100\pi \sin(3 \times 10^{10}\pi\, t - 100\pi\, z). \tag{E5-7.3}$$

Substituting (E5-7.3) in (E5-7.2), we obtain

$$\mu_0 \frac{\partial H_x}{\partial t} = \frac{\partial E_y}{\partial z} = 30 \times 100\pi \sin(3 \times 10^{10}\pi\, t - 100\pi\, z). \tag{E5-7.4}$$

Integrating (E5-7.4) with respect to t, we also obtain

$$\begin{aligned} H_x &= -\frac{30 \times 100\pi \cos(3 \times 10^{10}\pi\, t - 100\pi\, z)}{\mu_0 \times 3 \times 10^{10}\pi} \\ &= -\frac{30 \times 100\pi \cos(3 \times 10^{10}\pi\, t - 100\pi\, z)}{4\pi \times 10^{-7} \times 3 \times 10^{10}\pi} \\ &= -\frac{30 \times 100 \cos(3 \times 10^{10}\pi\, t - 100\pi\, z)}{4\pi \times 10^{-7} \times 3 \times 10^{10}} \\ &= -\frac{30 \times 100 \cos(3 \times 10^{10}\pi\, t - 100\pi\, z)}{4 \times 3.143 \times 10^{-7} \times 3 \times 10^{10}} \\ &= -\frac{\cos(3 \times 10^{10}\pi\, t - 100\pi\, z)}{4 \times 3.143} \\ &= -0.0795 \cos(3 \times 10^{10}\pi\, t - 100\pi\, z) \text{ A/m}. \end{aligned} \tag{E5-7.5}$$

Next, differentiating E_x given by (E5-7.1) with respect to z, we obtain

$$\frac{\partial E_x}{\partial z} = 20 \times 100\pi \sin(3\pi \times 10^{10} t - 100\pi\, z). \tag{E5-7.6}$$

Substituting (E5-6.6) in (E5-6.2), we then obtain

$$\mu_0 \frac{\partial H_y}{\partial t} = -\frac{\partial E_x}{\partial z} = -20 \times 100\pi \sin(3\pi \times 10^{10} t - 100\pi\, z). \tag{E5-7.7}$$

Further, integrating (E5-7.7) with respect to t, we obtain

$$H_y = \frac{20 \times 100\pi \cos(3\pi \times 10^{10} t - 100\pi\, z)}{\mu_0 3\pi \times 10^{10}}$$

$$= \frac{20 \times 100\pi \cos(3\pi \times 10^{10}t - 100\pi\, z)}{4\pi \times 10^{-7} \times 3\pi \times 10^{10}}$$

$$= \frac{\cos(3\pi \times 10^{10}t - 100\pi z)}{2 \times 3 \times 3.143}$$

$$= 0.053 \cos(3\pi \times 10^{10}t - 100\pi z) \text{ A/m}. \qquad \text{(E5-7.8)}$$

Using the expressions for H_x and H_y given by (E5-7.5) and (E5-7.8) respectively, we thus obtain the following expression for the magnetic field $\vec{H}$:

$$\vec{H} = H_x\vec{a}_x + H_y\vec{a}_y = (-0.0795\vec{a}_x + 0.053\vec{a}_y)\cos(3\pi \times 10^{10}t - 100\pi z) \text{ A/m}.$$

EXAMPLE E5-8

Show that the electric field in a cylindrical structure that supports waves with $\exp j(\omega t - \beta z)$ dependence and for which the axial electric field component $E_z = 0$ can be expressed in terms of the phase velocity of the wave v_{ph} and magnetic flux density components as $\vec{E} = v_{\text{ph}}(B_\theta\vec{a}_r - B_r\vec{a}_\theta)$.

SOLUTION

Let us express Maxwell's equation (5.78) in cylindrical coordinates with the help of (2.67) as follows:

$$\nabla \times \vec{E} = \left(\frac{1}{r}\frac{\partial E_z}{\partial \theta} - \frac{\partial E_\theta}{\partial z}\right)\vec{a}_r + \left(\frac{\partial E_r}{\partial z} - \frac{\partial E_z}{\partial r}\right)\vec{a}_\theta + \frac{1}{r}\left(\frac{\partial (rE_\theta)}{\partial r} - \frac{\partial E_r}{\partial \theta}\right)\vec{a}_z = -\frac{\partial \vec{B}}{\partial t}. \qquad \text{(E5-8.1)}$$

Since, here, $E_z = 0$ (given), we can write the r-component of (E5-8.1) as

$$-\frac{\partial E_\theta}{\partial z} = -\frac{\partial B_r}{\partial t}. \qquad \text{(E5-8.2)}$$

Similarly, the θ-component of (E5-8.1), remembering $E_z = 0$, gives

$$\frac{\partial E_r}{\partial z} = -\frac{\partial B_\theta}{\partial t}. \qquad \text{(E5-8.3)}$$

In view of the $\exp j(\omega t - \beta z)$ wave dependence, where β is the wave phase propagation constant, we can take $\partial/\partial z = -j\beta$ and $\partial/\partial t = j\omega$ to express (E5-8.2) and (E5-8.3) respectively as

$$\left.\begin{aligned} E_\theta &= -\frac{\omega}{\beta}B_r \\ E_r &= \frac{\omega}{\beta}B_\theta \end{aligned}\right\}. \qquad \text{(E5-8.4)}$$

However, the wave phase velocity v_{ph} is related to β as

$$v_{\text{ph}} = \frac{\omega}{\beta}. \tag{E5-8.5}$$

Plugging (E5-8.5) in (E5-8.4), we then get the relation

$$\left.\begin{aligned} E_\theta &= -v_{\text{ph}} B_r \\ E_r &= v_{\text{ph}} B_\theta \end{aligned}\right\}. \tag{E5-8.6}$$

In the present context, $E_z = 0$ (given), enabling us to write

$$\vec{E} = E_r \vec{a}_r + E_\theta \vec{a}_\theta. \tag{E5-8.7}$$

Substituting (E5-8.6) in (E5-8.7), we thus arrive at the desired relation

$$\vec{E} = v_{\text{ph}}(B_\theta \vec{a}_r - B_r \vec{a}_\theta).$$

EXAMPLE E5-9

Use the relevant Maxwell's equation to find the electric field $\vec{E}$ at a large distance r from a small current-loop antenna (see also Section 8.4) at an angle θ from the axis of the loop, in spherical polar coordinates (r, θ, ϕ), from the following expression for a given magnetic field $\vec{H}$ in free space:

$$\left.\begin{aligned} \vec{H} &= H_\theta \, \vec{a}_\theta \\ H_\theta &= -\frac{\omega \mu_0 p_m}{4\pi\eta_0} \frac{\exp(-j\beta\, r)}{r} \beta \sin\theta \end{aligned}\right\},$$

with time dependence $\exp(j\omega t)$. Here, p_m is the magnetic moment of the current loop treated as a magnetic dipole; $\beta = \omega(\mu_0\varepsilon_0)^{1/2}\beta = \omega(\mu_0\varepsilon_0)^{1/2}$ is the axial phase propagation constant, and $\eta_0 = (\mu_0/\varepsilon_0)^{1/2}\eta_0 = \mu_0/\varepsilon_0)^{1/2}$ is the intrinsic impedance of free space. Take the problem as azimuthally symmetric $(\partial/\partial\phi = 0)$.

SOLUTION

From the given magnetic field expression, we note that for the present problem, $H_r = 0$; $H_\theta \neq 0$; and $H_\phi = 0$. Also, $\partial/\partial\phi = 0$ in view of the given azimuthal symmetry of the problem $(\partial/\partial\phi = 0)$. From these magnetic field components, we can find the corresponding electric field components with the help of Maxwell's equation (5.79), which can be written, taking $\vec{J} = 0$ (source-free region) and in view of the time dependence $\exp(j\omega t)$ that enables one to take $\partial/\partial t = j\omega$, as

$$\nabla \times \vec{H} = \frac{\partial \vec{D}}{\partial t} = \varepsilon_0 \frac{\partial \vec{E}}{\partial t} = j\omega\varepsilon_0 \vec{E}. \tag{E5-9.1}$$

Expanding the left-hand side of (E5-9.1) with the help of the expression for curl obtainable from (2.68) and remembering that $\partial/\partial\phi = 0$, $H_r = H_\phi = 0$ (given), we can write

$$\frac{1}{r}\left(\frac{\partial(rH_\theta)}{\partial r}\right)\vec{a}_\phi = j\omega\varepsilon_0\vec{E},$$

that is,

$$\frac{1}{r}\left(\frac{\partial(rH_\theta)}{\partial r}\right) = j\omega\varepsilon_0 E_\phi. \tag{E5-9.2}$$

Putting in the above expression

$$H_\theta = -\frac{\omega\mu_0 p_m}{4\pi\eta_0}\frac{\exp(-j\beta r)}{r}\beta\sin\theta \quad \text{(given)},$$

we then obtain

$$-\frac{1}{r}\frac{\omega\mu_0 p_m}{4\pi\eta_0}\beta\sin\theta\left(\frac{\partial[\exp(-j\beta r)]}{\partial r}\right) = j\omega\varepsilon_0 E_\phi. \tag{E5-9.3}$$

Carrying out the differentiation in (E5-9.3), we can then write

$$-\frac{1}{r}\frac{\omega\mu_0 p_m}{4\pi\eta_0}\beta\sin\theta \times -j\beta\exp(-j\beta r) = j\omega\varepsilon_0 E_\phi,$$

whence we obtain

$$E_\phi = \frac{1}{r}\frac{\omega\mu_0 p_m}{4\pi\eta_0 j\omega\varepsilon_0}\beta^2\sin\theta \times j\exp(-j\beta r). \tag{E5-9.4}$$

We can simplify (E5-9.4) in view of the relations $\beta = \omega(\mu_0\varepsilon_0)^{1/2}$ and $\eta_0 = (\mu_0/\varepsilon_0)^{1/2}$ to obtain the following expression for the electric field component:

$$E_\phi = \frac{\omega\mu_0 p_m}{4\pi}\frac{\exp(-j\beta r)}{r}\beta\sin\theta,$$

which takes the vectorial form as follows:

$$\vec{E} = E_\phi\vec{a}_\phi = \frac{\omega\mu_0 p_m}{4\pi}\frac{\exp(-j\beta r)}{r}\beta\sin\theta\,\vec{a}_\phi.$$

5.5 Electric Scalar Potential and Magnetic Vector Potentials in Time-varying Fields

We have already developed the concept that led to a relation between the electric field and potential for static electric fields as (3.91), which can be rewritten for a ready reference as

$$\vec{E} = -\nabla V. \quad \text{(static fields)} \tag{5.101}$$

However, the concept that had led to (5.101) needs to be re-looked into for a time-varying situation. For this purpose, let us recall one of Maxwell's equations, namely (5.78), which can be rewritten for the sake of convenience as

$$\nabla \times \vec{E} = -\frac{\partial \vec{B}}{\partial t} . \tag{5.102}$$

Further, for the reason that would be understandable as we progress in the present analysis, let us express the vector $\vec{B}$ in terms of the curl of an arbitrarily chosen vector quantity $\vec{A}$, referred to as the magnetic vector potential, as

$$\vec{B} = \nabla \times \vec{A}. \tag{5.103}$$

Substituting (5.103) in (5.102), we can write

$$\nabla \times \vec{E} = -\frac{\partial(\nabla \times \vec{A})}{\partial t} = -\nabla \times \frac{\partial \vec{A}}{\partial t},$$

which can be rearranged as

$$\nabla \times \left(\vec{E} + \frac{\partial \vec{A}}{\partial t}\right) = 0. \tag{5.104}$$

At this stage, let us recall a well-known vector identity

$$\nabla \times \nabla \psi = 0, \tag{5.105}$$

where ψ is a scalar quantity. Now, by comparing (5.104) with (5.105), we can easily write

$$\vec{E} + \frac{\partial \vec{A}}{\partial t} = \nabla \psi,$$

which can be rearranged as

$$\vec{E} = \nabla \psi - \frac{\partial \vec{A}}{\partial t}. \tag{5.106}$$

We can read (5.106) for the special case of static fields by putting $\partial/\partial t = 0$ as

$$\vec{E} = \nabla \psi \quad \text{(static fields)}. \tag{5.107}$$

However, in order that (5.107) agrees with (5.101), we must identify ψ as

$$\psi = -V. \tag{5.108}$$

The substitution of (5.108) in (5.106) leads to the following relation for the time-varying situation:

$$\vec{E} = -\nabla V - \frac{\partial \vec{A}}{\partial t}. \tag{5.109}$$

Thus, in (5.109) we obtain for a time-varying situation a relation between the electric field and two potentials, namely, the scalar electric potential V and the magnetic vector potential $\vec{A}$.

Next, let us proceed to establish two wave equations—one in $\vec{A}$ and the other in V. For this purpose, take the curl of (5.103) as

$$\nabla \times \vec{B} = \nabla \times \nabla \times \vec{A},$$

which on expanding its right-hand side (with the help of the vector identity given at the end of the book following Chapter 11) can be written as

$$\nabla \times \vec{B} = \nabla(\nabla \cdot \vec{A}) - \nabla^2 \vec{A}. \tag{5.110}$$

Further, the left-hand side of (5.110) can be written with the help of (5.79) multiplied by μ_0 and remembering the relation $\vec{B} = \mu_0 \vec{H}$ as

$$\nabla \times \vec{B} = \mu_0 \nabla \times \vec{H} = \mu_0 \left(\vec{J} + \varepsilon_0 \frac{\partial \vec{E}}{\partial t} \right) = \mu_0 \vec{J} + \mu_0 \varepsilon_0 \frac{\partial \vec{E}}{\partial t}. \tag{5.111}$$

Now, we can compare the right-hand sides of (5.110) and (5.111) to write the relation

$$\nabla(\nabla \cdot \vec{A}) - \nabla^2 \vec{A} = \mu_0 \vec{J} + \mu_0 \varepsilon_0 \frac{\partial \vec{E}}{\partial t}. \tag{5.112}$$

Next, if we substitute (5.109) in (5.112), we obtain

$$\nabla(\nabla \cdot \vec{A}) - \nabla^2 \vec{A} = \mu_0 \vec{J} + \mu_0 \varepsilon_0 \frac{\partial}{\partial t} \left(-\nabla V - \frac{\partial \vec{A}}{\partial t} \right)$$

$$= \mu_0 \vec{J} - \mu_0 \varepsilon_0 \nabla \frac{\partial V}{\partial t} - \mu_0 \varepsilon_0 \frac{\partial^2 \vec{A}}{\partial t^2}. \tag{5.113}$$

Further, we can rearrange the terms of (5.113) to express it as

$$\nabla^2 \vec{A} - \mu_0 \varepsilon_0 \frac{\partial^2 \vec{A}}{\partial t^2} = -\mu_0 \vec{J} + \mu_0 \varepsilon_0 \frac{\partial}{\partial t} \nabla V + \nabla(\nabla \cdot \vec{A})$$

which can also be put as

$$\nabla^2 \vec{A} - \mu_0 \varepsilon_0 \frac{\partial^2 \vec{A}}{\partial t^2} = -\mu_0 \vec{J} + \nabla(\nabla \cdot \vec{A} + \mu_0 \varepsilon_0 \frac{\partial V}{\partial t}). \tag{5.114}$$

Recall that in (5.102) we had defined the curl of $\vec{A}$, giving us liberty to define the divergence of $\vec{A}$ ($= \nabla \cdot \vec{A}$) which is independent of its curl ($= \nabla \times \vec{A}$). We can thus define $\nabla \cdot \vec{A}$ such that it satisfies the following condition called Lorentz condition:

$$\nabla \cdot \vec{A} + \mu_0 \varepsilon_0 \frac{\partial V}{\partial t} = 0 \tag{5.115}$$

which is useful in decoupling the scalar potential V from the magnetic vector potential $\vec{A}$ to obtain the wave equation in $\vec{A}$ (devoid of V) that involves the source term, namely, the current density $\vec{J}$. This is also useful to obtain the wave equation in V (devoid of $\vec{A}$) that involves the source term, namely, the volume charge density ρ. Thus, the condition (5.115), when applied to (5.114), yields the following wave equation in the magnetic vector potential $\vec{A}$:

$$\nabla^2\vec{A} - \mu_0\varepsilon_0\frac{\partial^2\vec{A}}{\partial t^2} = -\mu_0\vec{J}. \quad (5.116)$$

Similarly, we can develop the wave equation in the electric scalar potential V. For this purpose, let us recall (5.76) and rewrite it using the relation $\vec{D} = \varepsilon_0\vec{E}$ as [see also (3.105)]

$$\nabla\cdot\vec{E} = \frac{\rho}{\varepsilon_0}. \quad (5.117)$$

Substituting (5.109) in (5.117), we can write

$$\nabla\cdot\left(-\nabla V - \frac{\partial\vec{A}}{\partial t}\right) = \frac{\rho}{\varepsilon_0},$$

which can also be read as

$$-\nabla^2 V - \frac{\partial}{\partial t}\nabla\cdot\vec{A} = \frac{\rho}{\varepsilon_0}. \quad (5.118)$$

Further, substituting the Lorentz condition (5.115) in (5.118), we can express the latter in the form of the following wave equation in the scalar potential V:

$$\nabla^2 V - \mu_0\varepsilon_0\frac{\partial^2 V}{\partial t^2} = -\frac{\rho}{\varepsilon_0}. \quad (5.119)$$

Thus, subject to the Lorentz condition (5.115), we have obtained the wave equations (5.119) and (5.116) in scalar and vector potentials, V and $\vec{A}$ respectively, in terms of the source quantities $\vec{J}$ and ρ, which can be put together as follows:

$$\left.\begin{aligned} \nabla^2\vec{A} - \mu_0\varepsilon_0\frac{\partial^2\vec{A}}{\partial t^2} &= -\mu_0\vec{J} \\ \nabla^2 V - \mu_0\varepsilon_0\frac{\partial^2 V}{\partial t^2} &= -\frac{\rho}{\varepsilon_0} \end{aligned}\right\}. \quad (5.120)$$

If we know the solutions of (5.120) for $\vec{A}$ and V for a time-varying problem (which we will elaborate later in this section), we can use these solutions to find $\vec{E}$ with the help of (5.109) in terms of $\vec{A}$ and V, as well as $\vec{B}$ $(= \mu_0\vec{H})$ with the help of (5.103) in terms of only $\vec{A}$. Alternatively, we can find $\vec{E}$ also in terms of only $\vec{A}$ with the help of (5.109), in which, however, V has to replaced by an expression in terms of $\vec{A}$. For this purpose, invoking the time dependence of V as $\exp(j\omega t)$, we can take $\partial V/\partial t = j\omega V$, which can be then substituted in Lorentz condition (5.115) to write

$$\nabla \cdot \vec{A} = -j\omega\mu_0\varepsilon_0 V. \tag{5.121}$$

Similarly, invoking the time dependence of $\vec{A}$ as $\exp(j\omega t)$, we can take $\partial\vec{A}/\partial t = j\omega\vec{A}$, which can be substituted in (5.109) to write

$$\vec{E} = -\nabla V - j\omega\,\vec{A}. \tag{5.122}$$

Substituting V obtainable from (5.121) in (5.122), we then have

$$\vec{E} = -j\nabla\frac{\nabla \cdot \vec{A}}{\omega\mu_0\varepsilon_0} - j\omega\vec{A},$$

which can be put as

$$\vec{E} = -j\omega\left(\vec{A} + \frac{\nabla\nabla \cdot \vec{A}}{\omega^2\mu_0\varepsilon_0}\right). \tag{5.123}$$

Thus, for a problem, if the vector potential $\vec{A}$ is known by solving its wave equation (5.116), we will be able to find $\vec{E}$ and $\vec{B}$ $(= \mu_0\vec{H})$ with the help of (5.123) and (5.103) respectively.

5.5.1 Retarded scalar and vector potentials

Now, let us proceed to appreciate how we can solve the wave equation (5.120) for the scalar potential V as well as for the vector potential $\vec{A}$. For this purpose, for the sake of convenience, let us reproduce (5.119)

$$\nabla^2 V - \mu_0\varepsilon_0\frac{\partial^2 V}{\partial t^2} = -\frac{\rho}{\varepsilon_0}. \tag{5.124}$$

To be more specific, let us discuss the solutions of (5.119) or (5.124) independently for the following two special cases:

$$\left.\begin{array}{ll} \rho = 0 & \text{(a)} \\ \dfrac{\partial}{\partial t} = 0 & \text{(b)} \end{array}\right\}. \text{ (two special cases)} \tag{5.125}$$

For the special case depicted by (5.125*a*), we can read (5.124) as follows:

$$\nabla^2 V - \mu_0\varepsilon_0\frac{\partial^2 V}{\partial t^2} = 0. \quad (\rho = 0,\ \partial/\partial t \neq 0) \tag{5.126}$$

The solution to the wave equation can be represented as (see Chapter 6):

$$V \propto \exp j(\omega t - \beta r), \tag{5.127}$$

where r is the distance of the point where the potential is observed from the source of potential. ω is the wave angular frequency, and β is the phase propagation constant of the wave, with $v_p(=\omega/\beta) = 1/(\mu_0\varepsilon_0)^{1/2}$ being the wave phase velocity.

Similarly, for the special case depicted by (5.125*b*), we can read (5.124) as

$$\nabla^2 V = -\frac{\rho}{\varepsilon_0}. \quad (\partial/\partial t = 0,\ \rho \neq 0) \tag{5.128}$$

Equation (5.128) can be seen as being identical to Poisson's equation (3.108) for electrostatics. The solution for potential at a point at a distant r from a point charge q is given by (3.80), which can be rewritten here as

$$V = \frac{q}{4\pi\varepsilon_0 r}. \tag{5.129}$$

In terms of a charge distribution distributed over a volume τ, we can interpret (5.129) as

$$V = \frac{1}{4\pi\varepsilon_0}\int_\tau \frac{\rho}{r} d\tau. \quad (\partial/\partial t = 0,\ \rho \neq 0) \tag{5.130}$$

Therefore, combining (5.127) and (5.130), we can write, for $\partial/\partial t = 0,\ \rho \neq 0$, the solution of (5.119) for the scalar potential V as

$$V = \frac{1}{4\pi\varepsilon_0}\int_\tau \frac{\rho}{r} \exp[j(\omega t - \beta r)] d\tau. \quad (\partial/\partial t \neq 0,\ \rho \neq 0) \tag{5.131}$$

In terms of the wave phase velocity $v_{\mathrm{ph}}(=\omega/\beta)$ as defined following (5.127), we can alternatively express (5.131) as

$$V = \frac{1}{4\pi\varepsilon_0}\int_\tau \frac{\rho}{r} \exp[j\omega(t - r/v_{\mathrm{ph}})] d\tau. \tag{5.132}$$

Similarly, following the same procedure used in obtaining the solution for the scalar potential V as (5.131) or (5.132) starting from (5.119), which is the wave equation in V, we can obtain the solution for the vector potential $\vec{A}$ starting from (5.116), which is the wave equation in $\vec{A}$, as follows:

$$\vec{A} = \frac{\mu_0}{4\pi}\int \frac{\vec{J}}{r} \exp[j(\omega t - \beta r)] d\tau \tag{5.133}$$

which, in terms of the wave phase velocity $v_p(=\omega/\beta)$ can be put as

$$\vec{A} = \frac{\mu_0}{4\pi}\int \frac{\vec{J}}{r} \exp[j\omega(t - r/v_p)] d\tau. \tag{5.134}$$

Further, we can define a current element $I d\vec{l}$, corresponding to a current I that flows through a conductor, say a wire of cross-sectional area α of an infinitesimal length element

dl, where $d\vec{l}$ is a length vector whose magnitude is dl and whose direction is that of the current flow through the conductor, being the same as the direction of the current density vector $\vec{J}$. For such a current element, we can then write, putting $d\tau = (\alpha)(dl)$,

$$Jd\tau = \frac{I}{\alpha}d\tau = \frac{I}{\alpha}\alpha\, dl = I\, dl$$

and, consequently,

$$\vec{J}\, d\tau = I\, d\vec{l}. \tag{5.135}$$

Therefore, substituting (5.135) in (5.133) and (5.134), we obtain the following alternative expressions for the vector potential $\vec{A}$:

$$\vec{A} = \frac{\mu_0}{4\pi}\int \frac{I}{r}\exp[j(\omega t - \beta r)]d\vec{l} \tag{5.136}$$

and

$$\vec{A} = \frac{\mu_0}{4\pi}\int \frac{I}{r}\exp[j\omega\,(t - r/v_p)]d\vec{l}. \tag{5.137}$$

You will be able to now appreciate from (5.132) and (5.136) or (5.137) that as the source quantities ρ and I vary as $\exp(j\omega t)$ at an instant of time t, the scalar and vector potentials V and $\vec{A}$ at the observation point at a distance r vary as $\exp(j\omega\, t')$, where $t' = t - r/v_p$ is a time earlier than t. In other words, these potentials are *retarded* in time ($t' = t - r/v_p$), meaning thereby that the disturbances in the source, here in the quantities ρ and I, take a time r/v_p to reach the observation point at a distance r while travelling with the velocity v_{ph}, and they manifest themselves in the potential quantities V and $\vec{A}$ respectively.

As a special case of a steady current for which we can take the exponential factor as unity in (5.137), the latter passes on to (4.79) that was earlier derived directly from Biot–Savart's law (see also Section 8.4). The solution (5.137) for the vector potential $\vec{A}$, in conjunction with the expressions (5.123) and (5.103) for $\vec{E}$ and $\vec{B}$ ($= \mu_0\vec{H}$) respectively, has considerable application in the analysis of a conduction current antenna such as a dipole antenna (see Section 8.4). For such applications, it is worth putting (5.123), (5.103) and (5.137) together as follows:

$$\left.\begin{aligned} \vec{E} &= -j\omega\,(\vec{A} + \frac{\nabla\nabla\cdot\vec{A}}{\omega^2\mu_0\varepsilon_0}) \\ \vec{H} &= \frac{\nabla\times\vec{A}}{\mu_0} \\ \vec{A} &= \frac{\mu_0}{4\pi}\int\frac{I}{r}\exp[j(\omega t - \beta r)]\, d\vec{l} \end{aligned}\right\}. \tag{5.138}$$

Appendix A5.1

Electric field inside the conducting sheet and the dielectric sheets placed in an electric field

Let us consider a conducting sheet placed in an electric field. The relaxation time of a good conductor is small. Therefore, when placed in an electric field, the bulk of the conducting sheet is not charged; the charges in the form of electrons freely move to the surface to form surface charges which totally cancel the electric field within the conductor (Fig. A5.1*a*). Gauss's law also corroborates this [see the note following (3.48)].

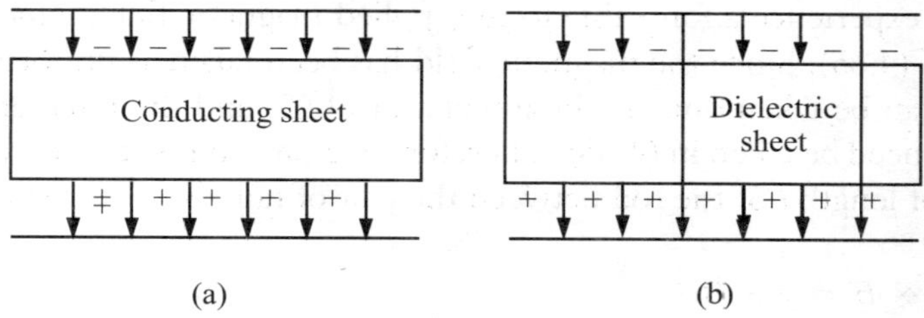

Fig. A5.1 *(a) Conducting and (b) dielectric sheets placed in an electric field, showing electric field lines outside the conducting and dielectric sheets. There are no electric field lines within the conducting sheet, while electric field lines exist within the dielectric sheet.*

On the contrary, a dielectric has a rather large relaxation time and when placed in an electric field, a dielectric sheet can be charged throughout its bulk. In fact, the electrons in a dielectric sheet are bound to their parent atoms and can only be displaced to a limited extent by the electric field. As a result only an insufficient surface charge is developed to totally cancel the electric field within the bulk of the sheet (Fig. A5.1*b*).

Appendix A5.2

Expression for the current in a circuit that is linked up with a magnetic flux density varying with time

Let us formulate the problem in rectangular coordinate system (x, y, z). We consider a closed circuit consisting of (i) a straight conducting rod placed in the direction of the X-axis, (ii) a pair of parallel conducting rails, each in the direction of Z-axis, over which the rod can freely slide, and (iii) a source of EMF connected across the rails such that the rod and the pair of rails lie on XZ plane. Further, we consider that the rod and the pair of rails are immersed in a uniform steady magnetic field applied in the positive direction of the Y-axis (Fig. A5.2).

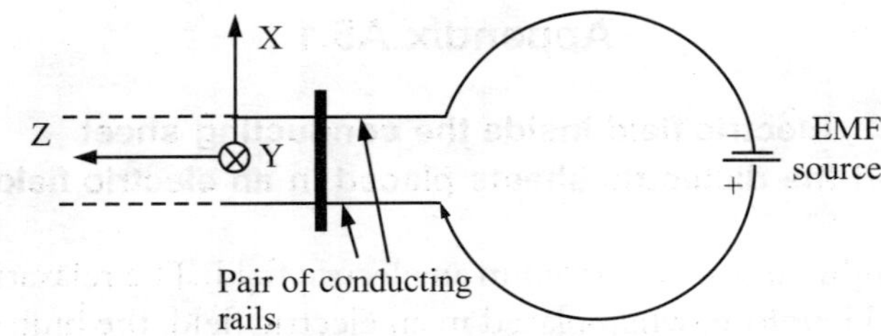

Fig. A5.2 *A conducting rod free to slide on a pair of transverse conducting rails, the whole immersed in a uniform steady normal magnetic field applied, through which a current passes in a circuit from a source of electromagnetic force.*

The rod will experience a force due to the applied magnetic field, which can be found with the help of (4.65). Since the magnetic field has been taken as uniform, the magnetic flux density $\vec{B}$ can be taken outside the integral of (4.65) and the integration of only the length vector $d\vec{l}$ need be taken in (4.65). Therefore, we can express the force $\vec{F}$ experienced by the portion of length $\vec{l}$ of the rod between the pair of rails as

$$\vec{F} = i\int d\vec{l} \times \vec{B} = i\vec{l} \times \vec{B}. \tag{A5.1}$$

The length vector $\vec{l}$, which has a magnitude l, will take up the direction of the current i in the rod, which, in turn, would depend on the polarity of the source of EMF. Thus, depending upon the polarity of the source, the length vector $\vec{l}$ will be either in the positive or the negative direction of x and accordingly, we can express $\vec{l}$ as

$$\vec{l} = \pm l\,\vec{a}_x \tag{A5.2}$$

where the upper and the lower signs correspond to the current in the rod being in the positive and the negative directions of x respectively (Fig. A5.2).

Since we have considered the magnetic field in the positive direction of y, we can express $\vec{B}$, which has the magnitude B, as

$$\vec{B} = B\,\vec{a}_y. \tag{A5.3}$$

Substituting (A5.2) and (A5.3) in (A5.1), we then have the expression for the force on the rod as

$$\vec{F} = \pm i\,l\,\vec{a}_x \times B\,\vec{a}_y = \pm i\,l\,B\,\vec{a}_x \times \vec{a}_y = \pm i\,l\,B\,\vec{a}_z \tag{A5.4}$$

where the explanation for the upper and the lower signs for the element of length vector as given following (A5.2) continues to hold good for the considered positive y direction of magnetic field, and hence that of magnetic flux density given by (A5.3). Thus, we can take the upper (plus) sign of (A5.4) if the direction of the current in the conducting rod is the positive x direction, which makes $\vec{l}$ have the positive x direction according to (A5.2), and hence the magnetic field $\vec{B}$ to have the positive y direction according to (A5.3). Further, we can easily appreciate that if we 'simultaneously' reverse both these directions of the current and the magnetic field, the upper (positive) sign of (A5.4) continues to hold good. On the other hand, we can take the lower (minus) sign of (A5.4) by changing either the direction of the current through the conducting rod or the direction of magnetic field, for instance, (i) if we take the current in the negative instead of the positive x direction, with the magnetic field continuing to be in the positive y direction or, (ii) if we take the magnetic field in the negative instead of the positive y direction, with the current continuing to be in the positive x direction.

If we take the upper sign of (A5.4), for instance, for (i) an appropriate polarity of the source of EMF that sends a current through the conducting rod in the positive x direction and (ii) the magnetic field linking with the circuit in the positive y direction, then the force on the conducting rod will be in the positive z direction as can be seen from (A5.2). Due to this force, the rod will move in the positive z direction, say, to an infinitesimal extent Δz in an infinitesimal time Δt, thereby causing an increase in the circuit area by an infinitesimal extent (Fig. A5.2)

$$\Delta S = l\,\Delta z. \tag{A5.5}$$

Consequently, due to this change in the circuit area, the flux of magnetic flux density $\vec{B}$ through the circuit will increase in an infinitesimal time Δt by an infinitesimal extent $\Delta\phi_B$ (see Section 3.4 for the definition of flux) given by

$$\Delta\phi_B = \vec{B}\cdot\vec{a}_n\Delta S. \tag{A5.6}$$

Now, if we substitute, in (A5.6), (i) $\vec{B} = B\,\vec{a}_y$ from (5.41) and (ii) $\vec{a}_n = \vec{a}_y$ (the unit vector in the direction of y) since the normal to the area ΔS has the y direction (Fig. A5.2), we can express the increase of magnetic flux through the circuit in time Δt as

$$\Delta\phi_B = B\,\vec{a}_y\cdot\vec{a}_y\,\Delta S = B\,\Delta S,$$

which, on putting $\Delta S = l\Delta z$ becomes

$$\Delta\phi_B = B\,l\,\Delta z. \tag{A5.7}$$

Next, let us find the element of work done ΔW by the source of EMF in maintaining the current i in time Δt. We can obtain ΔW by multiplying $\mathcal{E}$, the EMF of the source, by the element of charge ΔQ that is moved in time Δt corresponding to this current i as follows:

$$\Delta W = \mathcal{E}\,\Delta Q. \tag{A5.8}$$

ΔQ, in turn, is obtained by multiplying Δt by i, the time rate of the charge-flow as

$$\Delta Q = i\,\Delta t. \tag{A5.9}$$

Therefore, by substituting (A5.9) in (A5.8), we can obtain the element of work done by the source of EMF in maintaining the current i as

$$\Delta W = \mathcal{E}\, i\, \Delta t. \tag{A5.10}$$

A part $(\Delta W)_1$ of the element of work done ΔW, given by (A5.10), is used to overcome the resistance of the circuit R. This is interpreted as the resistive loss of energy and is obtained by multiplying the ohmic power loss $i^2 R$ (Joule's law)[4] by the element of time Δt. The remaining part $(\Delta W)_2$ of ΔW is spent in doing the work in moving the rod through the element of distance Δz. Thus, we get

$$\Delta W = (\Delta W)_1 + (\Delta W)_2,$$

which, in view of (A5.10) can be written as

$$\mathcal{E}\, i\, \Delta t = (\Delta W)_1 + (\Delta W)_2 \tag{A5.11}$$

where $(\Delta W)_1$ and $(\Delta W)_2$ are obtained as

$$(\Delta W)_1 = i^2 R\, \Delta t \tag{A5.12}$$

and

$$(\Delta W)_2 = \vec{F} \cdot \Delta z\, \vec{a}_z,$$

which, using (A5.4) with the upper sign on its right-hand side becomes

$$(\Delta W)_2 = i\, l\, B\, \vec{a}_z \cdot \Delta z\, \vec{a}_z = i\, l\, B\, \Delta z. \tag{A5.13}$$

Substituting (A5.10), (A5.12), and (A5.13) into (A5.11), we get then get

$$\mathcal{E}\, i\, \Delta t = i^2 R\, \Delta t + i\, l\, B\, \Delta z. \tag{A5.14}$$

Dividing (A5.14) by $i\, \Delta t$, we can simplify the former as

$$\mathcal{E} = iR + l\, B \frac{\Delta z}{\Delta t},$$

which, by the rearrangement of terms, becomes

$$i = \frac{\mathcal{E} - l\, B \frac{\Delta z}{\Delta t}}{R}. \tag{A5.15}$$

Further, dividing (A5.7) by Δt, we get the relation

$$\frac{\Delta \phi_B}{\Delta t} = lB \frac{\Delta z}{\Delta t}. \tag{A5.16}$$

Therefore, substituting (A5.16) in (A5.15), we can write

$$i = \frac{\mathcal{E} - \frac{\Delta \phi_B}{\Delta t}}{R},$$

[4]This can also be appreciated in Section 8.2.1 from field theory.

which, in the limit $\Delta t \to 0$ for the instantaneous current can be written as

$$i = \frac{\mathcal{E} - \dfrac{d\phi_B}{dt}}{R}. \tag{A5.17}$$

The expression (A5.17) is identical to (5.21) used in the text.

We have derived the expression for the circuit current (A5.17) here starting from the force equation (A5.4), with its upper (positive) sign corresponding to a typical situation considered with respect to the direction of the circuit current through the conducting rod and that of the magnetic field linking with the circuit as explained following (A5.4). Recall that we have taken the current through the conducting rod in the positive x direction and the magnetic field in the positive y direction.

Now, let us see what change, if any, would take place in the expression for the circuit current if we had taken in the formulation the force equation (A5.4) with its lower (negative) sign instead of its positive (plus) sign.

In the formulation with the lower sign of (A5.17), let us take, typically, the magnetic field reversed, that is, in the negative instead of the positive y direction, while taking the direction of the current through the conducting rod as unchanged, that is, in the positive x direction. Consequently, the expression for $(\Delta W)_2$ changes from (A5.13) to:

$$(\Delta W)_2 = -i\,l\,B\,\vec{a}_z \cdot \Delta z\,\vec{a}_z = -i\,l\,B\,\Delta z. \tag{A5.18}$$

However, the expression for $(\Delta W)_1$ continues to be that given by (A5.12). Therefore, substituting (A5.18) and (A5.12) in (A5.11), we obtain

$$\mathcal{E}\,i\,\Delta t = i^2 R\,\Delta t - i\,l\,B\,\Delta z. \tag{A5.19}$$

The rearrangement of terms of (A5.19) yields

$$i = \frac{\mathcal{E} + l\,B\frac{\Delta z}{\Delta t}}{R}, \tag{A5.20}$$

which is identical to (A5.15) except for a change in sign in the numerator. Substituting (A5.16) in (A5.20), we then obtain the circuit current expression as

$$i = \frac{\mathcal{E} + \frac{\Delta\phi_B}{\Delta t}}{R},$$

which in the limit $\Delta t \to 0$ for the instantaneous current becomes

$$i = \frac{\mathcal{E} + \frac{d\phi_B}{dt}}{R} \tag{A5.21}$$

which is identical to (A5.17) except for a change in the sign in the numerator.

Further, while deducing (A5.17) starting from the force expression (A5.4) with its upper (positive) sign, we have explained how, due to the magnetic field in the positive y direction, the conducting rod experiences a force in the positive z direction. Consequently, there is an increase in the area of the circuit and hence an increase in the magnetic flux. In view of this, we can express (A5.17) as

$$i = \frac{\mathcal{E} - \frac{d\phi_B}{dt}}{R},$$

which can be expressed as

$$i = \frac{\mathcal{E} - \left|\frac{d\phi_B}{dt}\right|}{R}, \tag{A5.22}$$

where $d\phi_B/dt$ represents the rate of increase in magnetic flux and $|d\phi_B/dt|$ represents the magnitude of the rate of change of magnetic flux with time. In view of this, we can also interpret expression (5.22) given in the text for the induced EMF (Section 5.2) as

$$\mathcal{E}_{\text{induced}} = -\frac{d\phi_B}{dt} = -\left|\frac{d\phi_B}{dt}\right|. \tag{A5.23}$$

Similarly, it is implied in the deduction of (A5.21), starting from the force expression (A5.4) with its lower (negative) sign, that, the conducting rod experiences a force in the negative z direction due to the magnetic field in the negative y direction, with a consequent decrease, instead of an increase, in the area of the circuit and hence, with a decrease in the magnetic flux. In view of this, $d\phi_B/dt$ represents here the rate of decrease in magnetic flux and, therefore, we can put $d\phi_B/dt = -\,|d\phi_B/dt|$ in (A5.21) to write the latter as

$$i = \frac{\mathcal{E} - \left|\frac{d\phi_B}{dt}\right|}{R} \tag{A5.24}$$

which is identical to (A5.22). Further, in view of (A5.24), the expression for the induced EMF (5.22) in the text (Section 5.2) can be interpreted as

$$\mathcal{E}_{\text{induced}} = -\frac{d\phi_B}{dt} = -\left|\frac{d\phi_B}{dt}\right| \tag{A5.25}$$

which is identical to (A5.23).

Thus, it is evident from (A5.22)-(A5.25) that, irrespective of whether the upper or the lower sign is taken in the starting expression (A5.4) for the force on the conducting rod in the formulation—with consequent increase or decrease of magnetic flux linking with the circuit with time—we deduce one and the same form of the circuit current and induced EMF expressions:

$$\left.\begin{aligned} i &= \frac{\mathcal{E} - \left|\frac{d\phi_B}{dt}\right|}{R} \\ \mathcal{E}_{\text{induced}} &= -\left|\frac{d\phi_B}{dt}\right| \end{aligned}\right\} \tag{A5.26}$$

in terms of the magnitude of the rate of change of magnetic flux with time $|d\phi_B/dt|$.

Appendix A5.3

Phasor representation of a time periodic quantity

We can represent a quantity say, E that varies periodically with time t according to sinusoidal or cosinusoidal function: $E = E_0 \sin\omega t$ or $E = E_0 \cos\omega t$, where E_o is the amplitude of the quantity; $\omega\,(= 2\pi/T = 2\pi f)$ is the angular frequency of variation, T and f being the time period and circular frequency of variation of the quantity. We can also represent the quantity E by the projections of a straight line representing the amplitude E_0 rotating anticlockwise with a uniform angular velocity ω on two perpendicular reference axes such that the reference $t = 0$ as well as $\omega t = 0$ is reckoned from one of these axes (Fig. A5.3*a*). Such a length changes its direction as it rotates and is called a rotating vector or a phasor. In such a phasor frame of reference, the instantaneous value of the quantity E at a time t can be represented as

$$E = E_0 \exp(j\omega t) \tag{A5.27}$$

where ωt can be called as the instantaneous phase of E at time t. We can expand $\exp(j\omega t)$ in terms of its real and imaginary parts in (A5.27) to express the latter as

$$E = E_0 \exp(j\omega t) = E_0(\cos\omega t + j\sin\omega t) = E_0\cos\omega t + jE_0\sin\omega t. \tag{A5.28}$$

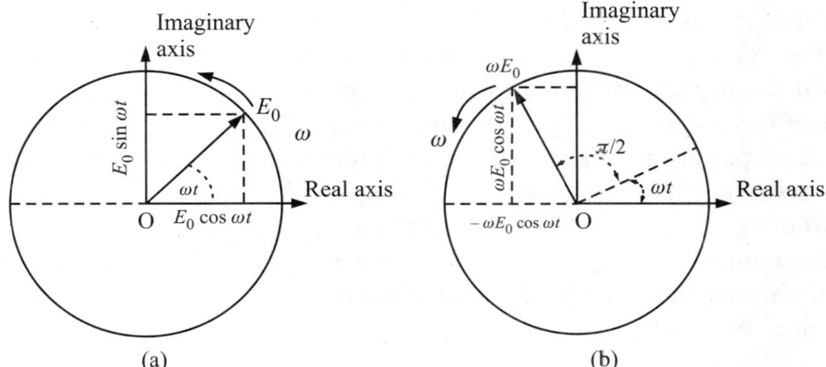

Fig. A5.3 *Phasor diagrams of (a) a time periodic quantity E, and (b) that of its time derivative dE/dt.*

In the phasor diagram, the instantaneous angle at time t traversed by the rotating vector of length E_0 (representing the amplitude of E) is ωt, measured from one of the two perpendicular reference axes to which we assign the reference $\omega t = 0$ (Fig. A5.3*a*). The projection of E_0 on this reference axis, as can be seen from the geometry of the phasor diagram, is clearly $E_0 \cos\omega t$. Further, this projection $E_0 \cos\omega t$ on the chosen reference axis is the real part of E given by (A5.28). This reference axis is also referred to as the real axis of the phasor diagram. Similarly, the projection of E_0 on the reference axis that is perpendicular to the real axis is $E_0 \cos\omega t$ (Fig. A5.3*a*), which can be identified as the imaginary part of E given by (A5.28). This reference axis is referred to as the imaginary axis of the phasor diagram.

The phasor representation of a time-periodic quantity, here E, can also be extended to its time derivative dE/dt, which we can write differentiating (A5.27) with t as:

$$\frac{dE}{dt} = j\omega\, E_0 \exp(j\omega t). \tag{A5.29}$$

Expanding $\exp(j\omega t)$ in terms of its real and imaginary parts in (A5.29), we can also write the latter as

$$\begin{aligned}\frac{dE}{dt} &= j\omega\, E_0 \exp(j\omega t) = j\omega\, E_0(\cos\omega\, t + j \sin\omega\, t)\\ &= -\omega\, E_0 \sin\omega\, t + j\omega\, E_0 \cos\omega\, t.\end{aligned} \tag{A5.30}$$

Further, if we substitute $\exp(j\pi/2)$ $[= \cos\pi/2 + j\sin\pi/2 = 0 + (j)(1)] = j$ in (A5.29), we obtain

$$\frac{dE}{dt} = \exp(j\pi/2)\omega\, E_0 \exp(j\omega t) = \omega\, E_0 \exp j(\omega\, t + \pi/2). \tag{A5.31}$$

In the phasor diagram (Fig. A5.3*b*), the length of the rotating vector representing dE/dt is $\omega\, E_0$, while the instantaneous phase of E is $\omega\, t$; the phase of dE/dt is $\omega\, t + \pi/2$. Thus, dE/dt leads E in phase by the angle $\pi/2$. Thus, the factor j as it appears in (A5.29) represents a phase lead of $\pi/2$ as indicated in (A5.31). Therefore, from the geometry of the phasor diagram (Fig. A5.3*b*), we can easily appreciate that the angle made by the instantaneous phasor with the negative direction of the real axis is $\pi - (\omega\, t + \pi/2) = \pi/2 - \omega\, t$. The projection of the instantaneous phasor representing dE/dt falls on the negative real axis and, therefore, putting the negative sign, the projection of the phasor on the real axis becomes $-\omega\, E_0 \cos(\pi/2 - \omega\, t) = -\omega\, E_0 \sin\omega\, t$ (Fig. A5.3*b*). This can be further identified as the real part of dE/dt, given by (A5.30). Furthermore, the angle between the instantaneous phasor representing dE/dt and the imaginary axis is $\omega\, t$. Therefore, the projection of the phasor with the imaginary axis is $\omega\, E_0 \cos\omega\, t$, which can be identified as the imaginary part of dE/dt, given by (A5.30).

Appendix A5.4

Phasor diagram depicting current density and electric field in an imperfect dielectric to find the loss tangent of the dielectric

We have already discussed how to represent a time periodic quantity and its derivative in phasor diagrams in Appendix A5.3. Let us now see how to represent the electric field and the current density in an imperfect dielectric ($\sigma \neq 0$) in a phasor diagram. For this purpose, let us recall the expression for the current through an imperfect dielectric given by (5.65) and rewrite it as follows:

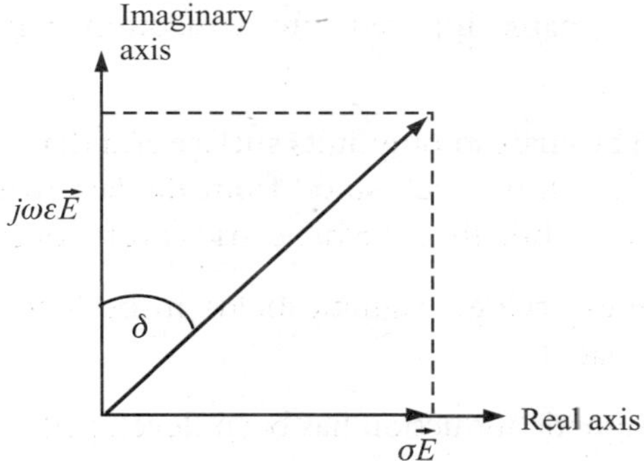

Fig. A5.4 *Phasor diagram for the electric field comprising conduction current density, displacement current density and the resultant of these two current densities, showing also the angle δ of the loss tangent tan δ.*

$$\vec{J} = \sigma\,\vec{E} + j\omega\,\varepsilon\,\vec{E}. \tag{A5.32}$$

The conduction current density $\sigma\,\vec{E}$ that appears in the first term of the right-hand side of (A5.32) is in phase with the electric field $\vec{E}$; these two quantities can be represented on the real axis of the phasor diagram (Fig. A5.4). As explained following (A5.31), the factor j, which appears in the second term of the right-hand side of (A5.32) in the displacement current density $j\omega\varepsilon\,\vec{E}$, represents a phase lead of $\pi/2$ over the electric field, and therefore, the displacement current density can be represented on the imaginary axis of the phasor diagram (Fig. A5.3). Thus, in the absence of conduction current density ($\sigma = 0$), the current density contributed only by the displacement current density, leads the electric field in phase by $\pi/2$. In the presence of the conduction current density ($\sigma \neq 0$), let the resultant electric current density, contributed by both the conduction and the displacement parts, make an angle δ with the displacement current density (Fig. A5.4). Thus, due to the finite conductivity of the dielectric ($\sigma \neq 0$), the resultant current density fails to have a phase lead of $\pi/2$ over the electric field by phase angle δ. From the geometry of the phasor diagram (Fig. A5.4), we can clearly see the following expression for $\tan\delta$, the loss tangent of the dielectric:

$$\tan\delta = \frac{\sigma\,E}{\omega\varepsilon\,E} = \frac{\sigma}{\omega\,\varepsilon}. \tag{A5.33}$$

The expression (A5.33) is found to be identical to (5.70).

SUMMARISING NOTES

- ✓ Continuity equation at a point relating the divergence of the current density to the time rate of the variation of volume charge density at the point has been derived starting from the conservation of charge in a region of charge flow.
- ✓ Continuity equation has been used to find the relaxation time of a medium, which

measures how long a charge injected into a medium will stay in the bulk of the medium.

- ✓ That a conductor can be charged only at its surface and that a dielectric can be charged throughout its volume can be understood from the fact that the relaxation time of a conductor is very short, while that of a dielectric is very long.
- ✓ Faraday's law for time-varying magnetic fields in both its integral and differential forms has been formulated.
- ✓ Concept of electromagnetic induction has been developed.
- ✓ That electric and magnetic fields are coupled in time-varying phenomena has been appreciated.
- ✓ Concept of displacement current has been developed with the help of the continuity equation, Poisson's equation and differential form of Ampere's circuital law.
- ✓ Dielectric medium, even though it is non-conducting, allows the displacement current to pass through it in a time-varying electric field.
- ✓ Capacitor, though it blocks a direct current, allows a time-varying current to pass through it in the form of displacement current.
- ✓ Loss tangent of a lossy dielectric in terms its conductivity and permittivity and the frequency of a time-periodic electric field has been derived.
- ✓ If the region between the conductors of a capacitor is filled up with a lossy dielectric, then the loss tangent of the dielectric measures the extent to which the current through the capacitor fails to lead the voltage across it by the phase angle of $\pi/2$.
- ✓ Maxwell's equations have been obtained in both integral and differential forms.
- ✓ Magnetic vector potential has been introduced defining it such that its curl represents the magnetic flux density.
- ✓ One of Maxwell's equations which relates the electric field with the time rate of variation of magnetic flux density as well as the relation between the magnetic flux density and vector potential has been used to find a relation between the electric field and the electric potential for a time-varying situation.
- ✓ Lorentz condition in terms of the time rate of scalar electric potential and the divergence of magnetic vector potential has been used to derive the wave equations in electric scalar and magnetic vector potentials.
- ✓ Solutions to the wave equations in electric scalar and magnetic vector potentials have been obtained keeping in view their potential applications (for instance, in the analysis of conduction current antennas).

Review Questions

Narrative type

5.1 How can you interpret the principle of conservation of charge in terms of the divergence of current density for a dc current?

5.2 Derive, from first principles, the continuity equation in terms of the divergence of current density and the time rate of change of volume charge density.

5.3 Write down the continuity equation in the curvilinear system of coordinates and interpret it for the rectangular, cylindrical and spherical polar coordinates.

5.4 Derive an expression for the relaxation time of a medium in terms of the medium conductivity and permittivity with the help of the continuity equation, Ohm's law and Poisson's equation. Explain the physical significance of relaxation time.

5.5 It is found that the relaxation time of copper is very small ($\sim 1.5 \times 10^{-19}$ s) and that of fused quartz is very high ($\sim$ 50 days). What does this finding physically signify?

5.6 How can you, by an appropriate physical reasoning, establish one of the forms of Faraday's law that states that the emf induced on a circuit is equal to the magnitude of the negative time rate of magnetic flux linked with the circuit?

5.7 What is Lenz's law? How can you explain the law with reference to a circuit that consists of a straight conducting rod which can freely slide on a pair of parallel conducting rails and which is immersed in a uniform magnetic field?

Hint: See the discussion following (5.22) and also Appendix A5.2.

5.8 Derive an expression for the voltage developed across a straight conducting rod of a given length that moves parallel to its length with a given velocity in a region of uniform magnetic field of a given flux density perpendicular to the plane of motion of the rod.

5.9 Give the theory of the Faraday disc generator that leads to an expression for the amount of voltage developed between the rim and the axle of the disc.

5.10 Deduce the integral form of Faraday's law starting from one of its forms that states that the emf induced on a circuit is equal to the magnitude of the negative time rate of magnetic flux linked with the circuit.

5.11 Develop the concept of displacement current density emerging out as the time rate change of electric displacement while deriving the differential or point from of Ampere's circuital law in order to make the latter consistent with the continuity equation.

5.12 A capacitor, in which a non-conducting dielectric constitutes the medium between its conducting parts, blocks a direct current flow through it and allows an alternating current to pass through it. Explain the phenomenon.

5.13 Verify that the magnitude of displacement current density in a parallel-plate capacitor connected to a time-varying voltage source is equal to the time rate of change of the magnitude of electric displacement in the capacitor.

5.14 Explain the physical significance of the loss tangent of a lossy dielectric medium.

5.15 Derive an expression for the loss tangent of a medium in terms of its conductivity and permittivity and operating frequency.

5.16 Explain with the help of relevant Maxwell's equations that a time-varying electric field is associated with a magnetic field and that a time-varying magnetic field is associated with an electric field.

5.17 Prove the differential form of Maxwell's equation $\nabla \times \vec{E} = -\partial\vec{B}/\partial t$ from its integral form $\oint_l \vec{E} \cdot d\vec{l} = -\int_S \partial\vec{B}/\partial t \cdot \vec{a}_n \, dS$.

5.18 Show that, while for a time-independent situation the electric field is related only to the scalar electric potential, for a time-varying situation the electric field is related to both scalar electric potential and vector magnetic potential.

5.19 Establish the wave equation in magnetic vector potential with the help of (i) the expression for electric field in terms of the scalar and vector potentials and (ii) one of Maxwell's equations that involves the conduction and displacement current densities, subject to Lorentz condition.

5.20 Establish the wave equation in electric scalar potential with the help of (i) the expression for the electric field in terms of the scalar and vector potentials and (ii) one of Maxwell's equations that involves the volume charge density, subject to Lorentz condition.

5.21 Solve the wave equations in electric scalar and magnetic vector potentials respectively and, hence, interpret the solutions of these equations to bring out the concept of 'retarded' scalar and vector potentials.

5.22 Discuss the solutions to the wave equations in electric scalar and magnetic vector potentials for time-dependent situation for the special case of time-independent situation.

5.23 For a time-varying situation, express (i) the magnetic field in terms of the magnetic vector potential $\vec{A}$; (ii) the electric field in terms of the scalar and vector potentials V and $\vec{A}$; and (iii) the electric field in terms of only the vector potential $\vec{A}$.

Review questions with solutions

5.24 Write down the differential or point forms of Maxwell's equations vis-à-vis the corresponding integral forms from which they are derived.

Answer:

Integral form	Differential or point form of Maxwell's equations
$\oint_S \vec{D}\cdot d\vec{S} = \oint_S \vec{D}\cdot\vec{a}_n\, dS = \int_\tau \rho\, d\tau$	$\nabla\cdot\vec{D} = \rho$
$\oint_S \vec{B}\cdot d\vec{S} = \oint_S \vec{B}\cdot\vec{a}_n\, dS = 0$	$\nabla\cdot\vec{B} = 0$
$\oint_l \vec{E}\cdot d\vec{l} = -\int_S \frac{\partial\vec{B}}{\partial t}\cdot\vec{a}_n\, dS$	$\nabla\times\vec{E} = -\frac{\partial\vec{B}}{\partial t}$
$\oint_l \vec{H}\cdot d\vec{l} = \int_S \left(\vec{J} + \frac{\partial\vec{D}}{\partial t}\right)\cdot\vec{a}_n\, dS$	$\nabla\times\vec{H} = \vec{J} + \frac{\partial\vec{D}}{\partial t}$

5-25 A large conducting plate of thickness t and conductivity σ is immersed in a dc magnetic field of flux density B perpendicular to the plate. The plate is moved parallel to the plane of the plate with a uniform velocity v. Show that the force developed per unit area of the plate is $B^2\sigma t v$.

Answer: Let us consider, for the sake of analysis, the geometry of the conducting plate as rectangular and let the plate be parallel to XY plane such that the thickness of the plate is t (given) along z. We consider the magnetic field to be directed along z. Let the dimension of the plate along x be l and that along y be b, which makes the area of the plate

$$\alpha = lb. \tag{1}$$

If the plate is moved parallel to the XY plane, a current i, say, will flow through the thickness of the plate due to electronic motion, the Lorentz force on an electron carrying charge $-|e|$ being $-|e|\,\vec{v}\times\vec{B}$. Since $\vec{v}$ is directed along y and $\vec{B}$ is directed along z, this Lorentz force will be along negative x, causing the electronic current along negative x and the conventional current i along positive x. We can write the following expression for i obtained by dividing the induced voltage Blv (see Example E5-1) by the resistance R of the plate as follows:

$$i = \frac{Blv}{R}. \tag{2}$$

Further, for the considered dimensions and conductivity of the plate, the resistance R can be expressed as

$$R = \frac{1}{\sigma}\frac{l}{\alpha'}, \tag{3}$$

where α' is the cross-sectional area of the plate across the thickness of the plate, given by

$$\alpha' = bt. \tag{4}$$

Substituting (4) in (3), we obtain

$$R = \frac{1}{\sigma}\frac{l}{bt}. \tag{5}$$

Substituting (5) in (2), we can write

$$i = Blv\frac{\sigma bt}{l} = Bv\sigma bt. \tag{6}$$

We can next find the force on the plate with the help of (4.65) as

$$\vec{F} = il\vec{a}_x \times B\vec{a}_z = -ilB\vec{a}_y. \tag{7}$$

It follows from (7) that the force on the plate is along y, that is, it is in a direction opposite to the direction of the motion of the plate and that the magnitude of the force F is

$$F = ilB. \tag{8}$$

Substituting (6) in (8), we obtain

$$F = B^2 v\sigma btl. \tag{9}$$

Dividing (9) by (1), we obtain

$$\text{Force per unit area of the plate} = \frac{B^2 v\sigma btl}{lb} = B^2\sigma tv.$$

5.26 Appreciate that the direction of the force on the moving plate immersed in a magnetic field found in Question 5-25 is in conformity with Lenz's law.

Answer: We have seen in the preceding problem that the voltage induced on the conducting plate is such that the direction of the force due to the consequent current through the plate immersed in magnetic field opposes the direction of motion of the plate that is the cause of the induced voltage. The finding is thus in conformity with Lenz's law.

5.27 An electron of charge e is accelerated from a planar electrode at a lower potential to another planar electrode at a higher potential. Show that a pulse of current of magnitude $|e|\,v/d$ is induced in the electrodes, where v is the magnitude of the velocity of the electron in flight at a point between the electrodes and d is the inter-electrode distance. Also, appreciate that at the instant of the arrival of the electron at the electrode at a higher potential, the induced current becomes zero.

Answer: Let us take the two planar electrodes, A and B, at $z = 0$ and $z = d$. Let the potential of A be $V = 0$ and that of B be $V = V_0$. If d is the distance of separation between the electrodes, we can write the magnitude of the uniform electric field E in the region between A and B as

$$E = \frac{V_0}{d}. \tag{1}$$

The electron will be then accelerated by this electric field. Let at any instant of time t the electron in flight between the electrodes be at a distance z from A. Further, at this instant, let the charge induced on the electrode B be q', so that we can write the energy required to induce this charge as

$$\text{Energy required to induce charge} = q'V_0. \qquad (2)$$

We can also write the following expression for the work done (= force×distance) in bringing the electron up to a point $z \leq d$, measured from the electrode A:

$$\text{Work done} = |e|\, Ez, \qquad (3)$$

the force on the electron due to the electric field being $|e|\, E$.

Now, if we plug (1) in (2), we can write

$$\text{Work done} = |e|\, \frac{V_0}{d} z. \qquad (4)$$

We can now equate (2) with (3) to write

$$q'V_0 = |e|\, \frac{V_0}{d} z,$$

which simplifies to

$$q' = |e|\, \frac{z}{d}. \qquad (5)$$

At this stage, note that as the electron moves from the electrode A to B, the induced charge q' on B will increase while the induced charge, say q, will decrease such that

$$q + q' = |e|\,.$$

Now, differentiating (1), we can find the induced current $i$on the electrode B as

$$i = \frac{dq'}{dt} = \frac{|e|}{d}\frac{dz}{dt} = \frac{|e|\, v}{d}.$$

Further, at the instant of the arrival of the electron at the electrode B (at a higher potential), the induced charge is $q' = |e|$ [which can be appreciated by putting $z = d$ in (4)]. Then the induced charge $|e|$ and the electron charge $-\,|e|$ will neutralize each other at the electrode and the induced current will become null.

5.28 Express the loss tangent of a medium in terms of the relaxation time of the medium.

Answer: $\tan\delta = 1/\omega \mathrm{T}$, where T is the relaxation time. [Recall (5.70) and (5.19)].

5.29 Prove the integral form of Maxwell's equation $\oint_l \vec{E} \cdot d\vec{l} = -\int_S \partial\vec{B}/\partial t \cdot \vec{a}_n\, dS$ from its differential form $\nabla \times \vec{E} = -\partial\vec{B}/\partial t$ using Stokes' theorem of vector calculus.

Answer: Stokes' theorem can be stated as

$$\oint_l \vec{G} \cdot d\vec{l} = \int_S (\nabla \times \vec{G}) \cdot d\vec{S} = \int_S (\nabla \times \vec{G}) \cdot \vec{a}_n dS \tag{1}$$

where $\vec{G}$ is a vector quantity. For the vector field $\vec{G} = \vec{E}$, we can write from (1)

$$\oint_l \vec{E} \cdot d\vec{l} = \int_S (\nabla \times \vec{E}) \cdot d\vec{S} = \int_S (\nabla \times \vec{E}) \cdot \vec{a}_n dS. \tag{2}$$

If we substitute the point form of Maxwell's equation $\nabla \times \vec{E} = -\partial \vec{B}/\partial t$ in (2), we get

$$\oint_l \vec{E} \cdot d\vec{l} = \int_S (\nabla \times \vec{E}) \cdot \vec{a}_n dS = -\int_S \frac{\partial \vec{B}}{\partial t} \cdot \vec{a}_n dS,$$

which is the integral form of Maxwell's equation [identical to (5.32)].

5.30 It is given that a uniform plane wave propagating along z in free-space with wave dependence $\exp(j\omega t - \beta z)$, where $\beta = \omega(\mu_0 \varepsilon_0)^{1/2}$ is the wave phase propagation constant, forms a standing wave represented by the electric field $\vec{E} = -2jE_0 \sin \beta z \exp(j\omega t)\, \vec{a}_x$. From this electric field, find the expression for the magnetic field in terms of the free-space intrinsic impedance $\eta_0 = (\mu_0/\varepsilon_0)^{1/2}$ using the relevant Maxwell's equation.

Answer: The given electric field $\vec{E} = -2jE_0 \sin \beta z \exp(j\omega\, t)\, \vec{a}_x$, which, can be written component-wise as

$$\left.\begin{aligned} E_x &= -2jE_0 \sin \beta z \exp(j\omega t) \\ E_y &= E_z = 0 \end{aligned}\right\}. \quad \text{(electric field components)} \tag{1}$$

We can write Maxwell's equation (5.78) with the help of (4.2) as

$$\nabla \times \vec{E} = \begin{vmatrix} \vec{a}_x & \vec{a}_y & \vec{a}_z \\ \dfrac{\partial}{\partial x} & \dfrac{\partial}{\partial y} & \dfrac{\partial}{\partial z} \\ E_x & E_y & E_z \end{vmatrix} = -\mu_0 \frac{\partial H_x}{\partial t} \vec{a}_x - \mu_0 \frac{\partial H_y}{\partial t} \vec{a}_y - \mu_0 \frac{\partial H_z}{\partial t} \vec{a}_z. \tag{2}$$

Substituting (1) in (2) and remembering $\partial/\partial x = \partial/\partial y = 0$ for uniform plane wave, we can write

$$\begin{vmatrix} \vec{a}_x & \vec{a}_y & \vec{a}_z \\ 0 & 0 & \dfrac{\partial}{\partial z} \\ -2jE_0 \sin \beta z \exp(j\omega t) & 0 & 0 \end{vmatrix} = -\mu_0 \frac{\partial H_x}{\partial t} \vec{a}_x - \mu_0 \frac{\partial H_y}{\partial t} \vec{a}_y - \mu_0 \frac{\partial H_z}{\partial t} \vec{a}_z. \tag{3}$$

Expanding the determinant of the left side of (3), we obtain

$$\frac{\partial}{\partial z}[(-2jE_0 \sin\beta\, z \exp(j\omega t))]\vec{a}_y = -\mu_0 \frac{\partial H_x}{\partial t}\vec{a}_x - \mu_0 \frac{\partial H_y}{\partial t}\vec{a}_y - \mu_0 \frac{\partial H_z}{\partial t}\vec{a}_z,$$

which simplifies, after carrying out the differentiation and invoking the RF time dependence as $\exp(j\omega t)$ that enables us to put $\partial/\partial t = j\omega$, to the following expression:

$$-2jE_0\beta \cos\beta\, z \exp(j\omega t)\,\vec{a}_y = -j\omega\mu_0 H_x \vec{a}_x - j\omega\mu_0 H_y \vec{a}_y - j\omega\mu_0 H_z \vec{a}_z. \tag{4}$$

Comparing the left- and right-hand sides of (4) and making use of the relation $\beta = \omega(\mu_0\varepsilon_0)^{1/2}$, we obtain

$$\left.\begin{aligned} H_x &= H_z = 0 \\ H_y &= \frac{2E_0\beta \cos\beta\, z \exp(j\omega t)}{\omega\mu_0} \\ &= \frac{2E_0(\mu_0\varepsilon_0)^{1/2} \cos\beta\, z \exp(j\omega t)}{\mu_0} \\ &= \frac{2E_0 \cos\beta\, z \exp(j\omega t)}{(\mu_0/\varepsilon_0)^{1/2}} \end{aligned}\right\}. \tag{5}$$

Expression (5) can be written in terms of the free-space intrinsic impedance $\eta_0 = (\mu_0/\varepsilon_0)^{1/2}$ as

$$\left.\begin{aligned} H_x &= H_z = 0 \\ H_y &= \frac{2E_0 \cos\beta\, z \exp(j\omega t)}{\eta_0} \end{aligned}\right\},$$

which in vectorial form becomes

$$\vec{H}(= H_y\vec{a}_y) = \frac{2E_0}{\eta_0}\cos\beta\, z \exp(j\omega t)\,\vec{a}_y.$$

Numerical type

5.31 Calculate the relaxation time of corn oil ($\sigma = 5 \times 10^{-4}$ mho/m and $\varepsilon_r = 3.1$).

Answer: 0.55 s

5.32 The value of loss tangent of bakelite ($\varepsilon_r = 4.74$, $\sigma = 10^{-9}$ mho/m) is given as $\tan\delta = 0.022$. What frequency does this value of $\tan\delta$ apply to?

Answer: 33.68 kHz

5.33 At what frequency can Earth ($\varepsilon_r = 8$, $\sigma = 5 \times 10^{-3}$ mho/m) be treated as a perfect dielectric?

Answer: Earth may be considered a perfect dielectric for $\sigma/\omega\varepsilon \sim 10^{-2}$, typically, which would correspond to $\sigma << \omega\varepsilon$ or $\omega\varepsilon >> \sigma$. Therefore, the frequency above which it can be treated as a perfect dielectric can be set as:

$$f = 100 \times \frac{\sigma}{2\pi\varepsilon} = 100 \times \frac{\sigma}{2\pi\varepsilon_0\varepsilon_r}$$

$$= 100 \times \frac{5 \times 10^{-3}}{2 \times 3.143 \times 8.854 \times 10^{-12} \times 8} = 1.12 \times 10^9 \text{ Hz} = 1.12 \text{ GHz}.$$

5.34 A circular loop of radius 0.15 m and resistance 10 ohm is immersed in a magnetic field of uniform flux density of magnitude $B = 0.25 \sin 10^3 t\vec{a}_z$T perpendicular to the plane of the loop. Find the current through the loop.

Answer: With the help of (5.22) and (5.30), we can get the magnitude of the induced emf as

$$|\text{induced emf}| = \int_S \frac{\partial}{\partial t}\vec{B} \cdot \vec{a}_n dS = \int_S \frac{\partial}{\partial t} B\vec{a}_z \cdot \vec{a}_z dS = \frac{\partial B}{\partial t} \int_S dS = \frac{\partial B}{\partial t} S, \qquad (1)$$

where S is the area of the loop.

In this problem,

$$\left.\begin{aligned} B &= 0.25 \sin 10^3 t \\ r &= 0.15 \end{aligned}\right\} . \text{ (given)} \qquad (2)$$

From (2) we can obtain

$$\left.\begin{aligned} \frac{\partial B}{\partial t} &= 0.25 \times 10^3 \cos 10^3 t \\ S &= \pi r^2 = \pi(0.15)^2 \end{aligned}\right\} . \qquad (3)$$

Substituting (3) in (1), we get the induced emf as

$$|\text{induced emf}| = \frac{\partial B}{\partial t} S = 0.25 \times 10^3 \cos 10^3 t \times \pi(0.15)^2 \text{V}. \qquad (4)$$

Since the resistance of the loop is 20 ohm (given), dividing (4) by 20, we obtain the current I through the loop as

$$I = \frac{0.25 \times 10^3 \times 3.143 \times 0.0225 \cos 10^3 t}{10} = 1.77 \cos 10^3 t \text{ A}.$$

5.35 A square loop lying on the plane $z = 0$ has its sides located at $x = \pm 0.5$ m and $y = \pm 0.5$ m. If the loop is placed in a region of magnetic flux density $\vec{B} = (0.3\vec{a}_x + 0.4\vec{a}_y + 0.5\vec{a}_z) \cos 50t$, what will be the current through the loop if the loop resistance is 5 kΩ?

Answer: In this problem, $\vec{B}$ is given as $\vec{B} = (0.3\vec{a}_x + 0.4\vec{a}_y + 0.5\vec{a}_z) \cos 50t$. This can be put in (7.29), in which we take $\vec{a}_n = \vec{a}_z$ since the loop is lying on the plane

$z = 0$, to write the following expression for the magnetic flux density through the loop:

$$\phi_B = \int_S \vec{B} \cdot \vec{a}_n \, dS = \int_S (0.3\vec{a}_x + 0.4\vec{a}_y + 0.5\vec{a}_z) \cdot \vec{a}_z \, dS \cos 50t,$$

which, putting $dS = dxdy$ becomes

$$\phi_B = \int_{x=-0.5}^{x=0.5} \int_{y=-0.5}^{y=0.5} (0.3\vec{a}_x + 0.4\vec{a}_y + 0.5\vec{a}_z) \cdot \vec{a}_z dxdy \cos 50t. \tag{1}$$

In view of the relations: $\vec{a}_x \cdot \vec{a}_z = \vec{a}_y \cdot \vec{a}_z = 0$; $\vec{a}_z \cdot \vec{a}_z = 1$, we can evaluate (1) as

$$\phi_B = 0.5 \cos 50t \int_{x=-0.5}^{x=0.5} \int_{y=-0.5}^{y=0.5} dxdy = 0.5 \cos 50t. \tag{2}$$

Taking the time derivative of (2), we get

$$\frac{d\phi_B}{dt} = -0.5 \times 50 \sin 50t = -25 \sin 50t. \tag{3}$$

Therefore, with the help of (3) and the expression for the induced emf given by (5.22), we obtain the following expression for the magnitude of induced emf:

$$|\text{induced emf}| = \left| \frac{d\phi_B}{dt} \right| = 25 \sin 50t. \tag{4}$$

Dividing (4) by the resistance 5 kΩ $= 5 \times 10^3 \Omega$, we get the current I through the loop as

$$I = \frac{25 \sin 50t}{5 \times 10^3} = 5 \times 10^{-3} \sin 50t \text{ A} = 5 \sin 50t \text{ mA}.$$

5.36 A straight conducting rod makes an oscillatory motion with a velocity $5 \cos 10^3 t$ m/s parallel to its length of 1 m over a plane perpendicular to a uniform magnetic field of flux density of 0.12 T. Calculate the emf developed across the length of the rod.

Answer: $0.6 \cos 10^3 t$ V (**Hint**: See Example E5-1.)

5.37 A coaxial cable consists of an inner conductor of radius 3 mm and an outer conductor of inner radius 6 mm with a dielectric filler of relative permittivity 2 between the conductors. The cable is connected to a voltage of $100 \sin 100t$. Calculate the capacitance and hence the radial displacement current over a length of 100 mm of the cable.

Answer: The capacitance C of the coaxial cable of length l can be written with the help of (E3-9.8) (see Example E3-9) as

$$C = \frac{2\pi\varepsilon_0\varepsilon_r}{\ln \frac{b}{a}} l, \tag{1}$$

where a is the radius of the inner conductor, b is the inner radius of the outer conductor , and ε_r is the relative permittivity of the medium between the conductors. Substituting the given data: $a = 3$ mm $= 3 \times 10^{-3}$ m, $b = 6$ mm $= 6 \times 10^{-3}$ m, $\varepsilon_r = 2$ and $l = 100$ mm$= 100 \times 10^{-3}$ m in (2), we can calculate

$$\begin{aligned} C &= \frac{2\pi\varepsilon_0\varepsilon_r l}{\ln \frac{b}{a}} = \frac{2 \times 3.143 \times 8.854 \times 10^{-12} \times 2 \times 100 \times 10^{-3}}{\ln \frac{6\times10^{-3}}{3\times10^{-3}}} \\ &= \frac{2 \times 3.143 \times 8.854 \times 10^{-12} \times 2 \times 100 \times 10^{-3}}{\ln 2} \qquad (2) \\ &= \frac{2 \times 3.143 \times 8.854 \times 10^{-12} \times 2 \times 10^{2} \times 10^{-3}}{0.693} \\ &= 10.66 \times 10^{-12}\,\text{F} = 10.66\ \text{pF} \end{aligned}$$

Since the cable is connected to a voltage $V = 100 \sin 100t$, we can find the current (displacement current) I through the capacitor using the value of the capacitance given by (2) as

$$\begin{aligned} I = C\frac{dV}{dt} &= 10.66 \times 10^{-12} \times \frac{d}{dt}(100 \sin 100t) = 10.66 \times 10^{-12} \times 100 \cos 100t\ \text{A} \\ &= 1.066 \times 10^{-9} \cos 100t\text{A} = 1.066 \cos 100t\ \text{nA} \end{aligned}$$

Multiple-choice questions

5.38 If the current density in a medium fails to lead the electric field by an angle $\theta = \pi/2$ by an amount greater than 0 but less than $\pi/2$, then the medium is a

(a) good conductor (b) bad conductor
(c) perfect dielectric (d) lossy dielectric.

5.39 Which one of the following expressions is incorrect?

(a) $\nabla \cdot \vec{E} = \rho$ (b) $\nabla \cdot \vec{B} = 0$
(c) $\nabla \times \vec{E} = -\frac{\partial \vec{B}}{\partial t}$ (d) $\oint_l \vec{H} \cdot d\vec{l} = \int_S \left(\vec{J} + \frac{\partial \vec{D}}{\partial t}\right) \cdot \vec{a}_n\, dS$

5.40 Which one of the following units of the magnetic vector potential is incorrect?

(a) Tm (b) Vs/m
(c) HA/m (d) HA

5.41 The unit of the expression $\oint_S \vec{D} \cdot d\vec{S} = \oint_S \vec{D} \cdot \vec{a}_n\, dS$ is

(a) C (b) C/m²
(c) V (d) V/m

5.42 $\nabla \cdot \vec{B} = 0$ means

(a) magnetic field vector is always irrotational and magnetic field lines originate from the north pole of a magnet and terminate on its south pole.
(b) magnetic field vector is always solenoidal and magnetic field lines originate from the north pole of a magnet and terminate on its south pole.
(c) magnetic field vector is always solenoidal and magnetic field lines are continuous.
(d) magnetic field vector is always irrotational and magnetic field lines are continuous.

5.43 Which one of the following expressions is incorrect?

(a) $\nabla \cdot \vec{A} = -\mu_0 \varepsilon_0 \frac{\partial V}{\partial t}$ (b) $\nabla^2 \vec{A} - \mu_0 \varepsilon_0 \frac{\partial^2 \vec{A}}{\partial t^2} = \mu_0 \vec{J}$

(c) $\vec{H} = \frac{\nabla \times \vec{A}}{\mu_0}$ (d) $\nabla \times \vec{B} = \mu_0 \vec{J} + \mu_0 \frac{\partial \vec{D}}{\partial t}$.

5.44 A conducting loop of area 0.5 m^2 is placed in a uniform magnetic field of flux density $2 \cos 100\pi t$ in a direction perpendicular to plane of the loop. The emf induced on the loop is

(a) $50\pi \sin 100\pi t$ V (b) $100\pi \sin 100\pi t$ V

(c) $150\pi \sin 100\pi t$ Wb/s (d) $200\pi \sin 100\pi t$ Wb/s. A

Answers

5.38 (d) 5.39 (a)
5.40 (d) (Recall that tesla or T is the unit of magnetic flux density and henry or H is the unit of inductance)
5.41 (a) 5.42 (c) 5.43 (b) 5.44 (b)

6

Wave Equation and Its Solution for a Wave Propagating through an Unbounded Medium

THE wave equations for electric and magnetic fields of an electromagnetic wave propagating through a medium are obtained by decoupling quantities that appear coupled in two of Maxwell's equations—one referring to the electric field due to the time-varying magnetic field, and the other to the magnetic field due to the time-varying electric field. Since we are going to deal with a wave and its propagation through a medium, it is worth recapitulating how a quantity associated with a wave could be represented in analytical problems (Section 6.1). The concepts developed in this chapter pertain to mediums unbounded in space: first we take a free-space medium (Section 6.2), and then a conducting medium (Section 6.3). Further, it will be seen that the concepts of the propagation of electromagnetic waves through a medium of finite resistivity lead us to concepts such as skin depth of a medium, surface resistance, and ac or RF resistance that are frequency dependent (Section 6.4).

6.1 Representation of a Quantity Associated with a Wave

Let us take a function $f_+(t, z)$ of the space coordinate z and the time coordinate t to represent a quantity associated with a 'forward wave' that propagates along the positive z with the progress of time t as:

$$f_+(t, z) = f_+\left(t - \frac{z}{v}\right) \tag{6.1}$$

where the subscript '+' stands for the forward wave. v is a quantity independent of z and t has the dimension of velocity, the physical meaning of which will become clear as we proceed further with this function. We may now interpret function (6.1) to represent the quantity associated with the forward wave when t and z each change by infinitesimal extents, Δt and Δz, to values $t + \Delta t$ and $z + \Delta z$ respectively, as follows:

$$f_+(t+\Delta t, z+\Delta z) = f_+\left(t+\Delta t - \frac{z+\Delta z}{v}\right) = f_+\left(t - \frac{z}{v} + \Delta t - \frac{\Delta z}{v}\right). \tag{6.2}$$

When subjected to the following condition

$$\Delta t - \frac{\Delta z}{v} = 0 \tag{6.3}$$

we get from (6.2)

$$f_+(t+\Delta t, z+\Delta z) = f_+\left(t - \frac{z}{v}\right)$$

which, in view of (6.1) becomes

$$f_+(t+\Delta t, z+\Delta z) = f_+(t, z). \tag{6.4}$$

The finding in (6.4) suggests that the function $f_+(t, z)$ representing a propagating wave at (t, z) is also realised at an infinitesimal time interval Δt later at an infinitesimal distance Δz forward along the positive z, that is, at $(t+\Delta t, z+\Delta z)$, provided condition (6.3) holds good. The latter is expressible as:

$$v = \frac{\Delta z}{\Delta t}$$

which, in the limit $\Delta t \to 0$ becomes

$$v = \underset{\Delta t \to 0}{Lt} \frac{\Delta z}{\Delta t} = \frac{dz}{dt}. \tag{6.5}$$

Thus the quantity v that appears in the function $f_+(t, z)$ given by (6.1) can be now identified as the velocity of propagation of the wave.

Further, you will appreciate that while the function $f_+(t, z) = f_+(t - z/v)$ given by (6.1) represents a quantity associated with a forward wave propagating along the positive z with a velocity v, the function

$$f_-(t, z) = f_-\left(t + \frac{z}{v}\right) \tag{6.6}$$

represents a quantity associated with a 'backward wave' propagating along the negative z with velocity v. The subscript '–' here stands for the backward wave. Using (6.6), we can then write the expression for the function at a time interval Δt later and at a distance Δz along the negative z, that is, at $(t+\Delta t, z-\Delta z)$ as

$$f_-(t+\Delta t, z-\Delta z) = f_-\left(t+\Delta t + \frac{z-\Delta z}{v}\right) = f_-\left(t + \frac{z}{v} + \Delta t - \frac{\Delta z}{v}\right). \tag{6.7}$$

In view of (6.6), we obtain from (6.7) the following relation for a wave propagating along the negative z:

$$f_{-}(t+\Delta t, z-\Delta z) = f_{-}\left(t+\frac{z}{v}\right) = f_{-}(t, z) \tag{6.8}$$

subject to the same condition as (6.3), which, in the limit $\Delta t \to 0$, gives the velocity of the wave as v given by (6.5). However, it is now with the interpretation that v represents the velocity of the wave along the negative z instead of the positive z. Thus, it becomes evident from (6.8) that the function $f_{-}(t, z)$ that is chosen to represent a backward wave at (t, z) is also realised at an infinitesimal time interval Δt later at an infinitesimal distance Δz along negative z, that is, at $(t+\Delta t, z-\Delta z)$.

However, we have yet to include the property of the periodicity of the wave in the functions $f_{+}(t, z) = f_{+}(t - z/v)$ and $f_{-}(t, z) = f(t + z/v)$ given by (6.1) and (6.6) respectively. Putting

$$t' = t - z/v \tag{6.9}$$

$$t'' = t + z/v \tag{6.10}$$

we can express the wave functions (6.1) and (6.6) as

$$f_{+}(t, z) = f_{+}(t') \tag{6.11}$$

$$f_{-}(t, z) = f_{-}(t''). \tag{6.12}$$

Now, let us consider the forward and the backward waves associated with a physical quantity such as electric field, volume charge density or velocity of a particle. With the help of (6.9) and (6.10) and using the phasor notation (See Section 5.3 and Appendix A5.3 in Chapter 5) for a time periodic quantity, we can represent the amplitude of electric field E_{+} associated with a forward wave and that associated with a backward wave E_{-} as

$$E_{+} = E_0 \exp(j\omega t') \tag{6.13}$$

$$E_{-} = E_0 \exp(j\omega t'') \tag{6.14}$$

where E_0 is the amplitude of the electric field and ω is the angular frequency of the wave. $\omega t'$ and $\omega t''$ are the phases of the forward and the backward waves respectively. Substituting (6.9) and (6.10) in (6.13) and (6.14) respectively, we obtain

$$E_{+} = E_0 \exp[j\omega (t - z/v)] \tag{6.15}$$

$$E_{-} = E_0 \exp[j\omega (t + z/v)] \tag{6.16}$$

which may also be expressed as

$$E_{+} = E_0 \exp[j(\omega t - \beta z)] \tag{6.17}$$

$$E_{-} = E_0 \exp[j(\omega t + \beta z)] \tag{6.18}$$

where $\omega t - \beta z \,(= \omega t')$ and $\omega t + \beta z \,(= \omega t'')$ can be identified as the phases of the wave with

$$\beta = \frac{\omega}{v} \tag{6.19}$$

being the phase propagation constant of the wave. The velocity v may be identified as the phase velocity of the wave, defined as the velocity with which the constant phase of the wave moves. This is because if we put, say, for a forward wave the phase

$$\omega t - \beta z = \text{constant}$$

and differentiate it with respect to t, we obtain $dz/dt = \omega/\beta$, the velocity of the constant phase of the wave, v_{ph}. The phase velocity of the wave, v_{ph}, is seen to be the same as v in (6.19). Thus, we obtain the relation $v_{ph} = \omega/\beta$.

Further, recalling the concept of phasor representation of a time periodic quantity (refer to Appendix A5.3 of Chapter 5), we can see that the real parts of (6.17) and (6.18) can be used to represent the electric fields associated with the forward and the backward waves respectively. For instance, if we take the forward wave at a fixed time, say, $t = 0$, the electric field quantity will be given by $E_{+} = E_0 \cos(\omega t - \beta z) = E_0 \cos(-\beta z) = E_0 \cos \beta z$, the expression giving thus the snapshot of the variation of the electric field with distance z. According to this variation, the electric field will take on the positive maximum value $E_{+} = E_0$ at $\beta z = 0,\ 2\pi,\ 4\pi,\ 6\pi,\ \ldots$, with two consecutive values of βz, being separated by a normalised (dimensionless) distance interval of $\beta |z|_{\text{interval}} = 2\pi$. The positive maximum value of the electric field represents a typical phase of the variation of electric field. The distance interval $|z|_{\text{interval}}$ between two consecutive points that have the same phase at a fixed time is the wavelength λ of the wave—here, typically, the phase that corresponds to the maximum value of the electric field quantity is considered. Thus, from the relation $\beta |z|_{\text{interval}} = 2\pi$, we obtain $\lambda = |z|_{\text{interval}} = 2\pi/\beta$. Similarly, considering again the forward wave, the electric field quantity at a given point at a fixed distance, say, at $z = 0$, will be given by $E_{+} = E_0 \cos(\omega t - \beta z) = E_0 \cos \omega t$, the value of which will take on the maximum value $E_{+} = E_0$ at $\omega t = 0,\ 2\pi,\ 4\pi,\ 6\pi,\ \ldots$, with two consecutive values of ωt separated by a normalised (dimensionless) time interval of $\omega |t|_{\text{interval}} = 2\pi$. From this, we obtain $T = |t|_{\text{interval}} = 2\pi/\omega$, the time interval between two consecutive phases at a fixed point, T called the time period of the wave. From this expression for T, we have the relation between the frequency at which a particular phase repeats in time, called the circular frequency f of the wave, and the angular frequency ω of the wave as $f = 1/T = \omega/2\pi$ or $\omega = 2\pi f$. Further, we can easily combine the relations $\lambda \,(= |z|_{\text{interval}}) = 2\pi/\beta$, $f(= 1/T) = \omega/2\pi$ and $v_{ph} = \omega/\beta$ to obtain the relation

$$v_{ph} = \omega/\beta = \frac{2\pi f}{2\pi/\lambda} = f\lambda.$$

6.2 Wave Propagation through a Free-Space Medium

In an electromagnetic wave, electric and the magnetic fields, both time-varying, are simultaneously present. Thus we see the electric field $\vec{E}$ and the magnetic field $\vec{H}$ as coupled in Maxwell's equations (5.78) and (5.79). We can, however, decouple these equations to write what is known as the wave equation in either of the field quantities, $\vec{E}$ or $\vec{H}$.

For the sake of the convenience of analysis, let us recall Maxwell's equations, (5.76)-(5.79), putting therein $\vec{D} = \varepsilon \vec{E}$ and $\vec{B} = \mu \vec{H}$ to write them as follows:

$$\nabla \cdot \vec{D} = \rho \tag{6.20}$$

$$\nabla \cdot \vec{H} = 0 \tag{6.21}$$

$$\nabla \times \vec{E} = -\mu \frac{\partial \vec{H}}{\partial t} \tag{6.22}$$

$$\nabla \times \vec{H} = \vec{J} + \varepsilon \frac{\partial \vec{E}}{\partial t} . \tag{6.23}$$

Taking the curl of (6.22)(refer to Section 2.7), we get

$$\nabla \times \nabla \times \vec{E} = -\mu \nabla \times \frac{\partial \vec{H}}{\partial t} . \tag{6.24}$$

The partial derivatives in the curl operation [see expressions (2.65)-(2.71)] and the partial derivative $\partial/\partial t$, both occurring on the right-hand side of (6.24), are interchangeable. Therefore, we can express (6.24) as

$$\nabla \times \nabla \times \vec{E} = -\mu \frac{\partial}{\partial t} \nabla \times \vec{H} . \tag{6.25}$$

Substituting (6.23) in (6.25), we then obtain

$$\nabla \times \nabla \times \vec{E} = -\mu \frac{\partial}{\partial t} \left(\vec{J} + \varepsilon \frac{\partial \vec{E}}{\partial t} \right) = -\mu \frac{\partial \vec{J}}{\partial t} - \mu\varepsilon \frac{\partial^2 \vec{E}}{\partial t^2} . \tag{6.26}$$

Further, substituting the vector identity

$$\nabla \times \nabla \times \vec{E} = \nabla(\nabla \cdot \vec{E}) - \nabla^2 \vec{E} \tag{6.27}$$

in (6.26), we get

$$\nabla(\nabla \cdot \vec{E}) - \nabla^2 \vec{E} = -\mu \frac{\partial}{\partial t} \left(\vec{J} + \varepsilon \frac{\partial \vec{E}}{\partial t} \right) = -\mu \frac{\partial \vec{J}}{\partial t} - \mu\varepsilon \frac{\partial^2 \vec{E}}{\partial t^2}$$

which, in view of (6.20) can be written as

$$\nabla \left(\frac{\rho}{\varepsilon} \right) - \nabla^2 \vec{E} = -\mu \frac{\partial \vec{J}}{\partial t} - \mu\varepsilon \frac{\partial^2 \vec{E}}{\partial t^2} . \tag{6.28}$$

We can rearrange (6.28) to express it as

$$\nabla^2\vec{E} - \mu\varepsilon\frac{\partial^2\vec{E}}{\partial t^2} = \nabla\left(\frac{\rho}{\varepsilon}\right) + \mu\frac{\partial\vec{J}}{\partial t}\,. \tag{6.29}$$

Equation (6.29) involves only electric field $\vec{E}$. Being devoid of magnetic field $\vec{H}$, it is known as the wave equation in electric field.

Going through the same procedure as that followed in deriving the wave equation in electric field (6.29) starting from the curl of (6.22), we can derive the wave equation in the magnetic field as well, however, now starting from the curl of (6.23):

$$\nabla\times\nabla\times\vec{H} = \nabla\times\vec{J} + \varepsilon\nabla\times\frac{\partial\vec{E}}{\partial t}\,. \tag{6.30}$$

Using the vector identity (6.27), applied now to $\vec{H}$ (instead of $\vec{E}$), and interchanging the derivatives in the second term of (6.30) as was done in obtaining (6.25) from (6.24), we obtain

$$\nabla(\nabla\cdot\vec{H}) - \nabla^2\vec{H} = \nabla\times\vec{J} + \varepsilon\frac{\partial}{\partial t}\nabla\times\vec{E}\,. \tag{6.31}$$

Substituting (6.22) in (6.31), we get

$$\nabla(\nabla\cdot\vec{H}) - \nabla^2\vec{H} = \nabla\times\vec{J} - \mu\varepsilon\frac{\partial^2\vec{H}}{\partial t^2}$$

which in view of (6.21), becomes

$$-\nabla^2\vec{H} = \nabla\times\vec{J} - \mu\varepsilon\frac{\partial^2\vec{H}}{\partial t^2}$$

which, after rearrangement of terms, becomes

$$\nabla^2\vec{H} - \mu\varepsilon\frac{\partial^2\vec{H}}{\partial t^2} = -\nabla\times\vec{J}\,. \tag{6.32}$$

In (6.32), we find the wave equation in magnetic field $\vec{H}$.

Now, let us take up as a special case the propagation of electromagnetic waves through a free-space medium. In free space, there are no charges. This makes $\rho = 0$ in free space. Moreover, in free space there is an absence of both conduction current density ($\sigma = 0$) and convection current density (there being no charges in the medium that can move to make a convection current). This makes $\vec{J} = \sigma\vec{E} = 0$ in free space. Therefore, for a free-space medium ($\varepsilon = \varepsilon_0, \mu = \mu_0$), since $\rho = 0$ and $\vec{J} = 0$, the wave equations (6.29) and (6.32) simplify to

$$\nabla^2\vec{E} - \mu_0\varepsilon_0\frac{\partial^2\vec{E}}{\partial t^2} = 0 \tag{6.33}$$

$$\nabla^2 \vec{H} - \mu_0 \varepsilon_0 \frac{\partial^2 \vec{H}}{\partial t^2} = 0 \,. \tag{6.34}$$

Let us formulate the problem by considering wave propagation along z in the rectangular coordinate system and read the vector equations (6.33) and (6.34) for their respective components as follows.

$$\left.\begin{aligned} \nabla^2 E_x &= \mu_0 \varepsilon_0 \frac{\partial^2 E_x}{\partial t^2} \quad \text{(a)} \\ \nabla^2 E_y &= \mu_0 \varepsilon_0 \frac{\partial^2 E_y}{\partial t^2} \quad \text{(b)} \\ \nabla^2 E_z &= \mu_0 \varepsilon_0 \frac{\partial^2 E_z}{\partial t^2} \quad \text{(c)} \end{aligned}\right\} \tag{6.35}$$

$$\left.\begin{aligned} \nabla^2 H_x &= \mu_0 \varepsilon_0 \frac{\partial^2 H_x}{\partial t^2} \quad \text{(a)} \\ \nabla^2 H_y &= \mu_0 \varepsilon_0 \frac{\partial^2 H_y}{\partial t^2} \quad \text{(b)} \\ \nabla^2 H_z &= \mu_0 \varepsilon_0 \frac{\partial^2 H_z}{\partial t^2} \quad \text{(c)} \end{aligned}\right\} . \tag{6.36}$$

Further, with the help of (2.75), we can expand the left-hand sides of (6.35) and (6.36) to write them respectively as

$$\left.\begin{aligned} \frac{\partial^2 E_x}{\partial x^2} + \frac{\partial^2 E_x}{\partial y^2} + \frac{\partial^2 E_x}{\partial z^2} &= \mu_0 \varepsilon_0 \frac{\partial^2 E_x}{\partial t^2} \quad \text{(a)} \\ \frac{\partial^2 E_y}{\partial x^2} + \frac{\partial^2 E_y}{\partial y^2} + \frac{\partial^2 E_y}{\partial z^2} &= \mu_0 \varepsilon_0 \frac{\partial^2 E_y}{\partial t^2} \quad \text{(b)} \\ \frac{\partial^2 E_z}{\partial x^2} + \frac{\partial^2 E_z}{\partial y^2} + \frac{\partial^2 E_z}{\partial z^2} &= \mu_0 \varepsilon_0 \frac{\partial^2 E_z}{\partial t^2} \quad \text{(c)} \end{aligned}\right\} \tag{6.37}$$

$$\left.\begin{aligned} \frac{\partial^2 H_x}{\partial x^2} + \frac{\partial^2 H_x}{\partial y^2} + \frac{\partial^2 H_x}{\partial z^2} &= \mu_0 \varepsilon_0 \frac{\partial^2 H_x}{\partial t^2} \quad \text{(a)} \\ \frac{\partial^2 H_y}{\partial x^2} + \frac{\partial^2 H_y}{\partial y^2} + \frac{\partial^2 H_y}{\partial z^2} &= \mu_0 \varepsilon_0 \frac{\partial^2 H_y}{\partial t^2} \quad \text{(b)} \\ \frac{\partial^2 H_z}{\partial x^2} + \frac{\partial^2 H_z}{\partial y^2} + \frac{\partial^2 H_z}{\partial z^2} &= \mu_0 \varepsilon_0 \frac{\partial^2 H_z}{\partial t^2} \quad \text{(c)} \end{aligned}\right\} . \tag{6.38}$$

Let us restrict our analysis to a uniform plane wave. For a plane wave, the wavefront is plane. A wavefront (equiphase surface) is defined as the surface over which the phase of the quantity associated with the wave remains constant (see Section 6.1 for the concept of phase). For a uniform plane wave, the amplitude of the quantity associated with the wave remains constant. For our analysis, we will consider the wavefront to be perpendicular to

z, the direction of propagation of the wave, so that the field quantities remain constant with respect to x and y. Hence, we can put

$$\frac{\partial}{\partial x} = \frac{\partial}{\partial y} = 0 \tag{6.39}$$

in (6.37) and (6.38) to express them as follows:

$$\left.\begin{aligned} \frac{\partial^2 E_x}{\partial z^2} &= \mu_0 \varepsilon_0 \frac{\partial^2 E_x}{\partial t^2} \quad \text{(a)} \\ \frac{\partial^2 E_y}{\partial z^2} &= \mu_0 \varepsilon_0 \frac{\partial^2 E_y}{\partial t^2} \quad \text{(b)} \\ \frac{\partial^2 E_z}{\partial z^2} &= \mu_0 \varepsilon_0 \frac{\partial^2 E_z}{\partial t^2} \quad \text{(c)} \end{aligned}\right\} \tag{6.40}$$

$$\left.\begin{aligned} \frac{\partial^2 H_x}{\partial z^2} &= \mu_0 \varepsilon_0 \frac{\partial^2 H_x}{\partial t^2} \quad \text{(a)} \\ \frac{\partial^2 H_y}{\partial z^2} &= \mu_0 \varepsilon_0 \frac{\partial^2 H_y}{\partial t^2} \quad \text{(b)} \\ \frac{\partial^2 H_z}{\partial z^2} &= \mu_0 \varepsilon_0 \frac{\partial^2 H_z}{\partial t^2} \quad \text{(c)} \end{aligned}\right\} . \tag{6.41}$$

Further, let us recall (6.20), putting therein $\rho = 0$ for the free-space medium considered, and (6.21). Reading them with the help of (2.47), we have

$$\frac{\partial E_x}{\partial x} + \frac{\partial E_y}{\partial y} + \frac{\partial E_z}{\partial z} = 0 \tag{6.42}$$

$$\frac{\partial H_x}{\partial x} + \frac{\partial H_y}{\partial y} + \frac{\partial H_z}{\partial z} = 0. \tag{6.43}$$

Invoking the condition (6.39) for a uniform plane wave, we can express (6.42) and (6.43) respectively as

$$\frac{\partial E_z}{\partial z} = 0 \tag{6.44}$$

$$\frac{\partial H_x}{\partial z} = 0. \tag{6.45}$$

In view of (6.44), we have $\partial^2 E_z/\partial z^2 = 0$, which, when substituted in (6.40*c*) gives

$$\frac{\partial^2 E_z}{\partial t^2} = 0. \tag{6.46}$$

The integration of (6.44) yields

$$E_z = A \tag{6.47}$$

where A is a constant with respect to z. Similarly, the integration of (6.46) yields

$$\frac{\partial E_z}{\partial t} = a$$

which, on another integration, gives

$$E_z = at + b \tag{6.48}$$

where a and b are constants with t.

Let us now examine if (6.47) and (6.48) conform to the propagating and periodic nature of a quantity associated with a forward wave in unbounded free-space medium that does not provide any scope for reflection for a backward wave to exist. Clearly, the constancy of E_z with z, given by (6.47), and the linear variation of E_z with t, given by (6.48), both with non-zero values of the constants A, a, and b, contradict with the dependence of the quantity as $\exp j(\omega t - \beta z)$ given by (6.17). This forces one to put $A = a = b = 0$ that makes

$$E_z = 0. \tag{6.49}$$

Proceeding exactly in a similar manner, starting from (6.41c) and (6.45) instead of (6.40c) and (6.44) respectively, we easily obtain

$$H_z = 0. \tag{6.50}$$

Thus, from (6.49) and (6.50) it is easy to appreciate that a uniform plane wave propagating through an unbounded free-space medium is essentially a transverse electromagnetic wave (TEM) with no electric and magnetic field components in the direction of propagation (along z) [refer to Fig. 6.1 and further elaboration given following (6.65)].

For such a transverse wave, let us see how the transverse components of the fields are related to one another. For this purpose, let us recall Maxwell's equations (6.22) and (6.23) and read them respectively with the help of (2.66) as

$$\left(\frac{\partial E_z}{\partial y} - \frac{\partial E_y}{\partial z}\right)\vec{a}_x + \left(\frac{\partial E_x}{\partial z} - \frac{\partial E_z}{\partial x}\right)\vec{a}_y + \left(\frac{\partial E_y}{\partial x} - \frac{\partial E_x}{\partial y}\right)\vec{a}_z = -\mu_0\frac{\partial \vec{H}}{\partial t} \tag{6.51}$$

$$\left(\frac{\partial E_z}{\partial y} - \frac{\partial E_y}{\partial z}\right)\vec{a}_x + \left(\frac{\partial E_x}{\partial z} - \frac{\partial E_z}{\partial x}\right)\vec{a}_y + \left(\frac{\partial E_y}{\partial x} - \frac{\partial E_x}{\partial y}\right)\vec{a}_z = \vec{J} + \varepsilon_0\frac{\partial \vec{E}}{\partial t}\,. \tag{6.52}$$

For a uniform plane wave, we have already obtained and $E_z = H_z = 0$ in (6.49) and (6.50), which enables us to take

$$\left.\begin{aligned} \vec{E} &= E_x\vec{a}_x + E_y\vec{a}_y + E_z\vec{a}_z = E_x\vec{a}_x + E_y\vec{a}_y \\ \vec{H} &= H_x\vec{a}_x + H_y\vec{a}_y + H_z\vec{a}_z = H_x\vec{a}_x + H_y\vec{a}_y \end{aligned}\right\}. \tag{6.53}$$

Moreover, since the medium is free space, we can take $\vec{J} = 0$ as discussed preceding (6.33). Therefore, putting $\vec{J} = 0$ and invoking (6.39) for a uniform plane wave, we can now write (6.51) and (6.52) as follows:

$$-\frac{\partial E_y}{\partial z}\vec{a}_x + \frac{\partial E_x}{\partial z}\vec{a}_y = -\mu_0 \frac{\partial}{\partial t}(H_x\vec{a}_x + H_y\vec{a}_y) \tag{6.54}$$

$$-\frac{\partial H_y}{\partial z}\vec{a}_x + \frac{\partial H_x}{\partial z}\vec{a}_y = \varepsilon_0 \frac{\partial}{\partial t}(E_x\vec{a}_x + E_y\vec{a}_y)\,. \tag{6.55}$$

Equating first the x-components and then the y-components of the left- and the right-hand sides of (6.54), we obtain

$$\frac{\partial E_y}{\partial z} = \mu_0 \frac{\partial H_x}{\partial t} \tag{6.56}$$

$$\frac{\partial E_x}{\partial z} = -\mu_0 \frac{\partial H_y}{\partial t}\,. \tag{6.57}$$

Similarly, the following expressions are obtained by equating first the x-components and then the y-components of the left and the right hand sides of (6.55):

$$-\frac{\partial H_y}{\partial z} = \varepsilon_0 \frac{\partial E_x}{\partial t} \tag{6.58}$$

$$\frac{\partial H_x}{\partial z} = \varepsilon_0 \frac{\partial E_y}{\partial t}\,. \tag{6.59}$$

Following (6.17), we can take the dependence of the field quantities as $\exp j(\omega t - \beta z)$, where ω is the angular frequency and β the phase propagation constant of the wave (Section 6.1). Thus, for instance, as in (6.17), we can take E_x in terms of its amplitude E_{x0} as

$$E_x = E_{x0} \exp j(\omega t - \beta z). \tag{6.60}$$

Differentiating (6.60), we get

$$\frac{\partial E_x}{\partial t} = j\omega\, E_{x0} \exp j(\omega t - \beta z) = j\omega\, E_x$$

$$\frac{\partial E_x}{\partial z} = -j\beta\, E_{x0} \exp j(\omega t - \beta z) = -j\beta\, E_x. \tag{6.61}$$

From (6.61), we get the following operators:

$$\left.\begin{aligned} \frac{\partial}{\partial t} &\equiv j\omega \\ \frac{\partial}{\partial z} &\equiv -j\beta \end{aligned}\right\}. \tag{6.62}$$

Making use of (6.62), we can express (6.56)-(6.59) as

$$\left.\begin{aligned} \beta\, E_y &= -\omega\mu_0 H_x \\ \beta\, E_x &= \omega\mu_0 H_y \\ \beta\, H_y &= \omega\varepsilon_0 E_x \\ \beta\, H_x &= -\omega\varepsilon_0 E_y \end{aligned}\right\}$$

which can be expressed as

$$\left.\begin{aligned} \frac{E_y}{H_x} &= \frac{-\omega\mu_0}{\beta} && \text{(a)} \\ \frac{E_x}{H_y} &= \frac{\omega\mu_0}{\beta} && \text{(b)} \\ \frac{E_x}{H_y} &= \frac{\beta}{\omega\varepsilon_0} && \text{(c)} \\ \frac{E_y}{H_x} &= \frac{-\beta}{\omega\varepsilon_0} && \text{(d)} \end{aligned}\right\}. \tag{6.63}$$

By comparing either (6.63*a*) with (6.63*b*) or (6.63*c*) with (6.63*d*), we obtain the relation

$$\frac{E_x}{H_y} = -\frac{E_y}{H_x} = \frac{\omega\mu_0}{\beta}$$

whence we then obtain

$$E_xH_x + E_yH_y = 0. \tag{6.64}$$

Further, since in the present context $E_z = H_z = 0$ [see (6.49) and (6.50)], the relation

$$\begin{aligned} \vec{E}\cdot\vec{H} &= (E_x\vec{a}_x + E_y\vec{a}_y + E_z\vec{a}_z)\cdot(H_x\vec{a}_x + H_y\vec{a}_y + H_z\vec{a}_z) \\ &= (E_x\vec{a}_x + E_y\vec{a}_y)\cdot(H_x\vec{a}_x + H_y\vec{a}_y) = E_xH_x + E_yH_y \end{aligned}$$

in view of (6.64) can be interpreted as

$$\vec{E}\cdot\vec{H} = 0. \tag{6.65}$$

The relation (6.65) indicates that the electric and magnetic field vectors, which have only the transverse components (x and y components), are at right angles to each other. Thus, for a wave propagating along z, if we choose the coordinate system such that the electric field is along x, then the magnetic field will be along y. A snapshot z of the electric and magnetic fields (variation with z at a fixed time) of the wave is obtainable from (6.60) and (6.63*b*) respectively (Fig. 6.1). Further, the electric and magnetic field vectors of the wave, each at different points along z, lie on the XZ and YZ planes respectively. Such a wave is also referred to as a plane polarised wave. Similarly, at a fixed point and at different instants of time, the tips of the electric and magnetic field vectors of the wave each lie on lines parallel to the X and Y axes respectively, and therefore, the wave is also called a linearly polarised wave. In more generalised terms, a wave could also be circularly or elliptically polarised (see Appendix A6.1 and also Examples E6-1 and E6-2).

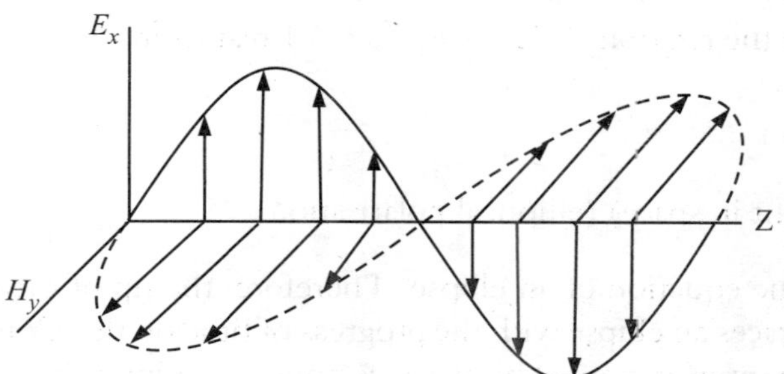

Fig. 6.1 *Snapshot of E_x (solid envelop) and H_y (broken envelop) for TEM wave propagating through an unbounded free-space medium in the z-direction.*

EXAMPLE E6-1

Show that for a wave propagating in the z-direction, two linearly polarised orthogonal components which the wave is composed of—one along x and the other along y, differing in a time phase angle of $\pi/2$—combine to yield elliptical polarisation.

SOLUTION

Let the two linearly polarised orthogonal components E_x and E_y, which the electric field vector $\vec{E} = E_x\vec{a}_x + E_y\vec{a}_y$ of the wave travelling along z is composed of, be represented as (see also Appendix A6.1):

$$\left.\begin{aligned} E_x &= E_{0x}\sin(\omega t - \beta z) \\ E_y &= E_{0y}\sin(\omega t - \beta z + \pi/2) = E_{0y}\cos(\omega t - \beta z) \end{aligned}\right\}. \tag{E6-1.1}$$

The subscript 0 in (E6-1.1) refers to amplitude. Also, here $\pi/2$ is the time-phase lead of E_y over E_x. However, at a fixed point in space, chosen as $z = 0$, expression (E6-1.1) becomes

$$\left.\begin{aligned} E_x &= E_{0x}\sin\omega t \\ E_y &= E_{0y}\cos\omega t \end{aligned}\right\}. \tag{E6-1.2}$$

From (E6-1.2) we obtain

$$\left.\begin{aligned} \sin\omega t &= \frac{E_x}{E_{0x}} \\ \cos\omega t &= \frac{E_y}{E_{0y}} \end{aligned}\right\}. \tag{E6-1.3}$$

Putting (E6-1.3) in the relation $\sin^2 \omega t + \cos^2 \omega t = 1$ results in

$$\frac{E_x^2}{E_{0x}^2} + \frac{E_y^2}{E_{0y}^2} = 1 \qquad \text{(E6-1.4)}$$

(at a fixed point in space) (elliptical polarisation)

which represents the equation of an ellipse. Therefore, the tip of the electric field vector $\vec{E} = E_x \vec{a}_x + E_y \vec{a}_y$ traces an ellipse with the progress of time on the XY plane, perpendicular to the direction of propagation. The wave is thus an elliptically polarised wave. (See Appendix A6.1).

EXAMPLE E6-2

Show that elliptical polarisation of Example E6-1 passes on to circular polarisation as a special case. Hence, also appreciate polarisations of the left-circular and right-circular types.

SOLUTION

Equation (E6-1.4), representing elliptical polarisation, is reproduced below for the sake of convenience of solving this problem.

$$\frac{E_x^2}{E_{0x}^2} + \frac{E_y^2}{E_{0y}^2} = 1. \quad \text{(Elliptical polarisation)} \qquad \text{(E6-2.1)}$$

In a special case of the amplitudes of the two orthogonal components being equal, that is, $E_{0x} = E_{0y} = E_0$, equation (E6-2.1) passes on to the following equation of a circle:

$$E_x^2 + E_y^2 = E_0^2 \qquad \text{(E6-2.2)}$$

(at a fixed point in space) (circular polarisation)

which represents the equation of a circle, thereby showing that elliptical polarisation passes on to circular polarisation. We can also write for this case ($E_{0x} = E_{0y} = E_0$), from (E6-1.2), the following field components:

$$\left.\begin{aligned} E_x &= E_0 \sin \omega t \\ E_y &= E_0 \cos \omega t \end{aligned}\right\}. \qquad \text{(E6-2.3)}$$

(at a fixed point in space) (circular polarisation)

From (E6-2.3), it follows that for circular polarisation, at $t = 0$ corresponding to $\omega t = 0$, the field components become $E_x = 0$ and $E_y = E_0$. Consequently, from the field expression $\vec{E} = E_x \vec{a}_x + E_y \vec{a}_y$, we obtain $\vec{E} = E_0 \vec{a}_y$, which shows that the electric field vector has

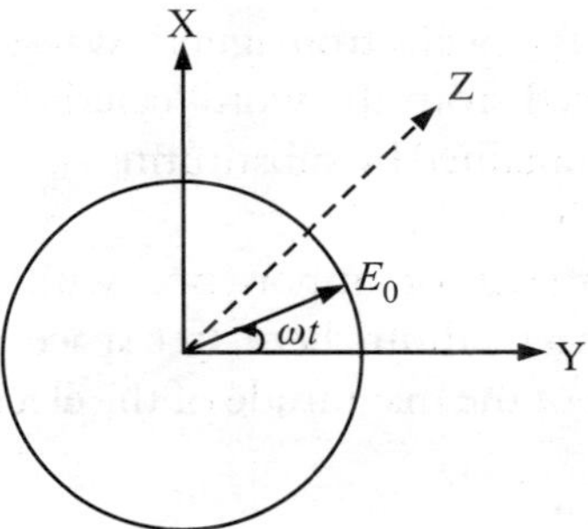

Fig. E6-2

magnitude E_0 and that it is directed along positive y. With the progress of time t, similarly, after a quarter of a cycle later, that is at $\omega t = \pi/2$, we obtain from (E6-2.3) the field components as $E_x = E_0$ and $E_y = 0$, giving $\vec{E} = E_x\vec{a}_x + E_y\vec{a}_y = E_0\vec{a}_x$, showing that the electric field vector has a magnitude of E_0 and that it is directed along positive x. Therefore, at fixed points in space, with the progress of time corresponding to the variations from $\omega t = 0$ to $\omega t = \pi/2$, the electric field vector of constant magnitude E_0 will rotate from the positive y to positive x (see Fig. E6-2, where the Z axis is shown as directed away from the reader, being in the direction of wave propagation).

You will observe such rotation of the electric field vector of constant magnitude E_0 as a 'clockwise rotation', viewing the wave as approaching you (as an observer), corresponding to the direction of positive z pointing towards you. According to IEEE convention (see Appendix A6.1), such a wave is referred to as a 'left-circularly polarised'. Correspondingly, according to the same convention, a 'right-circularly polarised wave' then refers to the 'anticlockwise' rotation of the electric field vector of constant magnitude E_0 for the wave viewing it as approaching you (as an observer) (Appendix A6.1). We have drawn this inference considering $\pi/2$ as the time-phase (ωt) 'lead' of E_y over E_x (see (E6-2.3) and Fig. E6-2). Thus, unlike a left-circularly polarised wave, a right-circularly polarised wave would correspond to the rotation of the electric field vector of constant magnitude E_0 from the positive x to positive y at fixed points in space with the progress of time for the variation of the time-phase from $\omega t = 0$ to $\omega t = \pi/2$.

Next, we can relate the phase propagation constant β with angular frequency ω of the wave by comparing either (6.63*a*) with (6.63*d*), or (6.63*b*) with (6.63*c*) as follows:

$$\beta = \omega\,(\mu_0\varepsilon_0)^{1/2}. \tag{6.66}$$

In an alternative approach, we can also obtain the same expression if we invoke relation (6.62) into any of the relations of (6.40).

Further, from (6.66), using (6.19), we obtain the following expression for the phase velocity v_{ph} $(= \omega/\beta)$ of the wave (see Section 6.1):

$$v_{\text{ph}}\left(= \frac{\omega}{\beta}\right) = \frac{1}{(\mu_0\varepsilon_0)^{1/2}}. \tag{6.67}$$

The value of the phase velocity v_{ph} of electromagnetic waves propagating through free space is assigned the symbol c (derived from the word 'celerity' , meaning swiftness or speed), and its value is $c = 3 \times 10^8$ m/s, obtained by substituting $\varepsilon_0 = 1/(36\pi) \times 10^{-9} = 8.854 \times 19^{-12}$ F/m and $\mu_0 = 4\pi \times 10^{-7}$ H/m in (6.67).

There is yet another parameter of importance with respect to the propagation of electromagnetic waves through a medium, here, free space, namely the intrinsic impedance η_0, obtained by taking the ratio of the magnitude of the electric field to that of the magnetic field:

$$\eta_0 = \frac{E}{H} = \frac{(E_x^2 + E_y^2)^{1/2}}{(H_x^2 + H_y^2)^{1/2}}. \tag{6.68}$$

Since the unit of E is V/m and that of H is A/m, you will appreciate that the unit of $\eta_0 = E/H$ is V/A or ohm (Ω). Now, substituting E_x from (6.63b) and E_y from (6.63a) in (6.68), we can express (6.68) as

$$\eta_0 = \frac{(E_x^2 + E_y^2)^{1/2}}{(H_x^2 + H_y^2)^{1/2}} = \frac{\left[\left(\frac{\omega\mu_0}{\beta}\right)^2 (H_x^2 + H_y^2)\right]^{1/2}}{(H_x^2 + H_y^2)^{1/2}} = \frac{\omega\mu_0}{\beta} \tag{6.69}$$

which, on substituting (6.66) becomes

$$\eta_0 = \frac{\omega\mu_0}{\beta} = \frac{\omega\mu_0}{\omega\,(\mu_0\varepsilon_0)^{1/2}} = \left(\frac{\mu_0}{\varepsilon_0}\right)^{1/2}. \tag{6.70}$$

Substituting $\varepsilon_0 = 1/(36\pi) \times 10^{-9} = 8.854 \times 19^{-12}$ F/m and $\mu_0 = 4\pi \times 10^{-7}$ H/m in (6.70), we then have the value of the intrinsic impedance η_0 of free space as $120\pi \cong 377$ohm (Ω).

EXAMPLE E6-3

A uniform plane wave propagates z in free space. If the amplitude of the electric field and frequency of the wave are $20\vec{a}_x + 30\vec{a}_y$ V/m and 15 GHz respectively, obtain the expressions for the instantaneous electric and magnetic fields.

SOLUTION

In view of the amplitude given, we can write the instantaneous electric field as

$$\vec{E} = (20\vec{a}_x + 30\vec{a}_y)\cos(\omega t - \beta z). \tag{E6-3.1}$$

In this example, the frequency is given as $f = 15$ GHz $= 15 \times 10^9$ Hz $= 1.5 \times 10^{10}$ Hz. So, we can calculate

$$\omega = 2\pi \times 1.5 \times 10^{10} = 3\pi \times 10^{10} \text{ rad/s} \tag{E6-3.2}$$

and

$$\beta = \frac{\omega}{c} = \frac{3\pi \times 10^{10}}{3 \times 10^8} = 100\pi\text{m}^{-1}. \tag{E6-3.3}$$

Substituting these values of ω and β from (E6-3.2) and (E6-3.3) in (E6-3.1), we obtain the following expression for the electric field:

$$\vec{E} = (20\vec{a}_x + 30\vec{a}_y)\cos(3\pi \times 10^{10}t - 100\pi\, z)\text{V/m}. \tag{E6-3.4}$$

Next, with the help of Maxwell's equation $\nabla \times \vec{E} = -\partial\vec{B}/\partial t$, taking $\partial/\partial x = \partial/\partial y = 0$ and $H_z = 0$ for a uniform plane wave propagating along z and using the expression (E6-3.4) for the electric field, we can obtain the expression for the magnetic field (following the approach in Example E5-7 in Chapter 5) as follows:

$$\vec{H} = (-0.0795\vec{a}_x + 0.053\vec{a}_y)\cos(3\pi \times 10^{10}t - 100\pi\, z)\ \text{A/m}.$$

EXAMPLE E6-4

From the electric and magnetic field expressions obtained in Example E6-3, show that the electric and magnetic fields are at right angles to each other for a uniform plane electromagnetic wave.

SOLUTION

We obtained the following field expressions in Example E6-3:

$$\left.\begin{aligned}\vec{E} &= (20\vec{a}_x + 30\vec{a}_y)\cos(3\pi \times 10^{10}t - 100\pi z)\ \text{V/m}\\ \vec{H} &= (-0.0795\vec{a}_x + 0.053\vec{a}_y)\cos(3\pi \times 10^{10}t - 100\pi z)\ \text{A/m}\end{aligned}\right\}$$

from which we identify

$$\left.\begin{aligned}E_x &= 20\cos(3\pi \times 10^{10}t - 100\pi z)\ \text{V/m}\\ E_y &= 30\cos(3\pi \times 10^{10}t - 100\pi z)\ \text{V/m}\\ H_x &= -0.0795\cos(3 \times 10^{10}\pi t - 100\pi z)\ \text{A/m}\\ H_y &= 0.053\cos(3\pi \times 10^{10}t - 100\pi z)\ \text{A/m}\end{aligned}\right\}. \tag{E6-4.1}$$

From the field components (E6-4.1), noting $H_z = 0$ for a uniform plane wave propagating along z, we can then calculate

$$\vec{E}.\vec{H} = E_xH_x + E_yH_y = [20(-0.0795) + 30 \times 0.053]\cos(3\pi \times 10^{10}t - 100\pi\, z)$$

$$\cong 0 \times \cos(3\pi \times 10^{10}t - 100\pi\, z) = 0$$

which shows that $\vec{E}$ and $\vec{H}$ are perpendicular to each other.

6.3 Wave Propagation through a Conducting Medium

We now consider the propagation of a uniform plane transverse electromagnetic wave (TEM) [$\partial/\partial x = \partial/\partial y = 0$ given by (6.39)] along z through an unbounded conducting medium using the same approach as followed in the analysis of the propagation of the wave through a free-space medium. A good conductor cannot be charged throughout its bulk—it can be charged only at its surface (Section 5.1.2). Therefore, we can put $\rho = 0$ in (6.29), the wave equation for the electric field for such a conductor, and write it as

$$\nabla^2 \vec{E} - \mu\varepsilon \frac{\partial^2 \vec{E}}{\partial t^2} = \mu \frac{\partial \vec{J}}{\partial t} \tag{6.71}$$

where

$$\vec{J} = \sigma \vec{E} \qquad \text{(5.11) rewritten} \tag{6.72}$$

is the conduction current density in the medium. Substituting (6.72) and the operator $\partial/\partial t = j\omega$ in (6.72) [see (6.62)], we obtain the wave equation for the electric field in a conducting medium, which can be expressed after rearranging terms as

$$\nabla^2 \vec{E} = j\omega\mu\ (\sigma + j\omega\varepsilon)\vec{E}. \tag{6.73}$$

We can expand the left-hand side of (6.73) and invoke the condition $\partial/\partial x = \partial/\partial y = 0$ for uniform plane wave $\partial/\partial x = \partial/\partial y = 0$ to express it for the transverse electric field components E_x and E_y as

$$\left.\begin{aligned} \frac{\partial^2 E_x}{\partial z^2} &= j\omega\mu\,(\sigma + j\omega\varepsilon)E_x \\ \frac{\partial^2 E_y}{\partial z^2} &= j\omega\mu\,(\sigma + j\omega\varepsilon)E_y \end{aligned}\right\}. \tag{6.74}$$

Putting

$$\gamma^2 = j\omega\mu\ (\sigma + j\omega\varepsilon) \tag{6.75}$$

in (6.74), we obtain the following wave equation for the transverse electric field components:

$$\left.\begin{aligned} \frac{\partial^2 E_x}{\partial z^2} &= \gamma^2 E_x \\ \frac{\partial^2 E_y}{\partial z^2} &= \gamma^2 E_y \end{aligned}\right\} \tag{6.76}$$

where γ is called the propagation constant of the medium. The following solution satisfies (6.76):

$$\left.\begin{aligned} E_x &= E_{x0+}\exp(-\gamma\, z) + E_{x0-}\exp(-\gamma\, z) \\ E_y &= E_{y0+}\exp(-\gamma\, z) + E_{y0-}\exp(-\gamma\, z) \end{aligned}\right\} \tag{6.77}$$

where E_{x0+} and E_{x0-} represent the amplitudes of the forward and the backward wave components respectively. Considering only the forward wave propagating along z, the following expression for the transverse electric field components is obtained by dropping the subscript + from (6.77):

$$\left.\begin{aligned} E_x &= E_{x0}\exp(-\gamma\, z) \\ E_y &= E_{y0}\exp(-\gamma\, z) \end{aligned}\right\}. \tag{6.78}$$

It may be noted that in expressions (6.77) or (6.78), the time dependence $\exp(j\omega\, t)$, which was absorbed by the operator $\partial/\partial t = j\omega$ as mentioned preceding (6.73), is understood.

The propagation constant γ is obtained from (6.75) as

$$\gamma = \sqrt{j\omega\mu(\sigma + j\omega\varepsilon)} = \alpha + j\beta \tag{6.79}$$

expressed in terms of its real and the imaginary parts. Therefore, in view of (6.79), we can put the field expression (6.78) as

$$\left.\begin{aligned} E_x &= E_{x0}\exp[-(\alpha + j\beta)z] = E_{x0}\exp(-\alpha\, z)\exp(-j\beta\, z) \\ E_y &= E_{y0}\exp[-(\alpha + j\beta)z] = E_{y0}\exp(-\alpha\, z)\exp(-j\beta\, z) \end{aligned}\right\}. \tag{6.80}$$

You can see from the field expression (6.80) that the amplitude of the electric field attenuates (or decays) exponentially as $\exp(-\alpha\, z)$, where α is called the attenuation constant of the wave. β continues to represent the phase propagation constant of the wave.

For the special case of free space ($\sigma = 0$; $\varepsilon = \varepsilon_0$, $\mu = \mu_0$), we get from (6.79)

$$\left.\begin{aligned} \gamma &= j\omega\sqrt{\mu_0\varepsilon_0} = j\beta \\ \beta &= \omega\sqrt{\mu_0\varepsilon_0} \\ \alpha &= 0 \end{aligned}\right\} \tag{6.81}$$

which agrees with (6.66) with respect to the phase propagation constant of the wave.

Now, let us take the special case of a conducting medium ($\mu = \mu_0$) with a good conductivity for which $\sigma >> \omega\varepsilon$, so that we can ignore $\omega\varepsilon$ in comparison with σ in (6.79). This gives

$$\gamma = \sqrt{j\omega\mu_0\sigma} = \alpha + j\beta. \tag{6.82}$$

You can easily separate the real and the imaginary parts of (6.82) to obtain α and β. For this purpose, if we take the square of (6.82), we get

$$j\omega\mu_0\,\sigma = (\alpha + j\beta)^2 = \alpha^2 + (j\beta)^2 + j2\alpha\beta = \alpha^2 - \beta^2 + j2\alpha\beta$$

from which equating the real and the imaginary parts we get

$$\alpha^2 - \beta^2 = 0 \tag{6.83}$$

$$2\alpha\beta = \omega\mu_0\sigma. \tag{6.84}$$

Substituting $\alpha = \beta$ obtainable from (6.83) in (6.84), we can then easily derive

$$\alpha = \beta = \sqrt{\frac{\omega\mu_0\sigma}{2}}. \tag{6.85}$$

The exponential field amplitude decay factor $\exp(-\alpha\, z)$ occurring in (6.80) can also be put as

$$\exp(-\alpha\, z) = \exp(-z/\delta),$$

where $\delta\ (= 1/\alpha)$ is called the 'skin depth'. Why it is so called becomes clear as we consider an electromagnetic wave entering from a free-space medium to a conducting medium extending up to infinity in the forward direction (semi-infinite medium). For such a wave penetrating the conducting medium, the amplitude of the field at the conducting surface (interface between the free-space and the conducting media) decays as $\exp(-z/\delta) = \exp(-\delta/\delta) = \exp(-1) = 1/e$, as at the surface or the 'skin' of the conductor, $z = 0$, as the wave penetrates through a distance or 'depth' equal to $z = \delta$. Hence, δ is called as the 'skin depth'. The concept of the skin depth will be further referred to as we develop the concept of ac or RF resistance in Section 6.4.

We can write expression for the skin depth $\delta (= 1/\alpha)$ with the help of (6.85) as

$$\delta = \sqrt{\frac{2}{\omega\mu_0\sigma}}. \tag{6.86}$$

Substituting $\omega = 2\pi f$ in (6.86), the expression for the skin depth δ in terms of the circular frequency $f\ (= \omega/2\pi)$ becomes:

$$\delta = \sqrt{\frac{1}{\pi f \mu_0 \sigma}}. \tag{6.87}$$

The skin depth, according to (6.87), has joint inverse proportionality with the square root of the wave frequency and the medium conductivity. Large conductivities and high wave frequencies result in small values of the skin depth, corresponding to significant values of the electric field only to small depths from the surface of the conductor. Typically, for copper ($\sigma = 5.8 \times 10^7$ mho/m), at power frequency 50 Hz, the value of the skin depth δ

calculated using (6.87) is found to be ~9.4 mm. Similarly, δ can be calculated as ~2.1 mm, ~66.1 μm, and ~2.1 μm for frequencies $f = 1$ kHz, 1 MHz, and 1 GHz, respectively. For $f = 10$ GHz, we find $\delta \sim 6.612 \times 10^{-7}$ m, which happens to be in the range of the wavelength of visible light.

EXAMPLE E6-5

Show that the skin depth of a good conductor is directly proportional to wavelength in the conductor.

SOLUTION

Let us recall (6.79) and rewrite it, putting $\mu = \mu_0$, as

$$\gamma = \sqrt{j\omega\mu_0(\sigma + j\omega\varepsilon)} = \sqrt{j\omega\mu_0\sigma\left(1 + \frac{j\omega\varepsilon}{\sigma}\right)} \tag{E6-5.1}$$

where γ is the wave propagation constant.

For a good conductor, taking $\sigma >> j\omega\varepsilon$, (E6-5.1) can be expressed as

$$\gamma = \sqrt{j\omega\mu_0\sigma} = \alpha + j\beta \tag{E6-5.2}$$

where α and β are the attenuation constant and phase propagation constant respectively. From (E6-5.2), we obtain by equating its real and imaginary parts (see 6.85):

$$\alpha = \beta = \sqrt{\frac{\omega\mu_0\sigma}{2}}. \qquad \text{(Good conductor)} \tag{E6-5.3}$$

The skin depth $\delta = 1/\alpha$ can be expressed with the help of (E6-5.3) as

$$\delta = \sqrt{\frac{2}{\omega\mu_0\sigma}} \tag{E6-5.4}$$

an expression that is identical to (6.85) (taking $\mu = \mu_0$).

In view of (E6-5.3), we can also express (E6-5.4) in terms of β as

$$\delta = \frac{1}{\beta}. \tag{E6-5.5}$$

Putting $\beta = 2\pi/\lambda$ in (E6-5.5), we then obtain

$$\delta = \frac{1}{\beta} = \frac{\lambda}{2\pi}. \qquad \text{(Good conductor)} \tag{E6-5.6}$$

Thus, we can see from (E6-5.6) that the skin depth of a good conductor is directly proportional to the wavelength in the conductor.

Example E6-6

Show that the skin depth of a poor conductor is independent of frequency.

Solution

Let us recall (6.79) to write it as (taking $\mu = \mu_0$)

$$\gamma = \sqrt{j\omega\mu_0(\sigma + j\omega\varepsilon)} = j\omega\sqrt{\mu_0\varepsilon}\left(1 + \frac{\sigma}{j\omega\varepsilon}\right)^{1/2} \tag{E6-6.1}$$

where γ is the wave propagation constant.

For a poor conductor, taking $\sigma << j\omega\varepsilon$, let us binomially expand (E6-6.1), retaining the second order terms and ignoring higher order terms to obtain

$$\gamma = j\omega\sqrt{\mu_0\varepsilon}\left(1 + \frac{1}{2}\frac{\sigma}{j\omega\varepsilon}\right) = j\omega\sqrt{\mu_0\varepsilon} + \frac{\sigma}{2}\sqrt{\frac{\mu_0}{\varepsilon}} = \alpha + j\beta \tag{E6-6.2}$$

where α and β are the attenuation constant and phase propagation constant respectively. By equating the real and the imaginary parts of (E6-6.2), we obtain the following expressions:

$$\left.\begin{array}{l}\alpha = \dfrac{\sigma}{2}\sqrt{\dfrac{\mu_0}{\varepsilon}} \\ \beta = \omega\sqrt{\mu_0\varepsilon}\end{array}\right\}. \qquad \text{(Poor conductor)} \tag{E6-6.3}$$

We can then obtain from (E6-6.3) the expression for the skin depth $\delta = 1/\alpha$ as

$$\delta = \frac{2}{\sigma}\sqrt{\frac{\varepsilon}{\mu_0}}. \qquad \text{(Poor conductor)} \tag{E6-6.4}$$

Thus, it can be seen that the skin depth of a poor conductor is independent of frequency.

Example E6-7

Deduce a general formula for the skin depth of a medium and interpret it for a poor conductor and a good conductor.

Solution

Recall the general expression (6.79) (taking $\mu = \mu_0$) for the propagation constant of a medium and rewrite it as

$$\alpha + j\beta = \sqrt{j\omega\mu_0(\sigma + j\omega\varepsilon)} \tag{E6-7.1}$$

where α is the attenuation constant and β is the phase propagation constant of the wave in a medium, the skin depth δ of the medium being the reciprocal of α. In order to separate (E6-7.1) into its real and imaginary parts, take the square of (E6-7.1) to get

$$\alpha^2 - \beta^2 + 2j\alpha\beta = j\omega\mu_0(\sigma + j\omega\varepsilon) = -\omega^2\mu_0\varepsilon + j\omega\mu_0\sigma. \tag{E6-7.2}$$

Equating the real and imaginary parts of (E6-7.2), we obtain respectively

$$\alpha^2 - \beta^2 = -\omega^2\mu_0\varepsilon \tag{E6-7.3}$$

and $$2\alpha\beta = \omega\mu_0\sigma. \tag{E6-7.4}$$

Further, let us introduce the following algebraic relation which will prove to be useful in finding α and β :

$$(\alpha^2 + \beta^2)^2 = (\alpha^2 - \beta^2)^2 + 4\alpha^2\beta^2. \tag{E6-7.5}$$

We can substitute (E6-7.3) and (E6-7.4) in (E6-7.5) to obtain

$$(\alpha^2 + \beta^2)^2 = \omega^4\mu_0^2\varepsilon^2 + \omega^2\mu_0^2\sigma^2$$

giving

$$\alpha^2 + \beta^2 = \pm\sqrt{\omega^4\mu_0^2\varepsilon^2 + \omega^2\mu_0^2\sigma^2}. \tag{E6-7.6}$$

Solving (E6-7.3) and (E6-7.6), we can easily get α^2 and β^2 and hence α and β as

$$\alpha = \sqrt{\frac{\pm\omega\mu_0\sqrt{\omega^2\varepsilon^2 + \sigma^2} - \omega^2\mu_0\varepsilon}{2}} \tag{E6-7.7}$$

and

$$\beta = \sqrt{\frac{\pm\omega\mu_0\sqrt{\omega^2\varepsilon^2 + \sigma^2} + \omega^2\mu_0\varepsilon}{2}}. \tag{E6-7.8}$$

The general formula for skin depth $\delta = 1/\alpha$ can be obtained by taking the reciprocal of (E6-7.7) as

$$\delta = \sqrt{\frac{2}{\pm\omega\mu_0\sqrt{\omega^2\varepsilon^2 + \sigma^2} - \omega^2\mu_0\varepsilon}}$$

which can be alternatively expressed, taking the admissible plus sign as

$$\delta = \frac{1}{\omega\sqrt{\mu_0\varepsilon}}\sqrt{\frac{2}{\sqrt{1 + \frac{\sigma^2}{\omega^2\varepsilon^2}} - 1}}. \quad \text{(General expression for skin depth)} \tag{E6-7.9}$$

In the special case of a good conductor, taking $\sigma >> \omega\varepsilon$, we can show from (E6-7.9)

$$\delta = \sqrt{\frac{2}{\omega\mu_0\sigma}} = \sqrt{\frac{1}{\pi f\mu_0\sigma}}. \quad \text{(Good conductor)} \tag{E6-7.10}$$

The expression (E6-7.10) agrees with (6.86) and (6.87).

Similarly, in another special case of a poor conductor, taking $\sigma << \omega\varepsilon$, we can expand binomially the denominator of the second square-root of (E6-7.9) while retaining only up to the second order term in the expansion to obtain

$$\delta = \frac{1}{\omega\sqrt{\mu_0\varepsilon}}\sqrt{\frac{2}{1+\frac{1}{2}\frac{\sigma^2}{\omega^2\varepsilon^2}-1}}$$

$$= \frac{1}{\omega\sqrt{\mu_0\varepsilon}}\sqrt{\frac{2}{\frac{1}{2}\frac{\sigma^2}{\omega^2\varepsilon^2}}} = \frac{1}{\omega\sqrt{\mu_0\varepsilon}}\frac{2}{\frac{\sigma}{\omega\varepsilon}},$$

which simplifies to

$$\delta = \frac{2}{\sigma}\sqrt{\frac{\varepsilon}{\mu_0}}. \qquad \text{(Poor conductor)} \qquad \text{(E6-7.11)}$$

EXAMPLE E6-8

Calculate the conductivity and skin depth of polythene of dielectric constant 2 and loss tangent 0.0002 at 100 MHz.

SOLUTION

In this problem, $\varepsilon = \varepsilon_r\varepsilon_0 = 2\varepsilon_0 = 2 \times 8.854 \times 10^{-12}$ F/m and $\tan\delta = 2 \times 10^{-4}$. The frequency $f = 100$ MHz , that is, $f = 100 \times 10^6 = 10^8$ Hz.

Using the data given, we can calculate using (5.70) the conductivity of the dielectric σ as

$$\sigma = \omega\varepsilon\tan\delta = 2\pi f\varepsilon\tan\delta$$

$$= 2 \times 3.143 \times 10^8 \times 2 \times 8.854 \times 10^{-12} \times 2 \times 10^{-4} \qquad \text{(E6-8.1)}$$

$$= 222.412 \times 10^{-8} \text{ mho/m}.$$

Since the value of $\sigma/\omega\varepsilon = (\tan\delta) = 2 \times 10^{-4} << 1$, we can use formula (E6-7.11) for the skin depth of a poor conductor:

$$\delta = \frac{2}{\sigma}\sqrt{\frac{\varepsilon}{\mu_0}} \qquad \text{(Poor conductor)}$$

which can be alternatively expressed as

$$\delta = \frac{2}{\sigma}\sqrt{\frac{\varepsilon_0\varepsilon_r}{\mu_0}} = \frac{2}{\sigma}\sqrt{\frac{\varepsilon_0}{\mu_0}}\sqrt{\varepsilon_r} = \frac{2\sqrt{\varepsilon_r}}{\sigma\eta_0} \qquad \text{(E6-8.2)}$$

where $\eta_0 = 120\pi = 377$ ohm is the intrinsic impedance of free space. Therefore, substituting $\varepsilon_r = 2$ (given) and σ from (E6-8.1) in (E6-8.2), we can calculate the skin depth

$$\delta = \frac{2 \times 1.414 \times 10^8}{222.412 \times 377} = 3.3727 \times 10^{-5} \times 10^8 \text{ m} = 3.3727 \text{ km}.$$

6.4 ac or RF Resistance

Let a uniform plane electromagnetic wave propagating along positive z penetrate a planar interface $z = 0$ (XY plane) between free-space and a semi-infinite conducting medium that extends from $z = 0$ to $z = \infty$ (Fig. 6.2). Let the electric field of the wave be directed along y and let us express its magnitude at depth z from the interface ($z = 0$) in terms of its exponentially decaying amplitude $E_0 \exp(-\alpha\, z)$ with the help of (6.80) as

$$E = E_0 \exp -(\alpha + j\beta)z) = E_0 \exp(-\alpha\, z) \exp(-j\beta\, z) \tag{6.88}$$

in which the time dependence $\exp(j\omega\, t)$ is understood. According to Ohm's law (6.72), the current density in the conducting medium due to the electric field will have the same direction as that of the electric field. We can then write, with the help of Ohm's law (6.72) the following expression for the magnitude of the current density along y at depth z from the interface ($z = 0$):

$$J = \sigma\, E_0 \exp -(\alpha + j\beta)z = \sigma\, E_0 \exp(-\alpha\, z) \exp(-j\beta\, z). \tag{6.89}$$

From (6.89), we see that $\sigma\, E_0 \exp(-\alpha\, z)$, the amplitude of the current density J, also attenuates with the depth z as does the amplitude of the electric field.

Let us next find the element of current $(\Delta I)_{(W)}$ through an infinitesimal strip of the conducting medium considered at depth z from the interface ($z = 0$). The strip has infinitesimal thickness dz, measured along z, and width W, measured along x. $(\Delta I)_{(W)}$ can be obtained by multiplying J, given by (6.89), by the infinitesimal area of the strip $W dz$ as follows (Fig. 6.2):

$$(\Delta I)_{(W)} = \sigma\, E_0 W \exp -(\alpha + j\beta)z\, dz. \tag{6.90}$$

The current $I_{(W)}$ through the width W of the strip for the entire conducting medium extending from $z = 0$ to $z = \infty$ is found by integrating (6.90) as follows:

$$I_{(W)} = \int_0^\infty \sigma\, E_0 W \exp -(\alpha + j\beta)\, z\, dz = \sigma\, E_0 W\, E_0 \int_0^\infty \exp -(\alpha + j\beta)\, z\, dz,$$

which, after working out the definite integral becomes

$$I_{(W)} = \sigma\, W\, E_0 \left[\frac{\exp[-(\alpha + j\beta)z]}{-(\alpha + j\beta)}\right]_0^\infty = \frac{\sigma\, W\, E_0}{\alpha + j\beta}. \tag{6.91}$$

For the uniform plane wave under consideration, we can take the electric field to be uniform [$\partial/\partial x = \partial/\partial y = 0$ according to (6.39)] over the surface of the conducting medium at $z = 0$ (XY plane). The amplitude of the electric field at the surface of the conducting medium $E_0 (= E_0 \exp(-\alpha z)$ corresponding to $z = 0$) can be interpreted as the potential drop per unit length at the surface of the conducting medium (which can be appreciated from the concept of electric field developed in terms of potential gradient in Section 3.2.2), the length being measured along y (Fig. 6.2). Therefore, the potential drop over the length l of the surface of the conducting medium measured along y, $V_{(l)}$, is obtained by multiplying E_0, the electric field amplitude at the surface, by the length l as

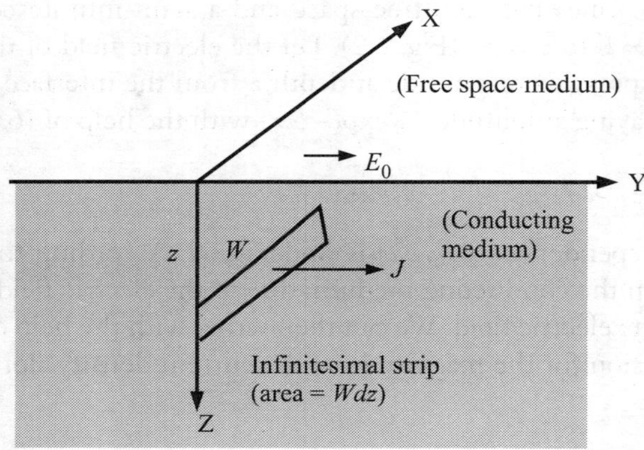

Fig. 6.2 *A uniform plane electromagnetic wave propagating along positive z penetrating the planar interface $z = 0$ (XY plane) between free-space and a semi-infinite conducting medium.*

$$V_{(l)} = E_0\, l. \tag{6.92}$$

We can then define the impedance of the conducting medium Z as

$$Z = \frac{V_{(l)}}{I_{(W)}}. \tag{6.93}$$

Substituting (6.91) and (6.92) into (6.93), we then obtain

$$Z = \frac{V_{(l)}}{I_{(W)}} = \frac{E_0 l}{\frac{\sigma\, W\, E_0}{\alpha + j\beta}} = \frac{l}{\sigma\, W}(\alpha + j\beta). \tag{6.94}$$

We can express (6.94) with the help of the relations $\delta = 1/\alpha$ and $\alpha = \beta$ (see (6.85) and the discussion following it) as:

$$Z = \frac{V_{(l)}}{I_{(W)}} = \frac{l}{\sigma\, W}\left(\frac{1}{\delta} + j\frac{1}{\delta}\right) = \frac{1}{\sigma\, \delta}\frac{l}{W}(1 + j). \tag{6.95}$$

At this stage, let us define a quantity called the surface impedance Z_S of the conducting medium as the ratio of the potential drop per unit length at the surface of the conducting medium to the current per unit width of the conducting medium. We can then easily obtain the surface impedance Z_S of the conducting medium by putting $l = W = 1$ in (6.95) as

$$Z_S = \frac{V_{(l=1)}}{I_{(W=1)}} = \frac{1}{\sigma}\left(\frac{1}{\delta} + j\frac{1}{\delta}\right) = \frac{1}{\sigma\,\delta}(1+j). \tag{6.96}$$

The surface impedance Z_S can be expressed in terms of its real part, called the surface resistance R_S, and the imaginary part, called the surface reactance X_S, as follows:

$$Z_S = R_S + jX_S. \tag{6.97}$$

Comparing (6.96) and (6.97), we then obtain following expression for the surface resistance R_S and the surface reactance:

$$R_S = X_S = \frac{1}{\sigma\delta} \tag{6.98}$$

which can be alternatively expressed by substituting (6.87) in (6.98) as

$$R_S = X_S = \sqrt{\frac{\pi\, f\mu_0}{\sigma}}. \tag{6.99}$$

The unit of the conductivity σ is mho/m and that of the skin depth δ is m. Therefore, the unit of $1/\sigma\delta$ is ohm, and since R_S and X_S are each equal to $1/\sigma\delta$, the unit of R_S and that of X_S and hence that of $Z_S(= R_S + jX_S)$ given by (6.97) are each ohm. This is also expressed as ohm per square or ohm/□, the square here referring to the unity length and unity width ($l = 1,\ W = 1$) implied in the definition (6.95) of Z_S.

As an example, let us find the resistance of a straight conducting wire of circular cross section. The resistance R_{dc} of the wire to a steady current is given by the well known relation[1]

$$R_{dc} = \frac{1}{\sigma}\frac{l}{\pi\, a^2} \tag{6.100}$$

where σ is the conductivity of the material of the wire, l the length and a the radius of the wire. However, we are now going to see how the expression for the resistance would change from (6.100) to R_{ac} for time-varying situations.

[1]The wire is aligned along z between $z = 0$ and $z = l$ at potentials $V = V$ and $V = 0$, respectively, sending current I from $z = 0$ to $z = l$. One-dimensional electric field $E = -\partial V/\partial z = -dV/dz$ (see (3.94), may be integrated as $\int_{V=V}^{V=0} dV = -E\int_{z=0}^{z=l} dz$ to give $E = V/l$. In addition, from Ohm's law $J = \sigma\, E$, we get $E = J/\sigma$. Equating the right-hand sides of these two expressions for E, we get $V/l = J/\sigma$, in which we can put the current per unit area or the current density $J = I/\pi\, a^2$ to obtain the familiar circuit theory expression for Ohm's law, $V = IR$, where $R = (1/\sigma)(l/\pi\, a^2)$ is the resistance of the wire, identical to (6.100).

We consider the skin depth of the material of the conducting medium to be rather small for good material conductivity and/or high wave frequencies. Such small skin depths correspond to the electric field, and hence to the current density becoming significant only to small depths from the surface of the conducting medium. The points at such small depths would fail to see the curvature of the wire of the circular cross section. Hence, this will enable us to use the expression for impedance derived for a planar conducting medium (6.95) by developing the circumference $2\pi a$ of the wire into the width corresponding to width W in the nomenclature of a planar conducting medium. Thus, putting $W = 2\pi a$ in (6.95), we obtain

$$Z = \frac{V_{(l)}}{I_{(W=2\,\pi\,a)}} = \frac{1}{\sigma\,\delta}\frac{l}{W}(1+j),$$

the real part of which yields R_{ac} as follows:

$$R_{\text{ac}} = \frac{1}{\sigma\,\delta}\frac{l}{2\pi\,a}. \tag{6.101}$$

The terminology ‘ac resistance’ is used at power frequencies, say, 50 or 60 Hz. One may use the terminology ‘RF resistance’ instead of ac resistance for higher frequencies, say, for frequencies $\sim$ a kilohertz (kHz) and above.

It would be of interest to obtain the ratio of the ac to the dc resistance of a wire, with the help of (6.100) and (6.101), as

$$\frac{R_{\text{ac}}}{R_{\text{dc}}} = \frac{\dfrac{1}{\sigma\,\delta}\dfrac{l}{2\pi\,a}}{\dfrac{1}{\sigma}\dfrac{l}{\pi\,a^2}} = \frac{a}{2\delta}. \tag{6.102}$$

Several practical situations of material conductivity and operating frequencies [see the text following (6.87)] correspond to $a >> \delta$, thereby making $R_{\text{ac}} > R_{\text{dc}}$.

EXAMPLE E6-9

Obtain an expression for the ac or RF resistance of a coaxial cable per unit length in terms of the skin depth of the conducting material of the cable and the geometry of the cable.

SOLUTION

Recall the expression (6.101) for the ac or RF resistance of a conducting wire of circular cross section of radius a:

$$R_{\text{ac}} = \frac{1}{\sigma\,\delta}\frac{l}{2\pi\,a} \tag{E6-9.1}$$

where l is the length of the wire and δ the skin depth of the material of the wire. δ is given by (6.86), which is reproduced here:

$$\delta = \sqrt{\frac{2}{\omega\mu_0\sigma}}. \tag{E6-9.2}$$

In the present context, interpreting expression (E6-9.1) as the ac or RF resistance of the coaxial cable contributed by the inner conductor of radius a of the cable, it can be written again as

$$R_{\mathrm{ac}} = \frac{1}{\sigma\,\delta}\frac{l}{2\pi\,a}. \qquad \text{(Contributed by inner conductor)} \tag{E6-9.3}$$

The resistance given by (E6-9.3) is attributed to the power going into the inner conductor, which is essentially an ohmic power loss. Similarly, the power going into the outer conductor of the cable will contribute to the ohmic power loss manifested in the ac or RF resistance of the outer conductor given by the same expression as (E6-9.1), though now the radius of the inner radius of the outer conductor b replaces a:

$$R_{\mathrm{ac}} = \frac{1}{\sigma\,\delta}\frac{l}{2\pi\,b} \qquad \text{(Contributed by outer conductor)} \tag{E6-9.4}$$

it being implied that the inner and the outer conductors are made of the same material. Therefore, in view of (E6-9.3) and (E6-9.4), we obtain the following expression for the ac or RF resistance of the coaxial cable:

$$R_{\mathrm{ac}} = \frac{1}{\sigma\,\delta}\frac{l}{2\pi\,a} + \frac{1}{\sigma\,\delta}\frac{l}{2\pi\,b} = \frac{1}{\sigma\,\delta}\left(\frac{1}{2\pi\,a} + \frac{1}{2\pi\,b}\right) l. \qquad \text{(Coaxial cable)} \tag{E6-9.5}$$

The expression for the ac or RF resistance per unit length of the coaxial cable (denoting it by R) can be obtained by dividing (E6-9.5) by l as:

$$\text{Resistance per unit length } R = \frac{1}{\sigma\,\delta}\left(\frac{1}{2\pi\,a} + \frac{1}{2\pi\,b}\right) = R_S\left(\frac{1}{2\pi\,a} + \frac{1}{2\pi\,b}\right),$$

where $R_S = 1/(\sigma\delta)$ is the surface resistance of the material of the conductors, assuming high conductivity and/or high operating frequencies.

EXAMPLE E6-10

Obtain an expression for the shunt conductance per unit length of a coaxial cable in terms of the capacitance per unit length of the cable and the loss tangent of the dielectric filling the region between the inner and outer conductors of the cable.

SOLUTION

Ohm's law (5.11) in the present one-dimensional problem, in which the potential varies only with r, being independent of θ and z in cylindrical coordinates (r, θ, z), can be written with the help of (3.91) in conjunction with (2.32) as

$$J = \sigma E = -\sigma \frac{dV}{dr} \qquad \text{(E6-10.1)}$$

where J is the current density through the lossy dielectric between the inner and outer conductors of the cable at the radial coordinate r off the axis ($r = 0$) of the cable (Fig. E6-10). If I is the current that flows, or in other words, is shunted between the inner and outer conductors over the length l of the coaxial cable through a area $2\pi rl$ due to the finite conductivity of the lossy dielectric filling the region between the conductors of the cable, then we can express J as

$$J = \frac{I}{2\pi rl} \, . \qquad \text{(E6-10.2)}$$

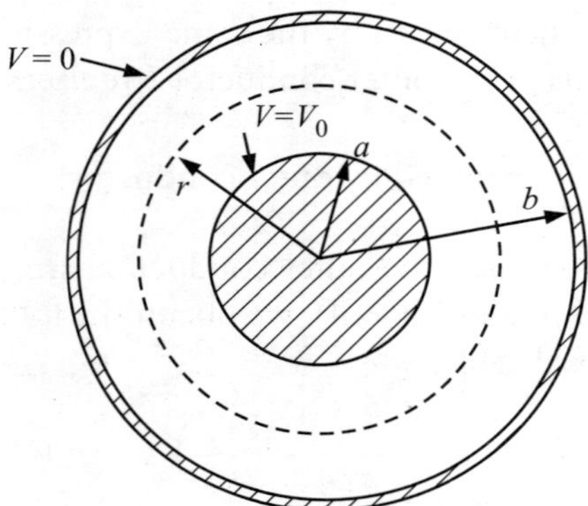

Fig. E6-10

Comparing (E6-10.1) and (E6-10.2), we can then write

$$-\sigma \frac{dV}{dr} = \frac{I}{2\pi rl}$$

which can be rearranged as

$$dV = \frac{-I}{2\pi\, l\sigma} \frac{dr}{r}. \qquad \text{(E6-10.3)}$$

Integrating (E6-10.3) over the limits $V = V_0$ on the inner conductor at $r = a$ and $V = 0$ on the outer conductor at $r = b$ (Fig. E6-10), we obtain

$$\int_0^{V_0} dV = \frac{-I}{2\pi l\sigma} \int_b^a \frac{dr}{r}. \qquad \text{(E6-10.4)}$$

Evaluating the integrals in (E6-10.4), we can write

$$[V]_0^{V_0} = \frac{-I}{2\pi l\sigma} \int_b^a \frac{dr}{r} = \frac{-I}{2\pi l\sigma} [\ln r]_b^a,$$

which simplifies to

$$V_0 = \frac{-I}{2\pi l\sigma}(\ln a - \ln b) = \frac{I}{2\pi l\sigma}(\ln b - \ln a) \tag{E6-10.5}$$

$$= \frac{I}{2\pi l\sigma}\ln\frac{b}{a}.$$

The shunt conductance of length l of the cable can be then obtained from (E6-10.5) as

$$\text{Shunt conductance} = \frac{I}{V_0} = \frac{2\pi l\sigma}{\ln\frac{b}{a}}.\ \text{(over the length } l) \tag{E6-10.6}$$

Dividing (E6-10.6) by length l, we get the shunt conductance per unit length G of the coaxial cable as

$$G\left(=\frac{1}{l}\frac{I}{V_0}\right) = \frac{2\pi\sigma}{\ln\frac{b}{a}}. \tag{E6-10.7}$$

Further, we have the relation (E3-9.8) for the capacitance per unit length C of the coaxial cable, which may be reproduced here as

$$C = \frac{2\pi\varepsilon}{\ln\frac{b}{a}} \tag{E6-10.8}$$

where ε is the permittivity of the dielectric filling the inner and outer conductors of the cable. We may also recall expression (5.70) for the loss tangent $\tan\delta$ as follows:

$$\tan\delta = \frac{\sigma}{\omega\varepsilon}. \tag{E6-10.9}$$

Dividing (E6-10.7) by (E6-10.8), we obtain

$$\frac{G}{C} = \frac{\sigma}{\varepsilon}. \tag{E6-10.10}$$

Combining (E6-10.9) with (E6-10.10) we thus derive the following expression for the shunt conductance per unit length of a coaxial cable in terms of the capacitance per unit length of the cable and the loss tangent

$$G = \omega C\tan\delta. \tag{E6-10.11}$$

6.5 Wave Propagation through Seawater

Let us now study the propagation of uniform plane electromagnetic waves propagating through an unbounded medium of weak conductivity. One such medium is seawater, whose conductivity σ is $\sim$ 4 mho/m. Further, for seawater, the value relative permittivity (dielectric constant) $\varepsilon_r = \varepsilon/\varepsilon_0$ is ~ 80, which makes its permittivity ε equal to $81\varepsilon_0$. We will see that the value of conductivity σ relative to that of $\omega\varepsilon$ vis-à-vis the wave frequency $f\ (= \omega/2\pi)$ will decide with how much attenuation the wave suffers as it propagates through seawater.

In order to appreciate the problem at lower frequencies, let us recall (6.79) and express it as

$$\gamma = \sqrt{j\omega\mu(\sigma + j\omega\varepsilon)} = \sqrt{j\omega\mu_0\sigma\left(1 + \frac{j\omega\varepsilon}{\sigma}\right)} = \alpha + j\beta. \tag{6.103}$$

For seawater, $\sigma = 4$ mho/m; $\varepsilon = 81\varepsilon_0 = 81 \times 8.854 \times 10^{-12}$ F/m. Taking the wave frequency typically as $f = 10$ kHz$= 10 \times 10^3$ Hz, a value close to frequencies used in underwater communication (such as in SONAR: sound navigation and ranging) compared to much higher wave frequencies ~ 10 GHz in microwave frequency range (used, for instance, in RADAR: radio detection and ranging) (to be considered later in this analysis), we obtain $\omega\varepsilon/\sigma = 2\pi f \varepsilon/\sigma \approx 1.1 \times 10^{-5}$. This enables us to ignore the second term under the radical in (6.103) and write the latter as

$$\gamma = \sqrt{j\omega\mu_0\sigma} = \alpha + j\beta \tag{6.104}$$

which is identical to (6.82). Therefore, we can continue to use (6.85) to express the real and the imaginary parts of (6.104) as

$$\alpha_{\text{lf}} = \beta_{\text{lf}} = \sqrt{\frac{\omega_{\text{lf}}\mu_0\sigma}{2}} \tag{6.105}$$

where we have used the subscript 'lf' to represent 'low' wave frequencies. With a view to treating higher frequencies, let us express (6.79) as

$$\gamma = \sqrt{j\omega\mu(\sigma + j\omega\varepsilon)} = j\omega\sqrt{\mu_0\varepsilon}\left(1 + \frac{\sigma}{j\omega\varepsilon}\right)^{1/2} = \alpha + j\beta. \tag{6.106}$$

Similarly, in order to appreciate the problem at higher frequencies, let us take the wave frequency as high ($f = 10$ GHz $= 10 \times 10^9$ Hz). For seawater ($\sigma = 4$ mho/m; $\varepsilon = 81\varepsilon_0 = 81 \times 8.854 \times 10^{-12}$ F/m) we then obtain $\sigma/\omega\varepsilon \approx 10^{-1}$. Now, we can no longer ignore the second term of the right-hand side of (6.106), and therefore, let us expand it in series and ignore higher order terms to write

$$\gamma = j\omega\sqrt{\mu_0\varepsilon}\left(1 + \frac{1}{2}\frac{\sigma}{j\omega\varepsilon}\right) = \alpha + j\beta. \tag{6.107}$$

By equating the real and the imaginary parts of (6.107), we obtain

$$\alpha_{\text{hf}} = \frac{\sigma}{2}\sqrt{\frac{\mu_0}{\varepsilon}} \tag{6.108}$$

$$\beta_{\text{hf}} = \omega\sqrt{\mu_0 \varepsilon} \tag{6.109}$$

where we have used the subscript 'hf' to represent 'high' wave frequencies, treating $f = 10$ GHz as much higher than $f = 10$ kHz for which we have deduced the corresponding expressions as (6.105).

With the help of (6.105) and (6.108), we can find the ratio of the low-to-high frequency attenuation constant as

$$\frac{\alpha_{\text{lf}}}{\alpha_{\text{hf}}} = \frac{\sqrt{\frac{\omega_{\text{lf}}\mu_0\sigma}{2}}}{\frac{\sigma}{2}\sqrt{\frac{\mu_0}{\varepsilon}}} = \sqrt{\frac{2\omega_{\text{lf}}\varepsilon}{\sigma}}. \tag{6.110}$$

For the typically chosen low-frequency value of $f = 10$ kHz $= 10 \times 10^3$ Hz, we can substitute $\omega_{\text{lf}} = 2\pi f = 2\pi \times 10 \times 10^3$ radian/s in (6.110) to obtain the ratio $\alpha_{\text{lf}}/\alpha_{\text{hf}}$ for seawater ($\sigma = 4$ mho/m; $\varepsilon = 81\varepsilon_0 = 81 \times 8.854 \times 10^{-12}$ F/m) as

$$\frac{\alpha_{\text{lf}}}{\alpha_{\text{hf}}} \approx 10^{-3}. \tag{6.111}$$

We can infer from (6.111) that, from the standpoint of attenuation, lower frequencies ($\sim$ 10 kHz) are preferable to higher frequencies ($\sim$ 10 GHz) for seawater communication. Further, it will be of interest to know what the wavelength of electromagnetic waves λ_{lf} in seawater will be for a typical lower frequency of $f =$ 10kHz= 10×10^3Hz. (The antenna size, for instance, may be typically of the order of $\lambda_{\text{lf}}/2$). We can find λ_{lf} from the following expression (see Section 6.1):

$$\lambda_{\text{lf}} = \frac{2\pi}{\beta_{\text{lf}}}$$

which, with the help of (6.105) and (6.86) can be written as

$$\lambda_{\text{lf}} = \frac{2\pi}{\beta_{\text{lf}}} = \frac{2\pi}{\sqrt{\frac{\omega_{\text{lf}}\mu_0\sigma}{2}}} = 2\pi\sqrt{\frac{2}{\omega_{\text{lf}}\mu_0\sigma}} = 2\pi\,\delta_{\text{skin}} \tag{6.112}$$

where δ_{skin} is the skin depth of seawater. Thus, with the help of (6.112) we can calculate the wavelength as $\lambda_{\text{lf}} \approx 15$ m for seawater ($\sigma = 4$ mho/m; $\mu_0 = 4\pi \times 10^{-7}$ H/m) at $f = 10$ kHz= 10×10^3 Hz.

6.6 Wave Propagation through a Medium of Charged Particles

Often, we need to study the behaviour of electromagnetic waves propagating through a medium of charge particles that may be either stationary, as in the ionosphere, or moving, as in an electron beam in many practical electronic devices.

The ionosphere comprises electrons and ions caused by the ionizing solar radiation (ultraviolet, soft X-ray) and ionizing particles (α-particle) emitted by the sun. It is located at heights of 50-300 km in earth's atmosphere and is differentiated into layers named D, E and F owing to the different physical properties and consequently different susceptibilities of the atmosphere to the solar radiations and ionizing particles at these heights. The ionosphere returns electromagnetic waves incident upon it from the earth; this phenomenon is utilised in the propagation of short radio waves for wave frequencies below a critical value. Each of the layers (D, E and F) of the ionosphere is characterised by a maximum electron number density corresponding to an optimum height of the ionospheric layer. The ionosphere has refractive index less than unity, and the return of electromagnetic waves from the ionosphere to the earth is attributed to total internal reflection. (See Example E6-11 to appreciate the dependence of refractive index of the ionosphere on charge density and wave frequency.)

Similarly, it is of interest to study the interaction of electromagnetic waves with a charge flow, for instance, a beam of electrons, which is a constituent of many practical electron devices such as vacuum electron devices like microwave tubes. These devices are based on space-charge waves, illustrated in Example E6-12, and cyclotron waves, illustrated in Example E6-13.

EXAMPLE E6-11

Show that the refractive index n of the ionosphere is given by $n = \sqrt{1 - \omega_p^2/\omega^2}$, where ω_p is the angular plasma frequency of the electron and ω is the wave angular frequency.

SOLUTION

The electron current density $\vec{J}$ in the ionosphere interacting with an electromagnetic wave is the sum of the convection current density $\rho_0\vec{v}$, obtainable from (3.124), and the displacement current density $\partial\vec{D}/\partial t$ (see the discussion following (5.47) and also the right-hand side of (6.23), remembering that $\vec{D} = \varepsilon_0\vec{E}$):

$$\vec{J} = \rho_0\vec{v} + \frac{\partial\vec{D}}{\partial t} = \rho_0\vec{v} + \varepsilon_0\frac{\partial\vec{E}}{\partial t} \qquad \text{(E6-11.1)}$$

where $\vec{E}$ and $\vec{D}$ are the wave electric field and displacement respectively; ρ_0 is the volume charge density of electrons and $\vec{v}$ is the velocity of electrons.

Invoking the wave's time-dependence, $\exp j\omega\, t$, we can now express $\vec{E}$ occurring in (E6-11.1) as

$$\vec{E} = \vec{E}_0 \exp(j\omega t) \qquad \text{(E6-11.2)}$$

which on differentiation gives

$$\frac{\partial\vec{E}}{\partial t} = j\omega\vec{E}_0 \exp(j\omega t). \qquad \text{(E6-11.3)}$$

In view of (E6-11.2) we can then express (E6-11.3) as

$$\frac{\partial \vec{E}}{\partial t} = j\omega \vec{E}$$

which can be substituted in (E6-11.1) to obtain

$$\vec{J} = \rho_0 \vec{v} + j\omega\varepsilon_0 \vec{E}. \tag{E6-11.4}$$

The electron velocity $\vec{v}$ appearing in this current density expression can be expressed in terms of $\vec{E}$ with the help of the force equation

$$m\frac{d\vec{v}}{dt} = e\vec{E}$$

whence, in view of the wave's time dependence $\exp(j\omega t)$ that would give $d\vec{v}/dt = j\omega\vec{v}$ [according to the explanation similar to that given following (E6-11.2)], we can write

$$\vec{v} = \frac{\eta \vec{E}}{j\omega} \tag{E6-11.5}$$

where e, m and $\eta = (e/m)$ are respectively the electronic charge, mass and charge-to-mass ratio (specific charge) of an electron.

Substituting (E6-11.5) in (E6-11.4), we obtain after a little algebra

$$\vec{J} = j\omega\varepsilon_0 \vec{E}\left[\rho_0 \frac{\eta \vec{E}}{j\omega}\frac{1}{j\omega} + 1\right] = j\omega\varepsilon_0 \vec{E}\left(1 - \frac{\eta\rho_0/\varepsilon_0}{\omega^2}\right). \tag{E6-11.6}$$

We can also express (E6-11.6) as

$$\vec{J} = j\omega\varepsilon_0 \vec{E}\left(1 - \frac{\omega_p^2}{\omega^2}\right) \tag{E6-11.7}$$

where

$$\omega_p = \left(\frac{\eta\rho_0}{\varepsilon_0}\right)^{1/2} = \left(\frac{|\eta|\,|\rho_0|}{\varepsilon_0}\right)^{1/2} \tag{E6-11.8}$$

noting that η and ρ_0 both carry the sign of the charge so that we can put $\eta = -|\eta|$ and $\rho_0 = -|\rho_0|$, enabling us to take $\eta\rho_0 = |\eta|\,|\rho_0|$.

ω_p is referred to as the electron angular plasma frequency, which was already introduced as (E3-6.5) in Example E3-6. Here, it is to be interpreted as the electron angular plasma frequency of the ionosphere.

Considering the ionosphere as a dielectric of an effective permittivity $\varepsilon_{\text{effective}}$, we can write, as an alternative to (E6-11.7), the expression for current density $\vec{J}$ in the ionosphere as the displacement current density

$$\vec{J} = \partial\vec{D}/\partial t = j\omega\varepsilon_{\text{effective}} \vec{E}. \tag{E6-11.9}$$

Equating the right-hand sides of these two alternative current density expressions (E6-11.7) and (E6-11.9), we obtain

$$j\omega\varepsilon_0\vec{E}\left(1-\frac{\omega_p^2}{\omega^2}\right)=j\omega\varepsilon_{\text{effective}}\,\vec{E},$$

whence we can write the expression for the effective permittivity of plasma as

$$\varepsilon_{\text{effective}}=\varepsilon_0\left(1-\frac{\omega_p^2}{\omega^2}\right). \tag{E6-11.10}$$

The effective relative permittivity of the ionosphere $\varepsilon_{r,\text{effective}}(=\varepsilon_{\text{effective}}/\varepsilon_0)$ then can be written from (E6-11.10) as

$$\varepsilon_{r,\text{ effective}}\,(=\varepsilon_{\text{effective}}/\varepsilon_0)=1-\frac{\omega_p^2}{\omega^2}. \tag{E6-11.11}$$

Taking the square root of $\varepsilon_{r,\text{effective}}$ given by (E6-11.11), we obtain the refractive index $n\,(=\sqrt{\varepsilon_{r,\text{effective}}})$ as

$$n=\sqrt{1-\frac{\omega_p^2}{\omega^2}}.$$

EXAMPLE E6-12

An electron beam of uniform dc velocity v_0 is perturbed longitudinally in motion. Consider the cross-sectional area of the beam to be large. Show that two space-charge waves are generated on the beam such that one of these waves, called the fast space-charge wave, has phase velocity $v_p > v_0$, while the other, called the slow space-charge wave, has phase velocity $v_p < v_0$, given by

$$v_p=\frac{\omega}{\omega\mp\omega_p}v_0,$$

the upper and the lower signs corresponding to the slow and the fast-waves respectively, where ω is the wave angular frequency and ω_p is the electron angular plasma frequency.

SOLUTION

This problem is an example that forms the basis for understanding the interaction between electromagnetic waves and a beam of electrons. Such interaction finds use in several electron devices in which some part of the energy of the beam of electrons is transferred to electromagnetic waves. The concept of space-charge waves developed through this problem has the scope of developing an understanding of vacuum electron devices such as travelling-wave tubes that find application in radar, electronic warfare, satellite communication, etc.

Since a practical vacuum electron device does not operate in perfect vacuum, it

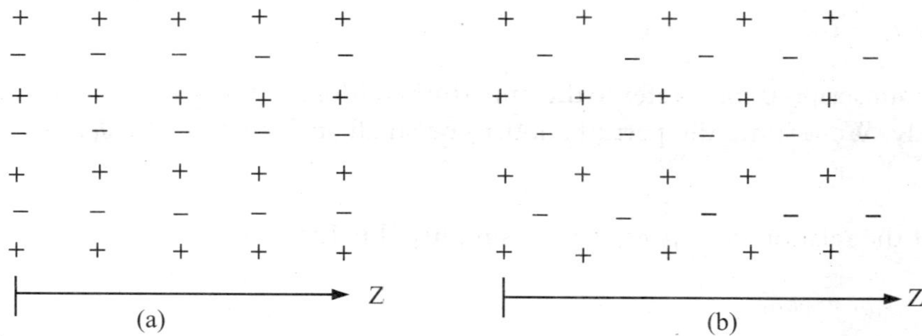

Fig. E6-12

can generate positive charges of ions by the process of ionisation, thereby providing a charge-neutralised background in the device. The flow of electrons would not be possible due to the repulsive force between the advancing and following electrons but for this charge-neutralised background in such devices. Hence, let us in the present problem consider a beam of electrons flowing in the longitudinal direction over a large cross-sectional area in such a charge-neutralised background. However, before proceeding further, let us consider the behaviour of a charge-neutralised region comprising negatively charged electrons and positively charged ions (Fig. E6-12). Now, in such a region, if the electrons are perturbed by an external force from their equilibrium positions with respect to the ions (which are relatively immobile due to their heavier mass as compared to the electrons), they will experience a restoring force due to the electric field, called the 'space-charge electric field' that is established between the electrons and ions. (Figure E6-12 depicts negatively charged electrons when they are in equilibrium positions with respect to positively charged ions (Fig. E6-12*a*) as well as their positions when they are perturbed (Fig. E6-12*b*).

In the charge-neutralised background, a beam of electrons if they are perturbed by an external force will also experience a restoring force due to such a space-charge electric field. Hence, let us treat the electron beam subject to this space-charge electric field in the direction of the flow of the beam of electrons, henceforth referred to as the axial direction of the beam. Further for the sake of simplicity of analysis of the problem, we treat the problem here as one-dimensional, considering the electron beam flow in the axial direction, z, say. We also take the transverse dimension of the beam to be large, that is, a large cross-sectional area the electron beam. This enables one to take, for an electron beam subject to the space-charge electric field perturbed by an external electric field, the beam velocity, charge density and current density to vary only in one dimension, here, in the longitudinal z-direction. This will correspond to $\partial/\partial x = \partial/\partial y = 0$ and $\partial/\partial z \neq 0$ in the framework of rectangular coordinates (x, y, z), and to $\partial/\partial r = \partial/\partial\theta = 0$ and $\partial/\partial z \neq 0$ in the framework of cylindrical coordinates (r, θ, z).

Now, let us recall the relation $J = \rho v$ given by (3.124) to write the following expression for the current density $J(= J_0 + J_1)$ in terms of the beam velocity $v(= v_0 + v_1)$ and volume charge density $\rho(= \rho_0 + \rho_1)$ of the electron beam considered as a charge fluid:

$$J_0 + J_1 = (\rho_0 + \rho_1)(v_0 + v_1) \tag{E6-12.1}$$

where the subscripts 0 and 1 refer to the unperturbed (dc) and the perturbed (ac) quantities, respectively. We assume the perturbation to be small such that $J_1 << J_0$, $\rho_1 << \rho_0$ and $v_1 << v_0$.

In view of the relation $J_0 = \rho_0 v_0$, we can simplify (E6-12.1) to

$$J_1 = \rho_0 v_1 + v_0 \rho_1 + \rho_1 v_1. \tag{E6-12.2}$$

However, we can ignore the third term $\rho_1 v_1$ of the right-hand side of (E6-12.2) compared to the remaining terms since $\rho_1 << \rho_0$ and $v_1 << v_0$, enabling us to write

$$J_1 = \rho_0 v_1 + v_0 \rho_1. \tag{E6-12.3}$$

Taking the partial derivative of (E6-12.3) with respect to z, we can write

$$\frac{\partial J_1}{\partial z} = \rho_0 \frac{\partial v_1}{\partial z} + v_0 \frac{\partial \rho_1}{\partial z}. \tag{E6-12.4}$$

Further, the continuity equation $\nabla . \vec{J} = -\partial \rho / \partial t$ given by (5.9) can be interpreted with the help of (2.47), to be read taking $\partial/\partial x = \partial/\partial y = 0$ as

$$\frac{\partial J_1}{\partial z} + \frac{\partial \rho_1}{\partial t} = 0. \tag{E6-12.5}$$

Let us next express (E6-12.4) with the help of (E6-12.5), as

$$\frac{\partial \rho_1}{\partial t} + v_0 \frac{\partial \rho_1}{\partial z} = -\rho_0 \frac{\partial v_1}{\partial z} \tag{E6-12.6}$$

Introducing a differential operator

$$\mathrm{D} = \frac{\partial}{\partial t} + v_0 \frac{\partial}{\partial z} \tag{E6.12.7}$$

we can also express (E6-12.6) as

$$\mathrm{D}\rho_1 = -\rho_0 \frac{\partial v_1}{\partial z}. \tag{E6-12.8}$$

Further, we can write the force equation for the electronic motion as

$$m \frac{dv_1}{dt} = eE_s \tag{E6-12.9}$$

where E_s represents the space-charge electric field in the perturbed electron beam.

The total derivative dv_1/dt occurring in (E6-12.9) can be put in terms of partial derivatives $\partial v_1/\partial t$ and $\partial v_1/\partial z$ as

$$\frac{dv_1}{dt} = \frac{\partial v_1}{\partial t} + \frac{dz}{dt}\frac{\partial v_1}{\partial z},$$

which can expressed putting $dz/dt = v$ as

$$\frac{dv_1}{dt} = \frac{\partial v_1}{\partial t} + v\frac{\partial v_1}{\partial z}. \qquad \text{(E6-12.10)}$$

Putting beam velocity $v = v_0 + v_1$ in (E6-12.10) we can then write

$$\frac{dv_1}{dt} = \frac{\partial v_1}{\partial t} + (v_0 + v_1)\frac{\partial v_1}{\partial z} = \frac{\partial v_1}{\partial t} + v_0\frac{\partial v_1}{\partial z} + v_1\frac{\partial v_1}{\partial z}. \qquad \text{(E6-12.11)}$$

Since $v_1 << v_0$, the third term of (E6-12.11) is much smaller than its second term, becoming of second order of importance. Therefore, ignoring the third term of (E6-12.11), we can write

$$\frac{dv_1}{dt} = \frac{\partial v_1}{\partial t} + v_0\frac{\partial v_1}{\partial z}. \qquad \text{(E6-12.12)}$$

If we now substitute (E6-12.12) in (E6-12.9), we obtain

$$m(\frac{\partial v_1}{\partial t} + v_0\frac{\partial v_1}{\partial z}) = eE_s. \qquad \text{(E6-12.13)}$$

Using the differential operator D given by (E6-12.7), we can also express (E6-12.10) as

$$\mathrm{D}v_1 = \eta E_s \qquad \text{(E6-12.14)}$$

where $\eta = (e/m)$ is the charge-to-mass ratio of an electron.

Now, if we operate D on (E6-12.8), we get

$$\mathrm{D}^2\rho_1 = -\rho_0\mathrm{D}\frac{\partial v_1}{\partial z}. \qquad \text{(E6-12.15)}$$

Since the operator D involves the partial derivatives and a constant according to (E6-12.7), we can interchange the operator D and the partial derivative $\partial v_1/\partial z$ on the right hand side of (E6-12.15) to write

$$\mathrm{D}^2\rho_1 = -\rho_0\frac{\partial}{\partial z}\mathrm{D}v_1. \qquad \text{(E6-12.16)}$$

We can then substitute (E6-12.14) in (E6-12.16) to obtain

$$\mathrm{D}^2\rho_1 = -\rho_0\frac{\partial}{\partial z}\eta E_s = -\eta\rho_0\frac{\partial E_s}{\partial z}. \qquad \text{(E6-12.17)}$$

Next, with the help of Poisson's equation $\nabla \cdot \vec{D} = \rho$, given by (3.105), interpreted with the help of (2.47) while taking $\partial/\partial x = \partial/\partial y = 0$, we can write

$$\frac{\partial E_s}{\partial z} = \frac{\rho_1}{\varepsilon_0},$$

which can be substituted in (E6-12.17) giving

$$D^2 \rho_1 = -\eta \rho_0 \frac{\rho_1}{\varepsilon_0} = -\frac{\eta \rho_0}{\varepsilon_0} \rho_1 = -\omega_p^2 \rho_1 \quad \text{(E6-12.18)}$$

where ω_p is the electron angular plasma frequency given by (E6-11.8). We thus have from (E6-12.18) the following expression for the operator D

$$D^2 = -\omega_p^2,$$

whence we get

$$D = \pm j\omega_p. \quad \text{(E6-12.19)}$$

Further, if we consider the wave quantities to vary as $\exp j(\omega t - \beta z)$, β being the phase propagation constant, we can take $\partial/\partial t = j\omega$ and $\partial/\partial z = -j\beta$, enabling us to write from (E6-12.7)

$$D = j\omega - j\beta v_0 = j(\omega - \beta v_0), \quad \text{(E6-12.20)}$$

Equating the right-hand sides of (E6-12.19) and (E6-12.20), we then obtain

$$\pm\omega_p = \omega - \beta v_0,$$

which by rearrangement of terms gives the following relation between the wave angular frequency ω and the wave phase propagation constant β or the dispersion relation of the space-charge waves:

$$\omega - \beta v_0 \mp \omega_p = 0. \quad \text{(E6-12.21)}$$

We can then derive from (E6-12.21) the expression for the wave phase propagation constant β as follows:

$$\beta = \frac{\omega \mp \omega_p}{v_0}. \quad \text{(E6-12.22)}$$

The wave phase velocity $v_p (= \omega/\beta)$ is then obtained with the help of (E6-12.22) as

$$v_p = \frac{\omega}{\omega \mp \omega_p} v_0. \quad \text{(E6-12.23)}$$

The two signs in the denominator of the right hand side of (E6-12.23) indicate that the electron beam supports two space-charge waves. The upper sign, as can be seen from (E6-12.23), makes $v_{ph} > v_0$, and the corresponding wave is referred to as the fast space-charge wave. Similarly, the lower sign makes $v_{ph} < v_0$ and the corresponding wave is referred to as the slow space-charge wave. The interaction of the slow space-charge wave on the electron beam (referred to in the problem) with electromagnetic wave is responsible for the operation of a device like the travelling-wave tube. The value of the plasma frequency ω_p plays a significant role in controlling the operating parameter of such a device.

EXAMPLE E6-13

An electron beam of large cross-sectional area and of uniform dc velocity v_0, immersed in a uniform dc magnetic field of flux density B along z in the axial direction of the beam, is perturbed in motion to a small extent in the transverse plane. Show that two waves are set up on the beam, one of which is called the fast cyclotron wave of phase velocity $v_{\text{ph}} > v_0$, while the other called the slow cyclotron wave has phase velocity $v_{\text{ph}} < v_0$, given by

$$v_{\text{ph}} = \frac{\omega}{\omega \mp \omega_c} v_0,$$

where ω is the wave angular frequency and $\omega_c = |\eta|\, B$ is the angular electron cyclotron frequency of the beam (see Example E4-11), $|\eta|$ being the magnitude of the charge-to-mass ratio of an electron.

SOLUTION

This problem is an example similar to that considered in Example E6-12 that forms the basis for understanding the interaction between electromagnetic waves and a beam of electrons in vacuum electron devices, for instance, the gyrotron that finds applications in areas such as in material processing. The concept of cyclotron waves developed through this problem forms the basis of developing this understanding.

Let us formulate the problem in the rectangular system of coordinates (x, y, z). The electrons in the beam will be perturbed in the transverse direction due to a longitudinal dc magnetic field along z such that $B_x = 0$; $B_y = 0$; $B_z = B$ ($\vec{B} = B_z \vec{a}_z$) (Fig. E6-13). We can then write, with the help of (4.58), the Lorentz force equations for electronic motion along x and y due to the magnetic field ($B_z = B$) as follows:

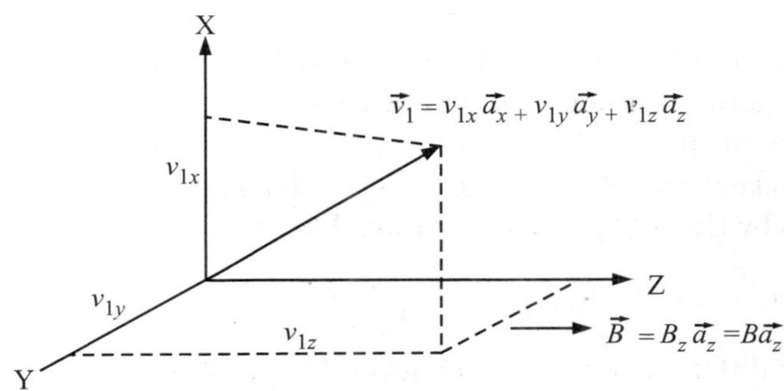

Fig. E6-13

$$m\frac{dv_{1x}}{dt} = e(\vec{v}_1 \times \vec{B})_x \tag{E6-13.1}$$

and

$$m\frac{dv_{1y}}{dt} = e(\vec{v}_1 \times \vec{B})_y \tag{E6-13.2}$$

where $\vec{v}_1 (= v_{1x}\vec{a}_x + v_{1y}\vec{a}_y + v_{1z}\vec{a}_z)$ represents the perturbed (ac) electron velocity (subscript 1 standing for a perturbed quantity) (Fig. E6-13).

For the sake of simplicity of analysis of the problem, let us follow the approach given in Example E6-12 to express the total derivative dv_{1x}/dt occurring in the left hand side of (E6-13.1) in terms of the partial derivatives $\partial v_{1x}/\partial t$ and $\partial v_{1x}/\partial z$ as

$$\frac{dv_{1x}}{dt} = \frac{\partial v_{1x}}{\partial t} + \frac{dz}{dt}\frac{\partial v_{1x}}{\partial z},$$

which can expressed, putting $dz/dt = v$, as

$$\frac{dv_{1x}}{dt} = \frac{\partial v_{1x}}{\partial t} + v\frac{\partial v_{1x}}{\partial z}. \tag{E6-13.3}$$

Putting beam velocity $v = v_0 + v_{1x}$ in (E6-13.3), we obtain

$$\frac{dv_{1x}}{dt} = \frac{\partial v_{1x}}{\partial t} + (v_0 + v_{1x})\frac{\partial v_{1x}}{\partial z} = \frac{\partial v_{1x}}{\partial t} + v_0\frac{\partial v v_{1x}}{\partial z} + v_{1x}\frac{\partial v_{1x}}{\partial z}. \tag{E6-13.4}$$

Ignoring the third term of the right-hand side of (E6-13.4) compared to its second term in view of $v_1 << v_0$ as discussed in Example E6-12 following (E6-12.11), we can then write (E6-13.4) as

$$\frac{dv_{1x}}{dt} = \frac{\partial v_{1x}}{\partial t} + v_0\frac{\partial v_{1x}}{\partial z} = (\frac{\partial}{\partial t} + v_0\frac{\partial}{\partial z})v_{1x} = \mathrm{D}v_{1x} \tag{E6-13.5}$$

where D is a differential operator given by

$$\mathrm{D} = \frac{\partial}{\partial t} + v_0\frac{\partial}{\partial z} \tag{E6-13.6}$$

That was introduced earlier in Example E6-12 as (E6-12.7) with a view to expressing the total derivative in terms of the partial derivatives ($d/dt = \partial/\partial t + v_0\partial/\partial z$). Further, if we consider the wave quantities to vary as $\exp j(\omega t - \beta z)$, β being the phase propagation constant, we can take $\partial/\partial t = j\omega$ and $\partial/\partial z = -j\beta$, which enables us to express the operator D, which is given by (E6-13.6), as (see Example E6-12)

$$\mathrm{D} = j\omega - j\beta v_0 = j(\omega - \beta v_0). \tag{E6-13.7}$$

Hence, using the differential operator D, given by (E6-13.6) or (E6-13.7), we can now express (E6-13.1) as

$$\mathrm{D}v_{1x} = \eta(\vec{v}_1 \times \vec{B})_x \tag{E6-13.8}$$

where $\eta(= e/m)$ is the charge-to-mass ratio of an electron.

Similarly, following the same approach as followed in deriving (E6-13.8) from (E6-13.1), we can also express (E6-13.2) as

$$\mathrm{D}v_{1y} = \eta(\vec{v}_1 \times \vec{B})_y \tag{E6-13.9}$$

Now, with the help of the vector cross-product relation

$$\vec{v}_1 \times \vec{B} = \begin{vmatrix} \vec{a}_x & \vec{a}_y & \vec{a}_z \\ v_{1x} & v_{1y} & v_{1z} \\ B_x & B_y & B_z \end{vmatrix}$$

in which $B_x = 0$, $B_y = 0$, $B_z = B$ (Fig. E6-13), we can find the cross-product components $(\vec{v}_1 \times \vec{B})_x$ and $(\vec{v}_1 \times \vec{B})_y$. Substituting these components in (E6-13.8) and (E6-13.9) respectively, we obtain

$$\mathrm{D}v_{1x} = \eta B v_{1y} = -\omega_c v_{1y} \tag{E6-13.10}$$

and $$\mathrm{D}v_{1y} = -\eta B v_{1x} = \omega_c v_{1x} \tag{E6-13.11}$$

where $\omega_c = -\eta B = |\eta|\, B$ is electron cyclotron frequency defined in Example E4-11.

Let us now operate D on (E6-13.10) to get

$$D^2 v_{1x} = -\omega_c D v_{1y}. \tag{E6-13.12}$$

Substitution of (E6-13.11) in (E6-13.12) yields

$$\mathrm{D}^2 v_{1x} = -\omega_c D v_{1y} = -\omega_c^2 v_{1x}$$

whence we get

$$\mathrm{D}^2 = -\omega_c^2 \tag{E6-13.13}$$

giving

$$\mathrm{D} = \pm j\omega_c. \tag{E6-13.14}$$

We can now see that (E6-13.14) (E6-13.10) is different from (E6-12.19) only in respect of the quantity ω_c cyclotron frequency replacing the plasma frequency ω_p (see Example E6-12). Therefore, following the same approach as in Example E6-12, we obtain the phase velocities of the cyclotron waves supported by the electron beam, by replacing ω_p in (E6-12.23) by ω_c as follows:

$$v_{\mathrm{ph}} = \frac{\omega}{\omega \mp \omega_c} v_0 \tag{E6-13.15}$$

where the upper sign in the denominator in the right-hand side refers to the fast cyclotron wave ($v_{ph} > v_0$) and the lower sign to the slow cyclotron wave ($v_{ph} < v_0$).

The interaction of the fast cyclotron wave on the electron beam (referred to in the problem) with electromagnetic wave is responsible for the operation of a device like the gyrotron. The value of the cyclotron frequency ω_c plays a significant role in controlling the operating parameter of such a device.

Appendix A6.1

Polarisation of a wave

A linearly polarised time-harmonic wave is characterised by the electric (or magnetic) field vector oriented in space along the same straight line at all instants of time. If the field vector has two components, then these components have to be at right angle to each other (orthogonal) in space and be either in phase or out of phase in time. Therefore, as a function of time, the tip of the electric field vector oscillates along the said straight line for the linearly polarised wave (see Fig. A6.1 showing the orientation of electric field vectors, typically at three instants of time: $t = 0$, $t = T/4$ and $t = T/2$, T being the wave time period).

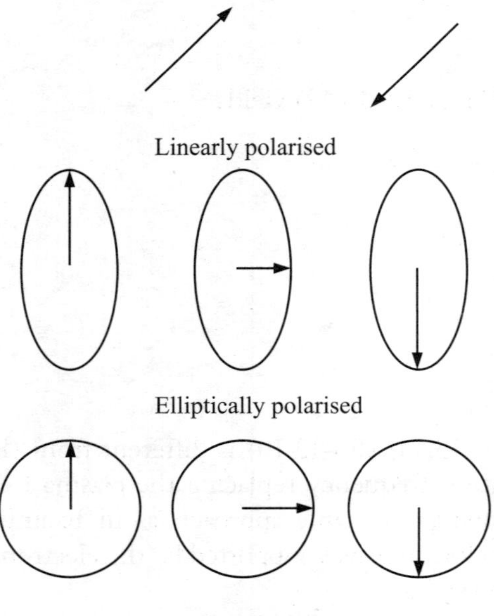

Fig. A6.1

However, if the electric (or magnetic) field vector of a time-harmonic wave is comprised of two orthogonal components in space such that one component leads the other in time phase by odd multiples of $\pi/2$, then as a function of time the tip of the field vector (resultant of the two orthogonal field components) traces a circle or an ellipse according to whether the amplitudes of the field components are the same or different. Accordingly, the wave is said to be circularly or elliptically polarised (see Fig. A6.1 showing the orientation of electric field vectors at $t = 0$, $t = T/4$ and $t = T/2$). Thus, a circularly polarised wave can be considered as a special case of an elliptically polarised wave. Further, the circular polarisation may be the left-circular or the right-circular types. According to IEEE convention, the wave is left-circularly or right-circularly polarised according to whether the electric field vector of the wave at a fixed point in space rotates clockwise or counterclockwise with time when viewing the wave as approaching the observer. (See also Examples E6-1 and E6-2 for an analytical appreciation of elliptical and circular polarisation.)

Summarising Notes

✓ Wave equations—one in electric field and the other in magnetic field—have been derived with the help of those two of Maxwell's equations in which these field quantities are coupled.

✓ Solutions to the wave equations in electric and magnetic fields have been obtained for uniform, plane electromagnetic waves propagating through an unbounded free-space medium.

✓ With the help of Maxwell's equations and the solutions to wave equations in electric and magnetic fields, it has been established, with reference to uniform, plane electromagnetic waves propagating through an unbounded free-space medium, that

- there exists no components of electric and magnetic fields in the direction of propagation;
- directions of the electric field, magnetic field and wave propagation are mutually perpendicular to one another;
- transverse electromagnetic (TEM) mode of propagation is supported by the unbounded medium;
- intrinsic impedance and phase velocity of the wave are each related to the permeability and the permittivity of free space, the wave phase velocity being the speed of light c.

✓ Concepts of skin depth, surface resistance and ac or RF resistance of a medium have been developed by studying the propagation of uniform, plane electromagnetic waves propagating through a semi-infinite conducting medium, considering such waves to be incident from a free-space region to a planar conducting surface extending to infinity.

- ✓ Ratio of the ac or RF resistance to dc resistance of a conducting wire of circular cross section becomes equal to the ratio of wire radius to twice skin depth, considering the wire to be of high conductivity and/or wave frequencies to be very high, such that the skin depth of the conductor becomes small enough compared to the wire radius to make a planar conductor approximation for the conducting wire of circular cross section.
- ✓ Lower frequencies, say ~10 kHz, is preferred to higher frequencies, say ~10 GHz, in view of comparatively lower attenuation of waves at such lower frequencies for sea-water communication as revealed by studying propagation through unbounded sea-water of finite conductivity and permittivity.

Review Questions

Narrative type

6.1 Explain why one should choose the functions $f(t - z/v)$ and $f(t + z/v)$ to represent the waves propagating with a velocity v in the positive z and negative z directions respectively.

6.2 Explain how can you put the physical quantities representing the forward and backward propagating waves as proportional to $\exp j(\omega t - \beta z)$ and $\exp j(\omega t + \beta z)$ respectively in the phasor notation, where β is the phase propagation constant of the wave.

6.3 From the phase part $\omega t - \beta z$ in the representation of a wave, show that the phase velocity of the wave is given by $v_{\text{ph}} = \omega/\beta$, where ω and β are the wave angular frequency and phase propagation constant respectively.

6.4 Develop the concept of (i) the frequency of a wave and (ii) the wavelength of a wave by examining the variation of a quantity representing the wave with (a) time at a fixed point in space and (b) distance in space at a fixed time that would provide a snapshot picture of variation respectively.

6.5 Establish the relation $v_{\text{ph}} = f\lambda$ for the wave phase velocity in terms of the wave frequency f and wavelength λ.

6.6 Starting from Maxwell's equation, establish the wave equation in electric field in a source-present (volume charge density $\rho \neq 0$ and current density $J \neq 0$) medium in terms of the volume charge density and current density.

6.7 Check if the following source-present (volume charge density $\rho \neq 0$ and current density $J \neq 0$)wave equation is dimensionally correct.

$$\nabla^2 \vec{E} - \mu\varepsilon \frac{\partial^2 \vec{E}}{\partial t^2} = \nabla\left(\frac{\rho}{\varepsilon}\right) + \mu \frac{\partial \vec{J}}{\partial t}.$$

Hint: Show that each of the terms of the wave equation has the dimension of V/m^3. Note that the dimension of $\mu\varepsilon$ is the dimension of the square of the reciprocal of velocity (m/s). Use the dimension of ρ as C/m^3 and that of ε as F/m. Recall that the dimension of μ is H/m and that of J is A/m^2. Further, take the dimension of the magnetic flux as Wb as well as HA (H standing for henry), and remember that Wb/s is synonymous with V.

6.8 Starting from Maxwell's equation, establish the wave equation in electric field in a source-free (volume charge density $\rho = 0$ and current density $J = 0$) medium.

6.9 Starting from Maxwell's equation, establish the wave equation in magnetic field in a source-present medium (volume charge density $\rho \neq 0$ and current density $J \neq 0$) in terms of a current density.

6.10 Starting from Maxwell's equation establish the wave equation in magnetic field in a source-free medium (volume charge density $\rho = 0$ and current density $J = 0$).

6.11 Write the wave equation in electric field in a free-space medium in generalised curvilinear coordinate system and expand and interpret it in rectangular, cylindrical and spherical coordinate systems.

6.12 How do you define a uniform plane electromagnetic wave?

6.13 Show that a uniform plane wave propagating through an unbounded free-space medium is essentially a transverse electromagnetic wave (TEM) with no electric and magnetic field components in the direction of propagation.

6.14 For a uniform plane TEM wave propagating through an unbounded free-space medium, prove the relation $\vec{E} \cdot \vec{H} = 0$, indicating thereby that the electric field ($\vec{E}$) and the magnetic field ($\vec{H}$) vectors are at right angles to each other,.

6.15 Discuss how a uniform plane wave on the one hand is plane-polarised, while on the other hand it is linearly polarised.

6.16 Show that an elliptically polarised wave results from the combination of two linearly polarised orthogonal components of a wave that have different amplitudes and differ in time phase angle by $\pi/2$.

6.17 Show that a circularly polarised wave results from the combination of two linearly polarised orthogonal wave components that have the same amplitude but differ in time phase angle by $\pi/2$.

6.18 What do you understand by left-circular and right-circular polarisations?

6.19 Prove that the phase velocity v_{ph} of a uniform plane wave propagating through free space is equal to $c = (\mu_0 \varepsilon_0)^{-1/2} = 3 \times 10^8$m/s.

6.20 Show that the intrinsic impedance of free space is $\eta_0 = (\mu_0/\varepsilon_0)^{1/2} = 120\pi \cong 377$ ohm (Ω).

6.21 Examining the wave equation in an electric field with reference to a uniform plane electromagnetic wave propagating through a conducting medium of conductivity σ, write the expression for the complex propagation constant of the medium and hence obtain the expression for the attenuation constant α and phase propagation constant β as $\alpha = \beta = (\omega\mu_0\sigma/2)^{1/2}$.

6.22 Define the skin depth of a conducting medium and derive its expression: $\delta = [1/(\pi f \mu_0 \sigma)]^{1/2}$ in terms of the conductivity σ of the medium and the wave frequency f.

6.23 Prove that the skin depth of a poor conductor is independent of the wave frequency, unlike that of a good conductor which is frequency dependent.

Hint: Derive the expression for the skin depth of a poor conductor $\delta = (2/\sigma)(\varepsilon/\mu_0)^{1/2}$, which is independent of frequency (see Example E6-6).

6.24 Considering the propagation of a uniform plane electromagnetic into a semi-infinite planar conducting medium, derive an expression for the surface impedance of the medium and hence show that the surface resistance and the surface reactance are each equal to $1/(\sigma\delta)$, where σ and $\delta = [1/(\pi f \mu_0 \sigma)]^{1/2}$ are the conductivity and skin depth of the medium respectively, and f is the wave frequency.

6.25 Derive the following expression for the ac resistance (R_{ac}) of a conducting wire of circular cross-section of radius a and conductivity σ at the operating frequency f:

$$R_{\text{ac}} = \frac{l}{2\pi a}\sqrt{\frac{\pi f \mu_0}{\sigma}}.$$

6.26 Prove that the ac-to-dc resistance of a wire of circular cross section is equal to the ratio of the wire radius to twice the skin depth for a good conductor and/or for high operating frequencies.

6.27 Obtain an expression for the ac resistance per unit length of a coaxial cable in terms of its geometrical dimensions and material conductivity.

Hint: See Example E6-9.

6.28 Deduce the expression for the shunt conductance per unit length of a coaxial cable $G = \omega C \tan\delta$, where C is the capacitance per unit length of the cable, $\tan\delta$ is the loss tangent of the dielectric region between the conductors of the cable, and ω is the wave angular frequency.

Hint: See Example E6-10.

6.29 Discuss, with relevant theory, why, for communication through seawater ($\sigma = 4$ mho/m; $\varepsilon = 81\varepsilon_0 = 81 \times 8.854 \times 10^{-12}$ F/m) you should prefer lower frequencies ~10 kHz to higher frequencies ~10 GHz.

6.30 Show that the attenuation constant α and the phase propagation constant β of a medium each become equal to $(\omega\mu_0\sigma/2)^{1/2}(= \alpha = \beta)$ if the operating frequency and the medium properties, that is, the values of σ and ε are such that $\omega\varepsilon/\sigma << 1$, enabling one to take $\omega\varepsilon/\sigma \approx 0$.

6.31 Show that if the operating frequency and the properties, namely the values of σ and ε of a medium are such that $\sigma << \omega\varepsilon$, enabling one to take $(\sigma/\omega\varepsilon)^2 \approx 0$, then the attenuation constant α and the phase propagation constant β of the medium can be expressed as $\alpha = (\sigma/2)(\mu_0/\varepsilon)^{1/2}$ and $\beta = \omega(\mu_0\varepsilon)^{1/2}$.

6.32 Prove that at lower frequencies, where $\omega\varepsilon/\sigma << 0$ such that one can take $\omega\varepsilon/\sigma \approx 0$, the wavelength in sea water becomes equal to the circumference of a circle of radius equal to the skin depth of seawater.

Hint: See (6.112).

6.33 Derive an expression for the refractive index of an ionospheric layer in terms of the wave frequency and the electron plasma frequency, and hence appreciate that a free-space region is denser than ionospheric layer.

6.34 Show that a beam of electrons supports two space-charge waves, one of which is slower while the other faster in phase velocity than the electron beam velocity.

6.35 Show that a beam of electrons supports two cyclotron waves in a magnetic field transverse to the electron beam velocity, one of which is slower while the other faster in phase velocity than the electron beam velocity.

Numerical type

6.36 A uniform plane wave propagating through a free-space region along z has an electric field of $\vec{E} = 60\pi \exp j(10^9 t - \beta z)\vec{a}_x$ V/m. Find the magnetic field and the wave phase propagation constant β. If the free-space region is replaced by a dielectric of relative permittivity 16, what will be the new value of β?

Answer

The electric field $\vec{E}$ is given as

$\vec{E} = 60\pi \exp j(10^9 t - \beta z)\vec{a}_x$ V/m. (1)

The magnitude of $\vec{E}$ from (1) is

$$E = 60\pi \exp j(10^9 t - \beta z). \tag{1}$$

With the help of (6.68), we can write the magnitude of magnetic field as

$$H = \frac{E}{\eta_0} \tag{2}$$

where η_0 is the intrinsic impedance of free space. Putting (2) in (3), we have

$$H = \frac{60\pi \exp j(10^9 t - \beta z)}{\eta_0}. \tag{3}$$

The direction of the magnetic field is along y since the given electric field is along x and the direction of wave propagation is along z. Therefore, with the help of (4) and remembering $\eta_0 = 120\pi$ ohm (Ω) [see the discussion following (6.70)], we obtain

$$\vec{H} = \frac{60\pi \exp j(10^9 t - \beta z)}{120\pi}\vec{a}_y = 0.5 \exp j(10^9 t - \beta z)\vec{a}_y \text{ A/s}. \tag{5}$$

Thus, in (6) we find the desired expression for the magnetic field.

Further, in the given problem, as can be seen from (1) or (5), the wave dependence $\exp j(\omega t - \beta z)$ is $\exp j(10^9 t - \beta z)$, whence we identify the wave angular frequency as

$$\omega = 10^9 \text{ rad/s}. \tag{6}$$

Then the phase propagation constant β in the free-space region can be found with the help of (7) and the relation (6.66) as follows [see the discussion following (6.67)]:

$$\beta = \omega\, (\mu_0 \varepsilon_0)^{1/2} = 10^9 (\mu_0 \varepsilon_0)^{1/2} = \frac{10^9}{3 \times 10^8} = \frac{10}{3} = 3.33 \text{ rad/m}. \tag{7}$$

If now the free-space region is replaced by a dielectric of relative permittivity $\varepsilon_r = 16$, then the new value of β is obtained by replacing ε_0 in (7) by $\varepsilon_r \varepsilon_0 = 16\varepsilon_0$ as:

$$\beta = \frac{10^9}{3 \times 10^8}\sqrt{16} = \frac{10^9 \times 4}{3 \times 10^8} = \frac{40}{3} = 13.33 \text{ rad/m}.$$

6.37 Show that the skin depth δ of copper ($\sigma = 5.8 \times 10^7$ mho/m) is given by $\delta = 0.066/\sqrt{f}$, and hence its value decreases from 66.1 to 2.1 mm as the wave frequency is increased from 1 Hz to 1 kHz.

6.38 Calculate the phase velocity of a uniform plane wave in copper ($\sigma = 5.8 \times 10^7$ mho/m) at 3 MHz.

Answer : 720 m/s.

6.39 Calculate the wave attenuation constant of copper ($\sigma = 5.8 \times 10^7$ mho/m) in dB/cm at 3 MHz.

Answer : 22.74×10^4 dB/m. (Recall 1 Np = 8.68 dB).

6.40 Show that the dc resistance of a rectangular conducting block of length l, breadth b and width equal to the skin depth δ of the material of the block at a given frequency is equal to the ac resistance of the block at the same frequency. Hence, also calculate the value of the ac resistance of the block of copper ($\sigma = 5.8 \times 10^7$ mho/m) of length $l = 4$ cm, breadth $b = 2$ cm and skin depth $\delta = 2.1\mu$ m at a frequency of 1 GHz.

Answer :

The dc resistance of the block R_{dc} of breadth b and width $h = \delta$ (and hence cross-sectional area $bh = b\delta$), where δ is the skin depth, can be written as

$$R_{DC} = \frac{1}{\sigma}\frac{l}{b\delta}\bigg|_{h=\delta}. \tag{1}$$

The ac resistance of the block can be written as

$$R_{\mathrm{ac}} = R_S\frac{l}{b}, \tag{2}$$

where R_S is the surface resistance of the block, which is given by (6.98) and which can be rewritten as

$$R_S = \frac{1}{\sigma\delta}. \tag{3}$$

Plugging (3) in (2) then we get

$$R_{\mathrm{ac}} = \frac{1}{\sigma\delta}\frac{l}{b},$$

which can be rearranged as

$$R_{\mathrm{ac}} = \frac{1}{\sigma}\frac{l}{b\delta}.$$

Comparing (4) with (1) we thus find

$$R_{\mathrm{ac}} = R_{\mathrm{dc}}|_{h=\delta}.$$

We can numerically appreciate the problem by substituting the given data: $l = 4$ cm, breadth $b = 2$ cm, $\sigma = 5.8 \times 10^7$ and skin depth $\delta = 2.1$ μm in (4) as follows:

$$R_{\mathrm{ac}} = \frac{1}{\sigma}\frac{l}{b\delta} = \frac{4 \times 10^{-2}}{5.8 \times 10^7 \times 2 \times 10^{-2} \times 2.1 \times 10^{-6}} = \frac{2}{5.8 \times 10 \times 2.1} = 0.016 \text{ ohm}.$$

6.41 A half-wave dipole antenna has a length of $\lambda/2$, that is, half the wavelength. Compare the length of the half-wave dipole in free space with that in seawater ($\sigma = 4$ mho/m; $\varepsilon = 81\varepsilon_0 = 81 \times 8.854 \times 10^{-12}$ F/m) at the operating frequency of 10 kHz, where one can consider the value of $\omega\varepsilon$ to be much smaller than that of σ.

Answer : The length of the antenna, which is half the wavelength in seawater, can be calculated with the help of (6.112) and is 7.5 m. The antenna length in free space

$= \lambda/2 = c/2f$ at the given frequency $f = 10$ kHz is 150 km.

Multiple-choice questions

6.42 In order to deduce the wave equation in electric field $\vec{E}$

$$\nabla^2 \vec{E} - \mu_0 \varepsilon_0 \frac{\partial^2 \vec{E}}{\partial t^2} = 0$$

for a free-space region, starting from Maxwell's equations, we need to take

(a) $\rho = 0$ (b) $\sigma = 0$
(c) $\rho = 0$ and $\sigma = 0$ (d) $\rho \neq 0$ and $\sigma \neq 0$.

6.43 In order to deduce the wave equation in magnetic field $\vec{H}$

$$\nabla^2 \vec{H} - \mu_0 \varepsilon_0 \frac{\partial^2 \vec{H}}{\partial t^2} = 0$$

for a free-space region, starting from Maxwell's equations, we need to take

(a) $\rho = 0$ (b) $\sigma = 0$
(c) $\rho = 0$ and $\sigma = 0$ (d) $\rho \neq 0$ and $\sigma \neq 0$.

6.44 We can double the skin depth of a conductor if we

(a) decrease the operating frequency four times.
(b) increase the operating frequency two times.
(c) decrease the operating frequency two times.
(d) increase the operating frequency four times.

6.45 The attenuation constant of seawater for relatively higher frequencies, where you take $(\sigma/\omega\varepsilon)^2 \approx 0$is

(a) proportional to frequency.
(b) inversely proportional to frequency.
(c) inversely proportional to frequency squared.
(d) independent of frequency.

6.46 The quantity E associated with a wave propagating along the negative z-direction can be represented by

(a) $E = E_0 \exp(\omega t - \beta z)$ (b) $E = E_0 \cos(\omega t + \beta z)$
(c) $E = E_0 \exp \omega(t - z/v_{\text{ph}})$ (d) $E = E_0 \cos(\omega t - \beta z)$,

where ω, β and v_{ph} are the wave angular frequency, phase propagation constant and phase velocity respectively.

6.47 For a uniform plane transverse electromagnetic wave propagating in the negative z-direction, the following relation holds good:

(a) $\dfrac{E_x}{H_y} = \dfrac{\omega\mu_0}{\beta}$ (b) $-\dfrac{E_y}{H_x} = \dfrac{\omega\mu_0}{\beta}$

(c) $\dfrac{E_y}{H_x} = \dfrac{\omega\mu_0}{\beta}$ (d) $\dfrac{E_x}{H_y} = \dfrac{v_{ph}}{\mu_0}$,

where ω, β and v_{ph} are the wave angular frequency, phase propagation constant and phase velocity respectively.

6.48 The unit of surface resistance is

(a) Ω (b) Ωm (c) Ω/m (d) Ω/m^2

6.49 The wave attenuation constant of a good conductor is

(a) inversely proportional to the wave frequency.
(b) directly proportional to the square root of the conductivity of the medium.
(c) inversely proportional to the square root of the wave frequency.
(d) directly proportional to the conductivity of the medium.

6.50 The resistance of a conducting wire of circular cross section can be reduced by

(a) decreasing its diameter without changing the operating frequency
(b) decreasing its diameter and decreasing the operating frequency
(c) increasing its diameter and decreasing the operating frequency
(d) increasing its diameter and increasing the operating frequency

6.51 Lower frequencies ~10 KHz is preferred to higher frequencies ~10 GHz for seawater communication because

(a) antenna-size is smaller in seawater than in free space while the wave attenuation becomes frequency-independent
(b) wave-attenuation is less at lower frequencies than at higher frequencies even though the antenna size is larger in seawater than in free space
(c) antenna-size is smaller in seawater than in free space even though the wave attenuation is more at lower frequencies than at higher frequencies
(d) wave-attenuation is less at lower frequencies than at higher frequencies and the antenna size is smaller in seawater than in free space.

Answers

6-42 (c) 6-43 (b) 6-44 (a) 6-45 (d) 6-46 (b) 6-47 (c) 6-48 (a) 6-49 (b)
6-50 (c) 6-51(d)

7

Electromagnetic Boundary Conditions

Field quantities for both steady (time-independent) and time-varying (time-dependent) situations will, in general, get modified when the medium is perturbed by the presence of another medium because of the abrupt change in the medium properties at the interface (common boundary) between the media. However, the field quantities would pass through a common set of electromagnetic boundary conditions at the interface or common boundary between the media. Consequently, subject to these boundary conditions, the propagation characteristics of electromagnetic waves change in such a system of bounded media. You will see in one of the subsequent chapters (Chapter 9), how these boundary conditions are used in obtaining the propagation characteristics of electromagnetic waves propagating through a hollow pipe known as the waveguide.

In this chapter, we will first deduce the electromagnetic boundary conditions at the interface between two media in general, without considering whether the media are made of dielectrics or conductors, irrespective of whether the situations are time-independent or time-dependent (Section 7.1). Next, we will interpret special cases of these general boundary conditions at a dielectric-dielectric interface (Section 7.2), dielectric-conductor interface (Section 7.3), and conductor-conductor interface (Section 7.4). We will also take up the application of these boundary conditions to study the phenomenon of reflection of an electromagnetic wave at a dielectric-conductor interface (Section 7.5) and that of reflection as well as refraction at a dielectric-dielectric interface (Section 7.6), it being implied that if one of the two media forming the interface is a dielectric, it could as well be free space whose permittivity $\varepsilon = \varepsilon_0$.

7.1 General Boundary Conditions

Let us consider two media with a boundary or an interface between them (Fig. 7.1), taking $\mu = \mu_1$, $\varepsilon = \varepsilon_1$ for medium 1 and $\mu = \mu_2$, $\varepsilon = \varepsilon_2$ for medium 2. Let us then, for the purpose of deriving the electromagnetic boundary conditions at a point P on the interface between these media, construct a parallelepiped volume element enclosing the point P (Gaussian volume; see Section 3.1.3). Let this volume element be say, in the form of a 'pillbox' , typically of rectangular cross section, with infinitesimal thickness dh and infinitesimal area dS for each of its bottom and top faces, AB and DC respectively, such that these faces lie in medium 1 and medium 2 respectively. The infinitesimal volume element is then $d\tau = dS\,dh$ (Fig. 7.1). (We could have taken the parallelepiped volume element to take some other shape, for instance, circular, instead of rectangular, such that the areas of its bottom and top faces each have the same infinitesimal area dS).

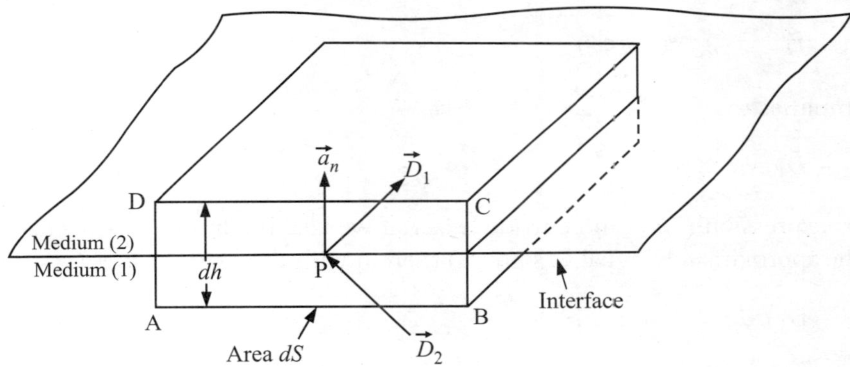

Fig. 7.1 *Pillbox Gaussian parallelepiped volume element enclosing the point P at the interface between medium 1 and medium 2 showing the electric displacement vectors $\vec{D}_1$ and $\vec{D}_2$ as well as the unit vector $\vec{a}_n$ directed from medium 1 to 2 at point P.*

By applying the integral form of Maxwell's equation (5.72)

$$\oint_S \vec{D} \cdot \vec{a}_n dS = \int_\tau \rho\, d\tau \qquad \text{(5.72) rewritten} \tag{7.1}$$

to the volume element $d\tau$, the following expression is obtained:

$$\vec{D}_2.\vec{a}_n dS + \vec{D}_1.(-\vec{a}_n) dS \cong \rho d\tau\,. \tag{7.2}$$

Here, $\vec{D}_1 (= \varepsilon_1 \vec{E}_1)$ and $\vec{D}_2 (= \varepsilon_2 \vec{E}_2)$ are the electric displacements in medium 1 and medium 2 respectively at point P, and $\vec{a}_n$ is the unit vector at P directed from medium 1 to 2. We have taken the unit vector normal to the top face as $\vec{a}_n$ and that normal to the bottom face as $-\vec{a}_n$. The first and the second terms of the left-hand side of (7.2) are the flux of electric displacement contributed by the top and bottom faces, DC and AB, of the Gaussian volume element. Furthermore, while writing (7.2), we have made the following approximations (Fig. 7.1):

1. The electric displacements $\vec{D}_1$ and $\vec{D}_2$ are each constant inside the volume element

$d\tau$, and hence also at the bottom and top faces of the pillbox volume element ABCD ($d\tau = dS\,dh$).

2. The volume charge density ρ is approximately constant within the volume element ABCD ($d\tau = dS\,dh$).

3. The contribution of the side wall to the surface integral on the left-hand side of (7.1) is ignored considering the volume element to be extremely thin ($dh \cong 0$).

In view of the above approximations, we have used the 'approximately equal' sign ($\simeq$) in (7.2). Putting $d\tau = dS\,dh$, we can now write the approximate expression (7.2) as follows:

$$(\vec{D}_2 - \vec{D}_1) . \vec{a}_n\, dS \cong \rho\, dS\, dh$$

which simplifies to

$$(\vec{D}_2 - \vec{D}_1) . \vec{a}_n \cong \rho\, dh. \tag{7.3}$$

However, expression (7.3) can be made 'exact' if we take the limit $dh \to 0$, enabling us to replace the approximately equal to sign ($\simeq$) therein with the equal to sign (=) as follows:

$$(\vec{D}_2 - \vec{D}_1) . \vec{a}_n = \lim_{dh \to 0} \rho\, dh. \tag{7.4}$$

You will appreciate that taking the limit $dh \to 0$ amounts to shrinking the pillbox volume element to a thin layer at the interface or boundary between the two media (Fig. 7.1). This is how we obtain in (7.4), one of the electromagnetic boundary conditions at a point on the interface between medium 1 and medium 2.

At this stage, let us define a quantity called *surface charge density* ρ as

$$\rho_s = \lim_{dh \to 0} \rho\, dh. \tag{7.5}$$

Since the dimension of the volume charge density ρ is C/m^3 and that of dh is m, we can see from (7.5) that the unit of the surface charge density is C/m^2. It is inherently assumed that as we decrease the thickness of the pillbox, that is, as $dh \to 0$, the volume charge density will increase, that is, $\rho \to \infty$, such that the surface charge density ρ_s given by (7.5) takes on a finite value.

Now, substituting (7.5) in (7.4), we obtain the electromagnetic boundary condition (7.4) as:

$$(\vec{D}_2 - \vec{D}_1) . \vec{a}_n = \rho_s. \tag{7.6}$$

Since $\vec{D}_2 . \vec{a}_n$ and $\vec{D}_1 . \vec{a}_n$ represent the normal components of $\vec{D}_2$ and $\vec{D}_1$ respectively, the electromagnetic boundary condition (7.6) states that, in general, the normal component of electric displacement is discontinuous at the interface between the two media and that the amount of discontinuity is equal to the surface charge density at the interface.

Now, if we start from the integral form of Maxwell's equation (5.73) involving magnetic flux density instead of the integral form involving electric displacement and use exactly the same method as followed while obtaining the boundary condition (7.6), then we will arrive at the following boundary condition:

$$(\vec{B}_2 - \vec{B}_1) \,.\, \vec{a}_n = 0. \tag{7.7}$$

Since $\vec{B}_2 . \vec{a}_n$ and $\vec{B}_1 . \vec{a}_n$ represent the normal components of $\vec{B}_2$ and $\vec{B}_1$ respectively, the electromagnetic boundary condition (7.7) states that, in general, the normal component of magnetic flux density is continuous at the interface between the two media (Fig. 7.1).

Next, let us consider a rectangle element of infinitesimal length dl and infinitesimal thickness dh enclosing the point P on the interface between medium 1 and medium 2 such that the bottom and top lengths of the rectangle lay in medium 1 and medium 2 respectively. With a view to deriving yet another electromagnetic boundary condition, now we have to apply Maxwell's equation in the integral form (5.75) to the closed path along the perimeter of the rectangle. For the sake of convenience, let us rewrite (5.75), replacing $\vec{a}_n$ therein by $\vec{n}$, as follows:

$$\oint_l \vec{H} \cdot d\vec{l} = \int_S \left(\vec{J} + \frac{\partial \vec{D}}{\partial t} \right) \cdot \vec{n} \, dS. \tag{7.8}$$

Using the symbol $\vec{J}_{\text{total}}$ to represent the total current density comprising the conduction current density $\vec{J}$ and the displacement current density $\partial \vec{D}/\partial t$,

$$\vec{J}_{\text{total}} = \vec{J} + \frac{\partial \vec{D}}{\partial t} \tag{7.9}$$

we can express (7.8) as

$$\oint_l \vec{H} . d\vec{l} = \int_S \vec{J}_{\text{total}} . \vec{n} \, dS. \tag{7.10}$$

We have used the symbol $\vec{n}$ instead of $\vec{a}_n$ in (7.10) to represent the unit vector perpendicular to the area of the rectangle element, since the symbol $\vec{a}_n$ has been already used here to represent the unit vector directed from the medium 1 to medium 2 at point P (Fig. 7.2). The sense of the direction of the unit vector $\vec{n}$ is taken as the direction of the linear motion that a screw would have if it were rotated along the sequence of the closed line integral of the left-hand side of (7.8), that is, from A to B; B to C; C to D; and then back to A from D (Fig. 7.2).

Let $\vec{H}_1$ and $\vec{H}_2$ be the magnetic fields in the media 1 and 2 respectively at the point P on the interface between them, $\vec{J}_{\text{total}}$ being the current density through the area element.

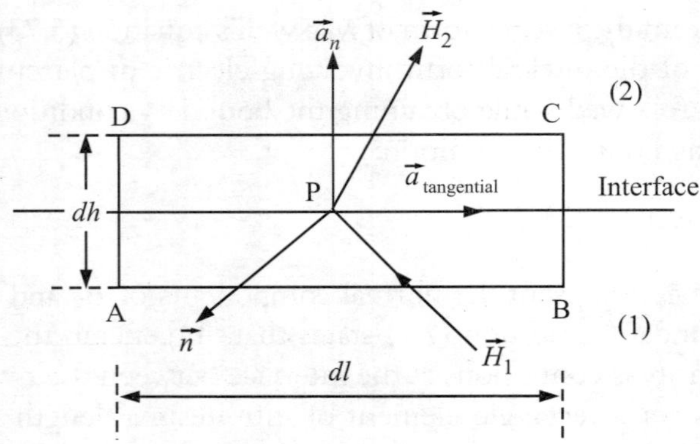

Fig. 7.2 *Rectangle element enclosing the point P on the interface between the media, showing the path along which the line integral of magnetic field is taken. Also shown are the unit vectors: (i) $\vec{n}$ normal to the area element, (ii) $\vec{a}_n$ directed from the medium 1 to 2 at the point P, and (iii) $\vec{a}_{tangential}$, tangential to the interface at point P such that the relation $\vec{a}_n \times \vec{n} = \vec{a}_{tangential}$ holds good.*

Also, let $\vec{a}_{\text{tangential}}$ be a unit vector tangential to the interface at the point P and parallel to the direction from A to B such that the relation

$$\vec{a}_n \times \vec{n} = \vec{a}_{\text{tangential}} \tag{7.11}$$

is satisfied (Fig. 7.2).

If we now apply (7.10) to the closed path, from A to B; B to C; C to D; and D to A of the rectangle element (Fig. 7.2), then we obtain

$$\vec{H}_1 . \vec{a}_{\text{tangential}}\, dl + \vec{H}_2 . (-\vec{a}_{\text{tangential}})\, dl \cong \vec{J}_{\text{total}} . \vec{n}\, dS$$

which, on putting $dS = dh\, dl$ can be written as

$$(\vec{H}_1 - \vec{H}_2) . \vec{a}_{\text{tangential}} \cong \vec{J}_{\text{total}} . \vec{n}\, dh. \tag{7.12}$$

The first and the second terms of the left-hand side of (7.12) are contributions to the line integral of magnetic field occurring in (7.10) by the bottom and the top lengths of the rectangle element, AB and CD, respectively.

Moreover, we have made the following approximations here while writing (7.11):

1. The magnetic fields $\vec{H}_1$ and $\vec{H}_1$ are each constant inside the area element dS, and hence also along the bottom and top lengths, AB and DC, respectively of the area element ABCD ($dS = dh\, dl$)
2. The current density $\vec{J}_{\text{total}}$ is approximately constant within the area element ABCD ($dS = dh\, dl$), and
3. The side wall contribution to the line integral on the left-hand side of (7.8) is ignored, considering the area element to be extremely thin ($dh \cong 0$).

We can now substitute (7.11) into the approximate expression (7.12) to express the latter as

$$(\vec{H}_1 - \vec{H}_2).\,\vec{a}_n \times \vec{n} \cong \vec{J}_{\text{total}}.\,\vec{n}\,dh. \tag{7.13}$$

Using the vector identity $\vec{A}\cdot(\vec{B}\times\vec{C}) = \vec{C}\cdot(\vec{A}\times\vec{B})$, in which we interpret $\vec{A} = \vec{H}_1 - \vec{H}_2$, $\vec{B} = \vec{a}_n$ and $\vec{C} = \vec{n}$, we can express the left-hand side of (7.13) as

$$(\vec{H}_1 - \vec{H}_2).\vec{a}_n \times \vec{n} = \vec{n}\cdot(\vec{H}_1 - \vec{H}_2)\times\vec{a}_n,$$

which can be now substituted in (7.13) to obtain

$$\vec{n}\,.\,(\vec{H}_1 - \vec{H}_2)\times\vec{a}_n \cong \vec{J}_{\text{total}}\,.\,\vec{n}\,dh,$$

which can be further expressed, by interchanging the dot product on the its right-hand side as

$$\vec{n}\,.\,(\vec{H}_1 - \vec{H}_2)\times\vec{a}_n \cong \vec{n}\,.\,\vec{J}_{\text{total}}\,dh. \tag{7.14}$$

By rearranging the terms of (7.14), we can express it as

$$\vec{n}\,.\,[(\vec{H}_1 - \vec{H}_2)\times\vec{a}_n - \vec{J}_{\text{total}}dh] \cong 0,$$

which can be put as

$$\vec{n}\cdot\vec{G} = (1)(G)\cos\theta = G\cos\theta \cong 0 \tag{7.15}$$

where

$$\vec{G} = (\vec{H}_1 - \vec{H}_2)\times\vec{a}_n - \vec{J}_{\text{total}}dh. \tag{7.16}$$

G being the magnitude of $\vec{G}$, and θ the angle between the unit vector $\vec{n}$ and $\vec{G}$.

However, the orientation of the rectangle element ABDC is arbitrary, which would, in turn, make $\vec{n}$, which is normal to the area of the rectangle, arbitrary as well. This would also make θ [defined following (7.16)] and hence $\cos\theta$ arbitrary. Therefore, irrespective of the orientation of the rectangle ABDC, and hence that of the value of $\cos\theta$, we can write from (7.15)

$$\vec{G} \cong 0. \tag{7.17}$$

Substituting (7.16) in (7.17), then we obtain

$$(\vec{H}_1 - \vec{H}_2)\times\vec{a}_n - \vec{J}_{\text{total}}dh \cong 0,$$

which can be rearranged as

$$(\vec{H}_1 - \vec{H}_2)\times\vec{a}_n \cong \vec{J}_{\text{total}}dh. \tag{7.18}$$

However, we can make the expression (7.18) 'exact' and replace the approximately equal sign ($\cong$) therein by the exactly equal sign ($=$) if we now take the limit $dh \to 0$ as follows:

$$(\vec{H}_1 - \vec{H}_2) \times \vec{a}_n = \lim_{dh \to 0} \vec{J}_{\text{total}} dh. \tag{7.19}$$

Substituting $\vec{J}_{\text{total}}$ from (7.9) into (7.19), we can write

$$(\vec{H}_1 - \vec{H}_2) \times \vec{a}_n = \lim_{dh \to 0} \left(\vec{J} + \frac{\partial \vec{D}}{\partial t}\right) dh = \lim_{dh \to 0} \vec{J} dh + \lim_{dh \to 0} \frac{\partial \vec{D}}{\partial t} dh. \tag{7.20}$$

We assume that as we decrease the thickness of the rectangle element, that is, as $dh \to 0$, the conduction current density will increase, that is, $\vec{J} \to \infty$, such that the first term of the right-hand side of (7.20) takes on a finite value called the surface current density $\vec{J}_S$ given by

$$\vec{J}_S = \lim_{dh \to 0} \vec{J} dh \quad . \tag{7.21}$$

However, the second term of (7.20) vanishes on taking the limit $dh \to 0$, since $\vec{D}$ and hence also its partial derivative $\partial \vec{D}/\partial t$ is finite in the region (ABDC). Therefore, taking into account the contribution only of the first term in (7.20), we obtain

$$(\vec{H}_1 - \vec{H}_2) \times \vec{a}_n = \lim_{dh \to 0} \vec{J} dh = \vec{J}_s \quad . \tag{7.22}$$

Notice from (7.22) that the dimension of the surface current density $\vec{J}_s$ is A/m since the dimension of $\vec{J}$ is A/m^2 and that of dh is m.

In view of the vector identity $\vec{A} \times \vec{B} = -\vec{B} \times \vec{A}$, we can express the left-hand side of (7.22) as

$$(\vec{H}_1 - \vec{H}_2) \times \vec{a}_n = -\vec{a}_n \times (\vec{H}_1 - \vec{H}_2) = \vec{a}_n \times (\vec{H}_2 - \vec{H}_1),$$

which, substituted in (7.22) gives

$$\vec{a}_n \times (\vec{H}_2 - \vec{H}_1) = \vec{J}_s. \tag{7.23}$$

In (7.23), we obtain yet another electromagnetic boundary condition. Since the directions of $\vec{a}_n \times \vec{H}_2$ that of $\vec{a}_n \times \vec{H}_1$ are each tangential to the surface at P, the boundary condition (7.23) states that the tangential component of the magnetic field is discontinuous at the interface between two media, the amount of discontinuity being equal to the magnitude of surface current density at the interface.

Next, let us start from the integral form of Maxwell's equation (5.74) involving the electric field instead of the integral form (5.75) involving the magnetic field, and use the exactly the same method as followed while obtaining the boundary condition (7.23). Now, instead of (7.20), we will easily arrive at

$$(\vec{E}_1 - \vec{E}_2) \times \vec{a}_n = \lim_{dh \to 0} \left(-\frac{\partial \vec{B}}{\partial t} dh\right),$$

which, in view of the vector identity $\vec{A} \times \vec{B} = -\vec{B} \times \vec{A} = (\vec{B}) \times (-\vec{A})$, in which we take $\vec{A} = \vec{E}_1 - \vec{E}_2$, that is, $-\vec{A} = \vec{E}_2 - \vec{E}_1$ and $\vec{B} = \vec{a}_n$, becomes

$$\vec{a}_n \times (\vec{E}_2 - \vec{E}_1) = \lim_{dh \to 0} \left(-\frac{\partial \vec{B}}{\partial t} dh \right). \tag{7.24}$$

However, since irrespective of the thickness of the the rectangle element, $\vec{B}$ is finite and hence $\partial \vec{B}/\partial t$ is finite as well. Hence the limit $dh \to 0$ will make the right-hand side of (7.21) zero, making it

$$\vec{a}_n \times (\vec{E}_2 - \vec{E}_1) = 0. \tag{7.25}$$

Thus, we derive in (7.25) another electromagnetic boundary condition at the interface between the media 1 and 2. Further, since the directions of $\vec{a}_n \times \vec{E}_2$ and $\vec{a}_n \times \vec{E}_1$ are each tangential to the surface at P, the boundary condition (7.25) states that the tangential component of the electric field is continuous at the interface between two media. Let us now put together all the general four electromagnetic boundary conditions for the sake of future reference, derived from the integral forms of Maxwell's equations in this section (Table 7.1).

Table 7.1 *General electromagnetic boundary conditions indicating the integral Maxwell's equations from which they are respectively derived**

General electromagnetic boundary condition	Originating integral Maxwell's equation	Remark
$(\vec{D}_2 - \vec{D}_1).\vec{a}_n = \rho_s$ (7.6)	$\oint_S \vec{D} \cdot \vec{a}_n dS = \int_\tau \rho \, d\tau$ (5.72)	Derivation is given in the text from first principles with all the steps of deduction.
$(\vec{B}_2 - \vec{B}_1).\vec{a}_n = 0$ (7.7)	$\oint_S \vec{B} \cdot d\vec{S} = \oint_S \vec{B} \cdot \vec{a}_n \, dS = 0$ (5.73)	Deduction steps are not given to avoid repetition, since they exactly follow those given for deriving (7.6).
$\vec{a}_n \times (\vec{H}_2 - \vec{H}_1) = \vec{J}_s$ (7.23)	$\oint_l \vec{H} \cdot d\vec{l} = \int_S \left(\vec{J} + \frac{\partial \vec{D}}{\partial t} \right) \cdot \vec{a}_n \, dS$ (5.75)	Derivation is given in the text from first principles with all the steps of deduction.
$\vec{a}_n \times (\vec{E}_2 - \vec{E}_1) = 0$ (7.24)	$\oint_l \vec{E} \cdot d\vec{l} = -\int_S \frac{\partial \vec{B}}{\partial t} \cdot \vec{a}_n \, dS$ (5.74)	Deduction steps are not given to avoid repetition, since they exactly follow those given for deriving (7.20).

*The cross product of two vector quantities being another vector quantity, it would be more appropriate to take the right-hand side of (7.25), rewritten in Table 7.1, as the null vector $\vec{0}$ instead of 0.

We can interpret these general electromagnetic boundary conditions for two special cases of a dielectric-dielectric (Section 7.1) and conductor-dielectric/free-space interfaces (Section 7.2). For this purpose, it is worth reviewing some of the basic behaviours of media—conductors and dielectrics—with regard to relaxation time, existence of a free charge in the bulk of the media, surface resistance, surface current density, and field (electric or magnetic) in the media (Table 7.2).

7.2 Boundary Conditions at Dielectric-Dielectric Interface

Let medium 1 ($\varepsilon = \varepsilon_1, \mu = \mu_0$) and medium 2 ($\varepsilon = \varepsilon_2, \mu = \mu_0$) both be dielectrics (Fig. 7.3a). It is possible to establish electric as well as magnetic field in a dielectric (Table 7.2) and, therefore we have $\vec{E}_{1,2} \neq 0$, $\vec{D}_{1,2} \neq 0$ and$\vec{H}_{1,2} \neq 0$, $\vec{B}_{1,2} \neq 0$, the subscripts 1 and 2 referring to media 1 and 2 respectively. Further, the relaxation time of a dielectric being rather large, the bulk of a dielectric can be charged with a finite volume charge density ($\rho \neq 0$) that makes the surface charge density $\rho_s = 0$ at the interface between the dielectrics, as can be seen by putting $\rho \neq 0$ in (7.5) (Section 5.1.2 and Table 7.2). Moreover, a dielectric does not conduct current, making the current density $\vec{J} = 0$, which in turn makes $\vec{J}_s = 0$ as can be seen by putting $\vec{J} = 0$ in (7.21) (Section 5.1.2 and Table 7.2). (See also the note following (3.48) and Appendix A5.4.)

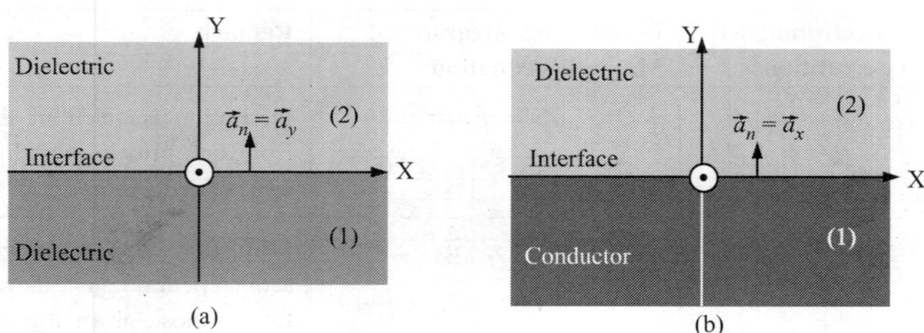

Fig. 7.3 *Interface between (a) medium 1 made of a dielectric ($\varepsilon = \varepsilon_1, \mu = \mu_0$), and medium 2 made of another dielectric ($\varepsilon = \varepsilon_2, \mu = \mu_0$); and (b) between medium 1, made of a conductor ($\varepsilon = \varepsilon_1 = \varepsilon_0, \mu = \mu_0$), and medium 2, made of a dielectric ($\varepsilon = \varepsilon_2, \mu = \mu_0$) (Z axis being perpendicular the plane of the paper and directed towards the reader, as indicated by a dot surrounded by a circle).*

Therefore, if we now substitute $\vec{E}_{1,2} \neq 0$, $\vec{D}_{1,2} \neq 0$ and $\vec{H}_{1,2} \neq 0$, $\vec{B}_{1,2} \neq 0$ as well as $\rho_s = 0$ and $\vec{J}_s = 0$ in the general electromagnetic boundary conditions (7.6), (7.7), (7.23), and (7.25) (Table 7.1), we obtain the following electromagnetic boundary conditions at the dielectric-dielectric interface for both time-independent and time-dependent situations (Table 7.3):

$$\left.\begin{array}{ll}(\vec{D}_2 - \vec{D}_1).\vec{a}_n = 0 & \text{(a)}\\ (\vec{B}_2 - \vec{B}_1).\vec{a}_n = 0 & \text{(b)}\\ \vec{a}_n \times (\vec{H}_2 - \vec{H}_1) = 0 & \text{(c)}\\ \vec{a}_n \times (\vec{E}_2 - \vec{E}_1) = 0 & \text{(d)}\end{array}\right\}. \tag{7.26}$$

Table 7.2 *Behaviour of a good conductor and a perfect dielectric*

Dielectric	**Conductor**
• The relaxation time of a dielectric is large. • The charge can stay longer inside the bulk of a dielectric without appearing at its surface. • The bulk of a dielectric can be electrically charged and a finite volume charge density can be established inside the bulk of a dielectric resulting in a zero surface charge density at the surface of a dielectric ($\rho_s = 0$). • The electric field and the electric displacement can be established inside a dielectric ($\vec{E} \neq 0$, $\vec{D} \neq 0$), for both *time-independent* and *time-dependent* situations. • A finite magnetic field or magnetic flux density can be established inside a dielectric ($\vec{H} \neq 0$, $\vec{B} \neq 0$), independent of electric field, for both *time-independent* and *time-dependent* situations. • No current flows through a dielectric and therefore, the surface current density becomes zero at the dielectric surface ($\vec{J}_s = 0$).	• The relaxation time of a good conductor is small. • The charge inside the bulk of a good conductor decays very fast to appear with a large volume charge density at the conductor surface. The volume charge density is concentrated over a thin layer on the surface of the conductor, resulting in a finite surface charge density on the conductor surface ($\rho_s \neq 0$) for both *time-independent* and *time-dependent* situations. • The bulk of a conductor cannot be electrically charged, resulting in no electric field or electric displacement inside the conductor ($\vec{E} = 0$, $\vec{D} = 0$) for both *time-independent* and *time-dependent* situations. • The magnetic field and the magnetic flux density each become zero ($\vec{H} = 0$, $\vec{B} = 0$) inside a conductor for *time-dependent* situations. • A finite magnetic field or magnetic flux density can be established ($\vec{H} \neq 0$, $\vec{B} \neq 0$) independent of an electric field for time-independent situations. • A finite current can be made to flow through the volume of a conductor for time-independent situations, resulting in zero surface current density ($\vec{J}_s = 0$) at the conductor surface for time-independent situations. • Large current density can be concentrated over a thin layer on the conductor surface, resulting in a finite surface current density ($J_s \neq 0$) for *time-dependent* situations.

Let us take up a very simple problem to illustrate the use of the boundary conditions at the interface between a dielectric of relative permittivity (or dielectric constant) ε_{r1} (medium 1) and another dielectric of relative permittivity ε_{r2} (medium 2). Now, suppose we know the electric field $\vec{E}_1$ in the first of these dielectrics as $\vec{E}_1 = l\vec{a}_x + m\vec{a}_y + n\vec{a}_z$ (given). How can we find the electric displacement $\vec{D}_1$ (unknown) in the first dielectric (medium 1) as well as the electric field $\vec{E}_2$ (unknown) and the electric displacement $\vec{D}_2$ (unknown) in the second dielectric (medium 2)?

Table 7.3 *Electromagnetic boundary conditions at the dielectric-dielectric and at the conductor-dielectric interfaces for both time-independent and time-dependent situations**

Dielectric 1–Dielectric 2 interface	**Conductor 1–Dielectric 2 interface**	
Both time-independent and time-dependent	Time-independent	Time-dependent
$(\vec{D}_2 - \vec{D}_1).\vec{a}_n = 0$ (a) $(\vec{B}_2 - \vec{B}_1).\vec{a}_n = 0$ (b) $\vec{a}_n \times (\vec{H}_2 - \vec{H}_1) = 0$ (c) $\vec{a}_n \times (\vec{E}_2 - \vec{E}_1) = 0$ (d) (7.26)	$\vec{D}_2.\vec{a}_n = \rho_s$ (a) $(\vec{B}_2 - \vec{B}_1).\vec{a}_n = 0$ (b) $\vec{a}_n \times (\vec{H}_2 - \vec{H}_1) = 0$ (c) $\vec{a}_n \times \vec{E}_2 = 0$ (d) (7.45)	$\vec{D}_2.\vec{a}_n = \rho_s$ (a) $\vec{B}_2.\vec{a}_n = 0$ (b) $\vec{a}_n \times \vec{H}_2 = \vec{J}_s$ (c) $\vec{a}_n \times \vec{E}_2 = 0$ (d) (7.46)

*The boundary conditions at a conductor-dilectric interface is analysed in section 7.3.

Let us take the interface on the XZ plane. The region $y < 0$ refers to medium 1 and the region $y > 0$ refers to medium 2 such that the unit vector directed from medium 1 to 2 is along y (Fig. 7.3a), that is:

$$\vec{a}_n = \vec{a}_y. \tag{7.27}$$

Thus, in this problem, let us find the components of

$$\vec{D}_1 = D_{1x}\vec{a}_x + D_{1y}\vec{a}_y + D_{1z}\vec{a}_z \tag{7.28}$$

$$\vec{D}_2 = D_{2x}\vec{a}_x + D_{2y}\vec{a}_y + D_{2z}\vec{a}_z \tag{7.29}$$

$$\vec{E}_2 = E_{2x}\vec{a}_x + E_{2y}\vec{a}_y + E_{2z}\vec{a}_z \tag{7.30}$$

in terms of the given components l, m, and n of the 'given' electric field in the first dielectric (medium 1)

$$\vec{E}_1 = l\vec{a}_x + m\vec{a}_y + n\vec{a}_z \tag{7.31}$$

where

$$\left.\begin{array}{ll} \vec{D}_1 = \varepsilon_1\vec{E}_1 = \varepsilon_0\varepsilon_{r1}\vec{E}_1 & \text{(a)} \\ \vec{D}_2 = \varepsilon_2\vec{E}_2 = \varepsilon_0\varepsilon_{r2}\vec{E}_2 & \text{(b)} \end{array}\right\}. \tag{7.32}$$

Substituting (7.31) in (7.32a), we get

$$\vec{D}_1 = \varepsilon_0 \varepsilon_{r1}(l\vec{a}_x + m\vec{a}_y + n\vec{a}_z) \tag{7.33}$$

which when compared with (7.28) immediately gives the following components of $\vec{D}_1$:

$$\left.\begin{array}{ll} D_{1x} = \varepsilon_0 \varepsilon_{r1} l & \text{(a)} \\ D_{1y} = \varepsilon_0 \varepsilon_{r1} m & \text{(b)} \\ D_{1z} = \varepsilon_0 \varepsilon_{r1} n & \text{(c)} \end{array}\right\}. \tag{7.34}$$

Now, if we substitute (7.28) and (7.29) into the following boundary condition at the interface between the two dielectrics (Table 7.3):

$$(\vec{D}_2 - \vec{D}_1)\,.\,\vec{a}_n = 0 \qquad \text{(7.26}a\text{) rewritten} \tag{7.35}$$

with the unit vector $\vec{a}_n$ interpreted as $\vec{a}_n = \vec{a}_y$ [see (7.27)], we get

$$[D_{2x}\vec{a}_x + D_{2y}\vec{a}_y + D_{2z}\vec{a}_z - (D_{1x}\vec{a}_x + D_{1y}\vec{a}_y + D_{1z}\vec{a}_z)] \cdot \vec{a}_y = 0. \tag{7.36}$$

Rearranging the terms of (7.36), we can write

$$[(D_{2x} - D_{1x})\vec{a}_x + (D_{2y} - D_{1y})\vec{a}_y + (D_{2z} - D_{1z})\vec{a}_z].\vec{a}_y = 0,$$

which, in view of the relations: $\vec{a}_x \cdot \vec{a}_y = 0$, $\vec{a}_y \cdot \vec{a}_y = 1$ and $\vec{a}_z \cdot \vec{a}_y = 0$ becomes

$$D_{2y} - D_{1y} = 0,$$

giving

$$D_{2y} = D_{1y}. \tag{7.37}$$

Substituting (7.34b) in (7.37), we then obtain

$$D_{2y} = \varepsilon_0 \varepsilon_{r1} m\,. \tag{7.38}$$

Next, let us substitute (7.30) and (7.31) into the following boundary condition (See Table 7.3)

$$\vec{a}_n \times (\vec{E}_2 - \vec{E}_1) = 0 \qquad \text{[(7.26}d\text{) rewritten]} \tag{7.39}$$

interpreting therein $\vec{a}_n = \vec{a}_y$ to write

$$\vec{a}_y \times [(E_{2x}\vec{a}_x + E_{2y}\vec{a}_y + E_{2z}\vec{a}_z) - (l\vec{a}_x + m\vec{a}_y + n\vec{a}_z)] = 0,$$

which can be rearranged as

$$\vec{a}_y \times [(E_{2x} - l)\vec{a}_x + (E_{2y} - m)\vec{a}_y + (E_{2z} - n)\vec{a}_z)] = 0. \tag{7.40}$$

In view of the relations $\vec{a}_y \times \vec{a}_x = -\vec{a}_z$, $\vec{a}_y \times \vec{a}_y = 0$ and $\vec{a}_y \times \vec{a}_z = \vec{a}_x$, we can express (7.40) as

$$-(E_{2x} - l)\vec{a}_z + (E_{2z} - n)\vec{a}_x = 0,$$

which gives (since the unit vectors $\vec{a}_x$ and $\vec{a}_z$ are unequal, though they are equal in magnitude, being equal to unity, yet different in directions being at right angle to each other):

$$\left.\begin{aligned} E_{2x} - l = 0 \\ E_{2z} - n = 0 \end{aligned}\right\},$$

whence we obtain

$$\left.\begin{aligned} E_{2x} = l \\ E_{2z} = n \end{aligned}\right\}. \tag{7.41}$$

With the help of (7.41), we can then write

$$\left.\begin{aligned} D_{2x} = \varepsilon_2 E_{2x} = \varepsilon_2 l = \varepsilon_0 \varepsilon_{r2} l \\ D_{2z} = \varepsilon_2 E_{2z} = \varepsilon_2 n = \varepsilon_0 \varepsilon_{r2} n \end{aligned}\right\}. \tag{7.42}$$

Substituting (7.38) and (7.42) into (7.29), we obtain

$$\vec{D}_2 = \varepsilon_0 \varepsilon_{r2} l\, \vec{a}_x + \varepsilon_0 \varepsilon_{r1} m\, \vec{a}_y + \varepsilon_0 \varepsilon_{r2} n\, \vec{a}_z \tag{7.43}$$

with the help of which we get

$$\begin{aligned} \vec{E}_2 = \frac{\vec{D}_2}{\varepsilon_2} &= \frac{\varepsilon_0 \varepsilon_{r2} l\, \vec{a}_x + \varepsilon_0 \varepsilon_{r1} m\, \vec{a}_y + \varepsilon_0 \varepsilon_{r2} n\, \vec{a}_z}{\varepsilon_0 \varepsilon_{r2}} \\ &= l\, \vec{a}_x + m \frac{\varepsilon_{r1}}{\varepsilon_{r1}} \vec{a}_y + n\, \vec{a}_z. \end{aligned} \tag{7.44}$$

The result of the problem is summarised in Table 7.4.

Table 7.4 *Results of a typical problem on dielectric-dielectric interface* $(y = 0)$

Quantity given		Quantity obtained with the help of the boundary conditions: $(\vec{D}_2 - \vec{D}_1).\vec{a}_n = 0$ (7.26*a*) and $\vec{a}_n \times (\vec{E}_2 - \vec{E}_1) = 0$ (7.26*d*)	
$\vec{E}_1 = l\vec{a}_x + m\vec{a}_y + n\vec{a}_z$	(7.31)	$\vec{D}_1 = \varepsilon_0 \varepsilon_{r1} (l\vec{a}_x + m\vec{a}_y + n\vec{a}_z)$	(7.33)
		$\vec{D}_2 = \varepsilon_0 \varepsilon_{r2} l\vec{a}_x + \varepsilon_0 \varepsilon_{r1} m\vec{a}_y + \varepsilon_0 \varepsilon_{r2} n\vec{a}_z$	(7.43)
		$\vec{E}_2 = l\vec{a}_x + m\frac{\varepsilon_{r1}}{\varepsilon_{r2}}\vec{a}_y + n\vec{a}_z$	(7.44)

EXAMPLE E7-1

The two dielectric regions, 1 and 2, of relative permittivity values $\varepsilon_{r1} = 5$ and $\varepsilon_{r2} = 2$ respectively, form an interface at the XY plane defined by $z = 0$. If $\vec{D}_1 = 2\vec{a}_x - 3\vec{a}_z$ C/m^2, find $\vec{D}_2$ and the angle θ that $\vec{D}_2$ makes with the normal to the interface.

SOLUTION

The approach is given following (7.26). The results given in Table 7.4 are valid for the dielectric-dielectric interface ($y = 0$). However, in the present problem, the interface is $z = 0$. Accordingly, you can modify the results of Table 7.4, to be more specific, (7.43), to write

$$\vec{D}_2 = \varepsilon_0\varepsilon_{r2} l\vec{a}_x + \varepsilon_0\varepsilon_{r2} m\vec{a}_y + \varepsilon_0\varepsilon_{r1} n\vec{a}_z. \tag{E7-1.1}$$

Similarly, modifying (7.44), we can write

$$\vec{E}_2 = l\vec{a}_x + m\vec{a}_y + n\frac{\varepsilon_{r1}}{\varepsilon_{r2}}\vec{a}_z. \tag{E7-1.2}$$

Further, we can rewrite (7.31) as

$$\vec{E}_1 = l\vec{a}_x + m\vec{a}_y + n\vec{a}_z. \tag{E7-1.3}$$

Similarly, (7.33) can be rewritten as

$$\vec{D}_1 = \varepsilon_0\varepsilon_{r1}(l\vec{a}_x + m\vec{a}_y + n\vec{a}_z). \tag{E7-1.4}$$

In the present problem,

$$\vec{D}_1 = 2\vec{a}_x - 3\vec{a}_z. \quad \text{(given)} \tag{E7-1.5}$$

Therefore, comparing (E7-1.4) and (E7-1.5), we get $\varepsilon_0\varepsilon_{r1} l = 2$, $m = 0$ and $\varepsilon_0\varepsilon_{r1} n = -3$, whence putting $\varepsilon_{r1} = 5$ and $\varepsilon_{r2} = 2$ (given), we obtain

$$\left.\begin{aligned} l &= \frac{2}{5\varepsilon_0} \\ m &= 0 \\ n &= -\frac{3}{5\varepsilon_0} \end{aligned}\right\}. \tag{E7-1.6}$$

Putting (E7-1.6) in (E7-1.1), and in view of $\varepsilon_{r1} = 5$ and $\varepsilon_{r2} = 2$ (given), we obtain

$$\vec{D}_2 = \varepsilon_0(2)\left(\frac{2}{5\varepsilon_0}\right)\vec{a}_x - \varepsilon_0(5)\left(\frac{3}{5\varepsilon_0}\right)\vec{a}_z,$$

which simplifies to

$$\vec{D}_2 = 0.8\vec{a}_x - 3\vec{a}_z \text{ nC/m}^2. \tag{E7-1.7}$$

In this problem, the unit vector normal to the interface $z = 0$ is $\vec{a}_z$. In order to find the angle θ that $\vec{D}_2$ makes with the normal to the interface, let us introduce the following expression:

$$\vec{D}_2 \cdot \vec{a}_z = D_2 \cos\theta. \tag{E7-1.8}$$

The magnitude D_2 of $\vec{D}_2$ can be calculated from (E7-1.7) as

$$D_2 = \sqrt{(0.8)^2 + 9} = \sqrt{0.64 + 9} = \sqrt{9.64} = 3.1. \quad \text{(E7-1.9)}$$

Further, putting (E7-1.7) in (E7-1.8) we obtain

$$\vec{D}_2 \cdot \vec{a}_z = D_2 \cos\theta = (0.8\vec{a}_x - 3\vec{a}_z) \cdot \vec{a}_z = -3. \quad \text{(E7-1.10)}$$

Substituting (E7-1.9) and (E7-1.10) in (E7-1.8), we then have

$$-3 = (3.1)\cos\theta,$$

whence we get

$$\cos\theta = -\frac{3}{3.1} = -0.968$$

giving $\theta = 165.47^0$.

EXAMPLE E7-2

Find the electric field $\vec{E}_1$ in dielectric region 1 ($\varepsilon_{r1} = 2$) if it forms an interface at the XY plane ($z = 0$) with another dielectric region 2 ($\varepsilon_{r2} = 4$), where the electric field is given as $\vec{E}_2 = 3\vec{a}_x + 4\vec{a}_y + 6\vec{a}_z$ V/m.

SOLUTION

The interface between the two dielectrics is the XY or the $z = 0$ plane. Here the unit vector $\vec{a}_n$, directed from region 1 to 2 is $\vec{a}_n = \vec{a}_z$.

Further, here,

$$\vec{E}_2 = 3\vec{a}_x + 4\vec{a}_y + 6\vec{a}_z \text{V/m.} \quad \text{(given)} \quad \text{(E7-2.1)}$$

Let

$$\vec{E}_1 = E_{1x}\vec{a}_x + E_{1y}\vec{a}_y + E_{1z}\vec{a}_z. \quad \text{(E7-2.2)}$$

If we substitute (E7-2.1) and (E7-2.2) in the boundary condition (7.26*d*) at the interface between the two dielectric regions 1 and 2 (remembering $\vec{a}_n = \vec{a}_z$), we obtain

$$\vec{a}_n \times (\vec{E}_2 - \vec{E}_1) = \vec{a}_z \times [(3 - E_{1x})\vec{a}_x + (4 - E_{1y})\vec{a}_y + (6 - E_{1z})\vec{a}_z] = 0$$

which becomes

$$(3 - E_{1x})\vec{a}_y - (4 - E_{1y})\vec{a}_x = 0,$$

whence we get [see the reasoning given while arriving at (7.41)]:

$$\left.\begin{aligned} E_{1x} &= 3 \\ E_{1y} &= 4 \end{aligned}\right\}. \tag{E7-2.3}$$

Next, let us take

$$\left.\begin{aligned} \vec{D}_1 &= D_{1x}\vec{a}_x + D_{1y}\vec{a}_y + D_{1z}\vec{a}_z \\ \vec{D}_2 &= D_{2x}\vec{a}_x + D_{2y}\vec{a}_y + D_{2z}\vec{a}_z \end{aligned}\right\} \tag{E7-2.4}$$

and put it in the boundary condition (7.26*a*) to write

$$(\vec{D}_2 - \vec{D}_1)\cdot\vec{a}_n = [(D_{2x} - D_{1x})\vec{a}_x + (D_{2y} - D_{1y})\vec{a}_y + (D_{2z} - D_{1z})\vec{a}_z]\cdot\vec{a}_z = 0,$$

giving

$$(D_{2z} - D_{1z})\vec{a}_z\cdot\vec{a}_z = D_{2z} - D_{1z} = 0,$$

whence we get

$$D_{2z} = D_{1z}. \tag{E7-2.5}$$

Further, from (E7-2.5) we can write

$$\varepsilon_0\varepsilon_{r2}E_{2z} = \varepsilon_0\varepsilon_{r1}E_{1z}$$

which gives

$$E_{1z} = \frac{\varepsilon_{r2}}{\varepsilon_{r1}}E_{2z},$$

which, in view of the values: $\varepsilon_{r1} = 2$ and $\varepsilon_{r2} = 4$ (given), yields

$$E_{1z} = 2E_{2z}. \tag{E7-2.6}$$

Now, from (E7-2.1) we can easily identify

$$E_{2z} = 6. \tag{E7-2.7}$$

Therefore, substituting (E7-2.7) in (E7-2.6), we get

$$E_{1z} = 12. \tag{E7-2.8}$$

In (E7-2.3) we have already obtained $E_{1x} = 3$ and $E_{1y} = 4$. Therefore, substituting these values of E_{1x} and E_{1y} as well as E_{1z} from (E7-2.8) in (E7-2.2), we then get

$$\vec{E}_1 = 3\vec{a}_x + 4\vec{a}_y + 12\vec{a}_z \text{ V/m}.$$

EXAMPLE E7-3

Two dielectric regions, 1 and 2, of relative permittivity values $\varepsilon_{r1} = 2$ and $\varepsilon_{r2} = 4$ respectively, form an interface at a plane that makes a slant angle $\pi/4$ with the negative

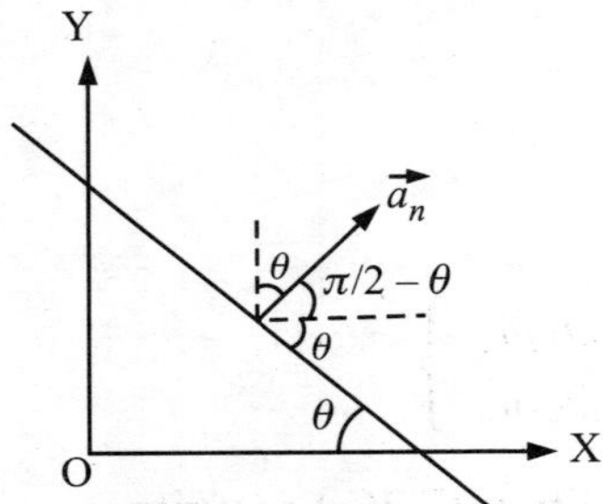

Fig. E-7.3.1

x direction. Find the electric field in region 2 if the electric field in region 1 is given as $\vec{E}_1 = 4\vec{a}_y + 6\vec{a}_z$ V/m.

SOLUTION

Let us represent the fields in the regions 1 and 2 by

$$\vec{E}_1 = E_{1x}\vec{a}_x + E_{1y}\vec{a}_y + E_{1z}\vec{a}_z \tag{E7-3.1}$$

and

$$\vec{E}_2 = E_{2x}\vec{a}_x + E_{2y}\vec{a}_y + E_{2z}\vec{a}_z. \tag{E7-3.2}$$

We now rewrite the boundary condition (7.26*d*) as

$$\vec{a}_n \times (\vec{E}_2 - \vec{E}_1) = 0 \tag{E7-3.3}$$

where $\vec{a}_n$ is the unit vector normal to the slanting interface directed from the region 1 to 2. You can express $\vec{a}_n$ in terms of the slant angle θ, the latter measured from the direction of negative X axis (Fig. E-7.3.1), as follows:

$$\vec{a}_n = \sin\theta\,\vec{a}_x + \cos\theta\,\vec{a}_y. \tag{E7-3.4}$$

Substituting (E7-3.1), (E7-3.2) and (E7-3.4) in (E7-3.3), we can show

$$\begin{aligned}\vec{a}_n &\times (\vec{E}_2 - \vec{E}_1)\\ &= (\sin\theta\,\vec{a}_x + \cos\theta\,\vec{a}_y) \times \left[E_{2x}\vec{a}_x + E_{2y}\vec{a}_y + E_{2z}\vec{a}_z - (E_{1x}\vec{a}_x + E_{1y}\vec{a}_y + E_{1z}\vec{a}_z)\right]\\ &= \cos\theta(E_{2z} - E_{1z})\vec{a}_x - \sin\theta(E_{2z} - E_{1z})\vec{a}_y\\ &\qquad + [\sin\theta(E_{2y} - E_{1y}) - \cos\theta(E_{2x} - E_{1x})]\,\vec{a}_z = 0.\end{aligned} \tag{E7-3.5}$$

In this problem, $\theta = \pi/4$ (given), which makes

$$\sin\theta = \cos\theta = \frac{1}{\sqrt{2}}. \tag{E7-3.6}$$

In view of (E7-3.6), we can write (E7-3.5) as

$$(E_{2z} - E_{1z})\vec{a}_x - (E_{2z} - E_{1z})\vec{a}_y + [(E_{2y} - E_{1y}) - (E_{2x} - E_{1x})]\vec{a}_z = 0. \quad \text{(E7-3.7)}$$

From (E7-3.7) we get

$$\left.\begin{array}{l} E_{2z} = E_{1z} \\ E_{2y} - E_{1y} = E_{2x} - E_{1x} \end{array}\right\}. \quad \text{(E7-3.8)}$$

Further, in this problem,

$\vec{E}_1 = 4\vec{a}_y + 6\vec{a}_z$, (given) which gives

$$\left.\begin{array}{l} E_{1x} = 0 \\ E_{1y} = 4 \\ E_{1z} = 6 \end{array}\right\}. \quad \text{(E7-3.9)}$$

Putting (E7-3.9) in (E7-3.8), we get

$$E_{2z} = 6 \quad \text{(E7-3.10)}$$

and

$$E_{2y} - 4 = E_{2x}. \quad \text{(E7-3.11)}$$

Next, let us recall the boundary condition (7.26*a*) and rewrite it as

$$(\vec{D}_2 - \vec{D}_1).\vec{a}_n = 0\,. \quad \text{(E7-3.12)}$$

Putting

$$\vec{D}_1 = D_{1x}\vec{a}_x + D_{1y}\vec{a}_y + D_{1z}\vec{a}_z$$

$$\vec{D}_2 = D_{2x}\vec{a}_x + D_{2y}\vec{a}_y + D_{2z}\vec{a}_z$$

as well as $\vec{a}_n = \cos\theta\,\vec{a}_x + \sin\theta\,\vec{a}_y$ from (E7-3.4) in (E7-3.12), we obtain

$$(D_{2x} - D_{1x})\cos\theta + (D_{2y} - D_{1y})\sin\theta = 0. \quad \text{(E7-3.13)}$$

Further, putting $D_{1x,y} = \varepsilon_{r1}\varepsilon_0 E_{1x,y}$ and $D_{2x,y} = \varepsilon_{r2}\varepsilon_0 E_{2x,y}$ in (E7-3.13), we can write

$$(\varepsilon_{r2}E_{2x} - \varepsilon_{r1}E_{1x})\cos\theta + (\varepsilon_{r2}E_{2y} - \varepsilon_{r1}E_{1y})\sin\theta = 0. \quad \text{(E7-3.14)}$$

In the present problem, $\varepsilon_{r1} = 2$ and $\varepsilon_{r2} = 4$; also, $\sin\theta = \cos\theta = 1/\sqrt{2}$ [given by (E7-3.6)], which makes (E7-3.14) as

$$4E_{2x} - 2E_{1x} + 4E_{2y} - 2E_{1y} = 0 \quad \text{(E7-3.15)}$$

Substituting (E7-3.9) in (E7-3.15), we obtain

$$E_{2x} + E_{2y} = 2. \tag{E7-3.16}$$

Similarly, we can put (E7-3.9) in (E7-3.8) to write

$$E_{2y} - 4 = E_{2x}. \tag{E7-3.17}$$

Simultaneous equations (E7-3.16) and (E7-3.17) may be solved to obtain

$$\left.\begin{aligned} E_{2x} &= -\frac{1}{2} \\ E_{2y} &= \frac{3}{2} \end{aligned}\right\}. \tag{E7-3.18}$$

Substituting (E7-3.10) and (E7-3.18) in (E7-3.2), we obtain the desired expression

$$\vec{E}_2 = -\frac{1}{2}\vec{a}_x + \frac{3}{2}\vec{a}_y + 3\vec{a}_z \text{ V/m}.$$

EXAMPLE E7-4

A region is divided in two parts: region 1 [$z \le 0$; $\mu_1 = 60\mu_0$ (iron)] and region 2 [$z \ge 0$; $\mu_2 = \mu_0$ (free space)], across the interface $z = 0$. Hence, determine $\vec{B}_2$ if $\vec{B}_1 = 6\vec{a}_x + 12\vec{a}_y$T.

SOLUTION

In the problem, we have

$$\vec{B}_1 = 6\vec{a}_x + 12\vec{a}_y. \quad \text{(given)} \tag{E7-4.1}$$

Let us put

$$\vec{B}_2 = B_{2x}\vec{a}_x + B_{2y}\vec{a}_y + B_{2z}\vec{a}_z. \tag{E7-4.2}$$

Recalling the boundary condition (7.26*b*), we can write:

$$(\vec{B}_2 - \vec{B}_1).\vec{a}_n = 0. \tag{E7-4.3}$$

Putting (E7-4.1) and (E7-4.2) as well as $\vec{a}_n = \vec{a}_z$ in view of the interface being ZX or the $y = 0$ plane in (E7-4.3), we obtain

$$[(B_{2x}\vec{a}_x + B_{2y}\vec{a}_y + B_{2z}\vec{a}_z) - (6\vec{a}_x + 12\vec{a}_y)].\vec{a}_z = 0,$$

whence we obtain

$$B_{2z} = 0. \tag{E7-4.4}$$

Next, let us rewrite the boundary condition (7.26*c*) as:

$$\vec{a}_n \times (\vec{H}_2 - \vec{H}_1) = 0. \tag{E7-4.5}$$

In view of the relations $\vec{B}_1 = \mu_1\vec{H}_1$ and $\vec{B}_2 = \mu_2\vec{H}_2$, where μ_1 and μ_2 are the permeability values of the regions 1 and 2 respectively, we can express (E7-45), making use of (E7-4.1) and (E7-4.2), as follows:

$$\vec{a}_z \times \left(\frac{B_{2x}\vec{a}_x + B_{2y}\vec{a}_y + B_{2z}\vec{a}_z}{\mu_2} - \frac{6\vec{a}_x + 12\vec{a}_y}{\mu_1}\right) = 0\,,$$

which simplifies to

$$\left(\frac{B_{2x}}{\mu_2} - \frac{6}{\mu_1}\right)\vec{a}_y - \left(\frac{B_{2y}}{\mu_2} - \frac{12}{\mu_1}\right)\vec{a}_x = 0\,. \tag{E7-4.6}$$

From (E7-4.6), we get

$$\frac{B_{2x}}{\mu_2} - \frac{6}{\mu_1} = 0$$

giving

$$B_{2x} = 6\frac{\mu_2}{\mu_1} \tag{E7-4.7}$$

and

$$\frac{B_{2y}}{\mu_2} - \frac{12}{\mu_1} = 0$$

giving

$$B_{2y} = 12\frac{\mu_2}{\mu_1}. \tag{E7-4.8}$$

Substitution of (E7-4.4), (E7-4.7) and (E7-4.8) in (E7-4.2) gives

$$\vec{B}_2 = 6\frac{\mu_2}{\mu_1}\vec{a}_x + 12\frac{\mu_2}{\mu_1}\vec{a}_y. \tag{E7-4.9}$$

Putting $\mu_1 = 60\mu_0$ and $\mu_2 = \mu_0$ in (E7-4.9), we then obtain the desired value

$$\vec{B}_2 = 0.1\vec{a}_x + 1.2\vec{a}_y\text{T}.$$

We will apply the concept of electromagnetic boundary conditions developed in this section for studying the phenomenon of reflection and refraction of electromagnetic waves at a dielectric-dielectric interface next.

7.3 Boundary Conditions at Conductor-Dielectric Interface

Let us now interpret the general electromagnetic boundary conditions (7.6), (7.7), (7.20), and (7.22) (Table 7.1) for the interface between a conductor, treated as medium 1

($\varepsilon = \varepsilon_0, \mu = \mu_0$), and a dielectric ($\varepsilon = \varepsilon_2, \mu = \mu_0$), treated as medium 2 (Fig. 7.3*b*). Since the bulk of a good conductor cannot be charged, the electric field and the electric displacement are each zero inside a good conductor, which makes $\vec{E}_1 = \vec{D}_1 = 0$ (see Section 5.1.2 and the text under the note immediately following (3.48) in Section 3.1.3) for both time-independent and time-independent situations. Further, for time-independent situations, it is possible to establish a finite magnetic field as well as a finite magnetic flux density ($\vec{H}_1 \neq 0,\ \vec{B}_1 \neq 0$) inside a conductor, independent of the electric field inside. However, for time-dependent situations, since the electric and the magnetic fields are coupled, an electric field would always be accompanied by a magnetic field. Therefore, for time-dependent situations, $\vec{E}_1 = \vec{D}_1 = 0$ makes $\vec{H}_1 = \vec{B}_1 = 0$ as well inside a conductor (see Section 5.1.2 and Table 7.2). In a dielectric, however, it is possible to establish magnetic field and the magnetic flux density ($\vec{H}_2 \neq 0,\ \vec{B}_2 \neq 0$) for both time-independent and time-dependent situations. Further, both for time-independent and time-dependent situations, charge that cannot stay longer inside the bulk of a conductor appears with large volume charge density ($\rho \to \infty$) over a thin layer ($dh \to 0$) of the conducting surface on the conductor-dielectric interface, which makes the surface charge density to have a finite value ($\rho_s \neq 0$). This can be appreciated by putting $\rho \to \infty$ and $dh \to 0$ in (7.5).

Further, for time-independent situations, a finite current density ($\vec{J} \neq 0$) can be made to flow through the volume of a conductor. This makes the surface current density $\vec{J}_s = 0$ at the conducting surface on the conductor-dielectric interface, which can be appreciated by putting $\vec{J} \neq 0$ and $dh \to 0$ in (7.21). However, for time-dependent situations, a large current density ($\vec{J} \to \infty$) can be established over a thin layer of the conducting surface on the interface ($dh \to 0$) such that the surface current density would become finite ($\vec{J}_s \neq 0$). This can be appreciated by putting $\vec{J} \to \infty$ and $dh \to 0$ in (7.21).

From the foregoing discussion, we can establish that : (i) for time-independent situations, $\vec{E}_1 = \vec{D}_1 = 0,\ \rho_s \neq 0,\ \vec{H}_1 \neq 0,\ \vec{B}_1 \neq 0,\ \vec{H}_2 \neq 0,\ \vec{B}_2 \neq 0$, and $\vec{J}_s = 0$, and (ii) for time-dependent situations, $\vec{E}_1 = \vec{D}_1 = 0,\ \rho_s \neq 0,\ \vec{H}_1 = \vec{B}_1 = 0,\ \vec{H}_2 \neq 0,\ \vec{B}_2 \neq 0$, and $\vec{J}_s \neq 0$, which when substituted in the general boundary conditions (7.26) to lead to the following boundary conations at a conductor-dielectric interface (Table 7.2):

For time-independent situations:

$$\left.\begin{array}{ll} \vec{D}_2.\vec{a}_n = \rho_s & \text{(a)} \\ (\vec{B}_2 - \vec{B}_1).\vec{a}_n = 0 & \text{(b)} \\ \vec{a}_n \times (\vec{H}_2 - \vec{H}_1) = 0 & \text{(c)} \\ \vec{a}_n \times \vec{E}_2 = 0 & \text{(d)} \end{array}\right\}. \tag{7.45}$$

For time-dependent situations:

$$\left.\begin{array}{ll} \vec{D}_2.\vec{a}_n = \rho_s & \text{(a)} \\ \vec{B}_2.\vec{a}_n = 0 & \text{(b)} \\ \vec{a}_n \times \vec{H}_2 = \vec{J}_s & \text{(c)} \\ \vec{a}_n \times \vec{E}_2 = 0 & \text{(d)} \end{array}\right\}. \tag{7.46}$$

The electromagnetic boundary conditions at the conductor-dielectric interface are used in studying the propagation characteristics of many physical structures supporting electromagnetic waves (for instance, in predicting the propagation characteristics of waveguides (Chapter 9). We will use these boundary conditions in the present chapter while studying the reflection of electromagnetic waves at a dielectric/free-space-conductor interface (Section 7.5). However, as a simple exercise, let us apply the relevant boundary conditions to find the surface charge density developed at the surface of a conductor in free space at a point adjacent to it where the electric field is given as $3\vec{a}_x - 4\vec{a}_y$ V/m.

Substituting the given electric field in the free-space region:

$$\vec{E}_2 = 3\vec{a}_x - 4\vec{a}_y = E_2\vec{a}_E \tag{7.47}$$

in the boundary condition (7.45*d*), we obtain

$$\vec{a}_n \times \vec{E}_2 = \vec{a}_n \times (E_2\vec{a}_E) = E_2(\vec{a}_n \times \vec{a}_E) = 0 \tag{7.48}$$

where subscript 2 refers to region 2, a dielectric, which, in the present problem, is interpreted as free space ($\varepsilon = \varepsilon_0$). $\vec{a}_n$ is the unit vector normal to the conducting surface where the electric field is given. $\vec{a}_E$ is the unit vector in the direction of $\vec{E}_2$, and E_2 is the magnitude of $\vec{E}_2$, which from (7.47) is

$$E_2 = \sqrt{(3)^2 + (-4)^2} = \sqrt{9+16} = \sqrt{25} = 5 \text{ V/m}\,. \tag{7.49}$$

In order for (7.48) to happen, since $E_2 \neq 0$, we must have

$$\vec{a}_n \times \vec{a}_E = 0. \tag{7.50}$$

It follows from (7.50) that the vectors $\vec{a}_n$ and $\vec{a}_E$ both of which have the same magnitude, namely, unity, must have the same direction as well; in other words, we must put

$$\vec{a}_n = \vec{a}_E. \tag{7.51}$$

Therefore, substituting (7.49) and (7.51) in (7.47), we get

$$\vec{E}_2 = E_2\vec{a}_E = 5\vec{a}_n. \tag{7.52}$$

Corresponding to (7.52), the electric displacement in free space is then given by

$$\vec{D} = \varepsilon_0\vec{E}_2 = 5\varepsilon_0\vec{a}_n \text{ C/m}^2. \tag{7.53}$$

Substituting (7.53) in (7.45*a*), we then obtain the required surface charge density developed on the conducting surface as

$$\rho_s = 5\varepsilon_0\vec{a}_n.\vec{a}_n = 5\varepsilon_0 \text{ C/m}^2.$$

EXAMPLE E7-5

Find the surface charge density developed on the conducting plate of a parallel-plate capacitor in terms of the potential difference between the plates and hence derive an expression for the capacitance of the capacitor in terms of the dimensions and permittivity of the dielectric filling the region between the plates of the capacitor.

SOLUTION

The problem enjoys rectangular symmetry, and for large plates we can treat the problem as one-dimensional, considering the potential to vary only along z ($\partial/\partial x = \partial/\partial y = 0$; $\partial/\partial z \neq 0$). Let the two plates be located at $z = 0$ and $z = d$ respectively. The plate at $z = 0$ is raised to the potential $V = V_0$ with respect to the potential $V = 0$ of the plate at $z = d$. Using (3.117), we can write the following expression for the electric field between the conducting plates:

$$\vec{E} = \frac{V_0}{d}\vec{a}_z. \tag{E7-5.1}$$

Multiplying (E7-5.1) by ε, the permittivity of the dielectric filling the region between the plates, we can write

$$\vec{D} = \varepsilon\frac{V_0}{d}\vec{a}_z. \tag{E7-5.2}$$

Apply the boundary condition (7.46a) at the conducting plate ($z = 0$) to

$$\vec{D}\cdot\vec{a}_n = \vec{D}\cdot\vec{a}_z = \rho_s \tag{E7-5.3}$$

where ρ_s is the surface charge density on the conducting plate ($z = 0$) . The surface charge density ρ_s is obtained by putting (E7-5.2) in (E7-5.3) as

$$\rho_s = \vec{D}\cdot\vec{a}_z = \varepsilon\frac{V_0}{d}\vec{a}_z\cdot\vec{a}_z = \frac{\varepsilon V_0}{d}. \tag{E7-5.4}$$

The expression for the capacitance C of the capacitor can then be obtained by dividing the charge on the plate $\rho_s A$ by V_0 (see Section 3.2.2) while making use of (E7-5.4), where A is the area of each plate.

$$C = \frac{\rho_s A}{V_0} = \frac{\frac{\varepsilon V_0}{d}A}{V_0} = \frac{\varepsilon A}{d}. \tag{E7-5.5}$$

The expression (E7-5.5) agrees with (3.100).

EXAMPLE E7-6

Find the surface charge density developed on the conducting surface of a long cylindrical capacitor comprising two coaxial long conducting cylinders in terms of the potential difference between the conducting cylinders and hence derive an expression for the capacitance per unit length of the capacitor in terms of the inner and outer radii of the capacitor and the permittivity of the dielectric filling the region between the conducting cylinders of the capacitor.

SOLUTION

The problem enjoys cylindrical symmetry, and for long coaxial conductors we may treat the problem as one-dimensional, considering the potential to vary only along r ($\partial/\partial\theta = \partial/\partial z = 0$; $\partial/\partial r \neq 0$). Let the inner and outer radii of the coaxial conductors be a and b respectively, and their respective potentials be $V = V_0$ at $r = a$ and $V = 0$ at $r = b$.

The following Laplace's equation holds good between the conducting cylinders, which can be written with the help of (3.109) in conjunction with the expression for the Laplacian of potential given by (2.76) in cylindrical coordinates in the context of the present problem: $\partial/\partial\theta = \partial/\partial z = 0$:

$$\nabla^2 V = \frac{1}{r}\frac{\partial}{\partial r}\left(r\frac{\partial V}{\partial r}\right) = 0,$$

whence we get, taking care to change the partial derivative to total derivative since V is a function of only r, the following expression:

$$\frac{d}{dr}\left(r\frac{dV}{dr}\right) = 0. \qquad \text{(E7-6.1)}$$

Integrating (E7-6.1), we obtain

$$r\frac{dV}{dr} = C_1,$$

which can be put as

$$\frac{dV}{dr} = \frac{C_1}{r} \qquad \text{(E7-6.2)}$$

where C_1 is a constant. Integrating (E7-6.2), we obtain

$$V = C_1 \ln r + C_2 \qquad \text{(E7-6.3)}$$

where C_2 is another constant. You can find C_1 and C_2 by recalling the potentials at the inner and outer conductors of the capacitor ($V = V_0$ at $r = a$ and $V = 0$ at $r = b$). Thus, by putting $V = V_0$ and $r = a$ in (E7-6.3) we get

$$V_0 = C_1 \ln a + C_2 \qquad \text{(E7-6.4)}$$

and, similarly, putting $V = 0$ and $r = b$ in (E7-6.3) we get

$$0 = C_1 \ln b + C_2. \qquad \text{(E7-6.5)}$$

From (E7-6.4) and (E7-6.5) we can easily get

$$\left.\begin{aligned} C_1 &= \frac{V_0}{\ln\left(\frac{a}{b}\right)} \\ C_2 &= -\frac{V_0}{\ln\left(\frac{a}{b}\right)}\ln b \end{aligned}\right\}. \tag{E7-6.6}$$

Substituting (E7-6.6) in (E7-6.3), we obtain

$$V = \frac{\ln\left(\frac{r}{b}\right)}{\ln\left(\frac{a}{b}\right)} V_0. \tag{E7-6.7}$$

We can find $\vec{E}$ in the region between the inner and outer conductors of the capacitor with the help of (3.91) in conjunction with (2.32), the latter, in the present context, to be read taking $\partial/\partial\theta = \partial/\partial z = 0$. Thus, we get

$$\vec{E} = -\nabla V = -\frac{\partial V}{\partial r}\vec{a}_r = -\frac{dV}{dr}\vec{a}_r. \tag{E7-6.8}$$

Differentiating (E7-6.7) and substituting the derivative in (E7-6.8), we obtain

$$\vec{E} = \frac{V_0}{\ln\left(\frac{b}{a}\right)}\frac{1}{r}\vec{a}_r,$$

whence we also get

$$\vec{D} = \varepsilon\vec{E} = \varepsilon\frac{V_0}{\ln\left(\frac{b}{a}\right)}\frac{1}{r}\vec{a}_r \tag{E7-6.9}$$

where ε is the permittivity of the dielectric filling the region between the conductors. We can read (E7-6.9) at the surface of the inner conductor by putting $r = a$ as follows:

$$\vec{D} = \varepsilon\frac{V_0}{\ln\left(\frac{b}{a}\right)}\frac{1}{a}\vec{a}_r. \quad (r = a) \tag{E7-6.10}$$

Applying the boundary condition (7.46*a*) at the surface of the inner conductor ($r = a$), we can then write, with the help of (E7-6.10),

$$\rho_s = \vec{D}\cdot\vec{a}_n = \vec{D}\cdot\vec{a}_r = \varepsilon\frac{V_0}{\ln\left(\frac{b}{a}\right)}\frac{1}{a}\vec{a}_r\cdot\vec{a}_r = \varepsilon\frac{V_0}{\ln\left(\frac{b}{a}\right)}\frac{1}{a}. \tag{E7-6.11}$$

The expression for the capacitance C of the capacitor can then be obtained by dividing the charge on the plate $\rho_s A$ by V_0 (see Section 3.2.2) while making use of (E7-6.11), where $A = 2\pi al$ is the area of the inner conductor and l the length of the capacitor. Thus, we obtain

$$C = \frac{\rho_s A}{V_0} = \frac{\varepsilon \dfrac{V_0}{\ln\left(\dfrac{b}{a}\right)} \dfrac{1}{a}}{V_0} 2\pi a l,$$

which simplifies to

$$C = \frac{2\pi\varepsilon}{\ln\left(\dfrac{b}{a}\right)} l. \tag{E7-6.12}$$

Dividing (E7-6.12) by the length of the capacitor l, we obtain

$$\text{Capacitance per unit length of the coaxial capacitor} = \frac{2\pi\varepsilon}{\ln\left(\dfrac{b}{a}\right)}.$$

The expression for the capacitance per unit length of a coaxial able agrees with (E3-9.8) deduced earlier in Chapter 3.

EXAMPLE E7-7

Find the surface charge density developed on the conducting surface of a spherical capacitor comprising two concentric spherical conductors in terms of the potential difference between the conductors, and hence derive an expression for the capacitance of the capacitor in terms of the inner and outer radii of the capacitor and the permittivity of the material filling the region between the inner and outer conductors of the capacitor.

SOLUTION

The problem enjoys spherical symmetry, and for large conductors we may treat the problem as one-dimensional considering the potential to vary only along r ($\partial/\partial\theta = \partial/\partial z = 0$; $\partial/\partial r \neq 0$). Let the inner and outer radii of the concentric conductors be a and b respectively, and their respective potentials be $V = V_0$ at $r = a$ and $V = 0$ at $r = b$.

We can write Laplace's equation in the region between the conducting spheres with the help of (3.109) in conjunction with the expression for the Laplacian of potential given by (2.77) in spherical coordinates, taking $\partial/\partial\theta = \partial/\partial\phi = 0$ in the context of the present problem.

$$\nabla^2 V = \frac{1}{r^2}\frac{\partial}{\partial r}\left(r^2 \frac{\partial V}{\partial r}\right) = 0. \tag{E7-7.1}$$

Since V is a function of only r, we can change the partial derivative to total derivative in (E7-7.1) to write:

$$\frac{d}{dr}\left(r^2 \frac{dV}{dr}\right) = 0. \tag{E7-7.2}$$

By integrating (E7-7.2), we obtain

$$r^2\frac{dV}{dr} = C_1,$$

whence we can write

$$\frac{dV}{dr} = C_1 r^{-2} \tag{E7-7.3}$$

where C_1 is a constant. We can then integrate (E7-7.3) to obtain

$$V = -\frac{C_1}{r} + C_2. \tag{E7-7.4}$$

Putting $V = V_0$ and $r = a$ in (E7-7.4), we obtain the following relation:

$$V_0 = -\frac{C_1}{a} + C_2. \tag{E7-7.5}$$

Similarly, putting $V = 0$ and $r = b$ in (E7-7.4) we obtain

$$C_2 = \frac{C_1}{b}. \tag{E7-7.6}$$

We can solve (E7-7.5) and (E7-7.6) to find the constants as

$$\left.\begin{aligned} C_1 &= \frac{ab}{a-b}V_0 \\ C_2 &= \frac{a}{a-b}V_0 \end{aligned}\right\}. \tag{E7-7.7}$$

Substituting (E7-7.7) in (E7-7.4), we can then show

$$V = \frac{r-b}{a-b}\frac{a}{r}V_0. \tag{E7-7.8}$$

We can find $\vec{E}$ in the region between the inner and outer conductors of the capacitor with the help of (3.91) in conjunction with (2.33), the latter, in the present context, to be read taking $\partial/\partial\theta = \partial/\partial z = 0$. Thus, we get

$$\vec{E} = -\nabla V = -\frac{\partial V}{\partial r}\vec{a}_r = -\frac{dV}{dr}\vec{a}_r. \tag{E7-7.9}$$

Substituting the derivative of (E7-7.8) in (E7-7.9), we get

$$\vec{E} = -\frac{dV}{dr}\vec{a}_r = \frac{ab}{b-a}\frac{V_0}{r^2}\vec{a}_r,$$

whence we get

$$\vec{D} = \varepsilon\vec{E} = \varepsilon\frac{ab}{b-a}\frac{V_0}{r^2}\vec{a}_r \tag{E7-7.10}$$

where ε is the permittivity of the dielectric filling the region between the conductors.

Reading (E7-7.10) at the surface of the inner conductor by putting $r = a$, we obtain

$$\vec{D} = \varepsilon \vec{E} = \varepsilon \frac{ab}{b-a} \frac{V_0}{a^2} \vec{a}_r . \qquad (r = a) \tag{E7-7.11}$$

Substituting (E7-7.11) in the boundary condition (7.46a) at the surface of the inner conductor ($r = a$), the surface charge density of the inner conductor is obtained as

$$\rho_s = \vec{D} \cdot \vec{a}_n = \vec{D} \cdot \vec{a}_r = \varepsilon \frac{ab}{b-a} \frac{V_0}{a^2} \vec{a}_r \cdot \vec{a}_r = \varepsilon \frac{ab}{b-a} \frac{V_0}{a^2} . \tag{E7-7.12}$$

Next, find the expression for the capacitance C of the capacitor by dividing the charge on the plate $\rho_s A$ by V_0 (see Section 3.2.2) while making use of (E7-7.12), where $A = 4\pi a^2$ is the area of the inner conductor of the capacitor. Thus, we obtain

$$C = \frac{\rho_s A}{V_0} = \frac{\varepsilon \dfrac{ab}{b-a} \dfrac{V_0}{a^2}}{V_0} 4\pi a^2 ,$$

which simplifies to

$$C = 4\pi\varepsilon \frac{ab}{b-a} = \frac{4\pi\varepsilon}{\dfrac{1}{a} - \dfrac{1}{b}} .$$

7.4 Boundary Conditions at Conductor-Conductor Interface for Time-independent Situations: Refraction of Currents

Let us consider dc currents through two dissimilar conducting media, 1 and 2, of conductivities σ_1 and σ_2 respectively (Fig. 7.4). Let the current densities $\vec{J}_1$ and $\vec{J}_2$ in the media 1 and 2 make the angles θ_1 and θ_2 respectively with the normal to the interface between the conductors at a point on the interface between the two conductors (Fig. 7.4). Under dc (time-independent) conditions ($\partial/\partial t = 0$) (Section 5.1.1), let us apply (5.2)

$$\oint_S \vec{J} . \vec{a}_n \, dS = 0 \qquad [(5.2) \text{ rewritten}] \tag{7.54}$$

to a pillbox volume element enclosure (Section 7.1) such that it encloses a point P at the interface between the conductors. The pillbox has an infinitesimal thickness dh across the interface, and its top and bottom faces have the same area, dS (Fig. 7.4). In the limit $dh \to 0$, which makes the volume of the pillbox shrink to a thin layer on the interface, the surface integral of (7.54) will be contributed to only by the top and the bottom faces of the element of the volume enclosure so that we can write (7.54), as applicable to the volume

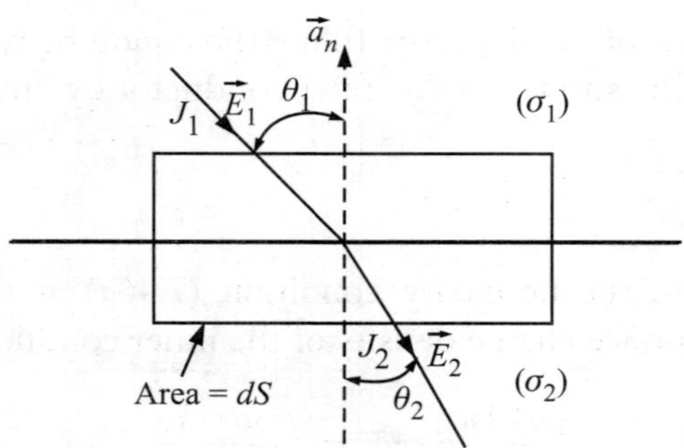

Fig. 7.4 *Refraction of current at the interface between two conductors.*

element considered, as (Fig. 7.4)

$$\int_{\text{bottom face}} \vec{J}_2.\vec{a}_n dS + \int_{\text{top face}} \vec{J}_1.(-\vec{a}_n)dS = 0. \tag{7.55}$$

Here, the unit vector that is outwardly normal to the bottom face in medium 2 is $\vec{a}_n$ and that outwardly normal to the top face in medium 1 is $-\vec{a}_n$, remembering the convention already mentioned following (7.2) that $\vec{a}_n$ is directed from medium 1 to 2 (Fig. 7.4).

Considering $\vec{J}_1$ and $\vec{J}_2$ to be uniform over the infinitesimal areas of the top and the bottom faces respectively, each being $dS = \Delta S$, say, we can interpret (7.55) as

$$\int_{\Delta S} \vec{J}_2.\vec{a}_n dS + \int_{\Delta S} \vec{J}_1.(-\vec{a}_n)dS = 0,$$

which gives

$$\vec{J}_2.\vec{a}_n \Delta S - \vec{J}_1.(\vec{a}_n)\Delta S = 0. \tag{7.56}$$

From (7.56), we get, since $\Delta S \neq 0$:

$$\vec{J}_2.\vec{a}_n - \vec{J}_1.(\vec{a}_n) = 0. \tag{7.57}$$

Evaluating the dot products of (7.57), we obtain

$$(J_2)(1)\cos\theta_2 - (J_1)(1)\cos\theta_1 = 0$$

which simplifies to

$$J_2 \cos\theta_2 = J_1 \cos\theta_1 . \tag{7.58}$$

However, J_1 and J_2 are related to E_1 and E_2 respectively by Ohm's law [see (5.11)] as follows:

$$\left.\begin{aligned} J_1 &= \sigma_1 E_1 \\ J_2 &= \sigma_2 E_2 \end{aligned}\right\}. \tag{7.59}$$

Substituting (7.59) in (7.58), we get

$$\sigma_2 E_2 \cos\theta_2 = \sigma_1 E_1 \cos\theta_1 . \tag{7.60}$$

Now, let us recall the boundary condition (7.25)

$$\vec{a}_n \times (\vec{E}_2 - \vec{E}_1) = 0 \qquad [(7.25) \text{ rewritten}] \tag{7.61}$$

and apply it to the interface between the conductors. Since $\vec{E}_1$ and $\vec{E}_2$ in the media 1 and 2 make angles θ_1 and θ_2 respectively with the normal to the interface between the conductors at point P, as do the current densities $\vec{J}_1$ and $\vec{J}_2$ (as mentioned in the beginning of the present section), the magnitude of $\vec{a}_n \times \vec{E}_1$ is $E_1 \sin\theta_1$ and that of $\vec{a}_n \times \vec{E}_2$ is $E_2 \sin\theta_2$. Hence, we can write the magnitude of the vector equation (7.61) as (Fig. 7.4)

$$E_2 \sin\theta_2 - E_1 \sin\theta_1 = 0,$$

which can be put as

$$E_2 \sin\theta_2 = E_1 \sin\theta_1 . \tag{7.62}$$

Then, dividing (7.62) by (7.60), we get

$$\frac{E_2 \sin\theta_2}{\sigma_2 E_2 \cos\theta_2} = \frac{E_1 \sin\theta_1}{\sigma_1 E_1 \cos\theta_1},$$

which simplifies to

$$\frac{\tan\theta_1}{\tan\theta_2} = \frac{\sigma_1}{\sigma_2}. \tag{7.63}$$

Furthermore, due to the discontinuity of the normal components of electric displacement at the interface between the dielectrics, a surface charge density develops at the conductor-conductor interface, which can be predicted with the help of the boundary condition (7.6) as

$$D_2 \cos\theta_2 - D_1 \cos\theta_1 = \rho_s . \tag{7.64}$$

$\vec{D}_1 = \varepsilon_0 \vec{E}_1$ and $\vec{D}_2 = \varepsilon_0 \vec{E}_2$ make angles θ_1 and θ_2 respectively with the normal to the interface between the conductors at point P, as do the current densities $\vec{E}_1$ and $\vec{E}_2$ [as mentioned following (7.61)].

Further, in terms of the normal components we can express (7.58) and (7.64) as

$$J_{n2} = J_{n1} = J_n, \text{ say} \tag{7.65}$$

and

$$D_{n2} - D_{n1} = \rho_s \tag{7.66}$$

where $J_{n1,2} = J_{1,2}\cos\theta_{1,2}$ and $D_{n1,2} = D_{1,2}\cos\theta_{1,2}$, with subscript n representing the normal components of the current densities and the electric displacements respectively at the interface (Fig. 7.4).

Further, we can express (7.66) in terms of the electric field components using the relations $D_{n1,2} = \varepsilon_0 E_{n1,2}$ and rearranging terms as follows:

$$\rho_s = \varepsilon_0(E_{n2} - E_{n1}). \tag{7.67}$$

Furthermore, the relation between the current density and the electric field given by Ohm's law, $J_{n1,2} = \sigma_{1,2}E_{n1,2}$, [see (5.11)] will enable (7.67) to be expressed as

$$\rho_s = \varepsilon_0\left(\frac{J_{n2}}{\sigma_2} - \frac{J_{n1}}{\sigma_1}\right),$$

which, in view of the relation (7.65) can be put as

$$\rho_s = \varepsilon_0\left(\frac{1}{\sigma_2} - \frac{1}{\sigma_1}\right)J_n. \tag{7.68}$$

Alternatively, we can also express (7.68) in terms of the relaxation times of the two conductors (media 1 and 2) , $T_1 = \varepsilon_0/\sigma_1$ and $T_2 = \varepsilon_0/\sigma_2$ respectively [see (5.19)] as

$$\rho_s = \left(\frac{\varepsilon_0}{\sigma_2} - \frac{\varepsilon_0}{\sigma_1}\right)J_n = (T_2 - T_1)J_n. \tag{7.69}$$

Thus, the boundary condition at the interface between the two conducting media gives us the law of refraction of current density in terms of the conductivities of the media as (7.63), as well as the surface charge density developed at the interface in terms of the relaxation times of the media as (7.69).

Now, as a corollary to the boundary condition at the interface between the conductors studied in this section, you will find it very interesting to study the circuit law of parallel resistances. For this purpose, consider two rectangular slabs of conductors of conductivities σ_1 and σ_2 and cross-sectional areas A_1 and A_2 respectively, each of length l, in contact, thereby forming a planar interface between them (Fig. 7.5). Let the dc currents

$$I_1 = J_1A_1 \tag{7.70}$$

and

$$I_2 = J_2A_2 \tag{7.71}$$

flow through these conductors (of conductivities, σ_1 and σ_2 and cross-sectional areas A_1 and A_2 respectively), parallel or tangential to the interface between them. Here, J_1 and J_2 are respectively the magnitudes of the current densities in the conductors, the direction of each being tangential to the interface (Fig. 7.5).

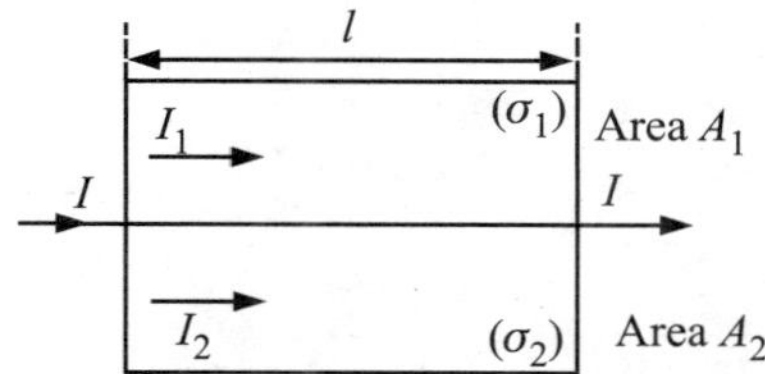

Fig. 7.5 *Two rectangular slabs of dissimilar conductivities forming a planar interface between them considered for the appreciation of the law of parallel resistances.*

The magnitudes of the current densities J_1 and J_2 tangential to the interface between the conductors are related to the corresponding tangential electric fields E_1 and E_2 by Ohm's law [see (5.11)] as:

$$J_1 = \sigma_1 E_1 \tag{7.72}$$

and

$$J_2 = \sigma_2 E_2 \, . \tag{7.73}$$

Substituting (7.72) and (7.73) in (7.70) and (7.71) respectively, we obtain

$$I_1 = \sigma_1 E_1 A_1$$

and

$$I_2 = \sigma_2 E_2 A_2,$$

which, by rearranging terms, can be put as

$$E_1 = \frac{I_1}{\sigma_1 A_1} \tag{7.74}$$

and

$$E_2 = \frac{I_2}{\sigma_2 A_2}. \tag{7.75}$$

Now, since the tangential component of the electric field is continuous at the interface between two media (see (7.25) and the text following it), we can equate E_1 to E_2, given by (7.74) and (7.75) respectively, to write

$$\frac{I_1}{\sigma_1 A_1} = \frac{I_2}{\sigma_2 A_2}. \tag{7.76}$$

Multiplying both sides of (7.76) by l, we get

$$I_1 \frac{l}{\sigma_1 A_1} = I_2 \frac{l}{\sigma_2 A_2}. \tag{7.77}$$

In (7.77), it is easy to identify

$$R_1 = \frac{l}{\sigma_1 A_1} \tag{7.78}$$

and

$$R_2 = \frac{l}{\sigma_2 A_2} \tag{7.79}$$

as the dc resistances of the two conducting slabs in contact. Therefore, substituting (7.78) and (7.79) in (7.77), we obtain

$$I_1 R_1 = I_2 R_2 = I R_{\text{eq}} \tag{7.80}$$

where I_1 and I_2 are the currents through the two slabs respectively, and $I(= I_1 + I_2)$ is the total current through the two conducting slabs in contact, taken together. R_{eq} is the equivalent resistance of the two resistances R_1 and R_2 connected in parallel (Fig. 7.5).

Further, we can easily obtain from (7.80)

$$\frac{I_1}{I_1 + I_2} = \frac{R_2}{R_2 + R_1}$$

which you we write, in terms of $I(= I_1 + I_2)$ as

$$I_1 = I \frac{R_2}{R_2 + R_1}. \tag{7.81}$$

In parallel, we can obtain from (7.80) the following expression:

$$I_1 = I \frac{R_{\text{eq}}}{R_1}. \tag{7.82}$$

Comparing (7.81) and (7.82) then you obtain

$$I \frac{R_2}{R_2 + R_1} = I \frac{R_{\text{eq}}}{R_1},$$

which, on rearranging terms, yields the following expression for the equivalent resistance R_{eq} of the two resistances R_1 and R_2 connected in parallel (Fig. 7.5):

$$R_{\text{eq}} = \frac{R_1 R_2}{R_2 + R_1} \tag{7.83}$$

which can also be expressed in the following well-known form known as the law of parallel resistances in circuit theory:

$$\frac{1}{R_{\text{eq}}} \left(= \frac{R_2 + R_1}{R_1 R_2} \right) = \frac{1}{R_1} + \frac{1}{R_2}. \tag{7.84}$$

7.5 Reflection of Electromagnetic Waves at the Interface between a Dielectric/Free-Space and a Conductor

Let a uniform plane electromagnetic wave propagating along z in a dielectric [$\partial/\partial x = \partial/\partial y = 0$; see (6.39)] be incident on a conducting surface (Fig. 7.6). Most practical cases, however, would involve a wave propagating through a free space rather than through a dielectric before impinging on a conductor. Therefore, without loss of generality, let us consider the medium to be free space, for which $\varepsilon = \varepsilon_0$. Further, let the incident wave propagate along z and let it be treated as the forward wave, for which we can take field quantities to vary as $\exp j(\omega t - \beta z)$ as discussed following (6.69). For the incident forward wave, therefore, recall the relation immediately preceding (6.64) and write it in the present context as follows:

$$\frac{E_x}{H_y} = \frac{\omega\mu_0}{\beta},$$

which, in view of the relation $\beta = \omega(\mu_0\varepsilon_0)^{1/2}$ from (6.66), becomes

$$\frac{E_x}{H_y} = \frac{\omega\mu_0}{\beta} = \frac{\omega\mu_0}{\omega(\mu_0\varepsilon_0)^{1/2}} = \left(\frac{\mu_0}{\varepsilon_0}\right)^{1/2} = \eta_0 \tag{7.85}$$

where the symbol η_0 has already been introduced earlier [see Chapter 6; (6.68)-(6.70)].

On the other hand, the reflected wave would essentially be a backward wave propagating in the negative z direction, For such a backward wave, we can take the field quantities to vary as $\exp j(\omega t + \beta z)$, with a change of sign of β as compared to the forward quantities of the incident wave for which the field quantities vary as $\exp j(\omega t - \beta z)$ [as mentioned preceding (7.85)]. Thus, incorporating a change of sign in β in the expression (7.85), we can write for the reflected backward wave the expression corresponding to (7.85) as

$$\frac{E_x}{H_y} = \frac{\omega\mu_0}{-\beta} = \frac{\omega\mu_0}{-\omega(\mu_0\varepsilon_0)^{1/2}} = -\left(\frac{\mu_0}{\varepsilon_0}\right)^{1/2} = -\eta_0. \tag{7.86}$$

Denoting the incident and the reflected field quantities by the subscripts i and r respectively, we can then put together the expressions (7.85) and (7.86) as follows:

$$\left.\begin{aligned} \frac{E_{xi}}{H_{yi}} &= \eta_0 \quad \text{(a)} \\ \frac{E_{xr}}{H_{yr}} &= -\eta_0 \quad \text{(b)} \end{aligned}\right\}. \tag{7.87}$$

You would have noticed that in the present context the free-space medium is occupying the negative z region while the conducting medium is occupying the positive z region (Fig. 7.6). At any point z in the free-space region, the electric field $\vec{E}$ comprises both the incident and the reflected components, and, therefore can be represented as

$$\vec{E} = \vec{E}_{0i} \exp j(\omega t - \beta z) + \vec{E}_{0r} \exp j(\omega t + \beta z) \tag{7.88}$$

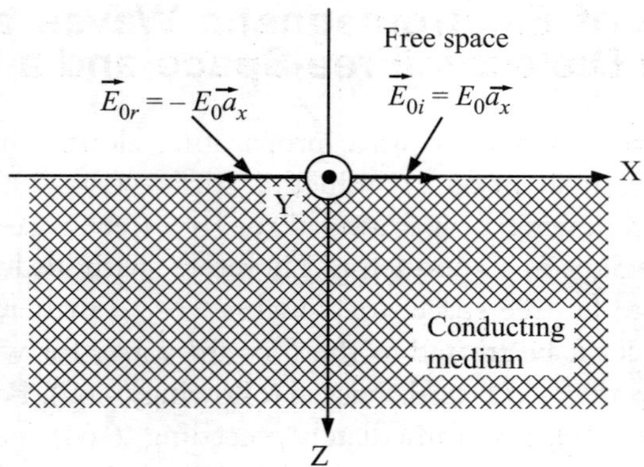

Fig. 7.6 *Uniform plane electromagnetic propagating along z, incident from a free-space negative-z region normally on a conducting surface and reflected back normally into the free-space region from a conducting surface at $z = 0$. The conducting medium occupies the positive z region. The incident wave propagates in the positive z direction and the reflected wave in the negative z direction. At the conducting surface ($z = 0$), the electric fields of the incident and reflected waves are in the positive and negative x directions respectively and, correspondingly, the magnetic field is in the positive y direction for both the waves shown by a dot surrounded by a circle, indicating the direction perpendicular to the plane of the paper and pointed towards the reader.*

where $\vec{E}_{0i}$ and $\vec{E}_{0r}$ are the amplitudes of the incident and the reflected electric field components respectively, and $\exp j(\omega t - \beta z)$ and $\exp j(\omega t + \beta z)$ are their respective phase factors.

Further, you can see from (7.88) that the quantities $\vec{E}_{0i}$ and $\vec{E}_{0r}$ represent the quantities at $z = 0$, the interface between the free-space (negative z) and the conducting (positive z) regions. They are related to each other by the electromagnetic boundary condition that the electric field at the interface between a conducting and a dielectric/free-space ($\varepsilon = \varepsilon_0$) medium is zero [see Table 7.3; (7.46*d*)]. Therefore, in view of this boundary condition, we can relate $\vec{E}_{0i}$ and $\vec{E}_{0r}$ as follows:

$$\vec{E}_{0i} + \vec{E}_{0r} = 0. \tag{7.89}$$

Substituting $\vec{E}_{0r} = -\vec{E}_{0i}$, obtainable from (7.89) in (7.88), we can express (7.88) in the following alternative form:

$$\begin{aligned}\vec{E} &= \vec{E}_{0i} \exp j(\omega t - \beta z) - \vec{E}_{0i} \exp j(\omega t + \beta z) \\ &= \vec{E}_{0i}[\exp j(\omega t - \beta z) - \exp j(\omega t + \beta z)] \\ &= \vec{E}_{0i} \exp j\omega t\,[\exp(-j\beta z) - \exp(j\beta z)],\end{aligned}$$

which, in view of the relation $\exp(\pm j\beta z) = \cos\beta z \pm \sin\beta z$ further simplifies to

$$\vec{E} = -2j\vec{E}_{0i} \sin\beta z \exp(j\omega t). \tag{7.90}$$

Further, in the present problem, we can express $\vec{E}_{0i}$ and $\vec{E}_{0r}$ in terms of their respective magnitudes as follows (see Fig. 7.6):

$$\vec{E}_{0i} = E_{0i}\vec{a}_x = E_0\vec{a}_x \tag{7.91}$$

and hence, in view of (7.89) and (7.91),

$$\vec{E}_{0r} = E_{0r}\vec{a}_x = -\vec{E}_{0i} = -E_0\vec{a}_x \tag{7.92}$$

where we have put $E_{0i} = E_0$ and, therefore, can identify that $E_{0r} = -E_0$.
Substituting (7.91) in (7.90), we can then write the expression for the electric field $\vec{E}$ at any point z in the free-space region as:

$$\vec{E} = -2jE_0 \sin\beta\, z \exp(j\omega\, t)\, \vec{a}_x. \tag{7.93}$$

Similarly, at any point z in the free-space region, the electric field $\vec{H}$ comprising the incident and the reflected components can be represented as

$$\vec{H} = \vec{H}_{0i} \exp j(\omega\, t - \beta\, z) + \vec{H}_{0r} \exp j(\omega\, t + \beta\, z) \tag{7.94}$$

where $\vec{H}_{0i}$ and $\vec{H}_{0r}$ represent the amplitudes of the incident and the reflected magnetic field components respectively and $\exp j(\omega t - \beta\, z)$ and $\exp j(\omega t + \beta\, z)$ their respective phase factors. You will be able to easily appreciate that for the forward wave incident on the conducting surface (Fig. 7.6) and propagating in the positive z direction, the direction of the magnetic field at the conducting surface $\vec{H}_{0i}$ would be along positive y corresponding to the incident electric field component along positive x [see (7.91)]. Similarly, the direction of $\vec{H}_{0r}$ would also be along positive y, corresponding to the reflected electric field component along negative x [see (7.92)]. Therefore, corresponding to the electric field expressions (7.91) and (7.92), we can respectively write the following magnetic field expressions

$$\vec{H}_{0i} = H_{0i}\vec{a}_y \tag{7.95}$$

and

$$\vec{H}_{0r} = H_{0r}\vec{a}_y. \tag{7.96}$$

Substituting (7.95) and (7.96) in (7.94), we then obtain

$$\vec{H} = [H_{0i} \exp j(\omega\, t - \beta\, z) + H_{0r} \exp j(\omega\, t + \beta\, z)]\, \vec{a}_y. \tag{7.97}$$

However, the magnitude of the incident and the reflected magnetic field components at the conducting surface, H_{0i} and H_{0r}, being the magnitudes of $\vec{H}_{0i}$ and $\vec{H}_{0r}$ occurring in the first and second terms of (7.97) respectively, are not independent of their corresponding electric field components, and they can be related with the help of (7.87). Thus, we can interpret (7.87) here to write

$$\left.\begin{aligned} H_{0i} &= \frac{E_{0i}}{\eta_0} \\ H_{0r} &= \frac{E_{0r}}{-\eta_0} \end{aligned}\right\}. \tag{7.98}$$

Substituting (7.98) in (7.97), we obtain

$$\vec{H} = \left[\frac{E_{0i}}{\eta_0} \exp j(\omega\, t - \beta\, z) + \frac{E_{0r}}{-\eta_0} \exp j(\omega\, t + \beta\, z)\right] \vec{a}_y,$$

which, in view of the relations $E_{0i} = E_0$ and $E_{0r} = -E_0$, as pointed out immediately following (7.92) becomes

$$\begin{aligned} \vec{H} &= \left[\frac{E_0}{\eta_0} \exp j(\omega\, t - \beta\, z) + \frac{-E_0}{-\eta_0} \exp j(\omega\, t + \beta\, z)\right] \vec{a}_y \\ &= \frac{E_0}{\eta_0} [\exp(-j\beta\, z) + \exp(j\beta\, z)] \exp(j\omega\, t)\, \vec{a}_y \end{aligned}$$

which, in view of the relation $\exp(\pm j\beta\, z) = \cos\beta\, z \pm \sin\beta\, z$ simplifies to

$$\vec{H} = \frac{2E_0}{\eta_0} \cos\beta\, z \exp(j\omega\, t)\, \vec{a}_y. \tag{7.99}$$

An alternative approach to obtaining the magnetic field expression (7.99) would be to start from the electric field expression (7.93) and use Maxwell's equations (see Appendix A7.1).

In order to appreciate how the electric and the magnetic fields at any point z are related with respect to phase, let us make use of the relation $\exp(-j\pi/2) = \cos(\pi/2) - j\sin(\pi/2) = 0 - (j)(1) = -j$ in (7.93) and express the electric field as

$$\vec{E} = 2E_0 \sin\beta\, z \exp j(\omega t - j\pi/2)\, \vec{a}_x. \tag{7.100}$$

Clearly, on examining the electric field expression (7.100) and the magnetic field expression (7.99), we can see that the electric field lags behind the magnetic field in phase by $\pi/2$.

Now, we are going to see that the propagating forward wave incident on a conducting surface combines with the propagating backward wave reflected from the conducting surface to form a resultant non-propagating wave. We will see that neither the phase factor $\exp j(\omega t - \beta\, z)$, characteristic of a quantity representing a forward wave (Section 6.1), nor the phase factor $\exp j(\omega t + \beta\, z)$, characteristic of a quantity representing a backward wave are present in the field expressions (7.93) or (7.10) and (7.99), each of which are obtained here by combining the forward and the backward wave components. These field expressions, in fact, represent the quantities associated with a standing wave formed by the combination of the forward and the backward waves [see (7.88) and (7.97)]. From (7.93), we will also see that the amplitude of the electric field of this standing wave is $-2E_0 \sin\beta z$, which takes on the maximum value of $2E_0$ at $\beta\, z = -\pi/2, -3\pi/2, -5\pi/2, \ldots$, that is, at distances $z = -\lambda/4,\ -3\lambda/4,\ -5\lambda/4,\ ...$ (using the relation $\beta = 2\pi/\lambda$; see the text at the end of Section 6.1), giving the distance between two consecutive maxima of electric field as $2\lambda/4 = \lambda/2$. These distances have negative values ($z = -\lambda/4,\ -3\lambda/4,\ -5\lambda/4,\ ...$) since the free-space region where the standing-wave field variation is observed in the present formulation is located in the negative z region (Fig. 7.6). Similarly, the amplitude of the magnetic field of the standing wave takes on maximum values at $\beta z = 0,\ -\pi,\ -2\pi,\ ...,$ as can be seen from (7.99), that is, at distances $z = 0,\ -\lambda/2,\ -\lambda,\ ...$ (using the relation $\beta = 2\pi/\lambda$), giving the distance between two consecutive maxima of the magnetic field as $\lambda/2$. This is the same distance obtained between two consecutive maxima of the electric field. Similarly, by inspecting the field expressions (7.93) or (7.100) and (7.99), you will find it interesting to see that for this standing-wave field pattern, the minima of the amplitudes of electric field coincides with the maxima of the amplitudes of magnetic field and vice versa[1].

EXAMPLE E7-8

The electric field at the XY planar interface at $z = 0$ between a dielectric of relative permittivity 4 and a conductor is $\vec{E} = 5 \sin \beta z \sin \omega t \vec{a}_x$ V/m for a uniform plane wave incident from the dielectric on the interface, forming a standing wave in the dielectric region. Calculate the peak surface current density developed at the interface.

SOLUTION

The given electric field is

$$\vec{E} = 5 \sin \beta z \sin \omega t \vec{a}_x. \tag{E7-8.1}$$

We can write from (E7-8. 1)

$$\left.\begin{aligned} E_x &= 5 \sin \beta\, z \sin \omega t \\ E_y &= 0 \\ E_z &= 0 \end{aligned}\right\}. \tag{E7-8.2}$$

Further, for a uniform plane wave propagating along z,

$$\frac{\partial}{\partial x} = \frac{\partial}{\partial y} = 0. \tag{E7-8.3}$$

Let us now recall Maxwell's equation (5.78) and write it with the help of (2.66) and in view of the relations $E_y = E_z = 0 = \partial/\partial x = \partial/\partial x = 0$ [see (E7-8. 2) and (E7-8. 3)] as

[1]It is easy to appreciate that there is no power flow in such a standing wave, since the quantity $1/2(\mathrm{Re}\,\vec{E} \times \vec{H}^*)$ representing the time-averaged power flow (see Section 8.22) becomes zero in view of the relations (7.93) and (7.99).

$$\frac{\partial E_x}{\partial z} = -j\omega B_y. \qquad \text{(E7-8.4)}$$

Substituting E_x from (E7-8.2) in (E7-8. 4) and carrying out the differentiation, we get

$$5\beta \cos \beta\, z \sin \omega t = -j\omega B_y. \qquad \text{(E7-8.5)}$$

In view of the relations $B_y = \mu_0 H_y$ and $\beta = \omega\sqrt{\mu_0 \varepsilon_0 \varepsilon_r}$, we can deduce from (E7-8.5) the following relation:

$$H_y = \frac{j5\sqrt{\varepsilon_r} \cos \beta\, z \sin \omega t}{\eta_0},$$

which can be read at the planar interface $z = 0$ between the dielectric and the conductor as

$$H_y = \frac{j5\sqrt{\varepsilon_r} \sin \omega t}{\eta_0} \qquad \text{(interface } z = 0) \qquad \text{(E7-8.6)}$$

where ε_r is the relative permittivity of the dielectric region and η_0 is the free-space intrinsic impedance. Putting $\eta_0 = 120\pi \cong 377\ \Omega$ and $\varepsilon_r = 4$ (given) in (E7-8.6), we get

$$H_y = \frac{j10 \sin \omega t}{377} = j0.0265 \sin \omega t \text{ A/m. (interface } z = 0) \qquad \text{(E7-8.7)}$$

Now, we can use (E7-8.7) to find the surface current density at the interface ($z = 0$). For this purpose, let us write (7.46*c*) interpreting therein $\vec{a}_n = \vec{a}_z$ as

$$\vec{J}_s = H_y(\vec{a}_z \times \vec{a}_y) = -H_y \vec{a}_x. \qquad \text{(E7-8.8)}$$

Substituting (E7-8.7) in (E7-8.8) we get

$$\vec{J}_s = -j0.0265 \sin \omega t\, \vec{a}_x \text{ A/m},$$

whence the peak surface current density comes out to be 0.0265 A/m or 26.5 mA/m.

7.6 Reflection and Refraction of Electromagnetic Waves at a Dielectric-Dielectric Interface

Let an electromagnetic wave propagating through a dielectric medium of permittivity ε_1 (medium 1) be obliquely incident on the surface of another dielectric medium of permittivity ε_2 (medium 2). We consider two cases, namely parallel polarisation and perpendicular polarisation of the electromagnetic wave. In the case of parallel polarisation, the electric field is parallel to the plane of incidence while the magnetic field is perpendicular to this plane (Fig. 7.7). In the case of perpendicular polarisation, the electric field is perpendicular to the plane of incidence while the magnetic field is parallel to this plane (Fig. 7.8), where the plane of incidence is defined as the plane containing the incident ray and the normal to the interface between the two media. In the present context, we will find it convenient to resort to ray diagram of optics, the ray representing the direction of propagation of electromagnetic waves. The incident ray will be reflected from as well as refracted through the interface between the two dielectric media (Figs. 7.7 and 7.8). Analytical results obtained in this section will bring about two well known concepts—Brewster's phenomenon with reference to parallel polarisation and total internal reflection with reference to both parallel and perpendicular polarisations.

The directions of the electric and the magnetic field components as well as the direction of propagation with reference to the incident, reflected and refracted rays for the cases of both parallel and perpendicular polarisations (Figs. 7.7 and 7.8) can be appreciated from the discussion following (6.65). Accordingly, for parallel polarisation, the direction of the magnetic field would be perpendicular to the plane of the paper and directed towards the reader (Fig. 7.7). Similarly, for perpendicular polarisation, it is the direction of the electric field that would be perpendicular to the plane of the paper and directed towards the reader (Fig. 7.8). This point will become clear further with the help of the Poynting vector $\vec{E} \times \vec{H}$ (to be introduced in Chapter 8), whose direction gives the direction of propagation. In other words, for a given direction of the Poynting vector $\vec{E} \times \vec{H}$, that is, for a given direction of a ray, if we assign the direction of any one of the electric and the magnetic fields, we can predict the direction of the other one.

7.6.1 Reflection and transmission coefficients for parallel polarisation

The tangential component of electric field is continuous at the interface [boundary condition (7.26*d*), understood with the help of the interpretation given following (7.25)]. We can therefore write the following expression for the electric field quantities associated with the incident, reflected and refracted (transmitted from medium (1) of permittivity ε_1 to medium (2) of permittivity ε_2) rays (Fig. 7.7):

$$E_i \cos\theta_1 - E_r \cos\theta_1 = E_t \cos\theta_2 \qquad (7.101)$$

where θ_1 is the angle of incidence, which is also equal to the angle of reflection, and θ_2 is the angle of refraction. The subscripts i, r and t refer to the components of the incident, reflected, and refracted rays respectively (Fig. 7.7).

Similarly, the tangential component of magnetic field is continuous at the interface [boundary condition (7.26*c*), understood with the help of the interpretation given following (7.25)], there being no current at the interface between the two dielectric media. We can therefore write the following expression for the magnetic field quantities associated with the incident, reflected and refracted rays (Fig. 7.7):

$$H_i + H_r = H_t \qquad (7.102)$$

where the subscripts i, r, and t have the same meaning as explained following (7.101). Here, H_i, H_r and H_t are all directed perpendicular to the plane of polarisation (pointed

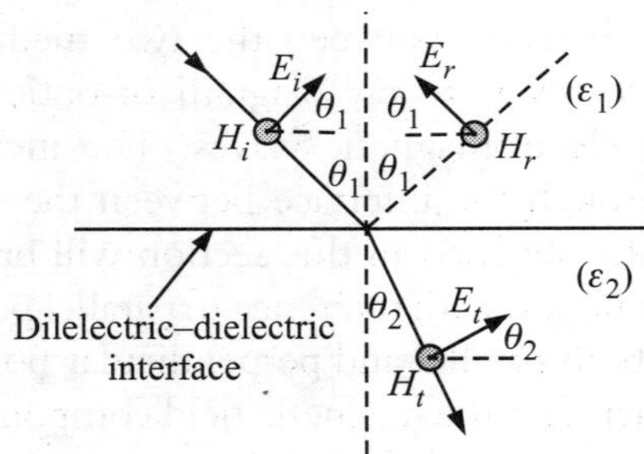

Fig. 7.7 *Electromagnetic wave with parallel polarisation obliquely incident from a dielectric medium on the surface of another dielectric medium.*

toward the reader in Fig. 7.7). Useful information can be derived with the help of (7.101) and (7.102) with reference to the reflection and refraction of electromagnetic waves at the interface between two dielectric media. However, for that purpose, we have to express (7.102) in terms of the electric field quantities with the help of the relations (6.63*b*) and (6.70) between the electric and magnetic fields in a medium involving the intrinsic impedance of the medium (see Section 6.2). Hence, in the present context, interpreting (6.63*b*) and (6.70), we can express H_i, H_r and H_t occurring in (7.102) in terms of E_i, E_r and E_t respectively as

$$\left.\begin{aligned} \frac{E_i}{H_i} &= \eta_1 \quad \text{(a)} \\ \frac{E_r}{H_r} &= \eta_1 \quad \text{(b)} \\ \frac{E_t}{H_t} &= \eta_2 \quad \text{(c)} \end{aligned}\right\} \tag{7.103}$$

where η_1 and η_2 are the intrinsic impedances of medium 1 of permittivity ε_1 and medium 2 of permittivity ε_2 respectively, given by [see (6.70)]

$$\left.\begin{aligned} \eta_1 &= (\mu_0/\varepsilon_1)^{1/2} \quad \text{(a)} \\ \eta_2 &= (\mu_0/\varepsilon_2)^{1/2} \quad \text{(b)} \end{aligned}\right\}. \tag{7.104}$$

We note that since the incident and the reflected rays both lie in medium 1, the intrinsic impedance η_1 of medium 1 occurs in both the relations (7.103*a*) and (7.103*b*), which are valid for the incident and the reflected rays respectively. Similarly, since the refracted ray (transmitted to medium 2) lies in medium 2, the intrinsic impedance η_2 of medium 2 occurs in the relation (7.103*c*) valid for the refracted ray. Therefore, with the help of (7.103), we can now write (7.102) as

$$\frac{E_i}{\eta_1} + \frac{E_r}{\eta_1} = \frac{E_t}{\eta_2}. \tag{7.105}$$

Now, by simple algebra we can eliminate E_t from (7.101) and (7.105) to obtain the following relation

$$\frac{E_r}{E_i} = \frac{-\eta_2 \cos\theta_2 + \eta_1 \cos\theta_1}{\eta_1 \cos\theta_1 + \eta_2 \cos\theta_2}. \tag{7.106}$$

Similarly, by eliminating E_r from (7.101) and (7.105), we can easily obtain the following relation

$$\frac{E_t}{E_i} = \frac{2\eta_2 \cos\theta_1}{\eta_1 \cos\theta_1 + \eta_2 \cos\theta_2}. \tag{7.107}$$

At this stage, with reference to the reflection for parallel polarisation at the interface between two dielectric media, let us define the reflection coefficient $\Gamma_{\|}$ for parallel incidence as

$$\Gamma_{\|} = -\frac{E_r}{E_i}. \tag{7.108}$$

A minus sign is deliberately put in the definition of the reflection coefficient (7.108) in its right-hand side, since, for such reflection the direction of the electric field in the incident ray and that in the reflected ray tend to oppose each other, more and more so as the angle of incidence (θ_1) is progressively decreased, approaching the situation of normal incidence (Fig. 7.7). Thus, the minus sign in the definition ensures that the value of the reflection coefficient $\Gamma_{\|}$ becomes a positive quantity. We can then obtain the following expression for the reflection coefficient for parallel incidence by substituting (7.107) in (7.108) and making use of the relation (7.104):

$$\Gamma_{\|} = -\frac{E_r}{E_i} = \frac{\eta_2 \cos\theta_2 - \eta_1 \cos\theta_1}{\eta_1 \cos\theta_1 + \eta_2 \cos\theta_2} = \frac{\sqrt{\varepsilon_1}\cos\theta_2 - \sqrt{\varepsilon_2}\cos\theta_1}{\sqrt{\varepsilon_1}\cos\theta_2 + \sqrt{\varepsilon_2}\cos\theta_1}. \tag{7.109}$$

Similarly, we can define the transmission coefficient for parallel incidence as

$$T_{\|} = \frac{E_t}{E_i} \tag{7.110}$$

in which now the minus sign need not be introduced since the electric field in the incident ray and that in the transmitted ray tend to be more and more in the same direction as the angle of incidence (θ_1) is progressively decreased when approaching normal incidence (Fig. 7.7). Thus, without the minus sign in the definition, the value of the transmission coefficient $T_{\|}$ becomes a positive quantity. We can then obtain the expression for the transmission coefficient $T_{\|}$ by substituting (7.107) in (7.110) and making use of the relation (7.104), as follows:

$$T_{\|} = \frac{E_t}{E_i} = \frac{2\eta_2 \cos\theta_1}{\eta_1 \cos\theta_1 + \eta_2 \cos\theta_2} = \frac{2\sqrt{\varepsilon_1}\cos\theta_1}{\sqrt{\varepsilon_1}\cos\theta_2 + \sqrt{\varepsilon_2}\cos\theta_1}. \tag{7.111}$$

Brewster's phenomenon

We can put the numerator of (7.119) equal to zero to obtain the following condition for nil reflection for parallel polarisation at the interface between two dielectric media:

$$\sqrt{\varepsilon_1}\cos\theta_2 - \sqrt{\varepsilon_2}\cos\theta_1 = 0,$$

which can be rearranged as

$$\sqrt{\varepsilon_2}\cos\theta_1 = \sqrt{\varepsilon_1}\cos\theta_2. \tag{7.112}$$

In order to know at what angle of incidence θ_1 the condition for nil reflection takes place, we have to combine (7.112) with Snell's law given by

$$\sqrt{\varepsilon_1}\sin\theta_1 = \sqrt{\varepsilon_2}\sin\theta_2. \tag{7.113}$$

Multiplying (7.112) with (7.113), we obtain

$$\sqrt{\varepsilon_1}\sqrt{\varepsilon_2}\sin\theta_1\cos\theta_1 = \sqrt{\varepsilon_1}\sqrt{\varepsilon_2}\sin\theta_2\cos\theta_2,$$

which simplifies to

$$\sin\theta_1\cos\theta_1 = \sin\theta_2\cos\theta_2. \tag{7.114}$$

Multiplying both sides of (7.114) by 2, we obtain

$$2\sin\theta_1\cos\theta_1 = 2\sin\theta_2\cos\theta_2$$

which, in view of the trigonometric relation $\sin 2\theta = 2\sin\theta\cos\theta$ can be put as

$$\sin 2\theta_1 = \sin 2\theta_2. \tag{7.115}$$

Thus, in compliance with (7.115), the following relation between θ_1 and θ_2 holds good

$$2\theta_2 = \pi - 2\theta_1,$$

which can be simplified to

$$\theta_2 = \frac{\pi}{2} - \theta_1. \tag{7.116}$$

Substituting (7.115) in (7.113), we can then write

$$\sqrt{\varepsilon_1}\sin\theta_1 = \sqrt{\varepsilon_2}\sin(\frac{\pi}{2} - \theta_1),$$

which, in view of the trigonometric relation $\sin(\pi/2 - \theta) = \cos\theta$, can be put as

$$\sqrt{\varepsilon_1}\sin\theta_1 = \sqrt{\varepsilon_2}\cos\theta_1,$$

which simplifies to

$$\tan\theta_1 = \sqrt{\frac{\varepsilon_2}{\varepsilon_1}}. \tag{7.117}$$

Thus, from the foregoing analysis we can see that for parallel polarisation the phenomenon of nil reflection takes place ($\Gamma_{||} = 0$) at an angle of incidence given by (7.117). This is known as Brewster's phenomenon. The angle of incidence, $\theta_1 = \theta_B$, called the Brewster's angle, can then be written from (7.117) as

$$\theta_B = \tan^{-1}\sqrt{\frac{\varepsilon_2}{\varepsilon_1}}. \tag{7.118}$$

We can numerically appreciate (7.118) by finding the Brewster's angle for a ray of light refracting from air, taken as medium 1 of refractive index $n_1 \approx 1$, to glass, taken as medium 2 of refractive index $n_2 \approx 1.5$, with the help of (7.118). However, we have to replace the ratio $\sqrt{\frac{\varepsilon_2}{\varepsilon_1}}$ by $\frac{n_2}{n_1}$ the ratio of the refractive indices of the two media. We can easily derive the foregoing relation, $\sqrt{\varepsilon_2/\varepsilon_1} = n_2/n_1$ by recalling (i) the relation (6.67) which gives the phase velocities of the media 2 and 1 as $v_{ph,2} = 1/\sqrt{\mu_0\varepsilon_2}$ and $v_{ph,1} = 1/\sqrt{\mu_0\varepsilon_2}$ respectively, and (ii) the relation between the refractive index and the phase velocity: $n = c/v_{ph}$, which gives $n_2 = c/v_{ph,2}$ and $n_1 = c/v_{ph,1}$, where c is the velocity of light. Thus, interpreting (7.118) as $\theta_B (= \tan^{-1}\sqrt{\varepsilon_2/\varepsilon_1}) = \tan^{-1} n_2/n_1$, we find, by taking $n_2 \approx 1.5$ and $n_1 \approx 1$ the value of the Brewster's angle as $\theta_B \approx 56.31^0$.

With the help of (7.119), in which, however, we have to take θ_2 in terms of θ_1 from Snell's law (7.113), it is easy to appreciate how the reflection coefficient $\Gamma_{||}$ varies with the angle of incidence θ_1. Thus, a plot $\Gamma_{||}$ against θ_1 will show that for $\varepsilon_1 < \varepsilon_2$, corresponding to incidence of a ray from a rarer to a denser medium, $\Gamma_{||}$ increases with θ_1 from a negative value of $\Gamma_{||} = (\sqrt{\varepsilon_1} - \sqrt{\varepsilon_2})/(\sqrt{\varepsilon_1} + \sqrt{\varepsilon_2})$ at $\theta_1 = 0$ to $\Gamma_{||} = 0$ at $\theta_1 = \theta_B$, the Brewster's angle. Further, beyond the Brewster's angle ($\theta_1 > \theta_B$), the value of $\Gamma_{||}$ becomes positive and continues to increase up to $\Gamma_{||} = 1$ at $\theta_1 = \pi/2$, the latter corresponding to normal incidence (Fig. 7.8).

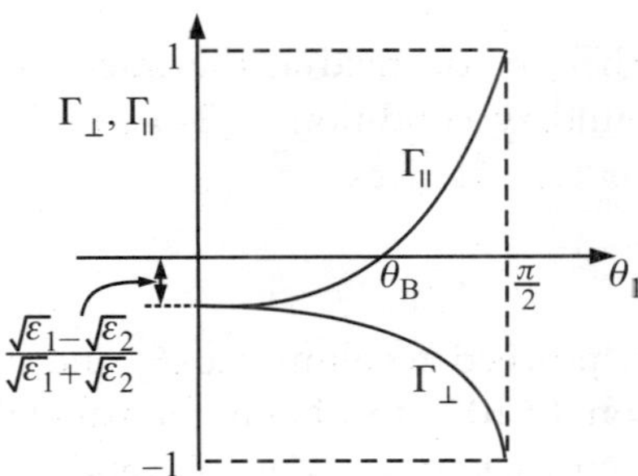

Fig. 7.8 *The nature of the variation of the reflection coefficients $\Gamma_{||}$ and $\Gamma_\perp$ for parallel and perpendicular polarisations with the angle of incidence θ_1, considering the reflection of electromagnetic ray, typically, from a rarer to denser medium ($\varepsilon_1 < \varepsilon_2$) at the interface between two dielectric media of permittivity values ε_1 and ε_2.*

7.6.2 Reflection and transmission coefficients for perpendicular polarisation

Now, let us proceed to the case of perpendicular polarisation, in which the electric field is perpendicular to the plane of incidence while the magnetic field is parallel to this plane (Fig. 7.9). The boundary condition of the continuity of the tangential component of magnetic field at the interface between the two dielectric media in this case gives (according to (7.26*c*) understood with the help of the interpretation given following (7.25), there being no current at the interface between the two dielectric media):

$$H_i \cos\theta_1 - H_r \cos\theta_1 = H_t \cos\theta_2. \tag{7.119}$$

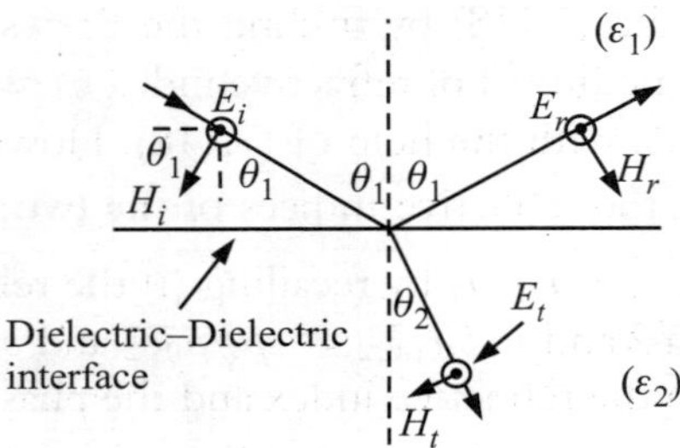

Fig. 7.9 *Electromagnetic wave with perpendicular polarisation obliquely incident from a dielectric medium on the surface of another dielectric medium.*

Therefore, with the help of (7.103), we can now write (7.119) as

$$\frac{E_i}{\eta_1}\cos\theta_1 - \frac{E_r}{\eta_1}\cos\theta_1 = \frac{E_t}{\eta_2}\cos\theta_2,$$

which can be simplified as

$$E_i - E_r = \frac{\cos\theta_2}{\cos\theta_1}\frac{\eta_1}{\eta_2}E_t. \tag{7.120}$$

Further, the continuity of the electric field at the interface between the two dielectric media [according to the boundary condition (7.26*d*), understood with the help of the interpretation given following (7.25)] gives

$$E_i + E_r = E_t. \tag{7.121}$$

We can, as in Section 7.6.1, proceed to eliminate E_t and E_r from (7.120) and (7.121) and make use of the relation (7.104) to obtain for parallel polarisation the following expressions for the reflection coefficient, $\Gamma_\perp = E_r/E_i$, and for the transmission coefficient, $T_\perp = E_t/E_i$:

$$\Gamma_\perp = \frac{E_r}{E_i} = \frac{\eta_2\cos\theta_1 - \eta_1\cos\theta_2}{\eta_2\cos\theta_1 + \eta_1\cos\theta_2} = \frac{\sqrt{\varepsilon_1}\cos\theta_1 - \sqrt{\varepsilon_2}\cos\theta_2}{\sqrt{\varepsilon_1}\cos\theta_1 + \sqrt{\varepsilon_2}\cos\theta_2} \tag{7.122}$$

$$T_{\perp} = \frac{E_t}{E_i} = \frac{2\eta_2 \cos\theta_1}{\eta_1 \cos\theta_1 + \eta_2 \cos\theta_2} = \frac{2\sqrt{\varepsilon_1}\cos\theta_1}{\sqrt{\varepsilon_1}\cos\theta_2 + \sqrt{\varepsilon_2}\cos\theta_1}. \tag{7.123}$$

Note that in the present case (perpendicular polarisation), no change in sign is required to be incorporated in the definition of the reflection coefficient or transmission coefficient in order to ensure that each of these coefficients takes on a positive value. This is because the field quantities, E_i, E_r, and E_t, each have the same direction, being perpendicular to the plane of polarisation (directed toward the reader in Fig. 7.9). Further, in this case, for no value of the angle of incidence θ_1 would the reflection coefficient $\Gamma_{\perp}$ become zero. This can be verified by putting $\Gamma_{\perp} = 0$, which will make the numerator of the right-hand side of (7.122) equal to zero giving

$$\sqrt{\varepsilon_1}\cos\theta_1 - \sqrt{\varepsilon_2}\cos\theta_2 = 0,$$

which can be rearranged as

$$\frac{\cos\theta_1}{\cos\theta_2} = \frac{\sqrt{\varepsilon_2}}{\sqrt{\varepsilon_1}} \text{ (for } \Gamma_{\perp} = 0),$$

a result that clearly contradicts Snell's law (7.113) according to which we have

$$\frac{\sin\theta_1}{\sin\theta_2} = \frac{\sqrt{\varepsilon_2}}{\sqrt{\varepsilon_1}}.$$

Therefore, in order to avoid a result that will contradict Snell's law, we cannot put $\Gamma_{\perp} = 0$ in the case of perpendicular polarisation,. In other words, in the case of perpendicular polarisation, Brewster's phenomenon would not take place, unlike in the case of parallel polarisation .

Furthermore, with the help of (7.122) and noting how θ_2 could be expressed in terms of θ_1 from Snell's law (7.113), we can see that for $\varepsilon_1 < \varepsilon_2$, corresponding to the incidence of a ray from a rarer to a denser medium, $\Gamma_{\perp}$ increases with θ_1 from a negative value of $\Gamma_{\perp} = (\sqrt{\varepsilon_1} - \sqrt{\varepsilon_2})/(\sqrt{\varepsilon_1} + \sqrt{\varepsilon_2})$ at $\theta_1 = 0$ to more and more negative values until the value $\Gamma_{\perp} = -1$ at $\theta_1 = \pi/2$ (normal incidence; see Fig. 7.8).

7.6.3 Total internal reflection

Irrespective of whether the polarisation of an electromagnetic wave is parallel or perpendicular, the wave incident from a denser medium to a rarer medium undergoes reflection at the interface between these two media with a magnitude of the reflection coefficient equal to unity when the angle of incidence is greater than a value called the 'critical angle'. The phenomenon is called total internal reflection.

Let us develop the understanding of the phenomenon of total internal reflection starting from Snell's law (7.113) expressed in terms of the refractive indices of the media at the interface. Thus, in view of the relation $\sqrt{\varepsilon_2/\varepsilon_1} = n_2/n_1$, we can express (7.113) as [refer to the discussions following (7.118)]

$$\frac{\sin\theta_1}{\sin\theta_2} = \frac{n_2}{n_1}. \tag{7.124}$$

It clearly follows from (7.124) that for an electromagnetic wave refracting from a denser to a rarer medium ($n_1 > n_2$), we obtain $\sin\theta_2 > \sin\theta_1$, that is, $\theta_2 > \theta_1$ corresponding to the refracting ray bending away from the normal. In the limit, we obtain the angle of refraction $\theta_2 = \pi/2$ corresponding to the angle of incidence $\theta_1 = \theta_c$, called the critical angle. Thus, at the critical angle of incidence θ_c, we obtain the electromagnetic ray grazing the interface between the two media ($\theta_2 = \pi/2$) (Fig. 7.10).

Further, if we put $\theta_2 = \pi/2$ in (7.124), we obtain the following relation for the critical angle $\theta_1 = \theta_c$

$$\frac{\sin\theta_c}{\sin\pi/2} = \frac{n_2}{n_1},$$

which can be put as (since $\sin\pi/2 = 1$):

$$\sin\theta_c = \frac{n_2}{n_1}. \tag{7.125}$$

It would be of interest to see how the magnitudes of the reflection coefficients $\Gamma_\parallel$ and $\Gamma_\perp$ for the parallel and the perpendicular polarisations respectively, each of which involves $\cos\theta_2$, behave for the angle of incidence $\theta_1 > \theta_c$. For this purpose, let us express $\cos\theta_2$ with the help of (7.124) as

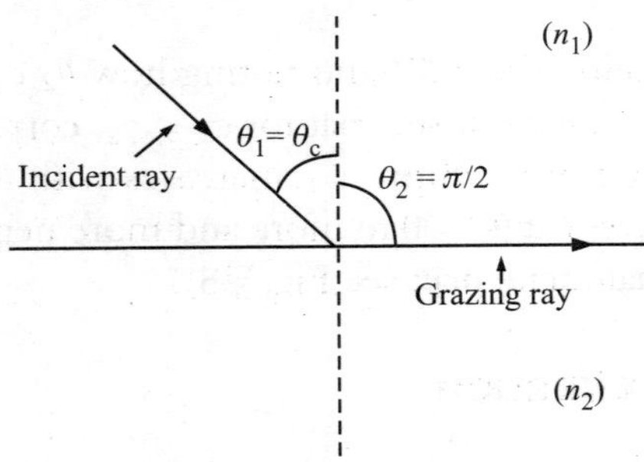

Fig. 7.10 *Ray diagram showing the incident and the grazing rays for an electromagnetic wave refracting from a denser to a rarer medium ($n_1 > n_2$) when the angle of refraction $\theta_2 = \pi/2$ for an angle of incidence $\theta_1 = \theta_c$ called the critical angle.*

$$\cos\theta_2 = \sqrt{1 - \sin^2\theta_2} = \sqrt{1 - (\frac{n_1}{n_2})^2 \sin^2\theta_1},$$

which, with the help of (7.125) can be written as

$$\cos\theta_2 = \sqrt{1 - \frac{\sin^2\theta_1}{\sin^2\theta_c}}. \tag{7.126}$$

For $\theta_1 > \theta_c$, which makes $\sin\theta_1 > \sin\theta_c$, that is, $\sin^2\theta_1/\sin^2\theta_c > 1$, we see from (7.126) that $\cos\theta_2$ becomes imaginary. We can then represent $\sqrt{\varepsilon_1}\cos\theta_2$ and $\sqrt{\varepsilon_2}\cos\theta_2$ occurring in the expressions (7.109) and (7.122) for $\Gamma_{\|}$ and $\Gamma_{\perp}$ respectively as the imaginary quantities:

$$\left.\begin{aligned} \sqrt{\varepsilon_1}\cos\theta_2 &= jB \\ \sqrt{\varepsilon_2}\cos\theta_2 &= jD \end{aligned}\right\} \tag{7.127}$$

where B and D are real quantities. We may further put the quantities $\sqrt{\varepsilon_2}\cos\theta_1$ and $\sqrt{\varepsilon_1}\cos\theta_1$ occurring in (7.119) and (7.122) respectively as

$$\left.\begin{aligned} \sqrt{\varepsilon_2}\cos\theta_1 &= A \\ \sqrt{\varepsilon_1}\cos\theta_1 &= C \end{aligned}\right\} \tag{7.128}$$

where A and C are real quantities. Therefore, using the expressions (7.127) and (7.128), we can express (7.1019) and (7.122) as

$$\left.\begin{aligned} \Gamma_{\|} &= \frac{jB - A}{jB + A} \\ \Gamma_{\perp} &= \frac{C - jD}{C + jD} \end{aligned}\right\}. \tag{7.129}$$

The magnitudes of the reflection coefficients $|\Gamma_{\|}|$ and $|\Gamma_{\perp}|$ for parallel and perpendicular polarisations are then obtained from (7.129) as

$$\left.\begin{aligned} |\Gamma_{\|}| &= \left|\frac{jB - A}{jB + A}\right| = \frac{\sqrt{B^2 + A^2}}{\sqrt{B^2 + A^2}} = 1 \\ |\Gamma_{\perp}| &= \left|\frac{C - jD}{C + jD}\right| = \frac{\sqrt{C^2 + D^2}}{\sqrt{C^2 + D^2}} = 1 \end{aligned}\right\}. \tag{7.130}$$

Thus by the deduction of (7.130) we see that for an electromagnetic wave refracting from a denser to a rarer medium, if the angle of incidence is greater that the critical angle, then 'total internal reflection' takes place, giving the magnitude of the reflection coefficient as unity, irrespective of whether the polarisation is parallel or perpendicular.

7.7 Boundary Conditions for a Structure Model

Electromagnetic boundary conditions can be easily developed for a simple physical structure. Take for example a straight hollow metal pipe of circular cross section. Assuming that the metal which the wall of the pipe is made of is a very good conductor and that the hollow region of the pipe is free space, we can directly apply electromagnetic boundary conditions to such a structure for analysis. However, the necessity of replacing an actual physical structure by its model for analysis arises if the structure geometry becomes rather complex. As a simple example of a complex physical structure, one can take a helix (see Fig. E7-9.1 in Example E7-9 below), which looks like a spring made of a conducting material.

EXAMPLE E7-9

A conducting wire or tape wound in the form of a helix of circular cross section in free space is modelled by a circular-cylindrical sheath of infinitesimal thickness. The sheath has infinite conductivity parallel to the winding direction and nil conductivity in the direction perpendicular to it (sheath-helix model). Derive the following four boundary conditions at the surface of the sheath in cylindrical coordinate system (r, θ, z):

$$\left.\begin{aligned} &E_{\theta_1} \cos\psi + E_{z1} \sin\psi = 0 \\ &E_{\theta 2} \cos\psi + E_{z2} \sin\psi = 0 \\ &H_{\theta_1} \cos\psi + H_{z1} \sin\psi = H_{\theta_2} \cos\psi + H_{z2} \sin\psi \\ &E_{z1} = E_{z2} \end{aligned}\right\}$$

where ψ is the helix pitch angle, and subscripts 1 and 2 represent the quantities inside and outside the winding radius of the helix respectively.

SOLUTION

It becomes convenient to analyse a conducting wire or tape wound in the form of a circular helix (Fig. E7-9.1) in a coordinate system taking three unit vectors $\vec{a}_{\|}$, $\vec{a}_{\perp}$ and $\vec{a}_r$ in three mutually perpendicular directions: parallel to the winding direction, perpendicular to the winding direction and in the outward radial direction respectively (see Fig. E7-9.2 in which the direction of $\vec{a}_r$ is shown perpendicular to the plane of the paper pointed towards the reader and ψ represents the pitch angle of the helix). The unit vectors so chosen in this system of coordinates will satisfy the relations $\vec{a}_{\|} \times \vec{a}_{\perp} = \vec{a}_r$, $\vec{a}_{\perp} \times \vec{a}_r = \vec{a}_{\|}$ and $\vec{a}_r \times \vec{a}_{\|} = \vec{a}_{\perp}$ (Fig. E7-9.2).

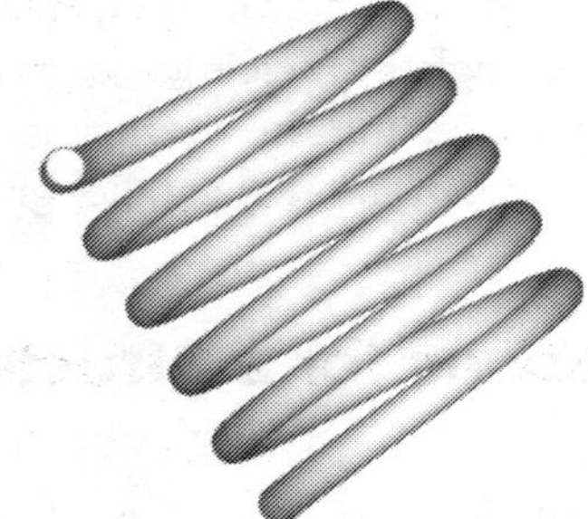

Fig. E7-9.1

We can then represent the electric and the magnetic fields respectively in this coordinate system as

$$\vec{E} = E_{\|}\vec{a}_{\|} + E_{\perp}\vec{a}_{\perp} + E\vec{a}_r \tag{E7-9.1}$$

and

$$\vec{H} = H_{\|}\vec{a}_{\|} + H_{\perp}\vec{a}_{\perp} + H\vec{a}_r. \tag{E7-9.2}$$

where the subscripts $\|$, $\perp$ and r represent the parallel, perpendicular and outward radial field components respectively (Fig. E7-9.2).

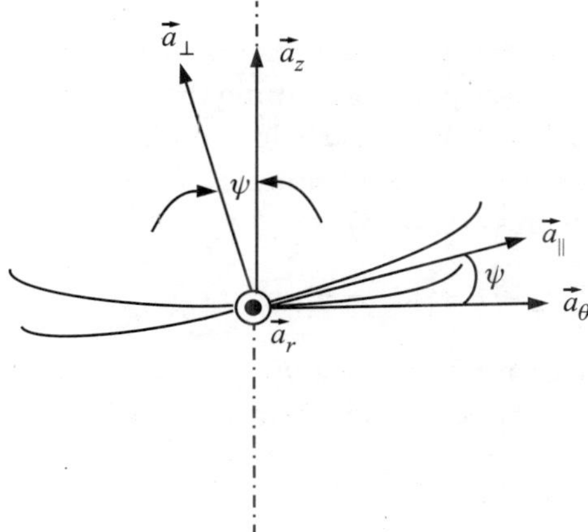

Fig. E7-9.2

We can use the field expressions (E7-9.1) and (E7-9.12) in the sheath-helix model defined in the given problem in which the actual helix for the purpose of analysis is replaced by a circular-cylindrical sheath of infinitesimal thickness that has infinite conductivity parallel to the winding direction and nil conductivity in the direction perpendicular to it (see Fig. E7-9.3 showing the cross section of the sheath divided in three regions of the model: region 1 representing the region inside the winding radius of the helix, region i representing the region inside the conducting medium of the helix, and region 2 representing the region outside the winding radius of the helix).

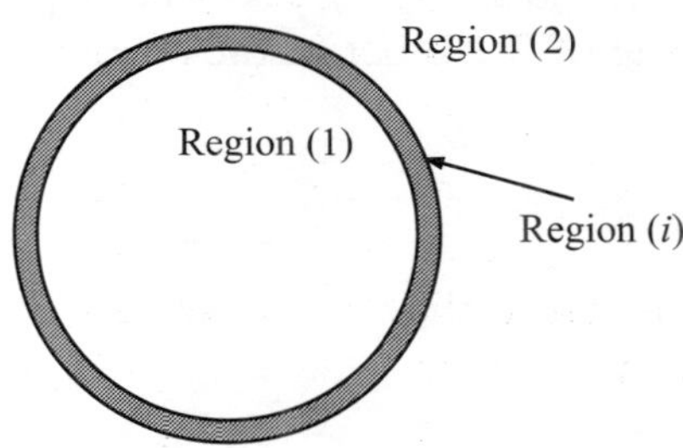

Fig. E7-9.3

Thus, the structure model (Fig. E7-9.3) has two interfaces. The first interface is the interface between the region inside the conducting medium of the helix (to be treated as the first region) and the region outside the helix winding radius (to be treated as the second region). The second interface is the interface between the region inside the helix winding radius (to be treated as the first region) and the region inside the conducting medium of the helix (to be treated as the second region). Therefore, with respect to the first interface we can interpret $\vec{E}_1 = \vec{E}_i$, $\vec{E}_2 = \vec{E}_2$ and $\vec{a}_n = \vec{a}_r$, and express the general boundary condition $\vec{a}_n \times (\vec{E}_2 - \vec{E}_1) = 0$ given by (7.25) while choosing to represent the electric field as in (E7-9.1) as follows:

$$\vec{a}_r \times [(E_{\parallel,2} - E_{\parallel,i})\vec{a}_\parallel + (E_{\perp,2} - E_{\perp,i})\vec{a}_\perp + (E_{r2} - E_{r,i})\vec{a}_r] = 0 \tag{E7-9.3}$$

the subscripts 1 and 2 representing the quantities inside and outside the helix winding radius respectively and the subscript i representing the field quantity inside the conducting medium of the helix (Fig. E7-9.3). With the help of the relations $\vec{a}_\parallel \times \vec{a}_\perp = \vec{a}_r$, $\vec{a}_\perp \times \vec{a}_r = \vec{a}_\parallel$ and $\vec{a}_r \times \vec{a}_\parallel = \vec{a}_\perp$(as mentioned in the beginning of the solution) as well as the relation $\vec{a}_r \times \vec{a}_r = 0$, we can then express (E7-9.3) as

$$(E_{\parallel,2} - E_{\parallel,i})\vec{a}_\perp - (E_{\perp,2} - E_{\perp,i})\vec{a}_\parallel = 0. \tag{E7-9.4}$$

From (E7-9.4) we then obtain the relations $E_{\parallel,2} - E_{\parallel,i} = 0$ and $E_{\perp,2} - E_{\perp,i} = 0$ ($\vec{a}_\parallel$ and $\vec{a}_\perp$ being perpendicular to each other and hence unequal), from the first of which we can write

$$E_{\parallel,2} - E_{\parallel,i} = 0. \tag{E7-9.5}$$

However, the electric field parallel to the winding direction being null in the sheath, we can put $E_{\parallel,i} = 0$ in (E7-9.5) to write

$$E_{\parallel,2} = 0. \tag{E7-9.6}$$

Similarly, with respect to the second interface, that is, at the interface between the region inside the helix winding radius (to be treated as the first region) and the region inside the conducting medium of the helix (to be treated as the second region), we can interpret $\vec{E}_1 = \vec{E}_1$, $\vec{E}_2 = \vec{E}_i$ and $\vec{a}_n = \vec{a}_r$, and following the same steps that led to the derivation of (E7-9.5), we can, with the help of (7.25) obtain the relation corresponding to (E7-9.5) as follows:

$$E_{\parallel,i} - E_{\parallel,1} = 0. \tag{E7-9.7}$$

From (E7-9.7), since $E_{\parallel,i} = 0$ (due to the electric field being null in the sheath parallel to the winding direction), we then obtain

$$E_{\parallel,1} = 0. \tag{E7-9.8}$$

Thus, starting from (7.25) we have derived two boundary conditions of the helix as (E7-9.8) and (E7-9.6).

Similarly, from the boundary condition $\vec{a}_n \times (\vec{H}_2 - \vec{H}_1) = \vec{J}_s$ given by (7.23), we can obtain

$$\vec{a}_r \times [(H_{\|,2} - H_{\|,i})\vec{a}_{\|} + (H_{\perp,2} - H_{\perp,i})\vec{a}_{\perp} \qquad \text{(E7-9.9)}$$

$$+(H_{r2} - H_{r,i})\vec{a}_r] = J_{s,\perp}\vec{a}_{\perp} + J_{s,\perp}\vec{a}_{\perp} + J_{sr}\vec{a}_r.$$

Now, with the help of the relation $\vec{a}_r \times \vec{a}_{\|} = \vec{a}_{\perp}$(as mentioned in the beginning of this appendix) the first term of the left-hand side of (A7.13) becomes equal to $(H_{\|,2} - H_{\|,i})\vec{a}_{\perp}$, which can also be equated to the terms $J_{s,\perp}\vec{a}_{\perp}$ occurring on the right-hand side of (E7-9.9). Therefore, equating these two terms each involving $\vec{a}_{\perp}$, we obtain

$$H_{\|,2} - H_{\|,i} = J_{s,\perp},$$

which, in turn, becomes the boundary condition:

$$H_{\|,2} = H_{\|,i} \qquad \text{(E7-9.10)}$$

since $J_{s,\perp} = 0$ in view of the nil conductivity of the sheath perpendicular to the winding direction assumed in the model. Now, we can easily interpret (E7-9.8), (E7-9.6) and (E7-9.10) to respectively obtain three boundary conditions of the helix:

$$E_{\theta_1} \cos\psi + E_{z1} \sin\psi = 0. \qquad \text{(E7-9.11)}$$

$$E_{\theta 2} \cos\psi + E_{z2} \sin\psi = 0 \qquad \text{(E7-9.12)}$$

$$H_{\theta_1} \cos\psi + H_{z1} \sin\psi = H_{\theta_2} \cos\psi + H_{z2} \sin\psi \qquad \text{(E7-9.13)}$$

where ψ is the helix pitch angle (Fig. E7-9.2).

As for the fourth boundary condition of the helix, we can, as before, start from the boundary condition (7.25), however, now in the usual cylindrical system of coordinates (r, θ, z) to obtain the relations

$$\left.\begin{aligned} E_{z1} &= E_{zi} \\ E_{zi} &= E_{z2} \end{aligned}\right\},$$

whence we get

$$E_{z1} = E_{z2}. \qquad \text{(E7-9.14)}$$

Thus, in (E7-9.11)–(E7-9.14) we have derived the desired four boundary conditions of the helix, which can be grouped together to be read as

$$\left.\begin{array}{l} E_{\theta_1}\cos\psi + E_{z1}\sin\psi = 0 \\ E_{\theta 2}\cos\psi + E_{z2}\sin\psi = 0 \\ H_{\theta_1}\cos\psi + H_{z1}\sin\psi = H_{\theta_2}\cos\psi + H_{z2}\sin\psi \\ E_{z1} = E_{z2} \end{array}\right\}.$$

Appendix A7.1

An alternative approach to deriving expression (7.100) for magnetic field from that for the electric field

The electric field at a distance z in the free-space region given by (7.93) can be rewritten for the sake of convenience as

$$\vec{E} = -2jE_0 \sin\beta\, z \exp(j\omega\, t\,)\vec{a}_x. \tag{A7.1}$$

We can write (A7.1) component-wise as

$$\left.\begin{array}{l} E_x = -2jE_0 \sin\beta z \exp j(\omega t) \\ E_y = E_z = 0 \end{array}\right\}. \tag{A7.2}$$

Moreover, for a uniform plane wave, we have

$$\frac{\partial}{\partial x} = \frac{\partial}{\partial y} = 0. \tag{A7.3}$$

We can then, with the help of Maxwell's equation (5.78), deduce (see Problem 5-6):

$$\vec{H}(= H_y\vec{a}_y) = \frac{2E_0}{\eta_0}\cos\beta\, z \exp j(\omega t)\,\vec{a}_y. \tag{A7.4}$$

Note that (A7.4) and (7.99) are identical.

SUMMARISING NOTES

✓ Electric and magnetic field quantities get modified due to an abrupt change of the medium properties at the interface between two media such that these quantities satisfy a set of electromagnetic boundary conditions at a point on the interface between the media.

✓ General electromagnetic boundary conditions at a point on the interface between two media have been deduced in vector form with the help of Maxwell's equations in integral form in terms of the field quantities and a unit vector directed from one medium to another.

✓ Surface charge density is defined as the product of the volume charge density and the infinitesimal thickness over which the charge is spread at the interface, in the limit of the infinitesimal thickness tending to zero. This definition emerges in course of the deduction of general electromagnetic boundary conditions.

✓ Surface current density is defined as the product of the current density and the infinitesimal thickness at the interface over which the current density is significant, in the limit of the infinitesimal thickness tending to zero. This definition emerges in course of the deduction of general electromagnetic boundary conditions.

✓ General boundary conditions can be interpreted for dielectric-dielectric and conductor-dielectric interfaces on the following findings.

- Relaxation time of a dielectric being quite large, the electric charge can stay longer inside the bulk of a dielectric and the bulk of a dielectric can be electrically charged with a finite volume charge density that makes the surface charge density nil at the dielectric surface, which can be appreciated from the definition of surface charge density. The finding is valid for both time-independent and time-dependent situations.
- Relaxation time of a good conductor being quite small, the charge inside the bulk of a good conductor decays very fast to appear with quite a large volume charge density at the conductor surface to get concentrated over a layer of infinitesimal thickness, which in turn renders a finite value of surface charge density according to its definition. The finding is valid for both time-independent and time-dependent situations.
- Electric field or electric displacement is absent in a good conductor since the bulk of the conductor cannot be electrically charged. The finding is valid for both time-independent and time-dependent situations.
- Finite magnetic field or magnetic flux density can be established inside a dielectric independently of electric field, for both time-independent and time-dependent situations.
- Magnetic field and the magnetic flux density, each become zero inside a conductor for time-dependent situations since for such situations the magnetic field is coupled to the electric field, which is absent in a conductor.
- Finite magnetic field or magnetic flux density can be established independently of an electric field in a conductor for time-independent situations.
- Surface current density at a dielectric surface is nil since a dielectric, supposedly perfect, does not conduct current. The finding is valid for both time-independent and time-dependent situations.

  - Surface current density at a conductor surface is nil for time-independent situations according to the definition of surface density, since a finite current can be made to flow through the bulk of the conductor for such situations.
  - Finite surface current density according to its definition can be established at the surface of a good conductor for time-dependent situations, since quite a large current density can be concentrated over a layer of infinitesimal thickness at the surface of the conductor.

✓ Circuit law of parallel resistances can be appreciated from the electromagnetic boundary condition that the tangential component of electric field is continuous at the interface between two media.

✓ Boundary conditions at the interface between two conducting media yield the law of refraction of current for time-independent situations.

✓ Understanding of the phenomenon of formation of a standing wave when a uniform plane electromagnetic wave is incident from the free-space region on the surface of a conducting medium has been developed with the help of the relevant electromagnetic boundary condition.

✓ Reflection and refraction of electromagnetic waves at a dielectric-dielectric interface have been studied with the help of the relevant electromagnetic boundary conditions for parallel and perpendicular polarisations.

✓ Brewster's phenomenon for parallel polarisation can be understood with the help of the relevant electromagnetic boundary conditions.

✓ Brewster's angle of incidence that corresponds to no reflection at a dielectric-dielectric interface has been deduced for parallel polarisation.

✓ Total internal reflection at a dielectric-dielectric interface for the angle of incidence greater than the critical angle has been understood for both parallel and perpendicular polarisations.

Review Questions

Narrative type

7.1 Define surface charge density and surface current density in terms of volume charge density and current density respectively.

Hint: See (7.5) and (7.21).

7.2 Write the units of volume charge density, surface charge density, current density.

7.3 Check the dimensional correctness of the expression (7.68) reproduced here as

$$\rho_s = \varepsilon_0(\frac{1}{\sigma_2} - \frac{1}{\sigma_1})J_n,$$

where ρ_s is the surface charge density and J_n the normal component of current density at the interface between two conducting media of conductivities σ_1 and σ_2 respectively.

7.4 Compare the behaviour of a good conducting medium with that of a perfect dielectric medium in regard to their relaxation times, charging of the medium bulk, appearance of surface charge, and the establishment of electric and magnetic fields in the media for both time-independent and time-dependent situations.

Hint: See Table 7.2.

7.5 State in words in a non-mathematical form the four general electromagnetic boundary conditions at a point on the interface between two media.

7.6 Prove the general electromagnetic boundary condition in vectorial form at a point on the interface between two media that involves the 'surface charge density' at the interface.

7.7 Prove the general electromagnetic boundary condition in vectorial form at a point on the interface between two media that refer to the 'normal components of magnetic flux densities' in the two media at the interface.

7.8 Prove the general electromagnetic boundary condition in vectorial form at a point on the interface between two media that involves the 'surface current density' at the interface.

7.9 Prove the general electromagnetic boundary condition in vectorial form at a point on the interface between two media that refer to the 'tangential components of electric fields' in the two media at the interface.

7.10 Interpret with physical grounds by giving reference to the concept of relaxation time in a dielectric medium (see Table 7.2) the four general electromagnetic boundary conditions at a point on a the interface between two dielectric media for (i) a 'time-independent' situation and (ii) 'time-dependent' situation.

7.11 Interpret, with physical grounds by giving reference to the concept of relaxation times in dielectric and conducting media (see Table 7.2), the four general electromagnetic boundary conditions at a point on a dielectric-conductor interface for (i) a 'time-independent' situation and (ii) 'time-dependent' situation.

7.12 Derive the law of refraction of direct current, $\tan\theta_1/\tan\theta_2 = \sigma_1/\sigma_2$, at the interface between two conducting media of conductivities σ_1 and σ_2 respectively, where θ_1 is the angle of incidence and θ_2 the angle of refraction.

7.13 Show that the surface charge density developed at a point on the interface between two media of relaxation times T_1 and T_2 respectively, in terms of the normal component of current density J_n at the point is equal to $(T_2 - T_1)J_n$.

7.14 Using the boundary condition that the tangential component of electric field is continuous at the interface between two media, and considering, typically, two conducting blocks making the interface, derive the circuit law of parallel resistances: $1/R_{\text{eq}} = 1/R_1 + 1/R_2$, where R_{eq} is the equivalent resistance of the two resistances R_1 and R_2 connected in parallel.

7.15 (i) Formulate the electric and magnetic fields with reference to a uniform plane electromagnetic wave in free space normally incident on a conducting surface forming a standing wave. (ii) Hence show that in such as standing wave, the electric field maxima falls on the magnetic field minima, and the distance between two consecutive maxima of electric field becomes the same as that between two consecutive maxima of magnetic field, being equal to half the wavelength. (iii) Also, state your observation on the power flow in such a standing wave.

7.16 With reference to the reflection and refraction of electromagnetic waves at a dielectric-dielectric interface, define 'parallel' and 'perpendicular' polarisations.

7.17 Derive an expression for the reflection coefficient for parallel polarisation (with reference to the reflection and refraction of electromagnetic waves at a dielectric-dielectric interface).

7.18 Derive an expression for the transmission coefficient for parallel polarisation (with reference to reflection and refraction of electromagnetic waves at a dielectric-dielectric interface).

7.19 For a wave incident from a dielectric medium of permittivity ε_1 to another dielectric medium of permittivity ε_2, develop the theory for Brewster's phenomenon with reference to parallel polarisation to show that the Brewster's angle is equal to $\tan^{-1}\sqrt{\varepsilon_2/\varepsilon_1}$.

7.20 Derive the expressions for the reflection and transmission coefficients for perpendicular polarisation (with reference to the reflection and refraction of electromagnetic waves at a dielectric-dielectric interface).

7.21 Prove that Brewster's phenomenon cannot take place for perpendicular polarisation (with reference to the reflection and refraction of electromagnetic waves at a dielectric-dielectric interface).

7.22 For a wave incident from a dielectric medium of permittivity ε_1 to another dielectric medium of permittivity ε_2 such that $\varepsilon_1 < \varepsilon_2$ (incidence from a rarer to a denser medium), show that the phenomenon of 'total internal reflection' takes place for an electromagnetic wave if the angle of incidence is greater that the critical angle, $\sin^{-1}\sqrt{\varepsilon_2/\varepsilon_1}$, irrespective of whether the polarisation is of the parallel or perpendicular type.

7.23 Obtain the electromagnetic boundary conditions of a conducting helix in the sheath-helix model in which the actual helix, for the sake of analysis, is replaced by an infinitesimally thin sheath that has infinite and zero conductivities in the directions parallel and perpendicular to the winding direction of the helix respectively.

Hint: See Example E7-9.

Review questions with solutions

7.24 Write down the four general electromagnetic boundary conditions in vectorial differential form at a point on the interface between two media vis-à-vis the corresponding integral form of Maxwell's equations from which these conditions can be derived.

Answer:

General electromagnetic boundary condition	Originating integral form of Maxwell's equation
$(\vec{D}_2 - \vec{D}_1).\vec{a}_n = \rho_s$	$\oint_S \vec{D} \cdot \vec{a}_n dS = \int_\tau \rho \, d\tau$
$(\vec{B}_2 - \vec{B}_1).\vec{a}_n = 0$	$\oint_S \vec{B} \cdot d\vec{S} = \int_S \vec{E} \cdot \vec{a}_n \, dS = 0$
$\vec{a}_n \times (\vec{H}_2 - \vec{H}_1) = \vec{J}_s$	$\oint_l \vec{H} \cdot d\vec{l} = \int_S \left(\vec{J} + \frac{\partial \vec{D}}{\partial t}\right) \cdot \vec{a}_n \, dS$
$\vec{a}_n \times (\vec{E}_2 - \vec{E}_1) = 0$	$\oint_l \vec{E} \cdot d\vec{l} = -\int_S \frac{\partial \vec{B}}{\partial t} \cdot \vec{a}_n \, dS$

The two regions are labelled as 1 and 2. The subscripts 1 and 2 refer to the regions 1 and 2, respectively. The unit vector $\vec{a}_n$ at the point on the interface is directed from region 1 to 2. (See Table 7.1).

7.25 One of the four general electromagnetic boundary conditions at a point on the interface between two media is stated as $(\vec{D}_2 - \vec{D}_1).\vec{a}_n = \rho_s$ in answer to question 7.1. How would this statement change if the unit vector $\vec{a}_n$ defined in answer to question 7.1 were directed from the region 2 to 1 instead of 1 to 2?

Answer: $(\vec{D}_1 - \vec{D}_2).\vec{a}_n = \rho_s$. (By replacing $\vec{a}_n$ by $-\vec{a}_n$ in the relevant boundary condition written in answer to question 7.1, you get the answer.)

7.26 Show how you can express the vectorial form of general boundary conditions

$$\left.\begin{array}{l} (\vec{D}_2 - \vec{D}_1).\vec{a}_n = \rho_s \\ (\vec{B}_2 - \vec{B}_1).\vec{a}_n = 0 \\ \vec{a}_n \times (\vec{H}_2 - \vec{H}_1) = \vec{J}_s \\ \vec{a}_n \times (\vec{E}_2 - \vec{E}_1) = 0 \end{array}\right\}$$

in terms of the tangential or normal components of relevant fields or flux densities.

Answer: Let us first take the following boundary condition involving the electric displacements or electric flux densities in the two regions 1 and 2 (see Fig. 1):

$$(\vec{D}_2 - \vec{D}_1).\vec{a}_n = \rho_s. \tag{1}$$

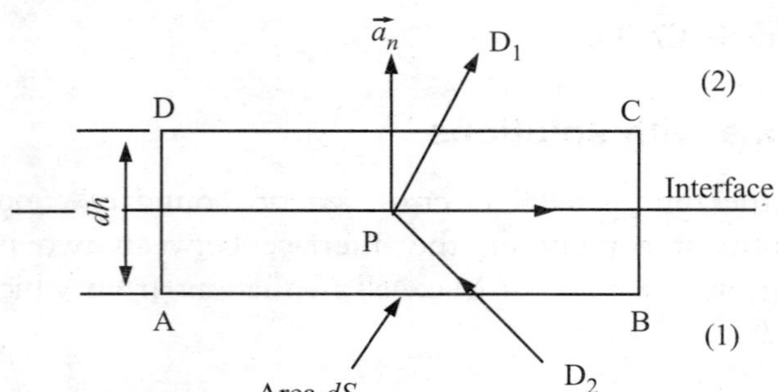

Fig. 1

Since $\vec{a}_n$ is the normal unit vector directed from the region 1 to 2 (see Fig.1), clearly, the dot products of $\vec{D}_1$ and $\vec{D}_2$ each with $\vec{a}_n$ will give respectively the normal components of these quantities: D_{nl} and D_{n2}. In view of this, we can then express (1) as

$$D_{\text{n2}} - D_{\text{n1}} = \rho_s. \tag{2}$$

Obviously then, proceeding from the boundary condition $(\vec{B}_2 - \vec{B}_1).\vec{a}_n = 0$ instead of (1), that is, $(\vec{D}_2 - \vec{D}_1).\vec{a}_n = \rho_s$, we obtain the expression corresponding to (2) as

$$B_{\text{n2}} - B_{\text{n1}} = 0. \tag{3}$$

Next, let us take the following boundary condition involving the current densities in the two regions 1 and 2 (see Fig. 2):

$$\vec{a}_n \times (\vec{H}_2 - \vec{H}_1) = \vec{J}_s. \tag{4}$$

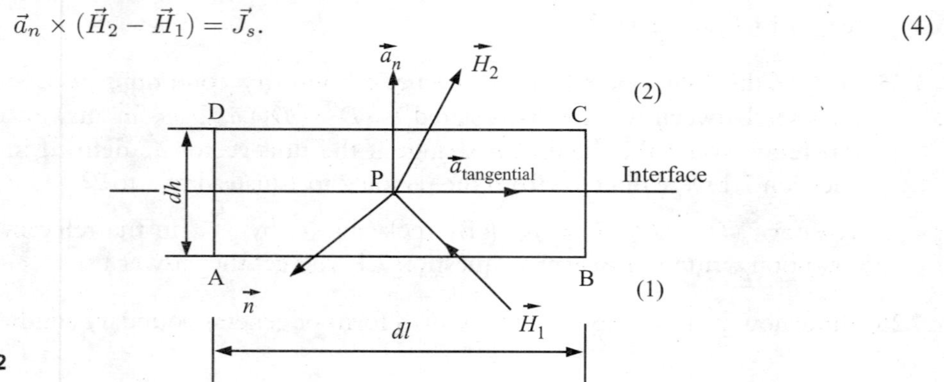

Fig. 2

In terms of the unit vectors $\vec{a}_n$, $\vec{n}$ and $\vec{a}_{\text{tangential}}$ defined preceding (7.12), let us represent $\vec{H}_1$ and $\vec{H}_2$ as

$$\left.\begin{aligned}\vec{H}_1 &= (H_{\text{tangential}})_1\vec{a}_{\text{tangential}} + (H_{a_n})_1\vec{a}_n + (H_n)_1\vec{n}\\ \vec{H}_2 &= (H_{\text{tangential}})_2\vec{a}_{\text{tangential}} + (H_{a_n})_2\vec{a}_n + (H_n)_2\vec{n}\end{aligned}\right\}. \tag{5}$$

Substituting (5) in (4) we get

$$\vec{a}_n \times \Big[\{(H_{\text{tangential}})_2 - (H_{\text{tangential}})_1\}\vec{a}_{\text{tangential}} + \{(H_{a_n})_2 - (H_{a_n})_1\}\vec{a}_n + \{(H_n)_2 - (H_n)_1\}\vec{n}\Big] = \vec{J}_s. \tag{6}$$

In view of the relations (see Fig. 2)

$$\left.\begin{aligned} \vec{a}_n \times \vec{a}_{\text{tangential}} &= -\vec{n} \\ \vec{a}_n \times \vec{a}_n &= 0 \\ \vec{a}_n \times \vec{n} &= \vec{a}_{\text{tangential}} \end{aligned}\right\},$$

we can then express (6) as

$$\Big[(H_{\text{tangential}})_2 - (H_{\text{tangential}})_1]\vec{n} + [(H_n)_2 - (H_n)_1\Big]\vec{a}_{\text{tangential}} = \vec{J}_S,$$

which can be rearranged as

$$\Big[(H_{\text{tangential}})_2\vec{n} + (H_n)_2\vec{a}_{\text{tangential}}] - [(H_{\text{tangential}})_1\vec{n} + (H_n)_1\vec{a}_{\text{tangential}}\Big] = \vec{J}_s. \tag{7}$$

The unit vectors $\vec{n}$ and $\vec{a}_{\text{tangential}}$ are both parallel to the interface between the two regions. Therefore, the first two terms of the left-hand side of (7)—one involving $\vec{n}$ and the other involving $\vec{a}_{\text{tangential}}$—put together will represent the tangential component of magnetic field H_{t2} in region 2 at the interface between the regions. Similarly, the remaining two terms of the left-hand side of (7) put together will represent the tangential component of magnetic field H_{t1} in region 1 at the interface between the regions. In view of this, therefore, we can express (7) in terms of the magnitude J_s of the surface current density as

$$H_{t2} - H_{t1} = J_s\,. \tag{8}$$

Obviously, therefore, starting from the boundary condition $\vec{a}_n \times (\vec{E}_2 - \vec{E}_1) = 0$ instead of $\vec{a}_n \times (\vec{H}_2 - \vec{H}_1) = \vec{J}_s$, we then obtain the expression corresponding to (8) as

$$E_{t2} - E_{t1} = 0. \tag{9}$$

Thus, we can put together (2), (3), (8) and (9) as follows.

Boundary conditions in vectorial form	**Boundary conditions in terms of components**
$(\vec{D}_2 - \vec{D}_1).\vec{a}_n = \rho_s$	$D_{n2} - D_{n1} = \rho_s$
$(\vec{B}_2 - \vec{B}_1).\vec{a}_n = 0$	$B_{n2} - B_{n1} = 0$
$\vec{a}_n \times (\vec{H}_2 - \vec{H}_1) = \vec{J}_s$	$H_{t2} - H_{t1} = J_s$
$\vec{a}_n \times (\vec{E}_2 - \vec{E}_1) = 0$	$E_{t2} - E_{t1} = 0$

7.27 Using the relevant boundary condition, deduce an expression for the surface charge density developed on the conducting plate of a parallel-plate capacitor in terms of the potential difference between the plates and hence derive an expression for the capacitance of the capacitor.

Answer:

(i) Surface charge density $\rho_s = \frac{\varepsilon V_0}{d}$

(ii) Capacitance $C = \frac{\varepsilon A}{d}$,

where V_0, d and ε are (i) the potential difference, (ii) the distance and (iii) the permittivity of the space between the plates of the capacitor (see Example E7-5).

7.28 Use the relevant boundary condition to find an expression for the surface charge density developed on the conducting surface of a cylindrical capacitor in terms of the potential difference between the inner and outer conductors of the capacitor and hence deduce an expression for its capacitance per unit length.

Answer:

(i) Surface charge density $\rho_s = \varepsilon \dfrac{V_0}{\ln\left(\frac{b}{a}\right)} \dfrac{1}{a}$

(ii) Capacitance per unit length $= \dfrac{2\pi\varepsilon}{\ln\left(\frac{b}{a}\right)}$,

where a and b are the inner and outer radii of the coaxial conductors of the capacitor and ε is the permittivity of the space between the conductors (see Example E7-6).

7.29 Deduce an expression for the surface charge density developed on the conducting surface of a spherical capacitor in terms of the potential difference between the inner and outer conductors of the capacitor using a relevant boundary condition, and hence deduce an expression for its capacitance.

Answer:

(i) Surface charge density $\rho_s = \vec{D} \cdot \vec{a}_n = \vec{D} \cdot \vec{a}_r = \varepsilon \dfrac{ab}{b-a} \dfrac{V_0}{a^2} \vec{a}_r \cdot \vec{a}_r = \varepsilon \dfrac{ab}{b-a} \dfrac{V_0}{a^2}$

(ii) Capacitance $C = \dfrac{4\pi\varepsilon}{\frac{1}{a} - \frac{1}{b}}$. (See Example E7-7)

Numerical type

7.30 Region 1: $x > 0$ is a dielectric medium of relative permittivity $\varepsilon_{r1} = 3$ and region 2: $x < 0$ is another dielectric medium of relative permittivity $\varepsilon_{r2} = 5$. If $\vec{E}_2 = 40\vec{a}_x + 60\vec{a}_y - 80\vec{a}_z$ V/m, find $\vec{D}_1$ and $\vec{D}_2$.

Answer: The given electric field in region 2 is

$$\vec{E}_2 = 40\vec{a}_x + 60\vec{a}_y - 80\vec{a}_z \text{ V/m}. \quad (1)$$

With the help of (1) we get

$$\vec{D}_2 = \varepsilon\vec{E}_2 = \varepsilon_0\varepsilon_{r2}\vec{E}_2$$
$$= 5 \times (40\vec{a}_x + 60\vec{a}_y - 80\vec{a}_z)\varepsilon_0 = (200\vec{a}_x + 300\vec{a}_y - 400\vec{a}_z)\varepsilon_0 \text{ C/m}^2. \qquad (2)$$

Since, in this problem, the unit vector is directed from the region 1 to 2 in the negative z direction, you can take $\vec{a}_n = -\vec{a}_x$ in the boundary condition $(\vec{D}_2 - \vec{D}_1).\vec{a}_n = 0$ (see Table 7.3) and write it in terms of the relevant vector components:

$$[(D_{x2} - D_{x1})\vec{a}_x + (D_{y2} - D_{y1})\vec{a}_y + (D_{z2} - D_{z1})\vec{a}_z] \cdot [-\vec{a}_x] = 0$$

to get

$$[(D_{x2} - D_{x1})\vec{a}_x + (D_{y2} - D_{y1})\vec{a}_y + (D_{z2} - D_{z1})\vec{a}_z] \cdot [-\vec{a}_x] = 0. \qquad (3)$$

We obtain from (3)

$$D_{x2} - D_{x1} = 0,$$

that is,

$$D_{x1} = D_{x2}. \qquad (4)$$

We can identify now from (1) that $D_{x2} = 200$ C/m^2 and put it in (4) to obtain

$$D_{x1} = 200\varepsilon_0 \text{ C/m}^2. \qquad (5)$$

In order to know the remaining components of $\vec{D}_1$, namely, D_{y1} and D_{z1} , proceed as follows. Recall the boundary condition $\vec{a}_n \times (\vec{E}_2 - \vec{E}_1) = 0$ (see Table 7.3). Putting therein $\vec{a}_n = -\vec{a}_x$ [as explained following (2)], write it in terms of vector components as follows:

$$-\vec{a}_x \times [(E_{x2} - E_{x1})\vec{a}_x + (E_{y2} - E_{y1})\vec{a}_y + (E_{z2} - E_z)\vec{a}_{z1}] = 0. \qquad (6)$$

We can simplify (6) as

$$[-(E_{y2} - E_{y2})\vec{a}_z + (E_{z2} - E_{z1})\vec{a}_y] = 0,$$

which gives $E_{y2} - E_{y1} = 0$ and $E_{z2} - E_{z1} = 0$, which, in turn, gives

$$\left.\begin{aligned} E_{y1} &= E_{y2} \\ E_{z1} &= E_{z2} \end{aligned}\right\}. \qquad (7)$$

From (1), we can identify

$$\left.\begin{aligned} E_{y2} &= 60 \\ E_{z2} &= -80 \end{aligned}\right\}. \text{ V/m} \qquad (8)$$

Therefore, putting (8) in (7) we obtain

$$\left.\begin{aligned} E_{y1} &= 60 \\ E_{z1} &= -80 \end{aligned}\right\}. \text{ V/m} \qquad (9)$$

We can then write with the help of (9)

$$\left.\begin{aligned} D_{y1} &= \varepsilon_1 60 = \varepsilon_0 \varepsilon_{r1} 60 = 3 \times 60 \varepsilon_0 = 180 \varepsilon_0 \\ D_{z1} &= -\varepsilon_1 80 = -\varepsilon_0 \varepsilon_{r1} 80 = -3 \times 80 \varepsilon_0 = -240 \varepsilon_0 \end{aligned}\right\} \text{C/m}^2. \tag{10}$$

Finally, with the help of (4) and (10), we obtain

$$\vec{D}_1 = D_{x1}\vec{a}_x + D_{y1}\vec{a}_y + D_{z1}\vec{a}_z = (200\vec{a}_x + 180\vec{a}_y - 240\vec{a}_z)\varepsilon_0 \text{ C/m}^2. \tag{11}$$

We can write the answer putting (11) and (2) together as follows:

$$\left.\begin{aligned} \vec{D}_1 &= (200\vec{a}_x + 180\vec{a}_y - 240\vec{a}_z)\varepsilon_0 \\ \vec{D}_2 &= (200\vec{a}_x + 300\vec{a}_y - 400\vec{a}_z)\varepsilon_0 \end{aligned}\right\} \text{C/m}^2.$$

7.31 If the electric field in region 2: $x < 0$ of relative permittivity ε_{r2} is $\vec{E}_2 = E_{2x}\vec{a}_x + E_{2y}\vec{a}_y + E_{2x}\vec{a}_z$ V/m, find the electric displacement $\vec{D}_1$ in the region 1: $x > 0$ of relative permittivity ε_{r1} with which the region 2 forms an interface at $x = 0$. Subscripts 1 and 2 refer to regions 1 and 2 respectively.

Answer: From the boundary condition $(\vec{D}_2 - \vec{D}_1).\vec{a}_n = 0$ [see (7.26*a*)], we can see that $D_{1x} = D_{2x} = \varepsilon_{r2}\varepsilon_0 E_{2x}$. Similarly, from the boundary condition $\vec{a}_n \times (\vec{E}_2 - \vec{E}_1) = 0$ [see (7.26*a*)], observe that (i) $E_{1y} = E_{2y}$ and (ii) $E_{1z} = E_{2z}$,. From the first of these, we obtain $D_{1y} = \varepsilon_{r1}\varepsilon_0 E_{2y}$, while from the second we obtain $D_{1z} = \varepsilon_{r1}\varepsilon_0 E_{2z}$. Therefore, substituting D_{1x}, D_{1y} and D_{1z} thus obtained in the expression $\vec{D}_1 = D_{1x}\,\vec{a}_x + D_{1y}\,\vec{a}_y + D_{1z}\,\vec{a}_z$, we get

$$\vec{D}_1 = \varepsilon_{r2}\varepsilon_0 E_{2x}\,\vec{a}_x + \varepsilon_{r1}\varepsilon_0 E_{2y}\vec{a}_y + \varepsilon_{r1}\varepsilon_0 E_{2z}\,\vec{a}_z \text{ C/m}^2.$$

7.32 If the electric field at a point on the surface of a conductor placed in free space is $6\vec{a}_x + 3\vec{a}_y + 2\vec{a}_z$ V/m, find the surface charge density developed on the point.

Answer: Let us write the given electric field in region 2 (free-space) as

$$\vec{E}_2 = 6\vec{a}_x + 3\vec{a}_y + 2\vec{a}_z = E_2\vec{a}_E \text{V/m} \tag{1}$$

where $\vec{a}_E$ is the unit vector in the direction of $\vec{E}_2$ and E_2 is the magnitude of $\vec{E}_2$. From (1) we get

$$E_2 = \sqrt{(6)^2 + (3)^2 + (2)^2} = \sqrt{36 + 9 + 4} = \sqrt{49} = 7 \text{ V/m}\,. \tag{2}$$

We can write the boundary condition (7.45*d*) as

$$\vec{a}_n \times \vec{E}_2 = \vec{a}_n \times (E_2\vec{a}_E) = E_2(\vec{a}_n \times \vec{a}_E) = 0\,. \tag{3}$$

Since, as can be seen from (2), $E_2 \neq 0$ for the compliance of (3), we must put

$$\vec{a}_n \times \vec{a}_E = 0$$

which then leads to the relation

$$\vec{a}_E = \vec{a}_n. \tag{4}$$

Putting (4) in (1), we can write

$$\vec{E}_2 = 6\vec{a}_x + 3\vec{a}_y + 2\vec{a}_z = E_2\vec{a}_n. \tag{5}$$

Substituting (2) in (5), we obtain

$$\vec{E}_2 = E_2\vec{a}_n = 7\vec{a}_n,$$

whence we get

$$\vec{D}_2 = \varepsilon_0\vec{E}_2 = 7\varepsilon_0\vec{a}_n. \tag{6}$$

Let us now recall the boundary condition (7.45a) and rewrite it as

$$\vec{D}_2.\vec{a}_n = \rho_S. \tag{7}$$

Substituting (6) in (7), we find the surface charge density as

$$\rho_S = 7\varepsilon_0\vec{a}_n.\vec{a}_n = 7\varepsilon_0 \text{ C/m}^2.$$

7.33 Calculate the angle of refraction of current and the surface charge density developed at the interface between brass and gold media of conductivities $\sigma = 4.1 \times 10^7$ and $\sigma = 1.1 \times 10^7$ mho/m respectively. A current density of 2 A/m^2 is getting refracted from the brass to gold medium at an angle of incidence of 30^0.

Answer: With the help of (7.53), we can find the angle of refraction θ_2 from the following given values: angle of incidence $\theta_1 = \pi/6$, $\sigma_1 = 1.1 \times 10^7$ mho/m and $\sigma_2 = 4.1 \times 10^7$ mho/m

$$\theta_2 = \tan^{-1}\left(\frac{\sigma_2}{\sigma_1}\tan\theta_1\right) = \tan^{-1}\left(\frac{4.1}{1.1}\tan\pi/6\right)$$

$$= \tan^{-1}\left(\frac{4.1}{1.1} \times 0.577\right) = \tan^{-1}(2.151) = 65.07^0. \tag{1}$$

From the given current density of 2 A/m^2 , we can find the normal component $J_n = 2\cos\pi/6 = 2 \times 0.866 = 1.732$ A/m^2, which can be substituted in (7.69) to find the surface charge density developed at the interface between the conducting media as

$$\rho_s = \left(\frac{\varepsilon_0}{\sigma_2} - \frac{\varepsilon_0}{\sigma_1}\right)J_n = \left(\frac{10^{-7}}{4.1} - \frac{10^{-7}}{1.1}\right) \times 1.732\varepsilon_0$$

$$= -\frac{3 \times 1.732 \times 10^{-7}}{4.1 \times 1.1}\varepsilon_0 = -1.152 \times 10^{-7}\varepsilon_0 \text{ C/m}^2. \tag{2}$$

From (1) and (2), we get the angle of refraction of current density and surface charge density respectively.

7.34 Calculate the Brewster's angle of pyrex glass $\varepsilon_r = 4$ in air ($\varepsilon_r \approx 1$).

Answer: 63.43^0.

7.35 Calculate the critical angle of glass $\varepsilon_r = 4$ in air ($\varepsilon_r \approx 1$) for a ray representing the transmission of light (electromagnetic wave) from glass to air.

Answer: 30^0.

7.36 Determine the reflection coefficient of both parallel and perpendicular polarisation types of electromagnetic wave normally incident from air (refractive index ≈ 1) to the surface of glass (refractive index≈ 1.5).

Answer: -0.2.

7.37 Show that for grazing incidence the reflection coefficients of parallel and perpendicular polarisation types of electromagnetic wave are 1 and -1 respectively.

Hint: See (7.109) and (7.122).

Multiple-choice questions

7.38 The circuit law of parallel resistances can be derived with the help of the following boundary conditions:

(a) The normal component of magnetic flux density is continuous at the interface between two media.

(b) The normal component of electric displacement is discontinuous at the interface between two media.

(c) The tangential component of electric field is continuous at the interface between two media.

(d) The tangential component of magnetic field is discontinuous at the interface between two media.

7.39 One of the following 'general' boundary conditions at the interface between two media is incorrect:

(a) $(\vec{D}_1 - \vec{D}_2).\vec{a}_n = \rho_s$ (b) $(\vec{B}_1 - \vec{B}_2).\vec{a}_n = 0$

(c) $\vec{a}_n \times (\vec{H}_1 - \vec{H}_2) = -\vec{J}_s$ (d) $\vec{a}_n \times (\vec{E}_1 - \vec{E}_2) = 0$,

where $\vec{a}_n$ is a unit vector directed from medium 1 to 2, the other symbols having their usual meaning given in the text.

7.40 One of the following boundary conditions at the dielectric-dielectric interface is incorrect for a 'time-independent' situation:

(a) $\vec{a}_n \times (\vec{H}_2 - \vec{H}_1) = 0$ (b) $\vec{a}_n \times (\vec{E}_2 - \vec{E}_1) = 0$

(c) $(\vec{D}_2 - \vec{D}_1).\vec{a}_n = \rho_s$ (d) $(\vec{B}_2 - \vec{B}_1).\vec{a}_n = 0$,

where $\vec{a}_n$ is a unit vector directed from medium 1 to 2, the other symbols having their usual meaning given in the text.

7.41 One of the following boundary conditions at the dielectric-dielectric interface is incorrect for a 'time-dependent' situation:

(a) $\vec{a}_n \times (\vec{E}_2 - \vec{E}_1) = 0$ (b) $\vec{a}_n \times (\vec{H}_2 - \vec{H}_1) = \vec{J}_S$

(c) $(\vec{D}_2 - \vec{D}_1).\vec{a}_n = 0$ (d) $(\vec{B}_2 - \vec{B}_1).\vec{a}_n = 0$,

where $\vec{a}_n$ is a unit vector directed from medium 1 to 2, the other symbols having their usual meaning given in the text.

7.42 One of the following boundary conditions at the conductor-dielectric interface is incorrect for a 'time-independent' situation:

(a) $\vec{D}_2.\vec{a}_n = \rho_S$ (b) $(\vec{B}_2 - \vec{B}_1).\vec{a}_n = 0$

(c) $\vec{a}_n \times \vec{H}_2 = 0$ (d) $\vec{a}_n \times \vec{E}_2 = 0$,

where $\vec{a}_n$ is a unit vector directed from medium 1 (conductor) to 2 (dielectric), the other symbols having their usual meaning given in the text.

7.43 One of the following boundary conditions at the conductor-dielectric interface is incorrect for a 'time-dependent' situation:

(a) $\vec{D}_2.\vec{a}_n = \rho_S$ (b) $(\vec{B}_2 - \vec{B}_1).\vec{a}_n = 0$

(c) $\vec{a}_n \times \vec{H}_2 = \vec{J}_s$ (d) $\vec{a}_n \times \vec{E}_2 = 0$,

where $\vec{a}_n$ is a unit vector directed from medium 1 (conductor) to 2 (dielectric), the other symbols having their usual meaning given in the text.

7.44 For wave incident from a dielectric medium of permittivity ε_1 and refractive index n_1 on another dielectric medium of permittivity ε_2 and refractive index n_2, Snell's law is stated in terms of the angle of incidence θ_1 and the angle of refraction θ_2 as

(a) $\sqrt{\varepsilon_1} \sin\theta_1 = \sqrt{\varepsilon_2} \sin\theta_2$

(b) $\sqrt{\varepsilon_1} \sin\theta_2 = \sqrt{\varepsilon_2} \sin\theta_1$

(c) $n_1 \sin\theta_2 = n_2 \sin\theta_1$

(d) $\sqrt{n_1} \sin\theta_1 = \sqrt{n_2} \sin\theta_2$.

7.45 One of the following statements is incorrect.

The phenomenon of total internal reflection takes place for

(a) a wave incidence from a denser to a rarer medium

(b) an angle of incidence greater than the critical angle

(c) parallel polarisation and not for perpendicular

(d) both parallel and perpendicular polarisation.

7.46 One of the following statements is incorrect with respect to Brewster's phenomenon for a wave incidence from a dielectric medium of permittivity ε_1 and refractive index n_1 to another dielectric medium of permittivity ε_2 and refractive index n_2.

(a) Brewster angle is $\theta_B = \tan^{-1} n_2/n_1$.

(b) Brewster angle is $\theta_B = \tan^{-1}(\sqrt{\varepsilon_1/\varepsilon_2})$.

(c) The reflection coefficient is zero at Brewster angle.

(d) The phenomenon takes place for parallel polarisation and not for perpendicular polarisation.

7.47 One of the following statements is incorrect.

(a) For parallel polarisation, the reflection coefficient is zero at Brewster's angle of incidence.

(b) For perpendicular polarisation, the reflection coefficient is zero at Brewster's angle of incidence.

(c) For parallel polarisation, the reflection coefficient is equal to 1 at grazing angle of incidence.

(d) For perpendicular polarisation, the reflection coefficient is –1 at grazing angle of incidence.

Answers

7.38 (c) 7.39 (a) 7.40(c) 7.41(b) 7.42 (c) 7.43 (b) 7.44 (a) 7.45 (c)
7.46 (b) 7.47 (b)

8

Electromagnetic Power Flow

Electrical and magnetic energies are associated with electric and magnetic fields respectively. For instance, there exists an electric field in the region between the plates of a parallel-plate capacitor (Section 3.2), and thus a capacitor stores electrical energy. Similarly, magnetic field is stored in a solenoid (Section 4.3). In this chapter, we will begin with static electric and magnetic fields and develop from first principles the concept of energy density stored in these fields (Section 8.1). We will then develop the concept of Poynting vector as a measure of the power density associated with electric and magnetic fields (Section 8.2). In the course of developing these concepts, a theorem called the Poynting theorem will be introduced, which will expose the various issues related to electromagnetic power flow. We will develop the general Poynting theorem involving instantaneous Poynting vector (Section 8.2.1) and then extend it to obtain the complex Poynting theorem for time-periodic electric and magnetic fields. We will also develop the concept of averaged electromagnetic power flow (Section 8.2.2), including the effect of the presence of an external source within the volume enclosure to which the theorem is applied (Section 8.2.3). We will also find the power loss at a conducting surface with reference to a uniform plane electromagnetic wave incident on a planar conducting surface (Section 8.3). The concept of power flow developed in this chapter will subsequently be used later in Chapter 9 to find the power transmitted through a waveguide (a wave-guiding physical structure) as well as the power attenuation in such a transmission due to the finite resistivity of the waveguide material.

8.1 Energy Stored in Electrostatic and Magnetostatic Fields

If we place a point charge in a region of a static electric field, the former will experience a force due to the latter (Chapter 3). Thus, free to move in free space, the point charge will move a distance under the influence of the electric field and, correspondingly, work will be done. The phenomenon may be interpreted as *storage of energy* in electric field. In order to find the density of energy stored in the electric field, let us first find the energy density of a given continuous charge distribution in a free-space region. This, in fact, can be interpreted as the work done in placing discrete point charges in the region.

Consider then that an electric field exists in the free-space region due to the distribution of electric point charges, say $Q_1, Q_2, Q_3, Q_4, \ldots, Q_n$. Let us find the work done when these point charges are placed, one by one, in the free-space region. Let point charge Q_1 be placed first, when there is no electric field in the region, followed by point charge Q_2, placed in the presence of the electric field in the region due to point charge Q_1. This is followed by point charge Q_3 placed in the presence of the electric field in the region due to point charges Q_1 and Q_2, and so on. Clearly, therefore, no work will be done in placing the point charge Q_1. The work done in placing the point charges, Q_1, Q_2, Q_3, and so on, can be expressed by the following relations:

$$\left.\begin{array}{ll}
\text{Work done in placing charge } Q_1 \text{ at its position} = 0; & \text{(a)} \\
\text{Work done in placing charge } Q_2 \text{ at its position } = Q_2V_{21}; & \text{(b)} \\
\text{Work done in placing charge } Q_3 \text{ at its position } = Q_3V_{31} + Q_3V_{32}; & \text{(c)} \\
\text{Work done in placing charge } Q_4 \text{ at its position} = Q_4V_{41} + Q_4V_{42} + Q_4V_{43}; & \text{(d)} \\
\text{and so on; and} & \\
\text{Work done in placing charge} Q_n \text{ at its position} & \\
\qquad = Q_nV_{n1} + Q_nV_{n2} + Q_nV_{n3} + \ldots + Q_nV_{n,n-1} & \text{(e)}
\end{array}\right\}. \tag{8.1}$$

Here, V_{21} is the potential at the location of point charge Q_2 due to point charge Q_1; V_{31} and V_{32} are the potentials at the location of point charge Q_3 due to the point charges Q_1 and Q_2 respectively, and so on; and $V_{n1}, V_{n2}, V_{n3}, \ldots$, and $V_{n,n-1}$ are the potentials at the location of point charge Q_n due to point charges Q_1, Q_2, Q_3, Q_4, ..., and Q_{n-1} respectively. Therefore, summing ($8.1a - e$), we obtain the following expression for the work done W_E in placing the electric point charges $Q_1,\ Q_2,\ Q_3,\ Q_4,\ \ldots, Q_n$ at their respective locations:

$$\begin{aligned} W_E &= 0 + Q_2V_{21} + Q_3V_{31} + Q_3V_{32} + Q_4V_{41} + Q_4V_{42} + Q_4V_{43} + \ldots \\ &\quad + Q_nV_{n1} + Q_nV_{n2} + Q_nV_{n3} + \ldots + Q_nV_{n,n-1}. \end{aligned} \tag{8.2}$$

Now, we can express with the help of (3.80) V_{21}, the potential at the location of point charge Q_2 due to point charge Q_1 as

$$V_{21} = \frac{Q_1}{4\pi\varepsilon_0 r_{21}} \tag{8.3}$$

where r_{21} is the distance of the location of Q_2 from that of Q_1. Similarly, V_{12}, the potential at the location of point charge Q_1 due to point charge Q_2 can be expressed as

$$V_{12} = \frac{Q_2}{4\pi\varepsilon_0 r_{12}}. \qquad (8.4)$$

Therefore, multiplying (8.3) by Q_2 and (8.4) by Q_1, we can respectively write

$$Q_2 V_{21} = \frac{Q_2 Q_1}{4\pi\varepsilon_0 r_{21}} \qquad (8.5)$$

$$Q_1 V_{12} = \frac{Q_1 Q_2}{4\pi\varepsilon_0 r_{12}}. \qquad (8.6)$$

Comparing (8.5) and (8.6), and since $r_{12} = r_{21}$, we obtain the following identity:

$$Q_2 V_{21} = Q_1 V_{12}. \qquad (8.7)$$

Similarly, the other identity relations may be written as

$$\left.\begin{aligned} Q_3 V_{31} &= Q_1 V_{13} \\ Q_3 V_{32} &= Q_2 V_{23} \\ Q_4 V_{41} &= Q_1 V_{14} \\ Q_4 V_{42} &= Q_2 V_{24} \\ Q_4 V_{43} &= Q_3 V_{34} \\ &\ldots.. \\ Q_n V_{n1} &= Q_1 V_{1n} \\ Q_n V_{n2} &= Q_2 V_{2n} \\ Q_n V_{n3} &= Q_3 V_{3n} \\ &\ldots.. \\ Q_n V_{n,n-1} &= Q_{n-1} V_{n-1,n} \end{aligned}\right\}. \qquad (8.8)$$

With the help of (8.7) and (8.8), we can alternatively express (8.2) as

$$W_E = 0 + Q_1V_{12} + Q_1V_{13} + Q_2V_{23} + Q_1V_{14} + Q_2V_{24} + Q_3V_{34} + \ldots..$$
$$+Q_1V_{1n} + Q_2V_{2n} + Q_3V_{3n} + \ldots. + Q_{n-1}V_{n-1,n}\,. \qquad (8.9)$$

Adding (8.2) and (8.9), and rearranging terms, we then obtain

$$2W_E = Q_1(V_{12} + V_{13} + V_{14} + \ldots..) + Q_2(V_{21} + V_{23} + V_{24}\ldots.) + Q_3(V_{31} + V_{32}$$
$$+V_{34}+\ldots..)+Q_4(V_{41}+V_{42}+V_{43}+\ldots..)+\ldots..+Q_n(V_{n1}+V_{n2}+V_{n3}+\ldots..V_{n,n-1})\,. \qquad (8.10)$$

We may alternatively express (8.10) as

$$2W_E = Q_1V_1 + Q_2V_2 + Q_3V_3 + Q_4V_4 + ... + Q_nV_n \tag{8.11}$$

where

$$\left.\begin{aligned} V_1 &= V_{12} + V_{13} + V_{14} + ...\\ V_2 &= V_{21} + V_{23} + V_{24} + ...\\ V_3 &= V_{31} + V_{32} + V_{34} + ...\\ V_4 &= V_{41} + V_{42} + V_{43} + ...\\ &.....\\ V_n &= V_{n1} + V_{n2} + V_{n3} + ... + V_{n,n-1} \end{aligned}\right\} \tag{8.12}$$

are the potentials at the locations of the point charges $Q_1, Q_2, Q_3, Q_4, ..., Q_n$ respectively, each due to the point charges located at the remaining other points. We can further put (8.11) using a summation notation as follows:

$$2W_E = \sum_{p=1}^{p=n} Q_pV_p,$$

whence we may write the expression for the work done W_E in placing the point charges $Q_1, Q_2, Q_3, Q_4, ..., Q_n$ at their respective locations as

$$W_E = \frac{1}{2}\sum_{p=1}^{p=n} Q_pV_p. \tag{8.13}$$

However, it will be of interest to us to smear out the discrete point charges $Q_1, Q_2, Q_3, Q_4, ..., Q_n$ into a continuous charge distribution within a volume enclosure. For this, we can write (8.13) in the following form to express the work done W_E in establishing such a distribution:

$$W_E = \frac{1}{2}\int_\tau V\rho\, d\tau \tag{8.14}$$

where τ is the volume over which the charge has been distributed, and V and ρ are the potential and the volume charge density respectively at a point inside the volume charge distribution.

The electric potential V and the volume charge density ρ, both occurring in (8.14), are related to the electric field through the gradient and the divergence relations (3.91) and (3.104) respectively. Therefore, it is implied that the work done W_E in establishing the charge distribution in the given volume, given by (8.14), will also be a measure of the energy stored in the electric field in the volume. Thus, it is of interest to us to express (8.14) in terms of the electric field $\vec{E}$. For this purpose, let us substitute ρ from the divergence relation (3.104) in (8.14) to write

$$W_E = \frac{1}{2}\int\limits_{\tau} V\rho\, d\tau = \frac{1}{2}\int\limits_{\tau} V\,\nabla.\vec{D}\, d\tau,$$

which, using the vector identity

$$\nabla.(V\vec{D}) = V\,\nabla.\,\vec{D} + \nabla V.\,\vec{D},$$

can be expressed as

$$W_E = \frac{1}{2}\int\limits_{\tau} V\,\nabla.\,\vec{D}\, d\tau = \frac{1}{2}\int\limits_{\tau} [(\nabla.(V\vec{D}) - \nabla V.\,\vec{D})\, d\tau] \tag{8.15}$$

where $\vec{D}$ is the electric displacement.

We can make further simplification by putting the gradient relation (3.91) in the second term of the right-hand side of (8.15) to write

$$W_E = \frac{1}{2}\left[\int\limits_{\tau} \nabla.(V\vec{D})\, d\tau + \int\limits_{\tau} \vec{E}.\,\vec{D}\, d\tau\right]. \tag{8.16}$$

In order to simplify (8.16), we make use of the Gauss's divergence theorem of vector calculus[1] to change the volume integral occurring in the first term of (8.16) into a closed surface integral over a closed area S enclosing the volume with reference to a vector quantity $\vec{A}$, say. This can be stated as

$$\int\limits_{\tau} \nabla\cdot\vec{A}d\tau = \oint\limits_{S} \vec{A}\cdot d\vec{S}. \tag{8.17}$$

In the present context, if we replace the vector $\vec{A}$ by $V\vec{D}$, which is also a vector quantity, we obtain

$$\int\limits_{\tau} \nabla\cdot V\vec{D}d\tau = \oint\limits_{S} V\vec{D}\cdot d\vec{S}. \tag{8.18}$$

Substituting (8.18) in (8.16) we obtain

[1] You can verify Gauss's divergence theorem with particular reference to the vector $\vec{D}$ by substituting ρ from Poisson's equation (3.104) in (3.64) to obtain

$$\int\limits_{\tau} \nabla\cdot\vec{D}d\tau = \oint\limits_{S} \vec{D}\cdot d\vec{S},$$

which is identical to the divergence theorem (8.17) if we replace vector $\vec{A}$ by $\vec{D}$.

$$W_E = \frac{1}{2}\left[\oint_S (V\vec{D}) \cdot d\vec{S} + \int_\tau \vec{E}.\vec{D}\, d\tau\right]. \tag{8.19}$$

However, at this stage the first surface integral term of the right-hand side of (8.19) needs to be examined further. For this purpose, let us consider a large volume enclosing all the charges of the given distribution such that the surface enclosing the volume is at a large distance $r \to \infty$ from the charge distribution. From such a surface enclosure, far from the charge distribution, an element of charge distribution can be regarded as a point charge. We can therefore appreciate that the potential V and the electric displacement $\vec{D}$ at the surface of such an enclosure [occurring in the first term of the right-hand side of (8.19)] decrease according to $V \propto 1/r$ (3.80) and $\vec{D} \propto 1/r^2$ (3.17) respectively. However, the magnitude of the area element dS on the surface of the enclosure increases as $dS \propto r^2$. In view of this dependence of V, $\vec{D}$ and $d\vec{S}$ on r, the integrand of the first term of the right-hand side of (8.19) decreases inversely as r, approaching zero as rapidly as $1/r$. The integral of this term thus becomes zero at a large distance $r \to \infty$ from the charge distribution. Consequently, we can write the following expression for the work done W_E in establishing the given volume electrostatic charge distribution, which is also the expression for the energy stored W_E in a static electric field due to such a charge distribution in the volume:

$$W_E = \frac{1}{2}\int_\tau \vec{E}.\vec{D}\, d\tau = \int_\tau \frac{1}{2}\vec{E}.\vec{D}\, d\tau. \tag{8.20}$$

We may alternatively express W_E in terms of the energy per unit volume, that is, the energy density U_E stored in a static electric field, as

$$W_E = \int_\tau U_E d\tau,$$

the right-hand side of which may be compared with that of (8.20) to identify U_E as follows:

$$U_E = \frac{1}{2}\vec{E}.\vec{D}. \tag{8.21}$$

Let us take the particular problem of finding the energy density stored in the electric field of a parallel-plate capacitor in order to exemplify the general expression (8.21) for energy density stored in a static electric field. In order to appreciate this, let us find from first principles the work done in charging a parallel-plate capacitor by a source of potential. We will consider the plate dimensions to be large compared to the distance between the plates of the capacitor. If V_0 be the potential difference between the plates of the capacitor when it is fully charged to the value of charge Q_0, then the quantities V_0 and Q_0 are related as (see Section 3.2)

$$Q_0 = CV_0 \tag{8.22}$$

where C is the capacitance of the capacitor. Similarly, if at any stage in the process of charging the capacitor the potential difference between the plates of the capacitor is V $(0 < V < V_0)$ and the charge of the capacitor is Q $(0 < Q < Q_0)$, then we can relate the quantities V and Q as

$$Q = CV. \tag{8.23}$$

The element of work done $dW_{\text{Capacitor}}$ to add an extra element of charge dQ to the capacitor when it is already raised to the potential V and has a charge Q established on it is given by (see Section 3.2)

$$dW_{\text{Capacitor}} = VdQ. \tag{8.24}$$

If we now substitute $V = Q/C$ from (8.23) in (8.24), we obtain

$$dW_{\text{Capacitor}} = \frac{Q}{C}dQ. \tag{8.25}$$

Integrating (8.25), we then obtain the expression for the work done $W_{\text{Capacitor}}$ in charging the capacitor to its full value of charge $Q = Q_0$ from $Q = 0$ as follows:

$$W_{\text{Capacitor}} = \int dW_{\text{Capacitor}} = \int_{Q=O}^{Q=Q_O} \frac{Q}{C}\, dQ = \frac{1}{C}\left[\frac{Q^2}{2}\right]_{Q=C}^{Q=Q_0} = \frac{1}{2}\frac{Q_0^2}{C},$$

into which let us substitute (8.22) to obtain

$$W_{\text{Capacitor}} = \frac{1}{2}\frac{Q_0^2}{C} = \frac{1}{2}\frac{C^2V_0^2}{C} = \frac{1}{2}CV_0^2. \tag{8.26}$$

For large plate dimensions compared to the distance between the plates, the electric field will be uniform between the plates, being directed from the plate at a higher potential to that at a lower potential (see Section 3.3). Further, we can express V_0 occurring in (8.26) in terms of the magnitude of the electric field E between the plates with the help of (3.117) as $V_0 = Ed$ and substitute it in (8.26) to write

$$W_{\text{Capacitor}} = \frac{1}{2}C(Ed)^2 \tag{8.27}$$

where d is the distance between the plates. Now, substituting C obtainable from (3.100) for a parallel-plate capacitor in (8.27), we may express (8.27) as

$$W_{\text{Capacitor}} = \frac{1}{2}\frac{\varepsilon A}{d}(Ed)^2 = \frac{1}{2}\varepsilon\,(Ad)E^2. \tag{8.28}$$

Dividing (8.28) by the volume Ad of the capacitor, we obtain the following expression for the energy per unit volume or energy density $U_{\text{Capacitor}}$ stored in the electric field of the capacitor:

$$U_{\text{Capacitor}} = \frac{1}{2}\varepsilon E^2. \tag{8.29}$$

The electric field in the capacitor is in the direction of the perpendicular to the plates of the capacitor. Therefore, if we assign this direction as the z direction, we can represent the electric field of the capacitor in the vectorial form as $\vec{E} = E\vec{a}_z$, where $\vec{a}_z$ is the unit vector in the z direction. Further, we can write $\vec{E} \cdot \vec{E} = E^2$, in view of which we can express (8.29) as

$$U_{\text{Capacitor}} = \frac{1}{2}\varepsilon E^2 = \frac{1}{2}\vec{E} \cdot \varepsilon \vec{E},$$

which, on putting $\vec{D} = \varepsilon \vec{E}$ becomes

$$U_{\text{Capacitor}} = \frac{1}{2}\vec{E}.\vec{D}. \tag{8.30}$$

We can see that the expression (8.30) for the energy density of a capacitor is identical to the general expression (8.21) for the energy density stored in a magnetic field.

EXAMPLE E8-1

Show that the magnitude of force per unit area (pressure) acting on each plate of a parallel-plate capacitor is the same as the energy density stored in the static electric field of the capacitor.

SOLUTION:

For large plate dimensions compared to the distance d between the plates considered, let us recall the expression (8.28) for the energy stored $W_{\text{Capacitor}}$ in the electric field of the capacitor as follows:

$$W_{\text{Capacitor}} = \frac{1}{2}\frac{\varepsilon A}{d}(Ed)^2 = \frac{1}{2}\varepsilon\,(Ad)E^2. \tag{E8-1.1}$$

If we move one of the plates of the capacitor against the force of attraction between them by an infinitesimal distance dz, thereby increasing the volume of the capacitor by an infinitesimal amount Adz, then there would be an infinitesimal change $\Delta W_{\text{Capacitor}}$ in the energy stored in electric field in the capacitor due to the change in the volume of the capacitor. The incremental change can then be written from (E8-1.1) as

$$\Delta W_{\text{Capacitor}} = \frac{1}{2}\varepsilon\, AdzE^2. \tag{E8-1.2}$$

The work done in moving the plates of the capacitor by the incremental distance dz against the force of attraction of magnitude F between the plates that are essentially oppositely charged can be written as

$$\text{Work} = Fdz. \qquad \text{(E8-1.3)}$$

Obviously, since the work done given by (E8-1.3) must be equal to $\Delta W_{\text{Capacitor}}$ given by (E8-1.2), we can equate the right-hand sides of (E8-1.2) and (E8-1.1) to write

$$Fdz = \frac{1}{2}\varepsilon\, AdzE^2,$$

whence we get

$$F = \frac{1}{2}\varepsilon\, AE^2. \qquad \text{(E8-1.4)}$$

Dividing (E8-1.4) by A, we then obtain the force per unit area

$$\frac{F}{A} = \frac{1}{2}\varepsilon\, E^2. \qquad \text{(E8-1.5)}$$

Now, let us recall expression (8.29) for the energy density $U_{\text{Capacitor}}$ stored in the electric field of a capacitor as follows:

$$U_{\text{Capacitor}} = \frac{1}{2}\varepsilon\, E^2. \qquad \text{(E8-1.6)}$$

Comparing (E8-1.5) with (E8-1.6) we obtain the force per unit area experienced by a plate of the capacitor as

$$\frac{F}{A} = U_{\text{Capacitor}},$$

this being the energy density stored in the capacitor.

Next, it will be of interest to us to find the expressions analogous to (8.20) and (8.21) for the energy and the energy density stored in a static magnetic field, noting, as explained following (4.5) (see also Table 4.1 in Chapter 4), that the quantities $\vec{D}$ and $\vec{E}$ in electrostatics are analogous to the quantities $\vec{H}$ and $\vec{B}$ respectively. Therefore, replacing $\vec{E}$ by $\vec{B}$ and $\vec{D}$ by $\vec{H}$ in both of the expressions (8.20) and (8.21), we can write the following expressions for the energy stored W_B and the energy density U_B stored in a static magnetic field:

$$W_B = \frac{1}{2}\int_{\tau} \vec{B}.\vec{H}\, d\tau \qquad (8.31)$$

$$U_B = \frac{1}{2}\vec{B}\cdot\vec{H}. \qquad (8.32)$$

Let us now proceed to the particular problem of finding the energy density stored in the magnetic field of a solenoid to exemplify the energy density stored in a static magnetic field. Similar to the capacitor problem, let us find the work done in establishing a current in a long solenoid by an induced electromotive force of magnitude $|E_{\text{induced}}|$, which, according to Faraday's law can be written as (see Section 5.2):

$$|E_{\text{induced}}| = \left|\frac{d\phi_B}{dt}\right| \tag{8.33}$$

where ϕ_B is the magnetic flux linked with the solenoid, which is related to the inductance L of the solenoid and the current i through the relation (4.48) reproduced below:

$$\phi_B = Li. \tag{8.34}$$

Let the induced electromotive force bring about an elemental increment of current di in the solenoid in an infinitesimal time dt, thereby adding an element of charge

$$dq = idt \tag{8.35}$$

to the solenoid in time dt. In doing so, an element of work dW_{Solenoid} is done. The induced electromagnetic force given by (8.33) can be interpreted as positive for the current building up with time, and correspondingly the magnetic flux ϕ_B increasing with time, thereby rendering $d\phi_B/dt$ positive. Hence we can remove the symbol representing its magnitude to write

$$|E_{\text{induced}}| = \frac{d\phi_B}{dt}. \tag{8.36}$$

Now, by multiplying this induced electromotive force by the charge element given by (8.35), we obtain

$$dW_{\text{Solenoid}} = \frac{d\phi_B}{dt} idt. \tag{8.37}$$

Differentiating (8.34) with time and substituting it in (8.37), we can write

$$dW_{\text{Solenoid}} = L\frac{di}{dt}idt = Lidi. \tag{8.38}$$

The work done W_{Solenoid} in building up current in the solenoid from $i = 0$ to $i = I_0$ is then obtained by integrating (8.38) as

$$W_{\text{Solenoid}} = \int dW_{\text{Solenoid}} = \int_{i=0}^{i=I_0} Lidi = L\left[\frac{i^2}{2}\right]_{i=0}^{i=I_0} = \frac{1}{2}LI_0^2. \tag{8.39}$$

If we substitute L from (4.48) in (8.39), we can express (8.39) as

$$W_{\text{Solenoid}} = \frac{1}{2}\mu_0 n^2 \alpha\, l\, I_0^2 = \frac{1}{2}\mu_0 (nI_0)^2 (\alpha\, l) \tag{8.40}$$

where l and α are the length and the cross-sectional area of the solenoid. Then with the help of the expression $H = ni$ (4.38) for the magnitude of the magnetic field, here to be interpreted for the solenoid current I_0, we can put (8.40) as

$$W_{\text{Solenoid}} = \frac{1}{2}\mu_0 H^2(\alpha\, l). \tag{8.41}$$

Dividing (8.41) by the volume αl of the solenoid, we obtain the following expression for the energy per unit volume or energy density U_{Solenoid} stored in the magnetic field of the solenoid:

$$U_{\text{Solenoid}} = \frac{1}{2}\mu_0 H^2. \tag{8.42}$$

The magnetic field in the solenoid is along the axis of the solenoid (Section 4.3). Therefore, if we assign this direction as the z direction, we can represent the magnetic field of the solenoid in the vectorial form as $\vec{H} = H\,\vec{a}_z$, where $\vec{a}_z$ is the unit vector in the z direction. Further, we can write $\vec{H} \cdot \vec{H} = H^2$, in view of which we can express (8.42) as

$$U_{\text{Solenoid}} = \frac{1}{2}\mu_0 \vec{H} \cdot \vec{H}$$

which, on putting $\vec{B} = \mu\,\vec{H}$ may be written as

$$U_{\text{Solenoid}} = \frac{1}{2}\vec{B} \cdot \vec{H}. \tag{8.43}$$

We can then easily appreciate that the energy density expression (8.43) of a solenoid in the typical example considered is identical to the general expression (8.32) for the energy density stored in a magnetic field.

EXAMPLE E8-2

Show that the same expression as that found in Example E4-10 for the inductance per unit length of a coaxial cable: $L = (\mu_0/2\pi)\ln(b/a)$ can be derived starting from the expression for the energy density stored in a magnetic field, where a is the radius of the inner conductor and b is the inner radius of the outer conductor of the cable.

SOLUTION:

Consider a cylindrical shell of radius r, thickness dr and length l considered between the inner and the outer conductors of the coaxial cable of radii a and b respectively ($a \le r \le b$). The element of volume of the shell is $2\pi\, r\, dr\, dl$. With the help of (8.32), putting $\vec{B} = \mu_0 \vec{H}$ therein, we can find the energy density $U_B = (1/2)\mu_0 H^2$. Multiplying U_B by the volume element $2\pi r\, dr\, dl$, we can find the element of magnetic energy stored ΔW_B in the element of volume as

$$\Delta W_B = \frac{1}{2}\mu_0 H^2 (2\pi r\, dr\, dl). \tag{E8-2.1}$$

The magnetic field H in the region between the conductors may be found with the help of Ampere's circuital law (4.23) as

$$H = \frac{I}{2\pi r}. \tag{E8-2.2}$$

We can substitute H from (E8-2.2) in (E8-2.1) to write

$$\Delta W_B = \frac{1}{2}\mu_0 \left(\frac{I}{2\pi r}\right)^2 (2\pi r\, dr\, dl). \tag{E8-2.3}$$

In order to find the energy stored in the coaxial cable W_B over a length l, let us integrate (E8-2.3) between the limits $r = a$ and b and $l = 0$ and l to obtain

$$W_B = \left[\int_{r=a}^{r=b} \frac{1}{2}\mu_0 \left(\frac{I}{2\pi r}\right)^2 2\pi r dr\right]\left[\int_0^l dl\right]$$

which simplifies to

$$W_B = \frac{\mu_0 I^2}{4\pi}\left[\int_{r=a}^{r=b} \frac{1}{r} dr\right]\left[\int_0^l dl\right] = \frac{\mu_0 I^2}{4\pi}[\ln r]_a^b [l]_0^l = \frac{\mu_0 I^2}{4\pi}\ln\left(\frac{b}{a}\right) l. \tag{E8-2.4}$$

Alternatively, we can also write the energy W_B of the cable of length l in terms of its inductance L', say, as

$$W_B = \frac{1}{2}L' I^2. \tag{E8-2.5}$$

While writing the expression (E8-2.5), it may be noted that, though the expression for the energy stored in terms of the inductance has been derived earlier for a solenoid in the form of (8.39), it is a general expression valid for other magnetic energy storing elements as well, including the coaxial cable. Equating the right-hand sides of (E8-2.4) and (E8-2.5), we get

$$\frac{1}{2}L' I^2 = \frac{\mu_0 I^2}{4\pi}\ln\left(\frac{b}{a}\right) l,$$

which gives the inductance per unit length of the coaxial cable $L (= L'/l)$ as

$$L = \frac{\mu_0}{2\pi}\ln\left(\frac{b}{a}\right),$$

This expression agrees with that derived in Example E4-10 with the help of Ampere's circuital law.

8.2 Poynting Theorem for Energy Balance in Electromagnetic Power Flow

The phenomenon of the storage, loss and flow of electromagnetic energy follows the basic energy balance principle well stated by the Poynting theorem. This will be formulated in this section from first principles in terms of a vector quantity $\vec{P} = \vec{E} \times \vec{H}$, called the Poynting vector, named after the British physicist John Henry Poynting. The dimension of the Poynting vector $\vec{P} = \vec{E} \times \vec{H}$ is that of (V/m) × (A/m), that is, W/m^2, and therefore the Poynting vector $\vec{P}$ can be called the power density vector as well.

8.2.1 Poynting theorem involving instantaneous Poynting vector

Very useful information about the energy in a closed volume can be obtained if we take the volume integral of the Poynting vector over the volume τ as follows:

$$\int_\tau \nabla\cdot\vec{P}d\tau = \int_\tau \nabla\cdot(\vec{E} \times \vec{H})d\tau\ ,$$

which, in view of the vector identity:

$$\nabla.(\vec{E} \times \vec{H}) = \vec{H}.\nabla \times \vec{E} - \vec{E}\cdot\nabla \times \vec{H}$$

can be expressed as

$$\int_\tau \nabla\cdot\vec{P}d\tau = \int_\tau (\vec{H}.\nabla \times \vec{E} - \vec{E}\cdot\nabla \times \vec{H})d\tau. \tag{8.44}$$

We can elaborate the right-hand side of (8.44) with the help of Maxwell's equations (5.38) and (5.45) and, for this purpose, let us recall and rewrite them respectively as follows:

$$\nabla \times \vec{E} = -\frac{\partial \vec{B}}{\partial t} \tag{8.45}$$

$$\nabla \times \vec{H} = \vec{J_c} + \frac{\partial \vec{D}}{\partial t} \tag{8.46}$$

where we have taken in (5.45) $\vec{J} = \vec{J_c}$, the conduction current density, assuming that the medium has a finite conductivity [as explained following (5.45)].

Thus, plugging Maxwell's equations (8.45) and (8.46) in (8.44), we can write

$$\int_{\tau} \nabla.\vec{P}\,d\tau = \int_{\tau} \left[\vec{H}.\left(-\frac{\partial \vec{B}}{\partial t}\right) - \vec{E}\cdot\left(\vec{J}_c + \frac{\partial \vec{D}}{\partial t}\right)\right] d\tau\,. \tag{8.47}$$

In view of the relations $\vec{B} = \mu\,\vec{H}$ [see (4.2)] and $\vec{D} = \varepsilon\,\vec{E}$ [see (3.16)], we can express (8.47) as

$$\int_{\tau} \nabla.\vec{P}\,d\tau = -\int_{\tau}\left(\mu\,H\frac{\partial H}{\partial t} + \vec{E}\cdot\vec{J}_c + \varepsilon\,E\frac{\partial E}{\partial t}\right)d\tau,$$

which making use of the relations:

$$\left.\begin{aligned}
\frac{\partial}{\partial t}\mu\,H^2 &= \mu\frac{\partial}{\partial t}(H^2) = \mu\left[\frac{\partial}{\partial H}(H^2)\right]\left(\frac{\partial H}{\partial t}\right) = \mu\,(2H)\left(\frac{\partial H}{\partial t}\right) = 2\mu\,H\frac{\partial H}{\partial t}\\
\frac{\partial}{\partial t}\varepsilon\,E^2 &= \varepsilon\frac{\partial}{\partial t}(E^2) = \varepsilon\left[\frac{\partial}{\partial E}(E^2)\right]\left(\frac{\partial E}{\partial t}\right) = \varepsilon\,(2E)\left(\frac{\partial E}{\partial t}\right) = 2\varepsilon\,E\frac{\partial E}{\partial t}
\end{aligned}\right\}.$$

can be further put as

$$\int_{\tau} \nabla.\vec{P}\,d\tau = -\int_{\tau}\left(\frac{1}{2}\frac{\partial}{\partial t}\mu\,H^2 + \frac{1}{2}\frac{\partial}{\partial t}\varepsilon\,E^2 + \vec{E}\cdot\vec{J}_c\right)d\tau\,. \tag{8.48}$$

If we now recall Gauss's divergence theorem (8.17) and use it by putting therein $\vec{A} = \vec{P}$, we can convert the volume integral of the left-hand side of (8.48) into a surface integral and express it as follows:

$$\oint_{S} \vec{P}\cdot d\vec{S} = -\int_{\tau}\left(\frac{1}{2}\frac{\partial}{\partial t}\mu\,H^2 + \frac{1}{2}\frac{\partial}{\partial t}\varepsilon\,E^2 + \vec{E}\cdot\vec{J}_c\right)d\tau \tag{8.49}$$

where S represents the area of the surface bounding the volume τ. Further, using the relations $\vec{B} = \mu\,\vec{H}$ [see (4.2)] and $\vec{D} = \varepsilon\,\vec{E}$ [see (3.16)], we obtain the relations $\mu\,H^2 = \mu\,\vec{H}.\vec{H} = \vec{B}.\vec{H}$ and $\varepsilon\,E^2 = \varepsilon\,\vec{E}\cdot\vec{E} = \vec{D}.\vec{E}$, which can then be used to express (8.49) after rearranging the terms as follows:

$$\oint_{S} \vec{P}\cdot d\vec{S} = -\left(\frac{\partial}{\partial t}\int_{\tau}\frac{1}{2}\vec{E}.\vec{D}\,d\tau + \frac{\partial}{\partial t}\int_{\tau}\frac{1}{2}\vec{B}.\vec{H}\,d\tau + \int_{\tau}\vec{E}\cdot\vec{J}_c\,d\tau\right). \tag{8.50}$$

It is easy to appreciate with the help of (8.20) and (8.31) that the first two terms within the parenthesis of the right-hand side of (8.50) represent the time rates of variation of energies W_E and W_B stored in electric and magnetic fields in the volume enclosure respectively. Therefore, we can express (8.50) as:

$$\oint_S \vec{P} \cdot d\vec{S} = -\left(\frac{\partial W_E}{\partial t} + \frac{\partial W_B}{\partial t} + \int_\tau \vec{E} \cdot \vec{J}_c \, d\tau\right). \tag{8.51}$$

With a change of sign, we can express (8.51) as

$$-\oint_S \vec{P} \cdot d\vec{S} = \frac{\partial W_E}{\partial t} + \frac{\partial W_B}{\partial t} + \int_\tau \vec{E} \cdot \vec{J}_c \, d\tau. \tag{8.52}$$

In (8.51) or (8.52), we thus obtain the integral statement of an energy balance theorem called the Poynting theorem. The left-hand sides of (8.51) and (8.52) each involve the closed surface integral of the instantaneous power density vector or instantaneous Poynting vector $\vec{P}$ or the flux of $\vec{P}$ through the surface area bounding a closed volume. (The meaning of the flux of a vector in terms of a surface integral has already been explained in Section 2.4). The flux of power density $\vec{P}$ can be outward or inward to the closed volume. The flux of power density $\vec{P}$ inwards the volume will (i) cause a change in the energy stored in the electric field, (ii) cause a change in the energy stored in the magnetic field and (iii) be dissipated in a conducting medium, if present, in the volume. Therefore, you will appreciate that the left-hand side of (8.52), that is, $-\oint_S \vec{P} \cdot d\vec{S}$, represents the instantaneous inward flux of $\vec{P}$ to a closed volume or the instantaneous power going into a closed volume, since the right-hand side of (8.52) comprises the following three terms: the first representing the time rate of increase of energy stored in electric field in the volume, the second representing the time rate of increase of energy stored in magnetic field in the volume, and the third representing the power lost in any conducting medium present in the medium. (That the third term the right-hand side of (8.52) represents the power lost in a conducting medium will be made amply clear with the help of a simple example of a straight wire of circular cross section carrying dc current ($\partial/\partial t = 0$) to follow.) Now that the negative surface integral $-\oint_S \vec{P} \cdot d\vec{S}$ represents the instantaneous inward flux of $\vec{P}$ to a volume enclosure or the instantaneous power going into a volume enclosure, as explained, we can conclude that the positive surface integral $\oint_S \vec{P} \cdot d\vec{S}$ represents the instantaneous outward flux of $\vec{P}$ from a volume enclosure or the instantaneous power going out of a volume enclosure. If there are free charges present in the volume, then, even in the absence of the conduction current density, $\vec{J}_c$ can be interpreted as the convection current density, and consequently the third term of the right-hand side of (8.52) will represent the energy spent in unit time (or power) in moving the charges.

For the case of the dc current-carrying straight wire considered within the volume enclosure τ, since $\partial/\partial t = 0$, the first two terms of the right-hand side of (8.52) each become null and only its third term becomes significant, enabling us to write (8.52) as follows:

$$-\oint_S \vec{P}\cdot d\vec{S} = \int_\tau \vec{E}\cdot\vec{J}_c\, d\tau\,. \tag{8.53}$$

Using the expression $\vec{J}_c = \sigma\,\vec{E}$ for the conduction current density [see (5.11)], σ being the conductivity of the material of the wire, we can write the integrand of the right-hand side of (8.53) as

$$\vec{E}\cdot\vec{J}_c = \vec{E}\cdot\sigma\,\vec{E} = \sigma\,E^2,$$

both $\vec{E}$ and $\vec{J}_c$ having the same direction of the axis of the wire, and thus express (8.53) as

$$-\oint_S \vec{P}\cdot d\vec{S} = \int_\tau \sigma\,E^2\, d\tau \tag{8.54}$$

where E is the magnitude of the electric field $\vec{E}$. There being no electric field existing outside the wire, the volume over which the integration of the right-hand side has to be taken is restricted to the region occupied by the wire. Therefore, carrying out the integration over the volume τ $(=\alpha\,l)$ occupied by the wire and considering the electric field to be uniform over the volume, and hence taking it outside the volume integral, we can put (8.54) as

$$-\oint_S \vec{P}\cdot d\vec{S} = \sigma\,E^2\int_\tau d\tau = \sigma\,E^2\tau = \sigma\,E^2\alpha\,l \tag{8.55}$$

where $\alpha = \pi a^2$ and l are respectively the cross-sectional area and the length of the wire, a being the radius of the wire (Fig. 8.1). Further, as explained in the footnote preceding (6.100), the electric field E can be related to the potential difference V between the ends of the wire as $E = V/l$, which can be now substituted in (8.55) to obtain

$$-\oint_S \vec{P}\cdot d\vec{S} = \sigma\left(\frac{V}{l}\right)^2 \alpha\,l = \sigma\frac{V^2}{l^2}\alpha\,l = \frac{V^2}{\dfrac{1}{\sigma}\dfrac{l}{\alpha}} = \frac{V^2}{\dfrac{1}{\sigma}\dfrac{l}{\pi a^2}}$$

in which we can plug (6.100) to obtain

$$-\oint_S \vec{P}\cdot d\vec{S} = = \frac{V^2}{R_{\mathrm{dc}}}$$

which can be expressed using the relation $V = IR_{\mathrm{dc}}$ as

$$-\oint_S \vec{P}\cdot d\vec{S} = I^2 R_{\mathrm{dc}} \tag{8.56}$$

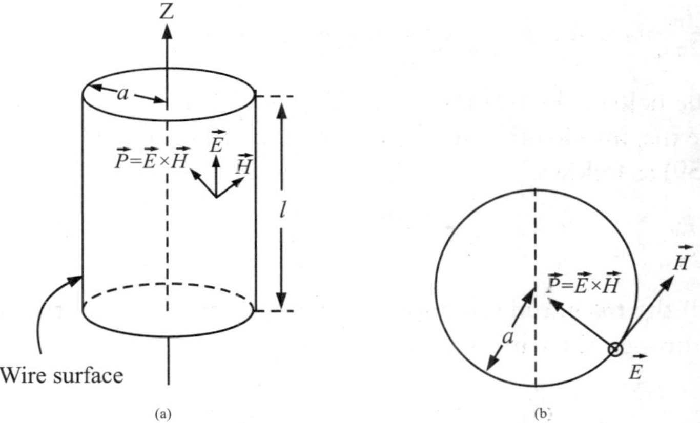

Fig. 8.1 *Straight wire of length l and radius a aligned along z: (a) perspective and (b) cross section, both showing Poynting vector $\vec{P}(=\vec{E}\times\vec{H})$ at the surface of the wire.*

where R_{dc} is the dc resistance of the wire. Thus the inward flux of the Poynting vector through the surface of the wire, represented by the left-hand side of (8.56), manifests itself as power loss in the wire—known as resistive loss or ohmic loss or I^2R-loss—in the form of heat, represented by the right-hand side of (8.56). This is known as Joule's law. Further, with the help of this example it becomes easy to appreciate that the third term $\int_\tau \vec{E}\cdot\vec{J}_c\,d\tau$ of the right-hand side of (8.52) represents the power lost in any conducting medium present in the volume as stated earlier. Consequently, the quantity $\vec{E}\cdot\vec{J}_c$ (integrand within the volume integral of the term), which also equals σE^2 as explained following (8.53), represents the power lost per unit volume within the volume.

To gain further insights, it will be interesting to have a re-look at the problem of power loss in a wire carrying a dc current starting from the power density or Poynting vector $\vec{P}\ (=\vec{E}\times\vec{H})$ at the surface of the wire of circular cross section (Fig. 8.1). With the help of Ampere's circuital law (4.27), we can write the azimuthal magnetic field at the surface of the wire due to the current through the wire along the length of the wire, supposedly in the axial z direction, as

$$\vec{H}=\frac{I_{\text{dc}}}{2\pi\,a}\vec{a}_\theta\,. \tag{8.57}$$

Let us use use (8.57) to write the Poynting vector $\vec{P}\ (=\vec{E}\times\vec{H})$ at the surface of the wire as

$$\vec{P}=\vec{E}\times\vec{H}=\vec{E}\times\frac{I_{\text{dc}}}{2\pi\,a}\vec{a}_\theta\,. \tag{8.58}$$

We can write the electric field inside the wire as $\vec{E}=E\vec{a}_z$, it being directed along z due to a potential difference across the wire that sends the current through the wire also in the z direction. Thus, using the relation $\vec{a}_z\times\vec{a}_\theta=-a_r$ (see Section 2.1) in (8.58), we obtain

$$\vec{P} = \vec{E} \times \frac{I_{\rm dc}}{2\pi\, a}\vec{a}_\theta = E\vec{a}_z \times \frac{I_{\rm dc}}{2\pi\, a}\vec{a}_\theta = E\frac{I_{\rm dc}}{2\pi\, a}\vec{a}_z \times \vec{a}_\theta = -E\frac{I_{\rm dc}}{2\pi\, a}\vec{a}_r. \quad (8.59)$$

Further, with the help of the relation $\vec{J}_c = \sigma\,\vec{E}$ [see (5.11)] and the relation $\vec{E} = E\vec{a}_z$, we can write J_c, the magnitude of $\vec{J}_c$, in terms of E, the magnitude of $\vec{E}$, as $J_c = \sigma\, E$ and use it to express (8.59) as follows:

$$\vec{P} = -\frac{J_c}{\sigma}\frac{I_{\rm dc}}{2\pi\, a}\vec{a}_r\ . \quad (8.60)$$

Putting in (8.60) the magnitude of current density (current per unit area) J_c in terms of the current $I_{\rm dc}$ through the wire as

$$J_c = \frac{I_{\rm dc}}{\pi\, a^2},$$

we then obtain the Poynting vector or power density vector at the surface of the wire as

$$\vec{P} = -\frac{I_{\rm dc}}{\pi\, a^2}\frac{I_{\rm dc}}{2\pi\, a\sigma}\vec{a}_r\ . \quad (8.61)$$

The negative sign of the right-hand side of (8.61) indicates that the power density vector $\vec{P}$ is directed radially inwards, suggesting that though the current flow is axial in the wire; the flow of power into the wire is in the radially inward direction. Further, we expect that the flux of power density vector into the volume of the wire that has a finite conductivity will be lost in the wire in the form of heat. We can then substitute (8.61) in the left-hand side of (8.56) to evaluate the inward flux through the surface bounding the wire as:

$$-\oint_S \vec{P}\cdot d\vec{S} = -\oint_S -\frac{I_{\rm dc}}{\pi a^2}\frac{I_{\rm dc}}{2\pi a\sigma}\vec{a}_r\cdot d\vec{S}\ ,$$

which on putting $d\vec{S} = dS\vec{a}_r$ becomes

$$-\oint_S \vec{P}\cdot d\vec{S} = \oint_S \frac{I_{\rm dc}}{\pi\, a^2}\frac{I_{\rm dc}}{2\pi\, a\sigma}\vec{a}_r\cdot\vec{a}_r dS = I_{\rm dc}^2\frac{1}{\sigma}\frac{1}{\pi\, a^2}\frac{1}{2\pi\, a}\int_S dS\ . \quad (8.62)$$

The integration of the area element on the right-hand side of (8.62) becomes the surface of the area of the wire as follows:

$$\int_S dS = 2\pi a l,$$

which substituted in (8.62) gives

$$-\oint_S \vec{P}\cdot d\vec{S} = I_{\rm dc}^2\frac{1}{\sigma}\frac{l}{\pi a^2}\ . \quad (8.63)$$

Plugging (6.100) in (8.63), we then obtain

$$-\oint_S \vec{P}\cdot d\vec{S} = I_{\mathrm{dc}}^2 R_{\mathrm{dc}}\,. \tag{8.64}$$

Thus, as expected, by evaluating the left-hand side of the Poynting theorem (8.52) we have now obtained the expression (8.64), which becomes exactly the same as the Joule's law (8.56) obtained earlier by evaluating the right-hand side of the Poynting theorem (8.52). Interestingly, if we reverse the direction of the wire current, in view of the relation $\vec{J}_c = \sigma\,\vec{E}$ [see (5.11)], it would correspond to a reversal of the electric field in the wire as well as to the reversal of the magnetic field too according to the relation (8.57). Consequently we can express for such a reversal of the direction of the electric field and that of the magnetic field: $\vec{E} = -E\vec{a}_z$ and $\vec{H} = -(I_{\mathrm{dc}}/2\pi a)\vec{a}_\theta$, and the Poynting vector as

$$\vec{P} = \vec{E}\times\vec{H} = (-E\vec{a}_z)\times\left(-\frac{I_{\mathrm{dc}}}{2\pi a}\right)\vec{a}_\theta = E\frac{I_{\mathrm{dc}}}{2\pi a}\vec{a}_z\times\vec{a}_\theta = -E\frac{I_{\mathrm{dc}}}{2\pi a}\vec{a}_r,$$

which is identical to (8.59). Thus the Poynting vector remains unchanged when there is a reversal of the direction of current in the wire. Consequently, starting from this Poynting vector expression and following the same steps as given following (8.59), we will arrive at the same Joule's law expression as (8.64) even if we reverse the direction of the wire current.

In the problem of the straight wire carrying a dc current, we have seen that the influx of power density vector is not stored; rather, it is lost in the form of resistive loss given by Joule's law (8.59) or (8.64). Now, let us apply Poynting theorem to the problem of a capacitor that stores electrical energy. To make life simple, let us take a parallel-plate capacitor with circular plates of a large radius a, supposedly large compared to the distance d between the plates (Fig. 8.2). The expression for the azimuthal magnetic field at the circular-cylindrical interface between the inside and the outside regions of the capacitor (Fig. 8.2) can be obtained by putting $r = a$ in the expression (5.84) as follows:

$$\vec{H} = \varepsilon\frac{dE}{dt}\frac{a}{2}\vec{a}_\theta \tag{8.65}$$

where a is the radius of the circular plate and ε is the permittivity of the dielectric filling the region between the plates of the capacitor (inside region of the capacitor) (Fig. 8.2). We can then find the Poynting vector $\vec{P}(=\vec{E}\times\vec{H})$ at the interface between the inside and outside regions of the capacitor with the help of (8.65) as

$$\vec{P} = \vec{E}\times\vec{H} = \vec{E}\times\varepsilon\frac{dE}{dt}\frac{a}{2}\vec{a}_\theta\,. \tag{8.66}$$

Considering the dimension of the plates of the capacitor to be large compared to the distance between the plates, we can take the electric field as $\vec{E} = E\vec{a}_z$ and substitute it into (8.66) to write

$$\vec{P} = \vec{E}\times\vec{H} = E\,\vec{a}_z\times\varepsilon\frac{dE}{dt}\frac{a}{2}\vec{a}_\theta = \varepsilon\,E\frac{dE}{dt}\frac{a}{2}\vec{a}_z\times\vec{a}_\theta = -\varepsilon\,E\frac{dE}{dt}\frac{a}{2}\vec{a}_r\,. \tag{8.67}$$

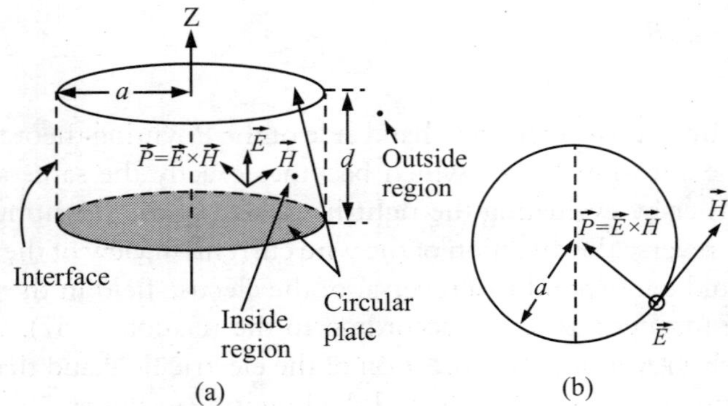

Fig. 8.2 *Parallel-plate capacitor with circular plates of radius* a *separated by a distance* d *along* z *showing: (a) perspective and (b) cross section, both showing Poynting vector* $\vec{P}$ $(= \vec{E} \times \vec{H})$ *at the interface between the inside and outside regions of the capacitor.*

The radially inward direction of the power density vector suggests that energy flows into the capacitor through the interface between the inside and outside regions of the capacitor (Fig. 8.2) and is stored in the electric field of the capacitor. Assuming that the inside region of the capacitor is either a free space or a dielectric which is not lossy (see Section 5.3), only the first term of the right-hand side of the Poynting theorem (8.52) representing the time rate of change of the stored energy in electric field will be significant in the problem of a capacitor, enabling us to write:

$$-\oint_S \vec{P} \cdot d\vec{S} = \frac{\partial W_E}{\partial t} . \tag{8.68}$$

Substituting (8.67) in (8.68), we then have

$$\oint_S \varepsilon\, E \frac{dE}{dt} \frac{a}{2} \vec{a}_r \cdot d\vec{S} = \frac{\partial W_E}{\partial t} ,$$

which on putting $d\vec{S} = dS\vec{a}_r$ becomes

$$\oint_S \varepsilon\, E \frac{dE}{dt} \frac{a}{2} dS\, \vec{a}_r \cdot \vec{a}_r = \oint_S \varepsilon\, E \frac{dE}{dt} \frac{a}{2} dS = \frac{\partial W_E}{\partial t} . \tag{8.69}$$

The term containing the electric field magnitude E and its time derivative, which is considered uniform inside the capacitor, can be taken outside the integral of (8.69) so that we can express (8.69) as

$$\varepsilon\, E \frac{dE}{dt} \frac{a}{2} \oint_S dS = \frac{\partial W_E}{\partial t} . \tag{8.70}$$

Putting in (8.70) $\oint_S dS = S = 2\pi ad$, the area of the circular-cylindrical interface between the inside and outside regions of the capacitor, (a being the radius of the circular plates of the capacitor and d the distance between the plates) (Fig. 8.2), we obtain

$$\frac{\partial W_E}{\partial t} = \varepsilon\, E\frac{dE}{dt}\frac{a}{2}\oint_S dS = \varepsilon\, E\frac{dE}{dt}\frac{a}{2}S$$

$$= \varepsilon\, E\frac{dE}{dt}\frac{a}{2}2\pi\, a\, d = \varepsilon\left(E\frac{dE}{dt}\right)(\pi a^2 d). \quad (8.71)$$

In view of

$$\frac{dE^2}{dt} = 2E\frac{dE}{dt},$$

we can also write (8.71) as

$$\frac{\partial W_E}{\partial t} = \frac{1}{2}\varepsilon\frac{dE^2}{dt}(\pi a^2 d) = \frac{d}{dt}\left(\frac{1}{2}\varepsilon\, E^2\right)(\pi a^2 d). \quad (8.72)$$

Integrating (8.72), we then obtain

$$W_E = \left(\frac{1}{2}\varepsilon\, E^2\right)(\pi a^2 d) \quad . \quad (8.73)$$

Dividing (8.73) by the volume of the capacitor $\pi a^2 d$, the following expression for the energy density U_E in the electric field of the capacitor is obtained:

$$U_E = \frac{1}{2}\varepsilon\, E^2 \quad (8.74)$$

which interestingly becomes the same as (8.29) obtained earlier by the circuit approach, in which the capacitor is assigned a capacitance corresponding to the charge developed on its plates because of the potential difference between them.

Next, let us take the case of a solenoid that stores energy in its magnetic field. The expression for the azimuthal electric field at the peripheral surface bounding the circular capacitor can be obtained by putting $r = a$ in expression (5.100) as follows:

$$\vec{E} = -\frac{dB}{dt}\frac{a}{2}\vec{a}_\theta \quad (8.75)$$

where a is the radius of the solenoid. In view of the relation $B = \mu H$, we can write (8.75) as:

$$\vec{E} = -\mu\frac{dH}{dt}\frac{a}{2}\vec{a}_\theta\, . \quad (8.76)$$

We will consider a long solenoid for which the magnetic field $\vec{H}$ is uniform inside the solenoid and is directed along the axial z direction. We may then take it as $\vec{H} = H\vec{a}_z$ and substitute it in (8.76) to obtain the Poynting vector $\vec{P}$ $(= \vec{E} \times \vec{H})$ as

$$\vec{P} = -\mu H \frac{dH}{dt} \frac{a}{2} \vec{a}_\theta \times \vec{a}_z = -\mu H \frac{dH}{dt} \frac{a}{2} \vec{a}_r \,. \tag{8.77}$$

As in the capacitor problem, in the present problem of the inductor too, the Poynting vector given by (8.77) is radially inward. In fact, the Poynting vector expressions for the capacitor and the inductor problems, (8.67) and (8.77) respectively, are similar in that that if we replace ε and E in (8.67) by μ and H respectively, we get (8.77). Therefore, if we proceed from (8.77) and follow the same method as that used to obtain the energy density expression (8.74) for the capacitor starting from (8.67), we will derive the following expression for the stored energy density U_B in the magnetic field of the solenoid

$$U_B = \frac{1}{2} \mu H^2 \tag{8.78}$$

which, with μ interpreted as μ_0 becomes identical to (8.42) obtained earlier by the circuit approach in which the solenoid was assigned an inductance corresponding to magnetic flux according to the current through the solenoid.

8.2.2 Complex Poynting vector theorem and time-averaged electromagnetic power flow

We come across numerous practical situations where the field intensities vary in a simple-harmonic manner. Such fields can be represented by a sine or a cosine variation with time, which, in fact, with proper interpretation of their initial phases represent one and the same variation. Further, in general, any other type of time-periodic variation of fields can also be considered to be a combination of such sine or cosine type variations. For time-periodic electric and magnetic fields, we need to time-average the power density or Poynting vector $\vec{P}$ $(= \vec{E} \times \vec{H})$ and also interpret the general Poynting theorem given by (8.51) or (8.52).

In order to understand the mechanism of power flow in problems involving time-varying electric and magnetic fields, let us first recall the simple example of a straight wire carrying a current (Fig. 8.1). In this example, we have already discussed following (8.64) how a reversal of current through the wire reverses the direction of the electric field and consequently that of the magnetic field as well. Thus the flux of the Poynting vector $\vec{P}$ $(= \vec{E} \times \vec{H})$ continues to be directed radially inwards in the wire, causing a power loss in the wire. This can be accounted for by using the Joule's law expression (8.64), which is essentially the same as Joule's law expression (8.59) obtained without reversing the current. Therefore, if we now consider the flow of a time-periodic current, say sinusoidal, through the wire and its associated time-periodic electric and the magnetic fields, then the direction of the flux of the instantaneous Poynting vector $\vec{P}$ $(= \vec{E} \times \vec{H})$ will continue to be radially inwards even with the reversal of wire current. This point can be further appreciated from the expression for the instantaneous Poynting vector that can be obtained by replacing I_{dc} in (8.61) by a time-periodic sinusoidal current $I = I_0 \sin \omega t$, where I_0 is the amplitude of the current, as follows:

$$\vec{P} = -\frac{I_0 \sin \omega t}{\pi a^2} \frac{I_0 \sin \omega t}{2\pi a \sigma} \vec{a}_r = -\frac{I_0^2 \sin^2 \omega t}{\pi a^2} \frac{1}{2\pi a \sigma} \vec{a}_r \, . \tag{8.79}$$

Since $\sin^2 \omega t$ is positive irrespective of the sign (positive or negative) of $\sin \omega t$ (that is, irrespective of the phase of the current), the direction of the instantaneous Poynting vector $\vec{P}$ remains inwards. The negative sign on the right-hand side of (8.79) indicates this, as mentioned preceding (8.79). Therefore, starting from (8.79) and using the method given following (8.61), we can easily obtain the expression for the instantaneous inward flux of $\vec{P}$ through the surface bounding the wire, or the instantaneous power flow into the wire, which is also the instantaneous power P absorbed or lost in the wire [see the discussion following (8.52)] as:

$$P = -\oint_S \vec{P} \cdot d\vec{S} = I_0^2 \sin^2 \omega t \, R_{dc} \, . \tag{8.80}$$

The average power flowing into the wire can then be found taking the average of (8.80) over a complete cycle as follows:

$$P_{\text{average}} = \left\langle I_0^2 \sin^2 \omega t \, R_{dc} \right\rangle_{\text{average}} = \frac{\int_{\omega t=0}^{2\pi} I_0^2 \sin^2 \omega t \, R_{dc} d(\omega t)}{\int_{\omega t=0}^{2\pi} d(\omega t)} = \frac{I_0^2 R_{dc} \int_{\omega t=0}^{2\pi} \sin^2 \omega t \, d(\omega t)}{\int_{\omega t=0}^{\omega t=2\pi} d(\omega t)}$$

$$= \frac{I_0^2 R_{dc} \int_{\omega t=0}^{2\pi} \frac{1-\cos 2\omega t}{2} d(\omega t)}{\int_{\omega t=0}^{2\pi} d(\omega t)}$$

$$= \frac{I_0^2 R_{dc} \left(\frac{1}{2} \int_{\omega t=0}^{2\pi} d(\omega t) - \frac{1}{2} \int_{\omega t=0}^{2\pi} \cos 2\omega t \, d(\omega t) \right)}{\int_{\omega t=0}^{2\pi} d(\omega t)}$$

$$= \frac{I_0^2 R_{dc} \left[\frac{1}{2} [\omega t]_0^{2\pi} - \frac{1}{2} \left[\frac{\sin 2\omega t}{2\omega} \right]_0^{2\pi} \right]}{[\omega t]_0^{2\pi}}$$

$$= \frac{I_0^2 R_{dc} \left[\frac{1}{2} (2\pi - 0) - \frac{1}{4\omega} (\sin 4\pi - \sin 0) \right]}{2\pi - 0}$$

$$= \frac{I_0^2 R_{\text{dc}} \left[\frac{1}{2} 2\pi - \frac{1}{4\omega} (0 - 0) \right]}{2\pi - 0} = \frac{1}{2} I_0^2 R_{\text{dc}}.$$

That is,

$$P_{\text{average}} = \frac{1}{2} I_0^2 R_{\text{dc}} \,. \tag{8.81}$$

We can also express P_{average} from (8.80) as the following time-averaged flux of the Poynting vector:

$$P_{\text{average}} = -\oint_S \vec{P}_{\text{average}} \cdot d\vec{S} \tag{8.82}$$

where

$$\vec{P}_{\text{average}} = \left\langle \vec{E} \times \vec{H} \right\rangle_{\text{average}} \tag{8.83}$$

is the time-averaged Poynting vector $\vec{P}$ $(= \vec{E} \times \vec{H})$. For a wire carrying a sinusoidal current, let the time-periodic electric and magnetic fields be

$$\left. \begin{aligned} \vec{E} &= E_0 \sin \omega t \, \vec{a}_E \quad \text{(a)} \\ \vec{H} &= H_0 \sin \omega t \, \vec{a}_H \quad \text{(b)} \end{aligned} \right\} \tag{8.84}$$

where E_0 and H_0 are the amplitudes of the electric and magnetic field magnitudes and $\vec{a}_E$ and $\vec{a}_H$ the unit vectors in these field directions respectively. We can then substitute (8.84) in (8.83) to obtain

$$\vec{P}_{\text{average}} = \left\langle \vec{E} \times \vec{H} \right\rangle_{\text{average}} = \langle E_0 \sin \omega t \, \vec{a}_E \times H_0 \sin \omega t \, \vec{a}_H \rangle_{\text{average}}$$

$$= \left\langle E_0 H_0 \sin^2 \omega t \right\rangle_{\text{average}} \vec{a}_E \times \vec{a}_H \,. \tag{8.85}$$

Let us now evaluate the average of the right-hand side of (8.85) as we did in obtaining (8.81) to obtain

$$\vec{P}_{\text{average}} = \frac{1}{2} E_0 H_0 \vec{a}_E \times \vec{a}_H \,. \tag{8.86}$$

It is easy to appreciate from (8.58) as well as from the discussion following it that $\vec{a}_E = \vec{a}_z$ and $\vec{a}_H = \vec{a}_\theta$ so that, in view of the relation $\vec{a}_z \times \vec{a}_\theta = -\vec{a}_r$, we can express (8.86) alternatively as

$$\vec{P}_{\text{average}} = \frac{1}{2} E_0 H_0 \vec{a}_z \times \vec{a}_\theta = -\frac{1}{2} E_0 H_0 \vec{a}_r \,. \tag{8.87}$$

Thus, as in (8.79), $\vec{P}_{\text{average}}$ predicted by (8.87) is directed radially inward to the wire. Now, substituting (8.87) in (8.82), we can find the average power P_{average} absorbed or lost in the wire as

$$P_{\text{average}} = -\oint_S -\frac{1}{2}E_0 H_0 \vec{a}_r \cdot d\vec{S} = \oint_S \frac{1}{2}E_0 H_0 \vec{a}_r \cdot d\vec{S}$$

which, on putting $\vec{H} = -(I_{\text{dc}}/2\pi a)\vec{a}_\theta$, as we did preceding (8.62), becomes

$$P_{\text{average}} = \oint_S \frac{1}{2}E_0 H_0 \vec{a}_r \cdot \vec{a}_r dS = \oint_S \frac{1}{2}E_0 H_0 dS\,. \tag{8.88}$$

Thus, in (8.81) and (8.88) we obtain two alternative expressions for the average power lost in a wire of finite conductivity.

Now, in order to gain a further insight into the power flow problem, let us return to the example of a capacitor with circular plates and put time-periodic electric field inside it. Hence the magnetic field is also time-periodic, which can be expressed with the help of (8.65) at the interface between the inside and outside regions of the capacitor (Fig. 8.2) $(r = a)$ as

$$\left.\begin{aligned} \vec{E} &= E_0 \sin\omega t\, \vec{a}_z = E_0 \sin\omega t\, \vec{a}_E \\ \vec{H} &= \varepsilon\frac{dE}{dt}\frac{a}{2}\vec{a}_\theta = \varepsilon\omega\cos\omega t\frac{a}{2}\vec{a}_\theta = H_0\cos\omega t\,\vec{a}_\theta = H_0\cos\omega t\,\vec{a}_H \end{aligned}\right\} \tag{8.89}$$

where $H_0 = \varepsilon\omega\, a/2$. With the help of (8.89), we can thus write the following expression for the average Poynting vector:

$$\vec{P}_{\text{average}} = \left\langle \vec{E}\times\vec{H}\right\rangle_{\text{average}} = \langle E_0 H_0 \sin\omega t\, \cos\omega t\, \vec{a}_E \times \vec{a}_H\rangle_{\text{average}}$$

$$= \frac{\int\limits_{\omega t=0}^{2\pi} E_0 H_0 \sin\omega t\, \cos\omega t\, d(\omega t)\, \vec{a}_E \times \vec{a}_H}{\int\limits_{\omega t=0}^{2\pi} d(\omega t)}$$

$$= \frac{E_0 H_0 \int\limits_{\omega t=0}^{2\pi} (\sin\omega t)\, d(\sin\omega t)\, \vec{a}_E \times \vec{a}_H}{\int\limits_{\omega t=0}^{2\pi} d(\omega t)}$$

$$= \frac{E_0 H_0 \left[\frac{\sin^2\omega t}{2}\right]_0^{2\pi} \vec{a}_E \times \vec{a}_H}{[\omega t]_0^{2\pi}}$$

$$= \frac{E_0 H_0 \frac{1}{2}(\sin^2 2\pi - \sin^2 0)\, \vec{a}_E \times \vec{a}_H}{2\pi - 0}$$

$$= \frac{E_0 H_0 \frac{1}{2}(0 - 0)\, \vec{a}_E \times \vec{a}_H}{2\pi - 0} = 0$$

That is,

$$\vec{P}_{\text{average}} = 0. \tag{8.90}$$

Further, since according to (8.90) the Poynting vector at the interface between the inside and outside regions of the capacitor (Fig. 8.2) ($r = a$) averages out to zero ($\vec{P}_{\text{average}} = 0$), the average power going into the capacitor, according to (8.82), will also become zero ($P_{\text{average}} = 0$).

You will get a clearer physical picture of the problem if you note the behaviour of the sine and the cosine functions involved in the electric and the magnetic field expressions (8.89) and (8.90) respectively. Studying the behaviour of the Poynting vector given by $\vec{P} = \vec{E} \times \vec{H} = E_0 H_0 \sin\omega t \cos\omega t\, \vec{a}_z \times \vec{a}_\theta = -E_0 H_0 \sin\omega t \cos\omega t\, \vec{a}_r$, we find that within the first quarter of a cycle $0 \le \omega t \le \pi/2$, $\sin\omega t$ and $\cos\omega t$ are each positive, and therefore the Poynting vector $\vec{P} = -E_0 H_0 \sin\omega t \cos\omega t\, \vec{a}_r$ will have the direction of $-\vec{a}_r$. That is, it will be directed radially inwards from the interface between the inner and outer regions of the capacitor ($r = a$) (Fig. 8.2). Within the second quarter of a cycle, $\pi/2 \le \omega t \le \pi$, since $\sin\omega t$ and $\cos\omega t$ are positive and negative respectively, $\sin\omega t \cos\omega t$ will have a negative value and consequently, $\vec{P} = -E_0 H_0 \sin\omega t \cos\omega t\, \vec{a}_r$ will have the direction of $\vec{a}_r$, that is, it will be directed radially outward from this surface. Within the third quarter of a cycle, $\pi \le \omega t \le 3\pi/2$, since $\sin\omega t$ and $\cos\omega t$ are each negative, $\sin\omega t \cos\omega t$ will have a positive value and consequently $\vec{P} = -E_0 H_0 \sin\omega t \cos\omega t\, \vec{a}_r$ will have the direction of $-\vec{a}_r$, that is, it will be directed radially inwards from the surface. During the fourth and last quarter of a cycle, $3\pi/2 \le \omega t \le 2\pi$, since $\sin\omega t$ and $\cos\omega t$ are negative and positive respectively, $\sin\omega t \cos\omega t$ will have a negative value and consequently $\vec{P} = -E_0 H_0 \sin\omega t \cos\omega t\, \vec{a}_r$ will have the direction of $\vec{a}_r$, that is, it will be be directed radially outward from the surface.

Thus, from the foregoing reasoning we can understand that the Poynting vector at the interface between the inner and outer regions of the capacitor (Fig. 8.2) becomes alternately radially inwards and outwards in quarter cycles, corresponding to a power alternately surging back and forth between the capacitor and the source, the latter presumably being used for charging the capacitor, thereby making zero the average power absorbed in the latter. We can draw a similar conclusion regarding the power flow with respect to the example of an inductor (solenoid, for instance) as well.

It becomes often convenient to describe the time-periodic variation of a quantity by phasor notation (see Appendix A5.3 in Chapter 5). Thus, we can represent the expression (8.89) for the time-periodic electric field in the capacitor by the phasor notation as

$$\vec{E} = E_0 \exp(j\omega t)\, \vec{a}_E\,. \tag{8.91}$$

Further, the time-periodic magnetic field expression $\vec{H} = H_0 \cos \omega t \, \vec{a}_H = H_0 \sin(\omega t + \pi/2) \, \vec{a}_H$ can be expressed by phasor notation as

$$\vec{H} = H_0 \exp j(\omega t + \pi/2) \, \vec{a}_H \, . \tag{8.92}$$

Thus, we can see that the magnetic field in a capacitor, given by (8.92), leads the electric field, given by (8.91), by a phase angle $\pi/2$. (In other words, the electric and the magnetic fields are in phase quadrature). However, if we somewhat generalise the problem by considering a lossy or imperfect dielectric filling the region between the capacitor plates (see Section 5.3), the magnetic field $\vec{H}$ will no longer differ in phase by $\pi/2$ from the electric field $\vec{E}$. Rather there will be a phase difference θ between them such that $\pi/2 < \theta < 0$. For this generalised case we can represent the electric and the magnetic fields as

$$\left.\begin{aligned} \vec{E} &= E_0 \exp(j\omega t) \vec{a}_E \qquad &\text{(a)} \\ \vec{H} &= H_0 \exp j(\omega t + \theta) \, \vec{a}_H \qquad &\text{(b)} \end{aligned}\right\} . \tag{8.93}$$

In the phasor formulation (see Appendix A5.3 of Chapter 5), the electric and the magnetic fields given by (8.84) correspond to $\theta = 0$ in (8.93) for the problem of a wire carrying a time-periodic current. We can thus represent these fields as

$$\left.\begin{aligned} \vec{E} &= E_0 \exp(j\omega t) \vec{a}_E \qquad &\text{(a)} \\ \vec{H} &= H_0 \exp(j\omega t) \vec{a}_H \qquad &\text{(b)} \end{aligned}\right\} . \tag{8.94}$$

Therefore, if we represent the electric field in the problem of a current-carrying wire (given by (8.94*a*), which has the amplitude E_0) on the real axis of the phasor diagram, then the in-phase ($\theta = 0$) magnetic field, given by (8.94*b*), will also be represented on the same real axis of the diagram. There will be no components of these fields on the imaginary axis. However, in the capacitor problem, according to (8.91) and (8.92), the electric and the magnetic fields are in quadrature phase corresponding to $\theta = \pi/2$ in (8.93). Therefore, if we represent the electric field on the real axis of the phasor diagram, then the magnetic field leading the electric field in phase by $\pi/2$ will be represented on the imaginary axis of the diagram. There will be no component of the magnetic field on the real axis. In the general case, considering the magnetic field to lead the electric field in phase by an angle θ, if we represent the electric field on the real axis of the phasor diagram, then the magnetic field will make an angle of θ with the real axis (Fig. 8.3). Further, in this diagram the component of the magnetic field, given by (8.94*b*), will have the magnitude $H_0 \cos\theta$ on the real axis, while its component on the imaginary axis has the magnitude $H_0 \sin\theta$. Therefore, since the electric field is represented on the real axis, we can call $H_0 \cos\theta$ and $H_0 \sin\theta$ as the magnitudes of the in-phase and the phase-quadrature components of the magnetic field respectively with respect to the electric field that has the magnitude E_0 (Fig. 8.1). Further, we can notice that while the electric and the magnetic fields are in phase ($\theta = 0$) in the problem of the current-carrying wire, contributing to a non-zero value of $\vec{P}_{\text{average}}$

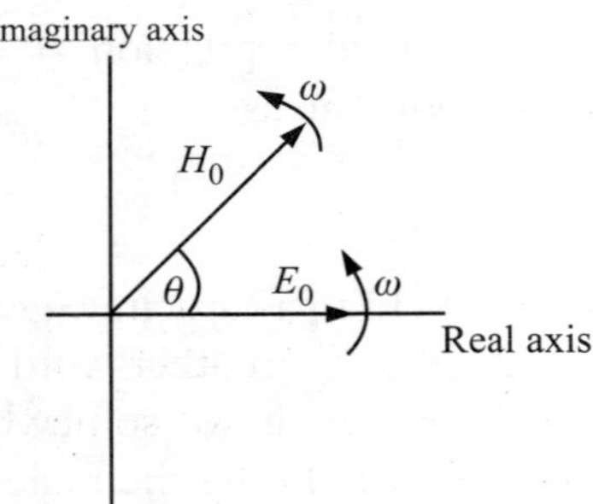

Fig. 8.3 *Phasor diagram showing rotating electric and magnetic field vectors, the latter leading the former by phase angle θ.*

given by (8.86), the electric and the magnetic fields are in phase quadrature ($\theta = \pi/2$) in the capacitor problem making $\vec{P}_{\text{average}}$ zero, as discussed immediately following (8.90). Therefore, if we now take the general case where the magnetic field leads the electric field by phase angle θ, only the in-phase component of magnetic field, namely, $H_0 \cos\theta$, will contribute to $\vec{P}_{\text{average}}$. We may then, by simply replacing H_0 by $H_0 \cos\theta$ in (8.86), write:

$$\vec{P}_{\text{average}} = \frac{1}{2} E_0 H_0 \cos\theta \, \vec{a}_E \times \vec{a}_H \quad . \tag{8.95}$$

It becomes useful while dealing with several time varying problems to express $\vec{P}_{\text{average}}$ given by (8.95) in terms of a complex Poynting vector $\vec{P}_{\text{complex}}$ defined as

$$\vec{P}_{\text{complex}} = \frac{1}{2} \vec{E} \times \vec{H}^* \tag{8.96}$$

where the asterisk represents the complex conjugate. In the case of the magnetic field $\vec{H}$ leading the electric field $\vec{E}$ by a phase of θ, as in the problem of capacitor with a lossy dielectric filling the region between the plates, these fields are given by (8.93). However, there may be a situation in which the magnetic field lags behind the electric field in phase by θ rather than lead it. In this case,while the electric field continues to be represented by (8.93*a*), the magnetic field will be represented by (8.93*b*) but with θ occurring there replaced by $-\theta$. Therefore, considering both the situations of the phase difference, $\pm\theta$ between the electric and the magnetic fields, we can represent these fields by

$$\left.\begin{aligned} \vec{E} &= E_0 \exp(j\omega t)\vec{a}_E && \text{(a)} \\ \vec{H} &= H_0 \exp j(\omega t \pm \theta)\, \vec{a}_H && \text{(b)} \end{aligned}\right\} . \tag{8.97}$$

Substituting $\vec{E} = E_0 \exp(j\omega t)\, \vec{a}_E$ from (8.97*a*) and the complex conjugate of (8.97*b*), that is, $\vec{H}^* = H_0 \exp -j(\omega t \pm \theta)\, \vec{a}_H$ in (8.96), we obtain

$$\vec{P}_{\text{complex}} = \frac{1}{2} \vec{E} \times \vec{H}^* = \frac{1}{2} E_0 H_0 \exp(j\omega t) \exp -j(\omega t \pm \theta)\, \vec{a}_E \times \vec{a}_H$$

$$= \frac{1}{2} E_0 H_0 \exp \mp (j\theta\ ) \vec{a}_E \times \vec{a}_H\ . \tag{8.98}$$

We can also express (8.98), separating its real and imaginary parts as

$$\vec{P}_{\text{complex}} = \frac{1}{2} \vec{E} \times \vec{H}^* = \frac{1}{2} E_0 H_0 \exp(\mp j\theta\ ) \, \vec{a}_E \times \vec{a}_H$$

$$= \frac{1}{2} E_0 H_0 (\cos\theta \mp j \sin\theta) \, \vec{a}_E \times \vec{a}_H\ . \tag{8.99}$$

Taking the real part of (8.99), we then have

$$\text{Re}\, \vec{P}_{\text{complex}} = \text{Re}\, \frac{1}{2} \vec{E} \times \vec{H}^* = \frac{1}{2} E_0 H_0 \cos\theta \, \vec{a}_E \times \vec{a}_H\ . \tag{8.100}$$

Similarly, taking the imaginary part of (8.99), we have

$$\text{Im}\, \vec{P}_{\text{complex}} = \text{Im}\, \frac{1}{2} \vec{E} \times \vec{H}^* = \mp \frac{1}{2} E_0 H_0 \sin\theta \, \vec{a}_E \times \vec{a}_H\ . \tag{8.101}$$

Therefore, comparing (8.100) with (8.95), we obtain the average Poynting vector as

$$\vec{P}_{\text{average}} = \text{Re}\, \vec{P}_{\text{complex}} = \frac{1}{2} \text{Re}\, \vec{E} \times \vec{H}^*. \tag{8.102}$$

The expression for the average Poynting vector (8.102) contributed by the real part of the complex Poynting vector (8.100) is used in the study of power flow problems in electromagnetic engineering. (We will elaborate this point further in Section 8.2.3.) The imaginary part of the complex Poynting vector given by (8.101) does not contribute to the average Poynting vector [as explained following (8.94)]; however, we can define the reactive power density vector $\vec{P}_{\text{reactive}}$ in terms of this imaginary part on the lines of (8.102) as follows:

$$\vec{P}_{\text{reactive}} = \text{Im} \vec{P}_{\text{complex}} = \frac{1}{2} \text{Im} \vec{E} \times \vec{H}^* = \mp \frac{1}{2} E_0 H_0 \sin\theta \, \vec{a}_E \times \vec{a}_H \quad . \tag{8.103}$$

We can depict the power density vectors in a triangle on a complex plane, in which the hypotenuse of the triangle represents the complex power density vector, also referred to as the virtual power density vector, the magnitude of which represents the apparent power density; the side of the triangle on the real axis represents the average power density vector or the real power density vector; and the side of the triangle on the imaginary axis represents the reactive power density vector (Fig. 8.2). Thus, we can see from this vector triangle that the following vector identity holds good (Fig. 8.4):

$$\overrightarrow{\text{AC}} = \overrightarrow{\text{AB}} + \overrightarrow{\text{BC}} \tag{8.104}$$

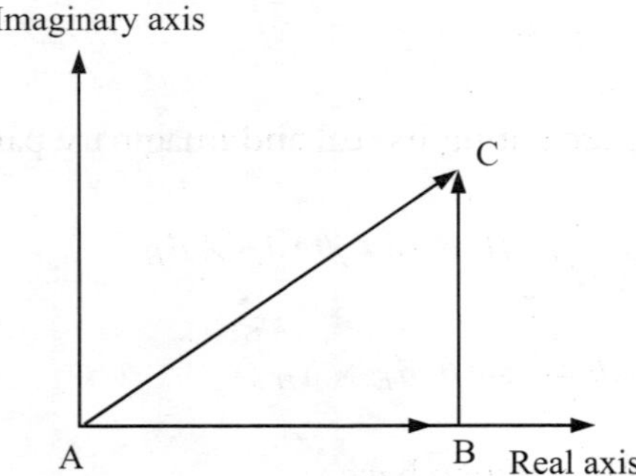

Fig. 8.4 *Power density triangle.*

where $\overrightarrow{AC}$ is the complex power density vector given by (8.99); $\overrightarrow{AB}$ is the average power density vector given by (8.95); and $\overrightarrow{BC}$ is the reactive power density vector given by (8.103).

Finally, since with reference to time-harmonic phenomena, the complex Poynting vector $\vec{P}_{\text{complex}}(= 1/2\vec{E} \times \vec{H}^*)$ is more relevant, let us now proceed to reformulate the concept of energy balance to derive the complex Poynting theorem. For this purpose, let us recall the following relation introduced preceding (8.44):

$$\nabla.(\vec{E} \times \vec{H}) = \vec{H}.\nabla \times \vec{E} - \vec{E} \cdot \nabla \times \vec{H},$$

which by replacing $\vec{H}$ by $\vec{H}^*$ can be written as

$$\nabla.(\vec{E} \times \vec{H}^*) = \vec{H}^*.\nabla \times \vec{E} - \vec{E} \cdot \nabla \times \vec{H}^*. \tag{8.105}$$

We can now express (8.105) with the help of (8.45) and the following complex conjugate of (8.46):

$$\nabla \times \vec{H}^* = \vec{J}_c^* + \frac{\partial \vec{D}^*}{\partial t},$$

which in view of the relation $\vec{D} = \varepsilon\vec{E}$ can be put as

$$\nabla \times \vec{H}^* = \vec{J}_c^* + \varepsilon\frac{\partial \vec{E}^*}{\partial t}. \tag{8.106}$$

Thus, substituting (8.45) and (8.106) in (8.105), we obtain

$$\nabla.(\vec{E} \times \vec{H}^*) = -\vec{H}^*.\frac{\partial \vec{B}}{\partial t} - \vec{E} \cdot \left(\vec{J}_c^* + \varepsilon\frac{\partial \vec{E}^*}{\partial t}\right),$$

which, in view of the relation $\vec{B} = \mu\vec{H}$, can be expressed as

$$\nabla.(\vec{E}\times\vec{H}^*) = -\mu\,\vec{H}^*.\frac{\partial\vec{H}}{\partial t} - \varepsilon\,\vec{E}\cdot\frac{\partial\vec{E}^*}{\partial t} - \vec{E}\cdot\vec{J}_c^*. \tag{8.107}$$

Taking the time dependence of the field quantities as $\exp(j\omega\,t)$, for instance, as in (8.97), we can write

$$\left.\begin{aligned}
&\vec{E} = E_0\exp(j\omega\,t)\,\vec{a}_E\\
&\vec{E}^* = E_0\exp(-j\omega\,t)\,\vec{a}_E\\
&\frac{\partial\vec{E}^*}{\partial t} = -j\omega\,E_0\exp(-j\omega\,t)\,\vec{a}_E = -j\omega\,\vec{E}^*\\
&\vec{H} = H_0\exp j(\omega\,t\pm\theta)\,\vec{a}_H\\
&\vec{H}^* = H_0\exp -j(\omega\,t\pm\theta)\,\vec{a}_H\\
&\frac{\partial\vec{H}}{\partial t} = j\omega\,H_0\exp j(\omega\,t\pm\theta)\,\vec{a}_H = j\omega\,\vec{H}]
\end{aligned}\right\}. \tag{8.108}$$

We can then express (8.107) with the help of (8.108) as

$$\begin{aligned}
\nabla.(\vec{E}\times\vec{H}^*) &= -j\omega\mu\,\vec{H}^*.\vec{H} + j\omega\varepsilon\,\vec{E}\cdot\vec{E}^* - \vec{E}\cdot\vec{J}_c^*\\
&= -j\omega(\mu\,\vec{H}^*.\vec{H} - \varepsilon\,\vec{E}\cdot\vec{E}^*) - \vec{E}\cdot\vec{J}_c^*.
\end{aligned} \tag{8.109}$$

Taking the volume integral of (8.109), we then obtain

$$\int_\tau \nabla.(\vec{E}\times\vec{H}^*)d\tau = -j\omega\int_\tau(\mu\,\vec{H}^*.\vec{H} - \varepsilon\,\vec{E}\cdot\vec{E}^*)d\tau - \int_\tau \vec{E}\cdot\vec{J}_c^* d\tau\,. \tag{8.110}$$

Now, applying Gauss's divergence theorem, we can convert the volume integral On the left-hand side of (8.110) into a surface integral [as we did earlier to convert the volume integral on the left-hand side of (8.48)] as follows:

$$\oint_S(\vec{E}\times\vec{H}^*)\cdot d\vec{S} = -j\omega\int_\tau(\mu\,\vec{H}^*\cdot\vec{H} - \varepsilon\,\vec{E}\cdot\vec{E}^*)d\tau - \int_\tau\vec{E}\cdot\vec{J}_c^* d\tau,$$

which, divided by 2 and making use of the relation (8.96) gives

$$\begin{aligned}
\oint_S\frac{1}{2}(\vec{E}\times\vec{H}^*)\cdot d\vec{S} &= \oint_S\vec{P}_{\text{complex}}\cdot d\vec{S}\\
&= -j\omega\int_\tau\frac{1}{2}(\mu\,\vec{H}^*\cdot\vec{H} - \varepsilon\,\vec{E}\cdot\vec{E}^*)d\tau - \int_\tau\frac{1}{2}\vec{E}\cdot\vec{J}_c^* d\tau\,.
\end{aligned} \tag{8.111}$$

Further, by changing the sign of (8.111) we can write

$$-\oint_S \frac{1}{2}(\vec{E}\times\vec{H}^*)\cdot d\vec{S} = -\oint_S \vec{P}_{\text{complex}}\cdot d\vec{S}$$

$$= j\omega\int_\tau \frac{1}{2}(\mu\,\vec{H}^*\cdot\vec{H} - \varepsilon\,\vec{E}\cdot\vec{E}^*)d\tau + \int_\tau \frac{1}{2}\vec{E}\cdot\vec{J}_c^*\,d\tau. \tag{8.112}$$

In (8.111) or (8.112), we thus obtain the complex Poynting vector theorem. Taking the real part of (8.112), we then obtain

$$-\oint_S \frac{1}{2}Re\,(\vec{E}\times\vec{H}^*)\cdot d\vec{S} = -\oint_S \text{Re}\,\vec{P}_{\text{complex}}\cdot d\vec{S} = \int_\tau \frac{1}{2}\vec{E}\cdot\vec{J}_c^*\,d\tau \tag{8.113}$$

which, making use of the relation (8.102) can also be put as

$$-\oint_S \vec{P}_{\text{average}}\cdot d\vec{S} = \int_\tau \frac{1}{2}\vec{E}\cdot\vec{J}_c^*\,d\tau \tag{8.114}$$

which is an expression for the balance of real power. We notice that for the case of time-harmonic fields considered, there cannot be an increase in energy stored in electric and magnetic fields, and, therefore, the energy storage terms are absent from (8.113). The left-hand side of (8.113) represents the average power entering thé volume enclosure over which the surface integral is taken. With a change of sign (from negative to positive), therefore, the left hand side of (8.113) will represent the average power going out of the volume enclosure. We use this result in numerous electromagnetic engineering problems, for instance, in finding radiation characteristics of an antenna (Section 8.4) and power transmitted through a waveguide.

Similarly, taking the imaginary part of (8.112), we obtain

$$-\oint_S \frac{1}{2}\text{Im}\,(\vec{E}\times\vec{H}^*)\cdot d\vec{S} = -\oint_S \text{Im}\,\vec{P}_{\text{complex}}\cdot d\vec{S} = \omega\int_\tau \frac{1}{2}(\mu\,\vec{H}^*\cdot\vec{H} - \varepsilon\,\vec{E}\cdot\vec{E}^*)d\tau \tag{8.115}$$

which, making use of the relation (8.103) can also be expressed as

$$-\oint_S \vec{P}_{\text{reactive}}\cdot d\vec{S} = \omega\int_\tau \frac{1}{2}(\mu\,\vec{H}^*\cdot\vec{H} - \varepsilon\,\vec{E}\cdot\vec{E}^*)d\tau.$$

A simple algebraic manipulation of the right-hand side of the above will give

$$\oint_S \vec{P}_{\text{reactive}}\cdot d\vec{S} = 2\omega\int_\tau \left[\frac{1}{2}\left(\frac{1}{2}\mu\,\vec{H}^*\cdot\vec{H}\right) - \frac{1}{2}\left(\frac{1}{2}\varepsilon\,\vec{E}\cdot\vec{E}^*\right)\right]d\tau \tag{8.116}$$

which essentially is an expression for the balance of reactive power.

Thus, in (8.114) and (8.116) put together we have an alternative form of the complex vector theorem stated by (8.112). The left-hand side of (8.114) represents the average power entering the volume enclosure. In order to appreciate the right-hand side of (8.114), let us proceed to substitute $\vec{E}$ by the time periodic function from (8.108) in the quantity $\frac{1}{2}\vec{E}\cdot\vec{J}_c^*$. We can write the quantity, in view of the ohmic law relation $\vec{J}_c = \sigma\vec{E}$ and hence its complex conjugate $\vec{J}_c^* = \sigma\vec{E}^*$, as:

$$\frac{1}{2}\vec{E}\cdot\vec{J}_c^* = \frac{1}{2}\vec{E}\cdot\sigma\vec{E}^* = \frac{1}{2}E_0\exp(j\omega t)\,\vec{a}_E\cdot\sigma E_0\exp(-j\omega t)\,\vec{a}_E$$
$$= \frac{1}{2}\sigma E_0^2\,\vec{a}_E\cdot\vec{a}_E = \frac{1}{2}\sigma E_0^2. \tag{8.117}$$

At this juncture, let us recall the discussion following (8.56) with reference to the problem of a wire carrying a dc current, in which we mentioned that the quantity σE^2 represents the power lost per unit volume of the wire (ohmic loss). For a time-periodic electric field, we need to evaluate the average of the power lost per unit volume of the quantity σE^2 over a cycle, that is, find $\langle\sigma E^2\rangle_{\text{average}}$, where the magnitude of the instantaneous time-periodic electric field is E. Representing the magnitude of such field as $E = E_0\sin\omega t$ [see for instance (8.84)], we can then find $\langle\sigma E^2\rangle_{\text{average}}$ as follows:

$$\langle\sigma E^2\rangle_{\text{average}} = \langle\sigma E_0^2\sin^2\omega t\rangle_{\text{average}} = \frac{\int_{\omega t=0}^{2\pi}\sigma E_0^2\sin^2\omega t\,d(\omega t)}{\int_{\omega t=0}^{2\pi}d(\omega t)} = \sigma E_0^2\frac{\int_{\omega t=0}^{2\pi}\sin^2\omega t\,d(\omega t)}{\int_{\omega t=0}^{2\pi}d(\omega t)}$$

$$= \sigma E_0^2\frac{\int_{\omega t=0}^{2\pi}\frac{1-\cos 2\omega t}{2}\,d(\omega t)}{\int_{\omega t=0}^{2\pi}d(\omega t)} = \sigma E_0^2\frac{\left[\frac{1}{2}\int_{\omega t=0}^{2\pi}d(\omega t) - \frac{1}{2}\int_{\omega t=0}^{2\pi}\cos 2\omega t\,d(\omega t)\right]}{\int_{\omega t=0}^{2\pi}d(\omega t)}$$

$$= \sigma E_0^2\frac{\frac{1}{2}[\omega t]_0^{2\pi} - \frac{1}{2}\left[\frac{\sin 2\omega t}{2\omega}\right]_0^{2\pi}}{[\omega t]_0^{2\pi}} = \sigma E_0^2\frac{\frac{1}{2}(2\pi-0) - \frac{1}{4\omega}(\sin 4\pi - \sin 0)}{2\pi - 0}$$

$$= \sigma E_0^2\frac{\left[\frac{1}{2}2\pi - \frac{1}{4\omega}(0-0)\right]}{2\pi - 0} = \frac{1}{2}\sigma E_0^2. \tag{8.118}$$

Thus, expression (8.118) for the average power lost per unit volume can be rewritten as

$$\langle\sigma E^2\rangle_{\text{average}} = \frac{1}{2}\sigma E_0^2. \tag{8.119}$$

Then comparing (8.117) with (8.119), we obtain

$$\frac{1}{2}\vec{E}\cdot\vec{J}_c^* = \frac{1}{2}\sigma E_0^2 = \langle\sigma E^2\rangle_{\text{average}} = \text{average power lost per unit volume.} \tag{8.120}$$

In view of (8.120), we can therefore easily appreciate that the right-hand side of (8.114) represents the average ohmic loss of power in the volume enclosure, while, as we have already stated, the left-hand side of (8.114) represents the average power entering the volume enclosure. Therefore, it follows from (8.114) that the average power entering a volume enclosure is equal to the average ohmic power loss in the volume.

Similarly, the left-hand side of (8.114) represents the average reactive power entering the volume enclosure. Let us next examine next the quantity $(1/2)\mu\,\vec{H}^*\cdot\vec{H}$ that occurs in the integrand of the volume integral of the right-hand side of (8.116). The quantity in phasor representation of time-harmonic magnetic field, with the help of (8.108) (taking $\theta = 0$ for the sake of simplicity of analytical understanding) can be expressed as

$$\frac{1}{2}\mu\,\vec{H}^*\cdot\vec{H} = \frac{1}{2}\mu\,H_0\exp(-j\omega\,t)\,\vec{a}_H\cdot H_0\exp(j\omega\,t)\,\vec{a}_H$$
$$= \frac{1}{2}\mu\,H_0^2\,\vec{a}_H\cdot\vec{a}_H = \frac{1}{2}\mu\,H_0^2\,. \tag{8.121}$$

Further, for a static magnetic field, recall (8.32) and rewrite it as follows:

$$U_B = \frac{1}{2}\vec{B}\cdot\vec{H},$$

which, in view of the relation $\vec{B} = \mu\,\vec{H}$, can also be put as

$$U_B = \frac{1}{2}\mu\,\vec{H}\cdot\vec{H} = \frac{1}{2}\mu\,H^2. \tag{8.122}$$

Correspondingly, for a time-harmonic magnetic field as such as that in (8.84), we can find the instantaneous energy density in magnetic field U_B by putting $H_0\sin\omega\,t$ in (8.122) as

$$U_B = \frac{1}{2}\mu\,H^2 = \frac{1}{2}\mu\,H_0^2\sin^2\omega t. \tag{8.123}$$

We can average (8.123) over a cycle as

$$\langle U_B\rangle_{\text{average}} = \left\langle\frac{1}{2}\mu\,H_0^2\sin^2\omega\,t\right\rangle_{\text{average}} = \frac{1}{2}\mu\,H_0^2\frac{\int\limits_{\omega t=0}^{2\pi}\sin^2\omega\,t\,d(\omega\,t)}{\int\limits_{\omega t=0}^{2\pi}d(\omega\,t)}$$

which can be evaluated, using exactly the same steps as those followed while deducing (8.118), as:

$$\langle U_B\rangle_{\text{average}} = \frac{1}{2}\frac{1}{2}\mu\,H_0^2. \tag{8.124}$$

Comparing (8.124) with (8.121), we can see that the following relation holds good:

$$\langle U_B \rangle_{\text{average}} = \frac{1}{2}\frac{1}{2}\mu\, \vec{H}^* \cdot \vec{H}. \tag{8.125}$$

Similarly, starting from the interpretation of the expression (8.21) for the energy density in electric field and making use of the relation $\vec{D} = \varepsilon\, \vec{E}$, we can deduce the relation corresponding to (8.125) for a time-harmonic electric field as follows:

$$\langle U_E \rangle_{\text{average}} = \frac{1}{2}\frac{1}{2}\varepsilon\, \vec{E} \cdot \vec{E}^*. \tag{8.126}$$

Plugging (8.125) and (8.126) in (8.116), we then obtain

$$-\oint_S \vec{P}_{\text{reactive}} \cdot d\vec{S} = 2\omega \int_\tau (\langle U_B \rangle_{\text{average}} - \langle U_E \rangle_{\text{average}})d\tau\,. \tag{8.127}$$

It follows from (8.127) that the reactive power flowing into a volume enclosure is equal to 2ω times the difference between the average energies stored in electric and magnetic fields in the volume. Thus, from the foregoing analysis we obtain, in addition to (8.111) or (8.112), two other forms of the complex Poynting theorem, namely (8.114) and (8.127) by separating the real and imaginary terms of (8.112).

8.2.3 Poynting theorem in the presence of an external source

Let us now review the Poynting theorem in the presence of external sources. We consider two situations: one in which the volume enclosure behaves as a 'transmitter' and the other in which it behaves as a 'receiver'.

Let us first take the case of the volume enclosure as a transmitter containing external sources. In order to take into account the effect of the external current due to this source, modify the third term of (8.52) by adding $\vec{J}_{\text{ext}}$ to $\vec{J}_c$ as follows:

$$-\oint_S \vec{P} \cdot d\vec{S} = \frac{\partial W_E}{\partial t} + \frac{\partial W_B}{\partial t} + \int_\tau \vec{E} \cdot (\vec{J}_c + \vec{J}_{\text{ext}})\, d\tau$$

which can be put as

$$-\oint_S \vec{P} \cdot d\vec{S} = \frac{\partial W_E}{\partial t} + \frac{\partial W_B}{\partial t} + \int_\tau \vec{E} \cdot \vec{J}_c\, d\tau + \int_\tau \vec{E} \cdot \vec{J}_{\text{ext}}\, d\tau \tag{8.128}$$

where $\vec{J}_{\text{ext}}$ is the external or the source current density. The said modification of the third term of (8.52) is due to one of the basic equations, namely (8.46) [used in the deduction of (8.52)] being modified by the addition of $\vec{J}_{\text{ext}}$ to $\vec{J}_c$.

Further, by rearranging the terms of (8.126) we can express it as

$$-\int_\tau \vec{E} \cdot \vec{J}_{\text{ext}}\, d\tau = \oint_S \vec{P} \cdot d\vec{S} + \frac{\partial W_E}{\partial t} + \frac{\partial W_B}{\partial t} + \int_\tau \vec{E} \cdot \vec{J}_c\, d\tau\,. \tag{8.129}$$

The left-hand side of (8.127) represents the 'source term'. We have already explained [see the explanation following (8.52)] that the first term of (8.129) of the right-hand side represents the instantaneous power flowing out of the volume enclosure; the second term the time rate of increase of energy stored in electric field in the volume, the third term the time rate of increase of energy stored in magnetic field in the volume, and the fourth term the power lost in any conducting medium present in the medium. All the terms of the right-hand side of (8.129) being each positive, the right-hand side of (8.129) is a positive quantity, which, therefore, also implies that the left-hand side of (8.129) representing the source term is a positive quantity meaning thereby that $\vec{E}$ and $\vec{J}_{\text{ext}}$ are oppositely directed.

Now, let us take the volume enclosure as a receiver containing no external sources such that in this case (8.52) continues to hold good and the left-hand side of (8.52) now stands for the source term representing the instantaneous power flow into the volume enclosure, all the terms of the right-hand side continuing to have the same interpretation, as explained following (8.52).

For time-periodic fields, it will be of interest to modify (8.114) considering the sources inside the volume enclosure as by adding $\vec{J}_{\text{ext}}$ to $\vec{J}_c$ as follows:

$$-\oint_S \vec{P}_{\text{average}} \cdot d\vec{S} = \int_\tau \frac{1}{2}\vec{E} \cdot (\vec{J}_c^* + \vec{J}_{\text{ext}}^*) d\tau = \frac{1}{2}\int_\tau \vec{E} \cdot \vec{J}_c^* d\tau + \int_\tau \frac{1}{2}\vec{E} \cdot \vec{J}_{\text{ext}}^* d\tau$$

which can be rearranged as

$$-\int_\tau \frac{1}{2}\vec{E} \cdot \vec{J}_{\text{ext}}^* d\tau = \oint_S \vec{P}_{\text{average}} \cdot d\vec{S} + \int_\tau \frac{1}{2}\vec{E} \cdot \vec{J}_c^* d\tau \tag{8.130}$$

where $\vec{P}_{\text{average}} = (1/2)\text{Re}\,\vec{E} \times \vec{H}^*$ [see (8.102)]. The left-hand side of (8.130), interpreted to have a positive value implying that $\vec{E}$ and $\vec{J}_{\text{ext}}$ are oppositely directed, represents the time-averaged source power within the volume enclosure. The first term of the right-hand side of (8.130) represents the average power flowing out of the volume enclosure and its second term represents the average ohmic power loss within the volume.

The power balance expression (8.130) finds extensive applications in electromagnetic problems. For instance, we can interpreting the left-hand side of the expression (8.130) as the power fed into an antenna, and we can use this expression to find the power radiated by an antenna in free space given by the first term of the right-hand side of (8.130), the second term (representing the ohmic power loss) of which becomes zero since the conduction current density $\vec{J}_c = 0$ in free space that does not have a conducting medium to support $\vec{J}_c$. In fact, in the chapter to follow (Chapter 9), we are going to use this expression to estimate the power transmitted through a hollow-pipe metal waveguide. However, in the section to follow (Section 8.3), we are going to develop an I^2R- type ohmic loss concept for the average power loss of electromagnetic waves at a conducting boundary.

8.2.4 Poynting theorem in different versions revisited

It worth putting together and summarizing the various versions of Poynting theorem obtained in this section for the benefit of readers as follows.

(i) Expression (8.52):

$$-\oint_S \vec{P} \cdot d\vec{S} = \frac{\partial W_E}{\partial t} + \frac{\partial W_B}{\partial t} + \int_\tau \vec{E} \cdot \vec{J}_c \, d\tau :$$

It is the Poynting theorem involving instantaneous Poynting vector. The left-hand side

$$-\oint_S \vec{P} \cdot d\vec{S}$$

represents the instantaneous power going into a closed volume. With a change of sign, $\oint_S \vec{P} \cdot d\vec{S}$ represents the instantaneous power going out of a closed volume [the term being identified as the left-hand side of (8.51)].

(ii) Expression (8.112):

$$-\oint_S \frac{1}{2}(\vec{E} \times \vec{H}^*) \cdot d\vec{S} = -\oint_S \vec{P}_{\text{complex}} \cdot d\vec{S} = j\omega \int_\tau \frac{1}{2}(\mu \vec{H}^* \cdot \vec{H} - \varepsilon \vec{E} \cdot \vec{E}^*) d\tau + \int_\tau \frac{1}{2} \vec{E} \cdot \vec{J}_c^* d\tau :$$

It is the complex Poynting vector theorem for time-periodic electric and magnetic fields. It has a valid interpretation for its real and imaginary parts.

(iii) Expression (8.113) or (8.114):

$$-\oint_S \frac{1}{2}\text{Re}\,(\vec{E} \times \vec{H}^*) \cdot d\vec{S} = -\oint_S \text{Re}\,\vec{P}_{\text{complex}} \cdot d\vec{S} = \int_\tau \frac{1}{2} \vec{E} \cdot \vec{J}_c^* d\tau$$

or

$$-\oint_S \vec{P}_{\text{average}} \cdot d\vec{S} = \int_\tau \frac{1}{2} \vec{E} \cdot \vec{J}_c^* d\tau :$$

It is the real part of the complex Poynting vector theorem for time-periodic electric and magnetic fields. It is the expression for the balance of real power. The left-hand side

$$-\oint_S 1/2\text{Re}\,(\vec{E} \times \vec{H}^*) \cdot d\vec{S} \text{ or } -\oint_S \vec{P}_{\text{average}} \cdot d\vec{S}$$

represents the average power entering the volume enclosure. With a change of sign, $\oint_S \vec{P}_{\text{average}} \cdot d\vec{S}$ represents the average power going out of the volume enclosure.

(iv) Expression (8.115) or (8.116):

$$-\oint_S \frac{1}{2}\mathrm{Im}\,(\vec{E}\times\vec{H}^*)\cdot d\vec{S} = -\oint_S \mathrm{Im}\,\vec{P}_{\text{complex}}\cdot d\vec{S} = \omega\int_\tau \frac{1}{2}(\mu\,\vec{H}^*\cdot\vec{H} - \varepsilon\,\vec{E}\cdot\vec{E}^*)d\tau$$

or

$$-\oint_S \vec{P}_{\text{reactive}}\cdot d\vec{S} = 2\omega\int_\tau \left[\frac{1}{2}\left(\frac{1}{2}\mu\,\vec{H}^*\cdot\vec{H}\right) - \frac{1}{2}\left(\frac{1}{2}\varepsilon\,\vec{E}\cdot\vec{E}^*\right)\right]d\tau :$$

It is the imaginary part of the complex Poynting vector theorem for time-periodic electric and magnetic fields. It is the expression for the balance of reactive power.

(v) Expression (8.129) in the presence of external sources:

$$-\int_\tau \vec{E}\cdot\vec{J}_{\text{ext}}\,d\tau = \oint_S \vec{P}\cdot d\vec{S} + \frac{\partial W_E}{\partial t} + \frac{\partial W_B}{\partial t} + \int_\tau \vec{E}\cdot\vec{J}_c\,d\tau :$$

It is the Poynting theorem involving the instantaneous Poynting vector for external sources within the volume enclosure behaving as a transmitter. The left-hand side

$$-\int_\tau \vec{E}\cdot\vec{J}_{\text{ext}}\,d\tau$$

is the source term. The first term of the right-hand side

$$\oint_S \vec{P}\cdot d\vec{S}$$

represents the instantaneous power flowing out of the volume enclosure.

(vi) Expression (8.52) for no external sources within the volume enclosure:

$$-\oint_S \vec{P}\cdot d\vec{S} = \frac{\partial W_E}{\partial t} + \frac{\partial W_B}{\partial t} + \int_\tau \vec{E}\cdot\vec{J}_c\,d\tau :$$

It is the Poynting theorem involving the instantaneous Poynting vector in the absence of external sources inside the volume enclosure, the latter behaving as a receiver. The left-hand side

$$-\oint_S \vec{P}\cdot d\vec{S}$$

stands for the source term representing the instantaneous power flow into the volume enclosure.

(vii) Expression (8.130):

$$-\int_{\tau}\frac{1}{2}\vec{E}\cdot\vec{J}^*_{\text{ext}}d\tau = \oint_{S}\vec{P}_{\text{average}}\cdot d\vec{S} + \int_{\tau}\frac{1}{2}\vec{E}\cdot\vec{J}^*_c d\tau \text{ (with } \vec{P}_{\text{average}} = \frac{1}{2}\text{Re}\vec{E}\times\vec{H}^*\text{):}$$

It is the real part of the complex Poynting vector theorem for time-periodic electric and magnetic fields for external sources within the volume enclosure, the latter behaving as a transmitter. The left-hand side

$$-\int_{\tau}\frac{1}{2}\vec{E}\cdot\vec{J}^*_{\text{ext}}d\tau$$

represents the time-averaged source power within the volume enclosure, the right-hand side representing the average power flowing out of the volume enclosure together with the average ohmic power loss, if any, within the volume.

8.3 Power Loss in Electromagnetic Wave Propagation at a Conducting Boundary

We have already developed the concept of the skin depth of a medium of finite resistivity while studying the propagation of uniform plane electromagnetic waves through such a medium (Section 6.3). We also introduced surface resistance of a conductor in terms of the medium conductivity and the wave frequency (Section 8.4). With this background, we now develop in this section the concept of power loss in electromagnetic wave propagation.

For this purpose, let us consider a uniform, plane electromagnetic wave propagating along z be normally incident from a free-space medium 2 on a semi-infinite planar conducting medium 1, the latter extending from $z = 0$ to $z = \infty$ (Fig. 8.5). (The meaning of a uniform, plane wave has been explained in Section 6.2). Let the electric field $\vec{E} = E_x\vec{a}_x$ and the magnetic field $\vec{H} = H_y\vec{a}_y$ be directed along x and y , respectively, such that the Poynting vector $\vec{E}\times\vec{H}(= E_xH_y\vec{a}_x\times\vec{a}_y = E_xH_y\vec{a}_z)$ is directed along z. We can then write the average power density or the average Poynting vector in the free-space (medium 2)at the interface between the regions 2 and 1, with the help of (8.102) as

$$\vec{P}_{\text{average}} = \frac{1}{2}\,\text{Re}\,\vec{E}\times\vec{H}^* = \frac{1}{2}\,\text{Re}\,E_xH_y^*\vec{a}_x\times\vec{a}_y = \frac{1}{2}\,\text{Re}\,E_xH_y^*\vec{a}_z. \tag{8.131}$$

In this problem we get another application of the power balance expression (8.130). Here, we can interpret the left-hand side of the expression (8.130) as the power fed into medium 2 (free space) and we can use this expression to find the average power P_{average} leaving the free-space medium 2 to enter the interface between medium 2 and medium 1 in the direction of z through a surface areaof the interface between the two media. P_{average} is given by the first term of the right-hand side of (8.130), the second term (representing the ohmic power loss) of which becomes zero since the conduction current density $\vec{J}_c = 0$ in free space (medium 2) that does not have a conducting medium to support $\vec{J}_c = 0$. Thus, we obtain:

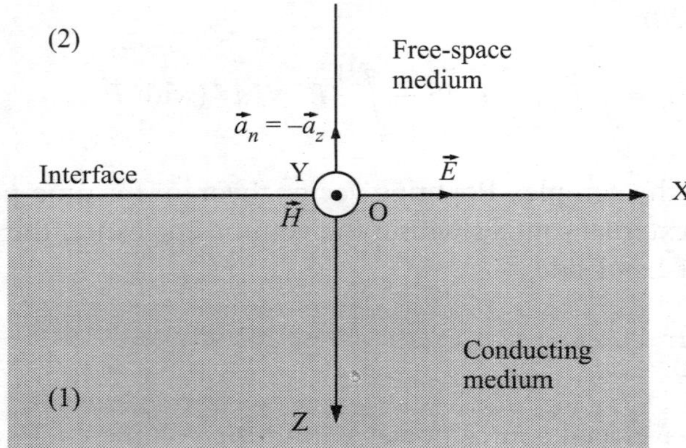

Fig. 8.5 *Incidence of a uniform plane wave, propagating along z direction, at the interface ($z = 0$) between semi-infinite conducting region 1 and a free-space region 2, showing the electric and magnetic fields along x and y directions, respectively, the unit vector $\vec{a}_n (= -a_z)$ taken as directed from the region 1 to 2 in the negative z direction.*

$$P_{\text{average}} = \int_S \vec{P}_{\text{average}} \cdot d\vec{S} = \int_S \vec{P}_{\text{average}} \cdot dS\,\vec{a}_z$$

in which we can substitute (8.1131) to obtain the following expression for the average power entering the conducting medium through the area S of the conducting surface:

$$P_{\text{average}} = \int_S \tfrac{1}{2}\text{Re}\, E_x H_y^* \vec{a}_z \cdot dS\,\vec{a}_z = \tfrac{1}{2}\text{Re}\, E_x H_y^* \int_S dS\,\vec{a}_z \cdot \vec{a}_z$$

$$= \frac{1}{2}\text{Re}\, E_x H_y^* \int_S dS = \frac{1}{2}\text{Re}\, E_x H_y^* S \tag{8.132}$$

where we have taken the field quantities as constants over a surface perpendicular to the direction of propagation z since uniform plane waves are considered (see Section 6.3) and taken outside the surface integral. Further, the power entering the conducting medium will be an ohmic power loss. We can find P_{LA}, the power loss per unit area of the surface perpendicular to the direction of propagation, by dividing (8.132) by the area S as

$$P_{\text{LA}} \left(= \frac{P_{\text{average}}}{S}\right) = \frac{1}{2}\text{Re}\, E_x H_y^* \,. \tag{8.133}$$

For a wave propagating through an unbounded medium, we can relate E_x to H_y through the intrinsic impedance of the medium. For a free-space medium, we have this relation as [see (6.70)]

$$\eta_0 = \frac{E_x}{H_y} = \left(\frac{\mu_0}{\varepsilon_0}\right)^{1/2} = \left(\frac{j\omega\mu_0}{j\omega\varepsilon_0}\right)^{1/2} \tag{8.134}$$

where η_0 is the intrinsic impedanceof free space and $\omega = 2\pi f$ is the wave angular frequency, f being the circular frequency. It may also be recalled that, while deducing (8.134) (which is (6.63*b*) rewritten), we had ignored the conduction current density and taken $\vec{J} = \sigma\vec{E} = 0$, since $\sigma = 0$ for free space (see Section 6.2). Further, by repeating the procedure already laid down in Section 6.2, however, now by taking $\sigma \neq 0$ for a conducting medium, we can easily modify (8.134) and deduce the following relation for the intrinsic impedance η of a conducting medium:

$$\eta = \frac{E_x}{H_y} = \left(\frac{j\omega\mu_0}{\sigma + j\omega\varepsilon_0}\right)^{1/2}. \tag{8.135}$$

For a good conductor, we can take $\sigma >> j\omega\mu_0$ in (8.135) and write it as

$$\eta = \frac{E_x}{H_y} = \left(\frac{j\omega\mu_0}{\sigma}\right)^{1/2}$$

which can be separated into its real and imaginary parts as follows (see Appendix A8.1):

$$\eta = \frac{E_x}{H_y} = \left(\frac{j\omega\mu_0}{\sigma}\right)^{1/2} = \left(\frac{\omega\mu_0}{2\sigma}\right)^{1/2} + j\left(\frac{\omega\mu_0}{2\sigma}\right)^{1/2}. \tag{8.136}$$

Further, recall (6.99) and rewrite it as

$$R_S = X_S = \sqrt{\frac{\pi f\mu_0}{\sigma}} = \sqrt{\frac{2\pi f\mu_0}{2\sigma}} = \sqrt{\frac{\omega\mu_0}{2\sigma}} \tag{8.137}$$

where R_S and X_S are respectively the surface resistance and the surface reactance of the conducting medium. Therefore, comparing (8.136) and (8.137) we obtain the relation

$$\eta = \frac{E_x}{H_y} = R_S + jX_S\,. \tag{8.138}$$

We can then substitute (8.138) in (8.133) to write the following expression for P_{LA}:

$$\begin{aligned} &= \frac{1}{2}\mathrm{Re}\; E_x H_y^* = \frac{1}{2}\mathrm{Re}\;\eta H_y H_y^* \\ &= \frac{1}{2}\mathrm{Re}\;(R_S + jX_S)H_y H_y^* = \frac{1}{2}R_S H_y H_y^* \end{aligned} \tag{8.139}$$

where the field H_y has to be interpreted as the quantity in free space at the conducting boundary between the regions 1 and 2.

We can find an alternative expression for P_{LA} in terms of the surface current density $\vec{J}_s$ at the conducting boundary, that is, at the interface between the free-space and the conducting media with the help of the electromagnetic boundary condition (7.46*c*), which can be rewritten as

$$\vec{a}_n \times \vec{H}_2 = \vec{J}_s \tag{8.140}$$

where $\vec{H}_2$ is the magnetic field in the region 2 (free-space) and $\vec{a}_n$ is a unit vector directed from the region 1 to 2 (Fig. 8.5). However, we have to read (8.140) in the present context, taking $\vec{a}_n = -\vec{a}_z$ and $\vec{H}_2 = H_y\vec{a}_y$ (Fig. 8.5) as

$$-\vec{a}_z \times H_y\vec{a}_y = -H_y\vec{a}_z \times \vec{a}_y = H_y\vec{a}_x = \vec{J}_s \,. \tag{8.141}$$

We can then use (8.141) to constitute the following relation

$$\vec{J}_s \cdot \vec{J}_s^* = H_y\vec{a}_x \cdot H_y^*\vec{a}_x = H_yH_y^*\vec{a}_x \cdot \vec{a}_x = H_yH_y^*$$

and use it to express (8.139) as

$$P_{\text{LA}} = \frac{1}{2}R_S\vec{J}_s \cdot \vec{J}_s^* \,. \tag{8.142}$$

The expression (8.142) for power loss per unit area of a conducting interface has been used in the chapter to follow (Chapter 9) to find the attenuation of electromagnetic waves propagating through a hollow-pipe metal waveguide due to the finite resistivity of the metal used in making the waveguide.

EXAMPLE E8-3

What is the incident power density and the power absorbed per unit area in a sheet of brass ($\sigma = 1.5 \times 10^7$mho/m) on which a uniform plane wave is incident with a peak electric field of 1 V/cm at 10 GHz?

SOLUTION:

For the electric field directed along x and the magnetic field along y, the amplitude of the electric field E_{x0} can be related to the amplitude of the magnetic field H_{y0} with the help of (8.134)

$$H_{y0} = \frac{E_{x0}}{\eta_0}, \tag{E8-3.1}$$

where $\eta_0 = 120\pi = 377$ ohm is the intrinsic impedance of free space.

Substituting $E_{x0} = 1$ V/cm (given) $= 100$ V/m in (E8.3.1), we then have

$$H_{y0} = \frac{100}{377} \text{ A/m.} \tag{E8-3.2}$$

The incident power density (see Section 8.2.2) can be written, with the help of (E8-3.1) and (E8-3.2) as

$$\text{Incident power density} = \frac{1}{2}E_{x0}H_{y0} = \frac{1}{2}\frac{(100)^2}{377} = 13.26 \text{ W/m}^2. \qquad \text{(E8-3.3)}$$

Next, with the help (8.137) we can find the power absorbed per unit area in the sheet as

$$\text{Power absorbed per unit area} = \frac{1}{2}R_S(H_{y0})^2, \qquad \text{(E8-3.4)}$$

where R_S is the surface resistance of the sheet given by (8.137), which can be rewritten as

$$R_S = \sqrt{\frac{\pi f \mu_0}{\sigma}}. \qquad \text{(E8-3.4)}$$

Substituting (E8-3.5) in (E8-3.4) we then obtain

$$\text{Power absorbed per unit area} = \frac{1}{2}\sqrt{\frac{\pi f \mu_0}{\sigma}}(H_{y0})^2. \text{ (8.216)} \qquad \text{(E8-3.6)}$$

Putting $f = 10$ GHz, that is, $10 \times 10^9 = 10^{10}$ Hz (given); $\sigma = 1.5 \times 10^7$ mho/m (given); $\mu_0 = 4\pi \times 10^{-7}$ H/m; and H_{y0}, given by (E8-3.2) in (E8-3.6), we obtain

$$\text{Power absorbed per unit area} = \frac{1}{2}\left(\sqrt{\frac{\pi \times 10^{10} \times 4\pi \times 10^{-7}}{1.5 \times 10^7}}\right)\left(\frac{100}{377}\right)^2$$

$$= 1.8 \times 10^{-3} \text{ W/m}^2 = 1.8 \text{ mW/ m}^2.$$

EXAMPLE E8-4

Calculate the average dielectric power loss per unit volume of a wave of electric field amplitude 4 V/m at a frequency of 4 GHz in a glass (pyrex) medium ($\varepsilon_r = 4$, $\tan\delta = 200 \times 10^{-4}$).

SOLUTION:

From (8.120) you know

$$\text{Average power lost per unit volume} = \frac{1}{2}\sigma E_0^2. \qquad \text{(E8-4.1)}$$

From the definition (5.70) we can write

$$\sigma = \omega\varepsilon \tan\delta. \qquad \text{(E8-4.2)}$$

Substituting (E8-4.2) in (E8-4.1) and, remembering $\varepsilon = \varepsilon_0\varepsilon_r$, we can write

$$\text{Average power lost per unit volume} = \frac{1}{2}\omega\varepsilon_0\varepsilon_r \tan\delta\, E_0^2. \qquad \text{(E8-4.3)}$$

Putting in (E8-4.3) the given values: $\omega = 2\pi f = 2\pi \times 4 \times 10^9$; $E_0 = 4$ V/m; $\varepsilon_r = 4$ and $\tan\delta = 200 \times 10^{-4}$, we can calculate as follows.

$$\text{Average power lost per unit volume} = \frac{1}{2}\omega\varepsilon_0\varepsilon_r \tan\delta\, E_0^2$$

$$= \pi \times 4 \times 10^9 \times 8.854 \times 10^{-12} \times 4 \times 200 \times 10^{-4} \times 4 \times 4 = 0.142 \text{ W/m}^3.$$

8.4 Radiated and Received Powers in Conduction Current Antennas

A conducting element through which an oscillating current passes radiates electromagnetic energy. You might have detected this energy as a noise in a radio receiver set when a nearby electrical appliance is either switched on or switched off. In a communication system, electromagnetic energy is made to propagate through a transmission line (see Section 10.1), say, in the form of a two-wire line, from one point to another such that no part of the line radiates out energy. It is implied that the separation between the wires of the line is small as compared to a wavelength such that the instantaneous magnetic field due to a current passing through an elemental portion of one of the wires is nullified by the oppositely directed instantaneous magnetic field caused by the current passing through the other wire in the opposite direction. However, we may be interested to radiate out energy into space either in many directions or in a specific direction at the terminating end of the transmission line. In order to implement the radiation of energy, the transmission line is terminated in a radiating system called the antenna. Further, at the receiving end as well we need to have a receiving antenna to receive electromagnetic energy from space and subsequently make it propagate through a transmission line in a receiving system. In this section, let us apply the concept of the vector potential (developed in Section 5.5) to discuss some of the fundamentals of conduction current antennas such as a dipole antenna. The displacement current antennas such as an antenna made out of an aperture of a wave-guiding structure for the radiation or reception of electromagnetic energy is kept outside the purview of the present section.

8.4.1 Infinitesimal dipole antenna

The infinitesimal dipole, also called the Hertzian dipole, is an oscillating filamentary current element Idl of infinitesimal length dl over which the amplitude of the current I remains uniform. Although the infinitesimal dipole is not a practical antenna, the results of its analysis are of immense significance in establishing a number of important concepts of practical antennas. Further, a finite-length dipole can be considered as constituted by a number of individual infinitesimal dipoles and therefore the results of analysis of an individual infinitesimal dipole can be integrated to obtain the results of a finite-length dipole antenna (see Section 8.4.3).

Let us analyse the infinitesimal dipole in spherical coordinate system (r, θ, ϕ) considering the element of length dl of the current element at the origin of coordinates $(0, 0, 0)$ (Fig.

8.6). However, for the sake of convenience, let us consider the current element as aligned along z in free space ($\mu = \mu_0$, $\varepsilon = \varepsilon_0$), in the framework of rectangular coordinate system. Hence, we can express the z-component of the element of vector potential A_z at the point $P(r, \theta, \phi)$, with the help of (5.137), as follows (Fig. 8.6):

$$A_z = \frac{\mu_0}{4\pi}\frac{Idl}{r}\exp(-j\beta r). \tag{8.143}$$

In (8.143) as well as in subsequent expressions such as field expressions to be derived from it to follow in this section, the time dependence $\exp(j\omega t)$ is understood. This is what is usually done to make analytical presentations more concise. The analysis, as we are going to see, will involve partial derivatives with respect to only r and not with respect to t. Consequently, the expressions that we are going to derive will each contain the factor $\exp(-j\beta r)$. However, it is implied that if we invoke the time dependence $\exp(j\omega t)$, which was otherwise kept understood, then these expressions will have the factor $\exp(j\omega t)\exp(-j\beta r) = \exp j(\omega t - \beta r)$ instead of $\exp(-j\beta r)$.

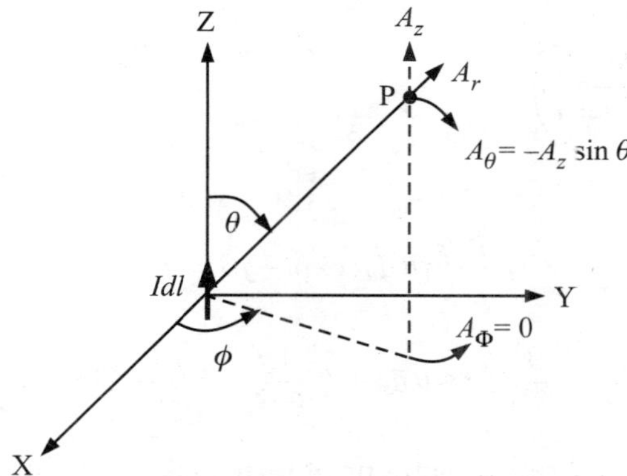

Fig. 8.6 *Vector potential components at the point $P(r, \theta, \phi)$ due to an infinitesimal dipole antenna at the origin of coordinates* $(0, 0, 0)$.

We can start from the expression (8.143) for A_z and proceed to find the electric and magnetic fields at the point $P(r, \theta, \phi)$. For this purpose, let us express with the help of (8.143) the components of vector potential at the point $P(r, \theta, \phi)$ in the spherical system as (Fig. 8.6)

$$\left.\begin{aligned} A_r &= A_z\cos\theta = \frac{\mu_0}{4\pi}\frac{Idl}{r}\cos\theta\exp(-j\beta r) \\ A_\theta &= -A_z\sin\theta = -\frac{\mu_0}{4\pi}\frac{Idl}{r}\sin\theta\exp(-j\beta r) \\ A_\phi &= 0 \end{aligned}\right\}. \tag{8.144}$$

The vector potential $\vec{A}$ at the point $P(r,\theta,\phi)$ then follows from (8.144) as

$$\vec{A} = A_r\vec{a}_r + A_\theta\vec{a}_\theta + A_\phi\vec{a}_\phi = \frac{\mu_0}{4\pi}\frac{Idl}{r}(\cos\theta\,\vec{a}_r - \sin\theta\,\vec{a}_\theta)\exp(-j\beta\,r). \tag{8.145}$$

In order to find $\vec{E}$, we have to find $\nabla\nabla\cdot\vec{A}$ with the help of (2.33) and (2.49) in which you have to take $\vec{A}$ from (8.145), and, subsequently, substitute $\nabla\nabla\cdot\vec{A}$ so obtained in (5.138).

In the first step, let us find $\nabla\cdot\vec{A}$ and $\nabla\nabla\cdot\vec{A}$ as (see Appendix A8.2)

$$\left.\begin{aligned}\nabla\cdot\vec{A} &= \frac{-\mu_0 Idl}{4\pi}\left(\frac{1}{r^2}+\frac{j\beta}{r}\right)\cos\theta\exp(-j\beta r)\\ \nabla\nabla\cdot\vec{A} &= \frac{\mu_0 Idl}{4\pi}\left[\left(\frac{2}{r^3}+\frac{2j\beta}{r^2}-\frac{\beta^2}{r}\right)\cos\theta\,\vec{a}_r+\left(\frac{1}{r^3}+\frac{j\beta}{r^2}\right)\sin\theta\,\vec{a}_\theta\right]\exp(-j\beta r)\end{aligned}\right\}. \tag{8.146}$$

Substituting the expression for $\vec{A}$ given by (8.145) and that for $\nabla\nabla\cdot\vec{A}$ given by (8.146) in the following expression obtainable from (5.138):

$$\vec{E} = -j\omega\left(\vec{A}+\frac{\nabla\nabla\cdot\vec{A}}{\omega^2\mu_0\varepsilon_0}\right)$$

we obtain

$$\vec{E} = -j\omega\frac{\mu_0}{4\pi}Idl\left(\frac{1}{r}(\cos\theta\,\vec{a}_r - \sin\theta\,\vec{a}_\theta)\exp(-j\beta\,r)\right.$$

$$\left.+\left(\left(\frac{2}{\beta^2r^3}+\frac{2j\beta}{\beta^2r^2}-\frac{\beta^2}{\beta^2r}\right)\cos\theta\,\vec{a}_r+\left(\frac{1}{\beta^2r^3}+\frac{j\beta}{\beta^2r^2}\right)\sin\theta\,\vec{a}_\theta\right)\right)\exp(-j\beta r),$$

which on simplification and rearrangement of terms takes the following form:

$$\vec{E} = -j\omega\frac{\mu_0}{4\pi}\,Idl\exp(-j\beta\,r)$$

$$\left(\left(\frac{2}{\beta^2r^3}+\frac{2j\beta}{\beta^2r^2}\right)\cos\theta\,\vec{a}_r+\left(-\frac{1}{r}+\frac{1}{\beta^2r^3}+\frac{j\beta}{\beta^2r^2}\right)\sin\theta\,\vec{a}_\theta\right). \tag{8.147}$$

Making use of the relations $\beta = \omega\sqrt{\mu_0\varepsilon_0}$ and $\eta_0 = (\mu_0/\varepsilon_0)^{1/2}$ for the propagation constant and intrinsic impedance respectively of free space, we can then easily identify from (8.147) the following components of $\vec{E} = E_r\vec{a}_r + E_\theta\vec{a}_\theta + E_\phi\vec{a}_\phi$:

$$\left.\begin{aligned} E_r &= -\frac{j\,I\,dl}{2\pi}\frac{\eta_0}{\beta}\cos\theta\left(\frac{1}{r^3}+\frac{j\beta}{r^2}\right)\exp(-j\beta\,r) \\ E_\theta &= -\frac{j\,I\,dl}{4\pi}\frac{\eta_0}{\beta}\sin\theta\left(\frac{-\beta^2}{r}+\frac{1}{r^3}+\frac{j\beta}{r^2}\right)\exp(-j\beta\,r) \\ E_\phi &= 0 \end{aligned}\right\}. \tag{8.148}$$

Similarly, we can find the magnetic field components with the help of the following expression obtainable from (5.138):

$$\vec{H} = \frac{\nabla\times\vec{A}}{\mu_0}. \tag{8.149}$$

We can express $\nabla\times\vec{A}$ occurring in (8.149) with the help of (2.68) as

$$\nabla\times\vec{A} = \frac{1}{r\sin\theta}\left(\frac{\partial(\sin\theta\,A_\phi)}{\partial\theta}-\frac{\partial A_\theta}{\partial\phi}\right)\vec{a}_r + \frac{1}{r}\left(\frac{1}{\sin\theta}\frac{\partial A_r}{\partial\phi}-\frac{\partial(r A_\phi)}{\partial r}\right)\vec{a}_\theta + \frac{1}{r}\left(\frac{\partial(r A_\theta)}{\partial r}-\frac{\partial A_r}{\partial\theta}\right)\vec{a}_\phi$$

which takes the following simple form remembering that in this problem $A_\phi = 0$ [see (8.144)] and $\partial/\partial\phi = 0$ (see Appendix A8.2):

$$\nabla\times\vec{A} = \frac{1}{r}\left(\frac{\partial(r A_\theta)}{\partial r}-\frac{\partial A_r}{\partial\theta}\right)\vec{a}_\phi,$$

which can be substituted in (8.149) to write, making use of the components of vector potential given by given by (8.145), the following expression for $\vec{H}$:

$$\vec{H} = \frac{1}{\mu_0 r}\frac{\mu_0}{4\pi}I dl\left(\frac{-\partial[\sin\theta\exp(-j\beta\,r)]}{\partial r}-\frac{\partial[\frac{1}{r}\cos\theta\exp(-j\beta\,r)]}{\partial\theta}\right)\vec{a}_\phi$$

which carrying out the partial derivatives takes the following form

$$H_\phi = \frac{I\,dl}{4\pi}\sin\theta\left(\frac{j\beta}{r}+\frac{1}{r^2}\right)\exp(-j\beta\,r). \tag{8.150}$$

We can then identify from (8.150) the following components of $\vec{H} = H_r\vec{a}_r + H_\theta\vec{a}_\theta + H_\phi\vec{a}_\phi$:

$$\left.\begin{aligned} H_r &= 0 \\ H_\theta &= 0 \\ H_\phi &= \frac{I\,dl}{4\pi}\sin\theta\left(\frac{j\beta}{r}+\frac{1}{r^2}\right)\exp(-j\beta\,r) \end{aligned}\right\}. \tag{8.151}$$

Far-field and near-field quantities of an infinitesimal dipole

At large distances $r$from the infinitesimal dipole, we can ignore the terms containing higher powers of r in the denominators of the field expressions (8.148) and (8.151) (see Appendix A8.3 for the more formal definition of the far-field region of an antenna). Accordingly, from (8.148) note that E_r is negligible compared to E_θ and only the first term in the expression for E_θ in (8.148) is significant. Similarly, for larger values of r, only the first term of the expression for H_ϕ in (8.151) is significant. Thus, the significant field quantities at large distances can be written from (8.148) and (8.151) as:

$$\left.\begin{aligned} E_\theta &= \frac{j\,\eta_0\beta}{4\pi\,r} I dl \sin\theta \exp(-j\beta\,r) \\ H_\phi &= \frac{j\beta}{4\pi\,r} I dl \sin\theta \exp(-j\beta\,r) \end{aligned}\right\}. \quad \text{(far-field quantities)} \tag{8.152}$$

It is easy to appreciate from (8.152) that the wave impedance of the infinitesimal dipole E_θ/H_ϕ becomes equal to the intrinsic impedance of free space η_0.

Thus, here we have presented a simple analysis to derive the expressions for the electric and maagnetic field components due to a Hertzian infinitesimal dipole. Further, one may treat a filamentary loop of current as an infinitesimal magnetic dipole antenna and deduce the expressions for fields due to it at a distant observation point from the analogous far-field expressions of a Hertzian infinitesimal dipole antenna (see Example E8-5).

EXAMPLE E8-5

Deduce the expressions for radiation fields due to a small circular turn or loop of current in the far-field region from the analogous field expressions of a Hertzian infinitesimal dipole.

SOLUTION:

You have already seen that electric and magnetic dipoles are analogous—the electric field of the electric dipole is analogous to the magnetic flux density of the magnetic dipole; the electric dipole moment $p = ql$ is analogous to the magnetic dipole moment $p_m(= i\alpha = i\pi a^2)$, where i and a represent the current and radius of a circular turn or loop making the magnetic dipole; and ε_0 is analogous to μ_0 respectively (see Examples 4-7 and 4-8).

Recall expression (8.152) in the far-field region for a Hertzian infinitesimal dipole antenna and put therein $Idl = (dq/dt)dl = j\omega q dl = j\omega p$ (introducing q and relating it to I so that one is able to express Idl in terms of electric dipole moment p), noting that for the short dipole, dl and l are synonymous and write it as:

$$E_\theta = \frac{j\,\eta_0\beta}{4\pi r} j\omega p \sin\theta \exp(-j\beta\,r). \tag{E8-5.1}$$

(Hertzian infinitesimal dipole antenna)

The expression for the magnetic flux density B_θ, analogous to E_θ, for the circular-loop antenna can be written from (E8-5.1) replacing p by the analogous p_m (see Example E4-8) as

$$B_\theta = \frac{j\,\eta_0\beta}{4\pi r} j\omega p_m \sin\theta \exp(-j\beta\,r). \tag{E8-5.2}$$

In order to get the expression for $H_\theta(= B_\theta/\mu_0)$, for the circular-loop antenna, we have to divide (E8-5.2) for B_θ by μ_0, the latter (μ_0), however, replaced by the analogous quantity $1/\varepsilon_0$ (see Example E4-8). Thus we obtain, for the circular-loop antenna, the expression for H_θ as

$$H_\theta = \frac{j\,\eta_0\beta}{4\pi\,r}\varepsilon_0 j\omega p_m \sin\theta \exp(-j\beta\,r) \quad , \tag{E8-5.3}$$

which, in view of the relation $\eta_0 = (\mu_0/\varepsilon_0)^{1/2}$, simplifies to

$$H_\theta = -\frac{\omega\mu_0 p_m}{4\pi\eta_0}\frac{\exp(-j\beta\,r)}{r}\beta\sin\theta. \quad \text{(circular-loop antenna)} \tag{8-5.4}$$

From the expression (E8-5.4) for H_θ, we can easily derive the expression for E_ϕ, using one of the Maxwell's equations (see Example 5-9 in Chapter 5), as follows:

$$E_\phi = \frac{\omega\mu_0 p_m}{4\pi}\frac{\exp(-j\beta\,r)}{r}\beta\sin\theta. \quad \text{(circular-loop antenna)} \tag{E8-5.5}$$

We can put together the field expressions (E8-5.5) and (E8-5.4) for the circular-loop antenna to write them as

$$\left.\begin{aligned} E_\phi &= \frac{\omega\mu_0 p_m}{4\pi}\frac{\exp(-j\beta\,r)}{r}\beta\sin\theta \\ H_\theta &= -\frac{\omega\mu_0 p_m}{4\pi\eta_0}\frac{\exp(-j\beta\,r)}{r}\beta\sin\theta \end{aligned}\right\}. \quad \text{(circular-loop antenna)} \tag{E8-5.6}$$

For distances r close to the infinitesimal dipole, similarly the significant field quantities are those involving higher powers of $r$in the denominators, which can be written from (8.148) and (8.151) as

$$\left.\begin{aligned} E_r &= -\frac{j\eta_0}{2\pi\beta r^3} I dl \cos\theta \exp(-j\beta\,r) \\ E_\theta &= -\frac{j\eta_0}{4\pi\beta r^3} I dl \sin\theta \exp(-j\beta\,r) \\ H_\phi &= \frac{1}{4\pi r^2} I\,dl \sin\theta\,\exp(-j\beta\,r) \end{aligned}\right\}. \quad \text{(near-field quantities)} \tag{8.153}$$

(See Appendix A8.2 for a more formal definition of the near-field region of an antenna).

Further, you will find it of interest to take up a problem to show that the near-field electric field amplitude due to an infinitesimal dipole given by (8.153), with an appropriate interpretation, becomes identical with the electric field due to a short electric dipole (see Example 8-6).

EXAMPLE E8-6

Show that, if appropriately interpreted, an infinitesimal dipole antenna is analogous to a short electric dipole as far as the electric field close to it is concerned.

SOLUTION:

The electric field close to an infinitesimal dipole can be written, with the help of (8.153), as

$$\vec{E} = E_r\vec{a}_r + E_\theta\vec{a}_\theta = \frac{-jIdl}{4\pi\omega\varepsilon_0 r^3}[\,2\cos\theta\,\vec{a}_r + \sin\theta\vec{a}_\theta]\exp(-j\beta\,r),$$

which, putting $I = dq/dt = j\omega q$ as the rate of charge flow, in view of the time dependence $\exp(j\omega t)$ understood, becomes:

$$\vec{E} = \frac{qdl}{4\pi\varepsilon_0 r^3}[\,2\cos\theta\,\vec{a}_r + \sin\theta\,\vec{a}_\theta]\exp(-j\beta\,r). \qquad \text{(E8-6.1)}$$

With $p = qdl$ interpreted as the dipole moment, the amplitude of electric field close to an infinitesimal dipole given by expression (E8-6.1) is identical to the static electric field due to a short electric dipole (see Example E3-4).

Reactive power associated with near-field quantities of an infinitesimal dipole

The complex Poynting vector $\vec{P}_{\text{complex}}$ defined by (8.96) for the near-field quantities (8.153) can be written as

$$\vec{P}_{\text{complex}} = \frac{1}{2}\vec{E}\times\vec{H}^* = \frac{1}{2}\begin{vmatrix} \vec{a}_r & \vec{a}_\theta & \vec{a}_\phi \\ E_r & E_\theta & 0 \\ 0 & 0 & H_\phi^* \end{vmatrix} = \frac{1}{2}(E_\theta H_\phi^*\vec{a}_r - E_r H_\phi^*\vec{a}_\theta). \qquad (8.154)$$

The quantities $E_\theta H_\phi^*$ and $E_r H_\phi^*$ occurring in (8.154) may be seen to be each purely imaginary as can be appreciated with the help of the field expressions (8.153) as follows:

$$\left.\begin{aligned} E_\theta H_\phi^* &= -j\left(\frac{\eta_0}{4\pi\beta r^3}\right)\left(\frac{1}{4\pi r^2}\right)(Idl)^2\sin^2\theta \\ E_r H_\phi^* &= -j\left(\frac{\eta_0}{2\pi\beta r^3}\right)\left(\frac{1}{4\pi r^2}\right)(Idl)^2\sin\theta\cos\theta \end{aligned}\right\}. \qquad (8.155)$$

In view of (8.155), we can appreciate from (8.154) that the quantity $(1/2)\vec{E}\times\vec{H}^*$ is purely imaginary. Consequently, therefore, the average complex Poynting $\vec{P}_{\text{average}} = (1/2)Re\vec{E}\times\vec{H}^*$ associated with the near-field quantities becomes zero. In other words, the near-field quantities do not correspond to the propagation of power. The power associated with these quantities is of non-radiative type which can also be referred to as the reactive power [see the discussion following (8.102)]. It is implied that the non-radiative power associated with the near-field quantities would be retuned to source feeding this power to the infinitesimal dipole.

Power radiated at large distances from an infinitesimal dipole

The complex Poynting vector $\vec{P}_{\text{complex}}$ defined by (8.96) for the far-field quantities (8.152) can be found as

$$\vec{P}_{\text{complex}} = \tfrac{1}{2}\vec{E}\times\vec{H}^* = \tfrac{1}{2}\begin{vmatrix} \vec{a}_r & \vec{a}_\theta & \vec{a}_\phi \\ 0 & E_\theta & 0 \\ 0 & H_\phi^* & \end{vmatrix} = \tfrac{1}{2}E_\theta H_\phi^* \vec{a}_r$$

$$= \frac{1}{2}\frac{\eta_0\beta^2}{(4\pi r)^2}(Idl)^2\sin^2\theta\vec{a}_r$$

whence we obtain the average complex Poynting vector $\vec{P}_{\text{average}} = (1/2)\text{Re}\,\vec{E}\times\vec{H}^*$ associated with the far-field quantities as

$$\vec{P}_{\text{average}} = \frac{1}{2}\text{Re}\,\vec{E}\times\vec{H}^* = \frac{1}{2}\frac{\eta_0\beta^2}{(4\pi r)^2}(Idl)^2\sin^2\theta\vec{a}_r\,. \tag{8.156}$$

Let us now find the power radiated out from the infinitesimal dipole in all directions. For this purpose, let us first find the power P propagating through a sphere of radius r first finding the element of power dP through an annular disc of element of area $dS = 2\pi r\sin\theta r d\theta$ at an angle θ as (Fig. 8.7):

$$dP = \frac{1}{2}\text{Re}\,(\vec{E}\times\vec{H}^*)\cdot dS\vec{a}_r = \frac{1}{2}\text{Re}\,(\vec{E}\times\vec{H}^*)\cdot\vec{a}_r 2\pi\, r\sin\theta\, r d\theta$$

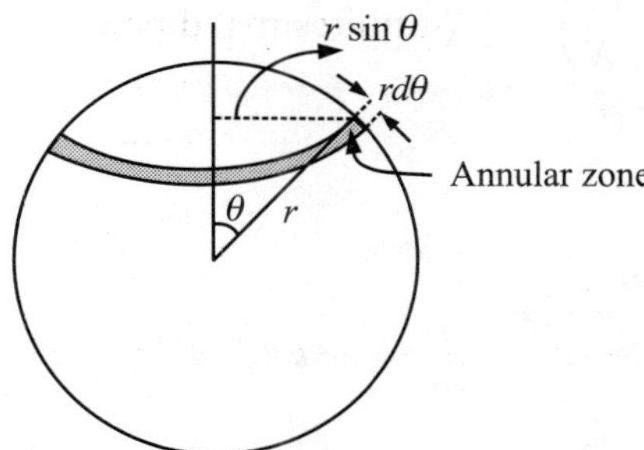

Fig. 8.7 *Annular element of disc of infinitesimal arc thickness $rd\theta$ and radius $r\sin\theta$ and hence of element of area $dS = 2\pi r\sin\theta r d\theta$ of the annular zone considered on a spherical area of radius r at an angle θ.*

which, using (8.156), can be written as

$$dP = \frac{1}{2}\frac{\eta_0\beta^2}{(4\pi\, r)^2}(Idl)^2\sin^2\theta\,\vec{a}_r\cdot\vec{a}_r 2\pi\, r\sin\theta\, r d\theta = \frac{\eta_0\beta^2}{16\pi}(Idl)^2\sin^3\theta\, d\theta\,. \tag{8.157}$$

Integrating (8.157) we then obtain

$$P = \int dP = \int_0^\pi \frac{\eta_0\beta^2}{16\pi}(Idl)^2\sin^3\theta d\theta = \frac{\eta_0\beta^2}{16\pi}(Idl)^2\int_0^\pi \sin^3\theta d\theta\,. \tag{8.158}$$

The definite integral occurring in (8.158) is[2]

$$\int_0^\pi \sin^3\theta d\theta = \frac{4}{3}$$

which can be substituted in (8.158) to obtain

$$P = \frac{\eta_0\beta^2}{16\pi}(Idl)^2\frac{4}{3},$$

which, putting $\eta_0 = \sqrt{\mu_0/\varepsilon_0} = 120\pi$ and $\beta = 2\pi/\lambda$ becomes

$$P = 40\pi^2 I^2\left(\frac{dl}{\lambda}\right)^2. \tag{8.159}$$

Clearly, the power P propagating through a sphere of radius r is the power radiated out from the infinitesimal dipole in all directions. The average power density, that is, the average power per unit area radiated out of the infinitesimal dipole W_0, is obtained by dividing P, given by (8.159), by the area $4\pi r^2$of the sphere of radius r as follows:

$$W_0 = \frac{P}{4\pi\, r^2} = \frac{10\pi I^2}{r^2}\left(\frac{dl}{\lambda}\right)^2. \quad \text{(infinitesimal dipole)} \tag{8.160}$$

---

[2] $$\int_0^\pi \sin^3\theta d\theta = \int_0^\pi \sin^2\theta\sin\theta d\theta d\theta = \int_0^\pi -\sin^2\theta d(\cos\theta)$$

$$= -\int_0^\pi (1-\cos^2\theta)d(\cos\theta) = -\int_0^\pi [d(\cos\theta) - \cos^2\theta)d(\cos\theta)]$$

$$= -\left[\int_0^\pi d(\cos\theta) - \int_0^\pi \cos^2\theta)d(\cos\theta)\right] = -\left[\cos\theta - \frac{\cos^3\theta}{3}\right]_0^\pi = \frac{4}{3}.$$

Let us at this stage define an isotropic radiator equivalent to the infinitesimal dipole as a hypothetical radiator that radiates power P, given by (8.159), however, now evenly distributed in all directions such that the isotropic radiator has the same power density in all directions, which is also its average power density. Thus, the infinitesimal dipole and its equivalent have the same average power density W_0 though they are different in that while the former has different power densities in different directions, the latter has same power density in all directions. Therefore, the average power density, that is, the power per unit area of the equivalent isotropic radiator, continues to be given by (8.160). This can be rewritten for the sake of convenience as

$$W_0 = \frac{10\pi I^2}{r^2}\left(\frac{dl}{\lambda}\right)^2. \quad \text{(isotropic radiator)} \tag{8.161}$$

Further, the power density W in the direction θ, given by the average complex Poynting vector $\vec{P}_{\text{average}}$ obtainable from (8.156), may be written as

$$W = \frac{1}{2}Re(\vec{E}\times\vec{H}^*) = \frac{1}{2}\frac{\eta_0\beta^2}{(4\pi r)^2}(Idl)^2\sin^2\theta,$$

which, putting $\eta_0 = \sqrt{\mu_0/\varepsilon_0} = 120\pi$ and $\beta = 2\pi/\lambda$ simplifies to

$$W = \frac{15\pi I^2}{r^2}\left(\frac{dl}{\lambda^2}\right)^2\sin^2\theta. \quad \text{(infinitesimal dipole)} \tag{8.162}$$

We can now find the directive gain D_g of the infinitesimal dipole antenna as the power density of the antenna relative to that of an isotropic radiator by dividing (8.162) by (8.161) as follows:

$$D_g = \frac{W}{W_0} = \frac{3}{2}\sin^2\theta. \quad \text{(infinitesimal dipole)} \tag{8.163}$$

The directive gain of the infinitesimal dipole according to (8.163) is proportional to $\sin^2\theta$. It has the maximum value $D_g = D_0$, called the directivity of the antenna in a direction perpendicular to the alignment of the dipole, that is, at an angle $\theta = \pi/2$. Putting $\theta = \pi/2$, that is, $\sin\theta = 1$ in (8.163), we obtain the directivity of the infinitesimal dipole D_0 as

$$D_0 = \frac{3}{2} = 1.5. \quad \text{(infinitesimal dipole)} \tag{8.164}$$

Radiation resistance of an infinitesimal dipole

The radiation resistance R_r of the infinitesimal dipole is the resistance of an equivalent resistor that consumes the same power as P, given by (8.159) in terms of the current amplitude I as follows:

$$P = 40\pi^2 I^2\left(\frac{dl}{\lambda}\right)^2 = \frac{1}{2}I^2 R_r. \tag{8.165}$$

The expression for R_r can be then obtained from (8.165) as

$$R_r = 80\pi^2 \left(\frac{dl}{\lambda}\right)^2. \quad \text{(infinitesimal dipole)} \tag{8.166}$$

It follows from (8.165) that for a given power P of an antenna to be radiated—here, the infinitesimal dipole—a lesser or higher current I will be required according to whether the radiation resistance of the antenna is larger or smaller.

Radiation pattern of an infinitesimal dipole

The radiation pattern of an antenna is the graphical representation of the radiation properties of the antenna with respect to related variables such as the field strength and power density as a function of space. Although the radiation pattern is three-dimensional, it is common to describe it in two planar patterns, obtained by making two slices through the three-dimensional pattern, namely the E-plane and H-plane patterns, where the E-plane is the plane containing the electric field vector and the direction of maximum radiation from the antenna, and the H-plane is the plane containing the magnetic field vector and the direction of maximum radiation from the antenna. Here, we are going to study these patterns with respect to an infinitesimal dipole (Fig. 8.8).

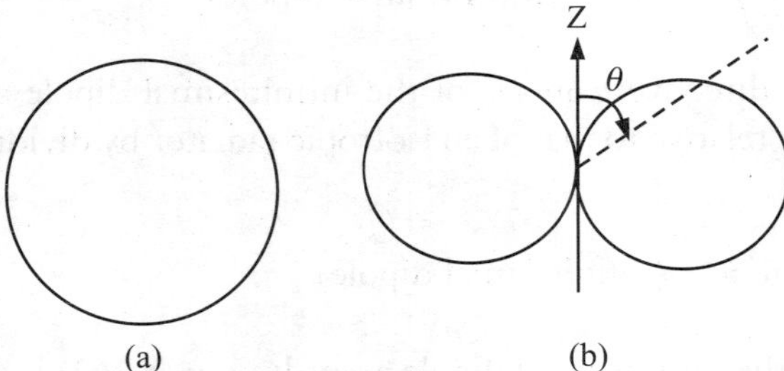

Fig. 8.8 *Patterns of an infinitesimal dipole aligned vertically: (a) horizontal (H-plane) and (b) vertical (E-plane).*

Referring to an infinitesimal dipole that is aligned in the vertical direction, the radiation pattern can be alternatively plotted as the horizontal pattern or the vertical pattern depending on whether the pattern is plotted on the (i) horizontal plane ($\theta = \pi/2$) on which the value of θ does not vary and remains constant at $\theta = \pi/2$ and the value of ϕ varies $\theta = \pi/2$ (H-plane pattern), or (ii) the vertical plane on which the value of ϕ remains constant (ϕ =constant) and the value of θ varies (E-plane pattern) (Fig. 8.8). In the light of this nomenclature of the radiation pattern, the horizontal pattern (H-plane pattern) ($\theta = \pi/2$) is a circle corresponding to the constant amplitude of $E_\theta[= (\eta_0\beta)/(4\pi\, rIdl)]$ obtained by putting $\theta = \pi/2$ in (8.152). On the other hand, the vertical pattern (E-plane pattern) (ϕ =constant) depends on the angle θ, the amplitude of E_θ, as can be again appreciated from (8.152), it becoming zero and maximum $[= (\eta_0\beta)/(4\pi\, rIdl)]$ for $\theta = 0$ and $\theta = \pi/2$ respectively. Thus, the shape of the vertical pattern (E-plane pattern) becomes the figure of infinity (∞) and remains azimuthally symmetric ($\partial/\partial\phi = 0$) about the axis of the dipole (Fig. 8.8).

The power density W in the direction θ is given by (8.162) and has a maximum value at $\theta = \pi/2$. At the angles $\theta = \pi/4$ and $\theta = 3\pi/4$, which are separated by $\theta = \pi/2$, the value of the power densitv W becomes half its maximum value. The separation between the values θ corresponding to the value of power density W becoming half its maximum value is called the half power bandwidth (HPBW) of an antenna, here, infinitesimal dipole. How much an antenna—here an infinitesimal dipole—is capable of beaming power in a specified direction can be estimated by HPBW. It is easy to appreciate this by putting $\sin^2\theta = \sin^2\pi/4 = \sin^2 3\pi/4 = (1/\sqrt{2})^2 = 1/2$ in (8.162). Thus, HPBW of the infinitesimal dipole becomes $\pi/2$ or 90^0, being equal to the angular separation $\theta = 3\pi/4 - \pi/4 = \pi/2$.

Effective aperture area of an infinitesimal dipole

The antenna is used not only to radiate power in the transmitting mode but also to receive power and deliver it to a load in the receiving mode. Interestingly, an antenna enjoys identical radiation and circuit characteristics in transmitting and receiving modes according to the reciprocity theorem of circuit theory[3].

Let a two-port equivalent network represent a system of two antennas, #1 and #2, separated by a distance r (Fig. 8.9). Let the input voltage and current at antenna #1 and at the corresponding inputs of its equivalent network be represented by V_1 and I_1 respectively, and the corresponding output quantities of antenna #2 and at its equivalent network by V_2 and I_2 respectively (Fig. 8.9). The circuit quantities can then be related by the circuit impedances Z as follows:

$$\left.\begin{aligned} V_1 &= Z_{11}I_1 + Z_{12}I_2 \\ V_2 &= Z_{21}I_1 + Z_{22}I_2 \end{aligned}\right\}. \tag{8.167}$$

If in the transmitting mode of antenna #1 we take antenna #2 far from antenna #1, then we may take the transfer impedance $Z_{12} = 0$ in (8.167), which enables us to write the following expression for antenna #1:

$$V_1 = Z_{11}I_1. \quad \text{(transmitting mode of antenna \#1).} \tag{8.168}$$

$Z_{11}(= V_1/I_1)$ can be identified as the self impedance of antenna #1. However, in the receiving mode of antenna #1, the transfer impedance $Z_{12} \neq 0$ provides the coupling between the antennas #1 and #2, enabling us to rewrite from (8.167):

$$V_1 = Z_{11}I_1 + Z_{12}I_2 \quad \text{(receiving mode of antenna \#1)} \tag{8.169}$$

[3]In any linear network containing bilateral linear impedances and energy sources, the ratio of the voltage on one mesh to the current in another mesh would remain unaltered if the voltage and the current were interchanged, other sources being removed.

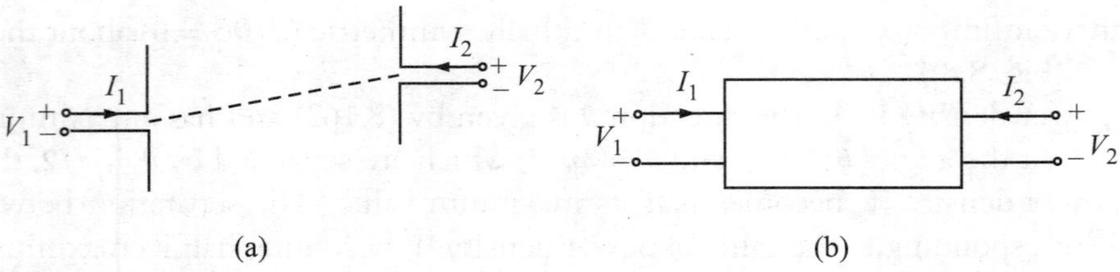

Fig. 8.9 *(a) Two antennas separated by a distance (b) and their equivalent network representation.*

to represent the voltage at the terminal of antenna #1, supposedly terminated in a load impedance Z_l. The second term $Z_{12}I_2$ may be interpreted as the emf of a constant voltage source since the value of $Z_{12}I_2$ remains constant in view of I_2 remaining unchanged with the change in the value of the load impedance Z_l (Fig. 8.10).

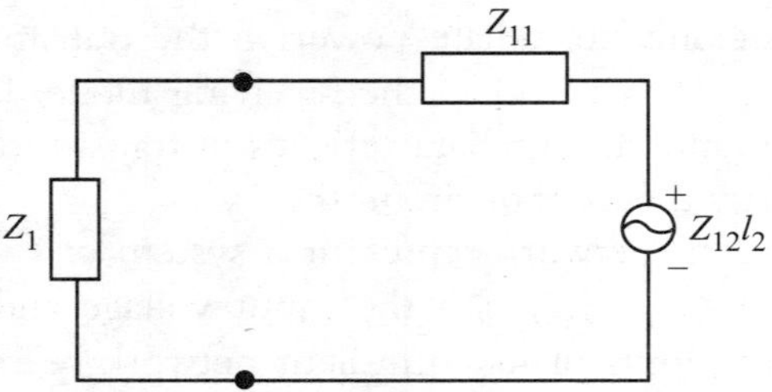

Fig. 8.10 *Equivalent network of antenna #1 in the receiving mode showing its self impedance Z_{11}, being equal to the internal impedance of the constant voltage source that has an emf $Z_{12}I_2$, and load impedance Z_l.*

Let us now replace antenna #1 (Fig. 8.10) by the infinitesimal dipole in its equivalent network representation (Fig. 8.11). Here, the emf of the constant voltage source $Z_{12}I_2$ (Fig. 8.10) is taken as the voltage induced on the infinitesimal dipole of length l due to electromagnetic energy received supposedly by the dipole from antenna #2 at a distance. The length l of the infinitesimal dipole being extremely small, we can take the potential gradient or the electric field induced to be uniform over its length l and therefore, take the voltage induced on the dipole as El (Fig. 8.11). Further, we can take that the self impedance Z_{11} comprises resistive and the reactive parts, R_{11} and X_{11} respectively, as

$$Z_{11} = R_{11} + jX_{11},$$

and subsequently write it by replacing R_{11} and X_{11} by the radiation resistance R_r and reactance X_A of the antenna as (Fig. 8.11):

$$Z_{11} = R_{11} + jX_{11} = R_r + jX_A. \tag{8.170}$$

Using (8.170) we can then write the amplitude of current I through the equivalent network as

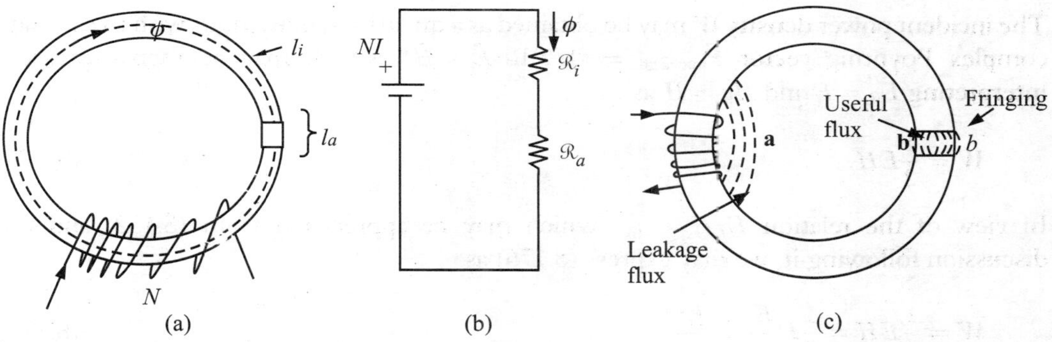

Fig. 8.11 *Equivalent network of the infinitesimal dipole in the receiving mode drawn from Fig. 8.10, interpreting antenna #1 as the infinitesimal dipole and replacing the emf $Z_{12}I_2$ of the constant voltage source by the voltage El induced on the length of the infinitesimal dipole.*

$$I = \frac{El}{\sqrt{(R_l + R_r)^2 + (X_l + X_A)^2}}. \tag{8.171}$$

where R_l and X_l are respectively the load resistance and reactance of the load impedance $Z_l = R_l + jX_l$ (Fig. 8.11). The power delivered to the load P_{load} may be then written with the help of (8.171) as

$$P_{\text{load}} = \frac{1}{2}I^2 R_l = \frac{1}{2}\left(\frac{El}{\sqrt{(R_l + R_r)^2 + (X_l + X_A)^2}}\right)^2 R_l. \tag{8.172}$$

Putting $R_l = R_r$ and $X_l + X_A$, the well-known maximum power transfer condition in (8.172), we can then find the expression for the maximum power transferred to the load $P_{\text{load, max}}$ as follows:

$$P_{\text{load,max}} = \frac{1}{2}\left(\frac{EI}{\sqrt{(2R_r)^2}}\right)^2 R_l = \frac{1}{2}\left(\frac{EI}{2R_r}\right)^2 R_r = \frac{E^2 l^2}{8R_r}. \tag{8.173}$$

Further, in the receiving mode, the power delivered to the load P_{load} may be related to the power density incident on the antenna W by a quantity called the effective aperture area A_e defined as the ratio

$$A_e = \frac{P_{\text{load}}}{W}. \tag{8.174}$$

Substituting $P_{\text{load}} = P_{\text{load, max}}$ in (8.174), we then obtain the expression for the maximum effective aperture area of an infinitesimal dipole as

$$A_{e,\,\text{max}} = \frac{P_{\text{load,max}}}{W}. \quad \text{(infinitesimal dipole)} \tag{8.175}$$

The incident power density W may be obtained as a quantity synonymous with the average complex Poynting vector $\vec{P}_{\text{average}} = (1/2)\text{Re}\,\vec{E} \times \vec{H}^*$, which may be then expressed, interpreting $E_\theta = E$ and $H_\theta = H$ as

$$W = \frac{1}{2}EH. \tag{8.176}$$

In view of the relation $E/H = \eta_0$, which may be appreciated from (8.152) and the discussion following it, we may express (8.176) as

$$W = \frac{1}{2}EH = \frac{1}{2}E\frac{E}{\eta_0} = \frac{E^2}{2\eta_0}. \tag{8.177}$$

The expression for the maximum effective aperture area of an infinitesimal dipole $A_{e,\,\text{max}}$ is then obtained by substituting (8.173) and (8.177) in (8.175) as follows:

$$A_{e,\,\text{max}} = \frac{P_{\text{load,max}}}{W} = \frac{\dfrac{E^2 l^2}{8R_r}}{\dfrac{E^2}{2\eta_0}},$$

which, putting $\eta_0 = 120\pi$ and substituting R_r from (8.166) where the length of the infinitesimal length is taken as l, may be written in the following form:

$$A_{e,\,\text{max}} = \frac{P_{\text{load, max}}}{W} = \frac{\dfrac{E^2 l^2}{8R_r}}{\dfrac{E^2}{2\eta_0}} = \frac{\dfrac{E^2 l^2}{8 \times 80\pi^2 (\frac{l}{\lambda})^2} \frac{E^2}{2\times 120\pi}}{=} \frac{3\lambda^2}{8\pi}. \quad \text{(infinitesimal dipole)} \tag{8.178}$$

Further, the following relation obtained by dividing (8.178) by (8.164) for the ratio of the maximum effective aperture area $A_{e,\,\text{max}}$ to the directivity D_0 of an infinitesimal dipole proves to be very useful in antenna theory:

$$\frac{A_{e,\,\text{max}}}{D_0} = \frac{\lambda^2}{4\pi}. \quad \text{(infinitesimal dipole)} \tag{8.179}$$

8.4.2 Constancy of the ratio of effective aperture area to directive gain of an antenna

Let us take two arbitrary antennas #1 and #2 separated by a distance r (Fig. 8.9a) and let D_{g1} and D_{g2} be their directive gains in the direction of antennas #2 and #1 respectively. If P_{t1} represent the power transmitted by antenna #1 in all directions, then we may relate the power density W_1 due to antenna #1 at the location of antenna #2 with D_{g1} [see (8.160)] as

$$D_{g1} = \frac{W_1}{\dfrac{P_{t1}}{4\pi\, r^2}} = \frac{W_1 4\pi\, r^2}{P_{t1}},$$

whence we may express W_1 as

$$W_1 = \frac{P_{t1} D_{g1}}{4\pi\, r^2}. \tag{8.180}$$

Now, let us express with the help of (8.174) the effective aperture A_{e2} of antenna #2 as

$$A_{e2} = \frac{P_{r2}}{W_1} \tag{8.181}$$

where P_{r2} is the power received by antenna #2 from the power transmitted by antenna #1, it being implied that the power so received will be transferred to a terminating load of antenna #2. Eliminating W_1 from (8.180) and (8.182), we than obtain the relation

$$\frac{P_{r2}}{P_{t1}} = \frac{D_{g1} A_{e2}}{4\pi r^2}. \tag{8.182}$$

We have deduced the relation (8.182) considering antennas #1 and #2 in the transmitting and receiving modes, respectively. On the other hand, if we now take antennas #2 and #1 in the transmitting and receiving modes respectively, we obtain the relation corresponding to (8.182) as

$$\frac{P_{r1}}{P_{t2}} = \frac{D_{g2} A_{e1}}{4\pi\, r^2}. \tag{8.183}$$

However, in the present context, the reciprocity theorem of circuit theory may be interpreted to write the relation:

$$\frac{P_{r2}}{P_{t1}} = \frac{P_{r1}}{P_{t2}} \tag{8.184}$$

which means that the power received by one of the antennas (#1 or #2) due to the power transmitted by the other would remain the same if the source for generating power and the instrument for measuring power are interchanged in the system of these two antennas in the transmitting and receiving modes. (Refer to footnote on page 399.)

Substituting (8.182) and (8.183) in (8.184), we then obtain the relation

$$\frac{A_{e1}}{D_{g1}} = \frac{A_{e2}}{D_{g2}}. \tag{8.185}$$

Note that the relation (8.185) continues to hold good when the effective aperture areas A_{e1} and A_{e2} are changed to their respective maxima $A_{e1,\max}$ and $A_{e2,\max}$, and similarly, when the directive gains D_{g1} and D_{g2} are changed to their respective maxima, that is, to directivities D_{01} and D_{02} respectively. Therefore, we can write from (8.185) the following relation

$$\frac{A_{e1,\,\max}}{D_{01}} = \frac{A_{e2,\,\max}}{D_{02}}. \tag{8.186}$$

Also notice that the relation (8.186) holds true for any two arbitrarily chosen antennas #1 and #2. Therefore, it follows from (8.186) that the maximum effective aperture of any antenna $A_{e,\,\max}$ bears a constant ratio to its directivity D_0 irrespective of the type of antenna:

$$\frac{A_{e,\,\max}}{D_0} = \text{constant.} \quad \text{(any type of antenna)} \tag{8.187}$$

Interestingly, we have already found the value of this constant ratio as $\lambda^2/4\pi$ in (8.179) for a particular antenna, namely, the infinitesimal dipole. Therefore, substituting this value of constant in (8.187), we obtain:

$$\frac{A_{e,\,\max}}{D_0} = \frac{\lambda^2}{4\pi}. \quad \text{(any type of antenna)} \tag{8.188}$$

Friis transmission equation

Let us now proceed to relate the maximum power received and delivered to a load by a receiving antenna to the power transmitted by a transmitting antenna at a distance through their respective power gains such that the receiving antenna is placed in the direction of the maximum power density radiated by the transmitting antenna and maximum power is delivered to the load. For this purpose, we need to first have the definition of the efficiency of an antenna.

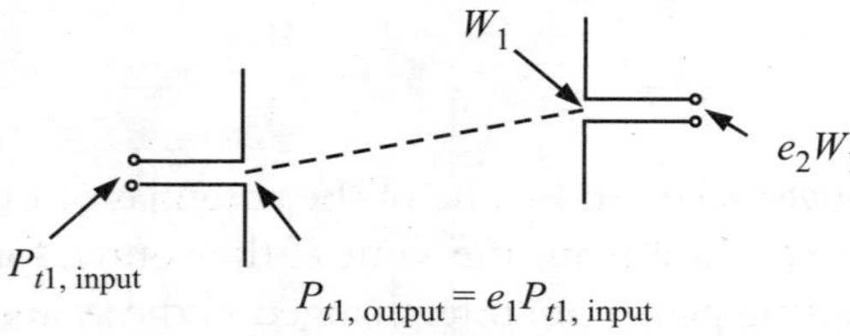

Fig. 8.12 *Transmitting and receiving antennas #1 and #2 showing the powers at the input and output of the transmission lines connected to the antennas #1 and #2 respectively.*

Let the powers at the input and output of a transmission line connected to a transmitting antenna be $P_{t1,\text{input}}$ and $P_{t1,\text{output}}$, respectively (Fig. 8.12). In general, in view of the loss in the transmission line of antenna #1, $P_{t1,\text{input}} \neq P_{t1,\text{output}}$ and we may relate $P_{t1,\text{input}}$ with $P_{t1,\text{output}}$ through the relation

$$P_{t1,\text{output}} = e_1 P_{t1,\text{input}} \tag{8.189}$$

where e_1 is defined as the efficiency of the transmitting antenna, the subscript 1 referring to the transmitting antenna taken as antenna #1.

Further, with the help of (8.180) and (8.189), let us now put the directive gain D_{g1} of antenna #1 as

$$D_{g1} = \frac{1}{e_1}\frac{W_1}{\dfrac{P_{t1,\text{input}}}{4\pi\, r^2}} \tag{8.190}$$

and define the power gain G_1 of antenna #1 with the help of (8.180) as

$$G_1 = \frac{W_1}{\dfrac{P_{t1,\text{input}}}{4\pi\, r^2}}. \tag{8.191}$$

Thus, with the help of (8.190) and (8.191) we can relate G_1 with D_{g1} as

$$G_1 = D_{g1}e_1. \tag{8.192}$$

Similarly, the power density appearing at the output of a transmission line connected to a receiving antenna, taken as antenna #2, will be different from that at its input due to the loss in the transmission line of antenna #2. Therefore, the power density at the output of the transmission line of antenna #2 may be put as e_2W_1, where W_1 is the power density due to antenna #1 appearing at the input of the transmission line of antenna #2. Thus, with the help of (8.174) we can relate the power density e_2W_1 at the output of the transmission line of antenna #2 to the power received $P_{r2,\text{output}}$ (the latter subsequently transferred to the load) as follows in terms of the effective aperture area A_{e2} of antenna #2, as

$$A_{e2} = \frac{P_{r2,\text{output}}}{e_2W_1}. \tag{8.193}$$

Substituting

$$W_1 = \frac{D_{g1}e_1P_{t1,\text{input}}}{4\pi\, r^2},$$

obtained from (8.190) in (8.193), we may write

$$A_{e2} = \frac{P_{r2,\text{output}}}{e_2}\frac{D_{g1}e_1P_{t1,\text{input}}}{4\pi\, r^2},$$

which simplifies to the relation

$$\frac{P_{r2,\text{output}}}{P_{t1,\text{input}}} = \frac{A_{e2}D_{g1}e_1e_2}{4\pi\, r^2}. \tag{8.194}$$

We may also express (8.194), using (8.192) as

$$\frac{P_{r2,\text{output}}}{P_{t1,\text{input}}} = \frac{G_1 A_{e2} e_2}{4\pi r^2}. \tag{8.195}$$

Further, substituting A_{e2} from (8.195) in (8.194) we obtain

$$\frac{P_{r2,\text{output}}}{P_{t1,\text{input}}} = \frac{G_1 A_{e1} D_{g2} e_2}{4\pi r^2 D_{g1}}. \tag{8.196}$$

In view of the relation $G_2 = D_{g2} e_2$ for antenna #2, corresponding to the relation (8.192) for antenna #1, we can express (8.196) as

$$\frac{P_{r2,\text{output}}}{P_{t1,\text{input}}} = \frac{G_1 G_2}{4\pi r^2} \frac{A_{e1}}{D_{g1}}. \tag{8.197}$$

However, in (8.197) we can replace A_{e1} and D_{g1} by $A_{e1,\,\text{max}}$ and D_{01} respectively, since, as mentioned following (8.188), in the present context the receiving antenna #2 is placed in the direction of maximum power density of the transmitting antenna #1, with the condition of maximum power transfer to the load connected to antenna #2. Therefore, we may express (8.197) as

$$\frac{P_{r2,\text{output}}}{P_{t1,\text{input}}} = \frac{G_1 G_2}{4\pi r^2} \frac{A_{e1,\,\text{max}}}{D_{01}}. \tag{8.198}$$

Invoking the relation (8.188) in (8.198), we then obtain the so-called Friis transmission equation as follows:

$$\frac{P_{r2,\text{output}}}{P_{t1,\text{input}}} = G_1 G_2 \left(\frac{\lambda}{4\pi r}\right)^2. \quad \text{(Friis transmission equation)} \tag{8.199}$$

Thus, in (8.199) we find a very useful formula to predict the power received by an antenna of known gain due to a given power transmitted by another antenna also of known gain, when the distance between the two antennas and the operating frequency ($= c/\lambda$) are also known.

EXAMPLE E8-7

You are given three antennas of unknown gains. Suggest a method of using Friis transmission equation to find the gain of each of these antennas if you can measure the power transmitted by an antenna and the power received by another for a given operating frequency.

SOLUTION

Let the gains of the three antennas #1, #2 and #3 be G_1, G_2 and G_3 respectively. Adjust antennas #1 and #2 to a known distance of separation r between them. Measure the power received by antenna #2 due to a measured power transmitted by antenna #1 at a known operating frequency f, which can be used to find the value of $\lambda(= c/f)$. We can now use

Friis transmission equation (8.199) to find the value of the product G_1G_2.

Proceeding similarly, we can independently find the values of the products G_1G_3 and G_2G_3. Now, from the values of the products G_1G_2, G_1G_3 and G_2G_3 thus obtained from the measurement of power transmitted from one antenna and that of power received by the other, we can calculate G_1, G_2 and G_3 from the following relation that can be obtained by simple algebra:

$$\left.\begin{aligned} G_1 &= \sqrt{\frac{(G_1G_2)(G_1G_3)}{G_2G_3}} \\ G_2 &= \sqrt{\frac{(G_2G_3)(G_1G_2)}{G_1G_3}} \\ G_3 &= \sqrt{\frac{(G_2G_3)(G_1G_3)}{G_1G_2}} \end{aligned}\right\}.$$

EXAMPLE E8-8

Two identical antennas having the same gain, one of which transmits power while the other receives it, have their significant dimension defined as 20 cm (see Appendix A8.3). The antennas are separated by a distance equal to 1.5 times the minimum distance prescribed for far-field zone (Appendix A8.3) in a measurement setup using a 10 GHz operating frequency. Calculate the antenna gain if the received power measured is 20 dB below the transmitted power.

SOLUTION

According to the given data, the power transmitted P_t and the power received P_r satisfy the relation

$$10\log_{10}\frac{P_t}{P_r} = 20,$$

which gives

$$\frac{P_t}{P_r} = 100. \tag{E8-8.1}$$

Since the operating frequency given is 10 GHz, we can find wavelength λ as

$$\lambda = \frac{c}{f} = \frac{3\times 10^{10}}{10\times 10^{9}}\,\text{cm} = 3\text{ cm}. \tag{E8-8.2}$$

Further, in view of the minimum distance prescribed for far-field zone by (A8.11) (see Appendix A8.3), and the given value $D = 20$ cm, we may write for the present problem

$$r = 1.5\times\frac{2D^2}{\lambda} = 1.5\times\frac{2(20)^2}{3} = (20)^2 = 400\text{ cm}. \tag{E8-8.3}$$

Next, putting $G_1 = G_2 = G$ as the gain G of each antenna, we can write with the help of (8.199):

$$G = \frac{4\pi r}{\lambda}\sqrt{\frac{P_r}{P_t}}. \tag{E8-8.4}$$

Let us then substitute $P_t/P_r = 100$ from (E8-8.1), $\lambda = 3$ cm from (E8-8.2), and $r = 400$ cm from (E8-8.3) in (E8-8.4) to get the gain of each antenna as

$$G = \frac{4\pi \times 400}{3}\sqrt{\frac{1}{100}} = \frac{4\pi \times 40}{3} = 167.47,$$

whence we obtain the gain in dB of each antenna as

$$\text{Gain} = 10\log_{10} 167.47 = 10 \times 2.239 = 22.39 \text{ dB}.$$

8.4.3 Finite-length dipole antenna

The infinitesimal dipole antenna proves to be impractical since the value of its radiation resistance given by (8.166) becomes very small in view of extremely small value of the length of the dipole compared to wavelength ($dl << \lambda$), calling for a very large amount of current I according to (8.165) for a reasonable amount of the radiation of power P from such an antenna. However, the analysis of the infinitesimal dipole (Section 8.4.1) continues to be useful in that that it can be easily extended to a more practical antenna providing a higher value of radiation resistance and consequently a lesser amount of current through it for a larger radiated power. Here, let us take up for analysis a centre-fed dipole of finite length.

For this purpose, let us consider two wires of an open-ended two-wire transmission line considered aligned horizontally (Fig. 8.13*a*), bent at points M and N at the same distance from their respective ends to make a centre-fed antenna of finite length l. Typically, for a half-wave dipole ($l = \lambda/2$) shown in Fig. 8.13*b*, the horizontal line portions at the ends MS and NT of the line (Fig. 8.13*a*) become the vertical lengths M′S′ and N′T′ of the antenna (Fig. 8.13*b*) respectively, each of length $l/2 = \lambda/4$. In other words, we have taken here MS= M′S′ $= l/2 = \lambda/4$ as well as M′S′= N′T′ $= l/2 = \lambda/4$. Since the line is open-ended, the current at the end of the line as well at the end points of the dipole antenna will be zero as shown in the standing-wave pattern[4] of current distribution shown by the dotted line in Fig. 8.13. Thus, for the typical half-wave dipole ($l = \lambda/2$), the current is maximum at the middle of the antenna near the feed point, and it is zero at both the ends of the dipole at S′ and T′ respectively (Fig. 8.13*b*). We can represent such a current distribution by the following cosine function:

[4](See Section 10.1.7.)

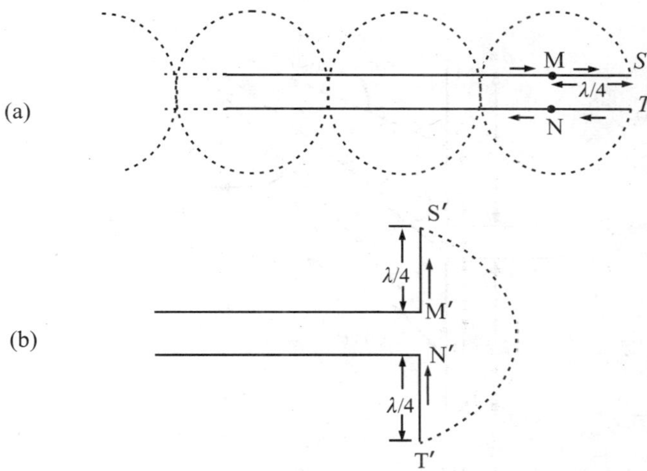

Fig. 8.13 *(a) Two wires of a two-wire transmission line, and (b) the same bent near their respective ends to develop a finite length, typically, a half-wave ($l = \lambda/2$), dipole antenna. The current distribution of the standing-wave pattern (dotted line) are shown in both (a) and (b). The points M and N, shown in (a), are the points where the wires of the line are bent to take the locations M' and N' respectively shown in (b). The end points S and T of the wire in (a) take the locations S' and T' respectively of the end points of the antenna. The distance between M' and N' shown exaggerated in (b) is to be taken much smaller than the length $l = S'T'$ of the antenna.*

$$I = I_0 \cos \beta z = I_0 \cos \frac{2\pi}{\lambda} z. \quad \text{[half-wave dipole: } -l/2(= -\lambda/4) < 0 < l/2(= \lambda/4)] \quad (8.200)$$

Let us take a dipole of finite length $MN = l$ aligned along the Z axis to be divided in two halves, OM= ON = l/2, and proceed to find the electric field due to it at angle θ at a distance OP= $r >> l$ from the middle of the dipole considered at the origin O ($z = 0$) (Fig. 8.14). For this purpose, let us take two infinitesimal current elements at the points A and B on the dipole at distances $z = |z|$ and $z = -|z|$, respectively, from the middle of the dipole (Fig. 8.14). You will appreciate that the distance of the point P from the current element at A is $r - |z| \cos \theta = r - z \cos \theta$, interpreting $z = |z|$ as positive for the half OM (positive z) of the antenna and the current element at B is $r + |z| \cos \theta = r - z \cos \theta$ interpreting $z = -|z|$ as negative for the half ON (negative z) of the antenna (Fig. 8.14). Therefore, with the help of (8.152) and the current distribution (8.200), we may find the contributions from current elements between O and M ($0 < z < l/2$) and those between O and N ($-l/2 < z < 0$) for a half-wave dipole ($l = \lambda/2$) as follows:

$$E_\theta = \int dE_\theta = \int_0^{\lambda/4} \frac{j\eta_0 \beta}{4\pi(r - z\cos\theta)} I_0 \cos \beta z \sin\theta \exp[-j\beta(r - z\cos\theta)] dz$$

$$+ \int_{\lambda/4}^{0} \frac{j\eta_0 \beta}{4\pi(r - z\cos\theta)} I_0 \cos \beta z \sin\theta \exp[-j\beta(r - z\cos\theta)] dz.$$

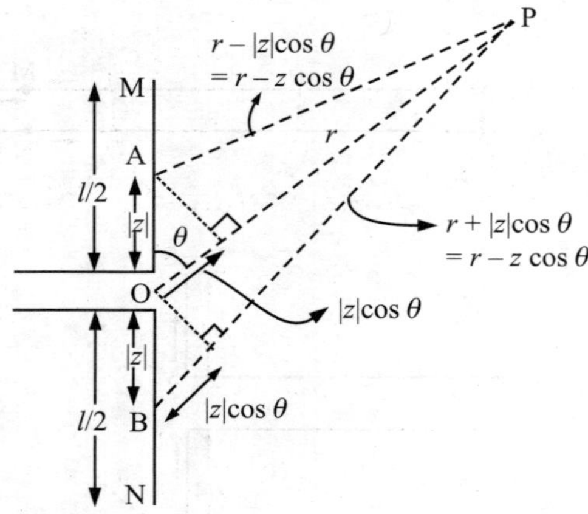

Fig. 8.14 *Finite-length dipole of length MN aligned vertically along z showing two infinitesimal current elements at the points A and B and a distant point P where the field of the dipole is sought.*

which may be simplified by taking $r - z\cos\theta \cong r$ in the denominator to

$$E_\theta = \frac{j\eta_0 \sin\theta\beta I_0}{4\pi r}\left[\int_0^{\lambda/4} \cos\beta z \exp[-j\beta(r - z\cos\theta)]dz + \int_{-\lambda/4}^{0} \cos\beta z \exp[-j\beta(r - z\cos\theta)]dz\right]$$

and subsequently by the rearrangement of terms to

$$E_\theta = \frac{j\eta_0 \sin\theta\beta I_0}{4\pi r}\exp(-j\beta r)\left[\int_0^{\lambda/4} \cos\beta z \exp(j\beta z\cos\theta)dz + \int_{-\lambda/4}^{0} \cos\beta z \exp(j\beta z\cos\theta)dz\right]. \tag{8.201}$$

The result of integration given below is worked out in Appendix A8.4, where a half-wave dipole has been dealt with as a special case of a finite-length dipole of length $l = \lambda/2$.

$$\int \cos\beta z \exp(j\beta z\cos\theta)dz = \frac{\exp(j\beta\cos\theta)z}{\beta\sin^2\theta}(j\cos\theta\cos\beta z + \sin\beta z).$$

We can then obtain from (8.201):

$$E_\theta = \frac{j\eta_0 \sin\theta\beta I_0}{4\pi r}\exp(-j\beta r)\left[\left(\frac{\exp(j\beta\cos\theta)z}{\beta\sin^2\theta}(j\cos\theta\cos\beta z + \sin\beta z)\right)_0^{\lambda/4}\right.$$

$$\left. + \left(\frac{\exp(j\beta\cos\theta)z}{\beta\sin^2\theta}(j\cos\theta\cos\beta z + \sin\beta z)\right)_{-\lambda/4}^{0}\right]. \tag{8.202}$$

Remembering $\beta = 2\pi/\lambda$, we can evaluate (8.202) as

$$E_\theta = \frac{j\eta_0 \sin\theta\beta I_0}{4\pi r}\exp(-j\beta r)\left(\frac{\exp j\frac{\pi}{2}\cos\theta - j\cos\theta}{\beta\sin^2\theta} + \frac{j\cos\theta + \exp -j\frac{\pi}{2}\cos\theta}{\beta\sin^2\theta}\right)$$

which simplifies to

$$E_\theta = \frac{j\eta_0 \sin\theta\beta I_0}{4\pi r}\exp(-j\beta r)\left(\frac{\exp j\frac{\pi}{2}\cos\theta + \exp -j\frac{\pi}{2}\cos\theta}{\beta\sin^2\theta}\right). \tag{8.203}$$

In view of the relation

$$\frac{\exp(j\psi) + \exp(-j\psi)}{2} = \cos\psi,$$

we can then express (8.203) as

$$E_\theta = \frac{j\eta_0 I_0}{4\pi r}\exp(-j\beta r)\frac{2\cos\left(\frac{\pi}{2}\cos\theta\right)}{\sin\theta}. \quad \text{(half-wave dipole)} \tag{8.204}$$

[See Appendix A8.4, where the field expression (8.204) has been derived as (A8.30) for the special case of a finite-length dipole of length $l = \lambda/2$ (half-wave dipole)].

Since E_θ and H_ϕ are related as $E_\theta/H_\phi = \eta_0$ [as already seen following (8.152)], we can write the expressions for H_ϕ and H_ϕ^* as

$$\left.\begin{aligned} H_\phi &= \frac{jI_0}{4\pi r}\exp(-j\beta r)\frac{2\cos\left(\frac{\pi}{2}\cos\theta\right)}{\sin\theta} \\ H_\phi^* &= \frac{-jI_0}{4\pi r}\exp(j\beta r)\frac{2\cos\left(\frac{\pi}{2}\cos\theta\right)}{\sin\theta} \end{aligned}\right\}. \quad \text{(half-wave dipole)} \tag{8.205}$$

To find P, as explained preceding (8.157), we first find the element of power dP through an annular disc of element of area $dS = 2\pi r\sin\theta r d\theta$ at an angle θ (Fig. 8.7), and subsequently integrate it to find the power P propagating through a sphere of radius r. The procedure, as given following (8.156), then leads to

$$P = \int dP = \int \frac{1}{2}\mathrm{Re}(\vec{E}\times\vec{H}^*)\cdot dS\vec{a}_r = \int \frac{1}{2}\mathrm{Re}\, E_\theta H_\phi^* \vec{a}_r \cdot \vec{a}_r 2\pi r\sin\theta r d\theta,$$

which, substituting the field expressions (8.204) and (8.205) takes the following simple form

$$P = \int_0^{\pi} \eta_0 \left(\frac{I_0}{4\pi r}\right)^2 \left(\frac{2\cos\left(\frac{\pi}{2}\cos\theta\right)}{\sin\theta}\right)^2 \pi r \sin\theta \, r d\theta. \tag{8.206}$$

Substituting $\eta_0 = 120\pi$ in (8.206), we obtain

$$P = 30 I_0^2 \int_0^{\pi} \frac{\cos^2\left(\frac{\pi}{2}\cos\theta\right)}{\sin\theta} d\theta \quad \text{(half-wave dipole)} \tag{8.207}$$

in which the integral can be evaluated numerically as

$$\int_0^{\pi} \frac{\cos^2\left(\frac{\pi}{2}\cos\theta\right)}{\sin\theta} d\theta = 1.218. \tag{8.208}$$

At this stage, we can define the radiation resistance of the finite-length dipole R_r as

$$P = \frac{1}{2} I_0^2 R_r. \tag{8.209}$$

Equating (8.206) and (8.208) and rearranging the terms, we then get the expression for R_r as

$$R_r = 60 \int_0^{\pi} \frac{\cos^2\left(\frac{\pi}{2}\cos\theta\right)}{\sin\theta} d\theta \quad \text{(half-wave dipole)} \tag{8.210}$$

which, on putting the value of the integral from (8.208) is evaluated as

$$R_r = 60 \times 1.218 = 73.08 \quad \Omega. \quad \text{(half-wave dipole)} \tag{8.211}$$

Further, as explained preceding (8.160), we can find the average power per unit area W_0 radiated out of a half-wave dipole, or the power density of an isotropic radiator equivalent to the half-wave dipole, by dividing P, given by (8.207) by the area $4\pi r^2$ of the sphere of radius r as

$$W_0 = \frac{30 I_0^2 \int_0^{\pi} \frac{\cos^2\left(\frac{\pi}{2}\cos\theta\right)}{\sin\theta} d\theta}{4\pi r^2}. \quad \text{(half-wave dipole)} \tag{8.212}$$

Further, the power density W of the half-wave dipole in the direction θ can be written with the help of (8.204) and (8.205) as

$$W = \frac{1}{2}\mathrm{Re}\,(\vec{E} \times \vec{H}^*) = \frac{1}{2}\mathrm{Re}\,(E_\theta H_\phi) = \frac{1}{2}\eta_0 I_0^2 \left(\frac{1}{4\pi r}\right)^2 \left(\frac{2\cos\left(\frac{\pi}{2}\cos\theta\right)}{\sin\theta}\right)^2,$$

which can be simplified and expressed by putting $\eta_0 = 120\pi$ as follows:

$$W = 15 I_0^2 \frac{1}{4\pi r^2} \left(\frac{2\cos\left(\frac{\pi}{2}\cos\theta\right)}{\sin\theta} \right)^2 . \quad \text{(half-wave dipole)} \tag{8.213}$$

Therefore, the directive gain of the half-wave dipole, as defined following (8.162), is obtained by dividing (8.213) by (8.212) as

$$D_g = \frac{2\left(\frac{\cos\left(\frac{\pi}{2}\cos\theta\right)}{\sin\theta} \right)^2}{\int\limits_0^\pi \frac{\cos^2\left(\frac{\pi}{2}\cos\theta\right)}{\sin\theta} d\theta} \quad \text{(half-wave dipole)} \tag{8.214}$$

which, substituting the value of the integral (8.208) becomes

$$D_g = \frac{2\left(\frac{\cos\left(\frac{\pi}{2}\cos\theta\right)}{\sin\theta} \right)^2}{1.218} . \quad \text{(half-wave dipole)} \tag{8.215}$$

The directivity of the half-wave dipole D_0 is obtained by maximising D_g given by (8.215) at the angle $\theta = \pi/2$

$$D_0 = \frac{2\left(\frac{\cos\left(\frac{\pi}{2}\cos\frac{\pi}{2}\right)}{\sin\frac{\pi}{2}} \right)^2}{1.218} = \frac{2\left(\frac{\cos(\frac{\pi}{2}\times 0)}{\sin\frac{\pi}{2}} \right)^2}{1.218},$$

which, substituting $\cos 0 = \sin\pi/2 = 1$, yields

$$D_0 = \frac{2}{1.218} = 1.64. \quad \text{(half-wave dipole)} \tag{8.216}$$

Thus, we can see by comparing (8.164) and (8.216) that a half-wave dipole is slightly more directional than an infinitesimal dipole (see Table 8.1). Next, as discussed following (8.166), we can find the radiation pattern of a half-wave dipole. As in the case of the infinitesimal dipole, the horizontal or H-plane pattern ($\theta = \pi/2$) is a circle since the amplitude of E_θ putting $\theta = \pi/2$ in (8.204) becomes uniform (independent of θ). Also, the vertical or E-plane pattern takes the shape of the figure of infinity (∞) since the amplitude of E_θ becomes zero and maximum for $\theta = 0$ and $\theta = \pi/2$ respectively. With the help of (8.204) we can find $(E_\theta)_{\text{amplitude}}$, the amplitude of E_θ at an angle θ, and its maximum value $(E_\theta)_{\text{amplitude}}\big|_{\max}$ at $\theta = \pi/2$ as

$$\left.\begin{aligned}(E_\theta)_{\text{amplitude}} &= \frac{\eta_0 I_0}{4\pi r}\frac{2\cos\left(\dfrac{\pi}{2}\cos\theta\right)}{\sin\theta}\\ (E_\theta)_{\text{amplitude}}\big|_{\max} &= \frac{\eta_0 I_0}{4\pi\, r}\frac{2\cos\left(\dfrac{\pi}{2}\cos\dfrac{\pi}{2}\right)}{\sin\dfrac{\pi}{2}} = \frac{2\eta_0 I_0}{4\pi\, r}\end{aligned}\right\},$$

whence we obtain the following expression for the relative electric field amplitude:

$$\frac{(E_\theta)_{\text{amplitude}}}{(E_\theta)_{\text{amplitude}}\big|_{\max}} = \frac{\cos\left(\frac{\pi}{2}\cos\theta\right)}{\sin\theta} \qquad \text{(half-wave dipole)}$$

the value of which may be put equal to $1/\sqrt{2}$ in order to find HPBW (remembering that the power density is proportional to the square of the field strength) as follows:

$$\frac{\cos\left(\frac{\pi}{2}\cos\theta\right)}{\sin\theta} = \frac{1}{\sqrt{2}},$$

which yields (see also Table 8.1):

$$\text{HPBW} = 78^\circ. \qquad \text{(half-wave dipole)} \tag{8.217}$$

Table 8.1 *Comparison between infinitesimal and half-wave dipoles*

Antenna type	Radiation resistance R_r	Directivity D_0	HPBW
Infinitesimal dipole	Insignificant	1.5	90^0
Half-wave dipole	73.08 Ω	1.64	78^0

EXAMPLE E8-9

Show that the maximum effective length of a half-wave dipole antenna is less than half a wavelength. For this purpose, define the effective length of the antenna as the length of a hypothetical equivalent antenna which carries over its length a uniform current amplitude equal to that of the actual antenna at its feed point such that it results in the same far-field as that of the actual antenna.

SOLUTION

Recall the expression for the far-field electric field component of a half-wave dipole from (8.204) (see also (A8.31) of Appendix A8.4) as

$$E_\theta = \frac{j\eta_0 \sin\theta\beta I_0}{4\pi\, r}\exp(-j\beta\, r)\frac{2\cos\left(\frac{\pi}{2}\cos\theta\right)}{\beta\sin^2\theta}. \qquad \text{(half-wave dipole)} \tag{E8-9.1}$$

In general, for a finite-length dipole (Fig. 8.14) we may write from (A8.22) (see Appendix A8.4)

$$I = I_0 \sin \beta(l/2 - z) \tag{E8-9.2}$$

where z is the distance of a point on the dipole measured from its middle (Fig. 8.14). Now, let us define $I(0)$ as the current I of the dipole at its middle $z = 0$. Then putting $z = 0$ in (E8-9.2) we get

$$I(0) = I_0 \sin \beta(l/2). \tag{E8-9.3}$$

Let us now define a hypothetical antenna equivalent to a finite-length dipole such that over its length the current remains uniform as in an infinitesimal dipole. Let the value of this uniform current be $I(0)$. Now, we need to find the effective length $l_{\text{effective}}$ of the equivalent antenna that gives the same far-field E_θ as that due to a half-wave dipole. Since the currents of both the equivalent antenna and infinitesimal dipole remain uniform over their lengths, we can use the same expression, namely (8.152) for both of them, however, interpreting $dl = l_{\text{effective}}$ and $I = I(0)$ for an equivalent dipole. With this interpretation, therefore, we can write the following far-field expression for an equivalent antenna $l_{\text{effective}}$ with the help of (8.152):

$$E_\theta = \frac{j\,\eta_0\beta}{4\pi\,r} I(0) l_{\text{effective}} \sin\theta \exp(-j\beta\,r). \tag{E8-9.4}$$

Equating the right-hand sides of (E8-9.1) and (E8-9.4), we can then write

$$\frac{j\eta_0 \sin\theta\beta I_0}{4\pi\,r} \exp(-j\beta\,r) \frac{2\cos\left(\frac{\pi}{2}\cos\theta\right)}{\beta\sin^2\theta} = \frac{j\,\eta_0\beta}{4\pi r} I(0) l_{\text{effective}} \sin\theta \exp(-j\beta\,r),$$

which simplifies to

$$l_{\text{effective}} = \frac{I_0}{I(0)} \frac{2\cos(\frac{\pi}{2}\cos\theta)}{\beta\sin^2\theta}. \tag{E8-9.5}$$

Further, for a half-wave dipole ($l/2$), we get from (E8-9.3)

$$I(0) = I_0 \sin\beta(l/2) = I_0 \sin[(2\pi/\lambda)(\lambda/4)] = I_0 \sin\pi/2 = I_0,$$

which substituted in (E8-9.5) gives

$$l_{\text{effective}} = \frac{2}{\beta} \frac{\cos\left(\frac{\pi}{2}\cos\theta\right)}{\sin^2\theta}. \tag{E8-9.6}$$

Notice that $l_{\text{effective}}$ given by (E8-9.6) for a half-wave dipole takes on the maximum value at $\theta = \pi/2$ as follows:

$$(l_{\text{effective}})_{\text{maximum}} = \frac{2}{\beta}\frac{\cos\left(\frac{\pi}{2}\cos\frac{\pi}{2}\right)}{\sin^2\frac{\pi}{2}} = \frac{2}{\beta}\frac{\cos\left(\frac{\pi}{2}\times 0\right)}{\sin^2\frac{\pi}{2}}$$

$$= \frac{2}{\beta}\frac{\cos 0}{\sin^2\frac{\pi}{2}} = \frac{2}{\beta} = \frac{2}{2\pi/\lambda} = \frac{\lambda}{\pi} = \frac{\lambda}{3.14} < \frac{\lambda}{2}. \quad \text{(E8-9.7)}$$

Thus, from (E8-9.7) we conclude that the maximum effective length of a half-wave dipole antenna is less than half the wavelength at the frequency of operation.

For the sake of the simplicity, we have presented the analysis here for a half-wave dipole taking $l = \lambda/2$. However, the generalised expressions for the radiation resistance and directive gain of a finite-length dipole of length l has been worked out in Appendix A8.4 following the same approach of analysis as used here for the half-wave dipole. For the finite length dipole,

$$R_r = 60\,\mathrm{I} \quad \text{[(A8.24) rewritten]} \quad (8.218)$$

$$D_g = \frac{2}{I}\left(\frac{\cos(\beta l\cos\theta) - \cos(\beta l/2)}{\sin\theta}\right) \quad \text{[(A8.25) rewritten]} \quad (8.219)$$

where I is the integral given by

$$\mathrm{I} = \int_0^{\pi} \frac{\left[\cos\left(\frac{\beta l}{2}\cos\theta\right) - \cos\frac{\beta l}{2}\right]^2}{\sin\theta} d\theta. \quad \text{[(A8.26) rewritten]} \quad (8.220)$$

We can find the far-field pattern using the same approach as explained following (8.166) for an infinitesimal dipole. For this purpose, however, instead of taking the amplitude of field from the expression (8.152) for an infinitesimal dipole, we take the expression preceding (8.217) for a finite-length dipole, for instance, $l = \lambda/2$ (half-wave dipole). Accordingly, we can plot, for instance, the vertical (E-plane) pattern of a half-wave dipole (Fig. 8.15*a*). For other antenna lengths, for instance, $l = \lambda$ (Fig. 8.15*b*), $l = (3/2)\lambda$ (Fig. 8.15*c*) and $l = 2\lambda$ (Fig. 8.15*d*), we take the generalised field expression derived in Appendix A8.4 (A8.29) for plotting the field pattern. Thus, we see that the shape of figure of infinity pattern of the infinitesimal dipole (Fig. 8.8) is retained for $l = \lambda/2$ (Fig. 8.15*a*) as well as for $l = \lambda$ (Fig. 8.15*b*). However, for further increase of the length of the dipole, for instance for $l = (3/2)\lambda$ (Fig. 8.15*c*) and $l = 2\lambda$ (Fig. 8.15*d*), side lobes appear, with the number of side lobes increasing and the antenna losing its directive property. The HPBW, which is 90° for an infinitesimal dipole ($l << \lambda$), decreases with the increase of l, thus becoming, for instance,78° and 48° for $l = \lambda/2$ and $l = \lambda$ respectively (Table 8.1).

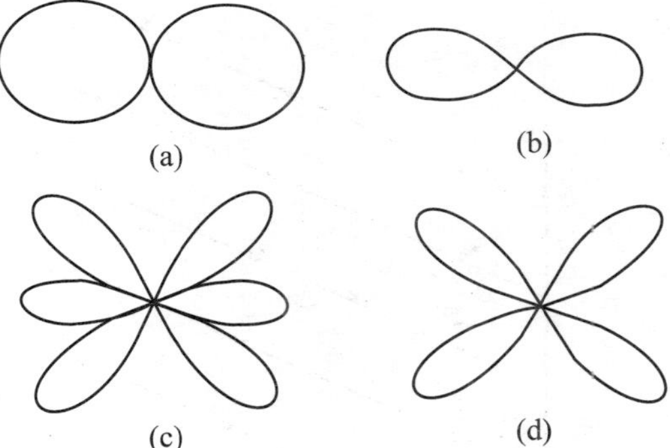

Fig. 8.15 *Vertical (E-plane) pattern of a fine-length dipole aligned vertically for typical lengths: (a) $l = \lambda/2$ (half-wave dipole), (b) $l = \lambda$ (full-wave dipole), (c) $l = (3/2)\lambda$ (one-and-half-wave dipole), and (d) $l = 2\lambda$ (two-wave dipole). (The angle θ, not shown in figure, is measured from the vertical axis).*

8.5 Array of Antennas

It often becomes necessary to beam power from an antenna in one specified direction. How much an antenna is capable of doing so can be estimated both by the antenna HPBW and the directivity, the latter rather more quantitatively. One can increase the directivity of an antenna by designing its geometry/shape and size. However, such an approach becomes somewhat difficult to implement and often leads to an inconvenient antenna geometry/shape and size for a practical antenna design. The alternative approach is to use linear (one-dimensional), planar (two-dimensional) or volume (three-dimensional) array of identical antenna elements. Moreover, such an array antenna provides means to steer the beam of the antenna electronically rather than mechanically, say, by physically rotating a bulky antenna.

For the sake of simplicity, let us consider here for analysis a uniform linear array of identical antenna elements (Fig. 8.16). For such an array, the spacing d and the arithmetically progressive phase between successive elements are considered the same throughout the array. The successive phase difference ψ between the electric fields due to the array elements at a large distance from the array thus becomes the excitation phase difference ξ plus the phase difference $\beta d \cos\theta$, corresponding to the path difference $d\cos\theta$, where $d\theta$ is the direction of the point where the electric field due to the array is sought from the direction of the line or axis of the array (Fig. 8.16):

$$\psi = \beta d \cos\theta + \xi \tag{8.221}$$

Therefore, taking, typically, four elements ($n = 4$) in an array, let antenna element #2 lead antenna element #1 by an excitation phase ξ, say, due to a corresponding lead of phase in the excitation current of the element. Similarly, antenna elements #3 and #4 each lead antenna elements #2 and #3 respectively by the same phase ξ (Fig. 8.17a). Later we will

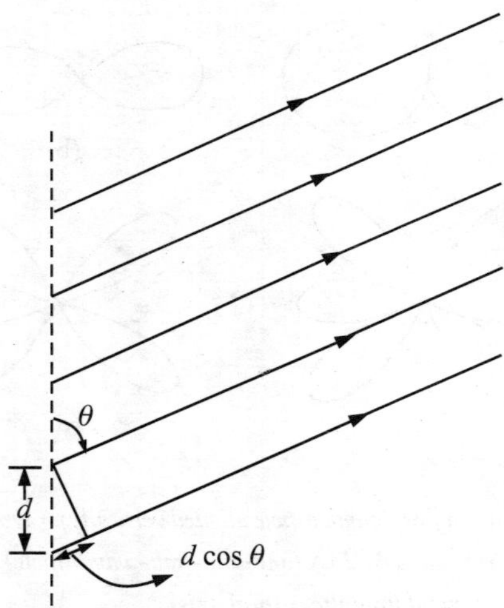

Fig. 8.16 *An array of elements showing the direction of a distant point where the electric field is sought with respect to the array alignment.*

extend the finding of the four-element array to the general case of an array of N number of elements.

With due consideration to this progressive phase difference ψ, we can then draw the vector diagram for the electric field due to each of these elements at a large distance from the array and, subsequently, the vector sum representing the resultant electric field (Fig. 8.17*a*):

$$\overrightarrow{A_1A_5} = \overrightarrow{A_1A_2} + \overrightarrow{A_2A_3} + \overrightarrow{A_3A_4} + \overrightarrow{A_4A_5} \tag{8.222}$$

where $\overrightarrow{A_1A_2}, \overrightarrow{A_2A_3}, \overrightarrow{A_3A_4}$ and $\overrightarrow{A_4A_5}$ represent the electric fields due to the antenna elements #1, #2, #3 and #4 respectively, and $\overrightarrow{A_1A_5}$ represents the resultant electric field of the four elements summed together. Let the successive phase difference ψ between the electric fields due to the array elements be (Fig. 8.17*a*)

$$\angle S_1A_2A_3 = \angle S_2A_3A_4 = \angle S_3A_4A_5 = \psi \tag{8.223}$$

such that (Fig. 8.17*a*)

$$\angle A_1A_2A_3 = \pi - \psi \tag{8.224}$$

and (Fig. 8.17a)

$$\angle A_1A_2O = \frac{\pi - \psi}{2}. \tag{8.225}$$

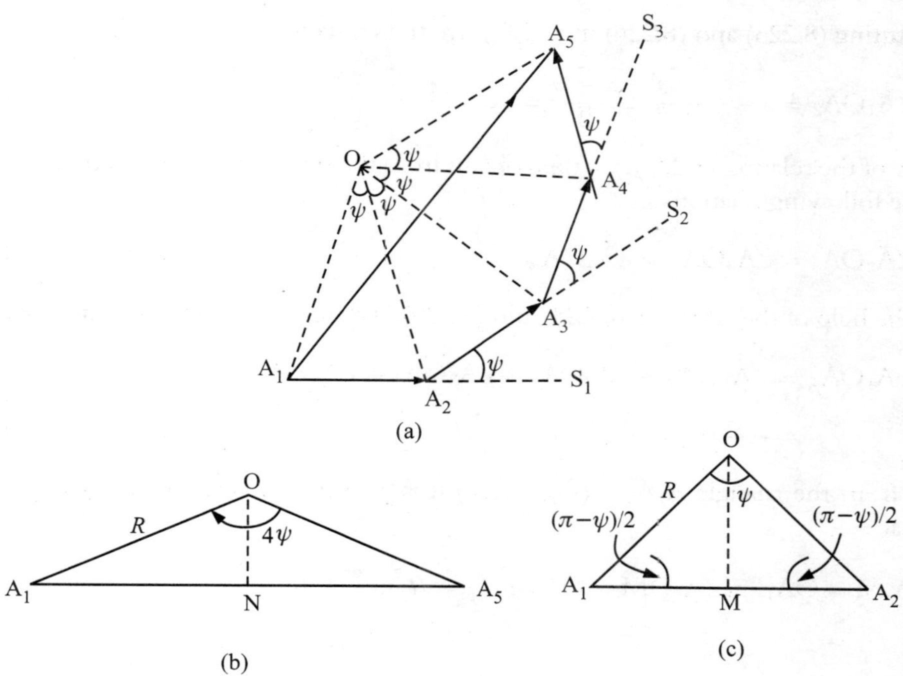

Fig. 8.17 *Vector diagram to represent the electric field due to an array of n elements taking, typically, n=4, at a large distance from the array: (a) the electric field vectors contributed by the individual elements as well as their vector sum making the resultant electric field , (b) the zoomed-out sections of (a) containing the electric field magnitude due to one of the antenna elements , and (c) the zoomed-out sections of (a) containing the electric field magnitude of the resultant electric field due to all the individual antenna elements vector-summed together .*

The geometry of the zoomed-out section of Fig. 8.17*c*, containing the electric field magnitude A_1A_2 due to one of the antenna elements—here, typically, the element #1—shows also angle $\angle A_1A_2O = (\pi - \psi)/2$ given by (8.225) . It is also clear from the symmetry of the geometry of the vector diagram (Fig. 8.17*c*) that

$$\angle A_2A_1O = \frac{\pi - \psi}{2}. \tag{8.226}$$

We can see that the triangle A_1A_2O satisfies the following relation between the angles (Fig. 8.17*c*):

$$\angle A_1A_2O + \angle A_2A_1O + \angle A_1OA_2 = \pi,$$

whence we get

$$\angle A_1OA_2 = \pi - \angle A_1A_2O - \angle A_2A_1O. \tag{8.227}$$

Substituting (8.225) and (8.226) in (8.227), we then obtain

$$\angle A_1OA_2 = \pi - \frac{\pi - \psi}{2} - \frac{\pi - \psi}{2} = \psi. \tag{8.228}$$

In view of the relation (8.228), we then obtain from the symmetry of geometry (Fig. 8.17*a*) also the following relations:

$$\angle A_2OA_3 = \angle A_3OA_4 = \angle A_4OA_5 = \psi\,. \tag{8.229}$$

With the help of the relations (8.228) and (8.229), we are able to further write (Fig. 8.17a)

$$\begin{aligned}\angle A_1OA_5 &= \angle A_1OA_2 + \angle A_2OA_3 + \angle A_3OA_4 + \angle A_4OA_5 \\ &= \psi + \psi + \psi + \psi = 4\psi.\end{aligned} \tag{8.230}$$

Now, from the triangle A_1A_2O (Fig. 8.17*c*) it is easy to appreciate that $\angle A_1OM = \psi/2$, and that

$$A_1M = OA_1 \sin \angle A_1OM = OA_1 \sin \frac{\psi}{2} = R \sin \frac{\psi}{2} \tag{8.231}$$

where

$$OA_1 = R. \tag{8.232}$$

Also, from the triangle A_1A_2O (Fig. 8.17*c*), putting $A_1A_2 = E_0$, we can write

$$E_0 = A_1A_2 = 2A_1M$$

which, with the help of (8.231), can be expressed as

$$E_0 = A_1A_2 = 2A_1M = 2R \sin \frac{\psi}{2}$$

whence we obtain

$$R = \frac{E_0}{2 \sin \frac{\psi}{2}}. \tag{8.233}$$

Now, from the geometry of the zoomed-out section of Fig. 8.17*a* containing the electric field magnitude A_1A_5 due to all the four antenna elements summed together (Fig. 8.15*b*), we can see that

$$A_1N = OA_1 \sin \angle A_1ON = OA_1 \sin \frac{4\psi}{2},$$

which, with the help of (8.232) may be expressed as

$$\mathrm{A_1N} = R\sin\frac{4\psi}{2}. \tag{8.234}$$

Also, putting $\mathrm{A_1A_5} = E_R$, we obtain (Fig. 8.15*b*)

$$E_R = \mathrm{A_1A_5} = 2\mathrm{A_1N},$$

which may be written with the help of (8.234) as

$$E_\mathrm{R} = 2R\sin\frac{4\psi}{2}. \tag{8.235}$$

Substituting R from (8.233) in (8.235), we then obtain the expression for the resultant field amplitude E_R as:

$$E_\mathrm{R} = E_0\frac{\sin\frac{4\psi}{2}}{\sin\frac{\psi}{2}}. \tag{8.236}$$

It becomes now quite simple to extend the expression (8.236), derived for four antenna elements ($N = 4$), to the following generalised form for N antenna elements [by replacing 4 by N in (8.235)]:

$$E_\mathrm{R} = E_0\frac{\sin\frac{N\psi}{2}}{\sin\frac{\psi}{2}}. \tag{8.237}$$

In terms of the array factor AF defined as

$$\mathrm{AF} = \frac{\sin\dfrac{N\psi}{2}}{\sin\dfrac{\psi}{2}} \tag{8.238}$$

we can now express (8.237) with the help of (8.238) as

$$E_R = (E_0)(\mathrm{AF}) \tag{8.239}$$

meaning thereby that the product of the array factor with the electric field amplitude due an individual array element gives the electric field amplitude due to the array of all the elements put together.

At this juncture, it is useful to define the normalised array factor as

$$(\mathrm{AF})_\mathrm{n} = \frac{\mathrm{AF}}{(\mathrm{AF})_{\max}} \tag{8.240}$$

where $(\mathrm{AF})_{\max}$ is the maximum value of AF obtained by maximising (8.238) as

$$(\mathrm{AF})_{\max} = N. \tag{8.241}$$

Therefore, substituting (8.238) and (8.241) in (8.240) we obtain the normalised array factor as

$$(\mathrm{AF})_{\mathrm{n}} = \frac{\mathrm{AF}}{(\mathrm{AF})_{\max}} = \frac{1}{N}\frac{\sin\dfrac{N\psi}{2}}{\sin\dfrac{\psi}{2}}. \tag{8.242}$$

For smaller values of the successive phase difference between the electric fields due to the elements ψ, we may take $\sin(\psi/2) \approx \psi/2$ in the denominator of the right-hand side of (8.242), enabling us to write the normalised array factor as

$$(\mathrm{AF})_{\mathrm{n}} \cong \frac{\sin\dfrac{N\psi}{2}}{\dfrac{N\psi}{2}}. \tag{8.243}$$

You can get an idea about how the normalised array factor $(\mathrm{AF})_{\mathrm{n}}$ varies with $N\psi/2$ from the approximate expression (8.243), which shows that $(\mathrm{AF})_{\mathrm{n}}$ obeys the well-known variation of the function $(\sin x)/x$, interpreting $x = N\psi/2$. Thus, the function $\sin x/x = \sin(N\psi/2)/(N\psi/2)$ varies periodically with $x(= N\psi/2)$, as does a cosine function. The values of the positive and negative maxima of $\sin x/x = \sin(N\psi/2)/(N\psi/2)$, however, unlike those of the cosine function, vary progressively with $x(= N\psi/2)$.

The minima of the function $\sin x/x = \sin(N\psi/2)/(N\psi/2)$ take place at $x(= N\psi/2) = \pi,\ 2\pi,\ 3\pi,\ 4\pi, 5\pi, 6\pi, \ldots\ldots.$, where the function becomes a null. The principal maximum of the function $\sin x/x = \sin(N\psi/2)/(N\psi/2)$ occurs at $x(= N\psi/2) = 0$ (which also corresponds to $\psi = 0$), while the first negative maximum occurs at $x(= N\psi/2) = 1.43\pi$ and the first positive maximum (secondary) at $x(= N\psi/2) = 2.46\pi$, and so on. However, depending on the value of the number of elements N, some of the minima of the normalised array factor$(\mathrm{AF})_{\mathrm{n}}$ predicted above at $N\psi/2 = \pi,\ 2\pi,\ 3\pi,\ 4\pi, 5\pi, 6\pi, \ldots\ldots.$ will be

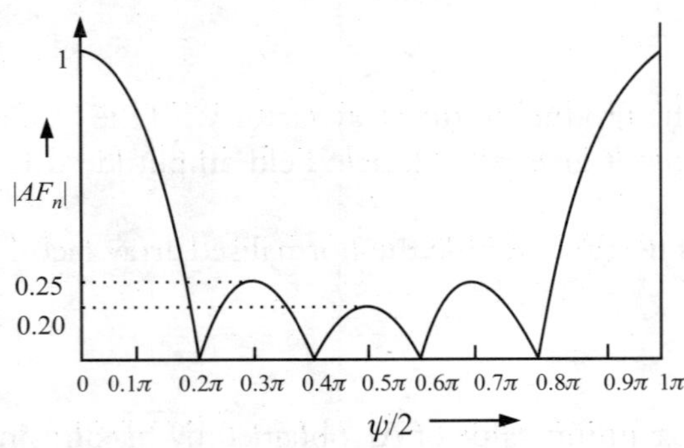

Fig. 8.18 *The magnitude of the normalised array factor* $|(AF)_{\mathrm{n}}|$ *of a uniform linear array of five identical antenna elements,* $N = 5$, *versus* $\psi/2$ *(half the successive phase difference between the electric fields due to the array elements).*

missing, being replaced by the principal maxima. For example, if we take the number of elements of the array as $N = 5$, then the Nth minima—here the fifth minima predicted as above—as $N\psi/2 = 5\pi$ or $\psi/2 = 5\pi/N = 5\pi/5 = \pi$ will be missing, being replaced by a principal maximum. We can appreciate this by noting that, at $\psi/2 = \pi$, the value of $(\text{AF})_n$ given by (8.242) takes on the maximum value of unity: $(\text{AF})_n = (1/N)(N) = 1$, which also corresponds to the principal maximum that occurs at $\psi/2 = 0$ as well (Fig. 8.18).

8.5.1 Broadside array

Let us design the linear array to make it a broadside array such that we obtain a principal maximum in a direction normal to the line of the array, that is, at $\theta = \pi/2$. The principal maximum occurs at $\psi = 0$, as explained following (8.243). Therefore, putting $\theta = \pi/2$ and $\psi = 0$ in (8.221), we obtain the required value of the excitation phase difference between successive elements as $\xi = 0$. Putting this in (8.221) gives the following expression for the successive phase difference ψ between the electric fields due to the array elements at a large distance from the broadside array:

$$\psi = \beta d \cos\theta. \quad \text{(broadside array)} \tag{8.244}$$

Let us now find, with the help of (8.244), the angle θ where the first minimum [which is zero (null)] in the field pattern will occur. As explained following (2.243), for the first null, we have the condition $N\psi/2 = \pi$, giving $\psi = 2\pi/N$, which substituted in (8.244) gives

$$\frac{2\pi}{N} = \beta d \cos\theta. \quad \text{(first null; broadside array)} \tag{8.245}$$

Putting $\beta = 2\pi/\lambda$ in (8.245) and after rearrangement of terms, we obtain the angle θ for the first null as

$$\frac{2\pi}{N} = \frac{2\pi}{\lambda} d \cos\theta,$$

which simplifies to give

$$\theta = \cos^{-1}\frac{\lambda}{Nd}. \quad \text{(broadside array: first null)} \tag{8.246}$$

Since the nulls in the field pattern take place symmetrically on both sides of the principal maximum, we can find the first null bandwidth FNBW as the separation between the first nulls on both sides of the principal maxima as twice the angular separation between one of these first nulls and the principal maximum at $\theta = \pi/2$. Thus, we obtain , with the help of (8.246) FNBW as

$$\text{FNBW} = 2 \times \left(\frac{\pi}{2} - \cos^{-1}\frac{\lambda}{Nd}\right) = \pi - 2\cos^{-1}\frac{\lambda}{Nd}. \quad \text{(broadside array)} \tag{8.247}$$

Similarly, we can find the half power bandwidth HPBW of the broadside array, for which as explained following (8.217), let us put in the present context, $(AF)_n = 1/\sqrt{2}$, which, with the help of (8.243) can be written as

$$\frac{\sin\dfrac{N\psi}{2}}{\dfrac{N\psi}{2}} = \frac{1}{\sqrt{2}}. \quad \text{(half-power condition)} \tag{8.248}$$

The value of $N\psi/2$ satisfying (8.248) is

$$\frac{N\psi}{2} = 1.39. \quad \text{(half-power condition)} \tag{8.249}$$

Putting (8.244) and remembering $\beta = 2\pi/\lambda$, we can express (8.249) as

$$\frac{N 2\pi d \cos\theta}{2\lambda} = 1.39,$$

which, in turn, gives

$$\theta = \cos^{-1}\frac{1.39\lambda}{N\pi d}. \quad \text{(broadside array: half-power location)} \tag{8.250}$$

Therefore, using the same approach as used to derive FNBW of the broadside array from (8.245), we can now obtain the expression for HPBW from (8.250) as follows:

$$\text{HPBW} = 2 \times \left(\frac{\pi}{2} - \cos^{-1}\frac{1.39\lambda}{N\pi d}\right) = \pi - 2\cos^{-1}\frac{1.39\lambda}{N\pi d}. \quad \text{(broadside array)} \tag{8.251}$$

Further, with the help of (8.242) we can find the locations of the other nulls besides the first null as well as those of the secondary maxima besides the principal maximum (at $\theta = \pi/2$) as well as the far-field electric field amplitudes of these secondary maxima relative to the principal maximum electric field amplitude.

Avoidance of the principal maximum along the line of the broadside array

In the broadside array, although there is no principal maximum along the line or axis of the array, a secondary maximum may exist along it for a typical situation. We can find the condition for the non-existence of the principal maximum along line of the array: $\theta = 0$. It follows from (8.242) that the normalised array factor $(AF)_n = 1$ for $\psi/2 = 0$ as well as for $\psi/2 = p\pi$, where $p = 1, 2, 3, \ldots$ is an integer. For $d \neq 0$, corresponding to $\psi/2 = 0$ (principal maximum), we get $\theta = \pi/2$, as can be appreciated from (8.244), which is the usual requirement of the broadside array. However, corresponding to $\psi/2 = p\pi$ (principal maximum), for $d \neq 0$, we obtain from (8.244) the condition:

$$\frac{\beta d \cos\theta}{2} = p\pi,$$

which, interpreted along the line of the array, $\theta = 0$, in view of the relation $\beta = 2\pi/\lambda$, becomes $d = p\lambda$. (existence of a principal maximum along the line of the arrow) (8.251)

Therefore, we obtain from (8.251) the condition for avoiding the principal maximum along the line of the broadside array as

$$d \neq p\lambda. \quad \text{(avoidance of a principal maximum along the line of the arrow)} \quad (8.252)$$

EXAMPLE E8-10

Consider a uniform broadside array of ten elements with the separation between the consecutive elements as quarter of a wavelength. Find the locations of the first and second nulls as well as those of the first and second secondary maxima. Find also the values of these two maxima relative to the principal maximum.

SOLUTION

Based on the discussion following (8.243), we can write the conditions for the first and second nulls as $N\psi/2 = \pi$ and $N\psi/2 = 2\pi$ respectively. Starting from $N\psi/2 = \pi$, we have found, with the help of (8.244), the location of the first null as $\theta = \cos^{-1}\lambda/(Nd)$ (the angle θ being measured from the axis of the array). Similarly, starting from $N\psi/2 = 2\pi$, the location of the second null is obtained as $\theta = \cos^{-1}2\lambda/(Nd)$. Also, based on the discussion following (8.243), we have the conditions for the first and second secondary maxima as $N\psi/2 \cong 1.5\pi$ and $N\psi/2 \cong 2.5\pi$ respectively, and, accordingly, we locate these secondary maxima at $\theta \cong \cos^{-1}1.5\lambda/(Nd)$ and $\theta \cong \cos^{-1}2.5\lambda/(Nd)$ respectively.

Therefore, in view of the locations of the nulls and secondary maxima as discussed above, putting $d = \lambda/4$ and $N = 10$, we can calculate the following:
First null location:

$$\theta = \cos^{-1}\frac{\lambda}{Nd} = \cos^{-1}\frac{\lambda}{10\lambda/4} = \cos^{-1}0.4 = 66.4^0$$

Second null location:

$$\theta = \cos^{-1}\frac{\lambda}{Nd} = \cos^{-1}\frac{2\lambda}{10\lambda/4} = \cos^{-1}2\times 0.4$$

$$= \cos^{-1}0.8 = 36.9^0$$

First secondary maximum location:

$$\theta \cong \cos^{-1}\frac{1.5\lambda}{Nd} = \cos^{-1}\frac{1.5\lambda}{10\lambda/4} = \cos^{-1}1.5\times 0.4$$

$$= \cos^{-1}0.6 = 53.1^0$$

Second secondary maximum location:

$$\theta \cong \cos^{-1}\frac{2.5\lambda}{Nd} = \cos^{-1}\frac{2.5\lambda}{10\lambda/4} = \cos^{-1} 2.5 \times 0.4$$

$$= \cos^{-1} 1.0 = 0^0 \text{ (along the axis of the arrow).}$$

Further, putting $N\psi/2 \cong 1.5\pi$ and $N\psi/2 \cong 2.5\pi$, as well as $N = 10$, we can find the relative values of the first and second secondary maxima with the help of the definition of the normalised array factor given by (8.242) as follows.

Relative value of the first secondary maximum:

$$20\log|(\mathrm{AF})_\mathrm{n}| = 20\log\left|\frac{1}{N}\frac{\sin\frac{N\psi}{2}}{\sin\frac{\psi}{2}}\right|$$

$$= 20\log\left|\frac{1}{10}\frac{\sin 1.5\pi}{\sin 0.15\pi}\right| = 20\log\left|\frac{1}{10}\frac{-1}{0.45}\right|$$

$$= 20\log 0.22 = -20 \times 0.66 = -13.2\,\mathrm{dB}$$

Relative value of the second secondary maximum:

$$|20\log(\mathrm{AF})_\mathrm{n}| = 20\log\left|\frac{1}{N}\frac{\sin\frac{N\psi}{2}}{\sin\frac{\psi}{2}}\right|$$

$$= 20\log\left|\frac{1}{10}\frac{\sin 2.5\pi}{\sin 0.25\pi}\right| = 20\log\left|\frac{1}{10}\frac{1}{0.707}\right|$$

$$= 20\log 0.14 = -20 \times 0.85 = -17\,\mathrm{dB}.$$

8.5.2 End-fire array

Let us design the linear array to make it the so-called end-fire array such that we obtain a principal maximum corresponding to $\psi = 0$ along the line or axis of the array, that is, at either $\theta = 0$ or $\theta = \pi$. Therefore, the required excitation phase difference between successive elements ξ is obtained by putting $\psi = 0$ and $\theta = 0$ in (8.221) as $\xi = -\beta d$ and $\xi = \beta d$ respectively. Substituting $\xi = -\beta d$ (for the end-fire array giving the principal maximum along its axis, $\theta = 0$) in (8.221), we then obtain

$$\psi = \beta d\cos\theta - \beta d = \beta d(\cos\theta - 1). \tag{8.253}$$

The first null, as explained following (2.243), will take place for $N\psi/2 = \pi$, that is, for $\psi = 2\pi/N$, which when put in (8.253) gives

$$\frac{2\pi}{N} = \beta d(\cos\theta - 1),$$

which, in view of the relation $\beta = 2\pi/\lambda$, may be expressed as

$$\cos\theta = 1 + \frac{\lambda}{Nd}. \tag{8.254}$$

However, since the value of $\cos\theta$ cannot be greater than unity, the expression is not admissible. However, the first null will also take place for $N\psi/2 = -\pi$, which we can appreciate recalling the explanation following (2.243) vis-à-vis the nature of the function $[\sin(N\psi/2)]/(N\psi/2)$. Therefore, taking $N\psi/2 = -\pi$ and going by the steps following (8.253) now yields the following admissible value of $\cos\theta$ (less than unity).

$$\cos\theta = 1 - \frac{\lambda}{Nd} \quad \text{(end-fire array)}$$

whence we obtain

$$\theta = \cos^{-1}\left(1 - \frac{\lambda}{Nd}\right). \quad \text{(end-fire array: first null)} \tag{8.255}$$

Since the first nulls occur on both sides of the principal maximum at $\theta = 0$ of the end-fire array, the first null bandwidth FNBW of the end-fire array is obtained as twice the value of θ given by (8.255), as follows:

$$\text{FNBW}\,(= 2\theta) = 2\cos^{-1}\left(1 - \frac{\lambda}{Nd}\right). \quad \text{(end-fire array)} \tag{8.256}$$

We can next find HPBW following the same procedure as used in Section 8.5.1 for the broadside array. However, for a valid solution now, instead of $N\psi/2 = 1.39$ [see the discussion following (8.246)], we have to take $N\psi/2 = -1.39$, that is, $\psi = -2 \times 1.39/N$. We can substitute this value of ψ in (8.253) and use the relation $\beta = 2\pi/\lambda$ to obtain

$$1 - \frac{1.39\lambda}{N\pi\, d} = \cos\theta,$$

which gives

$$\theta = \cos^{-1}\left(1 - \frac{1.39\lambda}{N\pi d}\right) \quad \text{(end-fire array: half-power location)} \tag{8.257}$$

Therefore, the half-power points being located on both sides of the principal maximum at $\theta = 0$ of the end-fire array, we obtain HPBW as as twice the value of θ given by (8.257), as follows:

$$\text{HPBW}\,(= 2\theta) = 2\cos^{-1}(1 - \frac{1.39\lambda}{N\pi d}). \quad \text{(end-fire array)} \tag{8.258}$$

Phased array

We have seen how the excitation phase difference between successive elements ξ need to be different for the broadside ($\xi = 0$) and end-fire arrays ($\xi = \mp\beta d$) that beam power normal to and along the axis of the array respectively. Therefore, by controlling the progressive phase excitation between the elements of the array, in principle, we can steer the beam of an antenna electronically rather than mechanically (by physically rotating the antenna). This is the principle of the phased-array scanning (which can be implemented by controlling the value of ξ by the control of the bias voltage of a hybrid-coupled varactor or the current of the coil wrapping a ferrite phase shifter).

Yagi-Uda antenna

We have, in a type of end-fire array antenna known as Yagi-Uda or fish-bone antenna, the means to transmit or receive a highly directive beam (Fig. 8.19). (You must have seen such an antenna installed on building rooftops for receiving television signals).

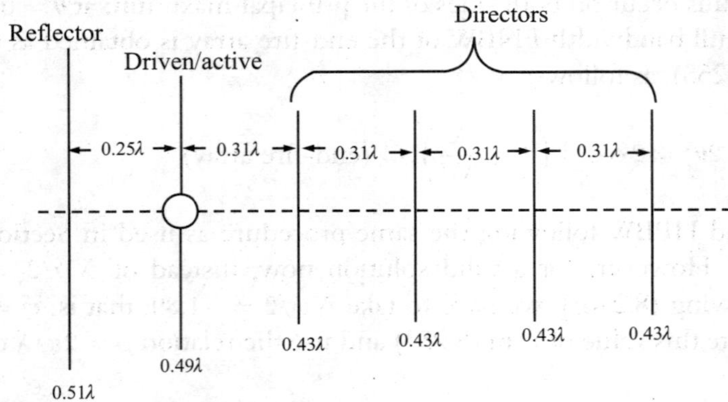

Fig. 8.19 *Yagi-Uda or fish-bone antenna.*

The Yagi-Uda antenna consisting of an array of parallel dipole elements directs power, in the transmitting mode, in the direction of the array axis as does an end-fire array. The thickness of each dipole is significantly less than a wavelength. However, whereas in the end-fire array all the elements are excited, active or driven, in the Yagi-Uda array only one of the elements is active and the remaining elements are all non-excited, passive or parasitic. All the parasitic elements except one are positioned on one side of the driven element along the axis of the array in the direction in which power is to be directed; they are called the 'directors'. The remaining parasitic element, called the 'reflector' is positioned on the opposite side of the driven element (Fig. 8.19). In the receiving mode the power is received only by only one of the elements, namely the driven element and is delivered to the load, although all the remaining elements intercept the incoming power.

Usually, the length of the driven dipole is slightly greater than a half wavelength; the dipole reflector is slightly longer and the dipole directors shorter than the driven dipole. The coupling between the array elements and hence the current distribution in these elements depend on the element lengths and the spacing between elements. In a typical simple example of a six-element Yagi-Uda array, the lengths of the reflector and driven dipoles are 0.51λ and 0.49λ respectively, while the lengths of the four directors are each 0.43λ. The spacing between the consecutive dipoles is 0.31λ throughout the array. The directivity of the antenna reported is 7.54 (8.77 dB) as compared to that of a half-wave dipole 1.64 (2.15 dB) [see (8.216)]. The foregoing simple design of the array is not the optimum for best performance. One must optimise the lengths of the dipole elements and the inter-element spacings, for instance, for attaining the maximum directivity of the Yagi-Uda array.

Appendix A8.1

Separation of $\eta = (j\omega\mu_0/\sigma)^{1/2}$ into its real and imaginary parts

Let us express the intrinsic impedance $\eta = (j\omega\mu_0/\sigma)^{1/2}$ of a conducting medium in terms of its real part A and imaginary part B as

$$\eta = \left(\frac{j\omega\mu_0}{\sigma}\right)^{1/2} = A + jB \tag{A8.1}$$

where A and B are the real quantities. The following expression is then obtained by squaring (A8.1):

$$\frac{j\omega\mu_0}{\sigma} = (A + jB)^2 = A^2 - B^2 + 2jAB. \tag{A8.2}$$

Equating the real and imaginary parts of (A8.2) we obtain

$$A^2 - B^2 = 0 \tag{A8.3}$$

and

$$2AB = \frac{\omega\mu_0}{\sigma}. \tag{A8.4}$$

From (A8.3) we obtain

$$A^2 = B^2$$

which gives

$$A = \pm B. \tag{A8.5}$$

If we take the minus sign in the right-hand side of (A8.5), that is, $A = -B$, and substitute it in (A8.4) we get

$$A^2 = -\frac{\omega\mu_0}{2\sigma}$$

giving

$$A = \pm\sqrt{-\frac{\omega\mu_0}{2\sigma}} = \pm j\sqrt{\frac{\omega\mu_0}{2\sigma}}. \tag{A8.6}$$

With the help of (A8.5) and (A8.6) we can then obtain

$$B = \mp A = \mp j\sqrt{\frac{\omega\mu_0}{2\sigma}}. \tag{A8.7}$$

We may note that A and B, given by (A8.6) and (A8.7) respectively, each become imaginary quantities and hence they are inadmissible since A and B each have to be real quantities as mentioned following (A8.1). However, if we take the plus sign instead of the minus sign in the right-hand side of (A8.5), that is, $A = B$, and proceed as we did to derive (A8.6) and (A 8.7) starting from (A8.1), we obtain A and B as admissible real quantities. Substituting the admissible quantities

$$A = B = \sqrt{\frac{\omega\mu_0}{2\sigma}},$$

thus obtained, in (A8.1) we then derive the following realistic expression for the intrinsic impedance of a conducting medium:

$$\eta = \left(\frac{j\omega\mu_0}{\sigma}\right)^{1/2} = A + jB = \sqrt{\frac{\omega\mu_0}{2\sigma}} + j\sqrt{\frac{\omega\mu_0}{2\sigma}}. \tag{A8.8}$$

The expression (A8.8) is identical to the expression (8.136).

Appendix A8.2

Finding $\nabla\nabla\cdot\vec{A}$ for $\vec{A}$ given by (8.145)

With the help of (2.49) we may write

$$\nabla\cdot\vec{A} = \frac{1}{r^2}\frac{\partial(r^2 A_r)}{\partial r} + \frac{1}{r\sin\theta}\frac{\partial(\sin\theta\, A_\theta)}{\partial\theta} + \frac{1}{r\sin\theta}\frac{\partial(A_\phi)}{\partial\phi}. \tag{A8.9}$$

Substituting (8.145) in (A8.9) we then get

$$\nabla \cdot \vec{A} = \frac{\mu_0 I dl}{4\pi r^2}\left(\cos\theta \frac{\partial[r\exp(-j\beta\,r)]}{\partial r} - \frac{\exp(-j\beta\,r)}{\sin\theta}\frac{\partial(\sin^2\theta)}{\partial\theta}\right). \quad \text{(A8.10)}$$

Carrying out the simple differentiations in (A8.10) , we then obtain

$$\nabla \cdot \vec{A} = \frac{\mu_0 I dl}{4\pi r^2}\left(\cos\theta\{\exp(-j\beta\,r) + (r)(-j\beta)\exp(-j\beta\,r)\} - \frac{\exp(-j\beta\,r)}{\sin\theta} 2\sin\theta\cos\theta\right)$$

which further simplifies to

$$\nabla \cdot \vec{A} = \frac{-\mu_0 I dl}{4\pi}\left(\frac{1}{r^2} + \frac{j\beta}{r}\right)\cos\theta\exp(-j\beta r). \quad \text{(A8.11)}$$

The expression (A8.11) is identical to that given in (8.146) in the text.

Next, we can obtain the following expression for $\nabla\nabla \cdot \vec{A}$ simply by replacing the scalar V in (2.33) by the right-hand side of (A8.11) and noting that the problem in question enjoys essentially azimuthal symmetry, $\partial/\partial\phi = 0$:

$$\nabla\nabla \cdot \vec{A} = \frac{-\mu_0 I dl}{4\pi}\left[\frac{\partial}{\partial r}\left\{\left(\frac{1}{r^2} + \frac{j\beta}{r}\right)\cos\theta\exp(-j\beta r)\right\}\vec{a}_r + \frac{1}{r}\frac{\partial}{\partial\theta}\left\{\left(\frac{1}{r^2} + \frac{j\beta}{r}\right)\cos\theta\exp(-j\beta r)\right\}\vec{a}_\theta\right]$$

which may be put as

$$\nabla\nabla \cdot \vec{A} = \frac{\mu_0 I dl}{4\pi}\left(-\cos\theta\frac{\partial}{\partial r}\{(\frac{1}{r^2} + \frac{j\beta}{r})\exp(-j\beta r)\}\vec{a}_r + (\frac{1}{r^3} + \frac{j\beta}{r^2})\exp(-j\beta r)\sin\theta\}\vec{a}_\theta\right)$$

$$\nabla\nabla \cdot \vec{A} = \frac{\mu_0 I dl}{4\pi}\left[-\cos\theta\frac{\partial}{\partial r}\left\{\left(\frac{1}{r^2} + \frac{j\beta}{r}\right)\exp(-j\beta r)\right\}\vec{a}_r + \left\{\left(\frac{1}{r^3} + \frac{j\beta}{r^2}\right)\exp(-j\beta r)\sin\theta\right\}\vec{a}_\theta\right] \quad \text{(A8.12)}$$

The differential coefficient occurring in the first term of the right-hand side of (A8.12) may be evaluated as

$$\frac{\partial}{\partial r}\left[\left(\frac{1}{r^2} + \frac{j\beta}{r}\right)\exp(-j\beta r)\right] = \exp(-j\beta r)\left(-\frac{2}{r^3} - \frac{j\beta}{r^2}\right) - j\beta\left(\frac{1}{r^2} + \frac{j\beta}{r}\right)\exp(-j\beta r)$$

which simplifies to

$$\frac{\partial}{\partial r}\left[\left(\frac{1}{r^2} + \frac{j\beta}{r}\right)\exp(-j\beta r)\right] = \exp(-j\beta r)\left(-\frac{2}{r^3} - \frac{2j\beta}{r^2} + \frac{\beta^2}{r}\right). \quad \text{(A8.13)}$$

Substituting (A8.13) in (A8.12), we then obtain

$$\nabla\nabla\cdot\vec{A} = \frac{\mu_0 Idl}{4\pi}\left[\left(\frac{2}{r^3}+\frac{2j\beta}{r^2}-\frac{\beta^2}{r}\right)\cos\theta\,\vec{a}_r + \left(\frac{1}{r^3}+\frac{j\beta}{r^2}\right)\sin\theta\,\vec{a}_\theta\right]\exp(-j\beta\, r). \qquad \text{(A8.14)}$$

The expression (A8.14) is identical with that given in (8.146) in the text.

Appendix A8.3

Far-field and near-field zones of an antenna

Let us consider an antenna with its significant linear dimension as D and an observation point P at a distance $r$from its mid-point O (Fig. A8.1). We can find the different ranges of an antenna depending upon the relative value of r with respect to the antenna dimension D and wavelengthλ. From the geometry of the figure (OP $= r$, AB $= D$, OA $=$ OB $= D/2$, AP $= R$, $\angle$AOP $= \theta$), we can write (Fig. A8.1):

$$\text{AP} = [(\text{OA})^2 + (\text{OP})^2 - 2(\text{OA})(\text{OP})\cos\angle\text{AOP}]^{1/2}$$

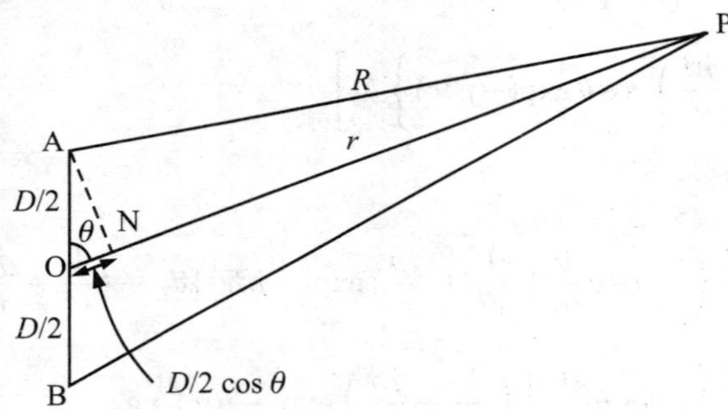

Fig. A8.1 *Observation point at a finite distance from an antenna*

which, in terms of symbols introduced, may be put as

$$R = \left(r^2 + \frac{D^2}{4} - 2r\frac{D}{2}\cos\theta\right)^{1/2}. \qquad \text{(A8.15)}$$

AP and OP tend to become parallel (Fig. A8.2) at very large distances r of the point P, while $r - R$ =the difference between the distances AP $= R$ and OP $= r$ becomes much less, tending to be equal to ON $=$ OA cos $\angle$AOP, that is, to ON $= D/2\cos\theta$:

$$r - R \cong \tfrac{D}{2}\cos\theta$$

which may be put as

$$R \cong r - \frac{D}{2}\cos\theta. \quad \text{(very large distances } r\text{)} \qquad \text{(A8.16)}$$

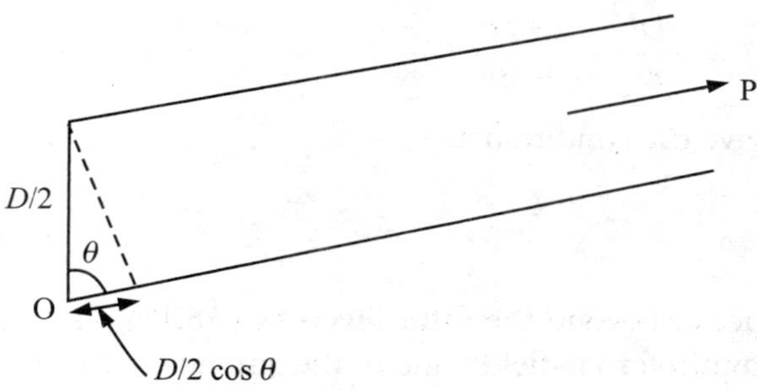

Fig. A8.2 *Observation point far from an antenna.*

In order to further elaborate on the relative importance of the various quantities in (A8.15), let us express the latter as

$$R = \left(r^2 + \frac{D^2}{4} - 2r\frac{D}{2}\cos\theta\right)^{1/2} = r\left(1 + \frac{D^2}{4r^2} - \frac{D\cos\theta}{r}\right)^{1/2}$$

$$= r\left[1 - \left(\frac{D\cos\theta}{r} - \frac{D^2}{4r^2}\right)\right]^{1/2}$$

which may be expanded as

$$R = \left(r - \frac{D}{2}\cos\theta\right) + \frac{D^2 r}{8r^2}(1 - \cos^2\theta) + \frac{D^3 r}{16r^3}\cos\theta(1 - \cos^2\theta) + ...$$

$$= \left(r - \frac{D}{2}\cos\theta\right) + \left(\frac{D^2 r}{8r^2}\sin^2\theta\right) + \left(\frac{D^3 r}{16r^3}\cos\theta\sin^2\theta\right) + ... \quad \text{(A8.17)}$$

If, for $r >> D$, if we ignore the second and third terms within parentheses of (A8.17), then the latter passes on to (A8.16). However, we need to quantitatively assign the values of these terms before their relative contributions in path length can be considered or ignored. Any difference in path length Δ (path) can be converted into difference in phase $\Delta\phi = \beta\Delta$ (path) in view of the RF dependence $\exp(-j\beta r)$. Thus, if we maximise the second term within parenthesis of (A8.17) by taking $\theta = \pi/2$ to the value $\Delta(\text{path}) = D^2/8r$, and ignore it compared to the first term within parenthesis of (A8.17), then the third term within parenthesis of (A8.17) will also be ignored, and consequently the expression (A8.17) approximates to

$$R \cong r - \frac{D}{2}\cos\theta \quad \text{(A8.18)}$$

which becomes identical with (A8.16). In order to implement this approximation, according to IRE standard, one may assign a phase difference $\Delta\phi = \beta\Delta$ (path) equal to $\pi/8$ as the condition, that is,

$$\Delta\phi = \beta\Delta(\text{path}) = \beta\frac{D^2}{8r} = \frac{2\pi}{\lambda}\frac{D^2}{8r} = \frac{\pi}{8}$$

which simplifies to give the condition as

$$r = \frac{2D^2}{\lambda}. \tag{A8.19}$$

Therefore, for distances r beyond the value given by (A8.19), that is, for $2D^2/\lambda < r < \infty$, referred to as the Fraunhofer far-field zone of the antenna, we can ignore the second and third terms and beyond within parentheses in (A8.17) and use the approximate relation (A8.18). Within this far-field zone, the power associated with electric and magnetic fields is essentially radiative, as elaborated in Section 8.4.1.

Similarly, one can find the condition for the third term and onwards within parenthesis of (A8.17) to be insignificant, the second term, however, remaining significant. This can be done by maximizing the third term within parenthesis of (A8.17) and then proceeding as discussed following (A8.17). The quantity $\cos\theta\sin^2\theta$ occurring in the third maximises to the value[5]

$$\cos\theta\sin^2\theta = \frac{2}{3\sqrt{3}}. \quad \text{(maximum value)} \tag{A8.20}$$

In view of (A8.20), the third term within parenthesis of (A8.17) maximizes to the value $\Delta(\text{path}) = (D^3/16r^2)(2/3\sqrt{3})$. Therefore, with reference to this value of $\Delta(\text{path})$, again assigning a phase difference $\Delta\phi = \beta\Delta(\text{path})$ equal to $\pi/8$ according to IRE standard, we obtain the following condition for only the first two terms within parentheses in (A8.17) becoming significant:

$$\Delta\phi = \beta\Delta(\text{path}) = \frac{2\pi}{\lambda}\frac{D^3}{16r^2}\frac{2}{3\sqrt{3}} = \frac{\pi}{8}\Delta\phi = \beta\Delta(\text{path}) = \beta\frac{D^2}{8r} = \frac{2\pi}{\lambda}\frac{D^2}{8r} = \frac{\pi}{8}$$

which simplifies to the condition

$$r = \left(\frac{D^3}{\lambda}\frac{2}{3\sqrt{3}}\right)^{1/2} = 0.62\sqrt{\frac{D^3}{\lambda}}. \tag{A8.21}$$

[5]The quantity $\cos\theta\sin^2\theta$ maximizes for

$$\frac{d}{d\theta}(\cos\theta\sin^2\theta) = 0$$

giving

$$(\cos\theta)(2\sin\theta\cos\theta) + (\sin^2\theta)(-\sin\theta) = 0$$

which simplifies to

$$\tan\theta = \sqrt{2}$$

whence we get

$$\left.\begin{aligned}\sin\theta &= \sqrt{\tfrac{2}{3}}\\ \cos\theta &= \tfrac{1}{\sqrt{3}}\end{aligned}\right\}$$

which, in turn, may be used to obtain (A8.12).

Therefore, according to (A8.21), we can define a range $0 < r < 0.62\sqrt{D^3/\lambda}$, referred to as the near-field zone of the antenna, for which only the first two terms within parentheses becomes significant while the remaining terms beyond insignificant in (A8.17). Within this near-field zone, the power associated with electric and magnetic fields is essentially reactive, as elaborated following (8.155).

Within the two zones $0 < r < 0.62\sqrt{D^3/\lambda}$ (reactive near-field zone) and $2D^2/\lambda < r < \infty$ (Fraunhofer radiative far-field zone), we can also now define an intermediate zone called the Fresnel radiative near-field zone: $0.62\sqrt{D^3/\lambda} < r < 2D^2/\lambda$, where the radiative power dominates the reactive power.

Thus, we may put together the three antenna zones as

1. Reactive near-field zone: $0 < r < 0.62\sqrt{D^3/\lambda}$,
2. Fresnel radiative near-field zone: $0.62\sqrt{D^3/\lambda} < r < 2D^2/\lambda$, and
3. Fraunhofer radiative far-field zone: $2D^2/\lambda < r < \infty$.

Appendix A8.4

Generalised expressions for the radiation resistance and directive gain of a finite-length dipole

For the general case, instead of the expression for (8.200) for current distribution for a half-wave dipole ($l = \lambda/2$), let us take the following expression for a finite-length antenna for both halves of the antenna length: $0 < z < l/2$ (positive z values) and $-l/2 < z < 0$ (negative z values) (Fig. 8.14)

$$I = I_0 \sin\beta(l/2 - z). \qquad \text{(A8.22)}$$

The current distribution assumed as (A8.22) ensures a null in the current distribution at the ends of the dipole: $z = l/2 I = I_0 \sin\beta(l/2 - z) = I_0 \sin\beta(l/2 - l/2) = 0$ at $z = l/2$. For the particular case of a half-wave dipole ($l = \lambda/2$), the current, as can be seen from (A8.22), becomes

$$I = I_0 \sin\beta(l/2 - z) = I_0 \sin\frac{2\pi}{\lambda}[(\lambda/2)/2 - z] = I_0 \sin(\frac{\pi}{2} - \frac{2\pi}{\lambda}z) = I_0 \cos\frac{2\pi}{\lambda}z$$

thus passing on to (8.200). $I = I_0 \sin\frac{2\pi}{\lambda}(l/2 - l/2) = 0$ The current, though it takes the maximum value I_0 at the middle of a half-wave dipole (z=0) (Fig. 8.13b), we can see from (A8.22) that for a general finite-length dipole the current has a finite value at its middle ($z = 0$): $I = I_0 \sin\beta(l/2) = I_0 \sin(2\pi/\lambda)(l/2) = I_0 \sin(\pi l/\lambda)$ and the maximum value $l = \lambda/2 I = I_0 \sin\beta(l/2 = I_0 \sin(2\pi/\lambda)(\lambda/4) = I_0 \sin(\pi/2) = I_0 I = I_0 \sin(\pi/2) = I_0$ at $z = \pi(l/\lambda - 1/2)$.

Similarly, for the general case, instead of (8.201), we will arrive at the following field expression:

$$E_\theta = \left(\frac{j\eta_0 \beta \sin\theta\, I_0}{4\pi r}\exp(-j\beta r)\right)(I_1 + I_2) \tag{A8.23}$$

$$\text{where } I_1 = \int_0^{l/2} \sin\beta\left(\frac{l}{2} - z\right)\exp(j\beta z\cos\theta)dz \tag{A8.24}$$

and

$$I_2 = \int_0^{l/2} \sin\beta\left(\frac{l}{2} + z\right)\exp(j\beta z\cos\theta)dz. \tag{A8.25}$$

In order to evaluate the integrals I_1 and I_2 , we may use the following result of integration:

$$\int \sin(bz + c)\exp azdz = \frac{\exp az}{a^2 + b^2}[a\sin(bz + c) - b\cos(bz + c)]. \tag{A8.26}$$

If we interpret $a = j\beta\cos\theta,\ b = -\beta,\ c = \beta l/2$, we can use the result (A8.26) to evaluate I_1, given by (A8.24) as

$$I_1 = \frac{\cos\left(\beta\frac{l}{2}\cos\theta\right) + j\sin\left(\beta\frac{l}{2}\cos\theta\right) - j\cos\theta\sin\beta\frac{l}{2} - \cos\beta\frac{l}{2}}{\beta\sin^2\theta}. \tag{A8.27}$$

Similarly, interpreting $a = j\beta\cos\theta,\ b = \beta,\ c = \beta l/2$, we can again use (A8.24) to evaluate I_2 given by (A8.17), as

$$I_2 = \frac{\cos\left(\beta\frac{l}{2}\cos\theta\right) - j\sin\left(\beta\frac{l}{2}\cos\theta\right) + j\cos\theta\sin\beta\frac{l}{2} - \cos\beta\frac{l}{2}}{\beta\sin^2\theta}. \tag{A8.28}$$

Adding (A8.27) and (A8.28) we obtain

$$I_1 + I_2 = \frac{2}{\beta\sin^2\theta}\left[\cos\left(\beta\frac{l}{2}\cos\theta\right) - \cos\beta\frac{l}{2}\right]. \tag{A8.29}$$

Substituting (A8.29) in (A8.23), we then obtain the following field expression

$$E_\theta = \frac{j\eta_0\beta\sin\theta\, I_0}{4\pi r}\exp(-j\beta r)\frac{2}{\beta\sin^2\theta}\left[\cos\left(\beta\frac{l}{2}\cos\theta\right) - \cos\beta\frac{l}{2}\right]. \tag{A8.30}$$

For a half-wave dipole ($l = \lambda/2$), putting $\beta l/2 = [(2\pi/\lambda)(\lambda/2)] = \pi/2$ in (A8.29), we obtain

$$E_\theta = \frac{j\eta_0\sin\theta\beta\, I_0}{4\pi r}\exp(-j\beta r)\frac{2\cos\left(\frac{\pi}{2}\cos\theta\right)}{\beta\sin^2\theta}. \quad \text{(half-wave dipole)} \tag{A8.31}$$

We can thus see that (A8.31) is identical to (8.204).

Further, starting from the field expression (A8.23) and using the method described in Section 8.4.3, next, instead of (8.207), (8.210), (8.214) and (8.216) for a half-wave dipole ($l = \lambda/2$), we can find respectively the following generalised expressions for P, R_r and D_g of a finite-length dipole:

$$P = 30 I_0^2\, I \tag{A8.32}$$

$$R_r = 60\, I \tag{A8.33}$$

$$D_g = \frac{2}{I}\left(\frac{\cos(\beta l \cos\theta) - \cos\frac{\beta l}{2}}{\sin\theta}\right) \tag{A8.34}$$

where I is the integral given by

$$\mathrm{I} = \int_0^{\pi} \frac{\left[\cos\left(\frac{\beta l}{2}\cos\theta\right) - \cos\frac{\beta l}{2}\right]^2}{\sin\theta}\, d\theta. \tag{A8.35}$$

As a special case of a half-wave dipole ($l = \lambda/2$), for which $\beta l/2 = (2\pi/\lambda)l/2 = \pi/2$, the integral I given by (A8.35) becomes

$$\mathrm{I} = \int_0^{\pi} \frac{\cos^2\left(\frac{\pi}{2}\cos\theta\right)}{\sin\theta}\, d\theta. \quad \text{(half-wave dipole)} \tag{A8.36}$$

Substituting $\mathrm{I}(= \mathrm{I}_1 + \mathrm{I}_2)$ from (A8.36) in (A8.32)-(A8.34), we obtain the expressions which are identical with (8.207), (8.210) and (8.214) respectively.

However, for the general case, the integral I when evaluated may be expressed as the following series form:

$$\mathrm{I} = C + \ln\beta l\text{-}\mathrm{Ci}(\beta l) + \frac{1}{2}\sin\beta l[\mathrm{Si}(2\beta l) - 2\mathrm{Si}(\beta l)]$$
$$+\frac{1}{2}\cos\beta l[C + \ln(\beta l/2) + \mathrm{Ci}(2\beta l) - 2\mathrm{Ci}(\beta l) \tag{A8.37}$$

where $C = \sum_{n=1}^{\infty} \frac{1}{n}\ln(1 + \frac{1}{n}) = 0.5772$ (Euler's constant)

and $\mathrm{Si}(x)$ and $\mathrm{Ci}(x)$ are sine and cosine integrals:

$$\left.\begin{aligned} \mathrm{Si}(x) &= \int_0^x \frac{\sin x}{x} dx \\ \mathrm{Ci}(x) &= \int_0^x \frac{\cos x}{x} dx \end{aligned}\right\}.$$

For a half-wave dipole ($l = \lambda/2$):$\beta l/2 = \pi/2, \beta l = \pi$;$\text{Ci}(\beta l) = \text{Ci}(\pi) = 0.08$, $\text{Ci}(2\beta l) = \text{Ci}(2\pi) = -0.01$; $\ln \beta l = \ln \pi = 1.114$, $\ln(\beta l/2) = \ln(\pi/2) = 0.45$; $\sin \beta l = \sin \pi = 0$ and $\cos \beta l = \cos \pi = -1$, which, in turn, yields from (A8.36):

$$I = 0.5772 + 1.114\text{-}0.08 + \frac{1}{2}(0.5772 + 0.45 - 0.01 - 2 \times 0.08) \cong 1.21. \quad \text{(half-wave dipole)} \quad \text{(A8.38)}$$

In view of (A8.34), we can evaluate the values of R_r, D_g and D_0 of a half-wave dipole using (A8.33) and (A8.34), which would agree to their values already obtained in Section 8.4.3.

SUMMARISING NOTES

- ✓ Expression for energy density in electric field has been derived in terms of the electric field and electric displacement (or electric flux density).
- ✓ Expression for the energy density in magnetic field analogous to the corresponding expression for energy density in electric field has been appreciated.
- ✓ Expression for the energy stored in a capacitor in terms of the capacitance of the capacitor and the charge of the capacitor or, alternatively, voltage across it has been obtained using the expression for the energy density in electric field..
- ✓ With reference to a parallel-plate capacitor, the expression for the energy density stored in electric field in terms of the electric field and electric displacement or electric flux density has been found to be valid.
- ✓ With reference to a solenoid, the expression for the energy density stored in magnetic field in terms of the magnetic field and magnetic flux density has been found to be valid.
- ✓ Poynting vector (power density vector) has been introduced.
- ✓ Energy balance theorem called the Poynting theorem has been derived involving instantaneous Poynting vector.
- ✓ Poynting theorem encapsulates the phenomenon of the storage, loss and flow of electromagnetic energy.
- ✓ Poynting theorem has been used to appreciate Joule's circuit law for the power loss in a wire of circular cross section and of finite resistance carrying a direct current.
- ✓ Poynting theorem has been used to derive the expression for energy density in electric field with reference to the problem of a parallel-plate capacitor of circular cross section.
- ✓ Poynting theorem has been applied to the problem of an inductor in the form of a solenoid of circular cross section and hence the expression for energy density in magnetic field has been derived,
- ✓ Complex Poynting theorem gives the concept of time averaged electromagnetic power flow.

✓ Complex form of the Poynting theorem has been derived that can be used to study time-averaged power flow through a bounded or an unbounded medium and associated power loss due to the presence of a lossy conducting medium.

✓ Complex Poynting vector has been identified as half the cross product of the electric field vector and the complex conjugate of magnetic field vector.

✓ Time averaged complex Poynting vector has been identified as the real part of the complex Poynting vector.

✓ Average power going out of a volume enclosure can be found as the outward flux of the time averaged complex Poynting vector through the enclosure.

✓ Concept of reactive power flowing into a volume enclosure and its relevance to average energies stored in electric and magnetic fields in the volume has been developed.

✓ Poynting theorem can be further generalised by studying it in the presence of an external source.

✓ Concepts of power flow developed finds extensive applications, for instance, in hollow-pipe waveguides (used for the transmission of power in the microwave frequency range) and antennas.

✓ Expression for the power loss per unit area in a conductor in terms of the surface resistance and the surface current density of the conductor has been deduced—an expression that has applications for instance in estimating the attenuation of power in a hollow-pipe waveguide.

✓ Concepts of power flow have been amply exemplified in their application to conduction current antennas.

✓ Starting from the fundamentals of Hertzian infinitesimal dipole, the antenna concepts have been developed such as

- directive gain;
- power gain;
- radiation resistance;
- effective length;
- effective aperture area; and
- Friis transmission equation relating the radiated power from a transmitting antenna to the power delivered to a load connected to a receiving antenna in terms of the antenna gains, distance between the antennas and wavelength.

✓ Study on antennas has characterised not only the infinitesimal dipole but also

- finite-length dipole;
- broad-side antenna array;

- end-fire antenna array;
- Yagi-Uda antenna; and
- loop-current antenna.

Review Questions

Narrative type

8.1 Derive from first principles the expression for energy density stored in a static electric field: $U_E = (1/2)\vec{E}.\vec{D}$.

8.2 Starting from the expression for the energy stored in a parallel-plate capacitor in terms of its capacitance and the expression for the capacitance in terms of the capacitor dimensions, verify that the energy density stored in the static electric field of the capacitor is $U_{\text{Capacitor}} = (1/2)\vec{E}.\vec{D}$.

8.3 Show that the pressure exerted on the oppositely charged plates of a parallel-plate capacitor is the same as the energy density stored in the static electric field of the capacitor.

8.4 Derive the expression for energy density stored in a static electric field: $U_E = (1/2)\vec{B}\cdot\vec{H}$.

8.5 Derive an expression for the energy density stored in a static magnetic field of a solenoid in terms of its inductance and hence verify that the energy density stored in the static magnetic field of the solenoid is $U_{\text{Solenoid}} = (1/2)\vec{B}\cdot\vec{H}$.

8.6 Derive an expression for the energy stored in the magnetic field of a coaxial cable and hence obtain an expression for the inductance per unit length of the cable in terms of the dimensions of its inner and outer conductors.

8.7 Derive Poynting theorem involving instantaneous Poynting vector as an integral statement of energy balance and explain the physical meaning of all the terms of the statement.

8.8 Derive the complex Poynting vector theorem for time-periodic electric and magnetic fields and interpret its real and imaginary parts.

8.9 Explain the significance of the real part of the complex Poynting vector theorem for time-periodic electric and magnetic fields and hence also how the average power going out of a volume enclosure can be estimated by the integral $1/2 \oint_S Re\,(\vec{E}\times\vec{H}^*)\cdot d\vec{S}$.

8.10 Explain how the imaginary part of the complex Poynting vector for time-periodic electric and magnetic fields expresses the balance of reactive power.

8.11 With the help of Poynting theorem show that the energy density U_E stored in the electric field of a circular parallel-plate capacitor is given by $U_E = (1/2)\varepsilon E^2$, where ε is the permittivity of the dielectric filling the gap between the plates of the capacitor.

8.12) With the help of Poynting theorem show that the energy density U_B stored in the magnetic field of a solenoid is given by $U_B = (1/2)\mu H^2$, where μ is the permeability of the medium of the solenoid.

8.13 For a uniform plane wave incident normally from a free-space medium on a planar conducting surface, derive the expression for the power loss per unit area of the surface P_{LA} as $P_{\text{LA}} = (1/2)R_S\vec{J}_s \cdot \vec{J}_s^*$ in terms of the surface resistance R_S and the surface current density $\vec{J}_s$, the superscript asterisk $(*)$ representing the complex conjugate.

8.14 Define the power density vector triangle and identify its three sides that represent the complex power density, the average power density and the reactive power density vectors respectively.

8.15 Develop a theoretical analysis leading to the definitions of (i) reactive near-field zone: $0 < r < 0.62\sqrt{D^3/\lambda}$, (ii) Fresnel radiative near-field zone: $0.62\sqrt{D^3/\lambda} < r < 2D^2/\lambda$, and (iii) Fraunhofer radiative far-field zone: $2D^2/\lambda < r < \infty$ of an antenna, where D is an antenna dimension of significance, λ is the free-space wavelength, and r is the distance of a point off from the antenna.

Hint: See Appendix A8.3.

8.16 Define directive gain, directivity, radiation resistance, efficiency, power gain and effective aperture area of an antenna.

8.17 What is an infinitesimal or Hertzian dipole? Discuss its role in establishing the concepts of practical antennas.

8.18 Derive the expressions for the far-field electric and magnetic fields with reference to an infinitesimal or Hertzian dipole.

8.19 Discuss the horizontal (H-plane) and vertical (E-plane) patterns of an infinitesimal dipole aligned vertically.

8.20 Show that, for large distances, the wave impedance of an infinitesimal dipole antenna is equal to the intrinsic impedance of free space.

Hint: The wave impedance E_θ/H_ϕ may be found with the help of (8.152).

8.21 Deduce the expressions for radiation fields due to a small circular loop conduction current antenna in the far-field region from the analogous field expressions of an infinitesimal dipole.

Hint: See Example E8-5.

8.22 Verify the dimensional correctness of the expression (8-5.6) in Example E8-5 with reference to a small circular-loop antenna:

$$H_\theta = -\frac{\omega\mu_0 p_m}{4\pi\eta_0}\frac{\exp(-j\beta r)}{r}\beta\sin\theta,$$

where the symbols have been given in Example E8-5.

Hint: In the left-hand side of the expression, the unit of magnetic field is H/m. In the right-hand side of the expression, the unit of p_m is Am^2 and that of η_0 is Ω; the unit of μ_0 is H/m; the unit of magnetic flux is (H)(A) or Wb, while we note that Wb/s (being the unit of the time rate of change of magnetic flux) is V.

8.23 Derive the expressions for the near-field electric and magnetic fields with reference to an infinitesimal or Hertzian dipole.

8.24 Justify that an infinitesimal dipole antenna is analogous to a short electric dipole as far as the electric field close to it is concerned.

8.25 Show that the average Poynting vector $(1/2)\vec{E} \times \vec{H}^*$ associated with the near-field quantities of an infinitesimal dipole antenna is zero. What conclusion regarding the radiative nature of the antenna for an observation point close to the antenna could be drawn from this finding?

8.26 Starting from the far-field expressions for an infinitesimal dipole antenna, derive an expression for its directive gain, directivity and radiation resistance.

8.27 Introducing the concept of the network equivalent of an antenna, prove that the maximum effective aperture area of an infinitesimal dipole is equal to $3\lambda^2/8\pi$.

8.28 Prove that the ratio of the maximum effective aperture area to the directivity of an infinitesimal dipole is equal to $\lambda^2/4\pi$.

8.29 Interpreting the reciprocity theorem of the circuit theory for an antenna system, show that the ratio of the effective aperture area of to its directive gain remains the same for all antennas.

8.30 Establish that the ratio of the maximum effective aperture area to the directivity of any type of antenna is equal to $\lambda^2/4\pi$, assuming a priori that the same ratio is valid with respect to infinitesimal dipole [see (8.179)].

8.31 Prove Friis transmission equation for the power received by an antenna of known power gain due to a given power transmitted by another antenna, also of known power gain, when the distance between the two antennas and the operating frequency $(= c/\lambda)$ are known as well.

Hint: Establish that the ratio of the power received by an antenna of known gain G_2 due to the power transmitted by another antenna of gain G_1 at a distance r is

$G_1G_2(\lambda/(4\pi r^2)$, where $\lambda(= c/f)$ is the wavelength corresponding to the operating frequency f. This ratio gives Friis transmission equation.

8.32 Describe the three-antenna method of finding the value of the power gain of an antenna if you are provided with the antenna power measurement facility as well as three antennas of unknown power gains including the one for which the gain is sought.

Hint: Use Friis transmission equation to find the product of the gains of any two antennas and hence proceed (see Example E8-7).

8.33 Explain why an infinitesimal Hertzian dipole is not suitable for practical applications calling for a finite-length dipole.

8.34 Define the effective length of an antenna and hence show that the maximum effective length of a half-wave dipole antenna is less than half a wave length.

Hint: See Example E8-9.

8.35 Discuss how the E-plane pattern of a fine-length dipole changes as the length l of the dipole is changed from $l = \lambda/2$ to $l = \lambda$ to $l = (3/2)\lambda$ to $l = 2\lambda$.

Hint: See Fig. 8.15.

8.36 Discuss the advantages of using an array of antenna elements over a single antenna element.

8.37 How do you arrive at the definitions of array factor and 'normalised' array factor of a uniform linear array antenna in terms of the number of array elements and the successive phase difference between the electric fields due to the array elements at a large distance from the array?

8.38 Deduce the expressions for FNBW and HPBW of a uniform broadside array in terms of the number of elements, the separation between the elements and the wavelength.

8.39 Deduce the expressions for FNBW and HPBW of a uniform end-fire array in terms of the number of elements, the separation between the elements, and the wavelength.

8.40 Describe Yagi-Uda antenna mentioning both its similarities and differences with a conventional end-fire array and comparing its directivity with that of a half-wave dipole.

Review questions with solutions

8.41 Two large parallel metallic plates separated by a distance d are immersed in a liquid of mass density ρ and permittivity ε. Show that, after applying a potential difference V between the plates, the level of the liquid surface occupied within the plates will rise above its original level by a height of $h = (\varepsilon - \varepsilon_0)V^2/(2d^2\rho g)$.

Answer: The level of the liquid within the plates rises by a height h after the application of the potential difference between the plates. The energy density within the height h of the liquid according to (8.29) is $(1/2)\varepsilon E^2$ after the application of the potential difference between the plates. The energy stored within the volume τ, say, occupying the height of h of the liquid is then $(1/2)\varepsilon E^2\tau$. However, before the application of this potential difference, the same region of height h was occupied by free space and the energy density within the region according to (8.29) was $(1/2)\varepsilon_0 E^2$ and the energy stored $(1/2)\varepsilon_0 E^2\tau$. Therefore, we can find the following expression for the change in the energy stored:

Change in stored energy occupying the height h within the plates

$$= \frac{1}{2}(\varepsilon - \varepsilon_0)\frac{V^2}{d^2}\tau. \qquad (1)$$

where τ is the volume of the portion of the liquid raised in level within the plates.

Next, let us write the following expression:

Work done in causing a rise in the height h of the liquid level between the plates

$$= mgh = \rho\tau gh \qquad (2)$$

where $m(= \rho\tau)$ is the mass of the portion of the liquid of height h within the plates.

We can then equate the right-hand sides of (1) and (2) to obtain

$$\frac{1}{2}(\varepsilon - \varepsilon_0)\frac{V^2}{d^2}\tau = \rho\tau gh,$$

which gives

$$h = \frac{(\varepsilon - \varepsilon_0)V^2}{2d^2\rho g}.$$

8.42 A magnetic field is produced in a helically wound coil (solenoid) by a 'start current pulse' by a high voltage pulse source supplied to one end of the solenoid; the other end is connected to a load. When the start current reaches a peak, a short is established at the input end of the coil thereby isolating the coil from the power source such that the short thus established moves towards the load end of the coil and the coil gets partly destroyed up to the beginning of the short in its movement until it reaches the load end when the coil gets completely destroyed. The situation is realised in a high-power microwave power source called the flux compression generator (FCG). Appreciate how a current multiplication takes place at the load end of the coil (with respect to the current at the input end of the coil) before its final disintegration in such an FCG.

Answer: Express with the help of (4.48) the inductance L of the coil in terms of the number of turns of the coil in the following form (Section 4.3): $L = \mu_0 \alpha\, n\, N$, where N is the number of turns, n is the number of turns per unit length, and α is the cross-sectional area of the coil. With the progress of the short towards the load end, while the values of α and n remain constant, the value of N decreases and hence, as can be seen from the above expression, the inductance of the coil L decreases, and, therefore, the current i through the coil increases since the energy of the coil: $(1/2)Li^2$ [elaborated in Section 8.1; see (8.39)] remains constant.

8.43 Assuming a coaxial cable to be made of a good conductor, show that the power carried by it due to a direct current flowing through it is given by the product $V_0 I_0$ of the potential difference V_0 between the inner and outer conductors of the cable and the current I_0 carried by it.

Answer: Applying Gauss' law, we can easily obtain by an expression for the electric field $\vec{E}$ in the region between the inner and outer conductors, identical with the expression (3.44) due to a long line charge distribution (derived in Chapter 3), as follows:

$$\vec{E} = \frac{\rho_L}{2\pi\varepsilon_0 r}\vec{a}_r \tag{1}$$

where ρ_L is the charge per unit length of the inner conductor of the cable. It is implied that, since the long conductors making the cable are made of a good conductor, the electric field tangential to the conducting surface is zero, thereby making the electric field to be essentially normal to the surface in the radial direction. Further, we can relate ρ_L to the potential difference between the conductors V_0 with the help of the relation (E3-9.6) as follows:

$$\rho_L = \frac{2\pi\varepsilon_0 V_0}{\ln\frac{b}{a}} \tag{2}$$

where a and b are the inner and outer conductor radii, respectively. Therefore, substituting (2) in (1), we get the expression for the electric field in the region between the conductors as

$$\vec{E} = \frac{V_0}{\ln\frac{b}{a}}\frac{1}{r}\vec{a}_r. \tag{3}$$

The magnetic field $\vec{H}$ the region between the conductors can be found using Ampere's circuital law as

$$\vec{H} = \tfrac{I_0}{2\pi r}\vec{a}_\theta, \tag{4}$$

I_0 representing the current of the inner conductor.

With the help of (3) and (4), we can find the Poynting vector $\vec{P}(= \vec{E}\times\vec{H})$ in the region between the conductors as

$$\vec{P} = \vec{E}\times\vec{H} = \frac{V_0}{\ln\dfrac{b}{a}}\frac{1}{r}\vec{a}_r \times \frac{I_0}{2\pi r}\vec{a}_\theta = \frac{V_0}{\ln\frac{b}{a}}\frac{1}{r}\frac{I_0}{2\pi r}\vec{a}_r\times\vec{a}_\theta = \frac{V_0}{\ln\frac{b}{a}}\frac{1}{r}\frac{I_0}{2\pi r}\vec{a}_z. \tag{5}$$

We can now find the element of power dP flowing through an element of area $dS = 2\pi r dr$ of an annular zone lying between r and $r + dr$ on the cross section of the cable as:

$$dP = \vec{P} \cdot d\vec{S} = \vec{P} \cdot 2\pi r dr \vec{a}_z \tag{6}$$

where we have put the element of area vector as $d\vec{S} = 2\pi r dr \vec{a}_z$. Substituting (2) in (1), we obtain

$$dP = \vec{P} \cdot 2\pi r dr \vec{a}_z = \frac{V_0}{\ln\frac{b}{a}} \frac{1}{r} \frac{I_0}{2\pi r} \vec{a}_z \cdot 2\pi r dr \vec{a}_z = \frac{V_0 I_0}{\ln\frac{b}{a}} \frac{dr}{r} \vec{a}_z \cdot \vec{a}_z = \frac{V_0 I_0}{\ln\frac{b}{a}} \frac{dr}{r}. \tag{7}$$

Integrating (7) between the inner and outer conductors of radii a and b respectively, we can find the power P transmitted through the region between the conductors of the cable as

$$P = \int_a^b \frac{V_0 I_0}{\ln\frac{b}{a}} \frac{dr}{r} = \frac{V_0 I_0}{\ln\frac{b}{a}} \int_a^b \frac{dr}{r} = \frac{V_0 I_0}{\ln\frac{b}{a}} \ln\frac{b}{a} = V_0 I_0.$$

8.44 Derive an expression for the radiation resistance of a half-wave dipole and hence find its value in view of the following definite integral value (obtainable numerically)

$$ds \int_0^\pi \frac{\cos^2\left(\frac{\pi}{2}\cos\theta\right)}{\sin\theta} d\theta = 1.218.$$

Answer: 73.08 Ω.

8.45 Derive an expression for the directivity of a half-wave dipole and hence find its value in view of the value of the definite integral given by (A8.38).

Answer: 1.64.

8.46 Show that the half-power bandwidth of a half-wave dipole is HPBW $= 78^0$.

8.47 Give a comparison between infinitesimal and half-wave dipoles with respect to radiation resistance, directivity and HPBW.

Answer: The radiation resistance increases from an insignificantly small value of an infinitesimal dipole to 73.08 Ω of a half-wave dipole. The directivity increases from the value 1.5 of an infinitesimal dipole to 1.64 of a half-wave dipole. HPBW decreases from the value 90^0 of an infinitesimal dipole to 78^0 of ı half-wave dipole. (See Table 8.1).

8.48 Write down the generalised expressions for the (a) radiation resistance and (b) directive gain of a finite-length dipole of length l in terms of a definite integral I defined as

$$\mathrm{I} = \int_0^{\pi} \frac{\left[\cos\left(\frac{\beta l}{2}\cos\theta\right) - \cos\frac{\beta l}{2}\right]^2}{\sin\theta} d\theta.$$

Answer:

(a) Radiation resistance $= 60\,\mathrm{I}$

(b) Directivity $= \dfrac{2}{\mathrm{I}}\left(\dfrac{\cos(\beta l\cos\theta) - \cos(\beta l/2)}{\sin\theta}\right)$.

8.49 With reference to a uniform linear array antenna, how can you define the successive phase difference between the electric fields due to the array elements at a large distance from the array ψ in terms of the excitation phase difference ξ, the separation d between the successive elements and the angle of the direction of the observation point from the line or axis of the array θ?

Answer:

$\psi = \beta d\cos\theta + \xi$, where $\beta(= 2\pi/\lambda)$ is the wave phase propagation constant.

8.50 What is a uniform broadside array? What phase condition of the array could implement this kind of array?

Answer:

$$\left.\begin{aligned} \xi &= 0 \\ \psi &= \beta d\cos\theta \end{aligned}\right\},$$

corresponding to the occurrence of the principal maximum at $\theta = \pi/2$, where $\beta(= 2\pi/\lambda)$ is the wave phase propagation constant and the remaining symbols as explained preceding (8.221).

8.51 What is a uniform end-fire array? What phase condition of the array could implement this kind of array?

Answer:

(i) $$\left.\begin{aligned} \xi &= -\beta d \\ \psi &= \beta d(\cos\theta - 1) \end{aligned}\right\},$$

corresponding to the occurrence of the principal maximum at $\theta = 0$,

and

(ii) $$\left.\begin{aligned} \xi &= \beta d \\ \psi &= \beta d(\cos\theta + 1) \end{aligned}\right\},$$

corresponding to the occurrence of the principal maximum at $\theta = \pi$, where $\beta(= 2\pi/\lambda)$ is the wave phase propagation constant and the remaining symbols as explained preceding (8.221).

Numerical type

8.52 Show that 1 electron volt (eV) $= 1.6 \times 10^{-19}$ J.

Answer: 1 electron volt or 1 eV is the kinetic energy an electron acquires when it falls through a potential energy as it is accelerated by a potential difference of 1 V. (Remember, a particle if released from a height from the ground, where it has a higher gravitational potential than the ground, falls towards the ground through a potential energy and acquires a kinetic energy by the attraction of the gravitational force). The potential V at a point, according to its definition, is the work required to bring a unit charge from infinity up to the point. Accordingly, the work required to move a charge of q against the electric field through a potential difference of V is qV, which is stored as the potential energy of the charge. Therefore, according to the principle of conversation of energy, the kinetic energy of qV will be acquired by a charge q when it is accelerated by a potential difference of V. Thus, an electron of charge $q = |e| = 1.6 \times 10^{-19}$C, when accelerated by a potential difference of $V = 1$ V, will acquire the kinetic energy of

$$qV = |e|\,V = (1.6 \times 10^{-19}) \times (1) = 1.6 \times 10^{-19}\,\text{J},$$

which, by definition, is equal to 1 eV. In other words, 1 eV $= 1.6 \times 10^{-19}$ J.

8.53 A uniform plane wave of peak electric field amplitude 1 V/m and frequency 1 GHz in free space is incident normally on a sheet of copper ($\sigma = 5.8 \times 10^{7}$ mho/m). What is the power density incident per unit area on the sheet?

Answer: 1.326 mW/m^2. (With the help of (8.102), you can find the quantity as $(1/2) \times (1) \times (1/377) = 0.001326$ W/m^2).

8.54 A uniform plane wave of angular frequency 10^{8} radian/s in free space is normally incident on a medium of relative permeability $\mu_r = 1000$ with power density 2 W/m^2. Calculate the average power loss density in the medium.

Answer:

In this problem,

Incident power density$= \frac{1}{2}\eta_0 H_{y0} H_{y0} = 2$ W/m^2 (given),

whence you can calculate

$$(H_{y0})^2 = \frac{4}{\eta} = \frac{4}{\sqrt{\dfrac{\mu}{\varepsilon_0}}} = \frac{4}{\sqrt{\dfrac{\mu_0}{\varepsilon_0}}} = \frac{4}{377} = \frac{4}{377} = 0.0106 \qquad (1)$$

which gives

$$H_{y0} = 0.103 \text{ A/m. } (2)$$

With the help (8.139) we can find the power absorbed per unit area in the medium as

$$\text{Power absorbed per unit area} = \frac{1}{2}R_S(H_{y0})^2 \text{ (3)}$$

where R_S is the surface resistance of the medium given by (8.137), which, in the present context, for the medium of relative permeability μ_r may be put as

$$R_S = \sqrt{\frac{\pi f \mu_0 \mu_r}{\sigma}}. \tag{4}$$

Substituting (4) in (3) you thus obtain

$$\text{Power absorbed per unit area} = \frac{1}{2}\sqrt{\frac{\pi f \mu_0 \mu_r}{\sigma}}(H_{y0})^2. \tag{5}$$

Putting $\mu_0 = 4\pi 10^{-7} = 4 \times 3.143 \times 10^{-7}$ H/m; $\pi f = 2\pi f/2 = \omega/2 = 10^8/2$; $\sigma = 2 \times 10^6$ mho/m; and $H_{y0} = 0.103$ A/m, given by (2), and $\mu_r = 1000$ in (5), we can calculate

$$\text{Power absorbed per unit area} = 0.0094\,\text{W/m}^2 = 9.4\,\text{mW/m}^2.$$

8.55 A direct current of $I = 1$mA passes through a conducting wire of circular cross section of radius $a = 1$cm and length $l = 10$ m, made of a material of conductivity 2×10^7 mho/m. Compare the power loss in the wire found by Joule's law and that found from the concept of Poynting theorem.

Answer: In view of the given values: $\sigma = 2 \times 10^7$ mho/m, $l = 10$ m and $a = 1$ cm $= 10^{-2}$ m, we can find the resistance R of the wire, as

$$R = \frac{1}{\sigma}\frac{l}{\pi a^2} = \frac{1}{2 \times 10^7} \times \frac{10}{3.143 \times (10^{-2})^2} = \frac{1}{2 \times 3.143} \times 10^{-7} \times 10 \times 10^4$$

$$= \frac{1}{2 \times 3.143} \times 10^{-2} = 0.159 \times 10^{-2}\ \Omega. \tag{1}$$

Using (1) and in view of the given value $I = 1\text{mA} = 10^{-3}$ A, we can find the power loss P_{Joule} according to Joule's law as

$$P_{\text{Joule}} = I^2 R = (1 \times 10^{-3})^2 \times 0.159 \times 10^{-2} = 0.159 \times 10^{-8}\ \text{W}. \tag{2}$$

For the current along the axial length, you can find with the help of Ohm's law, $\vec{J} = \sigma\vec{E}$, the axial electric field E as

$$E = \frac{J}{\sigma} = \frac{I}{\pi a^2 \sigma} = \frac{10^{-3}}{3.143 \times (10^{-2})^2 \times 2 \times 10^7}\ \text{V/m}. \tag{3}$$

Further, with the help of Ampere's law, you can find the azimuthal magnetic field H at the circumference of the wire as

$$H = \frac{I}{2\pi a}. \tag{4}$$

Next, we can find, with the help of (4), and using the Poynting theorem approach, the power going into the wire radially inward as the power loss P_{Poynting} as

$$P_{\text{Poynting}} = EH \times 2\pi al = E\frac{I}{2\pi a} \times 2\pi al = EIl. \qquad (5)$$

Therefore, substituting E from (3) and the given values, $I = 1$ mA $= 10^{-3}$ A and $l = 10$ m in (5), we obtain

$$P_{\text{Poynting}} = EIl = \frac{10^{-3}}{3.143 \times (10^{-2})^2 \times 2 \times 10^7} \times 10^{-3} \times 10 = 0.159 \times 10^{-8} \text{ W}. \qquad (6)$$

Thus, we find from (2) and (6) that the power loss in the wire found by Joule's law and that found from the concept of Poynting theorem are one and the same.

8.56 What would be the amount of error if you calculate the value of the radiation resistance of a half-wave dipole of length $l = \lambda/2$ if you use the formula for the radiation resistance of an infinitesimal dipole which is valid for the antenna length much less than λ?

Answer: If we put the value of $dl = \lambda/2$ in the formula (8.166) for the radiation resistance R_r of an infinitesimal dipole (which, in fact, should be applicable only for $dl << \lambda$), we get

$$R_r = 80\pi^2 \left(\frac{\lambda/2}{\lambda}\right)^2 = 80\pi^2 \times \frac{1}{4} = 20 \times (3.143)^2 = 197.57\ \Omega. \qquad (1)$$

However, as we have seen, the actual value of the radiation resistance is given by (8.211), which may be reproduced for comparison with the value found in (1) as

$$R_r = 73.08\ \Omega. \quad \text{(actual value of a half-wave dipole)} \qquad (2)$$

By comparing (1) and (2), we can see that the error is as high as ~170%.

8.57 The power received by an antenna from an identical transmitting antenna at a distance 2.54 m is 22 dB below the transmitted power. If the operating frequency is 10000 MHz, show that the gain of each antenna is 19.27 dB.

Answer: The ratio of the power transmitted to the power received, $P_{\text{rec}}/P_{\text{transm}}$, may be found with the help of Friis transmission equation (8.199) as

$$\frac{P_{\text{rec}}}{P_{\text{transm}}} = G^2 \left(\frac{\lambda}{4\pi r}\right)^2 \qquad (1)$$

where G is the gain of each antenna.

Therefore, we obtain G from (1) as

$$G = \frac{4\pi r}{\lambda}\sqrt{\frac{P_{\text{rec}}}{P_{\text{transm}}}}. \qquad (2)$$

In the present problem,

$$r = 2.54 \text{ m} \quad \text{(given)} \qquad (3)$$

$$\lambda = \frac{c}{f} = \frac{3 \times 10^8}{10000 \times 10^6} = \frac{3 \times 10^8}{100 \times 10^8} = 0.003 \text{ m (given)} \qquad (4)$$

$$10\log\frac{P_{\text{transm}}}{P_{\text{rec}}} = 22. \text{ (given)} \tag{5}$$

With the help of (4) and (5), we find

$$\frac{r}{\lambda} = \frac{2.54}{0.03} = 84.67. \tag{6}$$

From (5), we can find

$$\frac{P_{\text{transm}}}{P_{\text{rec}}} = \text{anti}\log 2.2 = 158.5,$$

whence we get

$$\frac{P_{\text{rec}}}{P_{\text{transm}}} = \frac{1}{158.5} = 0.0063. \tag{7}$$

Therefore, substituting (6) and (7) in (2), we find

$$G = 4\pi\frac{r}{\lambda}\sqrt{\frac{P_{\text{rec}}}{P_{\text{transm}}}} = 4\times 3.143\times 84.67\times\sqrt{0.0063}$$
$$= 4\times 3.143\times 84.67\times 0.0794 = 84.52. \tag{8}$$

The gain in decibel can be found from (8) as

$$G = 10\log 84.521 = 10\times 1.927 = 19.27 \text{ dB}.$$

Multiple-choice questions

8.58 The power flow through a straight wire of circular cross section aligned along z through which a direct current flows in the 'positive' z direction is

(a) along the positive z direction (b) along the negative z direction
(c) in the radially inward direction (d) in the radially outwardly direction.

8.59 The power flow through a straight wire of circular cross section aligned along z through which a direct current flows in the 'negative' z direction is

(a) along the positive z direction (b) along the negative z direction
(c) in the radially inward direction (d) in the radially outward direction.

8.60 For a direct current flow in the positive z direction through the inner conductor and in the negative z direction through the outer conductor of a coaxial cable made of a perfect conductor, the power flow is

(a) in the positive z direction in the region between the conductors
(b) radially inward into the inner or central conductor
(c) in the negative z direction in the region between the conductors
(d) radially outward into the outer conductor.

8.61 A capacitor is charged to a potential and thus an electrical energy is stored in the capacitor. Then the potential source is removed from the capacitor and the capacitor is connected across another identical capacitor. Consequently, the energy of the combination will

(a) remain unchanged from the original energy value
(b) become half the original energy value
(c) become two times the original energy value
(d) become four times the original energy value.

8.62 For a uniform plane wave incident on a conducting surface, the power loss in the conductor per unit area can be reduced by

(a) increasing the field amplitude of the incident wave
(b) decreasing the conductivity of the conductor
(c) decreasing the wave frequency
(d) increasing the resistivity of the conductor.

8.63 One of the following statements with reference to a Hertzian dipole of infinitesimal length *dl* is incorrect:

(a) Maximum effective aperture area $= \lambda^2/4\pi$
(b) Radiation resistance$= 80\pi^2(dl/\lambda)^2$
(c) Directivity $= 3/2$
(d) Far-field wave impedance $= \eta_0$.

8.64 Power received by an antenna from another distant transmitting antenna depends on the parameters: (i) transmitted power, (ii) gain of the transmitting antenna, (iii) gain of the receiving antenna, (iv) wavelength or operating frequency, and (v) distance between the antennas. The received power will remain unchanged if

(a) wavelength is increased two times and the distance between the antennas is reduced two times, the other parameters remaining unchanged
(b) gain of the receiving antenna is increased two times and the distance between the antennas is reduced two times, the other parameters remaining unchanged
(c) gain of the transmitting antenna is increased four times and the operating frequency is reduced two times, the other parameters remaining unchanged
(d) transmitted power is increased four times and the operating frequency is increased two times, the other parameters remaining unchanged.

Answers

8.58 (c) 8.59 (c) 8.60 (a) 8.61 (b) 8.62 (c) 8.63 (a) 8.64 (d)

9

Waveguides: Solution of the Wave Equation for a Wave in a Bounded Medium

A hollow metal pipe used for 'guiding' the transmission of electromagnetic 'waves' from one point to another—called a waveguide—is essentially a microwave-frequency counterpart of the lower-frequency connecting wire. The waveguide, thus, is essentially a type of transmission line, some other types being two-wire line, coaxial cable, parallel-plate line, stripline, and microstrip line. The waveguide is found to be more suitable than the other types of transmission lines for its capability of carrying relatively large powers from one point to another in the microwave frequency range.

We expect the electromagnetic waves propagating through space bounded by the conducting wall of the waveguide to behave differently from those propagating through an unbounded free-space medium (Chapter 6). We have already observed with respect to the unbounded free-space medium that (i) the transverse electromagnetic (TEM) mode of propagation is supported by the medium; (ii) the wave phase velocity v_{ph} is constant with the wave frequency, being equal to $1/(\mu_0\varepsilon_0)^{1/2}$ [see (6.67)]; and (iii) the medium intrinsic impedance η_0 is also constant with regard to frequency, being equal to $\eta_0 = (\mu_0/\varepsilon_0)^{1/2}$ [see (6.70)]. In this chapter we will see how the above parameters and their relationships are influenced by the presence of a conducting waveguide wall by solving the wave equation subject to the electromagnetic boundary conditions imposed by the conducting wall of the waveguide.

We will also see that a hollow-pipe waveguide, usually rectangular or circular in cross section and made of a single conductor, supports two types of modes: the transverse electric (TE) and the transverse magnetic (TM) modes for waveguide cross-sectional dimensions large compared to the wavelength at microwave frequencies. The TE mode is characterised by $H_z \neq 0$, $E_z = 0$ and the the TM mode by $E_z \neq 0$, $H_z = 0$, the direction of propagation being in the axial z direction of the waveguide. We will also see that the TEM mode (refer to Section 6.2) supported by the unbounded free-space medium is not supported by a hollow-pipe waveguide made of a single conductor, though a two-conductor structure like a coaxial waveguide supports it (Section 9.4). The TEM mode, moreover, is supported by a system conductors separated by less than a wavelength at lower frequencies below hundreds of MHz, for instance, by a transmission line such as the two-wire line and the coaxial cable. We will analyse a rectangular waveguide excited in the TE (Section 9.1) and the TM (Section 9.2) modes. The TE and the TM modes for both rectangular and cylindrical waveguides are assigned mode numbers that signify the nature of field variation over the cross-section of the structure perpendicular to the direction of propagation (Sections 9.3 and 9.4). With respect to a cylindrical waveguide, however, our analysis will be restricted to the TE mode of excitation (Section 9.4). The power flow through a waveguide and the associated power loss due to the finite conductivity of the waveguide material will be analysed to predict the attenuation constant of the waveguide (Section 9.5). Such approach of analysis to find the attenuation constant will be exemplified here with particular reference to a rectangular waveguide (Section 9.5).

9.1 Rectangular Waveguide in Transverse Electric Mode

Let us take the TE mode of propagation ($H_z \neq 0, E_z = 0$) for the present study and analyse a rectangular waveguide. This can be treated in the rectangular coordinate system of (x, y, z) in view of the rectangular symmetry of the problem. The TE mode is also known as the H mode since it is associated with a non-zero value of the axial magnetic field ($H_z \neq 0$). We will consider the TM mode later in this chapter (Section 9.2).

9.1.1 Wave equation

Let us recall the wave equation (6.38*c*) in the rectangular system for the non-zero axial magnetic field ($H_z \neq 0$) for the TE mode as:

$$\frac{\partial^2 H_z}{\partial x^2} + \frac{\partial^2 H_z}{\partial y^2} + \frac{\partial^2 H_z}{\partial z^2} = \mu_0 \varepsilon_0 \frac{\partial^2 H_z}{\partial t^2}. \quad \text{(TE mode)} \tag{9.1}$$

Taking the dependence of the field quantities with time and axial distance as $\exp j(\omega t - \beta z)$ for a forward wave propagating along the positive z direction, where ω is the angular frequency and β the axial phase propagation constant of the wave (Section 6.1), we can take the operators $\partial/\partial t \equiv j\omega$ and $\partial/\partial z \equiv -j\beta$ [according to (6.62)] so as to write

$$\left.\begin{aligned} \frac{\partial^2 H_z}{\partial z^2} &= (-j\beta)(-j\beta)H_z = j^2\beta^2 H_z = -\beta^2 H_z \\ \frac{\partial^2 H_z}{\partial t^2} &= (j\omega)(j\omega)H_z = j^2\omega^2 H_z = -\omega^2 H_z \end{aligned}\right\}. \tag{9.2}$$

In view of (9.2), we can then express (9.1) as

$$\frac{\partial^2 H_z}{\partial x^2} + \frac{\partial^2 H_z}{\partial y^2} - \beta^2 H_z = -\mu_0 \varepsilon_0 \omega^2 H_z,$$

which can be rearranged as

$$\frac{\partial^2 H_z}{\partial x^2} + \frac{\partial^2 H_z}{\partial y^2} + (k^2 - \beta^2) H_z = 0 \tag{9.3}$$

where

$$k = \omega(\mu_0 \varepsilon_0)^{1/2} = \frac{\omega}{c} \tag{9.4}$$

is the free-space propagation constant and

$$c = \frac{1}{(\mu_0 \varepsilon_0)^{1/2}} \tag{9.5}$$

is the velocity of light in free space [see (6.67)].

9.1.2 Field solutions

We can solve differential equation (9.3) representing the wave equation by the method of separation of variables, assuming a product solution leading to two ordinary differential equations. Let us try a solution of (9.3) as

$$H_z = XY, \tag{9.6}$$

where X is a function of only x and Y is a function of only y. Note that the factor of RF dependence, $\exp j(\omega t - \beta z)$, is absorbed in (9.3) in view of the interpretation of operators as $\partial/\partial t \equiv j\omega$ and $\partial/\partial z \equiv -j\beta$, and therefore, this factor representing the RF dependence with time t and axial distance z has to be understood in the trial field expression (9.6). The trial solution (9.6) has the partial derivative with respect to x as

$$\frac{\partial H_z}{\partial x} = Y \frac{\partial X}{\partial x},$$

which, on second partial differentiation with respect to x gives

$$\frac{\partial^2 H_z}{\partial x^2} = Y \frac{\partial^2 X}{\partial x^2}. \tag{9.7}$$

Similarly, the first partial differentiation followed by the second partial differentiation of (9.6) with respect to y will yield

$$\frac{\partial^2 H_z}{\partial y^2} = X \frac{\partial^2 Y}{\partial y^2}. \tag{9.8}$$

If we now substitute (9.6)-(9.8) in (9.3), we obtain

$$Y\frac{\partial^2 X}{\partial x^2} + X\frac{\partial^2 Y}{\partial y^2} + (k^2 - \beta^2)XY = 0. \tag{9.9}$$

The following expression is then obtained by dividing (9.9) by XY:

$$\frac{1}{X}\frac{\partial^2 X}{\partial x^2} + \frac{1}{Y}\frac{\partial^2 Y}{\partial y^2} + (k^2 - \beta^2) = 0,$$

which can be rearranged as

$$\frac{1}{X}\frac{\partial^2 X}{\partial x^2} = -\frac{1}{Y}\frac{\partial^2 Y}{\partial y^2} - (k^2 - \beta^2). \tag{9.10}$$

The expression (9.10) equates a function of x alone in its left-hand side with a function of y alone on its right-hand side. This can happen only when the individual functions each become equal to a constant. Putting this constant as $-A^2$, we can write (9.10), part-wise as

$$\frac{1}{X}\frac{\partial^2 X}{\partial x^2} = -A^2 \tag{9.11}$$

and

$$-\frac{1}{Y}\frac{\partial^2 Y}{\partial y^2} - (k^2 - \beta^2) = -A^2$$

which can be arranged as

$$\frac{1}{Y}\frac{\partial^2 Y}{\partial y^2} = A^2 - (k^2 - \beta^2). \tag{9.12}$$

If we put the right-hand side of (9.12) as

$$A^2 - (k^2 - \beta^2) = -B^2 \tag{9.13}$$

we can express (9.12) as

$$\frac{1}{Y}\frac{\partial^2 Y}{\partial y^2} = -B^2. \tag{9.14}$$

The differential equation (9.11) has the well-known solution

$$X = C_1 \cos Ax + C_2 \sin Ax \tag{9.15}$$

where C_1 and C_2 are constants. Similarly, we can write the solution of (9.14) as

$$Y = C_3 \cos By + C_4 \sin By \tag{9.16}$$

where C_3 and C_4 are constants. Therefore, substituting (9.15) and (9.16) in (9.6) and reintroducing the RF dependence factor $\exp j(\omega t - \beta z)$, we obtain the solution as

$$H_z = (C_1 \cos Ax + C_2 \sin Ax)(C_3 \cos By + C_4 \sin By) \exp j(\omega t - \beta z). \tag{9.17}$$

We can obtain the expressions for the transverse components of the electric field and magnetic field components E_x, E_y, H_x, and H_y with the help of (i) Maxwell's equations (5.78) and (5.79) and (ii) the axial component of the magnetic field given by (9.17) as follows. For this purpose, we will take the medium inside the waveguide as free space $(\mu = \mu_0, \varepsilon = \varepsilon_0)$ and the conduction current density $\vec{J}(= \sigma \vec{E})$ appearing in (5.79) to be zero. We can the respectively express (5.78) and (5.79) by expanding the curl in the rectangular coordinate system with the help of (2.66) and using the operator $\partial/\partial t \equiv j\omega$ [according to (6.62)].

$$\left.\begin{aligned}
&\left(\frac{\partial E_z}{\partial y} - \frac{\partial E_y}{\partial z}\right)\vec{a}_x + \left(\frac{\partial E_x}{\partial z} - \frac{\partial E_z}{\partial x}\right)\vec{a}_y + \left(\frac{\partial E_y}{\partial x} - \frac{\partial E_x}{\partial y}\right)\vec{a}_z \\
&\qquad = -j\omega\mu_0(H_x\vec{a}_x + H_y\vec{a}_y + H_z\vec{a}_z) \qquad \text{(a)} \\
&\left(\frac{\partial H_z}{\partial y} - \frac{\partial H_y}{\partial z}\right)\vec{a}_x + \left(\frac{\partial H_x}{\partial z} - \frac{\partial H_z}{\partial x}\right)\vec{a}_y + \left(\frac{\partial H_y}{\partial x} - \frac{\partial H_x}{\partial y}\right)\vec{a}_z \\
&\qquad = j\omega\varepsilon_0(E_x\vec{a}_x + E_y\vec{a}_y + E_z\vec{a}_z) \qquad \text{(b)}
\end{aligned}\right\}. \tag{9.18}$$

Let us take the x-component of (9.18*a*), putting therein $E_z = 0$ for the TE mode, as well as the y-component of (9.18*b*) as follows:

$$\left.\begin{aligned}
&-\frac{\partial E_y}{\partial z} = -j\omega\mu_0 H_x \qquad \text{(a)} \\
&\frac{\partial H_x}{\partial z} - \frac{\partial H_z}{\partial x} = j\omega\varepsilon_0 E_y \qquad \text{(b)}
\end{aligned}\right\}. \tag{9.19}$$

Using the operator $\partial/\partial z \equiv -j\beta$ [according to (6.62)], we can express (9.19) as

$$\left.\begin{aligned}
&j\beta E_y = -j\omega\mu_0 H_x \qquad \text{(a)} \\
&-j\beta H_x - \frac{\partial H_z}{\partial x} = j\omega\varepsilon_0 E_y \qquad \text{(b)}
\end{aligned}\right\}. \tag{9.20}$$

Substituting $H_x = -\beta E_y/(\omega\mu_0)$, obtainable from (9.20*a*) in (9.20*b*), we obtain

$$j\frac{\beta^2 E_y}{\omega\mu_0} - \frac{\partial H_z}{\partial x} = j\omega\varepsilon_0 E_y$$

which can be simplified by rearrangement of terms as

$$j\frac{\beta^2 - \omega^2\mu_0\varepsilon_0}{\omega\mu_0}E_y - \frac{\partial H_z}{\partial x} = 0\,. \tag{9.21}$$

In view of the relation (9.4), we can then write from (9.21) the following expression for the transverse component of the electric field E_y in terms of the rate of the variation of the axial magnetic field H_z with respect to x, one of the waveguide transverse coordinates:

$$E_y = \frac{j\omega\mu_0}{k^2 - \beta^2}\frac{\partial H_z}{\partial x}. \tag{9.22}$$

Similarly, we can find an expression for the electric field E_x. For this purpose, let us take the y-component of (9.18*a*), putting therein $E_z = 0$ for the TE mode, and the x-component of (9.18*b*) as follows:

$$\left.\begin{aligned} &\frac{\partial E_x}{\partial z} = -j\omega\mu_0 H_y && \text{(a)} \\ &\frac{\partial H_z}{\partial y} - \frac{\partial H_y}{\partial z} = j\omega\varepsilon_0 E_x && \text{(b)} \end{aligned}\right\}. \tag{9.23}$$

Using the operator $\partial/\partial z \equiv -j\beta$ [according to (6.62)], we can express (9.23) as

$$\left.\begin{aligned} &-j\beta E_x = -j\omega\mu_0 H_y && \text{(a)} \\ &\frac{\partial H_z}{\partial y} + j\beta H_y = j\omega\varepsilon_0 E_x && \text{(b)} \end{aligned}\right\}. \tag{9.24}$$

Now, substituting $H_y = \beta E_x/(\omega\mu_0)$, obtainable from (9.24*a*) in (9.24b), we obtain the following expression for the transverse component of the electric field E_x in terms of the rate of variation of the axial magnetic field H_z with respect to y, one of the waveguide transverse coordinates:

$$E_x = \frac{-j\omega\mu_0}{k^2 - \beta^2}\frac{\partial H_z}{\partial y}. \tag{9.25}$$

Thus, we obtain from (9.22) and (9.25) the following proportionality between the transverse components of electric field and the rate of the variation of the axial component of magnetic field:

$$\left.\begin{aligned} &E_y \propto \frac{\partial H_z}{\partial x} && \text{(a)} \\ &E_x \propto \frac{\partial H_z}{\partial y} && \text{(b)} \end{aligned}\right\}. \tag{9.26}$$

We will use this proportionality relation (9.26) and the solution for the axial magnetic field (9.17) in order to obtain the solutions for the other field components in the waveguide. For this purpose, we are going to apply the electromagnetic boundary condition (7.46*d*) at the interface between a free space and a conductor, the conducting medium here being each

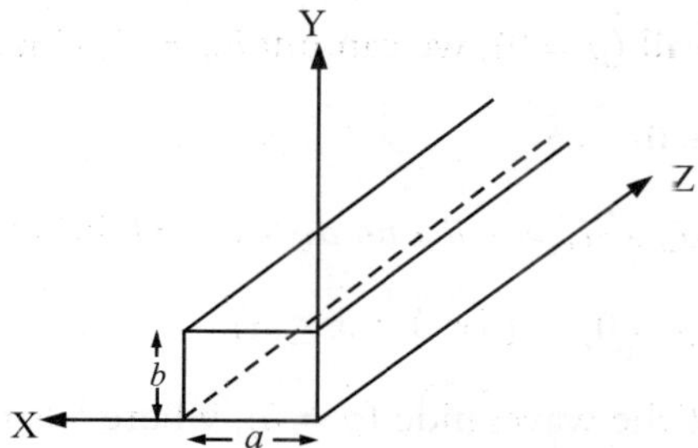

Fig. 9.1 *Rectangular waveguide represented in rectangular coordinate system with its walls located at $x = 0$ (right sidewall); $x = a$ (left sidewall); $y = 0$ (bottom wall); and $y = b$ (top wall).*

of the four waveguide walls (Fig. 9.1). Thus, taking the conducting wall of the waveguide as region 1 and the free-space medium inside the waveguide as region 2, we can write the electromagnetic boundary condition (7.46*d*) at the surface of any conducting waveguide wall as

$$\vec{a}_n \times \vec{E}_2 = \vec{a}_n \times (E_x\vec{a}_x + E_y\vec{a}_y + E_z\vec{a}_z) = 0,$$

which, taking $E_z = 0$ for the TE mode, becomes

$$\vec{a}_n \times \vec{E}_2 = \vec{a}_n \times (E_x\vec{a}_x + E_y\vec{a}_y) = 0\,. \tag{9.27}$$

Remember that the unit vector $\vec{a}_n$ is directed from region 1 (conductor) to region 2 (free space) (Section 7.3).

For the left sidewall of the waveguide ($x = a$), where a is the broad-wall dimension of the waveguide (Fig. 9.1), putting $\vec{a}_n = -\vec{a}_x$, we can then read (9.27) as:

$$-\vec{a}_x \times (E_x\vec{a}_x + E_y\vec{a}_y) = 0,$$

which can be simplified, in view of the relations: $\vec{a}_x \times \vec{a}_x = 0$; $\vec{a}_x \times \vec{a}_y = \vec{a}_z$, as $-E_y\vec{a}_z = 0$, giving the boundary condition

$$E_y = 0 \text{ (left sidewall } x = a) \quad (\text{for } 0 \le y \le b). \tag{9.28}$$

For the right sidewall ($x = 0$) (Fig. 9.1), similarly, putting $\vec{a}_n = \vec{a}_x$ we can read (9.27) as

$$\vec{a}_x \times (E_x\vec{a}_x + E_y\vec{a}_y) = 0,$$

which, using the relation

$$\vec{a}_x \times \vec{a}_x = 0;\ \vec{a}_x \times \vec{a}_y = \vec{a}_z, \text{ becomes } E_y\vec{a}_z = 0 \text{ giving}$$

$$E_y = 0 \text{ (right sidewall, } x = 0) \quad (\text{for } 0 \le y \le b). \tag{9.29}$$

Similarly, for the bottom wall ($y = 0$), we can put $\vec{a}_n = \vec{a}_y$ (Fig. 9.1) and read (9.27) as

$$\vec{a}_y \times (E_x\vec{a}_x + E_y\vec{a}_y) = 0$$

which, using the relations $\vec{a}_y \times \vec{a}_x = -\vec{a}_z and\ \vec{a}_y \times \vec{a}_y = 0$ becomes $-E_x\vec{a}_z = 0$, giving

$$E_x = 0 \text{ (bottom wall } y = 0) \quad (\text{for } 0 \le x \le a). \tag{9.30}$$

Further, for the top wall of the waveguide ($y = b$), where b is the narrow-wall dimension of the waveguide, we put $\vec{a}_n = -\vec{a}_y$ and read (9.27) as

$$-\vec{a}_y \times (E_x\vec{a}_x + E_y\vec{a}_y) = 0$$

which, using the relations $\vec{a}_y \times \vec{a}_x = -\vec{a}_z and\ \vec{a}_y \times \vec{a}_y = 0$ becomes $E_x\vec{a}_z = 0$ giving

$$E_x = 0 \text{ (top wall } y = b) \quad (\text{for } 0 \le x \le a)\ . \tag{9.31}$$

The quantities on the right-hand sides of (9.26) can be obtained by taking the partial differentiation of (9.17) as

$$\left.\begin{aligned} \frac{\partial H_z}{\partial x} &= (-C_1 A \sin Ax + C_2 A \cos Ax)(C_3 \cos By + C_4 \sin By) \quad \text{(a)} \\ \frac{\partial H_z}{\partial y} &= (C_1 \cos Ax + C_2 \sin Ax)(-C_3 B \sin By + C_4 B \cos By) \quad \text{(b)} \end{aligned}\right\}. \tag{9.32}$$

Therefore, applying (9.26*a*) ($E_y \propto \partial H_z/\partial x$) to the boundary condition (9.29) [$E_y = 0$ (right sidewall $x = 0$)], we obtain $\partial H_z/\partial x = 0$ (right sidewall $x = 0$), which, with the help of (9.32*a*), putting $x = 0$ therein gives:

$$\frac{\partial H_z}{\partial x} = (-C_1 A \sin Ax + C_2 A \cos Ax)(C_3 \cos By + C_4 \sin By)$$

$$C_2 A(C_3 \cos By + C_4 \sin By) = 0 \text{ (for all values of } y)$$

which can happen for two situations: either $C_2 = 0,\ A \ne 0$ (situation I) or $A = 0,\ C_2 \ne 0$ (situation II). For these two situations our findings would be as in the following:

For situation I, $C_2 = 0,\ A \ne 0$.

Applying $E_y \propto \partial H_z/\partial x$ (9.26*a*) to the boundary condition $E_y = 0$ (left sidewall $x = a$) (9.28), we obtain $\partial H_z/\partial x = 0$. Putting this value in (9.32*a*) and also letting $x = a$ therein, we get $-C_1 A \sin Aa + C_2 A \cos Aa = 0$ for all values of y ($0 \le y \le b$). Since for this situation $C_2 = 0$, we get

$$-C_1 A \sin Aa = 0. \tag{9.33}$$

Further, for situation I, $A \ne 0$. If $C_1 = 0$, together with $C_2 = 0$ (given for situation I), it makes $H_z = 0$ as can be seen from (9.17). This result violates the condition for the

TE mode ($H_z \neq 0$), and hence for situation I, both $A \neq 0$ and $C_1 \neq 0$. Therefore, from (9.34) we can infer that $\sin Aa = 0$, which is valid for $Aa = m\pi$ ($m = 1, 2, 3, \ldots$), giving the following values of A:

$$A = \frac{m\pi}{a} \; (m = 1, 2, 3, \ldots) \quad \text{(situation I)}. \tag{9.34}$$

Note that the condition $m = 0$ is excluded, since this condition, as can be seen from (9.34), will make $A = 0$, which is invalid for situation I. Therefore, for situation I ($C_2 = 0$, $A \neq 0$), we can express (9.17) as

$$H_z = (C_1 \cos Ax)(C_3 \cos By + C_4 \sin By) \exp j(\omega t - \beta z) \quad (m = 1, 2, 3, \ldots) \quad \text{(for all values of } y) \quad \text{(situation I)}. \tag{9.35}$$

Similarly, for situation II, $A = 0$, $C_2 \neq 0$, we can express (9.17) as

$$H_z = (C_1)(C_3 \cos By + C_4 \sin By) \exp j(\omega t - \beta z) \quad \text{(for all values of } y) \quad \text{(situation II)}. \tag{9.36}$$

Let us now tacitly express (9.36) for situation II as

$$H_z = (C_1 \cos Ax)(C_3 \cos By + C_4 \sin By) \exp j(\omega t - \beta z) \quad (m = 0) \quad \text{(for all values of } y) \quad \text{(situation II)}. \tag{9.37}$$

where

$$A = \frac{m\pi}{a} \; (m = 0) \quad \text{(situation II)}. \tag{9.38}$$

Thus, combining (9.35) and (9.37) encompassing both situations I ($m = 1, 2, 3, \ldots$) and situation II ($m = 0$), we obtain the following expression for the axial component of H_z:

$$H_z = (C_1 \cos Ax)(C_3 \cos By + C_4 \sin By) \exp j(\omega t - \beta z) \quad (m = 0, 1, 2, 3, \ldots) \quad \text{(for all values of } y), \tag{9.39}$$

where A satisfies both the relations (9.34) and (9.38) and hence can be expressed as

$$A = \frac{m\pi}{a} \; (m = 0, 1, 2, 3, \ldots). \tag{9.40}$$

Following the same approach as that used to obtain (9.39) and (9.40), we can start from the same expression as (9.17) but now use the proportionality relation (9.26*b*) instead of (9.26*a*), that is, use the relation $E_x \propto \partial H_z / \partial y$ to obtain, instead of (9.39) and (9.40) the following relations:

$$H_z = (C_3 \cos By)(C_1 \cos Ax + C_2 \sin Ax) \exp j(\omega t - \beta z) \quad (n = 0, 1, 2, 3, \ldots) \quad \text{(for all values of } x) \tag{9.41}$$

where

$$B = \frac{n\pi}{b} \quad (n = 0,\ 1,\ 2,\ 3,\ \ldots). \tag{9.42}$$

Comparing (9.39) and (9.40) with (9.41) and (9.42) respectively, we see that while the former two are valid for all values of y $(0 \le y \le b)$, the latter two are valid for all values of x $(0 \le x \le a)$. Further, the roles of A and m in the former have been taken up by B and n respectively in the latter. Therefore, if we combine the x dependence of the solution (9.39) for H_z, which is valid irrespective of the values of y $(0 \le y \le b)$ with the y dependence of the solution (9.41) for H_z, which is valid irrespective of the values of x $(0 \le x \le a)$, we can write the following expression for H_z valid for all values of y $(0 \le y \le b)$ and x $(0 \le x \le a)$:

$$H_z = (C_1 \cos Ax)(C_3 \cos By) \exp j(\omega t - \beta z)\,,$$

which can be expressed as

$$H_z = H_{z0} \cos Ax \cos By \exp j(\omega t - \beta z) \tag{9.43}$$

where $H_{z0}(= C_1C_3)$ is a field constant. Substituting (9.40) and (9.42) in (9.43), and including the ranges of m and n from (9.40) and (9.44) respectively, we finally obtain the expression for the axial magnetic field component for a rectangular waveguide excited in the TE mode as

$$H_z = H_{z0} \cos \frac{m\pi}{a} x \cos \frac{n\pi}{b} y \exp j(\omega t - \beta z) \quad (m, n = 0,\ 1,\ 2,\ 3,\ \ldots)\,. \quad \text{(TE mode)} \tag{9.44}$$

The transverse electric field components E_y and E_x can be obtained by substituting the partial derivatives of (9.44) with respect to x and y in (9.22) and (9.25) respectively. For this purpose, we can work out these partial derivatives of (9.44) as follows:

$$\left.\begin{aligned} \frac{\partial H_z}{\partial x} &= -\frac{m\pi}{a} H_{z0} \sin \frac{m\pi}{a} x \cos \frac{n\pi}{b} y \exp j(\omega t - \beta z) \quad \text{(a)} \\ \frac{\partial H_z}{\partial y} &= -\frac{n\pi}{b} H_{z0} \cos \frac{m\pi}{a} x \sin \frac{n\pi}{b} y \exp j(\omega t - \beta z) \quad \text{(b)} \end{aligned}\right\} \quad (m, n = 0,\ 1,\ 2,\ 3,\ \ldots). \tag{9.45}$$

Therefore, substituting (9.45*a*) and (9.45*b*) in (9.22) and (9.25) respectively, we obtain

$$\left.\begin{aligned} E_y &= -\frac{j\omega\mu_0}{k^2 - \beta^2} \frac{m\pi}{a} H_{z0} \sin \frac{m\pi}{a} x \cos \frac{n\pi}{b} y \exp j(\omega t - \beta z) \quad \text{(a)} \\ E_x &= \frac{j\omega\mu_0}{k^2 - \beta^2} \frac{n\pi}{b} H_{z0} \cos \frac{m\pi}{a} x \sin \frac{n\pi}{b} y \exp j(\omega t - \beta z) \quad \text{(b)} \end{aligned}\right\} \quad (m, n = 0,\ 1,\ 2,\ 3,\ \ldots). \tag{9.46}$$

Notice that the quantity $k^2 - \beta^2$ occurring in (9.46) can be expressed with the help of (9.13) as

$$k^2 - \beta^2 = A^2 + B^2,$$

which, with the help of (9.40) and (9.42) becomes

$$k^2 - \beta^2 = \left(\frac{m\pi}{a}\right)^2 + \left(\frac{n\pi}{b}\right)^2. \qquad \text{(TE mode)} \tag{9.47}$$

Substituting (9.47) in (9.46), the expressions for the transverse electric field components then take the following form:

$$\left.\begin{aligned}
E_y &= -\frac{j\omega\mu_0}{\left(\frac{m\pi}{a}\right)^2 + \left(\frac{n\pi}{b}\right)^2}\frac{m\pi}{a}H_{z0}\sin\frac{m\pi}{a}x\cos\frac{n\pi}{b}y\exp j(\omega t - \beta z) \quad \text{(a)}\\
E_x &= \frac{j\omega\mu_0}{\left(\frac{m\pi}{a}\right)^2 + \left(\frac{n\pi}{b}\right)^2}\frac{n\pi}{b}H_{z0}\cos\frac{m\pi}{a}x\sin\frac{n\pi}{b}y\exp j(\omega t - \beta z) \quad \text{(b)}
\end{aligned}\right\}. \qquad \text{(TE mode)} \tag{9.48}$$

The transverse components of magnetic field can then be obtained in terms of the transverse components of electric field with the help of (9.20*a*) and (9.24*a*) as follows:

$$\left.\begin{aligned}
H_x &= -\frac{\beta}{\omega\mu_0}E_y \quad \text{(a)}\\
H_y &= \frac{\beta}{\omega\mu_0}E_x \quad \text{(b)}
\end{aligned}\right\} \text{(TE mode)}$$

which can be then read with the help of (9.48) as

$$\left.\begin{aligned}
H_x &= \frac{j\beta}{\left(\frac{m\pi}{a}\right)^2 + \left(\frac{n\pi}{b}\right)^2}\frac{m\pi}{a}H_{z0}\sin\frac{m\pi}{a}x\cos\frac{n\pi}{b}y\exp j(\omega t - \beta z) \quad \text{(a)}\\
H_y &= \frac{j\beta}{\left(\frac{m\pi}{a}\right)^2 + \left(\frac{n\pi}{b}\right)^2}\frac{n\pi}{b}H_{z0}\cos\frac{m\pi}{a}x\sin\frac{n\pi}{b}y\exp j(\omega t - \beta z) \quad \text{(b)}
\end{aligned}\right\}. \tag{9.49}$$

Thus, from the foregoing analysis we have obtained the expressions for the two transverse electric field components E_x, E_y, given by (9.48). Remember that for the TE mode of excitation considered here, the axial electric field component $E_z = 0$. Further, while the expression for the axial magnetic field component H_z for the TE mode is given by (9.44), the expressions for the transverse components of magnetic field H_x and H_y are given by (9.49), the latter to be read with the help of (9.48).

To sum up, we have obtained here the field solutions for the TE mode ($H_z \neq 0, E_z = 0$) in a rectangular waveguide as follows:

(i) E_x (TE mode) : Equation (9.48*b*)

(ii) E_y (TE mode) : Equation (9.48*a*)

(iii) $E_z = 0$ (TE mode)

(iv) H_x (TE mode): Equation (9.49*a*)

(v) H_y (TE mode): Equation (9.49*b*)

(vi) H_z (TE mode): Equation (9.44)

The field pattern, that is, the variation of the electric field and that of the magnetic field with space coordinates, obtainable with the help of (i)-(vi) above for the TE mode, has been discussed later in this chapter (Section 9.3).

EXAMPLE E9-1

A wave propagates along z in a dielectric-filled rectangular waveguide of wider dimension $x = a$ and narrower dimension $y = b$. If the electric field in the waveguide is $\vec{E} = \vec{a}_z(1+j)\sin\frac{\pi}{a}x\sin\frac{\pi}{b}y\exp j(\omega t - \beta z)$, find the dielectric constant and loss tangent of the dielectric.

SOLUTION

In this problem, for the electric field in the waveguide we are given:

$$\left.\begin{aligned} &E_z = (1+j)\sin\frac{\pi}{a}x\sin\frac{\pi}{b}y\exp j(\omega t - \beta z) \\ &E_x = E_y = 0 \end{aligned}\right\}. \tag{E9-1.1}$$

In view of (E9-1.1), we can then find with the help of (2.66)

$$\nabla \times \vec{E} = \left(\frac{\partial E_z}{\partial y}\right)\vec{a}_x - \left(\frac{\partial E_z}{\partial x}\right)\vec{a}_y$$

and substitute it in Maxwell's equation

$$\nabla \times \vec{E} = -\frac{\partial \vec{B}}{\partial t} = -\mu_0\frac{\partial \vec{H}}{\partial t} = -j\omega\mu_0\vec{H}$$

to obtain

$$\frac{\pi}{b}(1+j)\sin\frac{\pi}{a}x\cos\frac{\pi}{b}y\vec{a}_x - \frac{\pi}{a}(1+j)\cos\frac{\pi}{a}x\sin\frac{\pi}{b}y\vec{a}_y = -j\omega\mu_0\vec{H},$$

which gives

$$\vec{H} = \frac{1+j}{-j\omega\mu_0}\left[\left(\frac{\pi}{b}\sin\frac{\pi}{a}x\cos\frac{\pi}{b}y\right)\vec{a}_x - \left(\frac{\pi}{a}\cos\frac{\pi}{a}x\sin\frac{\pi}{b}y\right)\vec{a}_y\right]. \tag{E9-1.2}$$

From (E9-1.2) we can therefore write

$$\left.\begin{aligned} H_z &= 0 \\ H_x &= -\frac{1+j}{j\omega\mu_0}\left(\frac{\pi}{b}\sin\frac{\pi}{a}x\cos\frac{\pi}{b}y\right) \\ H_y &= \frac{1+j}{j\omega\mu_0}\left(\frac{\pi}{a}\cos\frac{\pi}{a}x\sin\frac{\pi}{b}y\right) \end{aligned}\right\}. \tag{E9-1.3}$$

Next, recall Maxwell's equation

$$\nabla\times\vec{H} = \sigma\vec{E} + \varepsilon\frac{\partial\vec{E}}{\partial t} = (\sigma + j\omega\varepsilon)\vec{E}.$$

Since $\vec{E}$ has only the z-component, as can be seen from (E9-1.1), we can write

$$\begin{aligned} (\nabla\times\vec{H})_z &= (\sigma + j\omega\varepsilon)E_z \\ &= (\sigma + j\omega\varepsilon)(1+j)\sin\frac{\pi}{a}x\sin\frac{\pi}{b}y\exp j(\omega t - \beta z). \end{aligned} \tag{E9-1.4}$$

In view of (E9-1.3), read $(\nabla\times\vec{H})_z$ with the help of (2.66) as

$$(\nabla\times\vec{H})_z = \left(\frac{\partial H_y}{\partial x} - \frac{\partial H_x}{\partial y}\right)\vec{a}_z$$

and put it in (E9-1.4) to write

$$\frac{\partial H_y}{\partial x} - \frac{\partial H_x}{\partial y} = (\sigma + j\omega\varepsilon)(1+j)\sin\frac{\pi}{a}x\sin\frac{\pi}{b}y\exp j(\omega t - \beta z). \tag{E9-1.5}$$

Differentiating (E9-1.3), we get

$$\left.\begin{aligned} \frac{\partial H_y}{\partial x} &= -\frac{1+j}{j\omega\mu_0}\left(\frac{\pi}{a}\right)^2\left(\sin\frac{\pi}{a}x\sin\frac{\pi}{b}y\right)\exp j(\omega t - \beta z) \\ \frac{\partial H_x}{\partial y} &= \frac{1+j}{j\omega\mu_0}\left(\frac{\pi}{b}\right)^2\left(\sin\frac{\pi}{a}x\sin\frac{\pi}{b}y\right)\exp j(\omega t - \beta z) \end{aligned}\right\}. \tag{E9-1.6}$$

Substituting (E9-1.6) in (E9-1.5), the following simple expression results:

$$\frac{j}{\omega\mu_0}\left[\left(\frac{\pi}{a}\right)^2 + \left(\frac{\pi}{b}\right)^2\right] = (\sigma + j\omega\varepsilon). \tag{E9-1.7}$$

Equating the real part of (E9-1.7), we obtain $\sigma = 0$, whence the loss tangent $\tan\delta (= \sigma/\omega\varepsilon)$

is obtained as

$$\tan\delta = \frac{\sigma}{\omega\varepsilon} = 0.$$

Similarly, equating the imaginary part of (E9-1.7) we obtain

$$\omega\varepsilon = \frac{1}{\omega\mu_0}\left[\left(\frac{\pi}{a}\right)^2 + \left(\frac{\pi}{b}\right)^2\right]. \tag{E9-1.8}$$

Putting $\varepsilon = \varepsilon_r\varepsilon_0$ in (E9-1.8), where ε_r is the relative permittivity or dielectric constant of the dielectric filling the waveguide and in view of the relation $c = 1/\sqrt{\mu_0\varepsilon_0}$, we then derive:

$$\varepsilon_r = \frac{c^2}{\omega^2}\left[\left(\frac{\pi}{a}\right)^2 + \left(\frac{\pi}{b}\right)^2\right].$$

9.1.3 Characteristic equation or dispersion relation

We notice that the field components depend on the values of m, n $(= 0, 1, 2, 3, ...)$ (see 9.48). The integral numbers m and n are referred to as the mode numbers. Accordingly, the TE mode excited is referred to as the TE_{mn} waveguide mode. The propagation constant of the TE_{mn} waveguide mode depends on the mode numbers m and n according to (9.47), which can be alternatively expressed as

$$k^2 - \beta^2 = k_c^2 \tag{9.50}$$

putting

$$k_c = \left[\left(\frac{m\pi}{a}\right)^2 + \left(\frac{n\pi}{b}\right)^2\right]^{1/2} \quad (m, n = 0,\ 1,\ 2,\ 3,\ ...). \tag{9.51}$$

Substituting (9.4) in (9.50), we can express the latter as

$$\frac{\omega^2}{c^2} - \beta^2 = k_c^2$$

which can be simplified as

$$\omega^2 - \beta^2 c^2 - k_c^2 c^2 = 0, \tag{9.52}$$

which on putting

$$\omega_c = k_c c \tag{9.53}$$

becomes

$$\omega^2 - \beta^2 c^2 - \omega_c^2 = 0. \quad \text{(TE mode)} \tag{9.54}$$

We call $k_c (= \omega_c/c)$ as the cutoff wave number and ω_c as the cutoff angular frequency, the subscript c representing the cutoff quantities. Why these quantities are referred to as the 'cutoff' quantities is discussed later at the end of this subsection. We can obtain the expression for the angular cutoff frequency ω_c with the help of (9.51) and (9.53) along with the circular cutoff frequency $f_c (= \omega_c/2\pi)$ as well as that of the cutoff wavelength $(\lambda_c = c/f_c)$ as follows:

$$\left.\begin{aligned} \omega_c &= c\left[\left(\frac{m\pi}{a}\right)^2 + \left(\frac{n\pi}{b}\right)^2\right]^{1/2} \\ f_c &= \frac{c}{2}\left[\left(\frac{m}{a}\right)^2 + \left(\frac{n}{b}\right)^2\right]^{1/2} \\ \lambda_c\left(=\frac{c}{f_c}\right) &= \frac{2}{\left[\left(\frac{m}{a}\right)^2 + \left(\frac{n}{b}\right)^2\right]^{1/2}} \end{aligned}\right\} (m, n = 0, 1, 2, 3, ...). \text{ (TE mode)} \qquad (9.55)$$

In (9.54) we obtain a relation between ω and β. Such a relation is called the characteristic equation or the dispersion relation of the waveguide. The dispersion relation (9.54) derived here for the rectangular waveguide, with appropriate interpretation of the cutoff frequency, is valid for the cylindrical waveguide as well. We will see this later in Section 9.4. We can plot ω versus β or the dispersion characteristics of the waveguide with the help of (9.54) (Fig. 9.2). If, for a structure supporting an electromagnetic wave, ω varies linearly with β, that is, if the ω versus β plot is a straight line, or in other words if the phase velocity $v_{\text{ph}} (= \omega/\beta$; see Section 6.1) of the wave propagating through the waveguide remains constant with $f\ (= \omega/2\pi)$, then structure has no dispersion; otherwise, the structure is dispersive. The dispersion plot obtained using the dispersion relation (9.54) becomes a hyperbola (Fig. 9.2). The plot has makes an intercept with the ordinate, that is, with the ω-axis $(\beta = 0)$ at $\omega = \omega_c$, which is the value of ω corresponding to $\beta = 0$, as could be verified by putting $\beta = 0$ in (9.54).

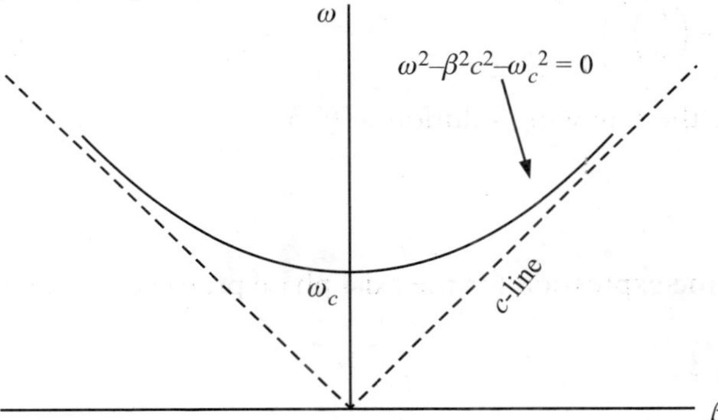

Fig. 9.2 *ω versus β dispersion characteristics of the waveguide which meets the c-line at $\omega \to \infty$, the latter shown as a dotted line having positive and negative slopes of magnitude equal to the velocity of light = c for the positive and the negative values of β respectively.*

It will be of interest to find a relation between the free-space wavelength λ, the guide wavelength λ_g (that is, the wavelength of the wave propagating through the waveguide), and the cutoff wavelength λ_c, defined by the following relation (see Section 6.1):

$$\left.\begin{aligned} k &= \frac{2\pi}{\lambda} \\ \beta &= \frac{2\pi}{\lambda_g} \\ k_c &= \frac{2\pi}{\lambda_c} \end{aligned}\right\}. \qquad (9.56)$$

The relation between λ, λ_g, and λ_c can be found by substituting (9.56) in (9.50) as follows:

$$\left(\frac{2\pi}{\lambda}\right)^2 - \left(\frac{2\pi}{\lambda_g}\right)^2 = \left(\frac{2\pi}{\lambda_c}\right)^2,$$

which simplifies to

$$\frac{1}{\lambda_c^2} = \frac{1}{\lambda^2} - \frac{1}{\lambda_g^2}. \qquad (9.57)$$

With the help of the relations (9.53), (9.55) and (9.56), we can easily obtain the following expression for the waveguide cutoff wavelength:

$$\lambda_c = \frac{2\pi}{k_c} = \frac{2\pi}{\omega_c} c = \frac{2\pi}{c\left[\left(\frac{m\pi}{a}\right)^2 + \left(\frac{n\pi}{b}\right)^2\right]^{1/2}} c$$

which simplifies to

$$\lambda_c = \frac{2}{\left[\left(\frac{m}{a}\right)^2 + \left(\frac{n}{b}\right)^2\right]^{1/2}}. \qquad (9.58)$$

Let us next examine the following solution of (9.54)

$$\beta^2 = \frac{\omega^2 - \omega_c^2}{c^2} \qquad (9.59)$$

from which we get the expression for the axial phase propagation constant β as

$$\beta = \pm \frac{(\omega^2 - \omega_c^2)^{1/2}}{c}. \qquad (9.60)$$

EXAMPLE E9-2

Show that the ionosphere behaves as a waveguide of cutoff frequency ω_p, taking the refractive index of the ionosphere as $n = \sqrt{1 - \omega_p^2/\omega^2}$, where ω is the wave frequency and ω_p the electron angular plasma frequency of the ionosphere (see Example E6-11).

SOLUTION

We obtain the given expression for the refractive index of the ionosphere (derived in Example E6-11) by taking the square root of $\varepsilon_{r,\text{effecive}}$ given by (E6-11.10) as follows:

$$n = \sqrt{1 - \frac{\omega_p^2}{\omega^2}} \tag{E9-2.1}$$

which is the given expression in the present example. We can then substitute the relation $n = c/v_p$ in (E9-2.1) to write

$$\frac{c}{v_p} = \sqrt{1 - \frac{\omega_p^2}{\omega^2}}\,. \tag{E9-2.2}$$

Substituting the relation $v_{\text{ph}} = \omega/\beta$ in (E9-2.2), we get

$$\frac{c\beta}{\omega} = \sqrt{1 - \frac{\omega_p^2}{\omega^2}} = \sqrt{\frac{\omega^2 - \omega_p^2}{\omega^2}}. \tag{E9-2.3}$$

Squaring (E9-2.3) and rearranging the terms, we obtain the following expression for the dispersion relation of the ionosphere:

$$\omega^2 - \beta^2 c^2 - \omega_p^2 = 0. \tag{E9-2.4}$$

Notice that (E9-2.4) is identical to the waveguide-mode dispersion relation of a waveguide $\omega^2 - \beta^2 c^2 - \omega_c^2 = 0$ given by (9.54), except that in the case of the ionosphere the electron angular plasma frequency ω_p has replaced the waveguide-mode cutoff frequency ω_c in the dispersion rélation. Hence we can infer that the ionosphere behaves as a waveguide of cutoff frequency $\omega_c = \omega_p$.

If we take the upper sign of (9.60), then for angular frequencies above ω_c, that is, for $\omega > \omega_c$, we obtain a positive real value of β occurring in the phase factor $\exp j(\omega t - \beta z)$. This also occurs in the expressions for the RF quantities associated with the wave propagating through the waveguide in the positive z direction, for instance, in the field expressions (9.48). (The plot in Fig. 9.2 encompasses both the positive and the negative values of β corresponding to the wave propagating along the positive and the negative directions of z respectively). For frequencies $\omega > \omega_c$, therefore, the waveguide supports a forward propagating wave along the positive z. As a corollary to this, the waveguide does not support a propagating wave for angular frequencies below ω_c, that is, for $\omega < \omega_c$. In other words, the propagating wave gets "cutoff" below the angular frequency ω_c, and, hence ω_c is called the waveguide cutoff angular frequency. Consequently, k_c, which is directly related to ω_c by (9.53), is called the cutoff wave number. Later in this section we will examine in details the cutoff behaviour of the waveguide in the case of $\omega < \omega_c$ (see Section 9.1.5).

9.1.4 Characteristic parameters

The wave phase velocity v_{ph} is related to the phase propagation constant β through the relation $v_{\text{ph}} = \omega/\beta$ (see (6.19) and the discussion following it), in which let us substitute (9.60) with its uppers sign to obtain

$$v_{\text{ph}} = \frac{\omega}{\beta} = \frac{\omega}{\dfrac{(\omega^2 - \omega_c^2)^{1/2}}{c}} = \frac{\omega c}{(\omega^2 - \omega_c^2)^{1/2}}$$

This can be rearranged in terms of the wave circular frequency $f = \omega/2\pi$ and the waveguide cutoff circular frequency $f_c = \omega_c/2\pi$ in the following normalised form:

$$\frac{v_{\text{ph}}}{c} = \frac{\omega}{(\omega^2 - \omega_c^2)^{1/2}} = \frac{1}{\left(1 - \dfrac{\omega_c^2}{\omega^2}\right)^{1/2}} = \frac{1}{\left(1 - \dfrac{f_c^2}{f^2}\right)^{1/2}}. \tag{9.61}$$

The guide wavelength $\lambda_g (= 2\pi/\beta)$ can also be expressed with the help of (9.56) in the following normalised form:

$$\frac{\lambda_g}{\lambda} = \frac{k}{\beta}$$

in which we can substitute $k = \omega/c$ (9.4) and $\beta = \omega/v_{\text{ph}}$ (see (6.19) and the discussion following it) to write

$$\frac{\lambda_g}{\lambda} = \frac{v_{\text{ph}}}{c},$$

which, with the help of (9.61) can be put as

$$\frac{\lambda_g}{\lambda} = \frac{1}{\left(1 - \dfrac{f_c^2}{f^2}\right)^{1/2}}. \tag{9.62}$$

Further, the group velocity v_g of the waves propagating through the waveguide can be found using the following relation (See Appendix A9.1):

$$v_g = \frac{\partial \omega}{\partial \beta}. \tag{9.63}$$

In order to find the right-hand side of (9.63), let us take the partial derivative of (9.54) with respect to β to obtain

$$2\omega\frac{\partial\omega}{\partial\beta} - c^2 2\beta = 0,$$

which can be simplified using the relation $v_{\text{ph}} = \omega/\beta$ (see (6.19) and the discussion following it) to write

$$\frac{\partial\omega}{\partial\beta} = \frac{c^2\beta}{\omega} = \frac{c^2}{\frac{\omega}{\beta}} = \frac{c^2}{v_{\text{ph}}}. \tag{9.64}$$

Substituting (9.64) in (9.63), we obtain the expression for the group velocity v_g as follows:

$$v_g = \frac{c^2}{v_{\text{ph}}}. \tag{9.65}$$

This can also be expressed as

$$v_{\text{ph}} v_g = c^2 \tag{9.66}$$

and in the normalised form as

$$\frac{v_g}{c} = \frac{c}{v_{\text{ph}}} = \frac{1}{\frac{v_{\text{ph}}}{c}},$$

which, with the help of (9.61) can be put as:

$$\frac{v_g}{c} = \left(1 - \frac{f_c^2}{f^2}\right)^{1/2}. \tag{9.67}$$

EXAMPLE E9-3

Establish a relation between the wave phase velocity v_{ph}, the wave group velocity v_g and the wave dispersion $\partial v_{\text{ph}}/\partial\omega$.

SOLUTION

We can start from the following two relations for the wave phase velocity v_{ph} (see (6.19) and the discussion following it) and group velocity [see (9.63)]:

$$v_{\text{ph}} = \frac{\omega}{\beta} \tag{R9-3.1}$$

where β is the wave propagation constant and

$$v_g = \frac{\partial\omega}{\partial\beta}. \qquad \text{(See Appendix 9.1.)} \tag{E9-3.2}$$

Writing (E9-3.1) as

$$\omega = v_{\rm ph}\beta$$

and differentiating it with respect to β, we get

$$\frac{\partial \omega}{\partial \beta} = v_{\rm ph} + \beta \frac{\partial v_{\rm ph}}{\partial \beta} = v_{\rm ph} + \beta \frac{\partial v_{\rm ph}}{\partial \omega}\frac{\partial \omega}{\partial \beta}. \tag{E9-3.3}$$

Plugging (E9-3.2) in (E9-3.3), we can then write

$$v_{\rm g} = v_{\rm ph} + \beta \frac{\partial v_{\rm ph}}{\partial \omega} v_{\rm g},$$

which can be rearranged to obtain the desired relation for the group velocity as

$$v_{\rm g} = \frac{v_{\rm ph}}{1 - \beta \frac{\partial v_{\rm ph}}{\partial \omega}},$$

where $\partial v_{\rm ph}/\partial \omega$ represents the wave dispersion, that is, the variation of the wave phase velocity with frequency.

EXAMPLE E9-4

The cutoff wavelength of the waveguide WR-137[1] is given as 69.70 mm. What should be the length of the guide in order to ensure that a signal at 8.6 GHz emerging out of the guide is delayed by 1 μs with respect to the signal that propagates outside the waveguide?

SOLUTION

The cutoff frequency f_c of the waveguide can be calculated as

$$f_c = \frac{c}{\lambda_c} = \frac{3 \times 10^8}{69.7 \times 10^{-3}} = 4.3 \times 10^9 \text{ Hz} = 4.3 \text{ GHz} \tag{E9-4.1}$$

where λ_c is the cutoff wavelength of the waveguide. The delay $\tau_{\rm delay}$ referred to in the problem is given by

$$\tau_{\rm delay} = \frac{l}{v_g} - \frac{l}{c} = l\left(\frac{1}{v_g} - \frac{1}{c}\right) \tag{E9-4.2}$$

where l is the length of the waveguide and v_g the group velocity of the signal through the waveguide (noting that the time of travel of the signal over the length l through the waveguide is l/v_g, while that outside the waveguide is l/c).

Now, with the help of (9.67) we can find the wave group velocity v_g as

$$v_g = \left(1 - \frac{f_c^2}{f^2}\right)^{1/2} \times 3 \times 10^8. \tag{E9-4.3}$$

[1] The commercial name of a waveguide (see also Table 9.1).

Putting $f_c = 4.3 \times 10^9$ Hz from (E9-4.1) and $f = 8.6$ GHz $= 8.6 \times 10^9$ Hz (given) in (E9-4.3), we can calculate $v_g = (1 - 0.25)^{1/2} \times 3 \times 10^8$ m/s, which together with $c = 3 \times 10^8$ m/s substituted in (E9-4.2) gives

$$\tau_{\text{delay}} = l\left(\frac{1}{v_g} - \frac{1}{c}\right) = l(0.3849 \times 10^{-8} - 0.3333 \times 10^{-8}) = l(0.0516 \times 10^{-8}),$$

whence we get

$$l = \frac{\tau_{\text{delay}}}{(0.0516) \times 10^{-8}}. \tag{E9-4.4}$$

Substituting the given value $\tau_{\text{delay}} = 10^{-6}$ s in (E9-4.4) we obtain the required length of the waveguide as

$$l = \frac{10^{-6}}{(0.0516) \times 10^{-8}} = 19.38 \times 10^2 \text{ m} = 1.938 \text{ km}.$$

For a waveguide, there is yet another parameter of relevance, namely, the wave impedance Z_W, which can be defined in the axial z direction as the ratio of the magnitude of the transverse electric field $E_{\text{transverse}}$ to the magnitude of the transverse magnetic field $H_{\text{transverse}}$ as follows:

$$Z_W = \frac{E_{\text{transverse}}}{H_{\text{transverse}}} = \frac{(E_x^2 + E_y^2)^{1/2}}{(H_x^2 + H_y^2)^{1/2}}. \tag{9.68}$$

If we now substitute (9.48) and (9.49) in (9.68), we obtain the following expression for the wave impedance:

$$Z_W = \frac{\omega\mu_0}{\beta}. \tag{9.69}$$

We can normalise the wave impedance (9.69) with respect to the intrinsic impedance of free space $\eta(= (\mu_0/\epsilon_0)^{1/2}$, given by (6.70) as

$$\frac{Z_W}{\eta_0} = \frac{\omega\mu_0}{\beta\eta_0} = \frac{\omega\mu_0}{\beta\left(\frac{\mu_0}{\eta_0}\right)^{1/2}} = \frac{\omega}{\beta}(\mu_0\epsilon_0)^{1/2}$$

which, in view of the relation $c = 1/(\mu_0\epsilon_0)^{1/2}$ given by (9.5) and the relation $v_{\text{ph}} = \omega/\beta$, can be expressed as

$$\frac{Z_W}{\eta_0} = \frac{v_{\text{ph}}}{c}$$

which, with the help of (9.61) can be put as

$$\frac{Z_W}{\eta_0} = \frac{1}{\left(1 - \frac{f_c^2}{f^2}\right)^{1/2}}. \tag{9.70}$$

It will be useful to put together the normalised expressions deduced for the wave phase velocity (9.61), guide wavelength (9.62), group velocity (9.67) and wave impedance (9.70) as follows for ready reference:

$$\left.\begin{aligned} \frac{v_{ph}}{c} &= \frac{1}{\left(1 - \frac{f_c^2}{f^2}\right)^{1/2}} \\ \frac{\lambda_g}{\lambda} &= \frac{1}{\left(1 - \frac{f_c^2}{f^2}\right)^{1/2}} \\ \frac{v_g}{c} &= \left(1 - \frac{f_c^2}{f^2}\right)^{1/2} \\ \frac{Z_W}{\eta_0} &= \frac{1}{\left(1 - \frac{f_c^2}{f^2}\right)^{1/2}} \end{aligned}\right\}. \tag{9.71}$$

The wave phase velocity versus frequency characteristics of a waveguide (Fig. 9.3) is an alternative to the dispersion characteristics represented by its $\omega - \beta$ plot (Fig. 9.2).

A point corresponding to an angular frequency $\omega(= 2\pi f)$ located on the $\omega - \beta$ dispersion plot (Fig. 9.2) represents a wave propagating through the waveguide with phase velocity equal to the slope of the line joining the origin ($\omega = \beta = 0$) with the point (Fig. 9.2). The slope of this line corresponding to a very high-frequency point ($\omega \to \infty$) represents the wave phase velocity tending to the velocity of light ($v_{\rm ph} \to c$) (Fig. 9.2). The slope of the tangent to a point on the dispersion characteristics represents the group velocity of the wave, $v_g = \partial\omega/\partial\beta$. This slope corresponding to a very high-frequency point ($\omega \to \infty$) represents wave group velocity tending to the velocity of light ($v_{\rm ph} \to c$). We can see this behaviour of the phase and the group velocities by putting $\omega = \infty$ or $f \to \infty$ in (9.71). Further, we can use (9.71) to plot not only the variation of the phase and the group velocities but also that of the guide wavelength as well as that of the wave impedance with frequency (Fig. 9.3). Thus let us appreciate from (9.71) as well as from these plots (Fig. 9.3) the values of these quantities, namely phase velocity, guide wavelength, group velocity and wave impedance at very high frequencies ($\omega = \infty$ or $f \to \infty$) as well as that at the waveguide cutoff frequency ($\omega = \omega_c$ or $f = f_c$)) as follows:

$$\left.\begin{aligned} v_{\rm ph} &= c \\ \lambda_g &= \lambda \\ v_g &= c \\ Z_W &= \eta_0 \end{aligned}\right\} \quad (f \to \infty) \tag{9.72}$$

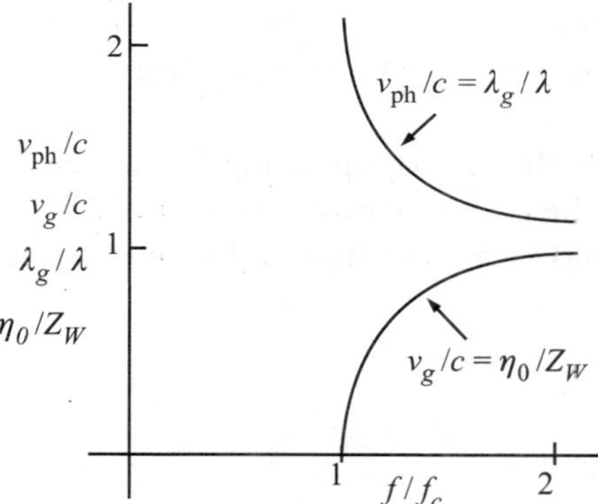

Fig. 9.3 *Variation of the normalised phase velocity, guide wavelength, group velocity and wave impedance with the normalised operating frequency, where the phase and the group velocities are each normalised with respect to the velocity of light; the guide wavelength is normalised with respect to the free-space wavelength; the wave impedance is normalised with respect to the free-space intrinsic impedance; and the operating frequency is normalised with respect to the cutoff frequency.*

and

$$\left.\begin{aligned} v_{ph} &= \infty \\ \lambda_g &= \infty \\ v_g &= 0 \\ Z_W &= 0 \end{aligned}\right\} \quad (f = f_c) \tag{9.73}$$

Thus, we can appreciate from (9.72) that as we deviate from the cutoff frequency to higher operating frequencies, the waveguide phase and group velocities each tend to approach the velocity of light such that the product of these velocities continues to remain the square of the velocity of light in accordance with (9.66). The guide wavelength and the wave impedance, on the other hand, tend to approach the free-space wavelength and the free-space intrinsic impedance respectively.

Similarly, it can be appreciated with the help of (9.73) that at operating frequencies close to the waveguide cutoff frequency, the wave phase velocity becomes high, tending to infinity, and the wave group velocity takes on small values, tending to zero. The product of these velocities continues to remain the square of the velocity of light in accordance with (9.66). Also, the values of the guide wavelength and wave impedance each tend to infinity.

9.1.5 Evanescent mode

So far we have considered propagation through the waveguide along the positive z above the waveguide cutoff frequency ($\omega > \omega_c$) corresponding to a positive real value of the phase propagation constant β occurring in the phase factor $\exp j(\omega t - \beta z)$, which, in turn, appears in the expressions for RF quantities pertaining to propagating waves (see the discussion on the above-the-cutoff situation $[(\omega > \omega_c)$ following (9.60)]. The upper sign of (9.60) gives such a positive real value of β—recall, for instance, that the factor $\exp j(\omega t - \beta z)$ appears in the field expressions (9.48).

Let us now turn towards the case of operating frequencies below the cutoff $(\omega < \omega_c)$. For this purpose, and for the sake of convenience of discussion, let us take a typical RF quantity, for instance E_y, and express its time and axial-distance dependence with the help of (9.48*a*) as

$$E_y = E_{y0} \exp[j(\omega t - \beta z)] \tag{9.74}$$

where

$$E_{y0} = -\frac{j\omega\mu_0}{\left(\frac{m\pi}{a}\right)^2 + \left(\frac{n\pi}{b}\right)^2} \frac{m\pi}{a} H_{z0} \sin \frac{m\pi}{a} x \cos \frac{n\pi}{b} y. \tag{9.75}$$

Choosing to put

$$\gamma = j\beta \tag{9.76}$$

we can then alternatively express (9.74) as

$$E_y = E_{y0} \exp(-\gamma z) \exp(j\omega t) \tag{9.77}$$

where the propagation constant γ can be put, with the help of (9.76) and (9.60), for the situation $\omega < \omega_c$ as

$$\gamma = j\beta = \pm j \frac{(\omega^2 - \omega_c^2)^{1/2}}{c} = \pm j \frac{[(-1)(\omega_c^2 - \omega^2)]^{1/2}}{c} = \pm (j)(j) \frac{(\omega_c^2 - \omega^2)^{1/2}}{c}$$
$$(\omega < \omega_c)$$

using the relation $j = (-1)^{1/2}$, which, on putting $(j)(j) = j^2 = -1$ becomes

$$\gamma = j\beta = \mp \frac{(\omega_c^2 - \omega^2)^{1/2}}{c} \qquad (\omega < \omega_c). \tag{9.78}$$

For a wave propagating along positive z, one can take the positive, lower sign for γ in (9.78) and express it as

$$\gamma = j\beta = \frac{(\omega_c^2 - \omega^2)^{1/2}}{c} \qquad (\omega < \omega_c) \tag{9.79}$$

and substitute it in (9.77) to obtain

$$E_y = E_{y0} \exp[-\frac{(\omega_c^2 - \omega^2)^{1/2}}{c}] z \exp(j\omega t) \qquad (\omega < \omega_c),$$

which by putting

$$\alpha = \frac{(\omega_c^2 - \omega^2)^{1/2}}{c} \qquad (\omega < \omega_c). \tag{9.80}$$

can be read as

$$E_y = E_{y0}\exp(-\alpha z)\exp(j\omega t). \qquad (\omega < \omega_c) \tag{9.81}$$

Similarly, with the help of (9.48*b*), we can express E_x as

$$E_x = E_{x0}\exp(-\alpha z)\exp(j\omega t) \qquad (\omega < \omega_c) \tag{9.82}$$

where E_{x0} is given by the following expression corresponding to (9.75):

$$E_{x0} = -\frac{j\omega\mu_0}{\left(\frac{m\pi}{a}\right)^2 + \left(\frac{n\pi}{b}\right)^2}\frac{n\pi}{b}H_{z0}\sin\frac{m\pi}{a}x\sin\frac{n\pi}{b}y. \tag{9.83}$$

Clearly, for frequencies $\omega < \omega_c$, the RF dependence $\exp(\omega t - \beta z)$ characteristic of a propagating wave is absent in the field expression (9.80) or (9.82). This shows that RF waves do not propagate through the waveguide below the cutoff frequency ω_c, or in other words, the wave is cut off below this frequency as discussed following (9.60). Further, below the cutoff ($\omega < \omega_c$), the value of α, as can be seen from (9.80) is positive and real. Consequently, the amplitude of the electric field $E_{y0}\exp(-\alpha z)$ [see (9.81)] decays or 'attenuates' exponentially with axial distance z without any change of phase. The factor α is hence called the attenuation constant. The non-propagating wave below the cutoff frequency ω_c is said to be supported by the waveguide in the so-called 'evanescent' mode of the waveguide. Thus, a waveguide behaves like a high-pass filter by allowing propagation of electromagnetic waves and associated power only above the cutoff frequency. Further, it can be noted that the attenuation in the evanescent mode of the wave launched by a source of energy to the waveguide does not correspond to the resistive or ohmic loss caused by the finite resistivity of the conducting material of the waveguide wall (the latter will be discussed in Section 9.5); rather, such attenuation corresponds to a return of energy from the waveguide to the source.

EXAMPLE E9-5

Show that the attenuation constant versus frequency plot of a waveguide in the evanescent mode is an ellipse.

SOLUTION

For the evanescent mode, let us rewrite the expression for the attenuation constant (9.81) as

$$\alpha = \frac{(\omega_c^2 - \omega^2)^{1/2}}{c} \qquad (\omega < \omega_c) \tag{E9-5.1}$$

which can also be expressed in the following form:

$$\alpha^2 c^2 + \omega^2 = \omega_c^2. \tag{E9-5.2}$$

The plot of α versus ω given by the expression (E9-5.2) becomes an ellipse. The intercept $\alpha_{\text{intercept}}$ of the ellipse on the α axis (ordinate) corresponding to $\omega = 0$ can be seen from (E9-5.2) to be $\alpha_{\text{intercept}} = \omega_c/c$. Similarly, the intercept $\omega_{\text{intercept}}$ of the ellipse on ω axis (abscissa) corresponding to $\alpha = 0$ is seen to be $\omega_{\text{intercept}} = \omega_c$.

It is of interest to examine the average Poynting vector $\vec{P}_{\text{average}} = \text{Re}\vec{P}_{\text{complex}} = 1/2\text{Re}\vec{E}\times\vec{H}^*$ (8.102) associated with the evanescent mode in a waveguide below cutoff ($\omega < \omega_c$)[see Section 8.2.2 and expressions (8.130) and (8.102)]. In fact, with reference to power propagating down the waveguide along z, we will be looking at the z-component of $\vec{P}_{\text{averge}}$, that is, $1/2\text{Re}(\vec{E}\times\vec{H}^*)_z$ below cutoff ($\omega < \omega_c$).

For this purpose, let us write with the help of (9.81) and (9.75) the expression for E_y and that of its complex conjugate as

$$\left.\begin{aligned} E_Y &= \frac{-j\omega\mu_0}{\left(\frac{m\pi}{a}\right)^2 + \left(\frac{n\pi}{b}\right)^2}\frac{m\pi}{a}H_{z0}\sin\frac{m\pi}{a}x\cos\frac{n\pi}{b}Y exp(-\alpha z)expj\omega t \\ E*_y &= \frac{-j\omega\mu_0}{\left(\frac{m\pi}{a}\right)^2 + \left(\frac{n\pi}{b}\right)^2}\frac{m\pi}{a}H_{z0}\sin\frac{m\pi}{a}x\cos\frac{n\pi}{b}Y exp(-\alpha z)expj\omega t \end{aligned}\right\}(\omega < \omega_c).$$

$$(\omega < \omega_c) \tag{9.84}$$

Similarly, with the help of (9.82) and (9.83), we can express E_x and its complex conjugate as

$$\left.\begin{aligned} E_x &= \frac{j\omega\mu_0}{\left(\frac{m\pi}{a}\right)^2 + \left(\frac{n\pi}{b}\right)^2}\frac{\pi}{b}H_{z0}\sin\frac{m\pi}{a}x\sin\frac{n\pi}{b}Y exp(-\alpha z)expj\omega t \\ E*_x &= \frac{-j\omega\mu_0}{\left(\frac{m\pi}{a}\right)^2 + \left(\frac{n\pi}{b}\right)^2}\frac{m\pi}{a}H_{z0}\sin\frac{m\pi}{a}x\sin\frac{n\pi}{b}Y exp(-\alpha z)exp(-j\omega t) \end{aligned}\right\}(\omega < \omega_c).$$

$$\tag{9.85}$$

Further, with the help of (9.49) and remembering $\gamma = j\beta$ [see (9.79)], that is, $\beta = -j\gamma$, we can write the following relation for the transverse components of magnetic field in terms of the transverse components of electric field:

$$\left.\begin{aligned} H_x &= \frac{j\gamma}{\omega\mu_0}E_y \\ H_y &= \frac{-j\gamma}{\omega\mu_0}E_x \end{aligned}\right\} (\omega < \omega_c). \tag{9.86}$$

We can then write with the help of (8.102) the z-component of the average Poynting vector as

$$\begin{aligned} (\vec{P}_{\text{average}})_{z-\text{component}} &= \text{Re}(\vec{P}_{\text{average}})_{z-\text{component}} \\ &= \frac{1}{2}\text{Re}\,(\vec{E} \times \vec{H}*)_z = \frac{1}{2}\text{Re}\,(E_x H *_y - E_y H *_y). \end{aligned} \tag{9.87}$$

The complex conjugates of the transverse magnetic field components (H_x^*, H_y^*) occurring in (9.87) can be obtained from (9.86) as

$$\left.\begin{aligned} H_x^* &= \frac{-j\gamma}{\omega\mu_0}E_y^* \\ H_y^* &= \frac{j\gamma}{\omega\mu_0}E_x^* \end{aligned}\right\} \quad (\omega < \omega_c). \tag{9.88}$$

Substituting (9.88) in (9.87), we then obtain

$$\begin{aligned} (\vec{P}_{\text{average}})_{z-\text{component}} &= \frac{1}{2}\text{Re}(E_x H_y^* - E_y H_y^*) \\ &= \frac{1}{2}\frac{\gamma}{\omega\mu_0}\text{Re}\,j(E_x E_x^* + E_y E^* E_y)(\omega < \omega_c). \end{aligned} \tag{9.89}$$

Further, with the help of (9.84) and (8.85), you can write the expressions $E_x E_x^*$ and $E_y E_y^*$ occurring in (9.89) as

$$\left.\begin{aligned} E_x E_x^* &= \left[\frac{\omega\mu_0}{\left(\frac{m\pi}{a}\right)^2 + \left(\frac{n\pi}{b}\right)^2}\frac{n\pi}{b}\right]^2 H_{z0}^2 \sin^2\frac{m\pi}{a}x \sin^2\frac{n\pi}{b}y \exp(-2\alpha z) \\ E_y E_y^* &= \left[\frac{\omega\mu_0}{\left(\frac{m\pi}{a}\right)^2 + \left(\frac{n\pi}{b}\right)^2}\frac{m\pi}{a}\right]^2 H_{z0}^2 \sin^2\frac{m\pi}{a}x \cos^2\frac{n\pi}{b}y \exp(-2\alpha z) \end{aligned}\right\} (\omega < \omega_c). \tag{9.90}$$

We can then substitute (9.90) in (9.89) to obtain

$$(\vec{P}_{\text{average}})_{z-\text{component}} = \frac{1}{2}\frac{\gamma}{\omega\mu_0}\text{Re}j\left[\frac{(\omega\mu_0)^2}{\left[\left(\frac{m\pi}{a}\right)^2+\left(\frac{n\pi}{b}\right)^2\right]^2}H_{z0}^2\sin^2\frac{m\pi}{a}x\exp(-2\alpha z)\right.$$
$$\left.\times\left[\left(\frac{n\pi}{b}\right)^2\sin^2\frac{n\pi}{b}y+\left(\frac{n\pi}{a}\right)^2\cos^2\frac{n\pi}{b}y\right]\right] = \frac{1}{2}\frac{Y}{\omega\mu_0}(0) = 0 \quad (\omega < \omega_c). \tag{9.91}$$

The average Poynting vector $(\vec{P}_{\text{average}})_{z-\text{component}}$ thus being zero according to (9.91), we infer that there will be no power flow in the waveguide in the evanescent mode below cutoff ($\omega < \omega_c$).

Let us next examine the nature of the waveguide quantities: phase velocity v_{ph}, guide wavelength λ_g, group velocity v_g, and wave impedance Z_W, which can be easily expressed for the evanescent mode below cutoff ($\omega < \omega_c$) with the help of (9.71) and by simple algebraic manipulation using the relation $j = (-1)^{1/2}$ as

$$\left.\begin{aligned}\frac{v_{\text{ph}}}{c} = \frac{\lambda_g}{\lambda} = \frac{Z_W}{\eta_0} &= -j\left(\frac{f_c^2}{f^2}-1\right)^{-1/2}\\ \frac{v_g}{c} &= j\left(\frac{f_c^2}{f^2}-1\right)^{-1/2}\end{aligned}\right\}(\omega < \omega_c). \tag{9.92}$$

From the above, we can see that the waveguide quantities v_{ph}, λ_g, v_g, and Z_W are each essentially the imaginary quantities in the evanescent mode as expected, since in this mode there is no power propagation. One important finding from this is that, in the non-propagating, evanescent mode below cutoff ($\omega < \omega_c$), the waveguide has imaginary wave impedance Z_W, which in other words means that guide sees a reactive load that does not consume power. Hence there will be no power flow in the waveguide, which corroborates the inference drawn from (9.91) and stated immediately thereafter.

9.1.6 Dimension-wise and mode-wise operating frequency criteria

Now that we have appreciated the evanescent waveguide mode below cutoff ($\omega < \omega_c$), let us look back at the propagating waveguide mode above cutoff ($\omega > \omega_c$) by continuing to consider the TE mode ($H_z \neq 0, E_z = 0$) for the rectangular waveguide. Later in this chapter (Section 9.2), we will touch upon the behaviour of the TM mode ($E_z \neq 0, H_z = 0$) as well.

From (9.55) it can be seen how for the TE mode considered ($H_z \neq 0, E_z = 0$) the value of the waveguide cutoff frequency f_c and correspondingly that of the waveguide cutoff wavelength λ_c each depend on the integers m and n. Further, we can also see the dependence of the RF field quantities on the integers m and n from (9.44), (9.48), and (9.49). The integers m and n are known as the waveguide mode numbers. The TE mode for a given set of values of m and n are referred to as the TE$_{mn}$ mode (written by subscripting the integers m and n). The broad dimension a of the rectangular waveguide is usually either equal to or greater than twice its narrow dimension b. The values of the cutoff wavelength $\lambda_c = 2a$ and cutoff frequency $f_c = c/2a$ for the TE_{10} mode ($m = 1$ and $n = 0$) for some of the commercial waveguides whose broad and narrow dimensions (a and b) are known are given in Table 9.1 with the help of (9.55).

We can see from the table that as the waveguide transverse dimensions b decreases, the value of its cutoff wavelength λ_c decreases and that of its cutoff frequency f_c increases. Let us examine the values of the cutoff quantities, namely λ_c and f_c for some typical lower order TE_{mn} waveguide modes corresponding to (a) $m = 1, n = 0$ (TE_{10} mode): (b) $m = 1, n = 1$ (TE_{11} mode); (c) $m = 2, n = 0$ (TE_{20} mode) and (d) $m = 0, n = 1$ (TE_{01} mode) respectively, which can easily be obtained from (9.55) as follows:

$$\left.\begin{aligned}
&TE_{10} : \lambda_c = 2a = 4bs;\ f_c = c/(4bs) && (a)\\
&TE_{11} : \lambda_c = 4bs\frac{1}{(1+4s^2)^{1/2}};\ f_c = (c/4bs)(1+4s^2)^{1/2} && (b)\\
&TE_{20} : \lambda_c = a = 2bs\ f_c = c/(2bs) && (c)\\
&TE_{01} : \lambda_c = 2b;\ f_c = c/2b && (d)
\end{aligned}\right\} \quad (9.93)$$

where we have put $a = 2bs$ with the factor $s \geq 1$. This is tantamount to writing $a \geq 2b$ such that if we take $s > 1$, we get $a > 2b$, while if we take $s = 1$, we get $a = 2b$.

In Table 9.2, we have taken the dimensions (a and b) of waveguides from Table 9.1 for different values of the $s(= a/2b)$ factor to find the value of λ_c and $f(c)$ with the help of (9.93) for some typical modes .

Table 9.1 *Typical waveguide dimensions ($a \geq b$) and the corresponding cutoff frequencies and cutoff wavelengths for the TE_{10} mode for some commercial waveguides*

Waveguide	a (inch)	b (inch)	$\lambda_c(= 2a)$(mm)	$f_c(= c/2a)$ (GHZ)
WR-2300	23	11.5	1168.4	0.257
WR-340	3.4	1.7	172.72	1.737
WR-284	2.84	1.34	144.27	2.08
WR-137	1.372	0.62	69.70	4.30
WR-112	1.122	0.497	57.0	5.26
WR-90	0.9	0.4	45.72	6.56
WR-3	0.034	0.017	1.73	173.7

Further, on examining (9.93c) and (9.93d), we notice that in the special case of (corresponding to $a = 2b$), the modes TE_{20} and TE_{01} of waveguides WR-2300 and WR-3 have the same value of λ_c as well as the same value of f_c, which, in other words, means that these modes coalesce or degenerate. Thus, typically, WR-2300 and WR-3, each corresponding to $s = 1$, have the coalescent or degenerate modes in TE_{20} and TE_{01} (Tables 9.2). Further, from the arrangement of the waveguide modes in the order of increasing cutoff frequency for the typical waveguide types considered (Table 9.3), we can see that the TE_{10} mode has the lowest cutoff frequency or the highest cutoff wavelength, irrespective of the waveguide type (Table 9.3). Also, that it is the waveguide type, that is, the waveguide transverse dimensions (a and b) (Table 9.1) on which will depend what will be the higher order mode next to the TE_{10} mode or in other words which mode will have the next higher cutoff frequency than the TE_{10} mode (Table 9.3). Therefore, placing the modes in increasing order of cutoff frequency (Table 9.3), we can see that typically, for the waveguide types WR-2300 and WR-3, for each of which $s = 1$ or $a = 2b$ (Tables 9.1 and 9.2), the next higher-order mode to the TE_{10} mode is the TE_{20} mode or the TE_{01} mode, the latter being coalescent or degenerate with the former (Table 9.3). However, for a waveguide corresponding to $s > 1$ or $a > 2b$, the next higher order mode to the TE_{10} mode will depend on the waveguide type (that is, on the waveguide transverse dimensions a and b) (Table 9.1). Thus, this mode will be the TE_{20} mode for the waveguide type WR-284, while for the WR-90 waveguide it will be the TE_{11} mode (Table 9.3).

Table 9.2 *Comparison between a few lower order modes for typical commercial waveguides with respect to the cutoff wavelength and the cutoff frequency*

Waveguide	s	**TE$_{10}$**		**TE$_{11}$**		**TE$_{20}$**		**TE$_{01}$**	
		λ_c (mm)	f_c (GHz)	λ_c (mm)	f_c (GHz)	λ_c (mm)	f_c (GHz)	λ_c (mm)	f_c (GHz)
WR-2300	1	1168.4	0.257	494.96	0.61	584.2	0.51	584.2	0.51
WR-284	1.06	144.27	2.08	61.57	4.87	72.16	4.16	68.1	4.41
WR-90	1.125	45.72	6.56	30.38	9.88	22.86	13.1	20.3	14.76
WR-3	1	1.73	173.7	0.77	388.6	0.86	347.4	0.86	347.4

If we choose the operating frequency of the waveguide above the cutoff frequency of TE_{10} mode (the lowest of the cutoff frequencies of all other modes), then TE_{10} mode will propagate through the waveguide along with any higher order mode whose cutoff frequency is lower than the operating frequency. For instance, consider 4.5 GHz as the operating frequency for the waveguide WR-284, which is higher than the cutoff frequencies 2.08 GHz, 4.16 GHz and 4.41 GHz of the modes TE_{10}, TE_{20}, and TE_{01} respectively, but lower than the cutoff frequency of the mode TE_{11} (Table 9.3). At this operating frequency, waveguide WR-284 will allow the propagation of modes TE_{10}, TE_{20} and TE_{01}, but will make mode TE_{11} evanescent and hence attenuate it (Section 9.1.5). Thus, TE_{10} mode which has the lowest cutoff frequency of all the modes will be excited in a rectangular waveguide if the operating frequency is higher than the cutoff frequency of the TE_{10} mode, irrespective of whether any other modes are exited or not. Thus, dominantly, TE_{10} mode is excited in a waveguide, and hence it is called the dominant mode of a rectangular waveguide. Further, again taking the example of waveguide WR-284, we can easily appreciate that if the operating frequency is chosen typically as 3 GHz, which is above the cutoff frequency 2.08 GHz of the dominant mode TE_{10} and at the same time below the cutoff frequency 4.16 GHz of the next higher order mode, namely TE_{20} (Table 9.3), it is only the dominant mode that will be excited in the rectangular waveguide. It turns out that the cutoff frequency of TE_{10} mode is the lowest of the cutoff frequencies of all the TE and TM modes (the latter treated in Section 9.2) taken together. (See also Section 9.3.) Thus, the TE_{10} mode continues to be called the dominant mode of a rectangular waveguide irrespective of whether the mode excited is of the TE or the TM type. Further, taking waveguides such as WR-2300 and WR-3 for which $s = 1$ or $a = 2b$ (Tables 9.1-9.3), the range of operating frequencies for the excitation of the dominant mode is prescribed as:

$$(1.15 - to - 1.20) f_c|_{TE_{10}} \leq f \leq (0.90 - to - 0.95) f_c|_{TE_{20}/TE_{01}}. \tag{9.94}$$

In view of (9.94) and making use of the cutoff frequency data (Table 9.3), therefore, one can choose the operating frequencies between 0.32 and 0.49 GHz, for waveguide WR-2300, and those between 220 and 325 GHz for waveguide WR-3.

Table 9.3 *Modes in order of increasing cutoff frequency for typical waveguide types*

Waveguide	Waveguide modes in order of increasing cutoff frequency f_c shown in parenthesis against the modes
WR-2300	TE_{10} (0.257 GHz) TE_{20}/ TE_{01} (0.51 GHz) TE_{11} (0.61 GHz)
WR-284	TE_{10} (2.08 GHz) TE_{20} (4.16 GHz) TE_{01} (4.41 GHz) TE_{11} (4.87 GHz)
WR-90	TE_{10} (6.56 GHz) TE_{11} (9.88 GHz) TE_{20} (13.1 GHz) TE_{01} (14.76 GHz)
WR-3	TE_{10} (173.71 (GHz) TE_{20}/ TE_{01} (347.4 GHz) TE_{11} (388.6 GHz)

Later, you are going to appreciate that while choosing the waveguide mode along with the operating frequency, waveguide dimensions and material of the waveguide, due consideration should also be given to minimise the attenuation constant of the wave (Section 9.5.3).

EXAMPLE E9-6

Show that a rectangular waveguide of cross-sectional dimensions 2.5 cm × 1 cm can support the propagating mode TE_{10} and the evanescent modes TE_{20} and TE_{01} at 10 GHz operating frequency.

SOLUTION

The cutoff frequency of a rectangular waveguide is given by (9.55), which can rewritten as

$$f_c = \frac{c}{2}\left[\left(\frac{m}{a}\right)^2 + \left(\frac{n}{b}\right)^2\right]^{1/2}. \tag{E9-6.1}$$

Putting $c = 3 \times 10^8$ m/s, $a = 2.5 \times 10^{-2}$ m and $b = 1.0 \times 10^{-2}$ m in (E9-6.1), we get

$$f_c = \frac{3 \times 10^8}{2}\left[\left(\frac{m}{2.5 \times 10^{-2}}\right)^2 + \left(\frac{n}{1.0 \times \times 10^{-2}}\right)^2\right]^{1/2}. \tag{E9-6.2}$$

For the TE_{10} mode, putting $m = 1, n = 0$ in (E9-6.2), we obtain

$$f_c = \frac{3 \times 10^8}{2 \times 2.5 \times 10^{-2}} = 6 \times 10^9 \text{ Hz} = 6 \text{ GHz}. \quad (TE_{10} \text{ mode})$$

Similarly, for the TE_{20} mode, putting $m = 2, n = 0$ in (E9-6.2), we obtain

$$f_c = \frac{3 \times 10^8}{2 \times 2.5 \times 10^{-2}} = 12 \times 10^9 \text{ Hz} = 12 \text{ GHz}. \quad (TE_{20} \text{ mode}) \tag{E9-6.3}$$

Also, for the TE_{01} mode, putting $m = 0, n = 1$ in (E9-6.2), we obtain

$$f_c = \frac{3 \times 10^8}{2 \times 10^{-2}} = 15 \times 10^9 \text{ Hz} = 15 \text{ GHz}. \quad (TE_{01} \text{ mode}) \tag{E9-6.4}$$

It follows from (E9-6.1) to (E9-6.4) that the 10 GHz operating frequency is above the TE_{10}-mode cutoff frequency of 6 GHz and below the TE_{20}-mode and TE_{01}-mode cutoff frequencies of 12 and 15 GHz respectively. Therefore, we can infer that in the present problem TE_{10} mode will be a propagating mode, while TE_{20} and TE_{01} modes will each be evanescent modes.

9.2 Rectangular Waveguide in Transverse Magnetic Mode

Let us now outline in brief the salient features of analysis of the TM mode ($E_z \neq 0, H_z = 0$) excited in a rectangular waveguide (Fig. 9.1). The TM mode is also known as the E mode since such a mode is associated with a non-zero value of the axial electric field ($E_z \neq 0$). The analysis of the TM mode closely follows that of the TE mode (Section 9.1) and therefore, we will skip several intermediate steps of derivation in order to avoid repetition.

Let us now start from the wave equation in E_z for the TM mode (instead of that in H_z for the TE mode) by replacing H_z by E_z in (9.1) as follows:

$$\frac{\partial^2 E_z}{\partial x^2} + \frac{\partial^2 E_z}{\partial Y^2} + \frac{\partial^2 E_z}{\partial xz^2} = \mu_0 \varepsilon_0 \frac{\partial^2 E_z}{\partial t^2}. \quad \text{(TM mode)} \tag{9.95}$$

We can then express (9.95) in the following form corresponding to (9.3) (see Section 9.1):

$$\frac{\partial^2 E_z}{\partial x^2} + \frac{\partial^2 E_z}{\partial y^2} + (k^2 - \beta^2)E_z = 0. \quad \text{(TM mode)} \tag{9.96}$$

Solving the differential equation (9.96) as we did for (9.3) for the TE mode (Section 9.1), we obtain the solution corresponding to (9.17) as

$$E_z = (C_1' \cos A'x + C_2' \sin A'x)(C_3' \cos B'y \sin B'y) \exp j(\omega t - \beta z) \tag{9.97}$$

where the primed quantities are all constants. In order to obtain an expression of (9.97) in a form corresponding to (9.44), now we have to use, instead of the boundary conditions (9.28)-(9.31), the following boundary conditions:

$$\left.\begin{array}{lll} E_z = 0 & \text{(left sidewall} & x = a) \text{ for all } y \\ E_z = 0 & \text{(right sidewall} & x = 0) \text{ for all } y \\ E_z = 0 & \text{(bottom wall} & y = 0) \text{ for all } x \\ E_z = 0 & \text{(left sidewall} & y = b) \text{ for all } x \end{array}\right\}. \quad \text{(TM mode)} \tag{9.98}$$

Thus, starting from (9.97), using the boundary conditions (9.98) and following the same procedure as laid down following (9.17), we obtain, with the help of (9.97) and (9.98), the TM mode solution for the axial electric field, corresponding to the TE field solution (9.43) as:

$$E_z = E_{z0} \sin A'x \sin B'y \exp j(\omega t - \beta z) \quad \text{(TM mode)} \tag{9.99}$$

with

$$\left.\begin{array}{l} A' = \dfrac{m\pi}{a} \\ B' = \dfrac{n\pi}{b} \end{array}\right\} \tag{9.100}$$

where E_{z0} is a field constant. Putting (9.100) in (9.99, we can then express the latter as

$$E_z = E_{z0} \sin \frac{m\pi}{a} x \sin \frac{n\pi}{b} y \exp j(\omega t - \beta z) \quad (m, n = 0, 1, 2, 3, ...). \quad \text{(TM mode)} \quad (9.101)$$

Further, while finding the solution of (9.96) by the method of separation of variables following the approach used for obtaining the TE-mode solution (Section 9.1.2), we obtain, corresponding to (9.13) the following expression for the TM mode:

$$A'^2 - (k^2 - \beta) = -B'^2 \quad \text{(TM mode)} \quad (9.102)$$

which, in view of (9.100) takes the form:

$$k^2 - \beta^2 = \left(\frac{m\pi}{a}\right)^2 + \left(\frac{n\pi}{b}\right)^2 \quad \text{(TM mode)} \quad (9.103)$$

which is found to be identical to (9.47) for a rectangular waveguide excited in the TE mode. Further, putting (9.51) in (9.103), we obtain a relationship identical to (9.50):

$$k^2 - \beta^2 = k_c^2 \quad \text{(TM mode)} \quad (9.104)$$

where

$$k_c = \left[\left(\frac{m\pi}{a}\right)^2 + \left(\frac{n\pi}{b}\right)^2\right]^{1/2} \quad (m, n = 0, 1, 2, 3, ...) \quad \text{(TM mode)} \quad (9.105)$$

is the cutoff wave number identical to (9.51). Putting $k = \omega/c$ from (9.4) and $k_c = \omega_c/C$ from (9.53), where ω_c is the cutoff frequency of the waveguide in (9.104) and rearranging the terms, we next obtain the following characteristic equation or the dispersion relation for a rectangular waveguide in the TM mode:

$$\omega^2 - \beta^2 c^2 - \omega_c^2 = 0. \quad \text{(TM mode)} \quad (9.106)$$

You can see that the dispersion relation (9.106) for the TM mode is identical to the dispersion relation (9.54) obtained earlier for the TE mode, with the same interpretation of the cutoff frequency ω_c or $f_c (= \omega_c/2\pi)$ and hence that of the related cutoff wavelength ($\lambda_c (= c/f_c)$ as (9.54). In other words, the expression (9.54) for the cutoff quantities continues to be valid for the TM mode as well. For the sake of convenience, let us rewrite this expression here for the TM mode:

$$\left.\begin{aligned}
\omega_c &= c\left[\left(\frac{m\pi}{a}\right)^2 + \left(\frac{n\pi}{b}\right)^2\right]^{1/2} \\
f_c &= \frac{c}{2}\left[\left(\frac{m}{a}\right)^2 + \left(\frac{n}{b}\right)^2\right]^{1/2} \\
\lambda_c\left(= \frac{c}{f_c}\right) &= \frac{2}{\left[\left(\frac{m}{a}\right)^2 + \left(\frac{n}{b}\right)^2\right]^{1/2}}
\end{aligned}\right\} \quad (m, n = 0, 1, 2, 3, \ldots) \quad \text{(TM mode)} \quad (9.107)$$

which can be seen to be identical to the corresponding TE-mode expression (9.55). Consequently, the dispersion characteristics depicted by Fig. 9.2 continues to be valid for the TM mode as well.

Next, starting from the solution for the axial electric field component E_z given by (9.101) for the TM mode, for which the axial component of the magnetic field is zero $(H_z = 0)$, and taking the help of Maxwell's equations (5.78) and (5.79) and following the same procedure as laid down in Section 9.1, we can easily derive the expressions for the transverse components of the electric and magnetic fields in the waveguide excited in the TM mode as follows:

$$\left.\begin{aligned} E_x &= \frac{-j\beta}{\left(\frac{m\pi}{a}\right)^2 + \left(\frac{n\pi}{b}\right)^2} \frac{m\pi}{a} E_{z0} \cos\frac{m\pi}{a}x \sin\frac{n\pi}{b}y \exp j(\omega t - \beta z) \quad \text{(a)} \\ E_y &= \frac{-j\beta}{\left(\frac{m\pi}{a}\right)^2 + \left(\frac{n\pi}{b}\right)^2} \frac{n\pi}{a} E_{z0} \sin\frac{m\pi}{a}x \cos\frac{n\pi}{b}y \exp j(\omega t - \beta z) \quad \text{(b)} \end{aligned}\right\}$$

$$\text{(TM mode)} \qquad (9.108)$$

$$\left.\begin{aligned} H_y &= \frac{-j\omega\epsilon_0}{\left(\frac{m\pi}{a}\right)^2 + \left(\frac{n\pi}{b}\right)^2} \frac{m\pi}{a} E_{z0} \cos\frac{m\pi}{a}x \sin\frac{n\pi}{b}y \exp j(\omega t - \beta z) \quad \text{(a)} \\ H_x &= \frac{-j\omega\varepsilon_0}{\left(\frac{m\pi}{a}\right)^2 + \left(\frac{n\pi}{b}\right)^2} \frac{n\pi}{a} E_{z0} \sin\frac{m\pi}{a}x \cos\frac{n\pi}{b}y \exp j(\omega t - \beta z) \quad \text{(b)} \end{aligned}\right\}.$$

$$\text{(TM mode)} \qquad (9.109)$$

Now, just as we summed up the field components for the waveguide excited in the TE mode at the end of Section 9.1.2, let us also do the same for the waveguide excited in the TM mode ($E_z \neq 0, H_z = 0$) as follows:

(i) E_x (TM mode): Equation (9.108*a*)

(ii) E_y (TM mode): Equation (9.108*b*)

(iii) $E_z = 0$ (TM mode): Equation (9.101)

(iv) H_x (TM mode): Equation (9.109*b*)

(v) H_y (TM mode): Equation (9.109*a*)

(vi) $H_z = 0$ (TM mode):

The field pattern for the TM mode obtainable with the help of (i)-(vi) above, together with that for the TE mode are discussed in Section 9.3 to follow.

9.3 Significance of Mode Numbers vis-à-vis Field Patterns

Let us now examine the field pattern, that is, the variation of the electric and magnetic fields with space coordinates with respect to some typical lower order modes first with respect to the TE_{mn} mode and then to the TM_{mn} mode for an electromagnetic wave propagating through a rectangular waveguide in the axial z direction. Later, we will examine the field pattern for cylindrical waveguides as well (Section 9.4).

For this purpose, let us first recall with the help of the field expressions that have been put together at the end of Section 9.1.2 the following with respect to the excitation of TE_{mn} mode in a rectangular waveguide ($H_z \neq 0, E_z = 0$):

(i) The TE_{00} ($m = 0, n = 0$) mode is associated with $H_z \neq 0$, $E_x = E_y = E_z = E_z = H_x = H_y = 0$ (TE_{00} mode). The expression for the non-zero magnetic field of this mode is obtained by putting $m = 0, n = 0$ in (9.44) as

$$H_z = H_{z0} \exp j(\omega t - \beta z). \qquad (\mathrm{TE}_{00}\ \text{mode}) \tag{9.110}$$

Thus, with only the axial magnetic field component H_z (which is independent of x and y) being present while all the remaining magnetic field components as well as all the electric field components being absent, the field solution as above for the TE_{00} mode is a trivial solution and, consequently, TE_{00} mode cannot be excited in the waveguide.

(ii) TE_{10} mode ($m = 1, n = 0$) mode is associated with $H_z \neq 0$, $E_x = 0, E_y \neq 0, E_z = 0, H_x \neq 0, H_y = 0$ (TE_{10} mode). The expressions for the non-zero electric and magnetic fields are obtained by putting $m = 1, n = 0$ in (9.44), (9.48a) and (9.49*a*) respectively as

$$\left.\begin{aligned} H_z &= H_{z0} \cos \frac{\pi}{a} x \exp j(\omega t - \beta z) \\ E_y &= -\frac{j\omega\mu_0 a}{\pi} H_{z0} \sin \frac{\pi}{a} x \exp j(\omega t - \beta z) \\ H_x &= \frac{j\beta a}{\pi} H_{z0} \sin \frac{\pi}{a} x \exp j(\omega t - \beta z) \end{aligned}\right\}. \quad (\mathrm{TE}_{10}\ \text{mode}) \tag{9.111}$$

Thus, the axial magnetic field component H_z as well as one transverse electric field component E_y and another transverse magnetic field component H_x being present, the field solution as above for the TE_{10} mode is a non-trivial solution. Therefore, TE_{10} mode can be excited in the waveguide.

(iii) TE_{01} mode ($m = 0, n = 1$) is associated with $H_z \neq 0$, $E_x \neq 0, E_y = 0, E_z = 0, H_z = 0, H_x = 0, H_y \neq 0$, ($\mathrm{TE}_{01}$ mode). The expressions for the non-zero electric and magnetic fields are obtained by putting $m = 0$, $n = 1$ in (9.44), (9.48*b*) and (9.49*b*) respectively as

$$\left.\begin{aligned} H_z &= H_{z0}\cos\frac{\pi}{b}y\exp j(\omega t-\beta z)\\ E_x &= \frac{j\omega\mu_0 b}{\pi}H_{z0}\sin\frac{\pi}{b}y\exp j(\omega t-\beta z)\\ H_y &= \frac{j\beta b}{\pi}H_{z0}\sin\frac{\pi}{b}y\exp j(\omega t-\beta z)\end{aligned}\right\}. \quad (\text{TE}_{01}\text{ mode}) \tag{9.112}$$

Thus, the axial magnetic field component H_z as well as one transverse electric field component E_x and another transverse magnetic field component H_y being present, the field solution as above for the TE_{01} mode is a non-trivial solution. Therefore, TE_{01} mode can be excited in the waveguide.

(iv) TE_{20} mode ($m = 2, n = 0$) is associated with $H_z \neq 0$, $E_x = 0, E_y \neq 0, E_z = 0; E_z = 0, H_x \neq 0, H_y = 0$ (TE_{20} mode). The expressions for the non-zero electric and magnetic fields are obtained by putting $m = 2, n = 0$ in (9.44), (9.48a) and (9.49a) respectively as:

$$\left.\begin{aligned} H_z &= H_{z0}\cos\frac{2\pi}{a}x\exp j(\omega t-\beta z)\\ E_y &= \frac{j\omega\mu_0 a}{2\pi}H_{z0}\sin\frac{2\pi}{a}x\exp j(\omega t-\beta z)\\ H_x &= \frac{j\beta a}{2\pi}H_{z0}\sin\frac{2\pi}{a}x\exp j(\omega t-\beta z)\end{aligned}\right\}. \quad (\text{TE}_{20}\text{ mode}) \tag{9.113}$$

Thus, the axial magnetic field component H_z as well as one transverse electric field component E_y and another transverse magnetic field component H_x being present, the field solution as above for the TE_{10} mode is a non-trivial solution. Therefore, TE_{10} mode can be excited in the waveguide.

Similarly, let us now take the case of the TM_{mn} mode ($E_z \neq 0, H_z = 0$). We can see, with the help of the field expressions put together at the end of Section 9.2 that for each of the typical lower order TM modes: (i) $\text{TM}_{00}(m = 0, n = 0)$, (ii) $\text{TM}_{10}(m = 1, n = 0)$ and (iii) ($m = 0, n = 1$), the following relation holds good:

$$E_x = E_y = E_z = E_z = H_x = H_y = H_z = 0 \quad (\text{TM}_{00}, \text{TM}_{10}, \text{TM}_{01}\text{ modes}).$$

Therefore, all the electric and magnetic field components being absent, the field solution as above for each of the modes TM_{00}, TM_{10} and TM_{01} is a trivial solution and therefore, none of these lower order modes can be excited in a rectangular waveguide. The cutoff frequency values of these non-excitable modes, obtainable with the help of (9.107), are $f_c = 0$ for the TM_{00} mode, $f_c = c/2a$ for the TM_{10} mode, and $f_c = c/2b$ for the TM_{01} mode, in order of increasing values (remembering $a > b$). The next higher order TM mode is the TM_{11} mode, whose cutoff frequency is obtained by putting $m = 1, n = 1$ in (9.107) as: $f_c = (c/2)[(1/a)^2 + (1+b)^2]^{1/2}$. Mode TM_{11} can be excited in the rectangular waveguide since it enjoys non-trivial field solutions as below, obtained by putting $m = 1$, $n = 1$ in the field expressions (9.101), (9.108), and (9.109) while remembering $H_z = 0$ (for the TM mode):

$$\left.\begin{aligned}
E_x &= \frac{-j\beta}{\left(\frac{\pi}{a}\right)^2+\left(\frac{\pi}{b}\right)^2}\frac{\pi}{a}E_{z0}\cos\frac{\pi}{a}x\sin\frac{\pi}{b}y\exp j(\omega t-\beta z)\\
E_y &= \frac{-j\beta}{\left(\frac{\pi}{a}\right)^2+\left(\frac{\pi}{b}\right)^2}\frac{\pi}{b}E_{z0}\cos\frac{\pi}{b}y\sin\frac{\pi}{a}x\cos\frac{\pi}{b}y\exp j(\omega t-\beta z)\\
E_z &= E_{z0}\sin\frac{\pi}{a}x\sin\frac{\pi}{b}y\exp j(\omega t-\beta z)
\end{aligned}\right\}\quad (\mathrm{TM}_{11}\text{ mode}) \tag{9.114}$$

$$\left.\begin{aligned}
H_x &= \frac{j\omega\varepsilon_0}{\left(\frac{\pi}{a}\right)^2+\left(\frac{\pi}{b}\right)^2}\frac{\pi}{b}E_{z0}\sin\frac{\pi}{a}x\cos\frac{\pi}{b}y\exp j(\omega t-\beta z)\\
H_y &= \frac{j\omega\varepsilon_0}{\left(\frac{\pi}{a}\right)^2+\left(\frac{\pi}{b}\right)^2}\frac{\pi}{a}E_{z0}\cos\frac{\pi}{a}x\sin\frac{\pi}{b}y\exp j(\omega t-\beta z)\\
H_z &= 0
\end{aligned}\right\}.\quad (\mathrm{TM}_{11}\text{ mode}) \tag{9.115}$$

Thus, we notice that the TM_{11} mode is the lowest order TM mode corresponding to non-trivial field solutions [given by (9.114) and (9.115)] that has the lowest value of the cutoff frequency. Still lower order TM modes, namely TM_{00}, TM_{10}, and TM_{01}, which each have a lower cutoff frequency value than the TM_{11} mode, correspond to trivial field solutions. Further, we can see that the cutoff frequency $f_c = (c/2)[(1/a)^2 + (1+b)^2]^{1/2}$ of the TM_{11} mode is greater than the cutoff frequency $f_c = c/2a$ of the lowest order TE mode, namely the TE_{10} mode that has non-trivial field solutions. Thus, the TE_{10} mode, which has the lowest cutoff frequency value of all the TE and the TM modes taken together, is called the dominant mode of the rectangular waveguide, as stated earlier. (See the discussion following Table 9.3.)

Let us plot the field pattern for a few typical lower order modes of a rectangular waveguide, for instance, the transverse electric modes TE_{10} (dominant mode) (Fig. 9.4) and TE_{20} (Fig. 9.6) with the help of the field expressions (9.111) and (9.113) respectively, and then that of the transverse magnetic TM_{11} mode (Fig. 9.7) with the help of the field expression (9.115). Note that for $a = 2b$, that is, waveguide broader dimension taken as twice its narrow dimension, although modes TE_{20} and TE_{01} are degenerate, having the same cutoff frequency as explained following Table 9.2, yet these two modes have different field patterns. [Here, the field pattern of only TE_{20} mode is plotted in Fig. 9.6, not that of the TE_{01} mode; the latter can, however, be easily done with the help of field expression (9.112)]. The field lines in these plots (Figs. 9.4 to 9.7) are drawn such that the direction of the cross product $\vec{E} \times \vec{H}$ representing the direction of the instantaneous Poynting vector indicates the direction of propagation z, the directions of $\vec{E}$ and $\vec{H}$ being represented by those of the electric and the magnetic field lines respectively.

The phase factor $\exp(-j\beta z)$ appearing in the field expressions (9.111) to (9.115), irrespective of modes considered will change sign at a phase interval of $\beta z = \pi$ since

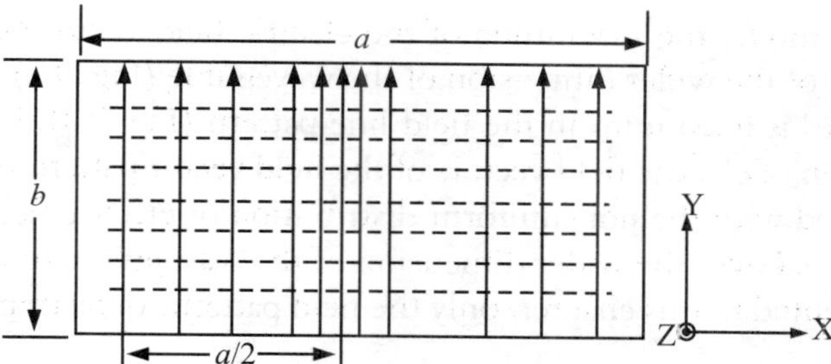

Fig. 9.4 *Field pattern of the* TE_{10} *mode of a rectangular waveguide showing electric and magnetic field lines indicated by solid and broken lines respectively.* Z *axis, shown by a dot surrounded by a circle, is perpendicular to the plane of the paper and directed towards the reader. The direction of propagation is along* z *being the direction of Poynting vector* $\vec{E} \times \vec{H}$.

$$\exp -j(\beta z + \pi) = \exp(-j\pi)\exp(-j\beta z) = (\cos\pi - j\sin\pi)\exp(-j\beta z) = -\exp(-j\beta z).$$

However, this phase interval of $\beta z = \pi$ corresponds to the distance interval of $z = \pi/\beta = \pi/(2\pi/\lambda) = \lambda/2$. This then means that the electric field lines will change its sign at a distance interval of $z = \lambda/2$ along the direction of propagation. Similarly, we can easily appreciate that the phase factor $\exp(-j\beta z)$ will not change its sign at a phase interval of $\beta z = 2\pi$ or at a distance interval of $z = \lambda$, which then means that the electric field lines will not change its sign at a distance interval of $z = \lambda$.

Further, for the TE_{10} mode, we notice with the help of (9.111) that the amplitude of E_y, the electric field along y, is proportional to the factor $\sin(\pi/a)x$, the latter being maximum and equal to 1 at $x = a/2\,(\sin(\pi/a)x = \sin\pi/2 = 1)$ and null both at $x = 0\,(\sin(\pi/a)x = \sin 0 = 0)$, and at $x = a(\sin(\pi/a)x = \sin(\pi/a)a = \sin\pi = 0)$. Thus, for the TE_{10} mode, you will see dense electric field lines parallel to the narrow dimension of the waveguide in the y direction at $x = a/2$, that is, in the middle of the broader dimension of the waveguide in the transverse XY-plane, and also see the absence of the electric field lines both at the ends $x = 0$ and $x = a$ (Fig. 9.4). Also, with the help of (9.111) you will see that the amplitude of H_x is independent of y, which means that the magnetic field lines which are parallel to the broader dimension of the waveguide will be uniformly distributed in the transverse XY-plane. Further, you will notice that the magnetic field lines have not been drawn up to the ends $x = 0$ and $x = a$ in the transverse plane in the field pattern of Fig. 9.4, since the amplitude of H_x, the magnetic field parallel to x, being proportional to $\sin(\pi/a)x$, becomes null at these ends (like the electric field lines as discussed above). However, the magnetic field lines are continuous and they do not end up at $x = 0$ and $x = a$; the continuity of these magnetic field lines can be seen in the field pattern the XZ-plane, which however is not drawn here.

There is an alternative approach to presenting the field pattern of a waveguide in which the variation of the magnitude of the field vector instead of the distribution of field lines, say, over the waveguide cross section is shown such that the length of the vector represents the field magnitude (Fig. 9.5). In such kind of field vector pattern which looks like a half wave for the TE_{10} mode, the magnitude of the electric field vector takes the maximum value at the middle of the wider dimension of the waveguide (Fig. 9.5), where the density of electric field lines is maximum in the field line pattern (Fig. 9.4). Please note that the uniform distribution of electric field vectors of the field vector pattern presented (Fig. 9.5) is not to be confused with the non-uniform distribution of electric field lines of the field line pattern of Fig. 9.4 over the wider dimension of the waveguide. In the rest of the field patterns to be presented in this chapter, only the field patterns depicting the field lines will be presented.

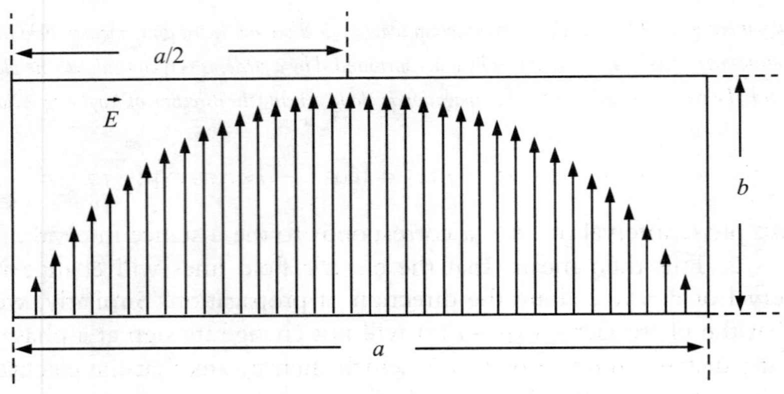

Fig. 9.5 *Field pattern of the TE_{10} mode of a rectangular waveguide showing the variation of the magnitude of the electric field vector over the wider dimension of the waveguide.*

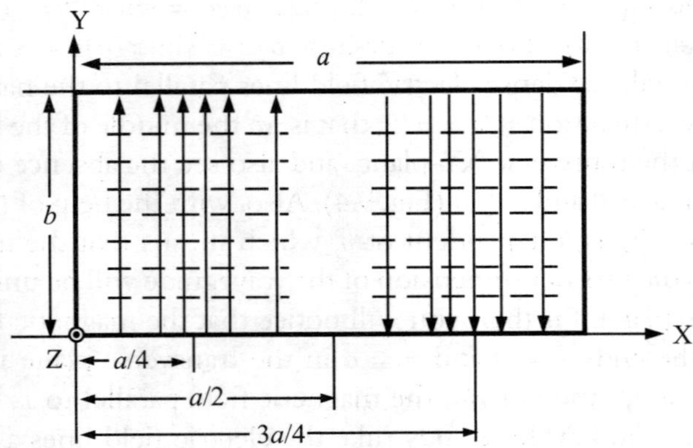

Fig. 9.6 *Field pattern of the TE_{20} mode of a rectangular waveguide showing electric and magnetic field lines indicated by solid and broken lines respectively. Z axis shown by a dot surrounded by a circle is perpendicular to the plane of the paper and directed towards the reader. The direction of propagation is along z being the direction of Poynting vector $\vec{E} \times \vec{H}$.*

Similarly, let us appreciate the field pattern of the rectangular waveguide excited in the TE_{20} mode (Fig. 9.6). For this mode TE_{20}, the amplitude of E_y, the electric field along

y, as can be seen from (9.113), is proportional to the factor $\sin(2\pi/a)x$, the latter being positive maximum and equal to 1 at $x = a/4$ (in view of $\sin(2\pi/a)x = \sin\pi/2 = 1$) and negative maximum and equal to -1 at $x = 3a/4$. Accordingly, the field pattern shows the concentration of electric field lines parallel to the y direction in these locations, that is, at the quarter and three quarter of the broader dimension of the waveguide in the transverse XY-plane respectively ($x = a/4$ and $x = 3a/4$). Note, however, the reversal of their directions in Fig. 9.6. Further, the factor $\sin(2\pi/a)x$ becomes null at the locations $x = 0, x = a/2$, and $x = a$ (in view of $\sin(2\pi/a)x = \sin 0 = 0$, $\sin(2\pi a)x = \sin\pi = 0$, and $\sin(2\pi/a)x = \sin 2\pi = 0$ respectively), which then corresponds to the absence of electric field lines at each of these locations. Further, we can also see from (9.113) that the amplitude of H_x, like that of E_y, is also proportional to $\sin(2\pi/a)x$. Therefore, like the electric field lines in the transverse XY-plane, the magnetic field also show their absence at the locations $x = 0, x = a/2$, and $x = a$; they will, like the electric field lines, show their reversal of directions at $x = a/4$ and $x = 3a/4$, noting that the magnetic field lines take on directions according to the directions of the electric field lines such that the direction of the corresponding cross product $\vec{E} \times \vec{H}$, representing the direction of the instantaneous Poynting vector, indicates the direction of propagation z (Fig. 9.6). Further, according to (9.113), the amplitude of H_x is independent of y, which means that the magnetic field lines are uniformly distributed in the transverse XY-plane (Fig. 9.6).

Following the same procedure above, plot the field patterns of the rectangular waveguide excited in the TE_{01} and the TE_{20} modes with the help of the field expressions (9.112) and (9.113) respectively. For instance, check with the help of (9.112) that for the TE_{01} mode, the electric field lines in the direction of x and parallel to the broad dimension of the waveguide become concentrated at $y = b/2$, that is, in the middle of the narrow dimension of the waveguide in the transverse XY-plane. You can also verify the absence of the electric field lines both at the ends $y = 0$ and $y = b$. Also, with the help of (9.112), appreciate that the amplitude of H_y is independent of x, which means that the magnetic field lines parallel to the narrow dimension of the waveguide will be uniformly distributed in the transverse XY-plane. We have however not presented the field pattern of the TE_{01} mode and leave the plotting of the field pattern to you as an exercise.

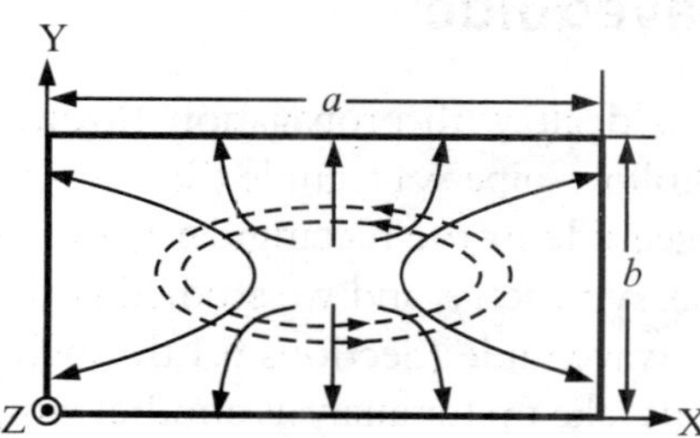

Fig. 9.7 *Field pattern of the TM_{11} mode of a rectangular waveguide showing electric and magnetic field lines indicated by solid and broken lines respectively. Z axis shown by a dot surrounded by a circle is perpendicular to the plane of the paper and directed towards the reader. The direction of propagation is along z being the direction of Poynting vector $\vec{E} \times \vec{H}$.*

Next, let us turn towards the field pattern of a rectangular waveguide excited in the TM_{11} mode. Unlike the TE_{10}, TE_{01} and TE_{20} modes for which there exist only three non-zero field components according to (9.111)-(9.113) respectively, in which the electric and magnetic field lines are everywhere parallel either to the broad or to the narrow waveguide dimension (see for instance Figs. 9.4 and 9.6), the TM_{11} mode is associated with five non-zero field components according to (9.114) and (9.115) taken together. Consequently, for the TM_{11} mode the field lines are no longer everywhere parallel either to the broad or the narrow waveguide dimension; in general they are curved (Fig. 9.7).

The significance of the TE_{mn} or TM_{mn} mode field patterns of a rectangular waveguide vis-à-vis the mode numbers m and n become apparent by examining the relevant patterns, for instance, Figs. 9.4 to 9.6. We can easily infer by examining such field patterns (including Figs. 9.4 to 9.6) that the mode number m, which is the first suffix of TE_{mn} or TM_{mn}, indicates the number of maxima of any field component, electric or magnetic, along the broad dimension of the waveguide, while the mode number n, which is the second suffix of TE_{mn} or TM_{mn}, indicates the number of maxima of any field component, electric or magnetic, along the narrow dimension of the waveguide. Alternatively, we can interpret m and n as the numbers of half-wave ($\lambda/2$) field patterns across the broad and narrow dimensions of the waveguide respectively. For example, we can see for the TE_{10} mode ($m = 1, n = 0$) that the first suffix being unity ($m = 1$), there is a single maximum of the field component along the waveguide broad dimension, and that the second suffix being zero ($n = 0$), there is no maximum of the field component along the waveguide narrow dimension (Fig. 9.4). Similarly, with reference to the TE_{20} mode ($m = 2$, $n = 0$) we can see that the first suffix being 2 ($n = 2$), there exist two maxima of the field component along the waveguide broad dimension and that the second suffix being zero ($n = 0$) there is no maximum of the field component along the waveguide narrow dimension (Fig. 9.6). We can make similar observations by examining the field pattern of the TM_{mn} mode, for instance, the TM_{11} mode as well (Fig. 9.7). Furthermore, in the section to follow, we are going also to make similar inferences on the significance of mode numbers vis-à-vis field pattern with respect to a cylindrical waveguide (Section 9.4).

9.4 Cylindrical Waveguide

The preceding three sections dealt with propagation through a bounded medium, the latter taken as a rectangular hollow-pipe waveguide (Sections 9.1 - 9.3). We formulated the analysis of a rectangular waveguide in the rectangular system of coordinates since such a waveguide enjoys rectangular symmetry, and we studied both the TE and the TM modes of excitation in a rectangular waveguide (Sections 9.1 through 9.3).

In the present section, let us take up for analysis another simple hollow-pipe waveguide, namely, a circular or cylindrical waveguide having a circular cross section (Fig. 9.8). Since a cylindrical waveguide enjoys cylindrical symmetry, we will now formulate the problem of this waveguide in the cylindrical coordinate system. We already seen that TE_{10} is the dominant mode in a rectangular waveguide, which is characterised by the lowest of the cutoff frequencies of all the TE_{mn} and the TM_{mn} modes of a rectangular waveguide taken together (Section 9.3). Similarly, for a cylindrical waveguide it turns out that the dominant mode is the TE_{11} mode, characterised by the lowest cutoff frequency. Further, with a view to putting a ceiling on the length of the present section, let us emphasize mainly the TE_{11} mode for discussion However, the analytical approach followed is rather general and is easily applicable to the analysis of other higher order TE_{mn} and TM_{mn} modes as well.

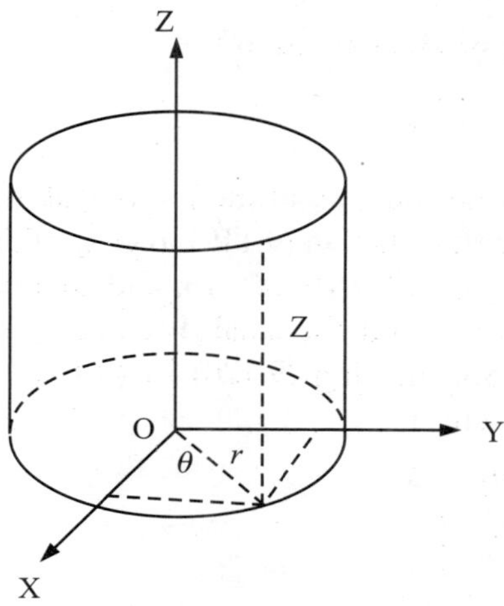

Fig. 9.8 *Geometry of the cylindrical waveguide analysed in the cylindrical coordinate system* (r, θ, z).

Let us start with the wave equation in the non-zero axial magnetic field component H_z, which can be written from (6.34) for the TE mode ($H_z \neq 0, E_z = 0$) excitation, as:

$$\nabla^2 H_z - \mu_0 \varepsilon_0 \frac{\partial^2 H_z}{\partial t^2} = 0. \tag{9.116}$$

Interpreting the Laplacian operator in the cylindrical coordinate system (r, θ, z) with the help of (2.76), we can express the wave equation (9.116) as:

$$\frac{1}{r}\frac{\partial}{\partial r}\left(r\frac{\partial H_z}{\partial r}\right) + \frac{1}{r^2}\frac{\partial^2 H_z}{\partial \theta^2} + \frac{\partial^2 H_z}{\partial z^2} - \mu_0 \varepsilon_0 \frac{\partial^2 H_z}{\partial t^2} = 0,$$

which can be written after expanding the differentiation of its first term as

$$\frac{\partial^2 H_z}{\partial r^2} + \frac{1}{r}\frac{\partial H_z}{\partial r} + \frac{\partial^2 H_z}{\partial z^2} - \mu_0 \varepsilon_0 \frac{\partial^2 H_z}{\partial t^2} + \frac{1}{r^2}\frac{\partial^2 H_z}{\partial \theta^2} = 0. \tag{9.117}$$

If we take $\partial/\partial t \equiv j\omega$ and $\partial/\partial z \equiv -j\beta$, taking the RF dependence with the axial distance z and with time t as $\exp j(\omega t - \beta z)$ (see Section 9.1.1), ω and β being the angular frequency and the axial phase propagation constant of the wave respectively, we can write the wave equation (9.117) as

$$\frac{\partial^2 H_z}{\partial r^2} + \frac{1}{r}\frac{\partial H_z}{\partial r} + (\omega^2 \mu_0 \varepsilon_0 - \beta^2) H_z + \frac{1}{r^2}\frac{\partial^2 H_z}{\partial \theta^2} = 0. \tag{9.118}$$

We will now follow the method of separation of variables (see Section 9.1.2) to solve the wave equation (9.118). For this purpose, let us rearrange the terms of (9.118) to write it as

$$\frac{\partial^2 H_z}{\partial r^2} + \frac{1}{r}\frac{\partial H_z}{\partial r} + (\omega^2 \mu_0 \varepsilon_0 - \beta^2) H_z = -\frac{1}{r^2}\frac{\partial^2 H_z}{\partial \theta^2}. \tag{9.119}$$

Let us try in this method the product solution

$$H_z = R\Theta, \tag{9.120}$$

where R is a function of r alone and Θ is a function of θ alone. Note that the factor of RF dependence, $\exp j(\omega t - \beta z)$, is absorbed in (9.119) in view of the interpretation of operators as $\partial/\partial t \equiv j\omega$ and $\partial/\partial z \equiv -j\beta$, and $\partial/\partial z \equiv -j\beta$, and therefore this factor (representing the RF dependence with time t and the axial distance z) has to be understood in the trial field solution (9.120). Substituting (9.120) in (9.119) and carrying out the partial differentiations involved therein as

$$\left.\begin{aligned} \frac{\partial H_z}{\partial r} &= \frac{\partial (R\Theta)}{\partial r} = \Theta\frac{\partial R}{\partial r} \\ \frac{\partial^2 H_z}{\partial r^2} &= \Theta\frac{\partial^2 R}{\partial r^2} \\ \frac{\partial^2 H_z}{\partial \theta^2} &= R\frac{\partial^2 \Theta}{\partial \theta^2} \end{aligned}\right\},$$

we can express (9.120) as follows:

$$\Theta\frac{\partial^2 R}{\partial r^2} + \frac{1}{r}\Theta\frac{\partial R}{\partial r} + (\omega^2 \mu_0 \varepsilon_0 - \beta^2) R\Theta = -\frac{1}{r^2}R\frac{\partial^2 \Theta}{\partial \theta^2}. \tag{9.121}$$

Next, dividing (9.121) by $R\Theta$, we can write

$$\frac{1}{R}\frac{\partial^2 R}{\partial r^2} + \frac{1}{r}\frac{1}{R}\frac{\partial R}{\partial r} + (\omega^2 \mu_0 \varepsilon_0 - \beta^2) = -\frac{1}{r^2}\frac{\partial^2 \Theta}{\partial \theta^2}\frac{1}{\Theta},$$

which, on putting

$$\gamma = (\omega^2 \mu_0 \varepsilon_0 - \beta^2)^{1/2} = (\omega^2/c^2 - \beta^2)^{1/2} = (k^2 - \beta^2)^{1/2} \tag{9.122}$$

$k(= \omega(\mu_0\varepsilon_0)^{1/2} = \omega/c)$ being the free-space propagation constant (9.4), becomes

$$\frac{1}{R}\frac{\partial^2 R}{\partial r^2} + \frac{1}{Rr}\frac{\partial R}{\partial r} + \gamma^2 = -\frac{1}{r^2}\frac{\partial^2 \Theta}{\partial \theta^2}\frac{1}{\Theta} \tag{9.123}$$

Further, multiplying (9.123) by r^2, we can express it as

$$\frac{r^2}{R}\frac{\partial^2 R}{\partial r^2} + \frac{r}{R}\frac{\partial R}{\partial r} + \gamma^2 r^2 = -\frac{\partial^2 \Theta}{\partial \theta^2}\frac{1}{\Theta}. \tag{9.124}$$

The above expression equates a function of r alone on its left-hand side with a function of θ alone on its right-hand side. This can hold good only when the individual functions each become equal to a constant. Putting this constant as m^2, we can then write (9.124) part-wise as

$$\frac{r^2}{R}\frac{\partial^2 R}{\partial r^2} + \frac{r}{R}\frac{\partial R}{\partial r} + \gamma^2 r^2 = m^2 \tag{9.125}$$

and

$$-\frac{\partial^2 \Theta}{\partial \theta^2}\frac{1}{\Theta} = m^2 \,. \tag{9.126}$$

Dividing (9.125) by r^2, we have

$$\frac{1}{R}\frac{\partial^2 R}{\partial r^2} + \frac{1}{Rr}\frac{\partial R}{\partial r} + \gamma^2 = \frac{m^2}{r^2}$$

which on multiplication by R followed by rearrangement of terms becomes

$$\frac{\partial^2 R}{\partial r^2} + \frac{1}{R}\frac{\partial R}{\partial r} + \left(\gamma^2 - \frac{m^2}{r^2}\right) = R = 0. \tag{9.127}$$

At this stage, let us define a dimensionless radial coordinate $x = \gamma r$ and express (9.127) as a differential equation in x. (Note that r has the dimension of distance; also, from (9.122) it can be seen that γ has the dimension of the reciprocal of distance so that the quantity becomes dimensionless). Therefore, substituting $r = x/\gamma$ (obtainable from the relation $x = \gamma r$) in (9.127) and rearranging the terms, we can easily write

$$\gamma^2 \frac{\partial^2 R}{\partial x^2} + \gamma^2 \frac{1}{x}\frac{\partial R}{\partial x} + \left(\gamma^2 - \frac{m^2}{r^2}\right) R = 0,$$

which, when divided by γ^2 while remembering the relation $x = \gamma r$ takes the following form

$$\frac{\partial^2 R}{\partial x^2} + \frac{1}{x}\frac{\partial R}{\partial x} + \left(1 - \frac{m^2}{x^2}\right) R = 0. \tag{9.128}$$

Let us next proceed to find the solutions of (9.126) and (9.128) for Θ and R respectively, which in turn can be substituted in (9.120) to obtain the solution for H_z. For this purpose, let us choose to express (9.126) as

$$\frac{\partial^2 \Theta}{\partial \theta^2} + m^2 \Theta = 0$$

which has the well-known solution of the form

$$\Theta = C\cos m\theta + D\sin m\theta, \tag{9.129}$$

where C and D are constants. Further, substituting

$$\left.\begin{aligned} C &= M\cos\psi \\ D &= M\sin\psi \end{aligned}\right\} \tag{9.130}$$

in (9.129), we obtain

$$\Theta = M(\cos m\theta\cos\psi + \sin m\theta\sin\psi), \tag{9.131}$$

where M and ψ are new constants which can be expressed with the help of (9.130) as

$$\left.\begin{aligned} M &= (C^2 + D^2)^{1/2} \\ \psi &= \tan^{-1}\tfrac{D}{C} \end{aligned}\right\}. \tag{9.132}$$

Further, using the trigonometric relation $\cos(A-B) = \cos A\cos B + \sin A\sin B$, we can express (9.31) as :

$$\Theta = M\cos(m\theta - \psi). \tag{9.133}$$

Notice that in the cross-sectional plane perpendicular to the axis of the cylindrical waveguide (Fig. 9.8), a point (r,θ) represented by an angle θ at given radial coordinate r will coincide with the point $(r,\theta+2\pi)$ represented by an angle $\theta+2\pi$ at the same radial coordinate r. Consequently, the value of Θ at angle θ would be the same as that at angle $\theta+2\pi$ (the radial coordinate r remaining unchanged). This can be mathematically stated with the help of (9.133) as

$$\cos(m\theta-\psi) = \cos[m(\theta+2\pi)-\psi] = \cos[(m\theta-\psi)+2\pi m]$$

a relation that holds good only for integral values of m. Thus, in the solution (9.133), the value of m has to be taken as $m = 0, 1, 2, 3, \ldots$. Further, the choice of the reference from which the value of θ has to be reckoned being arbitrary, we can choose it such that $\psi = 0$. This in turn corresponds to the maximum value of Θ at the reference $\theta = 0$, being equal to $\Theta = M$, which can be appreciated by putting $\psi = \theta = 0$ in (9.133). Therefore, putting $\psi = 0$ in (9.133), we can now write the solution for Θ as

$$\Theta = M\cos m\theta. \tag{9.134}$$

Similarly, let us proceed to solve (9.128) to obtain the solution for R, which however becomes rather easy since one can identify (9.128) as the well known ordinary Bessel equation that has the following solution (Appendix A9.2):

$$R = PJ_m(x) + QY_m(x) \tag{9.135}$$

where P and Q are arbitrary constants. $J_m(x)$ and $Y_m(x)$ are called the Bessel functions of order m of the first and second kinds respectively (Appendix A9.2). The function $Y_m(x)$ is known to shoot to infinity ($Y_m(x) \to \infty$) at $x = 0$, that is, at the axis of the cylindrical waveguide $r = 0$ (remembering the relation $x = \gamma r$) which was defined following (1.127). However, as can be seen from (9.135), $Y_m \to \infty$ would then make $R \to \infty$ at $r = 0$, which according to (9.120) would predict H_z shooting up to infinity at the waveguide axis ($r = 0$), which is however an impossibility. Therefore, in order to prevent this from happening, we have to put the constant $Q = 0$ in (9.135), enabling us to write

$$R = PJ_m(x) \tag{9.136}$$

which, in view of the relation $x = \gamma r$, can be read as

$$R = PJ_m(\gamma r). \tag{9.137}$$

Thus, in the present context, the values of the function $J_m(x)$ are of interest, here taking $x = \gamma r$. From literature on Bessel functions, we can note that the values of $J_m(x)$ are 'damped-sinusoidal-like' functions, alternating between positive and negative values with peak amplitudes gradually decreasing with the increase of the argument x of the function. The zeros of $J_0(x)$ are at $x = 2.408, 5.5201, 8.6537, \ldots$, which are nearly equi-spaced, the interval in x between successive zeros being close to the exactly equi-spaced interval $\pi = 3.14159$ of a sinusoidal function $\sin x$.

We can now obtain, by substituting (9.134) and (9.137) in (9.120), the following solution for the axial component of the magnetic field in terms of Bessel functions:

$$H_z = H_{z0} J_m(\gamma r) \cos m\theta,$$

which, after introducing the factor $\exp j(\omega t - \beta z)$ [understood in the field solution; see the discussion following (9.120)], can be read as

$$H_z = H_{z0} J_m(\gamma r) \cos m\theta \exp j(\omega t - \beta z). \tag{9.138}$$

Next, it will be of interest to find the expressions for the remaining field components, E_r, E_θ, H_r and H_θ, in terms of H_z, the latter given by (9.138) (remember that for the TE mode under consideration, $E_z = 0$). For this purpose, we take the Maxwell's equations (5.78) and (5.79), in which however we take the conduction current density $\vec{J} = 0$, the inside of the waveguide being a free-space region:

$$\nabla \times \vec{E} = -\frac{\partial \vec{B}}{\partial t}$$

and

$$\nabla \times \vec{H} = -\frac{\partial \vec{D}}{\partial t}.$$

In view of the relations $\vec{B} = \mu_0 \vec{H}$ and $\vec{D} = \varepsilon_0 \vec{E}$, these can be read as

$$\nabla \times \vec{E} = -\mu_0 \frac{\partial \vec{H}}{\partial t} \tag{9.139}$$

and $$\nabla \times \vec{H} = \varepsilon_0 \frac{\partial \vec{E}}{\partial t}. \tag{9.140}$$

We can read (9.139) and (9.140) by expanding $\nabla \times \vec{E}$ and $\nabla \times \vec{H}$ in cylindrical coordinate system with the help of (2.67) respectively as

$$\left(\frac{1}{r}\frac{\partial E_z}{\partial \theta} - \frac{\partial E_\theta}{\partial z}\right)\vec{a}_r + \left(\frac{\partial E_r}{\partial z} - \frac{\partial E_z}{\partial r}\right)\vec{a}_r + \frac{1}{r}\left(\frac{\partial (rE_\theta)}{\partial r} - \frac{\partial E_r}{\partial \theta}\right)\vec{a}_r$$

$$= -\mu_0\left(\frac{\partial H_r}{\partial t}\vec{a_r} + \frac{\partial H_\theta}{\partial t}\vec{a_0} + \frac{\partial H_z}{\partial t}\vec{a}_r\right) \tag{9.141}$$

and

$$\left(\frac{1}{r}\frac{\partial H_z}{\partial \theta} - \frac{\partial H_\theta}{\partial z}\right)\vec{a}_r + \left(\frac{\partial H_r}{\partial z} - \frac{\partial H_z}{\partial r}\right)\vec{a}_\theta + \frac{1}{r}\left(\frac{\partial (rH_\theta)}{\partial r} - \frac{\partial H_r}{\partial \theta}\right)\vec{a}_z$$

$$= \varepsilon_0\left(\frac{\partial E_r}{\partial t}\vec{a}_r + \frac{\partial E_\theta}{\partial t}\vec{a}_\theta + \frac{\partial E_z}{\partial t}\vec{a}_z\right). \tag{9.142}$$

We can then read the r and θ components of (9.141) respectively as

$$\frac{1}{r}\frac{\partial E_z}{\partial \theta} - \frac{\partial E_\theta}{\partial z} = -\mu_0 \frac{\partial H_r}{\partial t} \tag{9.143}$$

and $$\frac{\partial E_r}{\partial z} - \frac{\partial E_z}{\partial r} = -\mu_0 \frac{\partial H_\theta}{\partial t}. \tag{9.144}$$

Similarly, we can read the r and θ components of (9.141) respectively as

$$\frac{1}{r}\frac{\partial H_z}{\partial \theta} - \frac{\partial H_\theta}{\partial z} = -\varepsilon_0 \frac{\partial E_r}{\partial t} \tag{9.145}$$

and $$\frac{\partial H_r}{\partial z} - \frac{\partial H_z}{\partial r} = \varepsilon_0 \frac{\partial E_\theta}{\partial t}. \tag{9.146}$$

Further, we can take $\partial/\partial t = j\omega$ and $\partial/\partial z = -j\beta$ in view of the RF dependence $\exp j(\omega t - \beta z)$, as well as $E_z = 0$ for the TE mode of excitation of the waveguide, and hence express (9.143)-(9.146) respectively as

$$\beta E_\theta = -\omega \mu_0 H_r \tag{9.147}$$

$$\beta E_r = \omega\mu_0 H_\theta \tag{9.148}$$

$$\frac{1}{r}\frac{\partial H_z}{\partial\theta} + j\beta H_\theta = j\omega\varepsilon_0 E_r \tag{9.149}$$

$$-j\beta H_r - \frac{\partial H_z}{\partial r} = j\omega\varepsilon_0 E_\theta. \tag{9.150}$$

We can now easily eliminate H_r from (9.147) and (9.150) to obtain

$$E_\theta = \frac{j\omega\mu_0}{\gamma^2}\frac{\partial H_z}{\partial r} \tag{9.151}$$

where $\gamma = (\omega^2\mu_0\varepsilon 0 - \beta^2)^{1/2} = (k^2 - \beta^2)^{1/2}$ is defined by (9.122).

Substituting (9.151) in (9.147), we obtain

$$H_r = \frac{-j\beta}{\gamma^2}\frac{\partial H_z}{\partial r}. \tag{9.152}$$

Similarly, we can eliminate E_r from (9.148) and (9.149) to obtain

$$H_\theta = \frac{-j\beta}{\gamma^2 r}\frac{\partial H_z}{\partial\theta} \tag{9.153}$$

which can be substituted in (9.148) to obtain

$$E_r = \frac{-j\omega\mu_0}{\gamma^2 r}\frac{\partial H_z}{\partial\theta}. \tag{9.154}$$

It will be useful to group the field expressions (9.138) and (9.151)-(9.154) for the cylindrical waveguide excited in the TE mode ($E_z = 0$, $H_z \neq 0$), however, rearranging their order and adding the expression $E_z = 0$ (the mode excited being the TE mode) as follows:

$$\left.\begin{aligned}
E_r &= \frac{-j\omega\mu_0}{\gamma^2 r}\frac{\partial H_z}{\partial\theta}\\
E_\theta &= \frac{-j\omega\mu_0}{\gamma^2}\frac{\partial H_z}{\partial r}\\
E_z &= 0\\
H_r &= \frac{-j\beta}{\gamma^2}\frac{\partial H_z}{\partial r}\\
H_\theta &= \frac{-j\beta}{\gamma^2 r}\frac{\partial H_z}{\partial\theta}\\
H_z &= H_{z0}J_m(\gamma r)\cos m\theta \exp j(\omega t - \beta z)
\end{aligned}\right\}. \tag{155}$$

Further, differentiating (9.138) we obtain

$$\left.\begin{aligned}\frac{\partial H_z}{\partial \theta} &= -H_{z0} m J_m(\gamma r) \sin m\theta \exp j(\omega t - \beta z)\\ \frac{\partial H_z}{\partial r} &= H_{z0}\gamma J'_m \cos m\theta \exp j(\omega t - \beta z)\end{aligned}\right\} \tag{9.156}$$

where, $J'_m(\gamma r)$ represents the derivative of $J_m(\gamma r)$ with respect to γr.

Substitution of (9.156) in (9.155) yields the following expressions for the electric and the magnetic field components respectively for a cylindrical waveguide excited in TE mode:

$$\left.\begin{aligned}E_r &= \frac{j\omega\mu_0 m}{\gamma^2 r} H_{z0} J_m(\gamma r) \sin m\theta \exp j(\omega t - \beta z)\\ E_\theta &= \frac{j\omega\mu_0}{\gamma} H_{z0} J'_m(\gamma r) \cos m\theta \exp j(\omega t - \beta z)\\ E_z &= 0\end{aligned}\right\} \tag{9.157}$$

and

$$\left.\begin{aligned}H_r &= \frac{-j\beta}{\gamma} H_{z0} J'_m(\gamma r) \cos m\theta \exp j(\omega t - \beta z)\\ H_\theta &= \frac{j\beta}{\gamma^2 r} H_{z0} m J_m(\gamma r) \sin m\theta \exp j(\omega t - \beta z)\\ H_z &= H_{z0} J_m(\gamma r) \cos m\theta \exp j(\omega t - \beta z)\end{aligned}\right\}. \tag{9.158}$$

The quantity γ occurring in the field expressions (9.157) and (9.158), being related to the axial phase propagation constant β and the free-space propagation constant k by the relation (9.122), is called the radial or transverse propagation constant of the waveguide. This is because the quantity γ, by way of its appearance along with the radial coordinate r in the argument of the Bessel function and its derivative in (9.157) and (9.158), gives a measure of the variation of the field quantities with the radial coordinate r.

Note, however, that the field expressions (9.157) and (9.158) are subject to electromagnetic boundary conditions (7.46) at interface $(r = a)$ between the conducting region of the waveguide wall, interpreted as medium 1, and the free-space region inside the waveguide, interpreted as medium 2 (Fig. 9.8). Applying the boundary condition (7.46d) at the conducting surface of the waveguide wall, we can write

$$\vec{a}_n \times \vec{E}_2 = 0|_{r=a} \tag{9.159}$$

where $\vec{a}_n$ is the normal unit vector directed from the medium 1 (conducting region of the waveguide wall) to the medium 2 (free-space region inside the waveguide), the subscript 2 here referring to medium 2. Clearly, the direction of $\vec{a}_n$ is radially inwards, so that we can write

$$\vec{a}_n = -\vec{a}_r. \tag{9.160}$$

Further, we can interpret $\vec{E}_2$ as

$$\vec{E}_2 = E_r\vec{a}_r + E_\theta\vec{a}_\theta + E_z\vec{a}_z \tag{9.161}$$

where E_r, E_θ, and $E_z (= 0)$ are the electric field components given by (9.157). Therefore, substituting (9.160) and (9.161) in (9.159), and remembering $E_z = 0$, the electromagnetic boundary condition at $r = a$ can be written as

$$-\vec{a}_r \times (E_r\vec{a}_r + E_\theta\vec{a}_\theta) = 0|_{r=a}$$

which, in view of the relations $\vec{a}_r \times \vec{a}_r = 0$ and $\vec{a}_r \times \vec{a}_\theta = \vec{a}_z$ becomes $E_\theta\vec{a}_z = 0|_{r=a}$ giving the electromagnetic boundary condition at $r = a$ as

$$E_\theta = 0|_{r=a}. \tag{9.162}$$

Now, if we put $r = a$ in the expression for E_θ given by (9.157) and then substitute the expression so obtained in the boundary condition (9.162), we obtain the following condition:

$$J'_m(\gamma a) = 0. \tag{9.163}$$

This has a number of roots (or zeros) corresponding to its multiple solutions that can be put as

$$\gamma a = X_{mn} \tag{9.164}$$

corresponding to the nth zero, where X_{mn} is called the eigenvalue of the cylindrical waveguide excited in the mode TE$_{mn}$. From the easily available text on Bessel functions, one can list a few lower order values of the roots of (9.163), that is, the eigenvalues X_{mn} of the TE$_{mn}$ mode as follows:

$$\left.\begin{aligned} X_{01} &= 3.832, X_{02} = 7.016, X_{03} = 10.174 \\ X_{11} &= 1.841, X_{12} = 5.31, X_{13} = 8.536 \\ X_{21} &= 3.054, X_{22} = 6.706, X_{22} = 9.970 \end{aligned}\right\}. \quad (\text{TE}_{mn} \text{ mode}) \tag{9.165}$$

Obtaining the radial propagation constant Y in terms of the eigenvalue X_{mn} from (9.164) and substituting it in the field expressions (9.157) and (9.158), we can write the following field expressions for the cylindrical waveguide excited in the TE$_{mn}$ mode:

$$\left.\begin{aligned} E_r &= \frac{j\omega\mu_0 m}{\gamma^2 r} H_{z0} J_m\left(\frac{X_{mn}}{a}r\right) \sin m\theta \exp j(\omega t - \beta z) \\ E_\theta &= \frac{j\omega\mu_0}{\gamma} H_{z0} J'_m\left(\frac{X_{mn}}{a}r\right) \cos m\theta \exp j(\omega t - \beta z) \\ E_z &= 0 \end{aligned}\right\} \tag{9.166}$$

and

$$\left.\begin{aligned} H_r &= \frac{-j\beta}{\gamma} H_{z0} J'_m\left(\frac{X_{mn}}{a}r\right)\sin m\theta \exp j(\omega t - \beta z) \\ H_\theta &= \frac{j\beta}{\gamma^2} H_{z0} J_m\left(\frac{X_{mn}}{a}r\right)\cos m\theta \exp j(\omega t - \beta z) \\ E_z &= 0 \end{aligned}\right\}. \tag{9.167}$$

Let us next find the characteristic equation or the dispersion relation of a cylindrical waveguide with the help of (9.164) as we did with respect to a rectangular waveguide (Section 9.1.3). For this purpose, let us substitute (9.122) in (9.164) to write the latter as

$$(k^2 - \beta^2)^{1/2} a = X_{mn}$$

which upon squaring and rearranging terms gives the characteristic equation or dispersion relation of a cylindrical waveguide, which is similar to that of a rectangular waveguide (9.50), as:

$$k^2 - \beta^2 = k_c^2 \tag{9.168}$$

where

$$k_c = \frac{X_{mn}}{a}. \tag{9.169}$$

In view of the relation $k = \omega/c$, we can alternatively write the characteristic equation or dispersion relation (9.168) as:

$$\omega^2 - \beta^2 c^2 - \omega_c^2 = 0 \tag{9.170}$$

in terms of a quantity $\omega_c = k_c c$ (see (9.53). Using (9.168), ω_c can be expressed as

$$\omega_c = \frac{X_{mn} c}{a}. \tag{9.171}$$

EXAMPLE E9-7

Compare the cross-sectional area of a circular waveguide with that of a rectangular waveguide that has its wider dimension twice its narrower dimension and whose cutoff frequency is the same as that of the rectangular waveguide, taking both the waveguides excited in the dominant mode.

SOLUTION

From (9.93a) the cutoff frequency f_c of the rectangular waveguide excited in the dominant mode TE_{10} in terms of the wider dimension a is given by

$$f_c = \frac{c}{2a}$$

whence,

$$a = \frac{c}{2f_c}. \quad \text{(rectangular waveguide)} \tag{E9-7.1}$$

The narrower dimension of the rectangular waveguide being $a/2$, its cross-sectional area $(\text{area})_{\text{rect}}$ is given by

$$(\text{area})_{\text{rect}} = a \times \frac{a}{2} = \frac{a^2}{2}. \tag{E9-7.2}$$

Substituting (E9-7.1) in (E9-7.2), we get

$$(\text{area})_{\text{rect}} = \frac{c^2}{8f_c^2}. \tag{E9-7.3}$$

Similarly, we can find the area of a circular waveguide in terms of its cutoff frequency. (The cutoff frequency of the circular waveguide is the same as that of the rectangular waveguide in the given problem.) For this purpose, let us write, with the help of (9.171), the following expression for the circular waveguide excited in dominant mode TE_{11}:

$$2\pi f_c = \frac{X_{11}c}{a'},$$

which, with the help of (9.165) becomes

$$f_c = \frac{X_{11}c}{2\pi a'} = \frac{1.841c}{2\pi a'} \quad \text{(circular waveguide)} \tag{E9-7.4}$$

where a' is the radius of cross section of the circular waveguide. We then obtain

$$a' = \frac{1.841c}{2\pi f_c}. \tag{E9-7.5}$$

The cross-sectional area of the circular guide $(\text{area})_{\text{circular}} = \pi a'^2$ can then be written with the help of (E9-7.5) as

$$(\text{area})_{\text{circular}} = \pi a'^2 = \frac{(1.841)^2 c^2}{4\pi f_c^2}. \quad \text{(circular waveguide)} \tag{E9-7.6}$$

We can then find the ratio of the circular-to-rectangular waveguide cross-sectional area taking the same waveguide cutoff frequency f_c, by dividing (E9-7.6) by (E9-7.3) as

$$\frac{(\text{area})_{\text{circular}}}{(\text{area})_{\text{rect}}} = \frac{\dfrac{(1.841)^2 c^2}{4\pi f_c^2}}{\dfrac{c^2}{8f_c^2}} = \frac{2 \times (1.841)^2}{3.143} = 2.16.$$

It is of interest to note that the characteristic equation or the dispersion relation of a cylindrical waveguide (9.170) is identical to that of a rectangular waveguide, namely (9.54). Also, with reference to the cylindrical waveguide, the quantity ω_c given by (9.171) continues to have the same meaning as that of cutoff frequency and, similarly, $\omega - \beta$ dispersion characteristics depicted by Fig. 9.2 continues to hold good.

EXAMPLE E9-8

A cylindrical waveguide has a concentric dielectric rod insert at its axis and it is excited in the TE mode. Show that the waveguide would support only the azimuthally symmetric TE_{0n} mode, that is, the TE_{mn} mode corresponding to the waveguide mode number $m = 0$.

SOLUTION

The waveguide is divided in two regions: 1 $(0 \leq r \leq b)$ and $2(b \leq r \leq a)$, where a is the waveguide wall radius and b is the dielectric rod radius. Recalling (9.158), we can write the following field expression for the waveguide excited in the TE_{mn} mode:

$$H_{\theta 1} = \frac{j\beta}{\gamma^2 r} H_{z0} m J_m(\gamma r) \sin m\theta \exp j(\omega t - \beta z), \tag{E9-8.1}$$

where $\gamma = \sqrt{\omega^2 \mu_0 \varepsilon_1 - \beta^2}$ is the radial propagation constant, β the axial propagation constant; and subscript 1 refers to region 1 (dielectric rod).

Now, let our starting point be the condition $m \neq 0$. This condition, according to (E9-8.1), also corresponds to $H_{\theta 1} \neq 0$. In view of the boundary condition (7.26c), we can then write at the interface $r = a$:

$$H_{\theta 1} = H_{\theta 2}(r = a) \tag{E9-8.2}$$

subscript 2 referring to region 2 (dielectric-free, free-space region) of the waveguide. Also, you can interpret the boundary condition (7.26a) to write at the interface $r = a$:

$$D_{r1} = D_{r2}(r = a)$$

which, in view of the relations $D_{r1} = \varepsilon_1 E_{r1}$ and $D_{r2} = \varepsilon_0 E_{r2}$ can be written as

$$\varepsilon_1 E_{r1} = \varepsilon_0 E_{r2}. \quad (r = a) \tag{E9-8.3}$$

Further, with the help of (9.148) and interpreting $\beta = \omega\sqrt{\mu_0 \varepsilon_1}$ in region 1 and $\beta = \omega\sqrt{\mu_0 \varepsilon 0}$ in region 2, we can write the following relations:

$$\left.\begin{aligned} E_{r1} &= \sqrt{\frac{\mu_0}{\varepsilon_1}} H_{\theta 1} \\ E_{r2} &= \sqrt{\frac{\mu_0}{\varepsilon_0}} H_{\theta 2} \end{aligned}\right\}. \tag{E9-8.4}$$

Substituting (E9-8.4) in (E9-8.3), we obtain the following relation:

$$H_{\theta 1} = \frac{\sqrt{\varepsilon_0}}{\sqrt{\varepsilon_1}} H_{\theta 2}. \quad (r = a) \tag{E9-8.5}$$

You can clearly see that (E9-8.5) conflicts with (E9-8.2) and, therefore, the condition $m \neq 0$ as our starting point must be incorrect and we must put $m = 0$ as the right condition. Thus, we infer from this finding that the cylindrical waveguide with a coaxial dielectric rod supports the TE_{mn} mode only with the waveguide mode number $m = 0$, or in other words, it supports only the TE_{0n} mode.

The cutoff frequency of the cylindrical waveguide excited in the TE_{11} mode, which has the eigenvalue $X_{11} = 1.841$ [as given in (9.165)], is found to be the lowest of the cutoff frequencies of all the TE modes of excitation ($E_z = 0$, $H_z \neq 0$). In fact, had we analysed the waveguide starting from the wave equation in E_z instead of the wave equation (9.116) in H_z, we would have obtained the field expressions, dispersion relation and expression for the cutoff frequency of the waveguide excited in the TM mode of excitation ($H_z = 0$, $E_z \neq 0$). Such an analysis would also show that even the lowest cutoff frequency of the TM mode would be greater than the cutoff frequency of the TE_{11} mode. Therefore, the TE_{11} mode is the dominant mode of the cylindrical waveguide, which for a rectangular waveguide, you can remember, was identified as the TE_{10} mode (Section 9.1.6).

Let us now appreciate the field pattern of a cylindrical waveguide with the help of field expressions (9.166) and (9.167) with respect to a few lower order modes, typically, TE_{01}, TE_{02} and TE_{11} (Fig. 9.9). Thus, we can use (9.166) and (9.167) to write the following field expressions for the TE_{01} mode ($m = 0, n = 1$), while taking $X_{01} = 3.832$ from (9.165), as follows:

$$\left.\begin{aligned}
E_r &= 0 \\
E_\theta &= \frac{j\omega\mu_0}{\gamma} H_{z0} J_0'\left(\frac{3.832}{a} r\right) \exp j(\omega t - \beta z) \\
E_z &= 0 \\
H_r &= \frac{-j\beta}{\gamma} H_{z0} J_0'\left(\frac{3.832}{a} r\right) \exp j(\omega t - \beta z) \\
H_z &= H_{z0} J_0\left(\frac{3.832}{a} r\right) \exp j(\omega t - \beta z)
\end{aligned}\right\}. \ (\text{TE}_{01} \text{ mode}) \tag{9.172}$$

We can further express the field expression (9.172) for the TE_{01} mode in view of the relation $J_0'(x) = -J_1(x)$, which in turn follows from the recurrence relation available in literature: $J_m'(x) = (m/x)J_m(x) - J_{m+1}(x)$ as follows:

$$\left.\begin{aligned}
E_r &= 0\\
E_\theta &= \frac{j\omega\mu_0}{Y} H_{z0} J_0'\left(\frac{3.832}{a}r\right) \exp j(\omega t - \beta z)\\
E_z &= 0\\
H_r &= \frac{-j\beta}{Y} H_{z0} J_0'\left(\frac{3.832}{a}r\right) \exp j(\omega t - \beta z)\\
H_z &= H_{z0} J_0\left(\frac{3.832}{a}r\right) \exp j(\omega t - \beta z)
\end{aligned}\right\}. \text{ (TE}_{01}\text{ mode)} \tag{9.173}$$

where m with a Bessel function stands for the order of the function. For instance, $J_1(x)$ represents Bessel function of the first kind and order ($m =$)1.

Corresponding to the TE_{01} mode expression (9.173), we can similarly write using (9.166) and (9.167) and taking $X_{02} = 7.016$ from (9.165) the following field expression for the TE_{02} mode ($m = 0, n = 2$):

$$\left.\begin{aligned}
E_r &= 0\\
E_\theta &= \frac{j\omega\mu_0}{\gamma} H_{z0} J_1\left(\frac{7.016}{a}r\right) \exp j(\omega t - \beta z)\\
E_z &= 0\\
H_r &= \frac{-j\beta}{\gamma} H_{z0} J_1\left(\frac{7.016}{a}r\right) \exp j(\omega t - \beta z)\\
H_z &= H_{z0} J_0\left(\frac{7.016}{a}r\right) \exp j(\omega t - \beta z)
\end{aligned}\right\}. \text{ (TE}_{02}\text{ mode)} \tag{9.174}$$

Similarly, with the help of the field expressions (9.166) and (9.167), while taking $X_{11} = 1.841$ from (9.165), we can write for the TE_{11} mode ($m = 1, n = 1$) the following two expressions for the electric and the magnetic field components respectively:

$$\left.\begin{aligned}
E_r &= \frac{j\omega\mu_0}{\gamma^2 r} H_{z0} J_1\left(\frac{1.841}{a}r\right) \sin\theta \exp j(\omega t - \beta z)\\
E_\theta &= \frac{j\omega\mu_0}{\gamma^2 r} H_{z0} J_1'\left(\frac{1.841}{a}r\right) \cos\theta \exp j(\omega t - \beta z)\\
E_z &= 0
\end{aligned}\right\} \text{(TE}_{11}\text{ mode)} \tag{9.175}$$

and

$$\left.\begin{aligned}
H_r &= \frac{-j\beta}{\gamma} H_{z0} J_1'\left(\frac{1.841}{a}r\right) \cos\theta \exp j(\omega t - \beta z)\\
H_\theta &= \frac{j\beta}{\gamma^2 r} H_{z0} J_1\left(\frac{1.841}{a}r\right) \sin\theta \exp j(\omega t - \beta z)\\
H_z &= H_{z0} J_1\left(\frac{1.841}{a}r\right) \cos\theta \exp j(\omega t - \beta z)
\end{aligned}\right\} \text{(TE}_{11}\text{ mode)} \tag{9.176}$$

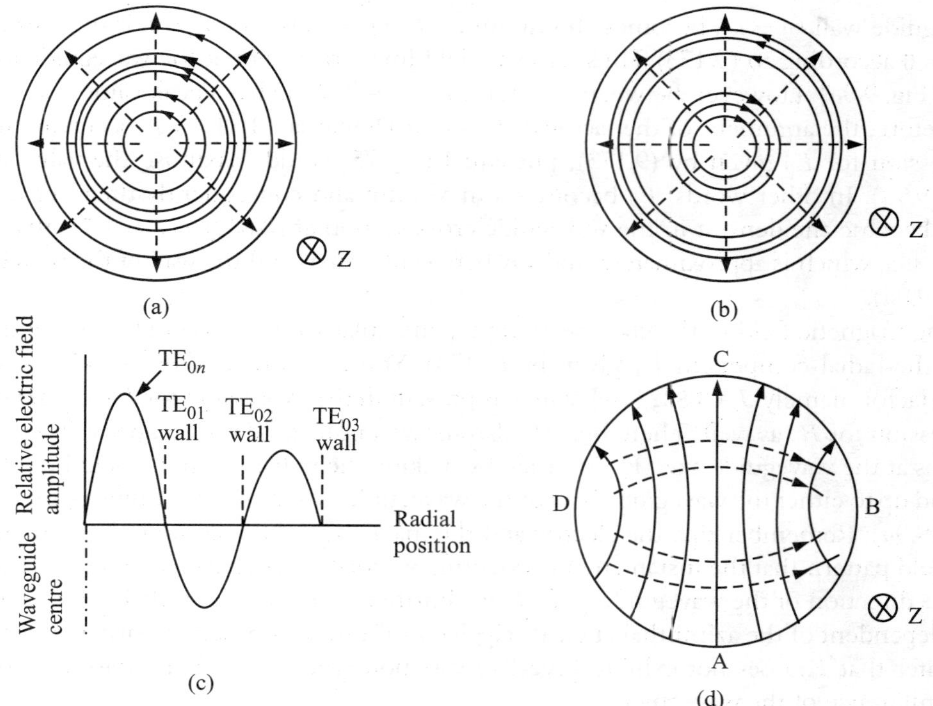

Fig. 9.9 *Field pattern/ variation of a few typical lower order modes of a cylindrical waveguide: (a)* TE_{01}*, (b)* TE_{02}*, (c)* $TE_{0n}(n = 1, 2, 3)$ *and (d)* TE_{11}*. Electric and magnetic field lines are indicated by solid and broken lines respectively. Z axis shown by a cross surrounded by a circle is perpendicular to the plane of the paper and directed away the reader. The direction of propagation is along z being the direction of Poynting vector* $\vec{E} \times \vec{H}$.

noting that the primed Bessel function involved in (9.175) and (9.176) can be expressed, with the help of the recurrence relation $J'_m(x) = (m/x)J_m(x) - J_{m+1}(x)$ as

$$J'_1(1.841r/a) = (a/1.841r)J_1(1.841r/a) - J_2(1.841r/a).$$

You will appreciate from (9.173) that the electric field in the waveguide excited in the TE_{01} mode comprises only the azimuthal component E_θ, which is independent of the azimuthal angle θ at a given value of r. This explains why the circular electric field lines are concentric with the centre of the cross-section of the cylindrical waveguide perpendicular to the waveguide axis (Fig. 9.9*a*). The nature of the variation of $J_1(x)$ with x (available in literature) shows that $J_1(x) = 0$ for $x = 3.832$, that is, $J_1(3.832) = 0$. At the waveguide wall $r = a$, the factor $J_1[(3.832/a)r]$ occurring in the expression for E_θ given by (9.173) therefore becomes $J_1(3.832) = 0$, which, in turn makes $E_\theta = 0$ at $r = a$ (waveguide wall), in compliance with the electromagnetic boundary condition (9.162) (Fig. 9.9*a*). Consequently, the electric field lines become fewer and fewer, and as we approach the waveguide wall ($r = a$), becomes absent. Since $J_1(x) = 0$ also for $x = 0$, it makes $E_\theta = 0$ at $r = 0$ according to (9.173).Thus, electric field lines are absent at the waveguide axis as well (Fig. 9.9a). However, between $x = 0$ and $x = 3.832$, $J_1(x)$ has a peak value at $x = 1.75$. Therefore, the argument of the factor $J_1[(3.832/a)r]$, that is, $(3.832)/a)r$ occurring in the expression for E_θ given by (9.173), put equal to 1.75 would maximise the value of E_θ (Fig. 9.9a). In other words, E_θ becomes a maximum and consequently the electric field lines become the densest in the waveguide cross section at $(3.832)/a)r = 1.75$, that is, at $r = 0.46a$, which is approximately midway between the axis and the wall of the waveguide (Fig. 9.9a).

The magnetic field in the cross section perpendicular to the waveguide axis comprises only the radial component H_r given by (9.173). You can also see from (9.173) that the same factor, namely $J_1[(3.832/a)r]$, which is present in the expression for E_θ, occurs in the expression for H_r as well. Therefore, H_r also shows a null at the waveguide wall ($r = a$) as well as at the waveguide axis ($r = 0$), thereby making the radial magnetic field lines fail to extend up to either the waveguide wall or the waveguide axis in the waveguide cross section (Fig. 9.9a). Remember that the electric and the magnetic field lines are to be so drawn in the field pattern that the instantaneous Poynting vector $\vec{E} \times \vec{H}$ remains directed along the axial z direction of the waveguide (Fig. 9.9a). Further, since the azimuthal component E_θ is independent of the azimuthal angle θ at a given value of r as observed earlier (Fig, 9.9a), we infer that E_θ does not exhibit wave-like variation around any part whatsoever of the circumference of the waveguide.

With a view to appreciating the significance of the mode number of the waveguide excited in the mode TE_{mn}, here TE_{01}, we note that the TE_{01} mode is characterised by nil half-wave field patterns around the half circumference. Corresponding to such characteristics one can assign the mode number $m = 0$ to the mode TE_{mn}, here TE_{01}. Further, for the same mode there exists one field maximum across a waveguide radius. Corresponding to such characteristics one can then assign the mode number $n = 1$ to the mode TE_{mn}, here TE_{01} (Fig. 9.9a).

With reference to the TE_{02} mode, as in the case of the TE_{01} mode, you will see from (9.174) that the electric field in the waveguide excited in the TE_0 mode comprises only the azimuthal component E_θ, which is independent of the azimuthal angle θ at a given value of r. This explains why the circular electric field lines are concentric with the centre of the cross section of the cylindrical waveguide perpendicular to the waveguide axis (Fig. 9.9). Furthermore, note that with the increase of x, the value of the function $J_1(x)$ increases from zero at $x = 0$ to become positive maximum at $x = 1.75$; then it decreases to become zero again at $x = 3.832$ (eigenvalue of the TE_{01} mode) and then it decreases further to reach the negative maximum at $x = 5.5$; and then it increases to become zero again at $x = 7.016$ (eigenvalue of the TE_{02} mode). Accordingly, therefore, interpreting x as the argument of the Bessel function occurring in the field expression (9.174), that is, $x = 7.016r/a$ for the TE_{02} mode, we can see that E_θ becomes positive maximum at $x = 7.016r/a = 1.75$, that is, at $r = 0.25a$, while increasing from zero at $x = 7.016r/a = 0$ that is, at $r = 0$. It becomes negative maximum at $x = 7.016r/a = 5.5$, that is, at $r = 0.78a$, passing through zero in between at $x = 7.016r/a = 3.832$, that is, at $r = 0.55a$ (Fig. 9.9*b*). Further, if the positive value of E_θ corresponds to clockwise electric field lines, its negative value will correspond to anticlockwise electric field line. Thus we can see the electric field lines reversing their direction from clockwise to anticlockwise across $r = 0.55a$, that is, as we move from the region $0 \le r \le 0.55$ to the region $0.55 \le r \le a$ (Fig. 9.9*b*). Correspondingly, since the radial dependence of H_r is the same as that of E_θ, as can be seen from (9.174), the radial magnetic field lines will also reverse their direction from radially inwards to radially outwards across $r = 0.55a$ while maintaining the direction of the instantaneous Poynting vector as the axial z direction of the waveguide (Fig. 9.9*b*).

Similarly, we can appreciate the field pattern for TE_{11} mode with the help of the field expressions (9.175) and (9.176). As mentioned with reference to the TE_{01} mode, in the case of the TE_{02} mode also, we have to draw the magnetic field lines in the field pattern in such a way that the instantaneous Poynting vector $\vec{E} \times \vec{H}$ remains directed along the axial z direction of the waveguide (Fig. 9.9*b*). As in the case of the TE_{01} mode, with a view to appreciating the significance of the mode number of the waveguide excited in the mode TE_{mn}, here TE_{02}, we note that the TE_{02} mode is characterised by nil half-wave field patterns around the half circumference, corresponding to the mode number $m = 0$, and two maxima across the waveguide radius, between the axis and the wall of the waveguide, corresponding to the mode number $n = 2$ such that if one of the maxima is positive, the other is negative (Fig. 9.9*b*). Further, though we have not drawn here the field pattern of the TE_{03} mode, we can see from the variation of the relative electric field amplitude with radius from the axis of the waveguide to its wall that the electric field amplitude passes through three maxima with diminishing amplitudes between the axis and the wall of the waveguide such that if we reckon the first maximum as the positive maximum, the second will be negative and the third positive (Fig. 9.9*c*).

Next, let us consider the cylindrical waveguide excited in the TE_{11} mode. Unlike in the case of the modes like TE_{01} and TE_{02} where the electric and the magnetic fields have azimuthal and radial components giving circular electric and radial magnetic field lines [as appreciated above with the help of (9.173) and (9.174)], in the case of the TE_{11} mode, the electric and the magnetic fields each have both azimuthal and radial components giving electric and magnetic field lines as depicted in Fig. 9.9*d* for the field pattern of the TE_{11} mode).

From the field pattern depicted for the TE_{11} mode of a cylindrical waveguide (Fig. 9.9*d*), we observe the following as we move anticlockwise around the circumference of the waveguide cross section starting from A and coming back to A via B, C and D: (i) at the extreme bottom point A of the waveguide, the upright electric field line has the tail of the arrow of the electric field line representing the maximum positive electric field there; (ii) at the extreme right end B of the waveguide, there are no upright electric field lines, representing zero electric field there; (iii) at the extreme top point C of the waveguide, the upright electric field line has the head of the arrow of the electric field line, representing maximum negative electric field there; (iv) at the extreme left end D of the waveguide, there are again no upright electric field lines, representing zero electric field there; and

(v) at point A we return to the point of maximum positive electric field. Therefore, we experience (a) electric field varying half-wave between the maximum positive value and the maximum negative value via zero value midway around half circumference [(i)-(iii)], and (b) electric field varying half-wave between the maximum negative value to the maximum positive value via zero value midway around the next half circumference [(iii)-(v)]. Thus, the TE_{11} mode is characterised by a single half-wave field pattern around the half circumference (or a single full-wave field pattern around the full circumference) of the waveguide corresponding to the mode number $m = 1$. Similarly, across the waveguide radius, there is a single maximum at the axis of the waveguide corresponding to the mode number $n = 1$ (Fig. 9.9*d*).

Based on the foregoing observation for the TE_{11} mode as well as on such observations for the TE_{01} and the TE_{02} modes already made, we, therefore, interpret the meaning of the mode numbers of a cylindrical waveguide excited in the TE_{mn} mode. The interpretation is:

m (azimuthal mode number) is the number of half-wave field patterns around the half waveguide circumference; and

n (radial mode number) is the number of maxima, positive and negative inclusive, across the waveguide radius.

Such interpretation is valid for the TM_{mn} mode as well. (You can compare this interpretation of the cylindrical waveguide mode numbers with that of the rectangular waveguide mode numbers, the latter discussed at the end of Section 9.3).

The inability of a hollow-pipe waveguide to support a TEM mode

We have, in the preceding sections, discussed the various aspects of the TE and the TM modes. We also mentioned in the beginning of this chapter that a hollow-pipe waveguide made of a single conductor does not support a TEM mode. The TEM mode is characterised by transverse electric and magnetic fields and no axial electric and magnetic field components ($E_z = H_z = 0$) (Chapter 6). However, for continuous magnetic field lines to exist in a hollow-pipe waveguide, there should be an axial current present in the waveguide in the form of either conduction or displacement current in the axial direction. The absence of a conductor does not allow such conduction current in the axial direction. Further, for the displacement current in the axial direction to exist in a waveguide, there should be an axial electric field, the time variation of which will be responsible for such current. However, the TEM mode does not permit the axial electric field. Therefore, a hollow-pipe waveguide cannot support TEM mode. However, a two-conductor structure comprising a hollow cylindrical waveguide with a conducting coaxial solid circular rod as insert, known as a coaxial waveguide, can support the TEM mode.

EXAMPLE E9-9

A coaxial waveguide comprises a conducting cylinder with a coaxial circular rod insert to support a pure TEM mode. For such a waveguide, the conducting material making the waveguide should be made of a perfect conducting material. Justify the statement.

SOLUTION

If the central rod of the waveguide is made of a perfect conductor, then the direction of the electric field is exactly normal to the surface of the rod, that is in the radial direction in conformity with the boundary condition (7.45*d*). Correspondingly, the magnetic field is azimuthal, thereby permitting power transmission through the region between the conductors of the coaxial waveguide in the axial direction. The mode supported is essentially TEM mode (Section 9.4). If the material of the rod is not perfectly conducting, then the electric field is not exactly normal to the surface of the rod, resulting in a finite value of the electric field in the axial direction which in turn provides a component of Poynting vector into the rod and consequently power loss. In such a situation, since the electric field is not exactly transverse to the rod, a pure TEM mode will no longer be supported by the waveguide. Similarly, if the outer conductor of the coaxial waveguide were not perfectly conducting, the waveguide would not support a pure TEM mode and there would be power loss in the outer conductor.

9.5 Power Flow and Power Loss in a Waveguide

We noted earlier in Section 9.1.5 that the wave is attenuated below the cutoff frequency of the waveguide in the evanescent mode. Such attenuation does not correspond to resistive or ohmic loss due to the finite resistivity of the conducting material of the waveguide wall. In this section, we are going to present an analytical approach to the analysis of power flow through a waveguide and power loss in it caused by the finite resistivity of the conducting waveguide wall, and interpret the power flow and the power loss for the attenuation constant of the wave in the waveguide. In order to illustrate the approach, let us take up here only a rectangular waveguide excited in the dominant mode TE_{10}. The approach followed is general and is easily applicable to a cylindrical waveguide as well, which, however, for the sake of brevity is excluded from the present section and left as an exercise for students.

9.5.1 Power flow

We can find the power P transmitted in the axial direction z of a rectangular waveguide by integrating the axial z-component of the average complex Poynting vector $\vec{P}_{\text{average}} = 1/2\text{Re}\vec{E} \times \vec{H}$, given by (8.102), over the waveguide cross-sectional area $(a \times b)$ transverse to the Z axis of the waveguide (Fig.9.1) as follows:

$$P = \int_{x=0}^{x=a} \int_{y=0}^{y=b} \frac{1}{2} \mathrm{Re}(\vec{E} \times \vec{H}^*)_z \; dxdy. \tag{9.177}$$

We can substitute the following expression for $(\vec{E} \times \vec{H}^*)_z$ (that is, the z-component of $\vec{E} \times \vec{H}^*$):

$$(\vec{E} \times \vec{H}^*)_z = E_x H_y^* - E_y H_x^*$$

in (9.177) to write

$$P = \int_{x=0}^{x=a} \int_{y=0}^{y=b} \frac{1}{2} \mathrm{Re}(E_x H_y^* - E_y H_x^*) dxdy. \tag{9.178}$$

Now, we have to put the expressions for the field components in (9.178) to evaluate P. For this purpose, let us, for the dominant TE_{10} mode considered, write the following field expressions putting $m = 1, n = 0$ in (9.48) and (9.49) respectively:

$$\left.\begin{aligned} E_y &= -\frac{j\omega\mu_0}{\pi/a} H_{z0} \sin\left(\frac{\pi}{a}\right) x \exp j(\omega t - \beta z) \quad &\text{(a)} \\ E_x &= 0 \quad &\text{(b)} \end{aligned}\right\} \tag{9.179}$$

$$\left.\begin{aligned} H_x &= -\frac{j\beta}{\pi/a} H_{z0} \sin\left(\frac{\pi}{a}\right) x \exp j(\omega t - \beta z) \quad &\text{(a)} \\ H_y &= 0 \quad &\text{(b)} \end{aligned}\right\} \tag{9.180}$$

Substituting $E_x = -0$ and $H_y = 0$ from (9.179*b*) and (9.180*b*) respectively in (9.178), we can then write

$$P = \int_{x=0}^{x=a} \int_{y=0}^{y=b} \frac{1}{2} \mathrm{Re}\,(-E_y H_x^*) dxdy. \tag{9.181}$$

Now, the expression for H_x^* occurring in (9.181) can be written with the help of (9.180*a*) as

$$H_x^* = \frac{-j\beta}{\pi/a} H_{z0} \sin\left(\frac{\pi}{a}\right) x \exp -j(\omega t - \beta z) \tag{9.182}$$

Therefore, substituting (9.179*a*) and (9.182) in (9.181), we easily obtain

$$P = \frac{1}{2} \frac{\omega\mu_0 a}{\pi} \frac{\beta a}{\pi} H_{z0}^2 \int_{x=0}^{x=a} \sin^2 \frac{\pi x}{a} dx \int_{y=0}^{y=b} dy$$

which, on evaluating the definite integrals, becomes

$$P = \frac{1}{2}\frac{\omega\mu_0 a}{\pi}\frac{\beta a}{\pi}H_{z0}^2\left(\frac{a}{2}\right)(b)$$

which can be rearranged as

$$P = \frac{1}{4\pi^2}\omega\mu_0\beta a^3 b H_{z0}^2. \quad (\text{TE}_{10} \text{ mode}) \tag{9.183}$$

From the expression for the power transmitted through the waveguide for the TE_{10} mode, we can easily deduce an expression for the maximum permissible power P_{maximum}, that is, the power handling capacity of the waveguide if we know the magnitude of the maximum electric field amplitude which the atmosphere of the inside of the waveguide can withstand before it breaks down. For this purpose, with respect to the TE_{10} mode, we can use (9.111) to write the following relation for H_{z0} that occurs in (9.183):

$$H_{z0} = \frac{\pi|E_y|_{\text{maximum}}}{\omega\mu_0 a},$$

in terms of the breakdown field of the medium $|E_y|_{\text{maximum}}$ inside the waveguide. Substituting this in (9.183), we obtain

$$P_{\text{maximum}} = \frac{1}{4\omega\mu_0}\beta ab(|E_y|_{\text{maximum}})^2,$$

which, in view of the relations $\omega/\beta = v_{\text{ph}}$ (see the beginning of Section 9.1.4) and $v_{\text{ph}}/c = 1/(1 - f_c^2/f^2)^{1/2}$ [see (9.71)] becomes

$$P_{\text{maximum}} = \frac{ab(1 - f_c^2/f^2)^{1/2}}{4\mu_0 c}(|E_y|_{\text{maximum}})^2. \quad (\text{TE}_{10} \text{ mode})$$

This relation can be used to find the maximum power handling capacity of a rectangular waveguide excited typically in the TE_{10} mode, provided we know the value of the breakdown field $|E_y|_{\text{maximum}}$, which, for instance, for dry air filling the waveguide can be taken as 29 kV/cm.

EXAMPLE E9-10

Calculate the maximum power handling capability of WR-340 rectangular waveguide excited in the dominant mode, operating at 3 GHz frequency, taking the breakdown limit of air as 29 kV/cm. See Table 9.1 to know the relevant WR-340 data.

SOLUTION

From Table 9.1, for the rectangular waveguide WR-340, we have $a = 3.4'' = 3.4\times2.54\times10^{-2}$ m; $b = 1.7'' = 1.7 \times 2.54 \times 10^{-2}$ m; $f_c = 1.737$ GHz (which you can independently find from the formula $f_c = c/2a$ for the dominant mode TE_{10}). In addition, the other data given in the problem: operating frequency $f = 3$ GHz (given); and breakdown field 29 kV/cm $= 29 \times 10^5$ (given).

We have already deduced the formula for the permissible maximum power P_{maximum} in terms of the breakdown field $|E_y|_{\text{maximum}}$ for a rectangular waveguide excited in the dominant mode:

$$P_{\text{maximum}} = \frac{ab(1 - f_c^2/f^2)^{1/2}}{4\mu_0 c}(|E_y|_{\text{maximum}})^2. \tag{E9-10.1}$$

From the data given, we can calculate the intermediate values to be used in (E9-10.1) as follows:

$$\left.\begin{aligned} ab &= 3.4 \times 2.54 \times 1.7 \times 2.54 \times 10^{-4} \\ &= 37.29 \times 10^{-4}\text{m}^2 \\ \left(\frac{f_c}{f}\right)^2 &= 0.335 \\ (|E_y|_{\text{maximum}})^2 &= (29 \times 10^5)^2(\text{V/m})^2 \end{aligned}\right\}. \tag{E9-10.2}$$

Putting (E9-10.2) in (E9-10.1), we can calculate P_{maximum} as follows:

$$\begin{aligned} P_{\text{maximum}} &= \frac{37.29 \times 0.8155 \times 29 \times 29}{4 \times 4 \times 3.143 \times 3} \times 10^{-4} \times 10^9\ \text{W} \\ &= 16.952 \times 10^6\ \text{W} = 16.952\ \text{MW}. \end{aligned}$$

9.5.2 Power loss per unit area and power loss per unit length

Let us find the power loss per unit area at each of the four waveguide walls at $x = 0$ (right sidewall), $x = a$ (left sidewall), $y = b$ (bottom wall), and $y = b$ (top wall) (Fig. 9.1). For this purpose, we will use expression (8.142) for power loss per unit area P_{LA} at a conducting interface which, for the sake of convenience, is rewritten as:

$$P_{\text{LA}} = \frac{1}{2}R_S\vec{J}_s.\vec{J}_s^* \tag{9.184}$$

where R_S is the surface resistance of the conducting boundary wall of the waveguide and $\vec{J}_s$ is the surface current density at the concerned waveguide wall. The surface current density $\vec{J}_s$, at a conducting boundary wall of the waveguide can be found from the electromagnetic boundary condition (7.46*c*) that states that the tangential component of magnetic field is discontinuous at the interface between a conducting medium and a dielectric/free-space ($\varepsilon = \varepsilon_0$) medium, the amount of discontinuity being equal to the surface current density $\vec{J}_s$ (Section 7.3). The said boundary condition (7.46*c*), is rewritten below:

$$\vec{a}_n \times \vec{H}_2 = \vec{J}_s \tag{9.185}$$

where $\vec{a}_n$ is the unit vector normal to the conducting boundary wall directed from the conducting region of the wall to the free-space region inside the waveguide, and subscript 2 refers to the free-space region inside the waveguide (Section 7.3). Therefore, $\vec{H}_2$ is interpreted as

$$\vec{H}_2 = H_x\vec{a}_x + H_y\vec{a}_y + H_z\vec{a}_z$$

which, in view of (9.180), becomes

$$\vec{H}_2 = \frac{j\beta a}{\pi} H_{z0} \sin\left(\frac{\pi}{a}\right) x \exp j(\omega t - \beta z)\vec{a_x} + H_z\vec{a}_z. \tag{9.186}$$

Further, the expression for H_z occurring in (9.186) can be written for the TE_{10} mode $(m = 1, n = 0)$ with the help of (9.44) as:

$$H_z = H_{z0} \cos\frac{\pi}{a} x \exp j(\omega t - \beta z). \tag{9.187}$$

Therefore, substituting (9.187) in (9.186), we obtain H_z as:

$$\vec{H}_z = \frac{j\beta a}{\pi} H_{z0} \sin\frac{\pi}{a} x \exp j(\omega t - \beta z)\vec{a}_x + H_{z0} \cos\left(\frac{\pi}{a}\right) x \exp j(\omega t - \beta z)\vec{a}_z. \tag{9.188}$$

Let us now use (9.188) to find the surface current density $\vec{J}_s$ at the right sidewall $(x = 0)$ of the waveguide, interpreting $\vec{a_n} = \vec{a_x}$ (since the normal unit vector $\vec{a_n}$ is is directed in the positive x direction from the conducting right sidewall to the free-space region inside the waveguide). Therefore, putting (9.188) as well as $\vec{a_n} = \vec{a}_x$, in (9.185), we obtain

$$\begin{aligned}\vec{J}_{s,\text{right sidewall}} = [\vec{a_x}] \times \Bigg[&\frac{j\beta a}{\pi} H_{z0} \sin\left(\frac{\pi}{a}\right) x \exp j(\omega t - \beta z)\vec{a}_x \\ &+ H_{z0} \cos\left(\frac{\pi}{a}\right) x \exp j(\omega t - \beta z)\vec{a_z}\Bigg]\end{aligned} \tag{9.189}$$

which, in view of the relations $\vec{a_x} \times \vec{a_x} = 0$ and $\vec{a}_x \times \vec{a}_z = -\vec{a}_y$, and remembering $x = 0$ at the right sidewall, becomes

$$\vec{J}_{\text{right sidewall}} = -H_{z0} \exp j(\omega t - \beta z)\vec{a_y}. \tag{9.190}$$

At the left sidewall $(x = a)$ of the waveguide, we can interpret $\vec{a_n} = -\vec{a_x}$ (since the normal unit vector $\vec{a_n}$ is directed in the negative x direction from the conducting right sidewall to the free-space region inside the waveguide) and accordingly, now we obtain $\vec{J}_s$ at the left sidewall $(x = a)$ of the waveguide, corresponding to (9.189), as

$$\begin{aligned}\vec{J}_{s\text{ left sidewall}} = [-\vec{a}_x] \times \Bigg[&\frac{j\beta a}{\pi} H_{z0} \sin\left(\frac{\pi}{a}\right) x \exp j(\omega t - \beta z)\vec{a}_x \\ &+ H_{z0} \sin\left(\frac{\pi}{a}\right) x \exp j(\omega t - \beta z)\vec{a_z}\Bigg]\end{aligned}$$

which, in view of the relations $\vec{a_x} \times \vec{a}_x = 0, \vec{a}_x \times \vec{a}_z = -\vec{a}_y$ and $\cos\pi = -1$, and remembering $x = a$ at the left sidewall, can be written as

$$\vec{J}_{r\text{ left sidewall}} = -H_{z0} \exp j(\omega t - \beta z)\vec{a}_y. \tag{9.191}$$

Similarly, let us now use (9.188) to find $\vec{J}_s$, at the bottom wall ($y = 0$) of the waveguide, interpreting $\vec{a}_n = \vec{a}_y$ (since the normal unit vector $\vec{a}_n$ is directed in the positive y direction from the conducting bottom wall to the free-space region inside the waveguide). Therefore, putting (9.188) as well as $\vec{a}_n = \vec{a}_y$ in (9.185), we obtain

$$\vec{J}_{s\text{ bottom wall}} = [\vec{a_y}] \times \left[\frac{j\beta a}{\pi} H_{z0} \sin\left(\frac{\pi}{a}\right) x \exp j(\omega t - \beta z)\vec{a}_x \right.$$
$$\left. + H_{z0} \cos\left(\frac{\pi}{a}\right) x \exp j(\omega t - \beta z)\vec{a}_z\right]$$

which, in view of the relations $\vec{a}_y \times \vec{a}_x = -\vec{a}_z$ and $\vec{a}_y \times \vec{a}_z = -\vec{a}_x$ simplifies to

$$\vec{J}_{s\text{ bottom wall}} = \frac{-j\beta a}{\pi} H_{z0} \sin\left(\frac{\pi}{a}\right) x \exp j(\omega t - \beta z)\vec{a}_z + H_{z0} \cos\left(\frac{\pi}{a}\right) x \exp j(\omega t - \beta z)\vec{a}_x. \tag{9.192}$$

At the top wall ($y = 0$) of the waveguide, $\vec{J}_s$ can be similarly found interpreting $\vec{a}_n = -\vec{a}_y$ (since the normal unit vector $\vec{a}_n$ is directed in the negative y direction from the conducting top wall to the free-space region inside the waveguide). Therefore, putting (9.188) as well $\vec{a}_n = -\vec{a}_y$ as in (9.185), we obtain

$$\vec{J}_{s\text{ top wall}} = [-\vec{a_y}] \times \left[\frac{j\beta a}{\pi} H_{z0} \sin\left(\frac{\pi}{a}\right) x \exp j(\omega t - \beta z)\vec{a}_x \right.$$
$$\left. + H_{z0} \cos\left(\frac{\pi}{a}\right) x \exp j(\omega t - \beta z)\vec{a}_z\right]$$

which, in view of the relations $\vec{a}_y \times \vec{a}_x = -\vec{a}_z$ and $\vec{a}_y \times \vec{a}_z = -\vec{a}_x$, simplifies to

$$\vec{J}_{s\text{ top wall}} = \frac{-j\beta a}{\pi} H_{z0} \sin\left(\frac{\pi}{a}\right) x \exp j(\omega t - \beta z)\vec{a}_z + H_{z0} \cos\left(\frac{\pi}{a}\right) x \exp j(\omega t - \beta z)\vec{a}_x. \tag{9.193}$$

In order to find power loss per unit area at a conducting waveguide wall with the help of (9.184), we need to know $\vec{J}_s$, which can be easily obtained from the complex conjugates of (9.190)-(9.193) as

$$\left.\begin{aligned}
\vec{J}_s^{*}{}_{\text{right sidewall}} &= -H_{z0}\exp -j(\omega t-\beta z)\vec{a}_y\\
\vec{J}_s^{*}{}_{\text{left sidewall}} &= -H_{z0}\exp -j(\omega t-\beta z)\vec{a}_y\\
\vec{J}_s^{*}{}_{\text{bottom sidewall}} &= \frac{j\beta a}{\pi}H_{z0}\sin\left(\frac{\pi}{a}\right)x\exp -j(\omega t-\beta z)\vec{a}_z + H_{z0}\cos\left(\frac{\pi}{a}\right)x\exp -j(\omega t-\beta z)\vec{a}_x\\
\vec{J}_s^{*}{}_{\text{top sidewall}} &= \frac{-j\beta a}{\pi}H_{z0}\sin\left(\frac{\pi}{a}\right)x\exp -j(\omega t-\beta z)\vec{a}_z + H_{z0}\cos\left(\frac{\pi}{a}\right)x\exp -j(\omega t-\beta z)\vec{a}_x
\end{aligned}\right\}. \tag{9.194}$$

Substituting (9.190)-(9.194) in (9.184), we obtain the following expression for power loss per unit area at the four waveguide walls:

$$\left.\begin{aligned}
P_{\text{LA,rightsidewall}} &= P_{\text{LA,leftsidewall}} = \tfrac{1}{2}R_sH_{z0}^2\\
P_{\text{LA,bottomwall}} &= P_{\text{LA,topwall}} = \frac{1}{2}R_sH_{z0}^2\left(\frac{\beta^2a^2}{\pi^2}\sin^2\left(\frac{\pi}{a}\right)x+\cos^2\left(\frac{\pi}{a}\right)x\right)
\end{aligned}\right\}. \tag{9.195}$$

Now that we have found the expressions for power loss per unit area at each of the waveguide walls, we can use them to find power loss per unit length P_{LL} at each of these walls, which, in turn would be required to find the wave attenuation in the waveguide (to be taken up in Section 9.5.3). In order to find power loss per unit length P_{LL} at the right side wall $(x=0)P_{\text{LL, right sidewall}}$, let us first find the element of power loss dP_{LL} over an element of strip of elemental thickness dz and elemental area bdz across the narrow dimension of the waveguide on its right side wall (Fig. 9.1). We can obtain dP_{LL} by multiplying the area element $dydz$ by power loss per unit area $P_{\text{LA,right sidewall}}$ given by (9.195) as follows:

$$dP_{\text{LL}} = \frac{1}{2}R_sH_{z0}^2dydz \quad \text{(right sidewall)}. \tag{9.196}$$

Integrating (9.196) over unit length across the narrow dimension of the waveguide, we then obtain power loss per unit length at the right sidewall $P_{\text{LL,right sidewall}}$ as follows:

$$P_{\text{LL,right sidewall}} = \int dP_{\text{LL}} = \frac{1}{2}R_sH_{z0}^2\int_{y=0}^{y=b}dy\int_{z=0}^{z=1}dz$$

which, on evaluating the definite integrals, becomes

$$P_{\text{LL,right sidewall}} = \frac{1}{2}R_sH_{z0}^2b. \tag{9.197}$$

Following exactly the same procedure as above, we can find the expression for power loss per unit length at the left sidewall $(x=a)$, $P_{\text{LL,left sidewall}}$, as follows:

$$P_{\text{LL,left sidewall}} = \frac{1}{2}R_sH_{z0}^2b. \tag{9.198}$$

In order to find power loss per unit length P_{LL} at the bottom wall $(y = 0)$, $P_{\mathrm{LL,bottom\ sidewall}}$, let us first find the element of power loss dP_{LL} over an element of strip of elemental thickness dz and elemental area $dzdx$ across the broad dimension of the waveguide on its bottom wall (Fig. 9.1). We can then obtain dP_{LL} by multiplying the area element $dzdx$ by power loss per unit area $P_{\mathrm{LL,bottom\ sidewall}}$, given by (9.195) as follows:

$$dP_{\mathrm{LL}} = \frac{1}{2}R_s H_{z0}^2 \left(\frac{\beta^2 a^2}{\pi^2} \sin^2\left(\frac{\pi}{a}\right) x + \cos^2\left(\frac{\pi}{a}\right) x \right) dzdx \quad \text{(bottom wall)}. \tag{9.199}$$

Integrating (9.199) over unit length across the broad dimension of the waveguide, we obtain

$$P_{\mathrm{LL\ bottom\ wall}} = \int dP_{LL} = \frac{1}{2}R_s H_{z0}^2 \int\limits_{y=0}^{y=a} \left(\frac{\beta^2 a^2}{\pi^2} \sin^2\left(\frac{\pi}{a}\right) x + \cos^2\left(\frac{\pi}{a}\right) x \right) dz \int\limits_{z=0}^{z=1} dz. \tag{9.200}$$

The definite integrals in (9.200) by can be easily evaluated by expanding trigonometric terms involved in the integrands using the following relations:

$$\left.\begin{aligned} \sin^2\left(\frac{\pi}{a}\right) x &= \frac{1 - \cos\frac{2\pi}{a}x}{2} \\ \cos^2\left(\frac{\pi}{a}\right) x &= \frac{1 + \cos\frac{2\pi}{a}x}{2} \end{aligned}\right\}.$$

The definite integrals in (9.200) thus evaluated yields

$$P_{\mathrm{LL,bottomwall}} = \frac{1}{2}R_s H_{z0}^2 \left(\frac{\beta^2 a^2}{\pi^2} \left(\frac{a}{2}\right) + \left(\frac{a}{2}\right) \right)$$

which can be simplified to write

$$P_{\mathrm{LL,bottomwall}} = \frac{1}{2}R_s H_{z0}^2 \left(\frac{\beta^2 a^2}{\pi^2} + 1 \right). \tag{9.201}$$

We can next find the expression for the power loss per unit length at the top wall $x = a$), $P_{\mathrm{LL,topwall}}$ by following exactly the same procedure as that used to derive (9.201):

$$P_{\mathrm{LL,topwall}} = \frac{1}{2}R_s H_{z0}^2 \left(\frac{\beta^2 a^2}{\pi^2} + 1 \right). \tag{9.202}$$

In order to find power loss per unit length of the waveguide P_{LL}, we have to sum up the contribution to it by all the four walls, or in other words, add (9.197), (9.198), (9.201) and (9.202) to obtain

$$P_{\mathrm{LL}} = \frac{1}{2}R_s H_{z0}^2 b + \frac{1}{2}R_s H_{z0}^2 b + \frac{1}{2}R_s H_{z0}^2 \frac{a}{2}\left(\frac{\beta^2 a^2}{\pi^2}+1\right) + \frac{1}{2}R_s H_{z0}^2 \frac{a}{2}\left(\frac{\beta^2 a^2}{\pi^2}+1\right)$$

which simplifies to

$$P_{LL} = R_s H_{z0}^2 \left[b + \frac{a}{2}\left(1+\frac{\beta^2 a^2}{\pi^2}\right)\right]. \quad (\mathrm{TE}_{10} \text{ mode}) \tag{9.203}$$

9.5.3 Attenuation constant from power flow and power loss per unit length

We have considered so far the RF dependence of field components as $\exp j(\omega t - \beta z)$. However, power loss caused by the finite resistivity of the waveguide wall can be accounted for by generalising the RF dependence of field components as $\exp(j\omega t - \gamma z) = \exp j\omega t \exp(-\alpha z)\exp(-j\beta z)$, where $\gamma = \alpha + j\beta$, α being the attenuation constant and β the phase propagation constant of the wave. Consequently, the amplitude of the field component no longer remains constant with axial distance z. Also, appreciate, for instance, from (9.181) that power transmitted through the waveguide varies with the axial distance z as $\exp(-2\alpha z)$. We can mathematically state this dependence of power as

$$P(z) = P(0)\exp(-2\alpha z). \tag{9.204}$$

If we differentiate (9.204), we obtain

$$\frac{dP(z)}{dz} = -2\alpha P(0)\exp(-2\alpha z)$$

which, in view of the relation (9.204) can be read as

$$\frac{dP(z)}{dz} = -2\alpha P(z). \tag{9.205}$$

The expression for the attenuation constant of the wave α then follows from (9.205) as

$$\alpha = \frac{-\dfrac{dP(z)}{dz}}{2P(z)}$$

which, interpreting the numerator with its negative sign as the power loss per unit length of the waveguide, and dropping the parenthesis z from $P(z)$ appearing in the denominator, can be expressed as

$$\alpha = \frac{P_{\mathrm{LL}}}{2P}. \tag{9.206}$$

Thus, expression (9.206) states that the wave attenuation constant α is the ratio of the power loss per unit length of the waveguide to twice the power transmitted through the waveguide. The dimension of the attenuation constant α is 1/m or m^{-1}, and it is expressed in the unit of Neper/m or Np/m. Note that 1 Np = 8.68 dB (see Appendix A9.3). We can now substitute (9.183) and (9.203) in (9.206) to obtain the following expression for the attenuation constant

$$\alpha = \frac{2\pi^2 R_s \left[b + \frac{a}{2}\left(1 + \frac{\beta^2 a^2}{\pi^2}\right)\right]}{\omega\mu_0\beta a^3 b}. \quad (\text{TE}_{10} \text{ mode}) \qquad (9.207)$$

Putting $R_s = 1/\sigma\delta$ from (6.98) in (9.207), we have an alternative form of the expression for the attenuation constant as

$$\alpha = \frac{2\pi^2 \left[b + \frac{a}{2}\left(1 + \frac{\beta^2 a^2}{\pi^2}\right)\right]}{\sigma\delta\omega\mu_0\beta a^3 b}. \quad (\text{TE}_{10} \text{ mode}) \qquad (9.208)$$

Further, with the help of the relation $\beta = \omega/v_{\text{ph}}$ (Section 9.1.4), to be read with the help of the relation (9.71) for v_{ph}/c, we can express (9.208), after a little algebraic rearrangement, as follows:

$$\alpha = \frac{1}{\sigma\delta\eta_0 b}\frac{\omega_c}{\omega}\left[\frac{\frac{2b}{a} + \frac{\omega^2}{\omega_c^2}}{\left(\frac{\omega^2}{\omega_c^2} - 1\right)^{1/2}}\right]. \quad (\text{TE}_{10} \text{ mode}) \qquad (9.209)$$

where $\omega_c = c\pi/a$ for the TE$_{10}$ mode (obtainable from (9.55), putting $m = 1, n = 0$ therein).

It is clear from (9.209) that the attenuation constant α of the waveguide can be reduced by increasing the conductivity σ of the material of the waveguide. A study of the expression (9.209) will further show that for a given ratio b/a, with the increase in the operating frequency $f(= \omega/2\pi)$, the attenuation constant α first decreases and then increases with f after passing through a minimum attenuation. For a rectangular waveguide of typical dimensions $a = 2.3$ cm, $b = 1$ cm excited in the TE$_{10}$ mode, the attenuation constant α of the waveguide reaches a minimum value < 0.1 dB/cm at around 13 GHz. (You can convert the attenuation constant in Np/m to dB by using the conversion: 1 Np = 8.68 dB; see Appendix A9.3). We have to take care while choosing the waveguide dimensions vis-a-vis the operating frequency and the frequency range of operation (Section 9.1.6) to see that the value of the attenuation constant does not get enhanced.

EXAMPLE E9-11

A rectangular waveguide ($a = 0.9''$; $b = 0.4''$) made of aluminum ($\sigma = 35.4 \times 10^4$ mho/m) has the cutoff frequency $f_c = 6.56$ GHz in the dominant mode. Calculate the attenuation constant α of the waveguide at the operating frequency $f = 9$ GHz.

SOLUTION

We can express (9.209) for the attenuation constant α of a rectangular waveguide excited in the dominant mode by replacing therein the value of $1/\sigma\delta = \sqrt{\pi f \mu_0/\sigma}$, the latter appreciated with the help of (6.98) and (6.99), as follows:

$$\alpha = \frac{1}{\eta_0 b}\sqrt{\frac{\pi f \mu_0}{\sigma}}\left[\frac{1+\dfrac{2b}{a}\dfrac{f_c^2}{f^2}}{\left(1-\dfrac{f_c^2}{f^2}\right)^{1/2}}\right]. \tag{E9-11.1}$$

The data of the problem given is:

$$\left.\begin{aligned} &a = 0.9'' = 0.9 \times 2.54 \times 10^{-2}\ \text{m} \\ &b = 0.9'' = 0.4 \times 2.54 \times 10^{-2}\ \text{m} \\ &f = 9 \times 10^9\ \text{Hz} \\ &f_c = 6.56 \times 10^9\ \text{Hz} \\ &\sigma = 35.4 \times 10^6\ \text{mho/m} \end{aligned}\right\}. \tag{E9-11.2}$$

The intermediate values as required to calculate α using (E9-11.1) can be found with the help of (E9-11.2) as

$$\left.\begin{aligned} &\left(1-\frac{f_c^2}{f^2}\right)^{1/2} = (1-0.5312)^{1/2} = \sqrt{0.4688} = 0.6847 \\ &1+\left[\frac{2b}{a}\frac{f_c^2}{f^2}\right] = 1 + 2 \times 0.44 \times 0.5312 = 1.467 \\ &\sqrt{\frac{\pi f \mu_0}{\sigma}} = 2\pi\sqrt{\frac{9 \times 10^9 \times 10^{-7}}{35.4 \times 10^6}} \\ &= 2 \times 3.143 \times 10^{-2} \times 0.504 = 3.17 \times 10^{-2}\ \text{ohm} \end{aligned}\right\}. \tag{E9-11.3}$$

remembering $\mu_0 = 4\pi \times 10^{-7}$ H/m. Substituting then (E9-11.3) in (E9-11.1), we get

$$\alpha = \frac{1}{\eta_0 b} \times 3.17 \times 10^{-2}\frac{1.467}{0.6847} = 0.0177\ \text{Np/m},$$

which, putting $\eta_0 = 377$ ohm and $b = 1.02 \times 10^{-2}$ m (given), results in

$$\alpha = \frac{1}{377 \times 1.02 \times 10^{-2}} \times 3.17 \times 10^{-2}\frac{1.467}{0.6847} = 0.0177\ \text{Np/m},$$

which, in view of the conversion 1 Np = 8.68 dB (see Appendix A9.3) becomes

$$\alpha = 0.0177\ \text{Np/m} = 8.68 \times 0.0177\ \text{dB/m} = 0.154\ \text{dB/m}.$$

EXAMPLE E9-12

Find the value of the wave frequency relative to the cutoff frequency of a rectangular waveguide excited in the dominant mode that results in the minimum attenuation due to the finite conductivity of the material of the waveguide in terms of the waveguide dimensions.

SOLUTION

The expression (9.209) for the attenuation constant α of a rectangular waveguide excited in the dominant mode can be reproduced here in the following form:

$$\alpha = \frac{1}{\sigma\delta\eta_0 b}\frac{f_c}{f}\left[\frac{\dfrac{2b}{a}+\dfrac{f^2}{f_c^2}}{\left(\dfrac{f^2}{f_c^2}-1\right)^{1/2}}\right]. \tag{E9.12.1}$$

We can identify the quantity $1/\sigma\delta$ occurring in (9-12.1) as the surface resistance of the waveguide, which can be expressed with the help of (6.98) and (6.99), as

$$\frac{1}{\sigma\delta} = \sqrt{\frac{\pi f \mu_0}{\sigma}}. \tag{E9-12.2}$$

Further, it is convenient to introduce a variable

$$\xi = \left(\frac{f_c}{f}\right). \tag{E9-12.3}$$

With the help of (E9-12.3), we can express (E9-12.2) as

$$\frac{1}{\sigma\delta} = \sqrt{\frac{\pi f \mu_0}{\sigma}} = \sqrt{\frac{f}{f_c}}\sqrt{\frac{\pi f_c \mu_0}{\sigma}} = \xi^{-1/4}\sqrt{\frac{\pi f_c \mu_0}{\sigma}}. \tag{E9-12.4}$$

Plugging (E9-12.4) in (E9-12.1), we can then express the attenuation constant as

$$\alpha = \frac{1}{\eta_0 b}\sqrt{\frac{\pi f_c \mu_0}{\sigma}}\xi^{-1/4}\frac{f_c}{f}\frac{\dfrac{2b}{a}+\left(\dfrac{f}{f_c}\right)^2}{\left(\left(\dfrac{f}{f_c}\right)^2-1\right)^{1/2}}. \tag{E9-12.5}$$

Let us now write with the help of (E9-12.3) the following quantities occurring in (E9-12.5):

$$\left.\begin{aligned}\frac{f_c}{f} &= \xi^{1/2}\\ \left(\frac{f}{f_c}\right)^2 &= \xi^{-1}\end{aligned}\right\}. \tag{E9-12.6}$$

Making use of (E9-12.6), we can then express (E9-12.5) after a little algebra as follows:

$$\alpha = G\left[\frac{\xi^{-1/4} + \frac{2b}{a}\xi^{3/4}}{(1-\xi)^{1/2}}\right] \tag{E9-12.7}$$

where G is a waveguide constant given by

$$G = \frac{1}{\eta_0 b}\sqrt{\frac{\pi\mu_0 f_c}{\sigma}}.$$

At the frequency where the minimum attenuation of the waveguide takes place, you can put

$$\frac{d\alpha}{df} = \frac{d\alpha}{d\xi}\frac{d\xi}{df} = 0. \tag{E9-12.8}$$

Therefore, substituting $d\alpha/d\xi$, obtained by differentiating (9-12.7) with respect to ξ, and $d\xi/df$, obtained by differentiating (9-12.3) with respect to f into (9-12.8), we can write the condition for the minimum attenuation of the waveguide as:

$$\frac{d\alpha}{df} = \frac{-2G}{f_c(1-\xi)^{1/2}}\left(\frac{-\frac{1-\xi}{2}\xi^{1/4} + \frac{3b}{a}(1-\xi)\xi^{5/4} + \xi^{5/4} + \frac{2b}{a}\xi^{9/4}}{2(1-\xi)}\right) = 0. \tag{E9-12.9}$$

The quantity outside the parenthesis of the right-hand side of (9-12.9) is not equal to zero for practical values of $\xi(=\omega_c^2/\omega^2)$. Therefore, we can put the quantity within the parenthesis equal to zero and rearrange terms to get

$$\frac{-\frac{1-\xi}{2}\xi^{1/4} + \frac{3b}{a}(1-\xi)\xi^{5/4} + \xi^{5/4} + \frac{2b}{a}\xi^{9/4}}{2(1-\xi)} = 0,$$

whence we obtain

$$\frac{3b}{a}(1-\xi)\xi^{5/4} + \xi^{5/4} + \frac{2b}{a}\xi^{9/4} = \frac{1-\xi}{2}\xi^{1/4}. \tag{E9-12.10}$$

Dividing (9-12.10) by $\xi^{1/4}$, we get

$$\frac{3b}{a}(1-\xi)\xi + \xi + \frac{2b}{a}\xi^2 = \frac{1-\xi}{2},$$

which can be rearranged to obtain the following quadratic relation for minimum attenuation:

$$2b\xi^2 - (3a+6b)\xi + a = 0. \tag{E9-12.11}$$

The solution of (9-12.11), in view of (9-12.3) can be read as

$$\xi = \left(\frac{f_c}{f}\right)^2 = \frac{(3a+6b) \pm \sqrt{(3a+6b)^2 - 8ab}}{4b},$$

whence we obtain the expression for the frequency where the minimum attenuation takes place as follows:

$$\frac{f}{f_c} = \sqrt{\frac{4b}{(3a+6b) \pm \sqrt{(3a+6b)^2 - 8ab}}}. \tag{E9-12.12}$$

We have restricted the analysis in this section to the problem of power loss and wave attenuation with reference to a rectangular waveguide excited in the dominant mode TE_{10}. The approach of analysis can be applied to higher orders of mode as well to show, for instance, that the attenuation constant would of higher for higher order modes than that for the lowest order TE_{10} mode for a rectangular waveguide. Further, as mentioned in the beginning of this section, the application of the approach followed here for a rectangular waveguide to find power loss and wave attenuation in a cylindrical waveguide is excluded here for the sake of brevity of analysis. This can be taken up as an exercise by a reader.

Appendix A9.1

Wave group velocity

We have already defined the phase velocity of a wave v_{ph} (see the discussion following (6.19) in Chapter 6). The phase velocity of the wave v_{ph} involves the phase propagation constant β, which, in turn, occurs in the representation of a wave associated with a physical quantity, for instance, the electric field E, as

$$E = E_0 \exp j(\omega t - \beta z). \tag{A9.1}$$

Further, we have identified the phase velocity of the wave as the velocity with which a plane of constant phase progresses. In order to find the phase velocity, we therefore put

$$\omega t - \beta z = \text{constant}$$

and differentiate it with respect to time t to write

$$\omega - \beta \frac{dz}{dt} = 0,$$

which can be rearranged to obtain the phase velocity v_{ph} as

$$v_{ph} = \frac{dz}{dt} = \frac{\omega}{\beta}. \tag{A9.2}$$

However, the phase velocity has a meaning only with reference to an infinite monochromatic wave train. In order to convey information, we take a group of wave trains at different frequencies which form a finite wave train or a wave packet, say, in the form of a modulated wave such that the modulation envelope contains the information to be conveyed. The velocity of the wave packet or group of waves is called the group velocity of the wave, which also represents the velocity with which energy is transported.

We can then find the group velocity as velocity of the amplitude of the group of waves in the wave packet. Through any medium which in general is dispersive, the wave components of the group travel with different phase velocities. We can develop further concepts on the group velocity of a wave by considering two components of the wave packet—one at $\omega - \Delta\omega$ and the other at $\omega + \Delta\omega$, separated in frequency by $2\Delta\omega$, with phase propagation constants $\beta - \Delta\beta$ and $\beta + \Delta\beta$ respectively. With the help of (A9.1), we can therefore represent the group comprising these two components associated with the electric field quantity of the wave as

$$E = E_0[\exp j[(\omega - \Delta\omega)t - (\beta - \Delta\beta)z] + E_0[\exp j[(\omega + \Delta\omega)t - (\beta + \Delta\beta)z]. \quad \text{(A9.3)}$$

For the sake of simplicity, we have considered the magnitudes of the two electric field components to be the same as E_0 in (A9.3). We can next put (A9.3) as

$$E = E_0 \exp j(\omega t - \beta z)[\exp -j[(\Delta\omega)t - (\Delta\beta)z] + \exp j[(\Delta\omega)t - (\Delta\beta)z],$$

which can be simplified using the relation $\exp \pm j\phi = \cos\phi \pm j\sin\phi$, taking $\phi(\Delta\omega)t - (\Delta\beta)z$ as

$$E = 2E_0 \cos(\Delta\omega t - \Delta\beta z) \exp j(\omega t - \beta z). \quad \text{(A9.4)}$$

The amplitude of the electric field of the combination of the wave components of the group represented by (A9.4) has wave-like variation with distance and time appearing in the cosine function: $\cos(\Delta\omega t - \Delta\beta z)$, where $\Delta\omega t - \Delta\beta z$ is the phase of this variation. In order to find the group velocity of the wave, which is the velocity with which a plane of constant phase of this wave-like variation progresses, we can put

$$\Delta\omega t - \Delta\beta z = \text{constant}$$

and differentiate it with respect to time t to write

$$\Delta\omega - \Delta\beta\frac{dz}{dt} = 0. \quad \text{(A9.5)}$$

We can rearrange (A9.5) to obtain

$$\frac{dz}{dt} = \frac{\Delta\omega}{\Delta\beta},$$

which for small values of $\Delta\omega$ in the limit yields the expression for the group velocity v_g as:

$$v_g = \frac{\partial\omega}{\partial\beta}. \tag{A9.6}$$

The expression (A9.6) is identical to (9.63) in the text.

Appendix A9.2

Bessel functions

A linear second-order differential equation in y as a function of x

$$x^2\frac{d^2y}{dx^2} + x\frac{dy}{dx} + (x^2 - m^2)y = 0 \tag{A9.7}$$

is known as the mth order Bessel's differential equation. Dividing (A9.7) by x^2, we obtain the Bessel equation in the following form:

$$\left(\frac{d^2y}{dx^2} + \frac{1}{x}\frac{dy}{dx}\right) + \left(1 - \frac{m^2}{x^2}\right)y = 0. \tag{A9.8}$$

The expression (A9.8) is of the form of (9.128) used in the text; the function R has replaced the function y in (9.128). The solution of the Bessel equation (A9.8) is given in literature as

$$y = PJ_m(x) + QY_m(x) \tag{A9.9}$$

where P and Q are constants. Interpreting $y = R$, we can identify (A9.9) as (9.135) used in the text. $J_m(x)$ and $Y_m(x)$ are called m^{th} order ordinary Bessel functions of argument x of the first and second kinds respectively, given by

$$J_n(x) = \sum_{s=0}^{\infty} \frac{(-1)^s (x/2)^{n+2s}}{s!(n+s)!}$$

and

$$Y_n(x) = \frac{2}{\pi}\left(\gamma + \ln\frac{x}{2}\right)J_n(x) - \frac{1}{n}\sum_{s=0}^{n-1}\frac{(n-s-1)!}{s!}\left(\frac{2}{x}\right)^{n+2s}$$

$$-\frac{1}{\pi}\sum_{s=0}^{\infty}\frac{(-1)^s(x/2)^{n+2s}}{s!(n+s)!}\left(1 + \frac{1}{2} + \frac{1}{3} + \ldots + \frac{1}{s} + 1 + \frac{1}{2} + \frac{1}{3} + \ldots\right),$$

where γ is Euler's constant defined by

$$\gamma = \lim_{n\to\infty}\left(1+\frac{1}{2}+\frac{1}{3}+\ldots+\frac{1}{n}-\log n\right) = 0.5772157.$$

Appendix A9.3

Relation between Neper and decibel

Let a power P_1 be lesser than another power P_2 by a factor of $\exp(2\alpha)$, where α is in Neper (Np). We can then state the relation between P_1 and P_2 as

$$\frac{P_2}{P_1} = \exp 2\alpha \tag{A9.10}$$

Taking the logarithm of (A9.10) to the exponential base e, we obtain

$$\log_e \frac{P_2}{P_1} = 2\alpha$$

which can be expressed as

$$\frac{\log_{10}\dfrac{P_2}{P_1}}{\log_{10} e} = 2\alpha. \tag{A9.11}$$

Putting $\log_{10} e = \log_{10} 2.718 = 0.435$ in (A9.11), we obtain

$$\frac{\log_{10}\dfrac{P_2}{P_1}}{0.435} = 2\alpha. \tag{A9.12}$$

We can then write with the help of (A9.12)

$$10\log_{10}\frac{P_2}{P_1} = (10)(2)(0.435)\alpha = 8.7\alpha. \tag{A9.13}$$

The left-hand side of (A9.13) represents the number of decibels (dB's) by which P_2 is greater than P_1. Therefore, the number of dB's corresponding to 1 Np is obtained by putting $\alpha = 1$ in (A9.13). Thus, we get

1 Np = 8.7 dB.

SUMMARISING NOTES

✓ Waveguide—a hollow pipe made of a conducting material—is extensively used for the transmission of power in the microwave frequency range.

✓ Waveguide can support transverse electric (TE) mode, which is characterised by non-zero axial magnetic field and zero axial electric field

✓ Waveguide can support transverse magnetic TM mode, which is characterised by non-zero axial electric field and zero axial magnetic field.

✓ Why a hollow-pipe waveguide cannot support transverse electromagnetic (TEM) mode, for which the axial electric field and the axial magnetic field are each nil, has been explained.

✓ Waveguide behaves as a high-pass filter supporting propagating waves above a cutoff frequency that is related to waveguide dimensions.

✓ TE-mode and TM-mode field solutions for both the rectangular and cylindrical waveguides have been obtained.

✓ Characteristic equation or dispersion relation of a waveguide can be found with the help of the field solutions and electromagnetic boundary condition that the tangential component of the electric field is nil at the conducting surface of the waveguide wall.

✓ One and the same dispersion relation is obtained between the wave angular frequency ω and phase propagation constant β of a waveguide for the TE and the TM modes involving their respective cutoff frequencies ω_c.

✓ Identical ω-β dispersion plots are generated for rectangular and cylindrical waveguides excited in TE or TM mode.

✓ Cutoff frequency of the waveguide is the frequency ω corresponding to zero value of phase propagation constant β (which can be identified as the point of intersection between the ω-β dispersion plot and the abscissa, that is, ω-axis of the plot).

✓ Cutoff frequency of the waveguide depends on the waveguide dimensions and the waveguide mode chosen.

✓ Characteristic parameters of a waveguide are guide wavelength, phase propagation constant, phase velocity, group velocity and wave impedance, each of them depending on the operating frequency relative to the cutoff frequency of the waveguide.

✓ Mode numbers m and n are subscripted in the nomenclatures TE_{mn} and the TM_{mn}, representing respectively the transverse electric and transverse magnetic modes of the waveguide.

✓ Dominant mode of a waveguide is characterised by the lowest value of the cutoff frequencies of all the TE_{mn} and the TM_{mn} modes of the waveguide.

✓ Dominant mode of a rectangular waveguide is the mode TE_{10}.

✓ Dominant mode of a cylindrical waveguide is the mode TE_{11}.

✓ Mode numbers m and n of the TE_{mn} and TM_{mn} modes of a rectangular or cylindrical waveguide can be correlated with their respective field patterns across the waveguide cross section.

✓ For a rectangular waveguide, the mode number m indicates the number of maxima (of any field component) along the broad dimension of the waveguide, while the mode number n indicates the number of maxima (of any field component) along the narrow dimension of the waveguide. Alternatively, you may interpret m and n as the numbers of half-wave field patterns across the broad and the narrow dimensions of the waveguide respectively.

✓ For a cylindrical waveguide, the mode number m indicates the number of half-wave field pattern around the half circumference and n indicates the number of positive or negative maxima across the waveguide radius.

✓ Evanescent mode is supported by a waveguide below its cutoff frequency associated with no component of the average Poynting vector in the direction of wave propagation, corresponding to no power flow in the waveguide.

✓ Expression for the power propagating through a rectangular waveguide above its cutoff frequency has been developed and hence the power handling capability of the waveguide has been found in terms of the breakdown voltage of the medium filling the hollow region of the waveguide for dominant-mode excitation of the waveguide.

✓ Expression for the power loss per unit length of the walls of a rectangular waveguide due to the finite resistivity of the material making the walls has been developed for dominant-mode excitation of the waveguide.

✓ Expression for the attenuation constant of a dominant-mode-excited rectangular waveguide has been developed using the expressions for propagating power and power loss per unit length of the waveguide.

✓ Attenuation constant of a waveguide depends on the operating frequency and the waveguide dimensions, which should be taken into consideration while choosing the waveguide mode and frequency for lower waveguide attenuation.

Review Questions

Narrative type

9.1 What do you understand by the excitation of the TE or H mode and that of the TM or E mode in a hollow-pipe waveguide?

9.2 Explain why it is impossible to excite the TEM mode in a hollow-pipe waveguide.

9.3 Set up the wave equation for the component of magnetic field in the direction of propagation of the TE-mode wave through a rectangular waveguide and solve the wave equation for this component by the method of separation of variables. Express the component in terms of the waveguide dimensions.

9.4 Starting from the expression for the component of magnetic field in the direction of propagation of wave through a rectangular waveguide excited in TE mode, in terms of the waveguide dimensions, and taking help of Maxwell's equations, derive the expressions for the remaining field components of the waveguide.

9.5 Write down the expressions for $E_{x,y,z}$ and $H_{x,y,z}$ in a rectangular waveguide excited in the TE mode ($H_z \neq 0, E_z = 0$).

9.6 Depict the field patterns of the TE_{10} and TE_{20} modes of a rectangular waveguide in a plane transverse to the direction of propagation.

9.7 Derive the characteristic equation or dispersion relation of a rectangular waveguide excited in the TE mode in terms of the waveguide cutoff frequency, the latter expressed in terms of the waveguide dimensions.

9.8 Deduce the following expression for the cutoff angular frequency ω_c of a rectangular waveguide excited in the TE_{mn} mode in terms of its broad and narrow dimensions a and b respectively: $\omega_c = c\left[(m\pi/a)^2 + (n\pi/b)^2\right]^{1/2}$.

9.9 The broader dimension a of a rectangular waveguide excited in the TE_{11} mode is s times the twice its narrow dimension b, that is, $a = 2bs$. Show that its cutoff wavelength is given by $\lambda_c = 4bs/(1 + 4s^2)^{1/2}$.

9.10 Plot $\omega - \beta$ dispersion characteristics of a waveguide indicating the velocity of light c line and the waveguide cutoff angular frequency.

9.11 Explain why a waveguide behaves as a high-pass filter.

9.12 Discuss how ionosphere behaves as a waveguide of cutoff frequency equal to the electron plasma frequency of the ionosphere.

Hint: See Example E9-2 as well as Example E6-11.

9.13 What do you understand by phase velocity v_{ph} and group velocity v_g of a wave? Establish from first principles the relations: $v_{ph} = \omega/\beta$ and $v_g = \partial\omega/\partial\beta$.

Hint: See Appendix A9.1.

9.14 Establish the following relation for a waveguide excited in the TE mode:

$$\frac{v_{ph}}{c} = \frac{\lambda_g}{\lambda} = \frac{c}{v_g} = \frac{Z_W}{\eta_0} = \frac{1}{(1 - f_c^2/f^2)^{1/2}},$$

where v_{ph} and v_g are the wave phase and group velocities respectively, Z_W is the wave impedance and λ_g the guide wavelength.

9.15 Show that the average Poynting vector in the evanescent mode below cutoff ($\omega < \omega_c$) of a waveguide is equal to zero. What conclusion out of this finding can be made with respect to power flow in the evanescent mode?

9.16 Show that an ellipse is generated if the attenuation constant α of a waveguide is plotted against the angular frequency ω in the evanescent mode of the waveguide.

Hint: See Example E9-6.

9.17 Show that the cutoff wavelength λ_c and the cutoff frequency f_c of a rectangular waveguide excited in the TE_{10} mode are given by $\lambda_c = 2a = 4bs$ and $f_c = c/(4bs)$ respectively, where $s = a/2b$.

9.18 Why is the TE_{10}mode called the dominant mode of a rectangular waveguide?

9.19 What do you understand by the degenerate modes? Show that for a rectangular waveguide, in which the broader dimension is twice the narrow dimension ($a = 2b$), modes TE_{20} and TE_{01} are degenerate.

9.20 Set up the wave equation for the component of electric field in the direction of propagation of the TM-mode wave through a rectangular waveguide and solve the wave equation for this component by the method of separation of variables. Express the component in terms of the waveguide dimensions.

9.21 Starting from the expression for the component of magnetic field in the direction of propagation of wave through a rectangular waveguide excited in the TM mode, derive the expressions for the remaining field components of the waveguide in terms of the waveguide dimensions, taking help of Maxwell's equations.

9.22 Write down for a rectangular waveguide excited in the TM mode ($E_z \neq 0, H_z = 0$) the expressions for $E_{x,y,z}$ and $H_{x,y,z}$.

9.23 Depict the field patterns of the TM_{11} mode of a rectangular waveguide in a plane transverse to the direction of propagation.

9.24 Write down the characteristic equation or dispersion relation of a rectangular waveguide excited in the TM mode in terms of (i) the waveguide cutoff frequency and (ii) the waveguide cutoff wavelength.

9.25 Deduce the expression for the cutoff angular frequency ω_c of a rectangular waveguide excited in the TM_{mn} mode in terms of its broad and narrow dimensions a and b respectively. Show that the expression is identical to that for a rectangular waveguide excited in the TE_{mn} mode: $\omega_c = c[(m\pi/a)^2 + (n\pi/b)^2]^{1/2}$.

9.26 Discuss with specific examples the significance of the mode numbers m and n of the TE_{mn} or TM_{mn} mode of a rectangular waveguide in terms of the number of maxima of the field relevant component along the broad and narrow dimensions of the guide transverse to the direction of propagation respectively, or, alternatively, in terms of the numbers of half-wave field patterns across the broad and narrow dimensions of the waveguide respectively.

9.27 Derive the characteristic equation or dispersion relation of a rectangular waveguide excited in the TM mode in terms of the waveguide cutoff frequency, the latter expressed in terms of the waveguide dimensions.

9.28 Show that the dispersion relation of a rectangular waveguide excited in the TM mode expressed in terms of the waveguide dimensions is identical to that of a rectangular waveguide excited in the TE mode.

9.29 Examining the field solutions of a rectangular waveguide excited in the TM mode, how can you conclude that the lowest order mode that can be excited in the guide is essentially the TM_{11} mode?

9.30 With reference to a rectangular waveguide, show that the cutoff frequency of the lowest order transverse magnetic mode, namely the TM_{11} mode is greater than that of the TE_{10} mode.

Hint: $(c/2)[(1/a)^2 + (1/b)^2]^{1/2} > c/2a$, the left-hand side being the cutoff frequency of the TM_{11} mode and the right-hand side that of the TE_{10} mode.

9.31 Set up the wave equation for the component of magnetic field in the direction of propagation of the TE-mode wave through a cylindrical waveguide and solve the wave equation for this component by the method of separation of variables. Express the component in terms of the waveguide dimensions.

9.32 Starting from the expression for the component of magnetic field in the direction of propagation of a wave through a cylindrical waveguide excited in the TE mode, derive the expressions for the remaining field components of the waveguide in terms of the waveguide dimensions, and taking help of Maxwell's equations.

9.33 Write down the expressions for $E_{r,\theta,z}$ and $H_{r,\theta,z}$ of a cylindrical waveguide excited in the TE mode $(H_z \neq 0, E_z = 0)$.

9.34 Depict the field patterns of the TE_{01}, TE_{02}, and TE_{11} modes of a cylindrical waveguide in a plane transverse to the direction of propagation.

9.35 Derive the dispersion relation of a cylindrical waveguide excited in the TE_{mn} mode as: $\omega^2 - \beta^2 c^2 - \omega_c^2 = 0$, where ω_c is the cutoff frequency of the waveguide given by $\omega_c = X_{mn} c/a$, X_{mn}, being the nth root of the function $J'_m(\gamma a) = 0$.

9.36 In what way will your analysis of a cylindrical waveguide excited in the TE_{mn} mode leading to its dispersion relation change if you were interested in deducing the relation for the TM_{mn} mode of excitation instead of the TE_{mn} mode?

Hint: Instead of the boundary condition (9.162), that is, $E_\theta = 0|_{r=a}$ used in the analysis of the TE_{mn}-mode excitation leading to the dispersion relation, you can take the boundary condition $E_\theta = 0|_{r=a}$. in the analysis of the TM_{mn}-mode excitation. You will then obtain the condition $J_m(\gamma a) = 0$ for the TM_{mn}-mode excitation instead of $J'_m(\gamma a) = 0$ obtained for the $\mathrm{TE_{mn}}$-mode excitation. Hence, proceed to derive the dispersion relation for the $\mathrm{TM_{mn}}$-mode excitation as: $\omega^2 - \beta^2 c^2 - \omega_c^2 = 0$, where ω_c is the cutoff frequency of the waveguide given by $\omega_c = X_{mn}c/a$, X_{mn}, being the nth root of the function $J_m(\gamma a) = 0$.

9.37 Show that the $\mathrm{TE_{11}}$ mode is the dominant mode of a cylindrical waveguide.

9.38 A cylindrical waveguide has a concentric dielectric rod insert at its axis and is excited in the TE mode. Using the appropriate boundary conditions, show that the waveguide would support only the azimuthally symmetric $\mathrm{TE_{0n}}$ mode, that is, the $\mathrm{TE_{mn}}$ mode corresponding to the waveguide mode number $m = 0$.

Hint: See Example E9-8.

9.39 Discuss the significance of the mode numbers m and n of a cylindrical waveguide excited in the $\mathrm{TE_{mn}}$ or $\mathrm{TM_{mn}}$ mode as the number of half-wave field patterns around the half waveguide circumference and the number of maxima—positive and negative inclusive—across the waveguide radius.

9.40 Why is the $\mathrm{TE_{11}}$ mode called the dominant mode of a cylindrical waveguide?

9.41 Obtain from first principles the expression for the power transmitted through a rectangular waveguide excited in the dominant mode in terms of the amplitude of a relevant field component. Hence predict the maximum power handling capacity of the waveguide in terms of the maximum allowable electric field component from the standpoint of the breakdown of the waveguide medium and waveguide dimensions.

9.42 Derive the expressions for the power lost per unit length in each of the walls of a rectangular waveguide excited in the dominant mode in terms of the surface resistance of the material of the waveguide wall and waveguide dimensions.

9.43 Obtain an expression for the attenuation constant of a rectangular waveguide excited in the dominant mode due to the finite conductivity of the material of the waveguide wall in terms of the conductivity of the material and the waveguide dimensions.

9.44 Give an analysis to predict the frequency relative to the cutoff frequency in terms of the waveguide dimensions, where the attenuation constant of a rectangular waveguide excited in the dominant mode takes on a minimum value.

Hint: See Example E9-12.

Review questions with solutions

9.45 Write down the dispersion relation or the characteristic equation of a rectangular waveguide excited in the TE mode in terms of the waveguide (i) cutoff frequency and (ii) cutoff wavelength.

Answer: (i) $\omega^2 - \beta^2 c^2 = 0$ and (ii) $l/\lambda_c^2 = 1/\lambda^2 - 1/\lambda_g^2$ are the wave angular frequency and axial phase propagation constant respectively; ω_c and λ_c are the waveguide cutoff frequency and the cutoff wavelength respectively; and λ and $\lambda_g (= 2\pi/\beta)$ are the free-space wavelength and guide wavelength respectively.

9.46 Find a relation for the wave dispersion $\partial v_{\text{ph}}/\partial\omega$ in terms of the wave phase velocity v_{ph} and group velocity v_g.

Answer: $\dfrac{\partial v_{\text{ph}}}{\partial W} = \dfrac{(1 - v_{\text{ph}}/v_{\text{g}})}{\beta}$. (See Example E9.3)

9.47 What is an evanescent mode in a waveguide? Find the attenuation constant of the wave in evanescent mode in terms of the wave frequency and cutoff frequency.

Answer: Attenuation constant $\alpha = \dfrac{\sqrt{\omega_c^2 - \omega^2}}{c} \quad (\omega < \omega_c)$.

9.48 (a) Write down the wave equations for the axial electric and magnetic field components for a wire or tape wound in the form of a helix in free space which supports slow electromagnetic waves of phase velocity $v_{\text{ph}} < c$. (b) Obtain the solutions of these equations for the axial and the azimuthal field components in the regions inside and outside a helix winding radius, treating the actual helix in the sheath-helix model. (c) Hence, derive the dispersion relation of the helix using the boundary conditions at the sheath representing the helix (see Problem 7-3) assuming that, subject to these boundary conditions, the helix supports both TE ($H_z \neq 0$) and TM ($E_z \neq 0$) modes, and that the modes are azimuthally symmetric ($\partial/\partial\theta = 0$).

Answer:

(a) Wave equation:

The wave equation (9.118) that we used for the analysis of a cylindrical waveguide continues to hold good for the present problem of the helix as well. This can be rewritten as

$$\frac{\partial^2(E_z, H_z)}{\partial r^2} + \frac{1}{r}\frac{\partial(E_z, H_z)}{\partial r} + (\omega^2\mu_0\varepsilon_0 - \beta^2)(E_z, H_z) + \frac{1}{r^2}\frac{\partial^2(E_z, H_z)}{\partial\theta^2} = 0,$$

which can further be expressed as

$$\frac{\partial^2(E_z, H_z)}{\partial r^2} + \frac{1}{r}\frac{\partial(E_z, H_z)}{\partial r} + \gamma^2(E_z, H_z) + \frac{1}{r^2}\frac{\partial^2(E_z, H_z)}{\partial\theta^2} = 0 \quad (1)$$

where $\gamma^2 = \omega^2\mu_0\varepsilon_0 - \beta^2$. Note that the cylindrical waveguide supports fast waves corresponding to $v_{\text{ph}} (= \omega/\beta) > c$, which makes $\omega^2\mu_0\varepsilon_0 > \beta^2$ in view of the relation $c = 1/\sqrt{\mu_0\varepsilon_0}$. Further, then this condition for fast waves will make $\gamma^2 (= \omega^2\mu_0\varepsilon_0 - \beta^2$ to have a positive value. However, in the present case of a helix that supports slow waves, we have to take $v_{\text{ph}} (= \omega/\beta) < c$, which makes $\omega^2\mu_0\varepsilon_0 < \beta^2$ which, in turn, makes $\gamma^2 (= (\omega^2\mu_0\varepsilon_0 - \beta^2)^{1/2})$ negative, that is, makes $\gamma\ (= j\gamma')$ a complex quantity, where $\gamma'(=(\beta^2-\omega^2\mu_0\varepsilon_0)^{1/2})$ is a real quantity. Thus, the solution to the above wave equation (1) (see Section 9.4) with reference to the present problem involves ordinary Bessel functions, however, with complex argument γr. Alternatively, we can express these solutions in terms of modified Bessel functions with real argument $\gamma' r$.

(b) Field solutions:

For the axial field components, the solutions of the wave equation (1) are:

$$\left.\begin{aligned} E_{z1} &= A_1 I_0(\gamma' r) B_1 K_0(\gamma' r) \\ H_{z1} &= C_1 I_0(\gamma' r) D_1 K_0(\gamma' r) \\ E_{z2} &= A_2 I_0(\gamma' r) B_2 K_0(\gamma' r) \\ H_{z2} &= C_2 I_0(\gamma' r) D_2 K_0(\gamma' r) \end{aligned}\right\} \tag{2}$$

where the subscripts 1 and 2 refer to regions 1 and 2 representing the inside and the outside of the helix winding radius respectively. However, in order to prevent fields from blowing up to infinity inside the helix winding radius at the helix axis $r=0$, put the two field constants $B_1=D_1=0$ since $K_0(\gamma' r)\to\infty$ as $\gamma' r\to 0$. Similarly, in order to prevent the fields from blowing up to infinity outside the helix winding radius at $r=\infty$, put the two field constants $A_2=C_2=0$ since $I_0(\gamma' r)\to\infty$ as $\gamma' r\to\infty$. Therefore, putting $B_1=D_1=0$ and $A_2=C_2=0$ in (2), the field expressions in terms of the four non-zero field constants A_1, C_1, B_2, D_2 works out as

$$\left.\begin{aligned} E_{z1} &= A_1 I_0(\gamma' r);\, H_{z1} = C_1 I_0(\gamma' r) \\ E_{z2} &= B_2 K_0(\gamma' r);\, H_{z2} = D_2 K_0(\gamma' r) \end{aligned}\right\}. \tag{3}$$

The following azimuthal field components can then be derived taking help of Maxwell's equations (5.78) and (5.79) (putting therein the current density $\vec{J}=0$) expanded in cylindrical system using (2.68), while assuming azimuthal symmetry of the problem ($\partial/\partial\theta=0$) and using the relations $I_0'(x)=I_1,(x)K_0'(x)=-K_1(x)$, the prime indicating the derivative of the function with respect to the argument of the function:

$$\left.\begin{aligned} E_{\theta 1} &= (-j\omega\mu_0/\gamma') C_1 I_1(\gamma' r) \\ H_{\theta 1} &= (j\omega\varepsilon_0/\gamma') A_1 I_1(\gamma' r) \\ E_{\theta 2} &= (j\omega\mu_0/\gamma') D_2 K_1(\gamma' r) \\ H_{\theta 2} &= (-j\omega\varepsilon_0/\gamma') B_2 K_1(\gamma' r) \end{aligned}\right\}. \tag{4}$$

(c) Dispersion relation:

We can then substitute the field expressions (3) and (4) into the boundary conditions at the sheath $r = a$ (see Problem 7-3)

$$\left.\begin{array}{l} E_{\theta 1}\cos\psi + E_{z1}\sin\psi = 0 \\ E_{\theta 2}\cos\psi + E_{z2}\sin\psi = 0 \\ H_{\theta 1}\cos\psi + H_{z1}\sin\psi = H_{\theta 2}\cos\psi + H_{z2}\sin\psi \\ E_{z1} = E_{z2} \end{array}\right\}$$

to obtain the following set of simultaneous equations in non-zero field constants

$$\left.\begin{array}{l} K_0(\gamma' a)\sin\psi B_2 + (j\omega\mu_0/\gamma')K_1(\gamma' a)\cos\psi D_2 = 0 \\ (j\omega\varepsilon_0/\gamma')I_1(\gamma' a)\cos\psi A_1 + I_0(\gamma' a)\sin\psi C_1 \\ \quad +(j\omega\varepsilon_0/\gamma')K_1(\gamma' a)\cos\psi B_2 - K_0(\gamma' a)\sin\psi D_2 = 0 \\ I_0(\gamma' a)\sin\psi A_1 - (j\omega\mu_0/\gamma')I_1(\gamma a)\cos\psi C_1 = 0 \\ I_0(\gamma' a)A_1 - K_0(\gamma' a)B_2 = 0 \end{array}\right\}. \tag{5}$$

We can express (5) as

$$\left.\begin{array}{l} a_{11}A_1 + a_{12}C_1 + a_{13}B_2 + a_{14}D_2 = 0 \\ a_{21}A_1 + a_{22}C_1 + a_{23}B_2 + a_{24}D_2 = 0 \\ (6)a_{31}A_1 + a_{32}C_1 + a_{33}B_2 + a_{34}D_2 = 0 \\ a_{41}A_1 + a_{42}C_1 + a_{33}B_2 + a_{44}D_2 = 0 \end{array}\right\} \tag{6}$$

where

$$\left.\begin{array}{l} a_{11} = 0,\ a_{12} = 0,\ a_{13} = K_0(\gamma' a)\sin\psi,\ a_{14} = (j\omega\mu_0/\gamma')K_1(\gamma' a)\cos\psi, \\ a_{21} = (j\omega\varepsilon_0/\gamma')I_1(\gamma' a)\cos\psi,\ a_{22} = I_0(\gamma' a)\sin\psi, \\ a_{23} = (j\omega\varepsilon_0/\gamma')K_1(\gamma' a)\cos\psi,\ a_{24} = -K_0(\gamma' a)\sin\psi, \\ a_{31} = I_0(\gamma' a)\sin\psi,\ a_{32} = -(j\omega\mu_0/\gamma')I_1(\gamma' a)\cos\psi,\ a_{33} = 0, a_{34} = 0, \\ a_{41} = I_0(\gamma' a),\ a_{42} = 0,\ a_{43} = -K_0(\gamma' a),\ a_{44} = 0. \end{array}\right\}. \tag{7}$$

For non-trivial solution of (6), the condition is that the determinant formed by the coefficients of the field constants occurring in (6) should be equal to zero.

$$\begin{vmatrix} a_{11} & a_{12} & a_{13} & a_{14} \\ a_{21} & a_{22} & a_{23} & a_{24} \\ a_{31} & a_{32} & a_{33} & a_{34} \\ a_{41} & a_{42} & a_{43} & a_{44} \end{vmatrix} = 0. \tag{8}$$

Condition (8) in conjunction with (7), making use of the relation $I_n(x)K_1(x) +$

$K_0(x)I_1(x) = 1/x$, simplifies to

$$\frac{k \cot \psi}{\gamma'} = \left[\frac{I_0(\gamma' a)K_0(\gamma' a)}{I_1(\gamma' a)K_1(\gamma' a)}\right]^{1/2}. \tag{9}$$

For slow waves ($v_{ph}(= \omega/\beta) < c$), we can appreciate from the discussion following (1) that the left-hand side of (9) is $k \cot \psi/\gamma' \approx (v_{ph}/c) \cot \psi$. Assigning a value to $\gamma' a$, we can calculate from the dispersion relation the value of $k \cot \psi/\gamma'$. Multiplying the latter by $\gamma' a$, we thus get the value of $ka \cot \psi$ $[= (k \cot \psi/\gamma')(\gamma' a)]$. Thus, we get a set of values $k \cot \psi/\gamma' \approx (v_{ph}/c) \cot \psi$, which is proportional to v_{ph} and $ka \cot \psi$, which is proportional to frequency $f(= \omega/2\pi)$. The relation (9) is therefore the dispersion relation of the helix in free space.

Numerical type

9.49 What will be the time delay for a signal to reach the end of a 20-m long rectangular 2 cm × 1.5 cm waveguide excited in the dominant mode at the operating frequency 10 GHz?

Answer:

With the help of (9.71), we can find the wave group velocity v_g as

$$v_g = \left(1 - \frac{f_c^2}{f^2}\right)^{1/2} \times 3 \times 10^8. \tag{1}$$

From (9.93a) the cutoff frequency f_c of the rectangular waveguide excited in the dominant TE_{10} mode in terms of the wider dimension a is given by

$$f_c = \frac{c}{2a},$$

which can be used to calculate

$$f_c = \frac{c}{2a} = \frac{3 \times 10^8}{2 \times 2 \times 10^{-2}} = 7.5 \times 10^9 \text{GHz} = 7.5 \text{ GHz}. \tag{2}$$

The operating frequency $f = 10$ GHz being given, we can calculate with the help of (2)

$$\left(\frac{f_c}{f}\right)^2 = (0.75)^2 = 0.5625,$$

which, in turn, can be put in (1) to find the group velocity v_g as

$$v_g = (1 - 0.5625)^{1/2} \times 3 \times 10^8 = (0.4375)^{1/2} \times 3 \times 10^8 = 1.98 \times 10^8 \text{ m/s}. \tag{3}$$

The time delay for the signal to reach the waveguide end is obtained by dividing the waveguide length by v_g, given by (3), as follows:

$$\text{Time delay} = \frac{20\text{m}}{1.98 \times 10^8 \text{m/s}} \approx 10^{-7}\text{s} = 100 \text{ ns}.$$

9.50 The broader and narrower dimensions of a rectangular waveguide are 3.4 and 1.7 inch respectively. Calculate the cutoff frequency and cutoff wavelength of the waveguide if it is excited in the TE_{10} mode.

Answer: 1.737 GHz and 172.72 mm.

9.51 Calculate the cutoff wave lengths of the modes TE_{01}, TE_{10}, TE_{20} and TE_{11} of the rectangular waveguide WR-284 with transverse dimensions: $a = 2.84$ inch, $b = 1.34$ inch and hence arrange the modes in the order of increasing cutoff wavelength.

Answer: TE_{11}, TE_{01}, TE_{20} and TE_{10}.

9.52 Verify that the use of the operating frequencies between 0.32 and 0.49 GHz ensures that only the dominant mode and not the next higher order modes are excited in the rectangular waveguide WR-2300 ($a = 23$ and $b = 11.5$ inch).

Hint: Calculate the cutoff frequency of the TE_{10} (dominant mode) as 0.257 GHz and that of the next higher order mode, in this case the TE_{20} and its degenerate TE_{01}, as 0.51 GHz.

9.53 Calculate the cutoff wavelengths of a rectangular waveguide ($a = 2$ cm and $b = 1$ cm) excited in the TE_{20} and TE_{01} modes and hence show that these modes are degenerate.

Answer: The calculation, with the help of (9.55), shows that both the modes TE_{20} and TE_{01} have the same cutoff wavelength, namely, 2cm. Hence, these modes will degenerate.

9.54 Calculate the cutoff wavelengths of a square waveguide ($a = b = 2$ cm) excited in the TE_{10} and TE_{01} modes and hence show that these modes are degenerate.

Answer: The calculation, with the help of (9.55), shows that both the modes TE_{10} and TE_{01} have the same cutoff wavelength, namely, 4 cm. Hence, these modes are degenerate.

9.55 A 20-GHz signal has to be transmitted through a cylindrical waveguide excited in the dominant mode. How will you choose the inside wall diameter of the guide in order to ensure that the signal frequency is 20% above the waveguide cutoff?

Answer:

The waveguide cutoff frequency is given by (9.171), which can be rewritten as

$$\omega_c = \frac{X_{mn}c}{a},$$

whence we get the wall radius as

$$a = \frac{X_{mn}c}{\omega_c}.$$

For the dominant mode TE_{11}, substituting $X_{11} = 1.841$ from (9.165) into (1), we get

$$a = \frac{1.841c}{2\pi f_c} = \frac{1.841 \times 3 \times 10^8}{2\pi f_c} \tag{2}$$

In order to ensure that the signal frequency is 20% above the waveguide cutoff, the cutoff frequency f_c should be 0.8 times the signal frequency 20 GHz, that is,

$$f_c = 0.8 \times 20 = 16 \text{ GHz} = 16 \times 10^9 \text{ Hz}. \tag{2}$$

Substituting (3) in (2), we obtain

$$a = \frac{1.841 \times 3 \times 10^8}{2\pi \times 16 \times 10^9} = \frac{1.841 \times 3 \times 10^8}{2 \times 3.143 \times 16 \times 10^9} = 0.00549 \text{ m}. \tag{4}$$

Therefore, from (4), the required waveguide diameter is $2a = 0.011$ m $= 1.1$ cm.

9.56 The broad dimension, $a = 0.9''$, and the narrow dimension, $b = 0.4''$, and the cutoff frequency, $f_c = 6.56$ GHz, of a rectangular waveguide excited in the dominant TE_{10} mode are given. Find the frequency of operation which would correspond to the minimum attenuation due to the finite conductivity of the wall of the waveguide if it is excited in the dominant mode TE_{10}.

Answer For the sake of convenience let us rewrite expression (9-12.9) for the frequency normalised with respect to the waveguide cutoff frequency that corresponds to the minimum attenuation of the waveguide excited in the dominant mode as

$$\frac{f}{f_c} = \sqrt{\frac{4b}{(3a+6b) \pm \sqrt{3a+6b)^2 - 8ab}}}. \tag{1}$$

From the data given in the problem, $a = 1.8b, 3a + 6b = 3 \times 1.8b + 6b = 11.4b$ and $8ab = 8 \times 1.8b \times b = 14.4b^2$, which when put in (1) gives

$$\frac{f}{f_c} = \sqrt{\frac{4b}{(11.4b) \pm \sqrt{(11.4)^2b^2 - 14.4b^2}}}. \tag{2}$$

Taking the minus sign in the denominator of the radical above, as it would give the realistic value of frequency above the cutoff frequency ($f > f_c$), we obtain

$$\frac{f}{f_c} = \sqrt{\frac{4}{11.4 - \sqrt{(11.4)^2 - 14.4}}} = 2.46. \tag{3}$$

Therefore, putting $f_c = 6.56$ GHz (given) in (3), we get the frequency where minimum attenuation takes place as $f = 6.56 \times 2.46 = 16.13$ GHz.

Multiple-choice questions

9.57 Above the cutoff frequency, the wave phase velocity in a waveguide is

(a) $< c$ (b) $=0$ (c) $> c$ (d) $= \infty$.

9.58 The product of the phase and group velocities of a rectangular waveguide excited in the TE mode is

(a) $v_{\text{ph}} v_g = c^2 \left(1 - \frac{f_c^2}{f^2}\right)^{-1}$ (b) $v_{\text{ph}} v_g = c^2$

(c) $v_{\text{ph}} v_g = c^2 \left(1 - \frac{f_c^2}{f^2}\right)^{1/2}$ (d) $v_{\text{ph}} v_g = \frac{c^2}{\left(1 - \frac{f_c^2}{f^2}\right)^{1/2}}$

9.59 A waveguide behaves as a filter of the following type:

(a) low-pass (b) high-pass (c) band-pass (d) band-elimination.

9.60 Power in a waveguide for operating frequencies below the cutoff frequency is

(a) reactive and is attenuated in the direction of propagation
(b) reactive and is not attenuated in the direction of propagation
(c) attenuated in the direction of propagation due to the wall loss due to its finite conductivity
(d) active and is radiated out of the waveguide.

9.61 At the cutoff frequency of a rectangular waveguide excited in the TE mode, the following hold true:

Choice	Phase velocity	Group velocity	Wave impedance	Guide wavelength
(a)	0	∞	0	∞
(b)	0	∞	∞	0
(c)	∞	0	∞	0
(d)	∞	0	0	∞

9.62 For frequencies tending to infinity above the cutoff frequency of a rectangular waveguide excited in the TE mode, the following hold true:

Choice	Phase velocity	Group velocity	Wave impedance	Guide wavelength
(a)	0	c	0	∞
(b)	0	c	η_0	λ
(c)	∞	0	η_0	∞
(d)	∞	0	η_0	λ

9.63 One of the following statements is incorrect:

(a) The dispersion relation of a waveguide is the relation between the wave angular frequency and the axial phase propagation constant (ω–β relationship) or the relation between the wave phase velocity and the wave frequency (v_{ph}–f relationship).

(b) The dispersion relation of a rectangular waveguide excited in the TE mode is the same as that of a rectangular waveguide excited in the TM mode with the same interpretation of the cutoff frequency in terms of the waveguide dimensions.

(c) The dispersion relation of a cylindrical waveguide excited in the TE mode is the same as that of the waveguide excited in the TM mode with the same dependence of the cutoff frequency on the waveguide dimensions.

(d) The dispersion relation of a rectangular waveguide is the same as that of a cylindrical waveguide, both excited in the TE mode, with appropriate definitions of cutoff frequencies in terms of waveguide dimensions.

9.64 The dominant mode of a rectangular waveguide is

(a) TE_{10} (b) TE_{01} (c) TE_{11} (d) TE_{21}.

9.65 The transverse electric field in the field pattern of a rectangular waveguide in the cross-sectional plane transverse to the direction of propagation of the TE mode becomes zero at any point in the middle of the broad dimension of the guide and reverses in direction on both sides of this point. The mode can be identified as

(a) TE_{01} (b) TE_{02} (c) TE_{10} (d) TE_{20}.

9.66 The dominant mode of a cylindrical waveguide is

(a) TE_{10} (b) TE_{01} (c) TE_{11} (d) TE_{21}.

9.67 The azimuthal electric field of the field pattern in the cross-sectional plane transverse to the direction of propagation of the TE mode which is along the axis of a cylindrical waveguide becomes zero at any point around midway between the axis and the wall of the waveguide and reverses in direction on both sides of this point. The mode can be identified as

(a) TE_{11} (b) TE_{21} (c) TE_{01} (d) TE_{02}.

Answer:

9.57: (c) 9.58:(b) 9.59: (b) 9.60: (a) 9.61: (d) 9.62: (b) 9.63: (c)
9.64:(a) 9.65: (d) 9.66: (c) 9.67: (d)

10

Waveguide Resonator: Analytic Appreciation by Equivalent Transmission Line Approach

A hollow metal pipe waveguide which supports an electromagnetic wave propagating through it and which is extensively used to transmit microwave power from one point to another (see Chapter 9) can be designed to store electromagnetic energy as well. Thus, for instance, if the two ends of a hollow metal pipe waveguide are closed by conducting metal walls to form a closed metal box, called waveguide resonator, it will serve as an energy storing component like an inductor or a capacitor or a combination thereof such as a tank circuit consisting of an an inductor in parallel with a capacitor. Note that at microwave frequencies circuit elements like an inductor or a capacitor are difficult to make because of their tiny sizes (at high frequencies). Moreover, at such frequencies the dimensions of these circuit elements become comparable to the operating wavelength, causing them to radiate electromagnetic energy that create electromagnetic interference in the neighbouring circuits and systems. Also, the radiation of energy at microwave frequencies is a loss of energy that is otherwise supposed to be stored in the resonant circuit. Furthermore, the effective resistance of circuit elements increases due to skin effect at high frequencies, causing an increased resistive loss. These radiative and resistive losses are reduced in a waveguide resonator made of a good conductor at microwave frequencies.

In this chapter we are going to appreciate that a specified length of a waveguide with both its ends open can also behave as a waveguide resonator. In fact, in the beginning of Chapter 9 we had described a waveguide as a transmission line, like the two-wire line or the coaxial cable. In this chapter we are going to use the transmission line theory to understand the behaviour of the waveguide as a resonator. For this purpose we will introduce the transmission line theory (Section 10.1) to find that length of a waveguide called the resonant length will make the waveguide behave as a resonator at a frequency called the resonant frequency. We will describe the waveguide as a short-circuited transmission line when it is closed at both its ends by a short, and as an open-circuited transmission line when it is open ended (Section 10.2). We will also deduce field solutions for a closed-end waveguide resonator (Section 10.3). Further, we will analyse the closed-end resonator for the energy stored and the power dissipated in it (Section 10.4) to derive the quality factor of the resonator (Section 10.5).

10.1 Basics of Transmission Line Theory

A transmission line consists of two or more parallel conductors. The common examples are the two-conductor, shielded two-conductor, coaxial cable and microstrip transmission lines used to transfer energy or information from one point to another.[1] One can extend the ac circuit theory to analyse the transmission line. This circuit approach of analysis of a transmission line is an alternative to the TEM-mode field theory approach that uses Maxwell's equations, wave equation, electromagnetic boundary conditions, Poynting theorem, etc. Though the transmission line circuit theory, unlike the field theory, cannot predict the field and mode patterns, yet it is found useful for predicting the propagation and impedance characteristics of a line. For instance, with the help of the transmission line theory you can find the phase velocity of waves propagating through the line and its variation with frequency, that is, the dispersion of the line.

In the transmission line circuit theory approach, the energies stored in electric and magnetic fields of the line are accounted for by the line series inductance per unit length L and shunt capacitance per unit length C respectively. Similarly, the power losses in the conductors and the dielectric, if present in the region between the conductors of the line, are accounted for by the line series resistance per unit length R and shunt conductance per unit length G respectively. L, C, R and G are called the 'distributed' circuit parameters of the transmission line. In the definition of these distributed line parameters, it is implied that the effect of the size of a circuit element relative to the operating wavelength is not ignored as can be done at low frequencies. For frequencies greater than a kilohertz, one needs to incorporate the effect of finite time of travel of the signal along the electric circuit while assigning electrical quantities such as voltage, current, resistance and capacitance to the signal. Thus, due to the transit-time effect, there will be a non-zero reactive drop for a lossless transmission line even though its resistive voltage drop is zero. Higher the operating frequencies, more pronounced is the manifestation of the transit-time effect in the reactance effect. Therefore, though a lumped circuit can represent a piece of wire passing a direct current (dc) through it since the current and the potential at all points on it can be regarded as constant, it is not valid for a time-varying situation since the current I and voltage V vary from point to point along the length of the wire, say, along z. Such a situation warrants the use of distributed circuit parameters expressed per unit length

[1]Readers are encouraged to think whether information transfer is synonymous with energy transfer (with a hint that while Shanon's information theory allows one to link information to entropy, thermodynamics allows one to link entropy to energy).

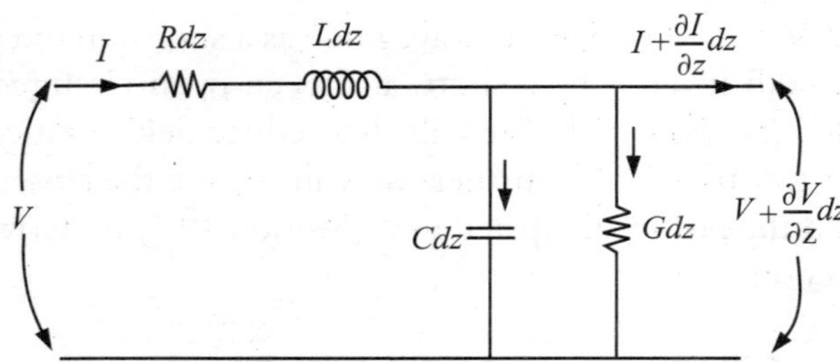

Fig. 10.1 *Representation of an infinitesimal length of an element of a transmission line by distributed line parameters.*

of the wire. Thus, while applying ac circuit theory to a transmission line, we take into consideration the continuous variation of V and I with z, and model the transmission line as a cascade of individual 'discrete' elemental line sections, each of infinitesimal length Δz, chosen small compared to the operating wavelength λ. By choosing $\Delta z \to 0$, we can obtain the results of the theory at all frequencies. This will allow us to treat individual elements of the line section of infinitesimal length Δz as lumped circuit elements and apply the laws of the circuit theory such as Kirchoff's laws for the analysis of the line.

Therefore, let us represent an element of length dz of a transmission line by discrete lumped parameters Rdz, Ldz, Cdz and Gdz as follows (Fig. 10.1):

1. Series resistance Rdz due to the finite conductivity of the conductor used in making the transmission line, accounting for ohmic power loss in the conductor;
2. Series inductance Ldz associated with the magnetic flux due to the current through the transmission line that links with the conductors making the line, which takes into account the energy stored in the magnetic field;
3. Shunt capacitance Cdz between the conductors of the transmission line because of the existence of an electric field between the conductors insulated by a dielectric medium;
4. Shunt conductance Gdz between the conductors of the transmission line, often referred to as leakage conductance or 'leakance', which manifests itself because of leakage current flowing between the conductors through the insulating dielectric if it is not perfect, thus accounting for the dielectric power loss in the insulator.

In the equivalent circuit representation of an element of length dz of a transmission line by the lumped parameters Rdz, Ldz, Cdz and Gdz shown in Fig. 10.1, it is implied that the four line parameters R, L, C and G are 'uniformly' distributed over the length of the line and are regarded as the 'primary line constants'. Further, because of the distributed nature, it is not possible to measure one line constant as a discrete parameter to the exclusion of the others without incurring an error. However, we can reduce the 'error of discretisation' by taking the element of line length $dz << \lambda$, thereby ensuring that the current through each element and the voltage across it can be regarded as practically constant over the line element even though they vary with z and t along the line. As mentioned earlier, the equivalent circuit representation of a transmission line of length dz by lumped parameters becomes more and more accurate as the length of the element is made smaller.

However, while applying the transmission line circuit theory to practical applications, one should be of aware of the domain of applicability of the simple theory developed in the present section. First, the present circuit theory is strictly valid only for axially uniform multi-conductor transmission lines that support TEM waves. Secondly, only if the line is lossless are the two successive cross-sections enclosing a segment not coupled, and on an infinite line one elementary segment indistinguishable from another. If there are losses in the line, the electric field must have a longitudinal component; then the successive line segments are not independent of each other. The derivation of the transmission line equations [(10.1) and (10.2) to follow] is strictly not valid for such cases. However, these equations are used with additional terms to account for the losses, and this is permissible provided the losses are small. Further, the analysis to follow in this section will consider the single-frequency, time-harmonic sources and signals. The limitation of the present analysis is that it cannot take into account the transients in situations where the sources and signals are not time harmonic and the steady-state conditions are not applicable, for instance, in computer networks processing digital signals or in power lines encountering sudden surges. Furthermore, we have also excluded from the purview of the present section the frequency dependence of the line parameters R, L, C and G, which is of practical relevance, for instance, in the analysis of microstrip lines. We have also ignored the effect of the return current generated by local fields and obviously excluded from the present simple theory the frequency dependence of the parameters of a transmission line with a ground return.

10.1.1 Telegrapher's equations of a transmission line

The potential at the output of an infinitesimal length dz of an element of a transmission line, which will change by an infinitesimal amount from the potential V at its input, can be represented by $V + (\partial V/\partial z)dz$ (Fig. 10.1). Therefore, the potential difference between the input and the output of the transmission line element can be written, estimating it as the sum of the potential drops across the series resistance and the series inductance Rdz and Ldz respectively as (Fig. 10.1):

$$V - \left(V + \frac{\partial V}{\partial z}dz\right) = I(Rdz) + (Ldz)\frac{\partial I}{\partial t},$$

where I is the current through the line, which can be simplified as

$$\frac{\partial V}{\partial z} = -\left(IR + L\frac{\partial I}{\partial t}\right). \tag{10.1}$$

We can now equate the current I that enters the input end of the line element with the sum of (i) the current that leaves its output end, which can be represented by $(I + \partial I/\partial z)dz$, (ii) the current that leaks through the shunt capacitance Cdz, which can be estimated as $\partial Q/\partial t = \partial(CdzV)/\partial t = (Cdz)(\partial V/\partial t)$, where $Q(= CdzV)$ is the charge on the capacitor element of capacitance Cdz, and (iii) the current $(Gdz)(V)$ that leaks through the shunt conductance Gdz (Fig. 10.1), to obtain (using Kirchoff's law)

$$I = I + \frac{\partial I}{\partial z}dz + Cdz\frac{\partial V}{\partial t} + GdzV,$$

which can be simplified as

$$\frac{\partial I}{\partial z} = -\left(GV + C\frac{\partial V}{\partial t}\right). \tag{10.2}$$

Taking the time dependence as $\exp(j\omega t)$, we can take $\partial/\partial t = j\omega$, enabling us to then express (10.1) and (10.2) respectively as

$$\frac{\partial V}{\partial z} = -(R + j\omega L)I \tag{10.3}$$

and

$$\frac{\partial I}{\partial z} = -(G + j\omega C)V. \tag{10.4}$$

In (10.1) and (10.2) or, alternatively, in (10.3) and (10.4), we obtain the so-called telegrapher's equations of a transmission line.

10.1.2 Circuit potential and current on a transmission line terminated in a load impedance

For a transmission line of a finite length that is terminated in a given load impedance, we can find an expression for the potential (voltage) $V(z)$ as a function of z, as well as that for the current $I(z)$ as a function of z, at a distance z from the input end of the line. For this purpose, let us first form a differential equation in V by taking the partial differentiation of (10.3) with respect to z and substituting therein (10.4) to obtain

$$\frac{\partial^2 V}{\partial^2 z} = -(R + j\omega L)\frac{\partial I}{\partial z} = (R + j\omega L)(G + j\omega C)V,$$

which can be expressed as

$$\frac{\partial^2 V}{\partial^2 z} = \gamma^2 V \tag{10.5}$$

where

$$\gamma^2 = (R + j\omega L)(G + j\omega C). \tag{10.6}$$

The differential equation (10.5) has the well-known solution

$$V = A\exp(-\gamma z) + B\exp(\gamma z) \tag{10.7}$$

which, invoking the time dependence $\exp(j\omega t)$ that is implied in the solution [see the statement preceding (10.3)] may be written as

$$V = [A\exp(-\gamma z) + B\exp\gamma z]\exp(j\omega t) \tag{10.8}$$

where A and B are constants and γ is called the propagation constant of the line. We can write the expression for the propagation constant γ of the transmission line by taking the square root of (10.6) as:

$$\gamma = [(R + j\omega L)(G + j\omega C)]^{1/2}. \tag{10.9}$$

It may be seen from (10.9) that γ is a complex quantity that can be expressed in terms of its real and imaginary parts, α and β respectively as

$$\gamma = \alpha + j\beta \tag{10.10}$$

so that now we can express the solution (10.8) with the help of (10.9) as[2]

$$V = A\exp(-\alpha z)\exp j(\omega t - \beta z) + B\exp(\alpha z)\exp j(\omega t + \beta z). \tag{10.11}$$

The phase parts, $\exp j(\omega t - \beta z)$ and $\exp j(\omega t + \beta z)$ of the first and the second terms of (10.11), refer respectively to the forward and the backward waves supported by the transmission line. β, which is the imaginary part of the propagation constant given by (10.10) and which can be seen to occur in the phase parts in (10.11), is called the phase propagation constant of the transmission line. You will appreciate from (10.11) that the amplitude of the forward wave $A\exp(-\alpha z)$ decreases or attenuates exponentially with the increase of z as the wave propagates forward, that is, along the positive z direction. Similarly, the amplitude of the backward wave $B\exp(\alpha z)$ also decreases or attenuates exponentially as the wave propagates backward along the negative z direction. α is called the attenuation constant of the wave. A forward wave can be launched on the line by a source at one end of the line called the input end of the line, while a backward wave is set up on the line due to reflection from load impedance connected to the other end of the line called the load end of the line.

In order to find the expressions for the attenuation constant α and phase propagation constant β, let us put (10.10) in (10.9) to write

$$\alpha + j\beta = [(R + j\omega L)(G + j\omega C)]^{1/2} \tag{10.12}$$

---

[2] Any linear superposition of solutions like (10.1) is also an acceptable solution. This leads us to conclude that discussing the details of a harmonic component of solution is all that is needed because a general solution can always be formed by taking a general linear superposition of such solutions.

which can also be expressed as

$$\alpha + j\beta = [(R^2 + \omega^2 L^2)^{1/2} \exp(j\theta_1)(G^2 + \omega^2 C^2)^{1/2} \exp(j\theta_2)]^{1/2}$$
$$= [(R^2 + \omega^2 L^2)(G^2 + \omega^2 C^2)]^{1/4} \exp j\left(\frac{\theta_1 + \theta_2}{2}\right) \qquad (10.13)$$

where

$$\left.\begin{aligned} \theta_1 &= \tan^{-1}\frac{\omega L}{R} \\ \theta_2 &= \tan^{-1}\frac{\omega C}{G} \end{aligned}\right\}.$$

Expanding the exponential term in the right-hand side of (10.13), we can then write

$$\alpha + j\beta = [(R^2 + \omega^2 L^2)(G^2 + \omega^2 C^2)]^{1/4}\left[\cos\left(\frac{\theta_1 + \theta_2}{2}\right) + j\sin\left(\frac{\theta_1 + \theta_2}{2}\right)\right]$$

whence equating the real and the imaginary parts we get

$$\alpha = [(R^2 + \omega^2 L^2)(G^2 + \omega^2 C^2)]^{1/4}\cos\left(\frac{\theta_1 + \theta_2}{2}\right) \qquad (10.14)$$

$$\beta = [(R^2 + \omega^2 L^2)(G^2 + \omega^2 C^2)]^{1/4}\sin\left(\frac{\theta_1 + \theta_2}{2}\right). \qquad (10.15)$$

Let us next find an expression for I. For this purpose, let us take the partial differentiation of (10.8) with respect to z and substitute it in (10.3) rearranged to obtain

$$I = -\frac{\dfrac{\partial V}{\partial z}}{R + j\omega L} = -\frac{-\gamma A\exp(j\omega t)\exp(-\gamma z) + \gamma B\exp(j\omega t)\exp(\gamma z)}{R + j\omega L} \qquad (10.16)$$

which simplifies to

$$I = \frac{\gamma}{R + j\omega L}[A\exp(-\gamma z) - B\exp(\gamma z)]\exp(j\omega t). \qquad (10.17)$$

Further, the expression for I takes the following form by substituting (10.9) in (10.17):

$$I = \frac{[A\exp(-\gamma z) - B\exp(\gamma z)]\exp(j\omega t)}{\left(\dfrac{R + j\omega L}{G + j\omega C}\right)^{1/2}},$$

which can be put as

$$I = \frac{A\exp(-\gamma z) - B\exp(\gamma z)}{Z_0}\exp(j\omega t) \qquad (10.18)$$

where

$$Z_0 = \left(\frac{R + j\omega L}{G + j\omega C}\right)^{1/2} \tag{10.19}$$

a quantity that has the dimension of impedance, called the characteristic impedance of the transmission line. The meaning of characteristic impedance will become clear after we develop later in this section the concept of the input impedance of the line (see Section 10.1.3).

10.1.3 Distortionless transmission line

The information or signal that we transmit through a transmission line is composed, in general, of a number of frequency components. For distortionless transmission, two conditions need to be satisfied:

(i) All the frequency components constituting the signal should simultaneously reach the same receiving point on the line, which also implies that the phase velocity of the wave ($= v_{ph} = \omega/\beta$) supported by the line should be constant with frequency, or in other words, the line should be 'dispersion-free' and

(ii) The attenuation constant of the line of the wave due to line losses, if any, should be the same for all the frequency components of the signal.

Let us examine if a lossless line ($R = G = 0$) is distortionless, satisfying the above two conditions (i) and (ii). Now, if we put $R = G = 0$ in the definitions of θ_1 and θ_2 given following (10.13), we get $\theta_1 = \theta_2 = \tan^{-1}\infty = \pi/2$. Therefore, putting $R = G = 0$ and $\theta_1 = \theta_2 = \pi/2$ and hence $\cos(\theta_1 + \theta_2)/2 = \cos\pi/2 = 0$ and $\sin(\theta_1 + \theta_2)/2 = \sin\pi/2 = 1$ for a lossless line, we obtain from (10.14) and (10.15) respectively,

$$\alpha = \omega\sqrt{LC}\cos\pi/2 = 0 \tag{10.20}$$

and

$$\beta = \omega\sqrt{LC}\sin\pi/2 = \omega\sqrt{LC} \tag{10.21}$$

the latter, in turn giving

$$v_{ph} = \frac{\omega}{\beta} = \frac{1}{\sqrt{LC}}. \tag{10.22}$$

Thus, for a lossless line we see from (10.20) and (10.22) that α and v_{ph} are both constant with frequency, satisfying the above distortionless conditions (i) and (ii) of the line. Hence we can infer that a lossless line is distortionless.

A line with losses ($R \neq 0$, $G \neq 0$) will not always satisfy the distortionless conditions (i) and (ii). We can appreciate this by examining expressions (10.14) and (10.15). However, it is a particular combination of the line parameters that obeys the relation

$$RC = LG \tag{10.23}$$

that needs to be examined with the help of (10.14) and (10.15) to see if it refers to the distortionless conditions (i) and (ii). In order to appreciate this, let us express (10.14) and (10.15) respectively as

$$\left.\begin{aligned} \alpha &= \sqrt{RG}\left[\left(1+\frac{\omega^2L^2}{R^2}\right)\left(1+\frac{\omega^2C^2}{G^2}\right)\right]^{1/4}\cos\left(\frac{\theta_1+\theta_2}{2}\right) \\ \beta &= \omega\sqrt{LC}\left[\left(1+\frac{R^2}{\omega^2L}\right)\left(1+\frac{G^2}{\omega^2C^2}\right)\right]^{1/4}\sin\left(\frac{\theta_1+\theta_2}{2}\right) \end{aligned}\right\}. \tag{10.24}$$

Now, from the relation (10.23), we can take $L/R = C/G$ and $R/L = G/C$ in the expressions for α and β given by (10.24) to write

$$\left.\begin{aligned} \alpha &= \sqrt{RG}\left(1+\frac{\omega^2L^2}{R^2}\right)^{1/2}\cos\left(\frac{\theta_1+\theta_2}{2}\right) \\ \beta &= \omega\sqrt{LC}\left(1+\frac{R^2}{\omega^2L}\right)^{1/2}\sin\left(\frac{\theta_1+\theta_2}{2}\right) \end{aligned}\right\}. \tag{10.25}$$

In view of the relation (10.23), we have from the definition of θ_1 and θ_2, following (10.13), the relations :

$$\left.\begin{aligned} \theta_1 = \theta_2 &= \tan^{-1}\left(\frac{\omega L}{R}\right) = \tan^{-1}\left(\frac{\omega C}{G}\right) \\ \tan\theta_1 &= \left(\frac{\omega L}{R}\right) \end{aligned}\right\} \quad (RC = LG) \tag{10.26}$$

from which we can also write

$$\left.\begin{aligned} \sin\theta_1 &= \frac{\omega L}{\sqrt{R^2+\omega^2L^2}} \\ \cos\theta_1 &= \frac{R}{\sqrt{R^2+\omega^2L^2}} \end{aligned}\right\}. \quad (RC = LG) \tag{10.27}$$

Making use of (10.26) and (10.27), we can then express (10.25) as

$$\begin{aligned} \alpha &= \sqrt{RG}\left(1+\frac{\omega^2L^2}{R^2}\right)^{1/2}\cos\left(\frac{\theta_1+\theta_2}{2}\right) = \sqrt{RG}\left(1+\frac{\omega^2L^2}{R^2}\right)^{1/2}\cos\theta_1 \\ &= \sqrt{RG}\left(1+\frac{\omega^2L^2}{R^2}\right)^{1/2}\frac{R}{\sqrt{R^2+\omega^2L^2}} = \sqrt{RG}\frac{\sqrt{R^2+\omega^2L^2}}{R}\frac{R}{\sqrt{R^2+\omega^2L^2}} \end{aligned}$$

which simplifies to

$$\alpha = \sqrt{RG}\ (RC = LG) \tag{10.28}$$

and

$$\beta = \omega\sqrt{LC}\left(1+\frac{R^2}{\omega^2 L}\right)^{1/2}\sin\left(\frac{\theta_1+\theta_2}{2}\right) = \omega\sqrt{LC}(1+\frac{R^2}{\omega^2 L})^{1/2}\sin\theta_1$$

$$= \omega\sqrt{LC}\left(1+\frac{R^2}{\omega^2 L}\right)^{1/2}\frac{\omega L}{\sqrt{R^2+\omega^2 L^2}} = \omega\sqrt{LC}\frac{\sqrt{R^2+\omega^2 L^2}}{\omega L}\frac{\omega L}{\sqrt{R^2+\omega^2 L^2}}$$

which simplifies to

$$\beta = \omega\sqrt{LC} \qquad (RC = LG) \tag{10.29}$$

whence we get

$$v_{\mathrm{ph}} = \frac{\omega}{\beta} = \frac{1}{\sqrt{LC}}. \qquad (RC = LG) \tag{10.30}$$

Thus, if the line parameters satisfy the relation $RC = LG$, then we see from (10.28) and (10.30) respectively that both α and v_{ph} are constant with frequency, thereby making the distortionless conditions of the line (i) and (ii) hold good. Hence we can infer that if the transmission line has losses, it can be made distortionless provided one can choose the line parameters to satisfy the relation $RC = LG$.

10.1.4 Reflection coefficient of the line

Let a transmission line of length l be terminated in load impedance Z_l (Fig. 10.2). If the location of the load is taken as $z = 0$ (output end or load end of the line), then the source becomes located at $z = -l$ (input end or source end of the line). Therefore, dividing (10.8) by (10.18), each interpreted at the load end $z = 0$, we obtain the load impedance Z_l as

$$Z_l = \left.\frac{V}{I}\right|_{z=0} = \frac{A+B}{\frac{A-B}{Z_0}} = \frac{A+B}{A-B}Z_0$$

whence we get

$$\frac{Z_l}{Z_0} = \frac{A+B}{A-B}. \tag{10.31}$$

The ratio B/A of the backward-wave to forward-wave voltages at the load end ($z = 0$) (see (10.11), putting therein $z = 0$) can be identified as the reflection coefficient Γ_l. We can then obtain by algebra, from (10.31), the following expression for the reflection coefficient $\Gamma_l (= B/A)$:

$$\Gamma_l = \frac{B}{A} = \frac{Z_l - Z_0}{Z_l + Z_0}. \tag{10.32}$$

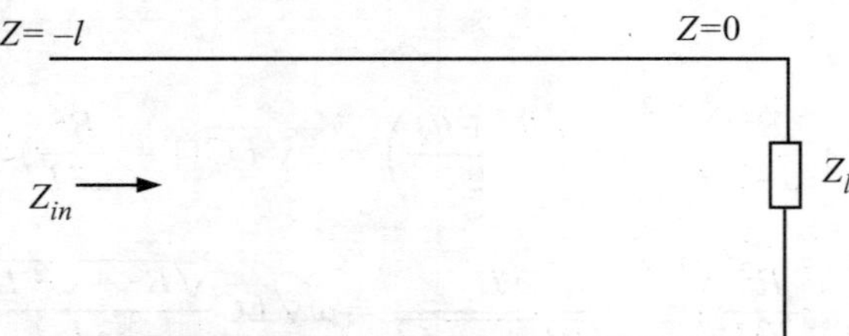

Fig. 10.2 *Representation of a transmission line of length l terminated in load impedance Z_l showing the output or load end at $z = 0$ and the input or source end at $z = -l$.*

10.1.5 Input impedance of the line

The input impedance Z_{in} of the line at $z = -l$ can be obtained by dividing (10.8) by (10.18) and putting therein $z = -l$ as

$$Z_{in} = \left.\frac{V}{I}\right|_{z=-l} = \frac{A\exp(-\gamma z) + B\exp(\gamma z)}{\dfrac{A\exp(-\gamma z) - B\exp(\gamma z)}{Z_0}} = Z_0\frac{A\exp(\gamma l) + B\exp(-\gamma l)}{A\exp(\gamma l) - B\exp(-\gamma l)}. \quad (10.33)$$

Dividing the numerator and the denominator of the right-hand side of (10.33) by A, we obtain

$$Z_{in} = Z_0\frac{\exp(\gamma l) + \dfrac{B}{A}\exp(-\gamma l)}{\exp(\gamma l) - \frac{B}{A}\exp(-\gamma l)},$$

in which we can substitute (10.32) to write

$$Z_{in} = Z_0\frac{\exp(\gamma l) + \dfrac{Z_l - Z_0}{Z_l + Z_0}\exp(-\gamma l)}{\exp(\gamma l) - \dfrac{Z_l - Z_0}{Z_l + Z_0}\exp(-\gamma l)}$$

$$= Z_0\frac{(Z_l + Z_0)\exp(\gamma l) + (Z_l - Z_0)\exp(-\gamma l)}{(Z_l + Z_0)\exp(\gamma l) - (Z_l - Z_0)\exp(-\gamma l)}. \quad (10.34)$$

We can then rearrange the terms of (10.34) to obtain

$$Z_{in} = Z_0\frac{Z_l[\exp(\gamma l) + \exp(-\gamma l)] + Z_0[\exp(\gamma l) - \exp(-\gamma l)]}{Z_l[\exp(\gamma l) - \exp(-\gamma l)] + Z_0[\exp(\gamma l) + \exp(-\gamma l)]} \quad (10.35)$$

which, in view of the relations

$$\left.\begin{aligned} \cosh(\gamma l) &= \frac{\exp(\gamma l) + \exp(-\gamma l)}{2} \\ \sinh(\gamma l) &= \frac{\exp(\gamma l) - \exp(-\gamma l)}{2} \end{aligned}\right\}$$

becomes

$$Z_{in} = Z_0 \frac{Z_l \cosh(\gamma l) + Z_0 \sinh(\gamma l)}{Z_l \sinh(\gamma l) + Z_0 \cosh(\gamma l)}.$$

An alternative form of the expression for the input impedance of the line is obtained by dividing the numerator and the denominator of (10.36) by $\cosh(\gamma l)$ and using the relation $\tanh(\gamma l) = \sinh(\gamma l)/\cosh(\gamma l)$ as follows:

$$Z_{in} = Z_0 \frac{Z_l + Z_0 \tanh(\gamma l)}{Z_l \tanh(\gamma l) + Z_0},$$

which after rearranging terms becomes

$$Z_{in} = Z_0 \frac{Z_l + Z_0 \tanh(\gamma l)}{Z_0 + Z_l \tanh(\gamma l)}. \tag{10.37}$$

For a lossless line, putting $R = G = 0$ in the expressions for θ_1, θ_2 given following (10.13), we get $\theta_1 = \theta_2 = \tan^{-1}\infty = \pi/2$, which, in turn, makes $(\theta_1 + \theta_2)/2 = \pi/2$ and which substituted in (10.14) gives $\alpha = 0$. Putting $\alpha = 0$ in (10.10), we get $\gamma = j\beta$, which when put in (10.37) gives the following expression for the input impedance of a line for a lossless line

$$Z_{in} = Z_0 \frac{Z_l + Z_0 \tanh j\beta l}{Z_0 + Z_l \tanh j\beta l},$$

which, invoking the relation $\tanh j\beta l = j\tan\beta l$ becomes

$$Z_{in} = Z_0 \frac{Z_l + jZ_0 \tan\beta l}{Z_0 + jZ_l \tan\beta l}. \quad \text{(lossless line)} \tag{10.38}$$

Putting $Z_l = 0$ in (10.38), we can write the expression for the input impedance of a lossless transmission line that is short-circuited at the load end as

$$Z_{in} = jZ_0 \tan\beta l. \quad \text{(lossless line short-circuited at the load end)} \tag{10.39}$$

Similarly, first dividing the numerator and the denominator of (10.38) by Z_l and then putting $Z_l = \infty$, we obtain the following expression for the input impedance of a lossless transmission line that is open-circuited at the load end:

$$Z_{in} = -jZ_0 \cot\beta l. \quad \text{(lossless line open-circuited at the load end)} \tag{10.40}$$

EXAMPLE E10-1

Show that the input impedance of a half-wave lossless transmission line is the same as the terminating load impedance of the line and hence find the thickness of a radome (used to protect the antenna of a radar), treated as a transmission line, made of a fibre glass of dielectric constant 4.9. Take the operating frequency as 3 GHz. Assume the incidence of uniform plane electromagnetic waves on the radome.

SOLUTION

The input impedance of a half-wave lossless transmission line Z_{in} of given characteristic impedance Z_0 and length $l(=\lambda/2)$ when it is terminated in a known impedance Z_l, can be written by putting $\beta = 2\pi/\lambda$ and $l = \lambda/2$ in (10.38) as

$$\begin{aligned} Z_{in} &= Z_0 \frac{Z_l + jZ_0 \tan \beta l}{Z_0 + jZ_l \tan \beta l} = Z_0 \frac{Z_l + jZ_0 \tan\left(\frac{2\pi}{\lambda}\frac{\lambda}{2}\right)}{Z_0 + jZ_l \tan\left(\frac{2\pi}{\lambda}\frac{\lambda}{2}\right)} \\ &= Z_0 \frac{Z_l + jZ_0 \tan \pi}{Z_0 + +jZ_l \tan \pi} = Z_0 \frac{Z_l}{Z_0} = Z_l \end{aligned} \quad \text{(E10-1.1)}$$

Thus, we see in (E10-1.1) that the input impedance of the half-wave line is the same as the load impedance. This concept can be used to find the thickness of the radome, which is a cover used for protecting a radar antenna from environmental disturbances. In the present context, if we take the thickness of the radome, treated as a transmission line of length $l = \lambda/2$, we will realise that $Z_{in} = Z_l = \eta_0$, the intrinsic impedance of free space, the input and the load ends of the radome both being free space. From the physical point of view, taking the thickness of the radome as $\lambda/2$ will ensure a path difference of the wave reflected from the radome as λ, thereby making it in anti-phase with the incident wave and hence causing cancellation of the incident and the reflected waves resulting in the matching of the radome to the incident wave. The wavelength λ in the radome material has to be calculated using the following relation:

$$\begin{aligned} \lambda &= \frac{2\pi}{\beta} = \frac{2\pi}{\omega\sqrt{\mu_0 \varepsilon}} = \frac{2\pi}{\omega\sqrt{\mu_0 \varepsilon_0 \varepsilon_r}} = \frac{2\pi}{2\pi f \sqrt{\mu_0 \varepsilon_0 \varepsilon_r}} \\ &= \frac{1}{f\sqrt{\varepsilon_r}} \frac{1}{\sqrt{\mu_0 \varepsilon_0}} = \frac{c}{f} \frac{1}{\sqrt{\varepsilon_r}} \end{aligned} \quad \text{(E10-1.2)}$$

where β is the phase propagation of the wave and ε_r is the relative permittivity (dielectric constant) of the radome material. Substituting $f = 3$ GHz$= 3 \times 10^9$ Hz, $c = 3 \times 10^8$ m/s, and $\varepsilon_r = 4.9$ in (E10-1.2), we calculate

$$\lambda = \frac{c}{f} \frac{1}{\sqrt{\varepsilon_r}} = \frac{3 \times 10^8}{3 \times 10^9} \frac{1}{\sqrt{4.9}} = 0.04517 \text{ m} = 45.17 \text{ mm},$$

whence the thickness of the radome comes out to be $\lambda/2 = 45.17/2 = 22.59$ mm.

EXAMPLE E10-2

A 3-cm long lossless transmission line of characteristic resistance 50 Ω which operates at

3.2 GHz is terminated in load impedance $Z_l = 25 - j15\ \Omega$. Calculate the input impedance of the line.

SOLUTION

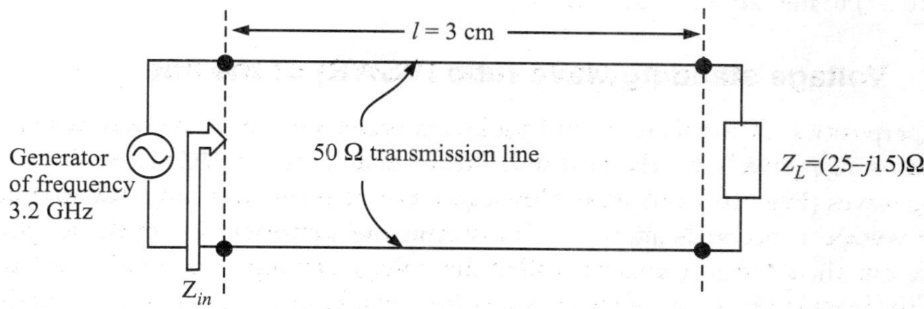

Fig. E10-2

The given quantities are: Z_0 =50 Ω; $Z_l = 25 - j15\ \Omega$; frequency $f = 3.2$ GHz= 3.2×10^9 Hz; and length $l = 3$ cm (Fig. E10-2). We can calculate the wavelength:

$$\left.\begin{aligned} \lambda &= c/f = (3 \times 10^{10})/(3.2 \times 10^9) = 9.375 \text{ cm} \\ \beta l &= (2\pi/\lambda)l = (2\pi/9.375)(3) = 2.01 \end{aligned}\right\}. \tag{E10-2.1}$$

Now, recall (10.38) as follows:

$$Z_{in} = Z_0 \frac{Z_l + jZ_0 \tan \beta l}{Z_0 + jZ_l \tan \beta l}$$

and substitute therein Z_0 =50 Ω (given); $Z_l = 25 - j15\ \Omega$ (given) and $\beta l = 2.01$ from (E10-2.1) to find the input impedance as

$$Z_{in} = 50 \frac{25 - j15 + j(50)(\tan 2.01)}{50 + j(25 - j15)(\tan 2.01)}\ \Omega. \tag{E10-2.2}$$

The separation of the real and imaginary parts of (E10-2.2) gives the value of the input impedance as: $Z_{in} = 109.5 - j13.6\ \Omega$.

10.1.6 Characteristic impedance of the line

The characteristic impedance of a transmission line mentioned following (10.19) can be identified as:

(i) The input impedance of the line of infinite length ($l = \infty$), which can be directly appreciated by putting large values of l in (10.33) that would make only the first term of each of the numerator and the denominator significant (the second term tending to zero for large values of l), thereby yielding from (10.33): $Z_{in} = Z_0$.

or

(ii) The terminating load impedance of the line that makes the input impedance of the line equal to the load impedance, which can be appreciated by putting $Z_l = Z_0$ in (10.37) to obtain: $Z_{in} = Z_l = Z_0$.

10.1.7 Voltage standing-wave ratio (VSWR) of the line

The superposition of the forward and backward waves on a transmission, which can be represented respectively by the first and the second terms of (10.11), will give rise to standing waves (Figs. 10.3 and 10.4). Consequent to such superposition, the magnitude of the line voltage V becomes alternately maximum and minimum down the length of the line. We can then define a quantity called the voltage standing-wave ratio VSWR as the ratio of the magnitude $|V_{\max}|$ of the circuit voltage maximum $V_{\max}$ to the magnitude $|V_{\min}|$ of the circuit voltage minimum $V_{\min}$ as

$$\text{VSWR} = |V_{\max}| / |V_{\min}|.$$

To make life simple, let us take the line to be lossless, for which we can put $\alpha = 0$ in (10.11) and hence interpret $\gamma = \alpha + j\beta = j\beta$ from (10.10) to write

$$V = [A\exp(-j\beta\, z) + B\exp(j\beta\, z)]\exp(j\omega t). \tag{10.41}$$

Dividing (10.41) by $A\exp(-j\beta\, z)$, we get

$$\frac{V}{A\exp(-j\beta z)} = \left[1 + \frac{B\exp(j\beta\, z)}{A\exp(-j\beta\, z)}\right]\exp(j\omega t),$$

which, by rearranging terms, takes the following normalised (dimensionless) form:

$$\frac{V}{A} = \left[1 + \frac{B}{A}\exp(j2\beta\, z)\right]\exp j(\omega t - \beta\, z). \tag{10.42}$$

If we define the load end of the line as the reference point $z = 0$, we can express the voltage at any point on the line at a distance d from the load end towards the source end, that is, at $z = -d$, with the help of (10.42) in the following normalised form:

$$\frac{V}{A} = [1 + \frac{B}{A}\exp(-j2\beta\, d)]\exp j(\omega t + \beta d). \tag{10.43}$$

In view of the relation $\Gamma_l = B/A$ [see (10.32)], we can express (10.43) in terms of the reflection coefficient Γ_l as

$$\frac{V}{A} = [1 + \Gamma_l\exp(-j2\beta\, d)]\exp j(\omega t + \beta d). \tag{10.44}$$

Expressing Γ_l occurring in (10.44) in terms of its magnitude $|\Gamma_l|$ and phase θ as $\Gamma_l = |\Gamma_l|\exp(j\theta)$, we can express (10.44) as

$$\frac{V}{A} = [1 + |\Gamma_l| \exp j(\theta - 2\beta\, d)] \exp j(\omega t + \beta d). \tag{10.45}$$

We can write from (10.45) the magnitude of the normalised voltage amplitude as

$$\frac{|V|}{A} = 1 + |\Gamma_l| \exp j(\theta - 2\beta\, d). \tag{10.46}$$

Note that the right-hand side of (10.46) is in general a complex quantity, and it assumes real values only for specific values of the argument of the exponential function. In fact, we are going discuss later at what values of the argument the right-hand side will take on maximum and minimum real values. Further, it is clear from (10.46) that the normalised voltage quantity $|V|/A$ at a point on the line varies with the position of the point on the line measured by the distance d of the point from the load end. We can elaborate on this further by representing $|V|/A$ in a vector diagram (Fig. 10.3) by a vector $\overrightarrow{AP}$, which is the sum of (i) the length vector $\overrightarrow{AO}$ of magnitude unity representing the first term 1 of (10.46), and (ii) another length vector $\overrightarrow{OP}$ of magnitude $|\Gamma_l|$ and phase $(\theta - 2\beta\, d)$ representing the second term of (10.46), namely $|\Gamma_l| \exp j(\theta - 2\beta\, d)$. The angular position of $\overrightarrow{OP}$ depends on the point on the line where the voltage is sought (Fig. 10.3). In order to find this angular position, we define a reference zero angle in line with AO (Fig. 10.3) and (i) first rotate the vector $\overrightarrow{OP}$ of length $|\Gamma_l|$ fixing its base on the tip O of the vector $\overrightarrow{AO}$ anticlockwise through an angle θ corresponding to the first term of the phase $(\theta - 2\beta\, d)$ around a circle to reach the point P_L representing the load position on the line and (ii) then rotate it clockwise through an angle $2\beta\, d$, corresponding to the second term $(-2\beta\, d)$ of the phase $(\theta - 2\beta\, d)$, around the same circle to reach the point P (Fig. 10.3). Thus, we get $\overrightarrow{AP} (= \overrightarrow{AO} + \overrightarrow{OP})$ to represent the quantity $1 + |\Gamma_l| \exp j(\theta - 2\beta\, d)$, it being the normalised voltage on the line at a distance d from the load end towards the source (Fig. 10.3).

The length of the vector $\overrightarrow{AP}$ takes on (i) the maximum value $\mathrm{AM} = 1 + |\Gamma_l|$ for $\theta - 2\beta\, d = 0, 2\pi, 4\pi, ...$; and (ii) the minimum value $\mathrm{AM}' = 1 - |\Gamma_l|$ for $\theta - 2\beta\, d = \pi, 3\pi, 5\pi, ...$ (Fig. 10.3) such that the distance Δd between two consecutive maxima and that between two consecutive minima each corresponds to $2\beta\Delta\, d = 2\pi$, which gives $\Delta\, d = \pi/\beta = \pi/(2\pi/\lambda) = \lambda/2$ in view of the relation $\beta = 2\pi/\lambda$, where λ is the wavelength of the wave supported by the transmission line (see Section 6.1).

Further, since $\overrightarrow{AP}$ is proportional to $|V|$, as can be seen from (10.45), the maximum and minimum values of $|V|$, namely $|V_{\max}|$ and $|V_{\min}|$ will be proportional to the maximum and minimum values of the length of the vector $\overrightarrow{AP}$, that is, to $\mathrm{AM} = 1+|\Gamma_l|$ and $\mathrm{AM}' = 1-|\Gamma_l|$ respectively. In other words, we get $|V_{\max}| \propto 1 + |\Gamma_l|$ and $|V_{\min}| \propto 1 - |\Gamma_l|$, and hence the voltage standing-wave ratio $\mathrm{VSWR} = |V_{\max}| / |V_{\min}|$ (defined earlier in this section) is

$$\mathrm{VSWR} = \frac{|V_{\max}|}{|V_{\min}|} = \frac{1 + |\Gamma_l|}{1 - |\Gamma_l|} \tag{10.47}$$

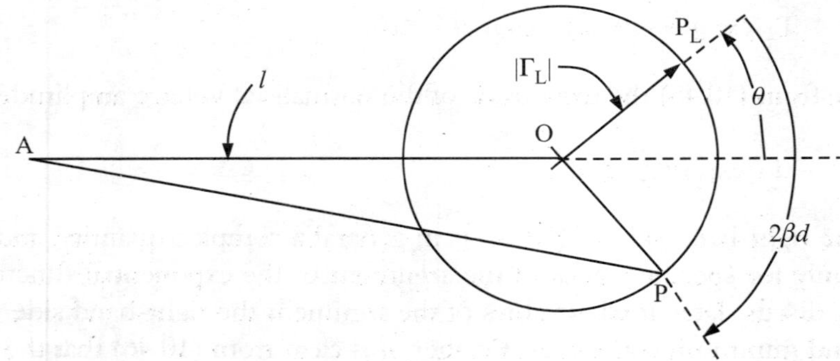

Fig. 10.3 *Vector diagram to represent the normalised voltage on a transmission line comprising a vector of a fixed length and another vector that can be rotated in a circular path with its base fixed on the tip of the vector of the fixed length.*

We can also appreciate this from the analytical expressions directly obtainable from (10.46) as

$$\frac{|V_{\max}|}{A} = 1 + |\Gamma_l| \; (\theta - 2\beta\, d = 0, 2\pi, 4\pi, ...) \tag{10.48}$$

and

$$\frac{|V_{\min}|}{A} = 1 - |\Gamma_l|. \; (\theta - 2\beta\, d = \pi, 3\pi, 5\pi, ...) \tag{10.49}$$

The division of (10.48) by (10.49) yields

$$\frac{|V_{\max}|}{|V_{\min}|} = \frac{1 + |\Gamma_l|}{1 - |\Gamma_l|}. \tag{10.50}$$

Note that (10.50) is identical to the expression for VSWR given by (10.47). Thus, we can find VSWR of the line from (10.47) if we know the magnitude $|\Gamma_l|$ of the reflection coefficient, $\Gamma_l = |\Gamma_l| \exp(j\theta)$, as defined preceding (10.45). And we can find Γ_l if we know the load impedance Z_l with the help of the expression: $\Gamma_l = (Z_l - Z_0)/(Z_l + Z_0)$ [see (10.32)].

We can also choose to have a definition of VSWR similar to (10.47) but in terms of the maximum and minimum values of $|I|$, namely, $|I_{\max}|$ and $|I_{\min}|$. For this purpose, let us take $\gamma = \alpha + j\beta = j\beta$, as explained preceding (10.41) for a lossless line, and substitute it in (10.18) to obtain the following expression for the current on the line

$$I = \frac{A \exp(-j\beta\, z) - B \exp(j\beta\, z)}{Z_0} \exp(j\omega\, t),$$

which can be rearranged as

$$\frac{I}{\dfrac{A}{Z_0}} = \left[1 - \frac{B}{A}\exp(j2\beta\, z)\right] \exp j(\omega\, t - \beta z). \qquad (10.51)$$

Since A has the dimension of voltage, as can be seen from (10.8), the denominator of the left-hand side of (10.51) has the dimension of current and therefore, in (10.51) we thus obtain an expression for a normalised current on the transmission line. Further, reckoning the load end as $z = 0$, as has been done preceding (10.43), we can then write from (10.51) the expression for the normalised current on the line at a point distant d from the load end towards the source end, that is, at $z = -d$ as follows:

$$\frac{I}{\dfrac{A}{Z_0}} = \left[1 - \frac{B}{A}\exp(-j2\beta\, d)\right] \exp j(\omega\, t + \beta d),$$

which, in view of the relation $\Gamma_l = B/A$ and $\Gamma_l = |\Gamma_l| \exp j\theta$ (as explained preceding (10.44) and (10.45) respectively) becomes

$$\frac{I}{\dfrac{A}{Z_0}} = [1 - |\Gamma_l| \exp j(\theta - 2\beta\, d)] \exp j(\omega\, t + \beta d). \qquad (10.52)$$

We can then write the amplitude of the normalised current from (10.52) as

$$\frac{|I|}{\dfrac{A}{Z_0}} = [1 - |\Gamma_l| \exp j(\theta - 2\beta\, d)]\,. \qquad (10.53)$$

From (10.53), we then obtain

$$\frac{|I_{\max}|}{\dfrac{A}{Z_0}} = 1 + |\Gamma_l|\; (\theta - 2\beta\, d = \pi, 3\pi, 5\pi, ...) \qquad (10.54)$$

and

$$\frac{|I_{\min}|}{\dfrac{A}{Z_0}} = 1 - |\Gamma_l|\,.\; (\theta - 2\beta\, d = 0, 2\pi, 4\pi, ...) \qquad (10.55)$$

Dividing (10.54) by (10.55), we get

$$\frac{|I_{\max}|}{|I_{\min}|} = \frac{1 + |\Gamma_l|}{1 - |\Gamma_l|}\,. \qquad (10.56)$$

Thus, we can summarise the findings (10.47), (10.50) and (10.56) as

$$\text{VSWR} = \frac{|V_{\max}|}{|V_{\min}|} = \frac{|I_{\max}|}{|I_{\min}|} = \frac{1 + |\Gamma_l|}{1 - |\Gamma_l|}\,. \qquad (10.57)$$

We can see from (10.48), (10.49), (10.54) and (10.55) that the conditions for the maxima and those for the minima in the standing-wave voltage or current patterns of a transmission line are essentially interchanged. Thus, you can see that the condition for the minimum current amplitude on the line: $\theta - 2\beta d = 0, 2\pi, 4\pi, ...$given in (10.55) is the same as the condition for the maximum voltage amplitude on the line given in (10.48). Similarly, the condition for the current maximum on the line: $\theta - 2\beta d = \pi, 3\pi, 5\pi, ...$ given in (10.54) is the same as the condition for the minimum voltage amplitude on the line given in (10.49). As a corollary we can then see that the conditions for the maximum and minimum 'impedance' amplitudes on the line will be the same as those for the maximum and minimum voltage amplitudes on the line respectively.[3]

We can also easily show that the impedance Z_{max} at the voltage maxima and the impedance Z_{min} at the voltage minimum are simply obtained by multiplying and dividing the characteristic impedance by the voltage standing-wave ratio:

$$Z_{max} = (Z_0)(\text{VSWR}) \text{ and}$$

$$Z_{min} = (Z_0)/(\text{VSWR})$$

respectively [Appendix A10.1; see (A10.1) and (A10.2)]. Also, we can see that the condition: $\theta - 2\beta d = 0, 2\pi, 4\pi, ...$ given by (10.48) for voltage amplitude maxima on the line (or by (10.55) for current amplitude minima on the line) is shifted by an amount π with respect to the condition: $\theta - 2\beta d = \pi, 3\pi, 5\pi, ...$ given by (10.49) for voltage amplitude minima (or by (10.54) for current amplitude maxima). Further, the positions of voltage amplitude maxima will coincide with those of current amplitude minima on the line and vice versa. We can also then appreciate from either of the conditions $\theta - 2\beta d = 0, 2\pi, 4\pi, ...$ or $\theta - 2\beta d = \pi, 3\pi, 5\pi, ...$ that the interval in the phase $\theta - 2\beta d$ between two consecutive voltage (or current) maxima and that between two consecutive voltage (or current) minima are each equal to 2π , which corresponds to $2\beta\Delta d = 2\pi$, giving $\Delta d = \lambda/2$ as the distance between two consecutive voltage or between two consecutive current amplitude maxima or minima on the line. This is also discussed earlier following (10.46) while explaining the vector diagram representing the voltage on a transmission line. Further, it may be noted that, while studying experimentally the standing-wave pattern of a transmission line, the voltage minima are brought within the purview of measurement rather than the voltage maxima as the former are sharper than the latter (see Fig. 10.4).

[3]The impedances at different locations on the line are different, which you can appreciate as follows. The conventional positive direction of the current is in the direction of propagation of the wave. [See the phase factor of the right-hand side of (10.51)]. So the positive sign of current changes according to the direction of propagation but the voltage does not change sign with the direction of propagation. Consequently, a forward wave and a backward wave add to the voltage in phase but they add to the current out of phase.

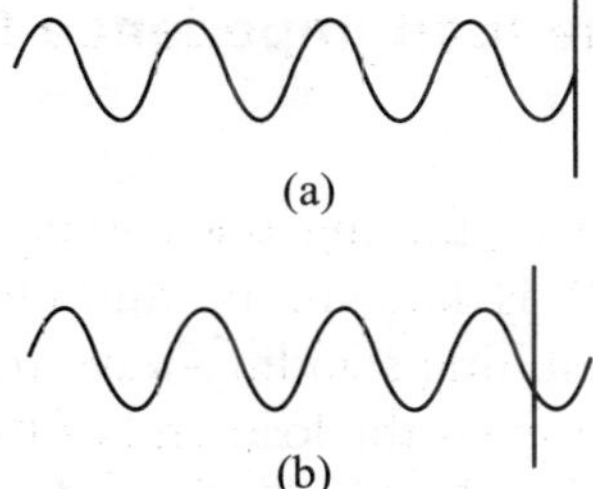

Fig. 10.4 *Standing-wave voltage maxima and minima on a transmission line for the situations in which the first extremum from the load end is (a) a minimum and (b) a maximum respectively.*

EXAMPLE E10-3

If a transmission line of characteristic resistance 50 Ω is terminated in complex impedance $25 + j100$ Ω, what will be the reflection coefficient and the VSWR of the line?

SOLUTION

The theory of the problem is dealt with in Sections 10.1.4, 10.1.7 and 10.1.8. Here, the given quantities are: $Z_l = 25 + j100\ \Omega$ and $Z_0 = 50\ \Omega$. Substituting these quantities in (10.32), we get

$$\Gamma_l = \frac{25 + j100 - 50}{25 + j100 + 50} = \frac{-25 + j100}{75 + j100}. \tag{E10-3.1}$$

We can separate the real and imaginary parts of Γ_l of (E10-3.1) by multiplying its numerator and denominator each by $75 - j100$ to obtain

$$\begin{aligned}\Gamma_l &= \frac{-25 + j100}{75 + j100} \times \frac{75 - j100}{75 - j100} \\ &= \frac{-25 \times 75 + j25 \times 100 + j100 \times 75 + 100 \times 100}{(75)^2 + (100)^2}.\end{aligned} \tag{E10-3.2}$$

We can simplify (E10-3.2) to obtain the required value of the reflection coefficient Γ_l as:

$$\Gamma_l = 0.8 + j0.36. \tag{E10-3.3}$$

From the value of Γ_l given by (E10-3.3), we can find its magnitude as:

$$|\Gamma_l| = \sqrt{(0.8)^2 + (0.36)^2} = 0.877,$$

which can be next substituted in (10.57) to obtain the required value of VSWR as:

$$\text{VSWR} = \frac{1 + |\Gamma_l|}{1 - |\Gamma_l|} = \frac{1 + 0.877}{1 - 0.877} = 15.3.$$

10.1.7.A Prediction of the load impedance from the standing-wave patterns

There is a well-known method of finding the complex load impedance of a lossless transmission line whose VSWR is known. In this method, the transmission line is terminated in a short and the resulting standing-wave pattern is noted. The spatial shift Δd of the first voltage minimum from the load end of the standing-wave pattern is then interpreted as the complex load impedance. The formula for the terminating complex load impedance of the line Z_l in terms of Δd (see Appendix A10.2 for its derivation) is:

$$\left.\begin{aligned} Z_l &= \frac{1+\Gamma_l}{1-\Gamma_l} Z_0 \\ \Gamma_l &= -\left(\frac{\text{VSWR}-1}{\text{VSWR}+1}\right)(\cos 2\beta\Delta d + j\sin 2\beta\Delta d) \end{aligned}\right\}. \qquad [(\text{A10.14}) \text{ rewritten}]$$

In this method, whether the first voltage amplitude extremum from the load end of the standing-wave pattern is a maximum or minimum when a complex load terminates the load end will depend on the value of the complex impedance. Also, whether the voltage standing-wave pattern will shift towards the load end or the source end on the length of the transmission line when the short—which is also one of the locations of the voltage minima—replaces the complex load is determined by the value of the complex impedance. Obviously, the distance of the first voltage minimum from the load end when the line is terminated in the complex load will be equal to Δd, the spatial shift of the voltage minimum of the standing-wave pattern. In fact, since it is somewhat difficult to exactly locate the load end and hence also the position of the first voltage minimum from the load end, we resort to finding the load impedance from the shift in the standing-wave pattern when the load is replaced by a short (Fig. 10.4). Further, in this formula for the load impedance Z_l which is rewritten above as (A10.14), Δd is interpreted as positive or negative according to whether the voltage standing-wave pattern shifts towards the load end (capacitive load) or source end (inductive load) as the short replaces the complex load at the terminating end. The method of finding the value of the complex impedance terminating a transmission line can also be implemented in the graphical method using the Smith chart (Section 10.1.8.H).

EXAMPLE E10-4

A transmission line of characteristic impedance 50 Ω terminated in a complex load impedance exhibits VSWR = 1.5. If the first extremum of the voltage standing-wave pattern of the line is a minimum that is located at a distance of 0.21λ from the load end, calculate (i) the value of the reflection coefficient and (ii) that of the load impedance.

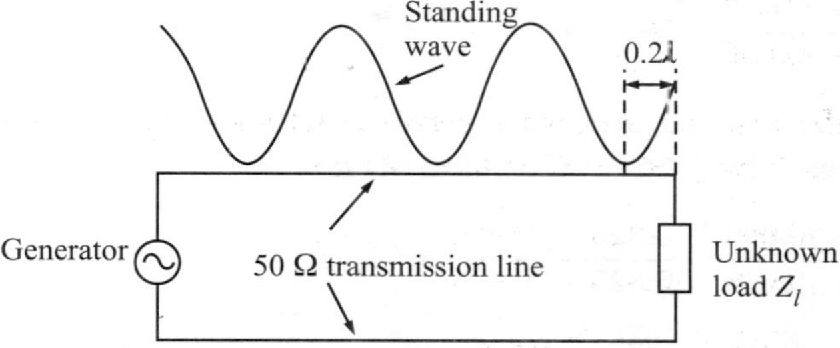

Fig. E10-4

SOLUTION

The theory of the problem is dealt with in Section 10.1.7.A with due reference to Appendix A10.2. In this problem, $Z_0 = 50\ \Omega$, VSWR $= 1.5$ and $\Delta d = 0.21\lambda$. Hence, let us apply (10.32). Further, since the first voltage amplitude extremum is a minimum from the load, it is implied, if the load were replaced by a short at the terminating end, then the standing-wave pattern would shift towards the load, thereby ensuring that a voltage minimum is formed at the terminating short end. From the location of the first minimum, we can take $\Delta d = 0.21\lambda$ and calculate the quantity (Fig. E10-4):

$$2\beta\Delta d = (2)(2\pi/\lambda)(0.21\lambda) = 0.84\pi,$$

which, in turn, can be used to calculate

$$\left.\begin{aligned}\cos 2\beta\Delta d &= \cos 0.84\pi = \cos 151.2^0 = -0.8763\\ \sin 2\beta\Delta d &= \sin 0.84\pi = \sin 151.2^0 = 0.4818\end{aligned}\right\}. \tag{E10-4.1}$$

Substituting the values of $\cos 2\beta\Delta d$ and $\sin 2\beta\Delta d$ from (E10-4.1) in the expression for Γ_l given in (A10.14), we obtain

$$\begin{aligned}\Gamma_l &= -\left(\frac{VSWR-1}{VSWR+1}\right)(\cos 2\beta\Delta d + j\sin 2\beta\Delta d)\\ &= -\left(\frac{1.5-1}{1.5+1}\right)(-0.8763 + j0.4818)\end{aligned} \tag{E10-4.2}$$

Simplifying (E10-4.2) we get the required value of the reflection coefficient:

$$\Gamma_l = 0.175 - j0.096. \tag{E10-4.3}$$

Putting now the value of Γ_l from (E10-4.3) as well as $Z_0 = 50$ (given) in the expression for Z_l obtainable from (10.32) (obtained as (A10.3) in Appendix A10.2), we can then calculate

$$Z_l = \frac{1+\Gamma_l}{1-\Gamma_l}Z_0 = \frac{1+(0.175-j0.096)}{1-(0.175-j0.096)} \times 50 = \frac{1.175-j0.096}{0.825+j0.096} \times 50$$

$$= \frac{1.175 - j0.096}{0.825 + j0.096} \times 50. \tag{E10-4.4}$$

We can separate the real and imaginary parts of (E10-4.4) by multiplying its numerator and denominator each by $(0.825 - j0.096)$ to obtain

$$Z_l = \frac{1.175 - j0.096}{0.825 + j0.096} \times \frac{0.825 - j0.096}{0.825 - j0.096} \times 50$$

$$= \frac{(1.175 - j0.096)(0.825 - j0.096)}{(0.825)^2 + (0.096)^2} \times 50\ \Omega,$$

which simplifies giving the required value

$$Z_l = \frac{0.9602 - j0.1920}{0.6898} \times 50 \doteq 69.6 - j13.9\ \Omega.$$

10.1.8 Smith chart

The Smith chart due to Phillip H. Smith, which is an elegant graphical tool for high frequency circuit applications, is popularly used to solve transmission line problems (Fig. 10.5). It charts complex impedances in coordinates defined by the complex reflection coefficient of a transmission line. The chart is similar to a slide rule used for carrying out calculations involving addition, subtraction, multiplication, division, logarithmic and trigonometric functions, etc. From the Smith chart, for instance, we can easily state the value of the VSWR of a transmission line if we know the value of the complex load impedance in which the line is terminated, it being implied that the line is lossless. Similarly, we can use the Smith chart to find easily the input impedance of a lossless transmission line of known characteristic impedance of a given length terminated in a given complex load impedance. An extended Smith chart required for studying a lossy transmission line characterised by complex characteristic impedance is kept outside the scope of the theory of the Smith chart developed in what follows. However, in a later section (Section 10.1.8.I), a method to estimate the attenuation constant of a lossy line using the conventional Smith chart for a lossless line is given.

10.1.8.A Reflection coefficient and the constant-$|\Gamma_l|$ circle

For a lossless transmission line ($R = G = 0$), the characteristic impedance of the line Z_0 becomes purely resistive and is thus to be taken as the characteristic resistance R_0 , which is given by (10.19) as $R_0 = (L/C)^{1/2}$. Let us start from (10.31), which can be rewritten in terms of $z_l = (Z_l/R_0)$, the complex load impedance normalised with respect to the characteristic resistance R_0 of the line, as follows:

$$z_l = \frac{Z_l}{R_0} = \frac{A + B}{A - B}. \tag{10.58}$$

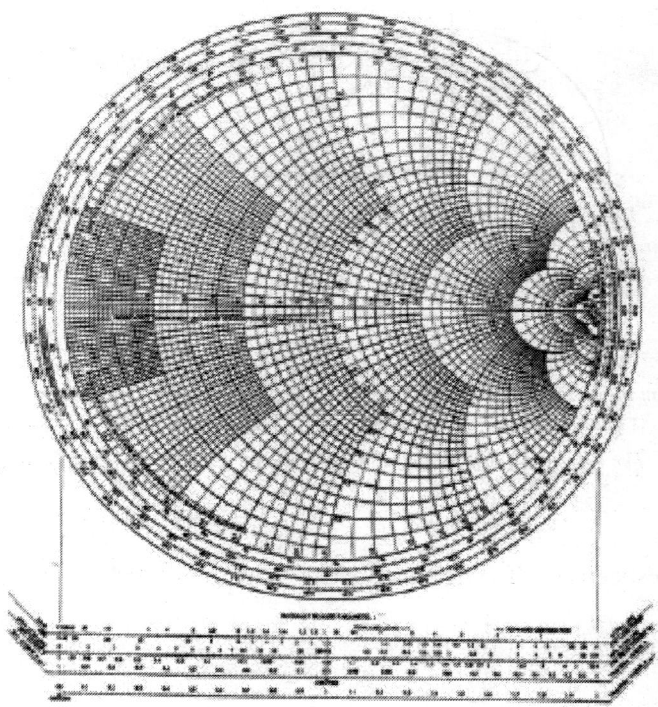

Fig. 10.5 *Smith chart (Source: Emeloid Company, Hillside, New Jersey).*

We can read (10.58) by dividing its numerator and denominator by A as

$$z_l = \frac{1 + (B/A)}{1 - (B/A)}. \tag{10.59}$$

Recall that while deriving (10.52) we had used the relation

$$\frac{B}{A} = \Gamma_l = |\Gamma_l| \exp j\theta. \tag{10.60}$$

Therefore, putting (10.60) in (10.59), we can obtain the following expression for the normalised load impedance:

$$z_l = \frac{1 + |\Gamma_l| \exp(j\theta)}{1 - |\Gamma_l| \exp(j\theta)} = r + jx, \text{ say} \tag{10.61}$$

where r and x are the real and imaginary parts of the complex load impedance z_l, to be interpreted here as the normalised load resistance (real part) and normalised load reactance

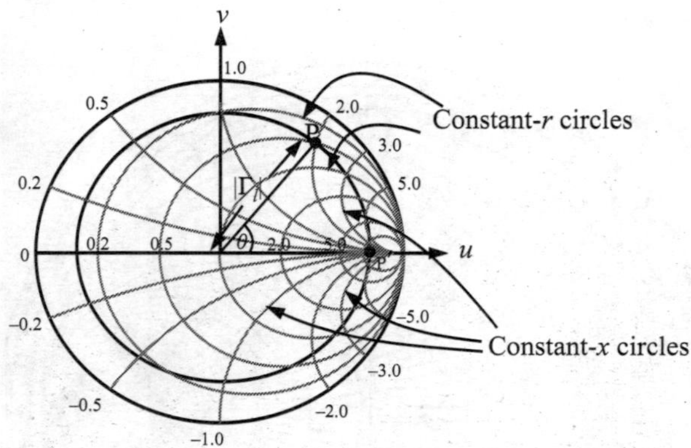

Fig. 10.6 *A point P on the reflection coefficient $\Gamma_l (= |\Gamma_l| \exp j\theta)$ plane of the Smith chart in the domain $|\Gamma_l| \leq 1$ indicating the magnitude $|\Gamma_l|$ represented by the length OP and the phase θ represented by the angle between OP and u axis (reference $\theta = 0$). The value of $|\Gamma_l|$ can be read on the horizontal scale provided at the bottom of the chart (Fig. 10.5).*

(imaginary part) respectively. Let us now choose to express (10.61) as

$$\frac{1 + |\Gamma_l| \exp(j\theta)}{1 - |\Gamma_l| \exp(j\theta)} = \frac{r + jx}{1}$$

from which by algebraic manipulation we can write

$$\frac{[1 + |\Gamma_l| \exp(j\theta)] - [1 - |\Gamma_l| \exp(j\theta)]}{[1 + |\Gamma_l| \exp(j\theta)] + [1 - |\Gamma_l| \exp(j\theta)]} = \frac{(r + jx) - 1}{(r + jx) + 1}$$

which simplifies to

$$|\Gamma_l| \exp j\theta = \frac{r + jx - 1}{r + jx + 1}. \tag{10.62}$$

Identifying the left-hand side of (10.62) as the complex reflection coefficient Γ_l as defined by (10.62), let us then express (10.60) separating its real and imaginary parts as

$$\Gamma_l = |\Gamma_l| \exp j\theta = \frac{r + jx - 1}{r + jx + 1} = u + jv. \tag{10.63}$$

On the Smith chart, we locate a point in the plot to represent the value of the reflection coefficient Γ_l in rectangular coordinates, taking the abscissa u to represent the real part and the ordinate v to represent the imaginary part of the reflection coefficient $\Gamma_l uv$ respectively (see (10.63) and Fig. 10.6). The plane on which this chart is generated may be thus called the reflection coefficient Γ_l plane or VSWR plane, since VSWR is directly related to Γ_l by the relation VSWR $= (1 + |\Gamma_l|)/(1 - |\Gamma_l|)$ [see (10.57)]. Alternatively, we can use polar coordinates to locate a point P on this chart (Fig. 10.6) such that the distance of the point from the origin O ($u = 0, v = 0$), the length OP, measures the magnitude $|\Gamma_l|$ (OP$= |\Gamma_l|$), while the angle between the real u-axis and the direction of the point from the origin measures the phase θ of the reflection coefficient $\Gamma_l = |\Gamma_l| \exp(j\theta)$. We can draw a circle (constant-$|\Gamma_l|$circle) centered at the origin O ($u = v = 0$) that passes through P to cut the positive u axis at a point P$'$ such that OP$'$ = OP = $|\Gamma_l|$ (Fig. 10.6). Thus, the length OP$'$($= |\Gamma_l|$), which could be read on a scale provided at the bottom of a commercial Smith chart (Fig. 10.5), measures the value of $|\Gamma_l|$. The angle θ can be read on a scale provided on the periphery of the chart. The constant-$|\Gamma_l|$ circles are not drawn on the 'bare' commercial Smith chart; they are to be drawn, for instance, as in Fig. 10.6, on the chart by its users for solving transmission line problems.

10.1.8.B Constant-r and constant-x circles

In order to find u and v each in terms of r and x , let us separate the real and imaginary parts of (10.62) by multiplying the numerator and denominator each of the left-hand side of it by $r + 1 - jx$ to obtain

$$\frac{r - 1 + jx}{r + 1 + jx} \times \frac{r + 1 - jx}{r + 1 - jx} = u + jv$$

which, after a little algebra can be simplified to the following form:

$$\frac{r^2 + r - jxr - r - 1 + jx + jxr + jx + x^2}{(r+1)^2 + x^2} = \frac{r^2 - 1 + 2jx + x^2}{(r+1)^2 + x^2}$$
$$= \frac{r^2 - 1 + x^2}{(r+1)^2 + x^2} + j\frac{2x}{(r+1)^2 + x^2} = u + jv$$

whence by equating its real and imaginary parts we obtain

$$u = \frac{r^2 - 1 + x^2}{(r+1)^2 + x^2} \tag{10.64}$$

and

$$v = \frac{2x}{(r+1)^2 + x^2}. \tag{10.65}$$

Similarly, we can choose to express r and x each in terms of u and v. For this purpose, let us write (10.63) as

$$\frac{u + jv}{1} = \frac{r + jx - 1}{r + jx + 1},$$

which by algebraic manipulation can be put as

$$\frac{1 + (u + jv)}{1 - (u + jv)} = \frac{(r + jx + 1) + (r + jx - 1)}{(r + jx + 1) - (r + jx - 1)}$$

which simplifies to

$$r + jx = \frac{1 + u + jv}{1 - u - jv}. \tag{10.66}$$

In order to separate the real and imaginary parts of the right-hand side of (10.66) with a view to getting the expressions for r and x, let us next multiply the numerator and denomination of the right-hand side of (10.66) by $1 - u + jv$ to obtain

$$r + jx = \frac{1 + u + jv}{1 - u - jv} \times \frac{1 - u + jv}{1 - u + jv},$$

which can be easily simplified by algebra to

$$r + jx = \frac{1 - u^2 - v^2 + 2jv}{(1 - u)^2 + v^2}. \tag{10.67}$$

The expressions for r and x are then obtained by equating the real and imaginary parts of (10.67) as follow:

$$r = \frac{1 - u^2 - v^2}{(1 - u)^2 + v^2} \tag{10.68}$$

and

$$x = \frac{2v}{(1 - u)^2 + v^2}. \tag{10.69}$$

With a little algebra, we can show that the expressions (10.68) and (10.69) represent the equations of the constant-r and constant-x circles respectively, by expressing them in the following form (algebraic steps detailed in Appendix A10.3):

$$\left(u - \frac{r}{r + 1}\right)^2 + v^2 = \frac{1}{(r + 1)^2} \quad \text{(constant-}r\text{ circle)} \tag{10.70}$$

$$(u - 1)^2 + \left(v - \frac{1}{x}\right)^2 = \frac{1}{x^2} \quad \text{(constant-}x\text{ circle)}\,. \tag{10.71}$$

The plot of v versus u for a constant value of r obtainable from (10.70) is a constant-r circle of the Smith chart. Thus, we can generate a number of constant-r circles of the Smith chart for different r values with the help of (10.70). Hence we can also find the values of the radius of the constant-r circles as well as the values of the intercepts of these circles on the real (u) and imaginary (v) axes (Table 10.1) (Figs. 10.5 and 10.6).

The radius of the constant-r circle is obtained by simply taking the square root of the right-hand side of (10.70). We can obtain the intercepts of the constant-r circle on the u and v axes, $u_{\text{intercept}}$ and $v_{\text{intercept}}$, simply by putting $v = 0$ and $u = 0$ respectively in (10.70), as follows (Table 10.1):

Table 10.1 *Constant-r circles (typical) showing their radii and intercepts on the reflection coefficient axes*

r	Radius $= \pm\frac{1}{r+1}$ (positive value)	$u_{\text{intercept}} = 1,\ \frac{r-1}{r+1}$ (corresponding to $v = 0$)	$v_{\text{intercept}} = \sqrt{\frac{1-r}{1+r}},\ -\sqrt{\frac{1-r}{1+r}}$ (corresponding to $u = 0$)
0	1	$1,\ -1$	$1,\ -1$
$\frac{1}{2}$	$\frac{2}{3}$	$1,\ -\frac{1}{3}$	$\frac{1}{\sqrt{3}},\ -\frac{1}{\sqrt{3}}$
1	$\frac{1}{2}$	$1,\ 0$	$0,\ 0$
$1\frac{1}{2}$	$\frac{2}{5}$	$1,\ \frac{1}{5}$	$j\sqrt{\frac{1}{5}}$
2	$\frac{1}{3}$	$1,\ \frac{1}{3}$	$j\sqrt{\frac{2}{3}},\ -j\sqrt{\frac{2}{3}}$ (both complex; no intercept)
∞	0	$1,\ 1$	$j,\ -j$ (both complex; no intercept)

$$\left.\begin{array}{l} \text{radius} = \pm\dfrac{1}{r+1} \\ u_{\text{intercept}} = 1,\ \dfrac{r-1}{r+1};\ v_{\text{intercept}} = \sqrt{\dfrac{1-r}{1+r}},\ -\sqrt{\dfrac{1-r}{1+r}} \end{array}\right\}.\ \text{(constant-}r\text{ circle)} \qquad (10.72)$$

The values of r of the constant-r circles are indicated at the points of intersection between the constant-r circles and the real u axis on the Smith chart (Figs. 10.5 and 10.7). Further, from (10.72), we can easily bring out the following information about the constant-r circles within the domain of the reflection coefficient $|\Gamma_l| \leq 1$ (Tables 10.1 and 10.2; Figs. 10.5 and 10.6).

1. The values of r of the constant-r circles are displayed on the real u axis of the chart at one of the two intercepts of each circle with the real u axis on the left-hand side; the other intercept on the right-hand side is the extreme end point of the chart irrespective of the circle.

2. The outermost constant-r circle is the constant $r = 0$ circle; it is the largest of the constant-r circles. All the points of relevance in the chart are located within this constant $r = 0$ circle. Further, the constant $r = 0$ circle coincides with the constant $|\Gamma_l| = 1$ circle.

3. The centre of the outermost constant-r circle (constant $r = 0$ circle; constant $|\Gamma_l| = 1$ circle) is the origin ($u = 0, v = 0$) of the v versus u plot; it is located on the real u axis at the point where the constant $r = 1$ circle intersects with the axis.

4. The outermost constant-r circle (constant $r = 0$ circle; constant $|\Gamma_l| = 1$ circle)

is of radius = 1 with intercepts ±1 on each of the real u and imaginary v axes ($u_{\text{intercept}} = \pm 1$; $v_{\text{intercept}} = \pm 1$).

5. All the constant-r circles meet the real u axis at a common point at the extreme right point of the real u axis corresponding to $u_{\text{intercept}} = 1$.
6. The constant $r = 0$ and $r = \infty$ circles pass through the extreme left and right points of the real u axis, respectively, the constant $r = \infty$ shrinking to a point.
7. The radius of the constant-r circle decreases with the increase of the value of r, in the limit $r \to \infty$, its value shrinking to zero at the extreme right point of the chart on the real u axis ($u_{\text{intercept}} = 1$) intersected by $r = \infty$ constant-r circle, the latter shrinking to a point.
8. The VSWR of the transmission line is the value of r of the constant-r circle (read on the real u axis) passing through the positive intercept of the constant-$|\Gamma_l|$ circle with the real u axis of the chart.
9. The values of the impedance maxima and minima are the values of r of the constant-r circles (read on the real u axis) intersecting with the extreme right and left intercepts of the constant-$|\Gamma_l|$ circle with the real u axis.
10. The extreme left and right points of the $|\Gamma_l| = 1$ circle (constant $r = 0$ circle) represent the short circuit and open circuits at the load end respectively.
11. The VSWR of the points on the chart representing the short and open circuits at the load end (see item (10) above) is infinity according to item (viii), noting that $|\Gamma_l| = 1$ holds good for each of these terminating loads.
12. The constant-r circles with negative r values are of no relevance to the Smith chart since they correspond to inadmissible values of both $u_{\text{intercept}}$ and $v_{\text{intercept}}$, each falling outside the domain of the reflection coefficient $|\Gamma_l| \le 1$.

Similarly, we can generate a constant-x circle of the Smith chart with the help of (10.71) by plotting v against u for a constant value of x. Thus, we find the values of the radius of the constant-x circles as well as the values of the intercepts of these circles on the real (u) and imaginary (v) axes (Table 10.2) (Figs. 10.5 and 10.6).

The square root of the right-hand side of (10.71) gives the radius of the constant-x circle. The intercepts of the constant-x circle on the u and v axes, $u_{\text{intercept}}$ and $v_{\text{intercept}}$, are obtained simply by putting $v = 0$ and $u = 0$ respectively in (10.70), enabling us to write (Table 10.2):

$$\left.\begin{aligned} &\text{radius} = \pm\frac{1}{x} \\ &u_{\text{intercept}} = 1;\ v_{\text{intercept}} = \sqrt{\frac{1}{x^2} - 1} + \frac{1}{x},\ -\sqrt{\frac{1}{x^2} - 1} + \frac{1}{x} \end{aligned}\right\} \text{(constant-x circle)} \qquad (10.73)$$

Table 10.2 *Constant-x circles (typical) showing their radii and intercepts on reflection coefficient axes*

x	Radius $= \pm\frac{1}{x}$ (positive value)	$u_{\text{intercept}} = 1$ (independent of x) (corresponding to $v = 0$)	$v_{\text{intercept}} = \sqrt{\frac{1}{x^2} - 1} + \frac{1}{x}, -\sqrt{\frac{1}{x^2} - 1} + \frac{1}{x}$ (corresponding to $u = 0$)
$-\infty$	0	1	$j, -j$ (both complex; no intercept)
-2	$\frac{1}{2}$	1	$j\frac{\sqrt{3}}{2} - \frac{1}{2}, -j\frac{\sqrt{3}}{2} - \frac{1}{2}$ (both complex; no intercept)
$-1\frac{1}{2}$	$\frac{2}{3}$	1	$j\frac{\sqrt{5}}{2} - \frac{2}{3}, -j\frac{\sqrt{5}}{2} - \frac{2}{3}$ (both complex; no intercept)
-1	1	1	$-1, \quad -1$
$-\frac{1}{2}$		1	$\sqrt{3} - 2$, $-2 - \sqrt{3}$ (< -1; inadmissible outside the domain $\lvert\Gamma_l\rvert \le 1$)
0	∞	1	∞(> 1; inadmissible outside the domain $\lvert\Gamma_l\rvert \le 1$), 0
$\frac{1}{2}$	2	1	$2 + \sqrt{3}$ (> 1; inadmissible outside the domain $\lvert\Gamma_l\rvert \le 1$), $2 - \sqrt{3}$
1	1	1	$1, \quad 1$
$1\frac{1}{2}$	$\frac{2}{3}$	1	$j\frac{\sqrt{5}}{2} + \frac{2}{3}, -j\frac{\sqrt{5}}{2} + \frac{2}{3}$ (both complex; no intercept)
2	$\frac{1}{2}$	1	$j\frac{\sqrt{3}}{2} + \frac{1}{2}, -j\frac{\sqrt{3}}{2} + \frac{1}{2}$ (both complex; no intercept)
∞	0	1	$j, -j$ (both complex; no intercept)

The x values of the constant-x circles are displayed on the Smith chart at the periphery of the outermost constant-r circle of the chart (Figs. 10.5 and 10.7). We then have the following information about the *constant-x circles within the domain of the reflection coefficient* $|\Gamma_l| \le 1$ obtainable from (10.73) (Table 10.2; Figs. 10.5 and 10.6).

1. The values of the constant-x circles are displayed near these circles at the periphery of the outermost ($r = 0$) constant-r circle.
2. The radius of the constant-x circle increases with the decrease of the value of x, becoming infinite for the constant $x = 0$ circle, making the radius develop into a straight line coinciding with the real u axis of the chart.
3. All the constant-x circles meet the real u axis of the chart at a common point which is the extreme right point of the real u axis ($u_{\text{intercept}} = 1$).
4. The radius of the constant-x circle decreases with the increase of the value of x, shrinking to zero for the constant $x = \infty$ circle at the extreme right point on the real u axis of the chart ($u_{\text{intercept}} = 1$). The constant $x = \infty$ circle, in the limit, shrinks to a point.
5. Those constant-x circles of relevance to the Smith chart correspond to the admissible values of $u_{\text{intercept}}$ and $v_{\text{intercept}}$ falling within the domain of the reflection coefficient $|\Gamma_l| \leq 1$. This in fact, contains only the family of 'arcs' rather than the full circles.
6. The constant-x circles for the positive and negative x values lie respectively on the positive and negative imaginary v regions of the chart.
7. Two constant-x circles of the Smith chart, which are assigned the same values of x but of opposite signs, are essentially the mirror images of each other across the real u axis of the chart.

10.1.8.C Location of load impedance on the chart

In order to locate the load impedance on the chart, normalise the impedance by dividing it by the characteristic resistance to obtain its real and imaginary parts, which are the values of the normalised load resistance and normalised load reactance r and x respectively. The load impedance is then located on the Smith chart as the point P, the intersection between the constant-r and constant-x circles corresponding to the values of the load resistance r and the load reactance x respectively. (See Fig. 10.6 in which the point P is shown typically as the point of intersection between the constant $r = 0.5$ and $x = 2.0$ circles.) Recall that while the values of r of the constant-r circle can be directly read on the u axis, we can read the values of x displayed on the periphery of the outermost constant-r circle.

Representation of purely resistive, purely inductive and purely capacitive, short-circuit, open-circuit, and matched loads

A purely resistive load is represented by a point on the constant $x = 0$ circle, or in other words on the real u axis, since, in the limit $x = 0$, the radius of the constant-x circle becomes a straight line coinciding with the real u axis. The value of r, the load resistance normalised with respect to the characteristic resistance, can be read on the real u axis at the point of intersection (A, for instance in Fig. 10.7) of the constant-r circle with the axis. A purely inductive load is represented on the outermost constant-r circle by a point (B, for instance in Fig. 10.7) that is intersected by the constant-x circle where x has a positive value representing the positive inductive load reactance normalised with respect to the characteristic resistance. Similarly, a purely capacitive load is represented on the outermost constant-r circle by a point (C, for instance in Fig. 10.7) that is intersected by the constant-x circle where x has a negative value representing the negative capacitive load reactance normalised with respect to the characteristic resistance.

The point representing a short circuit at the load end is the point of intersection between the constant $r = 0$ circle (outermost constant-r circle) and the constant $x = 0$ circle (the latter, in the limit $x = 0$ becoming a straight line coinciding with the real u axis). Thus the point representing a short circuit at the load end can be easily identified as the extreme left end point of the chart on the real u axis (P_{SC} in Fig. 10.7). Similarly, the point representing an open circuit at the load end is the point of intersection between the constant $r = \infty$ circle (the latter, in the limit $r = \infty$ shrinking to a point at the extreme right end point of the chart on the real u axis) and the constant $x = \infty$ circle (the latter also, in the limit $x = \infty$ shrinking to a point at the extreme right end point of the chart on the real u axis). Thus, the point representing an open circuit at the load end can easily be identified as the extreme right end point of the chart on the real u axis (P_{OC} in Fig. 10.7). Further, the matched load $Z_l = Z_0$ corresponds to $|\Gamma_l| = 0$ [see (10.32)], which can be identified as the origin 'O' ($u = 0$, $v = 0$) (Fig. 10.7) (see Section 10.1.8.A), the point where the constant-$|\Gamma_l|$ circle shrinks to zero, which can be identified as the point $r = 1$ on the real u axis.

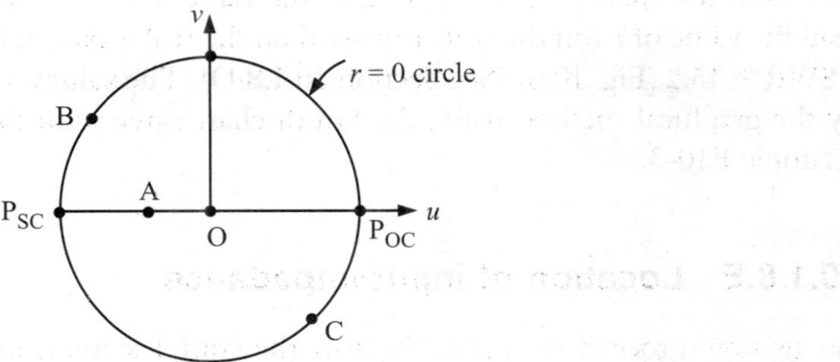

Fig. 10.7 *Typical points on the Smith chart representing purely resistive (A), purely inductive (B), purely capacitive (C), matched (O), short-circuited (P_{SC}), and open-circuited (P_{OC}) loads at the terminating end of the line.*

10.1.8.D Reading of VSWR

Immediately following (10.63), we had discussed how to represent a value of the reflection coefficient Γ_l in rectangular coordinates by a point P on the chart. We had also located point P′ on the positive u axis through which crosses a circle that is centered at the origin O ($u = v = 0$) and passes through the point P (constant-$|\Gamma_l|$ circle). Hence we identified $OP' = |\Gamma_l|$ (Fig. 10.6). At the point P′ on the positive u axis, let us then directly read the value of r of the constant-r circle and relate it with $u_{intercept} = OP' = |\Gamma_l|$. Thus with the help of (10.72), which gives $u_{intercept} = (r - 1)/(r + 1)$, and (10.57), which gives

VSWR= $(1+|\Gamma_l|)/(1-|\Gamma_l|)$, we obtain the following expression by simple algebra:

$$r = \frac{1+u_{\text{intercept}}}{1-u_{\text{intercept}}} = \frac{1+|\Gamma_l|}{1-|\Gamma_l|} = \text{VSWR}. \quad \text{(constant-r circle through P}')\qquad (10.74)$$

Thus, in view of (10.74), the value of VSWR can be read as the value of r at the positive intercept of the constant-$|\Gamma_l|$ circle with the real u axis.

EXAMPLE E10-5

Apply the Smith chart to find the reflection coefficient and the VSWR of the transmission line for solving the problem in Example E10-3.

SOLUTION

Divide $Z_l = 25 + j100\ \Omega$ by $Z_0 = 50\ \Omega$ to obtain the normalised the load impedance $z_l(= Z_l/Z_0) = (25 + j100)/50 = 0.5 + j2.0$. Locate the point P on the Smith chart as the point of intersection between the constant $r = 0.5$ and $x = 2.0$ circles (Fig. 10.6). Draw the constant-$|\Gamma_l|$ circle with the origin O ($u = 0$, $v = 0$) as the centre and locate the point of intersection P$'$ of this circle with the positive u axis (Fig. 10.6). Measure the length OP = OP$'$ by superposing it on the scale [supposed to be provided at the bottom of Fig. 10.6 as in the chart (Fig. 10.5)] to give the value $|\Gamma_l| \cong 0.87$ (see Section 10.1.8.A). Next, read the value of r (on the scale provided on the real u axis) at the point P$'$ to give the value VSWR $\cong$ 15.2 (Fig. 10.6) (See Section 10.1.8.D). The values of $|\Gamma_l|$ and VSWR so obtained by the graphical method using the Smith chart agree with those obtained analytically in Example E10-3.

10.1.8.E Location of input impedance

Let us now proceed to appreciate how the complex input impedance of a transmission line when it is terminated in a complex load can be represented on the Smith chart. For this purpose, let us recall (10.33) and interpret the latter, putting $\gamma = (\alpha + j\beta) = j\beta$ for a lossless line ($\alpha = 0$), as

$$z_{\text{in}} = \frac{Z_{\text{in}}}{Z_0} = \frac{A\exp(j\beta l) + B\exp(-j\beta l)}{A\exp(j\beta l) - B\exp(-j\beta l)}, \qquad (10.75)$$

where $z_{\text{in}} = (Z_{\text{in}}/Z_0)$ is the input impedance of the line of length l normalised with respect to the characteristic impedance. Dividing the numerator and denominator of the right-hand side of (10.75) by $A\exp(j\beta l)$, we then obtain

$$z_{\text{in}} = \frac{1 + \dfrac{B}{A}\exp(-2j\beta l)}{1 - \dfrac{B}{A}\exp(-2j\beta l)} = \frac{1 + \Gamma_l \exp(-2j\beta l)}{1 - \Gamma_l \exp(-2j\beta l)},$$

which, in view of (10.60), becomes

$$z_{\rm in} = \frac{1 + |\Gamma_l| \exp(j\theta) \exp(-j2\beta\, l)}{1 - |\Gamma_l| \exp(j\theta) \exp(-j2\beta\, l)} = \frac{1 + |\Gamma_l| \exp j(\theta - 2\beta\, l)}{1 - |\Gamma_l| \exp j(\theta - 2\beta\, l)}. \tag{10.76}$$

We can see that the expression (10.76) for the normalised input impedance $z_{\rm in}$ is identical to the expression (10.61) for the normalised load impedance z_l but for the phase factor $\exp(-j2\beta\, l)$. Therefore, as we could locate the normalised load impedance z_l on the Γ_l plane (Fig. 10.6), we can then locate the input impedance z_{in} on this plane as well, however, with due care to include this phase factor $\exp(-j2\beta l)$. Recall that we had located the reflection coefficient Γ_l at the point P on the chart at an angle θ measured 'anticlockwise' from the direction of the u axis ($\theta = 0$) around the circle centered at the origin O ($u = v = 0$)(constant-$|\Gamma_l|$ circle) that passes through P to take into account the phase factor $\exp(j\theta)$ in the reflection coefficient $\Gamma_l = |\Gamma_l| \exp(j\theta)$ (see (10.60) and (Fig. 10.6). We had also located the normalised load impedance z_l at the point P as the intersection between constant-r and constant-x circles, the values of r and x being the real and imaginary parts of z_l (Fig. 10.6). In order to take into account the additional phase factor $\exp -j2\beta\, l$ [see (10.72)], therefore, we have to locate the input impedance $z_{\rm in}$ at the point Q by moving 'clockwise' through an angle $2\beta\, l$ from the point P around the constant-$|\Gamma_l|$ circle centred at the origin O ($u = v = 0$) passing through P (Fig. 10.6). Thus, we can see that the movement of a point on the line 'towards the input or source or generator end' corresponds to the 'clockwise movement around the circle' and, correspondingly, the movement of a point on the line 'towards the load end' corresponds to the 'anticlockwise movement around the circle'. Note that in view of the relation $\beta = 2\pi/\lambda$, we can express the angle $2\beta\, l$ as $2\beta\, l = (4\pi/\lambda)l$. Further, the line length l can be expressed in terms of the number, p, of wavelengths as $l = p\lambda$, and therefore, $2\beta\, l = 4\pi p$. As mentioned earlier, in the commercially available Smith chart, an angle is displayed around the periphery of the outermost constant-r circle, allowing the angle $4\pi p$ to be read on the chart. One rotation through angle 2π around the circle corresponds to the distance $\lambda/2$ ($p = 1/2$) on the length of the transmission ($p = 1/2$; $2\beta\, l = 4\pi p = 2\pi$). The distance in wavelengths is indicated on the periphery of the outermost constant-r circle (constant $r = 0$ circle). The extreme left and right points of the circle on the real u axis of the chart respectively refer to 0 and 0.25 times the wavelength on the distance scale of the length of the transmission line, with the reference 0 being the same for both 'clockwise motion towards generator' and 'anticlockwise motion towards load'.

EXAMPLE E10-6

Apply the Smith chart to the problem in Example E10-2 to find the input impedance of the transmission line.

SOLUTION

The quantities with respect to the transmission line given are: characteristic impedance

$Z_0 = 50\ \Omega$; $Z_l = 25 - j15\ \Omega$; frequency $f = 3.2$ GHz $= 3.2 \times 10^9$ Hz; and length $l = 3$ cm. We can calculate the normalised load impedance as $z_l = Z/Z_0 = (25 - j15)/50 = 0.5 - j0.3$. Let us locate the point of intersection between the constant $r = 0.5$ and $x = -0.3$ circles (Fig. E10-6) as point A to represent z_l. Further, the wavelength in the present context can be calculated as $\lambda = 9.375$ cm [see Example E10-2: expression (E10-2.1)].

Therefore, the length of the line in number of wavelengths becomes $l/\lambda = 3/9.375 = 0.32$. Next, move 0.32 wavelengths clockwise towards generator around the constant-$|\Gamma_l|$ circle through the point A to reach the point B. (We can read the movement in wavelengths on scale provides around the periphery of the constant $r = 0$). Identify the constant-r and constant-x circles passing through the point B and note their constant values as $r = 2.2$ and $x = -0.3$. Thus, we find the normalised input impedance $z_{\text{in}}(= Z_{in}/Z_0) = 2.2 - j0.3$, and hence the input impedance $Z_{\text{in}} = (2.2 - j0.3)(Z_0) = (2.2 - j0.3)(50) = 110 - j15\ \Omega$, a value that is close to the value obtained analytically in Example E10-2.

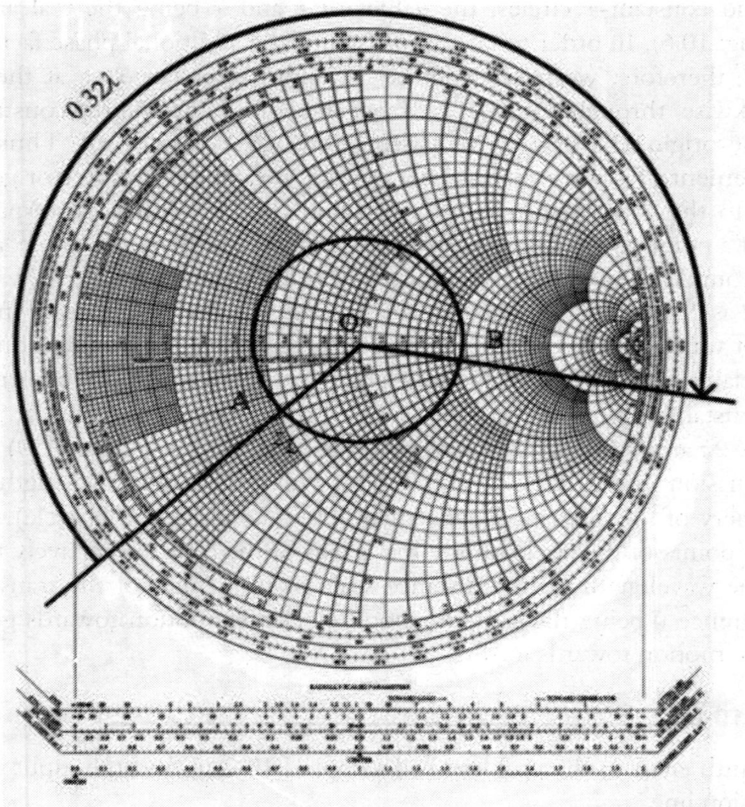

Fig. E10-6

10.1.8.F Location of admittance

It often becomes necessary to locate on the Smith chart the admittance instead of impedance of a point on a transmission line, for instance, in problems in which two or more lines are connected in parallel, for which the admittances of the parallel combination can be obtained simply by adding the admittances of the individual lines. For this purpose, let us, therefore, write the expression for the load admittance $Y_l = 1/Z_l$, normalised with respect to the characteristic admittance $Y_0 = 1/Z_0$ with the help of (10.61) as

$$y_l = \frac{Y_l}{Y_0} = \frac{Z_0}{Z_l} = \frac{1}{\frac{Z_l}{Z_0}} = \frac{1}{z_l} = \frac{1 - |\Gamma_l| \exp(j\theta)}{1 + |\Gamma_l| \exp(j\theta)}. \tag{10.77}$$

In view of the relation $\exp(j\pi) = \cos\pi + j\sin\pi = -1$, let us next choose to express (10.77) as

$$y_l = \frac{1 - |\Gamma_l| \exp(j\theta)}{1 + |\Gamma_l| \exp(j\theta)} = \frac{1 + |\Gamma_l| \exp(j\pi \exp j\theta)}{1 - |\Gamma_l| \exp(j\pi \exp j\theta)} = \frac{1 + |\Gamma_l| \exp j(\theta + \pi)}{1 - |\Gamma_l| \exp j(\theta + \pi)} = \frac{1 + |\Gamma_l| \exp(j\theta')}{1 - |\Gamma_l| \exp(j\theta')} \tag{10.78}$$

where $\theta' = \theta + \pi$. Thus, we can see that the expression (10.78) for normalised load admittance $y_l$is identical to the expression (10.61) for normalised load impedance except for the phase θ' of the reflection coefficient in the expression for y_l being different from the corresponding phase θ in the expression for z_l by an amount π. Therefore, on a constant-$|\Gamma_l|$ circle, a point representing the normalised load admittance y_l will be diametrically opposite to the point representing the normalised load impedance on the Smith chart. Thus, for instance, the extreme left and right end points of the constant-$|\Gamma_l| = 1$ circle on the real u axis in the impedance framework representing a short circuit and an open circuit at the load end respectively, will, in the admittance framework, represent an open circuit and a short circuit at the load end respectively. Correspondingly, the constant-r (normalised resistance) and constant-x (normalised reactance) circles intersecting a point representing the admittance at a point on the line will now have to be interpreted as the constant-g (normalised conductance) and constant-b (normalised susceptance) circles (g and b representing the real and imaginary parts of y_l respectively).

Similarly, it can also be seen that, on a constant-$|\Gamma_l|$ circle, a point representing the normalised input admittance y_{in} will be diametrically opposite to the point representing the normalised input impedance on the Smith chart. For instance, let us proceed to locate on the Smith chart the input admittance y_{in} normalised with respect to the characteristic admittance $Y_0 (= 1/Z_0)$ of a line of given characteristic impedance Z_0 open-circuited at the load end, with length expressed in number of wavelengths.

For this purpose, first locate the open-circuited load at the extreme right end point P_{OC} of the outermost constant-$|\Gamma_l| = 1$ circle (constant-$r = 0$ circle) on the real u axis of the chart in the impedance framework (Fig. 10.8). Moving round the periphery of the constant-$r = 0$ circle clockwise towards the generator through the number of wavelengths representing the given line length, we then reach point A, say, on the impedance chart (Fig. 10.8). This point represents the input impedance normalised with respect to the given characteristic impedance Z_0. Diametrically opposite to this point would be the point B representing the load admittance normalised with respect to the characteristic admittance $Y_0(=1/Z_0)$ (Fig. 10.8).

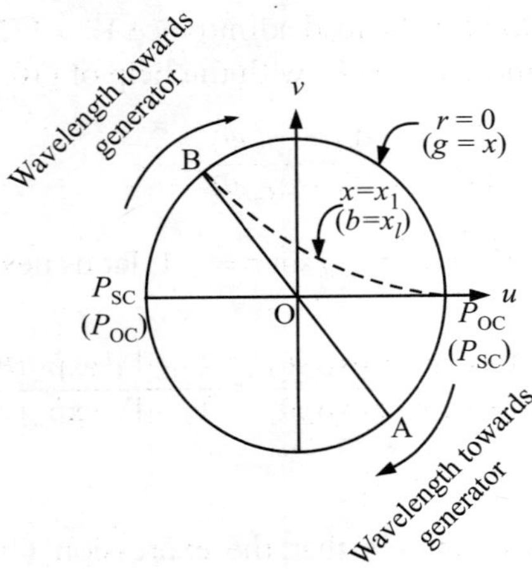

Fig. 10.8 *Smith chart applied to find the input admittance of a line open-circuited at the load end.*

Alternatively, we can locate at the extreme left end point of the constant-$r = 0$ circle on the real u axis of the chart in the admittance framework the point P_{OC} representing the open-circuited load (which is point P_{SC} in the impedance framework). Then, move from this point round the circle clockwise towards the generator through the number of wavelengths representing the line length to reach the same point as B to represent the load admittance normalised with respect to the characteristic admittance $Y_0(=1/Z_0)$. Note that the arc $P_{OC}A =$ the arc $P_{SC}B$ (Fig. 10.8). Then, as discussed above, we can find the value of the load admittance from the values of r and x of those constant-r and constant-x circles (interpreted as the constant-g and constant-b circles) which intersect with point B found by any one of the above two alternative ways. Thus, the value $r = 0$ of the constant-$r = 0$ circle intersecting with point B in the impedance framework is to be interpreted as $g = 0$ of the constant-g circle ($r = g = 0$). Similarly, the value $x = x_1$, say, of the constant-x circle read in the impedance framework intersecting with point B has to be interpreted as the value $b = b_1$ of the constant-b circle in the admittance framework (with the interpretation: $b_1 = x_1$) (Fig. 10.8). Thus we can find the input admittance Y_{in} normalised with respect to the characteristic admittance Y_0 as $y_{in}(= Y_{in}/Y_0)$ as

$$y_{in}(= Y_{in}/Y_0) = g + jb = 0 + jb_1 = 0 + jx_1 = jx_1$$

whence we obtain the input admittance Y_{in} as

$Y_{in} = y_{in} Y_0 = jx_1 Y_0 = j\frac{x_1}{Z_0}$.

10.1.8.G Impedance maximum and minimum in the standing-wave pattern

We have seen that when the transmission line is terminated in a load impedance other than the characteristic impedance, a standing wave pattern is generated on the line and that the points of voltage maxima and minima of the pattern are essentially the same as those of impedance maxima and minima (see Section 10.1.7). Consequently, the constant-$|\Gamma_l|$ circle for a given load impedance terminating the line will have positive and negative intercepts $u_{intercept}$ on the real u axis. The values of r of the constant-r circles passing through the positive and negative values of $u_{intercept}$, both directly read on the real u axis, will then give the values of the impedance maxima and minima respectively. We can verify that these values are the same as those obtained by multiplying and dividing the characteristic impedance respectively by VSWR of the line (see Section 10.1.7).

10.1.8.H Load impedance from the standing-wave patterns

We have shown how by analytical method we can find a complex load impedance from the distance Δd of the first voltage minimum from the load end when the line is terminated in the load or from the shift Δd in the standing-wave pattern when a short replaces the complex impedance at the terminating load end (Section 10.1.7A). The method involves a formula that has been derived by an involved analysis (developed in Appendix A10.1). However, the use of the Smith chart makes it simple to implement the method graphically (Fig. 10.9). The method, which assumes that the quantities Δd expressed in number of wavelengths, the characteristic resistance R_0 and the VSWR of the line (Section 10.1.1.A and Appendix A10.1) are known, involves the following simple steps to find the value of the terminating load impedance (Fig. 10.9).

(i) Locate the point P_{VSWR} to represent the VSWR of the line as the point of intersection of the constant-(r = VSWR) circle with the real u axis (see Section 10.1.8.D).

(ii) Draw the constant-$|\Gamma_l|$ circle passing through P_{VSWR}.

(iii) Locate point P to represent the point of impedance minimum as the extreme left end on the real u axis of the constant-$|\Gamma_l|$ circle passing through P_{VSWR} (see Section 10.1.8.G).

(iv) Locate point P′ on the outermost constant-r circle ($r = 0$) such that the arc length P_{SC} P′ measures the given value of Δd in number of wavelengths.

(v) Identify the point P as the point of intersection of the line OP′ with the constant-$|\Gamma_l|$ circle obtained in (ii) such that the angle of the arc PP becomes the same as the angle of the arc P_{SC} P′, each representing Δd expressed in number of wavelengths.

(vi) The point P, as obtained in (v), is identified as the point representing the terminating complex load.

(vi) Identify the constant-r and the constant-x circles which will intersect with the point P and note down the values of r and x of these circles. These will give the load resistance and reactance respectively, each normalised with respect to the characteristic resistance R_0 (Fig. 10.9). Multiplying these values of r and x each by the characteristic resistance R_0, we finally obtain the load impedance $(= (r + jx)R_0)$.

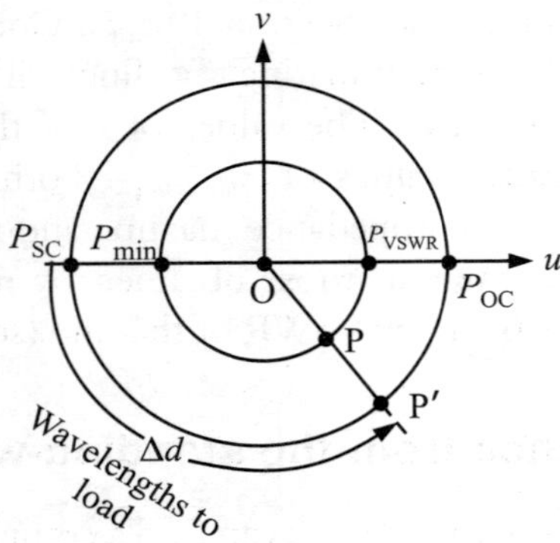

Fig. 10.9 *The use of the Smith chart to find the load impedance from the position of the first minimum of the standing-wave pattern or from the shift in the standing-wave pattern when the load is replaced by a short.*

10.1.8.I Attenuation constant of a lossy transmission line

The Smith chart is suitable for working out problems on lossless transmission lines. However, with care we can use it to estimate the attenuation constant of a lossy line as well by revisiting the concept of locating the input impedance on the Smith chart taking into account the effect of line losses. For this purpose, let us recall that had we started from (10.33) considering the line losses ($\alpha \neq 0$) as we did in Section 10.1.8.E, we would have obtained instead of (10.76) the expression for the input impedance z_{in} of length l in terms of the propagation constant $\gamma\ (= \alpha + j\beta)$ [see (10.10)] as follows:

$$z_{in} = \frac{1 + |\Gamma_l| \exp j(\theta - 2(\alpha + j\beta)\, l)}{1 - |\Gamma_l| \exp j(\theta - 2\beta\, l)} = \frac{1 + |\Gamma_l| \exp -2\alpha\, l \exp j(\theta - 2\beta l)}{1 - |\Gamma_l| \exp -2\alpha\, l \exp j(\theta - 2\beta l)},$$

which may be put as

$$z_{\text{in}} = \frac{1 + |\Gamma_l'| \exp j(\theta - 2\beta l)}{1 - |\Gamma_l'| \exp j(\theta - 2\beta l)} \tag{10.79}$$

where

$$|\Gamma_l'| = |\Gamma_l| \exp(-2\alpha\, l). \tag{10.80}$$

From (10.80), we can also write

$$\exp(2\alpha\, l) = \frac{|\Gamma_l|}{|\Gamma_l'|},$$

whence taking the logarithm of both its sides and by simple algebra we get

$$\alpha = \frac{1}{2l} \ln \frac{|\Gamma_l|}{|\Gamma_l'|}. \tag{10.81}$$

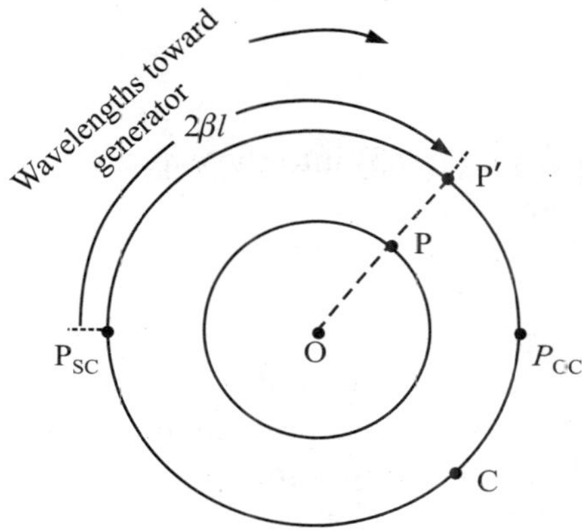

Fig. 10.10 *Use of the Smith chart to find the attenuation constant of a lossy transmission line.*

Therefore, it follows that, in view of (10.79)-(10.81), the method outlined in Section 10.1.8.E of finding the input impedance from the load impedance using the Smith chart for the lossless line needs to be modified by auxiliary calculation to estimate the value of the attenuation constant $\alpha$of a line with losses. In order to appreciate this modification, let us take up the problem of finding the value of α and that of the phase propagation constant β of a lossy line of known characteristic impedance and known length l for which the input impedance when it is short-circuited at the load end is known.

The input impedance of the 'lossless' transmission line that is short-circuited at the load end is purely reactive (see (10.39); Section 10.1.5) and can be located on the (outermost) constant-$r = 0$ or constant-$|\Gamma_l| = 1$ circle. Let us now represent the given input impedance of the 'lossy' line, normalised with respect to the given characteristic impedance, at point P and draw the constant-$|\Gamma_l|$ circle of radius OP $= |\Gamma_l'|$ passing through it, thus interpreting the circle as the constant-$|\Gamma_l'|$ circle (Fig. 10.10). Let us extend now the line OP to meet the outermost constant-$r = 0$ or constant-$|\Gamma_l| = 1$ circle at point P$'$, getting OP$' = |\Gamma_l| = 1$ (since P$'$ lies on the constant-$|\Gamma_l| = 1$ circle). Therefore, with the help of (10.81) we obtain

$$\alpha = \frac{1}{2l} \ln \frac{|\Gamma_l|}{|\Gamma_l'|} = \frac{1}{2l} \ln \frac{\mathrm{OP}'}{\mathrm{OP}} = \frac{1}{2l} \ln \frac{1}{\mathrm{OP}} \tag{10.82}$$

since $\mathrm{OP}' = |\Gamma_l| = 1$, the point P′ lying on the constant-$|\Gamma_l| = 1$ circle. We can read the value of OP (Fig. 10.10) by supersposing the length OP on the horizontal scale provided at the bottom of the Smith chart (see Fig. 10.5), and thus find the value of the attenuation constant of the line α in Np/m with the help of (10.82).

Further, from the arc length $\mathrm{P_{SC}P}' = p\lambda$ expressed in number p of wavelengths (Fig. 10.10), where the value of p that can be read from the scale provided on the periphery of the outermost constant-r circle of the chart (Fig. 10.5), we can find the distance of the input end P′ from the short-circuit termination of the line $\mathrm{P_{SC}}$, that is, the length l , as follows:

$$\frac{l}{\lambda} = p. \tag{10.83}$$

Therefore, putting $\lambda = l/p$ from (10.83) into the expression $\beta = 2\pi/\lambda$ for the axial phase propagation constant, we obtain

$$\beta = \frac{2\pi}{l}p. \tag{10.84}$$

Thus, knowing the value of p on the chart as discussed above and from the given value of l, we can find the value of the axial phase propagation constant of the line β in radian/m with the help of (10.84). Thus, we can find the values of both α and β of a lossy line with the help of (10.82) and (110.84) respectively.

10.1.9 Impedance transformer

A transmission line can be used as an impedance transformer since the input impedance of the line, in general, is different from its load impedance. One of the simplest examples is the quarter-wave transformer (Section 10.1.9.A) which has practical applications, for instance, in a radar duplexer. There is yet another example in the single-stub/double-stub transformer in the implementation of which the Smith chart is found to be very useful (Section 10.1.9.B).

10.1.9.A Quarter-wave transformer

You can make the so-called quarter-wave transformer out of a lossless transmission line by taking its length l equal to quarter of a wavelength: $l = \lambda/4$. The quarter-wave transformer can be used to match a resistive load with a lossless transmission line, it being implied that by such impedance transformation the line will be effectively connected to a load impedance Z_l equal to the characteristic impedance of the line Z_0 for which the reflection coefficient of the line Γ_l would become zero ($\Gamma_l = 0$) (see by putting $Z_l = Z_0$ in the expression (10.32) for Γ_l). For this purpose, one has to connect a section of a lossless transmission line (impedance transformer) of characteristic impedance Z_0', say, between the given transmission line of characteristic impedance $Z_0 = R_0$ and the given resistive load $Z_l (= R_l)$ such that the input impedance of the impedance transformer line Z'_{in} looking at the load end becomes equal to $Z_0(= R_0)$ $(Z'_{\text{in}} = Z_0)$. This effectively makes the reflection coefficient of the line with the transformer connected as $\Gamma_l = 0$ (see (10.32) where you have to put $Z_l = Z_0$ to appreciate this). With the help of (10.38), taking $l = \lambda/4$, we can then write

$$Z'_{\text{in}} = Z'_0 \frac{Z_l + jZ'_0 \tan \beta l}{Z'_0 + jZ_l \tan \beta l} = Z'_0 \frac{Z_l + jZ'_0 \tan \beta(\lambda/4)}{Z'_0 + jZ_l \tan \beta(\lambda/4)}$$

which, after dividing the numerator and denominator each of its right-hand side by $\tan \beta l$ becomes

$$Z'_{\text{in}} = Z'_0 \frac{\dfrac{Z_l}{\tan \beta(\lambda/4)} + jZ'_0}{\dfrac{Z'_0}{\tan \beta(\lambda/4)} + jZ_l}. \tag{10.85}$$

Putting $\beta = 2\pi/\lambda$ and hence $\tan \beta(\lambda/4) = \tan(2\pi/\lambda)(\lambda/4) = \tan(\pi/2) = \infty$ in (10.85), we have

$$Z'_{\text{in}} = Z'_0 \frac{\dfrac{Z_l}{\infty} + jZ'_0}{\dfrac{Z'_0}{\infty} + jZ_l} = Z'_0 \frac{Z'_0}{Z_l} = \frac{Z'^2_0}{Z_l}. \tag{10.86}$$

For a matched line, one has to realise $Z'_{in} = Z_0$, as explained already above, so that by putting $Z'_{in} = Z_0$ in (10.86) we then obtain

$$Z_0 = \frac{Z'^2_0}{Z_l}$$

whence we obtain the following expression for the characteristic impedance Z'_0 of the impedance transformer line:

$$Z'_0 = (Z_0 Z_l)^{1/2}. \tag{10.87}$$

Putting $Z_0 = R_0$ (characteristic resistance) for the lossless transmission line to be matched with a resistive load, as well as $Z_l = R_l$ for the resistive load in (10.86), we have the following relation:

$$Z'_0 = (R_0 R_l)^{1/2}. \tag{10.88}$$

Thus, in (10.88) we have the expression for the characteristic impedance (resistance) of a lossless quarter-wave ($\lambda/4$) impedance transmission line transformer. This can be connected between the resistive load and the lossless line for matching the load resistance and the characteristic resistance of the line (Fig. 10.11).

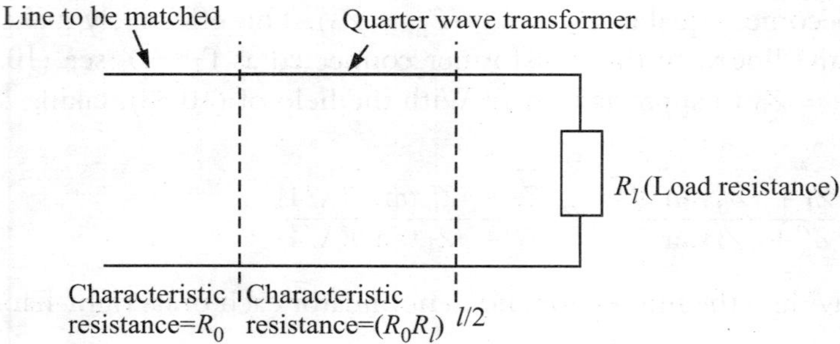

Fig. 10.11 *Quarter-wave impedance transformer for matching a resistive load with a lossless transmission line.*

Radar duplexer exemplifying the use of a quarter-wave transformer

Let us look at the use of a quarter-wave ($\lambda/4$) transmission line in a duplexer of radar (radio detection and ranging). In a radar system, the signal power is sent to a target in pulses and the echo signal is received between pulses. The duplexer in the radar permits the use of a single antenna in both transmitting and receiving modes of the radar. In the transmitting mode the duplexer allows the transmitter to send signal power to the antenna for radiation while protecting the receiver from the transmitted power and, in the receiving mode allows the signal power to be received by the receiver.

Let us consider the branch-type duplexer (Fig. 10.12) which uses the concept that the input impedance of a quarter-wave ($\lambda/4$) lossless transmission line terminated in a short circuit is 'infinity', and that terminated in an open circuit is 'zero'. This can be directly appreciated by putting $\beta = 2\pi/\lambda$ and $l = \lambda/4$ in (10.39) and(10.40) respectively. In this type of duplexer, the main transmission line is connected to the transmitter and the antenna at its two ends; it is also connected to two quarter-wave ($\lambda/4$) branch transmission lines terminated in the transmit–receive (TR) and the anti-transmit–receive (ATR) switches (typically, gas discharge types) respectively (Fig. 10.11). These switches are turned on (by gas ionisation discharge) when the radar sends signal power pulses and turned off (by gas deionisation) between pulses. The TR and the ATR switches primarily disconnect the receiver and the transmitter in the transmitting and receiving modes respectively.

In the transmitting mode, the turned-on TR switch short-circuits the transmitted power, thereby preventing it from entering the receiver connected to the branch line containing the TR switch (Fig. 10.12). Further, in the transmitting mode the $\lambda/4$ branch line with its terminating end shorted by the turned-on TR switch presents infinite impedance and hence an open circuit at the main transmission line, and thus does not impede power flow in the main transmission line to the antenna (Fig. 10.12). Similarly, the turned-on ATR switch also presents an open circuit at the main transmission line and has no effect on the latter. In the receiving mode, the TR and ATR switches are both turned off, thereby making both the $\lambda/4$ branch lines terminate in an open circuit. Therefore, in the receiving mode, the open-ended $\lambda/4$ branch line connected to the tuned-off ATR switch, that is, terminated in an open circuit will effectively short-circuit the main line which, in turn, would present infinite impedance at a distance $\lambda/4$ away on the main line where it is connected to the $\lambda/4$ branch line connected to the tuned-off TR switch. As a result, the transmitter will be effectively disconnected from the main line, and the signal power will be directed to the receiver through the lower impedance (short circuit) path presented by the $\lambda/4$ branch line terminated in the turned-off TR switch (open circuit) (Fig. 10.12). Thus, the branch type radar duplexer (Fig. 10.12) here exemplifies how the the concept of impedance transformation in a transmission line can be implemented in a practical application.

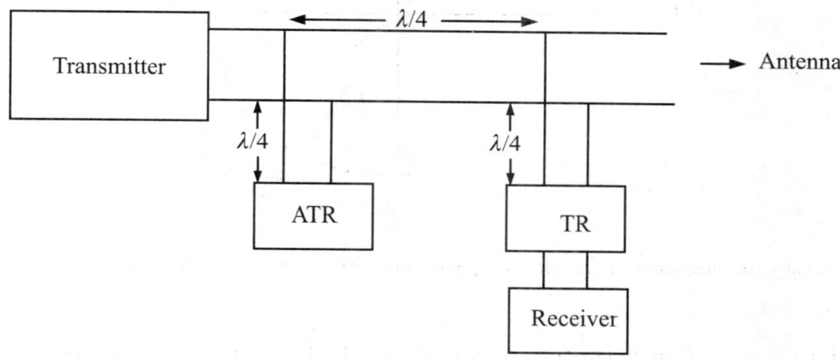

Fig. 10.12 *Branch-type radar duplexer.*

10.1.9.B Single-stub and double-stub impedance transformer

We have seen that a quarter-wave transformer is suitable for matching a resistive load (Section 10.1.9.A). However, when a complex load terminates a transmission line (main line), we can connect in parallel with it another transmission line which has the same characteristic resistance as the main line and which is terminated in a short circuit to form a single-stub matching network (Fig. 10.13*a*). The stub makes the main line effectively see a load resistance equal to the characteristic resistance and thus behave as a matched line. For the same cause, in another configuration, two such branch lines separated by a distance are used to form a double-stub matching network (Fig. 10.14*a*). It makes life simple to use the Smith chart (see Section 10.1.8) and design the single-stub (Fig. 10.13*b*) and double-stub (Fig. 10.14*b*) matching networks.

Single-stub matching

In order to implement single-stub matching (Fig. 10.13*a*), let us find the distance *d* on the main line from the complex load end at AA, where the real part of the line impedance is the characteristic resistance and the imaginary part, in general, a non-zero value representing a normalised reactance: $r = 1$, $x \neq 0$. In the admittance format (Section 10.1.8.F), this would

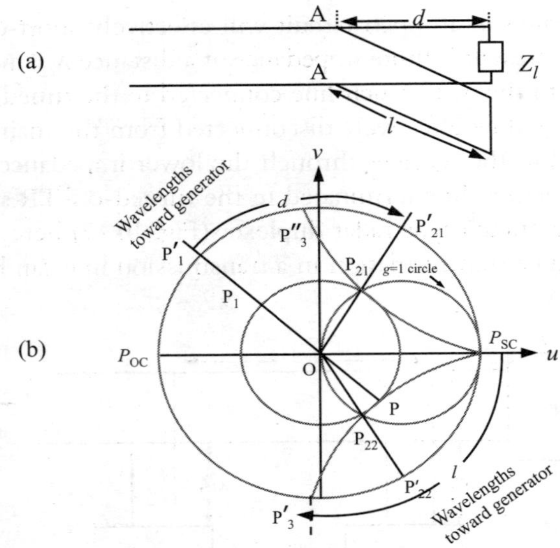

Fig. 10.13 *Single-stub matching: (a) circuit arrangement and (b) use of the Smith chart for the implementation of matching.*

correspond to $g = 1$, $b \neq 0$ (non-zero suseptance). Therefore, let the distance d be so adjusted as to make the normalised admittance at AA becomes equal $y = 1 + jb$ (Fig. 10.13 *a*). We may next connect the branch transmission line or stub terminated in a short circuit across AA (Fig. 10.13 *a*) which would provide at its input end a purely reactive impedance and hence a purely susceptive admittance, explicitly evident from the relation (10.39). More specifically, let us find the length l of the stub needed to provide a normalised susceptance equal to $y = -jb$ at its input at AA such that the parallel combination of the two normalised admittances, $y = 1 + jb$ and $y = -jb$ at AA provides an equivalent normalised admittance at AA equal to:$y = (1 + jb) + (-jb) = 1$, the value required for matching the line (Fig. 10.13*a*). However, it becomes somewhat complex by an analytical method to find the values of d and l to implement the above scheme. Alternatively, it easy to find the values of d and l graphically using the Smith chart in the following steps (Fig. 10.13*b*):

(i) Locate the point P on the chart to represent the complex load as the point of intersection between the constant-r and constant-x circles, where r and x refer to the normalised load impedance.

(ii) Draw the constant-$|\Gamma_l|$ circle of radius OP through point P, where O is the origin.

(iii) Locate a point P_1 that is diametrically opposite to P on the constant-$|\Gamma_l|$ circle obtained in (ii) and identify it as the point representing the load in the admittance chart (Section 10.1.8.F).

(iv) Draw a line from O through P_1 to meet it at point P'_1 on the outermost constant-$|\Gamma_l|$ or constant-$r = 0$ circle.

(v) The point representing the short circuit at the load end of a line is represented by the point P_{sc} at the extreme right end on the real u axis of the admittance chart (Section 10.1.8.F).

(vi) Draw the constant-$g = 1$ circle that would obviously pass through O and P_{sc}.

(vii) Identify the two points of intersection, P_{21} and P_{22}, between the constant-$|\Gamma_l|$ circle (which passes through P and P_1) and the constant-$g = 1$ circle.

(viii) Draw a line from O through P_{21} and P_{22} to meet the outermost constant-$|\Gamma_l|$ or constant-$r = 0$ circle at the points P'_{21} and P''_{22}, respectively.

(ix) Locate the points P'_3 and P''_3 on the outermost constant-$|\Gamma_l|$ or constant-$r = 0$ circle which are interested by the constant-x circles passing through P_{21} and P_{21}, respectively.

Now onwards, we have two alternative solutions to find the values of d and l—one starting from the points (a) P_1, P'_1, P_{21}, P'_{21}, P'_3 and P_{SC} and the other starting from (b) P_1, P'_1, P_{22}, P''_{22}, P''_3 and P_{SC} (Fig. 10.13*b*). If we choose to start from (a), let us move around the constant-$|\Gamma_l|$circle from P_1 to P_{21} clockwise and read the distance moved on the wavelength scale 'towards the generator' from P'_1 to P'_{21} (arc $P'_1P'_{21}$) to obtain the desired value of d. Thus, we obtain at AA the admittance $y = 1 + jb$ without the stub. The length l providing the admittance $y = -jb$ at AA due only to the short-circuited stub can be read on the wavelength scale 'towards the generator' from P_{SC} to P'_3 (arc $P_{SC}P'_3$). Therefore, the line now with the short-circuited stub will provide the desired value $y = (1 + jb) + (-jb) = 1$ for the matching of the line according to the conceived scheme. Similarly, we can implement the scheme starting from (b) as well (left to you as an exercise).

Double-stub matching

In the scheme of single-stub matching, the distance d of the stub from the load end (Fig. 10.13*a*) varies with change in the load impedance for a given frequency of operation. From the mechanical standpoint, this poses the difficulty of connecting the stub with the main line every time you change the load impedance, more so when a junction has to be constructed, for instance, by drilling a hole into a coaxial cable transmission line at some undesirable position of the line. The difficulty is circumvented in the scheme of the double-stub matching (Fig. 10.14*a*) in which we use two short-circuited stubs or branch lines connected to the main line keeping the distance d_0 between the branch lines or stubs the same irrespective of the load impedance to be matched. We are then supposed to adjust the lengths l_1 and l_2 of the short-circuited stubs, labelled as stub-1 and stub-2, connected at the terminals 1-1 and 2-2 respectively, according to the load to be matched. For the sake of simplicity, we consider here the situation in which the load and stub-1 are both connected across the terminal 1-1 (Fig. 10.14*a*), although one could have taken a general situation in which only stub-1 is connected to the terminal 1-1, while shifting stub-1 towards the source by a fixed distance from the load that is connected to the terminating end of the main line (figure not shown).

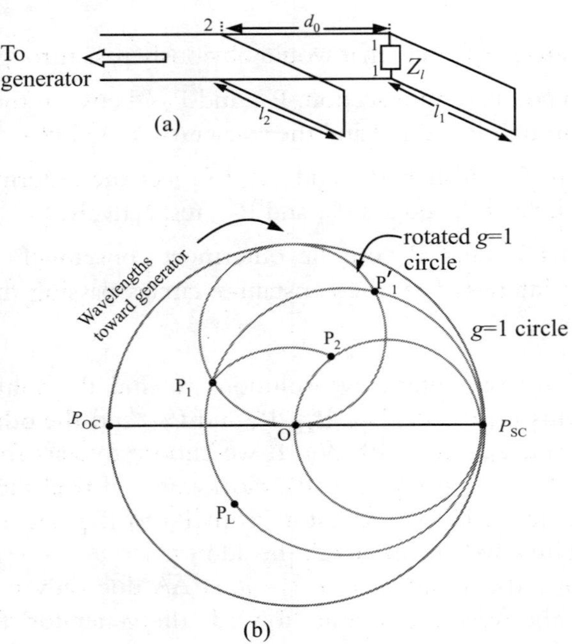

Fig. 10.14 *Double-stub matching: (a) circuit arrangement and (b) use of the Smith chart for the implementation of matching.*

Therefore, for the situation under study (Fig. 10.14*a*), in order to match the line, let us realise at the terminal 2-2 unity normalised admittance, that is, $1 = y_{s2} + y_2$, where y_{s2} is the input admittance of the short-circuited stub-2, which is purely susceptive, and y_2 is the admittance at the terminal 2-2 to be caused by the portion of the main line towards the load side along with the load connected to it. Let us design y_2 such that it obeys the relation $y_2 = 1 + jb_{P2}$, b_{P2} being the susceptive part. This relation, $y_2 = 1 + jb_{P2}$, when put in the unity admittance requirement: $1 = y_{s2} + y_2$ for an impedance-matched line, yields the condition $1 = y_{s2} + 1 + jb_{P2}$, which, in turn, gives the required input admittance of the short-circuited stub-2 alone as $y_{s2} = -jb_{P2}$. Notice that y_{s2} is purely susceptive, as expected of a short-circuited line. According to the design relation: $y_2 = 1 + jb_{P2}$, therefore, the point P_2 representing y_2 on the admittance chart would lie on the $g = 1$ circle (Fig. 10.14*b*). Next, let us translate this requirement at terminal 2-2 to that at terminal 1-1 'towards the load' on the 'rotated' $g = 1$ circle such that the rotation is 'anticlockwise towards the load', and the amount of rotation is $2\beta d_0$ (Fig. 10.14*b*). Taking the separation between stub-1 and stub-2 as $d_0 = \lambda/8$, typically, the required rotation of the $g = 1$ circle comes out to be $2\beta d_0 = 2(\beta)(\lambda/8) = 2(2\pi/\lambda)(\lambda/8) = \pi/2$. The rotated $g = 1$ circle is of relevance to locating the point P_1 representing the impedance y_1 at the terminal 1-1 (which is separated from the terminal 2-2 by the distance d_0). Thus, we can locate P_1 as one of the two points of intersection between the rotated $g = 1$ circle and the $g = g_l$ circle, where g_l is the conductive part of the normalised load admittance $y_l = g_l + jb_l$, b_l being the susceptive parts of y_l (Fig. 10.14*b*). Therefore, from the circuit standpoint, we may express y_1 as $y_1 = y_{s1} + y_l = y_{s1} + g_l + jb_l$ (substituting $y_l = g_l + jb_l$), where y_{s1} is the input admittance of the short-circuited stub-1, which is purely susceptive. Also, y_1 being represented by P_1 on the admittance chart, can also be expressed as $y_1 = g_l + jb_{p1}$, b_{P1} being the susceptive part, where g_l is the value of g of the constant-g circle passing through the points P_1 as well as P_L, the latter representing the load admittance $y_l (= g_l + jb_l)$. Therefore, equating these two expressions for y_1, we can write $(y_1 =)\, y_{s1} + g_l + jb_l = g_l + jb_{p1}$, whence by rearranging terms we obtain the input admittance of the short-circuited stub-1 as $y_{s1} = j(b_{p1} - b_l)$, which is purely susceptive as expected of a short-circuited line, as we had obtained above for the short-circuited stub-2 as $y_{s2} = -jb_{P2}$.

From the above obtained the expression for the short-circuited stub-1: $y_{s1} = j(b_{p1} - b_l)$ and that for the short-circuited stub-2: $y_{s2} = -jb_{P2}$, we can find the respective lengths l_1 and l_2 of these stubs in the following steps (Fig. 10.14*b*).

(i) Locate the point P_L representing the load admittance $y_l = g_l + jb_l$ as the point of intersection between the constant $g = g_l$ and the constant $b = b_l$ circles.

(ii) Draw the constant $g = 1$ circle.

(iii) Draw the constant $g = 1$ circle rotated by an angle $\pi/2$ 'anticlockwise' (towards the load).

(iv) Locate the point P_1 as one of the two points of intersection between the $g = g_l$ and the 'rotated' $g = 1$ circles.

(v) Taking the origin O ($r = 1$ on the impedance chart on the real u axis) as the centre, draw an arc clockwise (towards generator) staring from the point P_1 and end up at the point of intersection with the $g = 1$ circle. Identify this point of intersection as point P_2.

(vi) Read the value of b of the constant $b = b_l$ circle passing through the point P_L and the value of b of the constant b_{p1} circle passing through the point P_1 , thereby getting the values of b_l and b_{p1}, and hence the value of $b_{p1} - b_l$.

(vi) Identify the constant $b = b_{p1} - b_l$ circle.

(vii) Locate the point on the outermost constant-r circle ($r = 0$) (in the impedance framework) that is intersected by the constant $b = b_{p1} - b_l$ circle and read 'anticlockwise towards generator'(on the scale at periphery of the outermost constant-r circle) the distance of this point of intersection in terms of the number of wavelengths from the point P_{sc} (shown at the extreme end of the admittance chart) representing the short-circuit load of stub-1.

(viii) Locate the point on the outermost constant-r circle ($r = 0$) that is intersected by the constant $b = b_{p2}$ circle and read 'anticlockwise towards generator' the distance of this point of intersection in terms of the number of wavelengths from the point P_{sc}.

(ix) The distances as found in (vii) and (viii) are the lengths l_1 and l_2 of the short-circuited stub-1 and stub-2 respectively (Fig. 10.14*b*).

Note step (iv) in which the point P_1 was identified as one of the two points of intersection between the $g = g_l$ and the 'rotated' $g = 1$ circles. However, had we taken the other point of intersection, namely, P_1', instead of P_1 (Fig. 10.14*b*), we would have obtained the second set of the values of the lengths l_1 and l_2 of the short-circuited stub-1 and stub-2, respectively. (This may be taken as an interesting exercise to work out). Furthermore, you will notice that the rotated $g = 1$ circle does not intersect with the $g = g_l$ circle for the value of g_l beyond 2, which is a limitation of the double-stub matching method. This is thus to be treated as the 'forbidden' region, since in this region it is not possible to locate neither P_1 nor P_1' to proceed further after steps (iv) onwards for implementing double-stub matching.

10.2 Resonant Length and Resonant Frequency of a Waveguide Resonator

As mentioned in the beginning of this chapter, a waveguide which supports propagating wave, if properly designed with its both ends either closed by conducting metal walls to form a closed metal box, or kept open, makes a cavity resonator. Such a cavity resonator can be used as a tank circuit at microwave frequencies like their low-frequency counterparts to store electromagnetic energy, for instance, in a microwave oscillator. The device using a cavity resonator oscillates with greater amplitude at the resonant frequency of the resonator than at other frequencies and the frequency of oscillation of the oscillator is controlled by the resonant frequency which in turn depends on the dimension of the resonator.

We will see in this section how to use the transmission line theory (Section 10.1) to find the resonant length of the waveguide that makes it behave as a resonator at the resonant frequency. We take two situations: one in which the waveguide is closed by conducting walls and the other in which it is open at both its ends $z = 0$ and $z = l$ respectively.

For a closed-ended resonator (Fig. 10.15), we can use the relation (10.38) for the short-circuited line at the load end ($Z_l = 0$) corresponding to the waveguide closed by a conducting wall at $z = l$, in which, further, we have to take $Z_{in} = 0$ corresponding to the waveguide closed or short-ended by a conducting wall at $z = 0$. Therefore, putting $Z_{in} = 0$ in (10.39), we obtain

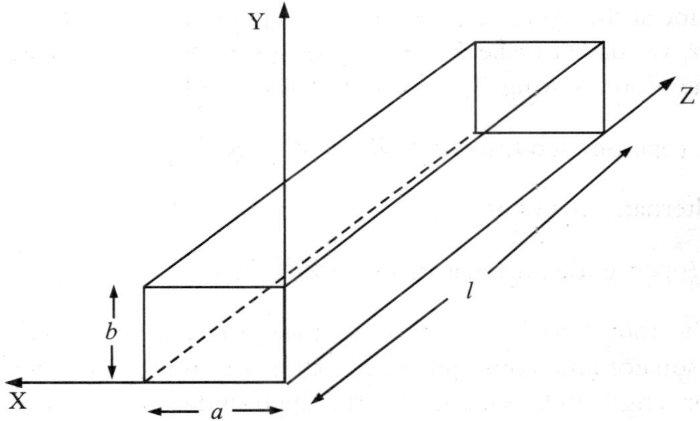

Fig. 10.15 *Closed-ended rectangular waveguide resonator* $(a \times b \times l)$, *which is a rectangular waveguide of broad dimension* a *and narrow dimension* b *and whose length is* l *and which has six conducting walls located at* $x = 0, a;\ y = 0, b;\ z = 0, l.$

$$\tan \beta l = 0. \quad \text{(closed-ended resonator: } Z_{in} = Z_l = 0) \tag{10.89}$$

In the context of the waveguide resonator, we have to interpret β as the propagation constant of the waveguide (Section 9.1). The length of the resonator that satisfies the trigonometric relation (10.8) is given by

$$\beta l = p\pi \quad (p = 1,\ 2,\ 3,\ \ldots). \tag{10.90}$$

Further, we may recall from (9.54) and (9.106) that for both the TE and the TM modes of excitation the same dispersion relation as follows holds good:

$$\omega^2 - \beta^2 c^2 - \omega_c^2 = 0 \quad \text{(relation (9.54) or (9.106) rewritten)} \tag{10.91}$$

where ω_c is the waveguide cutoff frequency. Therefore, substituting β from (10.90) as well as ω_c from (9.55) for a rectangular waveguide into (10.91), we obtain the expression for the length l of the resonator resonating at the frequency $\omega = \omega_r$, satisfying the condition $Z_{\text{in}} = Z_l = 0$ or, in other words, for the resonant frequency $\omega = \omega_r$ corresponding to the resonator length l, as follows:

$$\omega_r^2 - \left(\frac{p\pi}{l}\right)^2 c^2 - \left[\left(\frac{m\pi}{a}\right)^2 + \left(\frac{n\pi}{b}\right)^2\right] c^2 = 0$$

which may be rearranged as

$$\omega_r = \left[\left(\frac{m\pi}{a}\right)^2 + \left(\frac{n\pi}{b}\right)^2 + \left(\frac{p\pi}{l}\right)^2\right]^{1/2} c. \tag{10.92}$$

Similarly, for an open-ended resonator, we may use the relation (10.40) for the open-circuited line at the load end ($Z_l = \infty$) corresponding to the waveguide open-ended at $z = l$. Further, we have to take $Z_{\text{in}} = \infty$ corresponding to the waveguide open-ended also at $z = 0$. Therefore, putting $Z_{\text{in}} = \infty$ in (10.40), we obtain

$$\cot \beta l = \infty \text{ (open-ended resonator: } Z_{in} = Z_l = \infty),$$

which may be alternatively put as

$$\tan \beta l = 0 \text{ (open-ended resonator: } Z_{in} = Z_l = \infty) \tag{10.93}$$

a relation that is found to be identical to the corresponding relation (10.89) for a closed-ended resonator and, consequently, (10.90) continues to be valid as the condition for the resonator length to be satisfied by the open-ended resonator as well. Therefore, the same procedure as that used following (10.89) would lead us to the expression for the resonant frequency of an open-ended resonator identical to expression (10.92) for the closed-ended resonator.

EXAMPLE E10-7

A device (gyrotron) uses an open-ended cylindrical waveguide cavity in the fundamental axial mode. It operates close to the cutoff at a frequency ν% above the cutoff frequency ($\nu << 100$) of the cavity in the fundamental axial mode.

(i) Show that the length of the cavity l to be chosen for device operation is:

$$l = \frac{\lambda_c}{2\sqrt{(1 + \frac{\nu}{100})^2 - 1}},$$

where λ_c is the waveguide cutoff wavelength.

(ii) Calculate both the radius and the length of the cavity, taking the device to operate at a frequency 0.3% above the cutoff frequency. In one case, take the device operating frequency as 42 GHz for the waveguide mode TE_{03} , for which the eigenvalue is 10.173. In another case, take the device operating frequency as 94 GHz for the waveguide mode $TE_{10,4}$, for which the eigenvalue is 23.76.

SOLUTION

Let us rewrite (10.90) as

$$\beta l = p\pi \quad (p = 1,\ 2,\ 3,\ \ldots)$$

which for the fundamental mode ($p = 1$) (given) becomes

$$\beta l = \pi \tag{E10-7.1}$$

where l is the length of the resonator. Next, let us recall the waveguide-mode dispersion relation (10.91) as

$$\omega^2 - \beta^2 c^2 - \omega_c^2 = 0 \tag{E10-7.2}$$

where ω_c is the waveguide cutoff angular frequency.

If we substitute $\beta = \pi/l$ from (E10-7.1) in (E10-7.2), we obtain the following expression using the relations $\omega = 2\pi f$ and $\omega_c = 2\pi f_c$, where f_c is the waveguide cutoff circular frequency:

$$4\pi^2 f^2 - \frac{\pi^2}{l^2}c^2 - 4\pi^2 f_c^2 = 0. \tag{E10-7.3}$$

Since the device operates at a frequency ν% above the cutoff frequency we can write

$$f = f_c + \frac{\nu}{100} f_c = f_c\left(1 + \frac{\nu}{100}\right). \tag{E10-7.4}$$

Substituting (E10-7.4) in (E10-7.3), we then obtain the relation

$$\frac{1}{l^2}\frac{c^2}{f_c^2} = 4\left(1 + \frac{\nu}{100}\right)^2 - 4,$$

which, in view of the relation $f_c = c/\lambda_c$, where λ_c is the waveguide cutoff wavelength, becomes

$$\frac{1}{l^2}\lambda_c^2 = 4(1 + \frac{\nu}{100})^2 - 4. \tag{E10-7.5}$$

The length of the resonator l is then obtained by simplifying (E10-7.5) as

$$l = \frac{\lambda_c}{2\sqrt{(1 + \frac{\nu}{100})^2 - 1}}. \tag{E10-7.6}$$

Further, with the help of (9.171) we can write the waveguide radius a as

$$a = X_{mn} c/\omega_c \cong X_{mn} c/\omega \tag{E10-7.7}$$

taking $\omega \cong \omega_c$, where X_{mn} is the eigenvalue of the waveguide mode TE_{mn}.

In the given problem, with reference to 42-GHz device excited in the TE_{03}-mode ($m = 0$, $n = 3$), the operating frequency $f = 42$ GHz, cavity radius $a = 11.56$ mm and cavity length $l = 46.05$ mm (all given); and with reference to the 94-GHz excited in the $TE_{10,4}$-mode ($m = 10$, $n = 4$) gyrotron, the operating frequency $f = 94$ GHz gyrotron, cavity radius $a = 12.07$ mm and cavity length $l = 20.59$ mm (all given). Therefore, we can calculate the cavity length and radius with the help of (E10-7.6) and (E10-7.7) respectively and put the result of calculation as:

Frequency	Cavity mode	Eigenvalue	Cavity length	Cavity radius
42 GHz	$TE_{0,3}$	10.173	46.05 mm	11.56 mm
94 GHz	$TE_{10,4}$	23.76	20.59 mm	12.07 mm

10.3 Fields in a Waveguide Resonator

Let us take a waveguide resonator with its two ends at $z = 0$ and $z = l$ respectively, closed by conducting walls. To make life simple, take the dominant mode of the waveguide excited in the TE_{10} mode for which the field expressions are given by (9.111). However, here, we have to consider both forward and backward waves due to reflections at the two resonator conducting walls ($z = 0$ and $z = l$), which would form standing waves in the resonator. Therefore, considering forward and backward waves, we may express with the help of (9.111) the axial magnetic field component H_z as

$$H_z = [S\exp(-j\beta\, z) + S'\exp(j\beta\, z)]\cos\left(\frac{\pi}{a}x\right)\exp(j\omega\, t) \tag{10.94}$$

where S and S' are the field constants and the phase factors $\exp(-j\beta\, z)$ and $\exp(j\beta\, z)$ refer to the forward and the backward waves, respectively We will see later that S and S' can be related through the electromagnetic boundary condition of the problem. For this purpose, we may use the boundary condition that the tangential component of the electric field at the conducting boundary at $z = 0$ of the cavity (Fig. 10.15) is zero given by:

$$\vec{a}_n \times \vec{E}_2 = 0 \tag{10.95}$$

which is (7.46*d*) rewritten interpreting therein:

$$\left.\begin{aligned} \vec{a}_n &= \vec{a}_z \\ \vec{E}_2 &= E_y\vec{a}_y \end{aligned}\right\} \tag{10.96}$$

noting that, for the TE_{10} mode excited in the rectangular waveguide considered, the electric field comprises only the y-component (see (9.111) and the discussion preceding it). Substituting (10.96) in (10.95), we then have the boundary condition at $z = 0$:

$$\vec{a}_z \times E_y\vec{a}_y = 0 \ \ (z = 0)$$

which in view of the relation $\vec{a}_z \times \vec{a}_y = -\vec{a}_x$ may be read as

$$-\vec{a}_x E_y = 0 \ (z = 0)$$

whence we get

$$E_y|_{z=0} = 0. \tag{10.97}$$

Further, in view of the relation (9.26*a*), we may interpret (10.97) as

$$\left.\frac{\partial H_z}{\partial x}\right|_{z=0} = 0 \tag{10.98}$$

for all values of x in the resonator (Fig. 10.15). In order to interpret (10.98) further, let us take the partial derivative of (10.98) to write

$$\frac{\partial H_z}{\partial x} = -\frac{\pi}{a}\sin\frac{\pi}{a}x[S\exp(-j\beta\, z) + S'\exp(j\beta\, z)]$$

and evaluate it at $z = 0$ and substitute it into (10.98) to obtain

$$0 = S + S'. \tag{10.99}$$

Substituting $S'(= -S)$ from (10.99) in (10.94), we get

$$H_z = S[\exp(-j\beta\, z) - \exp(j\beta\, z)]\cos\left(\frac{\pi}{a}x\right)\exp(j\omega\, t),$$

which, using the trigonometric relation $\exp(\pm j\phi) = \cos\phi \pm j\sin\phi$, simplifies to

$$H_z = H_0\cos\left(\frac{\pi}{a}x\right)\sin\beta\, z\exp(j\omega\, t) \tag{10.100}$$

where $H_0 = -2jS$. Taking the partial derivative of (10.100) with respect to x and substituting it in (9.22), we obtain the following expression for E_y:

$$E_y = -\frac{\pi}{a}\frac{j\omega\mu_0}{k^2 - \beta^2}H_0\sin\left(\frac{\pi}{a}x\right)\sin\beta\, z\exp(j\omega\, t)$$

which, making use of the relation $k^2 - \beta^2 = (\pi/a)^2$ obtainable from (9.47) putting $m = 1, n = 0$ for the TE_{10} waveguide mode, takes the following form:

$$E_y = -\frac{j\omega\mu_0 a}{\pi}H_0\sin\left(\frac{\pi}{a}x\right)\sin\beta\, z\exp j\omega\, t. \tag{10.101}$$

Now, at the conducting boundary at $z = l$ of the cavity (Fig. 10.15), the boundary condition (10.98) has to be read taking $\vec{a}_n = -\vec{a}_z$, instead of $\vec{a}_n = \vec{a}_z$ [see (10.96)] and consequently we now obtain the condition corresponding to (10.97) as

$$E_y|_{z=l} = 0 \tag{10.102}$$

for all values of x. Therefore, putting $z = l$ in (10.101) and substituting it in (10.102) we obtain the condition

$$\sin\beta l = 0 \tag{10.103}$$

which is satisfied by

$$\beta\, l = p\pi \quad (p = 1,\ 2,\ 3,\ ...). \tag{10.104}$$

It is of interest to note that condition (10.104) derived in this section by field analysis becomes identical to condition (10.105) which was obtained earlier by the transmission line theory. Now, putting $\beta = 2\pi/\lambda_g$ from (9.56) in (10.104), we can then express the length of the resonator l in terms of the guide wavelength λ_g as follows:

$$l = \frac{p\pi}{\beta} = p\frac{\lambda_g}{2} \quad (p = 1,\ 2,\ 3,\ \ldots.). \tag{10.105}$$

This suggests that a short-ended waveguide has to be assigned a length equal to an integral number of half the guide wavelength $\lambda_g/2$ for the guide to resonate as the waveguide resonator. We note that the length of an open-ended waveguide resonator will also obey the relation (10.105) for the resonator length for a closed-ended waveguide resonator since the condition (10.104) (for a closed-ended waveguide resonator) from which the relation (10.105) has been obtained is valid for an open-ended resonator as well [see the discussion following (10.93)]. It is also worth noting at this juncture that though these two types of resonators (closed-ended and open-ended) enjoy identical relation (10.105) for the resonator length, they have different kinds of applications. For instance, a closed-ended resonator is used in an instrument called the 'wavemetre' for measuring microwave frequency and an open-ended resonator is used in a vacuum electron device such as the gyrotron for millimetre-wave frequency generation.

Now, substituting $\beta = p\pi/l$ from (10.104) in (10.100), we obtain the following expression for H_z in the resonator excited in the $\mathrm{TE}_{10\mathrm{p}}$ mode:

$$H_z = H_0 \cos\left(\frac{\pi}{a}x\right) \sin\left(\frac{p\pi}{l}z\right) \exp(j\omega\, t) \qquad (\mathrm{TE}_{10\mathrm{p}}\ \text{mode}) \tag{10.106}$$

where an extra mode number p is assigned to the resonator corresponding to the waveguide excited in the TE_{10} mode that is now bounded in the extra dimension along its length limited to the one given by (10.105). A similar substitution $\beta = p\pi/l$ in (10.101) results in the following expression for the resonator excited in the $\mathrm{TE}_{10\mathrm{p}}$ mode

$$E_y = -\frac{j\omega\mu_0 a}{\pi} H_0 \sin\left(\frac{\pi}{a}x\right) \sin\left(\frac{p\pi}{l}z\right) \exp(j\omega\, t). \quad (\mathrm{TE}_{10\mathrm{p}}\ \text{mode}) \tag{10.107}$$

Now, recall that H_x is related to E_y by the relation (9.19a) which, for the sake of convenience, can be put here as

$$\frac{\partial E_y}{\partial z} = j\omega\mu_0 H_x. \tag{10.108}$$

Taking the derivative of (10.107), we obtain

$$\frac{\partial E_y}{\partial z} = -\frac{j\omega\mu_0 a}{\pi}\frac{p\pi}{l} H_0 \sin\left(\frac{\pi}{a}x\right) \cos\left(\frac{p\pi}{l}z\right) \exp(j\omega\, t)$$

which substituted in (10.108) gives after rearrangement of terms the following expression for H_x for the resonator excited in the $\mathrm{TE}_{10\mathrm{p}}$ mode:

$$H_x = -p\frac{a}{l}H_0 \sin\left(\frac{\pi}{a}x\right)\cos\left(\frac{p\pi}{l}z\right)\exp(j\omega t). \quad (\text{TE}_{10\text{p}}\text{ mode}) \tag{10.109}$$

Finally, it is useful to put together the field expressions (10.106), (10.107) and (10.109) for a waveguide resonator excited in the TE_{101} mode, taking $p = 1$, as follows:

$$\left.\begin{aligned} H_z &= H_0 \cos\left(\frac{\pi}{a}x\right)\sin\left(\frac{\pi}{l}z\right)\exp(j\omega t) \\ E_y &= -\frac{j\omega\mu_0 a}{\pi}H_0 \sin\left(\frac{\pi}{a}x\right)\sin\left(\frac{\pi}{l}z\right)\exp(j\omega t) \\ H_x &= -\frac{a}{l}H_0 \sin\left(\frac{\pi}{a}x\right)\cos\left(\frac{\pi}{l}z\right)\exp(j\omega t) \end{aligned}\right\}. \quad (\text{TE}_{101}\text{ mode}) \tag{10.110}$$

We are going to use the field expression (10.110) for finding energy stored and power loss in a cavity resonator and hence study an important characteristics of a cavity resonator, namely, the quality factor (Section 10.4).

10.4 Energy Stored and Power Dissipated in a Waveguide Resonator

Let us consider a close-ended rectangular waveguide resonator excited in the TE_{101} mode for which we can use the field expression (10.110). First we will find the average energy stored by the resonator (Section 10.4.1), then the power loss due to the finite conductivity of the six conducting walls of the resonator (Section 10.4.2) and finally its quality factor to be defined in terms of energy stored and power loss (Section 10.4.2).

10.4.1 Energy stored

We have in (8.20) and (8.32) the expressions for the energies stored in static electric and magnetic fields respectively, which for the sake of convenience can be rewritten by putting $\vec{D} = \varepsilon_0\vec{E}$ and $\vec{B} = \mu_0\vec{H}$ as:

$$W_E = \frac{1}{2}\varepsilon_0 \oint_\tau \vec{E}.\vec{E}\,d\tau \quad (\text{static field}) \tag{10.111}$$

and

$$W_B = \frac{1}{2}\mu_0 \oint_\tau \vec{H}.\vec{H}\,d\tau. \quad (\text{static field}) \tag{10.112}$$

In the present context of time-varying fields for a rectangular closed-ended resonator in the TE_{101} mode, if we substitute (10.110) in (10.111) and then time-average the latter, we obtain

$$W_E = \frac{1}{2}\varepsilon_0(\oint_\tau \vec{E}.\vec{E})_{\text{time-average}}d\tau = \frac{1}{2}\varepsilon_0(\oint_\tau E_y\vec{a}_y.E_y\vec{a}_y)_{\text{time-average}}d\tau$$

$$= \frac{1}{2}\varepsilon_0(\oint_\tau E_y^2)_{\text{time-average}}d\tau \tag{10.113}$$

where $\tau = abl$ is the volume of the rectangular resonator (Fig. 10.15). If we now put E_y from (10.110) in (10.113) and carry out the time-averaging, an extra factor of $1/2$ will appear (due to time-averaging) and thus we obtain

$$W_E = \frac{1}{2}\varepsilon_0(\oint_\tau E_y^2)_{\text{time-average}}d\tau = \frac{1}{4}\varepsilon_0(\oint_\tau |E_y|^2\, d\tau \tag{10.114}$$

where $|E_y|^2$, obtainable from (10.110), is

$$|E_y|^2 = \frac{\omega^2\mu_0^2a^2}{\pi^2}H_0^2\sin^2\left(\frac{\pi}{a}x\right)\sin^2\left(\frac{\pi}{l}z\right),$$

which, when substituted in (10.114) while taking the volume element as $d\tau = dxdydz$ gives

$$W_E = \frac{\mu_0^2\varepsilon_0\omega^2a^2}{4\pi^2}H_0^2\int_0^a \sin^2\left(\frac{\pi}{a}x\right)dx\int_0^b dy\int_0^l \sin^2\left(\frac{\pi}{l}z\right)dz,$$

which, on evaluating the definite integrals becomes

$$W_E = \frac{\mu_0^2\varepsilon_0\omega^2a^2}{4\pi^2}H_0^2\left(\frac{a}{2}\right)(b)\left(\frac{l}{2}\right). \text{ (TE}_{101}\text{ mode)} \tag{10.115}$$

(One can easily carry out the definite integrals to obtain (10.115) using the trigonometric relation $\sin^2\phi = \frac{1}{2} - \frac{\cos 2\phi}{2}$), thereby getting

$$\int_0^a \sin^2\left(\frac{\pi}{a}x\right)dx = \frac{a}{2},\quad \int_0^b dy = b, \text{ and } \int_0^l \sin^2\left(\frac{\pi}{l}z\right) = \frac{l}{2}).$$

Substituting $\omega = 2\pi f$ in (10.115), we can then obtain in terms of the wave frequency f the following expression for the average magnetic energy stored in the rectangular waveguide resonator ($a \times b \times l$):

$$W_E = \frac{1}{4}\mu_0^2\varepsilon_0a^3bl\,f^2H_0^2. \text{ (TE}_{101}\text{ mode)} \tag{10.116}$$

Similarly, starting from (10.112) and using (10.110), we can easily find the time-averaged energy stored in the magnetic field as

$$W_B = (\frac{1}{2}\mu_0 \oint_\tau (H_x\vec{a}_x + H_y\vec{a}_y)\cdot(H_x\vec{a}_x + H_y\vec{a}_y)\,d\tau)_{\text{time-average}}$$

$$[\quad = \frac{1}{2}\mu_0 \oint_\tau (H_x^2 + H_y^2)_{\text{time-average}}\,d\tau. \tag{10.117}$$

Substituting (10.110) in (10.117) and carrying out time averaging, we then obtain an expression for the energy stored in magnetic field (corresponding to (10.114) for the stored energy in electric field) as

$$W_B = \frac{1}{4}\mu_0 \oint_\tau (|H_x|^2 + |H_y|^2)_{\text{time-average}}\,d\tau \tag{10.118}$$

where $|H_x|^2$ and $|H_z|^2$, obtainable from (10.110), are

$$\left.\begin{aligned} H_x^2 &= \left(\frac{a^2}{l^2}\right) H_0^2 \sin^2\left(\frac{\pi}{a}x\right)\cos^2\left(\frac{\pi}{l}z\right) \\ H_z^2 &= H_0^2 \cos^2\left(\frac{\pi}{a}x\right)\sin^2\left(\frac{\pi}{l}z\right) \end{aligned}\right\}. \tag{10.119}$$

If we substitute (10.119) in (10.118), where we take the volume element as $d\tau = dxdydz$, we obtain

$$W_B = \frac{1}{4}\mu_0 H_0^2\left[\left(\frac{a^2}{l^2}\right)\int_0^a \sin^2\left(\frac{\pi}{a}x\right)dx\int_0^b dy\int_0^l \cos^2\left(\frac{\pi}{l}z\right) + \int_0^a \cos^2\left(\frac{\pi}{a}x\right)dx\int_0^b dy\int_0^l \sin^2\left(\frac{\pi}{l}z\right)\right]$$

which can be simplified carrying out the definite integrals as

$$W_B = \frac{\mu_0}{16}abl\left(\frac{a^2}{l^2}+1\right)H_0^2. \text{ (TE}_{101}\text{ mode)} \tag{10.120}$$

(One can easily carry out the definite integrals to obtain (10.120), using the trigonometric relations $\sin^2\phi = \frac{1}{2} - \frac{\cos 2\phi}{2}$ and $\cos^2\phi = \frac{1}{2} + \frac{\cos 2\phi}{2}$, thereby getting

$$\int_0^a \sin^2\left(\frac{\pi}{a}x\right)dx = \int_0^a \cos^2\left(\frac{\pi}{a}x\right)dx = \frac{a}{2},\ \int_0^b dy = b, \text{ and}$$

$$\int_0^l \sin^2\left(\frac{\pi}{l}z\right)dz = \int_0^l \cos^2\left(\frac{\pi}{l}z\right)dz = \frac{l}{2}).$$

Thus, we have now two expressions, namely, (10.116) and (10.120) for the time-averaged energies stored in electric and magnetic fields respectively. Let us compare these two energies at the resonant frequency $f = f_r (= \omega_r/2\pi)$, the latter obtainable from (10.92), putting $m = 1, n = 0, p = 1$ for the TE_{101} mode as

$$f_r = \frac{1}{2}\left(\frac{1}{a^2} + \frac{1}{l^2}\right)^{1/2} c = \frac{1}{2(\mu_0 \varepsilon_0)^{1/2}}\left(\frac{1}{a^2} + \frac{1}{l^2}\right)^{1/2}. \ (\text{TE}_{101} \text{ mode}) \tag{10.121}$$

Substituting (10.121) for $f(= f_r)$ in (10.116), we obtain the following expression, after a little algebra:

$$W_E = \frac{\mu_0}{16} abl \left(1 + \frac{a^2}{l^2}\right) H_0^2 \ (\text{TE}_{101} \text{ mode}) \ (\text{at resonance } f = f_r) \tag{10.122}$$

By comparing (10.120) and (10.122) the following equality at the resonant frequency ($f = f_r$) can be noted:

$$W_E = W_B = \frac{\mu_0}{16} abl \left(1 + \frac{a^2}{l^2}\right) H_0^2. \tag{10.123}$$

In view of (10.123), therefore, we can express the total time-averaged energy W stored in electric and magnetic fields put together at resonance as

$$W = 2W_E = 2W_B = \frac{\mu_0}{8} abl \left(1 + \frac{a^2}{l^2}\right) H_0^2. \tag{10.124}$$

10.4.2 Power dissipation

We can find the power loss per unit area at each of the six walls of a rectangular waveguide resonator made of a conductor of finite conductivity by the method that was already used by us to find the power loss per unit area at each of the four conducting walls of a rectangular waveguide in Section 9.5.2. For this purpose, let us recall (9.184) and (9.185) and, for the sake of convenience, rewrite them respectively as follows:

$$P_{\text{LA}} = \frac{1}{2} R_S \vec{J}_s \cdot \vec{J}_s^* \tag{10.125}$$

and

$$\vec{a}_n \times \vec{H}_2 = \vec{J}_s \tag{10.126}$$

where P_{LA} and $\vec{J}_s$ are respectively the power loss per unit area and the surface current density at the conducting wall of the resonator. R_S is the surface resistance of the material of the conducting wall of the resonator. $\vec{a}_n$ is the unit vector directed from the concerned conducting wall to the free-space region and $\vec{H}_2$ is the magnetic field in the free-space region at the concerned wall. We can write the expressions for normal unit vector $\vec{a}_n$ at the six walls of the resonator as (refer to Fig. 10.15 and Section 9.5.2):

$$\left.\begin{array}{lll} \vec{a}_n|_{x=0} = \vec{a}_x & \text{(right wall)} \\ \vec{a}_n|_{x=a} = -\vec{a}_x & \text{(left wall)} \\ \vec{a}_n|_{y=0} = \vec{a}_y & \text{(bottom wall)} \\ \vec{a}_n|_{y=b} = -\vec{a}_y & \text{(top wall)} \\ \vec{a}_n|_{z=0} = \vec{a}_z & \text{(front wall)} \\ \vec{a}_n|_{z=l} = -\vec{a}_z & \text{(back wall)} \end{array}\right\}. \tag{10.127}$$

For the TE$_{101}$-mode-excited waveguide resonator, we can see from (9.44) and (9.49) and remembering $m = 1$, $n = 0$ for the mode that $\vec{H}_2 f$ has the x and the z components, respectively, and accordingly we can express it as

$$\vec{H}_2 = H_x\vec{a}_x + H_z\vec{a}_z. \tag{10.128}$$

The expressions for surface current density $\vec{J}_s$ at the six walls of the resonator can be then written substituting (10.27) and (10.128) in (10.126) as (Fig. 10.15):

$$\left.\begin{array}{lll} \vec{J}_s\Big|_{x=0} = \vec{a}_x \times (H_x\vec{a}_x + H_z\vec{a}_z) = -H_z\vec{a}_y & \text{(right wall)} \\ \vec{J}_s\Big|_{x=a} = -\vec{a}_x \times (H_x\vec{a}_x + H_z\vec{a}_z) = H_z\vec{a}_y & \text{(left wall)} \\ \vec{J}_s\Big|_{y=0} = \vec{a}_y \times (H_x\vec{a}_x + H_z\vec{a}_z) = -H_x\vec{a}_z + H_z\vec{a}_x & \text{(bottom wall)} \\ \vec{J}_s\Big|_{y=b} = -\vec{a}_y \times (H_x\vec{a}_x + H_z\vec{a}_z) = H_x\vec{a}_z - H_z\vec{a}_x & \text{(top wall)} \\ \vec{J}_s\Big|_{z=0} = \vec{a}_z \times (H_x\vec{a}_x + H_z\vec{a}_z) = H_x\,\vec{a}_y & \text{(front wall)} \\ \vec{J}_s\Big|_{z=l} = -\vec{a}_z \times (H_x\vec{a}_x + H_z\vec{a}_z) = -H_x\,\vec{a}_y & \text{(back wall)} \end{array}\right\}. \tag{10.129}$$

Now, the magnetic field components at the resonator walls can be written with the help of (10.110) as

$$
\left.\begin{array}{ll}
\left.\begin{array}{l}
H_z = H_0 \sin\left(\frac{\pi}{l}z\right)\exp(j\omega\, t) \\
H_x = 0
\end{array}\right\} & \text{(right wall; } x=0) \\
\left.\begin{array}{l}
H_z = -H_0 \sin\left(\frac{\pi}{l}z\right)\exp(j\omega\, t) \\
H_x = 0
\end{array}\right\} & \text{(left wall; } x=a) \\
\left.\begin{array}{l}
H_z = H_0 \cos\left(\frac{\pi}{a}x\right)\sin\left(\frac{\pi}{l}z\right)\exp(j\omega\, t) \\
H_x = -\frac{a}{l} H_0 \sin\left(\frac{\pi}{a}x\right)\cos\left(\frac{\pi}{l}z\right)\exp(j\omega\, t)
\end{array}\right\} & \text{(bottom wall; } y=0) \\
\left.\begin{array}{l}
H_z = H_0 \cos\left(\frac{\pi}{a}x\right)\sin\left(\frac{\pi}{l}z\right)\exp(j\omega\, t) \\
H_x = -\frac{a}{l} H_0 \sin\left(\frac{\pi}{a}x\right)\cos\left(\frac{\pi}{l}z\right)\exp(j\omega\, t)
\end{array}\right\} & \text{(top wall; } y=b) \\
\left.\begin{array}{l}
H_z = 0 \\
H_x = -\frac{a}{l} H_0 \sin\left(\frac{\pi}{a}x\right)\exp(j\omega\, t)
\end{array}\right\} & \text{(front wall; } z=0) \\
\left.\begin{array}{l}
H_z = 0 \\
H_x = \frac{a}{l} H_0 \sin\left(\frac{\pi}{a}x\right)\exp(j\omega\, t)
\end{array}\right\} & \text{(back wall; } z=l)
\end{array}\right\}. \qquad (10.130)
$$

Substituting (10.130) in (10.129) we then have

$$
\left.\begin{array}{ll}
\vec{J}_s\Big|_{x=0} = -\,H_0 \sin\left(\frac{\pi}{l}z\right)\exp(j\omega\, t)\vec{a}_y & \text{(right wall)} \\
\vec{J}_s\Big|_{x=a} = -H_0 \sin\left(\frac{\pi}{l}z\right)\exp(j\omega\, t)\vec{a}_y & \text{(left wall)} \\
\vec{J}_s\Big|_{y=0} = \frac{a}{l} H_0 \sin\left(\frac{\pi}{a}x\right)\cos\left(\frac{\pi}{l}z\right)\exp(j\omega\, t)\vec{a}_z & \\
\qquad + H_0 \cos\left(\frac{\pi}{a}x\right)\sin\left(\frac{\pi}{l}z\right)\exp(j\omega\, t)\vec{a}_x & \text{(bottom wall)} \\
\vec{J}_s\Big|_{y=b} = -\frac{a}{l} H_0 \sin\left(\frac{\pi}{a}x\right)\cos\left(\frac{\pi}{l}z\right)\exp(j\omega\, t)\vec{a}_z & \\
\qquad - H_0 \cos\left(\frac{\pi}{a}x\right)\sin\left(\frac{\pi}{l}z\right)\exp(j\omega\, t)\vec{a}_x & \text{(top wall)} \\
\vec{J}_s\Big|_{z=0} = -\frac{a}{l} H_0 \sin\left(\frac{\pi}{a}x\right)\exp(j\omega\, t)\,\vec{a}_y & \text{(front wall)} \\
\vec{J}_s\Big|_{z=l} = -\frac{a}{l} H_0 \sin\left(\frac{\pi}{a}x\right)\exp(j\omega\, t)\,\vec{a}_y & \text{(back wall)}
\end{array}\right\}. \qquad (10.131)
$$

The expressions for the complex conjugate $\vec{J}_s^*$ of $\vec{J}_s$ as required for deriving the expressions for the power loss per unit area P_{LA} at the resonator walls with the help of (10.125) are the same as (10.131), but for the factor $\exp(j\omega\, t)$ replaced by $\exp(-j\omega t)$. Therefore, we obtain the following expressions for the dot product of $\vec{J}_s$ and $\vec{J}_s^*$:

$$\left.\begin{aligned}
&(\vec{J}_s\cdot\vec{J}_s^*)\Big|_{x=0}\text{(right wall)} = (\vec{J}_s\cdot\vec{J}_s^*)\Big|_{x=a}\text{(left wall)}\\
&\quad = H_0^2\sin^2\left(\frac{\pi}{l}z\right)\\
&(\vec{J}_s\cdot\vec{J}_s^*)\Big|_{y=0}\text{(bottom wall)} = (\vec{J}_s\cdot\vec{J}_s^*)\Big|_{y=b}\text{(top wall)}\\
&\quad = \left(\frac{a}{l}\right)^2 H_0^2\sin^2\left(\frac{\pi}{a}x\right)\cos^2\left(\frac{\pi}{l}z\right) + H_0^2\cos^2\left(\frac{\pi}{a}x\right)\sin^2\left(\frac{\pi}{l}z\right)\\
&(\vec{J}_s\cdot\vec{J}_s^*)\Big|_{z=0}\text{(front wall)} = (\vec{J}_s\cdot\vec{J}_s^*)\Big|_{z=l}\text{(back wall)}\\
&\quad = \left(\tfrac{a}{l}\right)^2 H_0^2\sin^2\left(\tfrac{\pi}{a}x\right)
\end{aligned}\right\}. \tag{10.132}$$

We then substitute (10.132) in (10.125) to obtain the following expressions for power loss per unit area at the cavity walls:

$$\left.\begin{aligned}
&P_{\mathrm{LA}}\text{(right wall)} = P_{\mathrm{LA}}\text{(left wall)}\\
&\quad = \frac{1}{2}R_S H_0^2\sin^2\left(\frac{\pi}{l}z\right)\\
&P_{\mathrm{LA}}\text{(bottom wall)} = P_{\mathrm{LA}}\text{(top wall)}\\
&\quad = \frac{1}{2}R_S\left[\left(\frac{a}{l}\right)^2 H_0^2\sin^2\left(\frac{\pi}{a}x\right)\cos^2\left(\frac{\pi}{l}z\right) + H_0^2\cos^2\left(\frac{\pi}{a}x\right)\sin^2\left(\frac{\pi}{l}z\right)\right]\\
&P_{\mathrm{LA}}\text{(front wall)} = P_{\mathrm{LA}}\text{(back wall)}P_{\mathrm{LA}}\text{(front wall)} = P_{\mathrm{LA}}\text{(back wall)}\\
&\quad = \frac{1}{2}R_S\left(\frac{a}{l}\right)^2 H_0^2\sin^2\left(\frac{\pi}{a}x\right)
\end{aligned}\right\}. \tag{10.133}$$

The power loss through the right wall P_{L} (right wall) can be found by integrating the power loss through an infinitesimal area $dydz$ over the area $b\times l$ as follows:

$$P_{\mathrm{L}}\text{(right wall)} = \int\limits_{\text{right wall}} P_{\mathrm{LA}}\text{(right wall)}dydz$$

which can be written with the help of (10.133) as

$$P_{\mathrm{L}}\text{(right wall)} = \frac{1}{2}R_S H_0^2\int_0^b dy\int_0^l \sin^2\left(\frac{\pi}{l}z\right)dz,$$

which, on evaluating the definite integrals [as discussed following (10.120)] becomes

$$P_{\mathrm{L}}\text{(right wall)} = \frac{1}{2}R_S H_0^2(b)\left(\frac{l}{2}\right). \tag{10.134}$$

A similar procedure will then lead to the power loss through the left wall P_L (left wall) as follows:

$$P_L(\text{left wall}) = \frac{1}{2} R_S H_0^2(b)\left(\frac{l}{2}\right). \tag{10.135}$$

Similarly, the power loss through the bottom wall P_L(bottom wall) can be found by integrating the power loss through an infinitesimal area $dxdz$ over the area $a \times l$ as follows:

$$P_L(\text{bottom wall}) = \int\limits_{\text{bottom wall}} P_{LA}(\text{bottom wall})\, dxdz$$

which using (10.133) can be written as

$$P_L(\text{bottom wall}) = \frac{1}{2} R_S H_0^2 \left[\left(\frac{a}{l}\right)^2 \int_0^a \sin^2\left(\frac{\pi}{a}x\right) dx \int_0^l \cos^2\left(\frac{\pi}{l}z\right) dz + \int_0^a \cos^2\left(\frac{\pi}{a}x\right) dx \int_0^l \sin^2\left(\frac{\pi}{l}z\right) dz\right]. \tag{10.136}$$

Evaluating the definite integrals [as discussed following (10.120)], we can express (10.136) as

$$P_L(\text{bottom wall}) = \frac{1}{2} R_S H_0^2 \left[\left(\frac{a}{l}\right)^2 \left(\frac{a}{l}\right)\left(\frac{l}{2}\right) + \left(\frac{a}{l}\right)\left(\frac{l}{2}\right)\right],$$

which simplifies to

$$P_L(\text{bottom wall}) = \frac{1}{2} R_S H_0^2 \left(\frac{a}{l}\right)\left(\frac{l}{2}\right)\left[\left(\frac{a}{l}\right)^2 + 1\right]. \tag{10.137}$$

Proceeding in a similar way, we then obtain an expression identical to (10.137) for the power loss through the top wall P_L(top wall) as follows:

$$P_L(\text{top wall}) = \frac{1}{2} R_S H_0^2 \left(\frac{a}{l}\right)\left(\frac{l}{2}\right)\left[\left(\frac{a}{l}\right)^2 + 1\right]. \tag{10.138}$$

Next, let us find the power loss through the front wall P_L (front wall) by integrating the power loss through an infinitesimal area $dxdy$ over the area $a \times b$ as follows:

$$P_L(\text{front wall}) = \int\limits_{\text{front wall}} P_{LA}(\text{front wall}) dxdy$$

which, using (10.133) can be written as

$$P_{\rm L}(\text{front wall}) = \frac{1}{2} R_S H_0^2 \left(\frac{a}{l}\right)^2 \int_0^a \sin^2\left(\frac{\pi}{a}x\right) dx \int_0^b dy$$

which, on evaluating the definite integrals [as discussed following (10.120) (10.61)], becomes

$$P_{\rm L}(\text{front wall}) = \frac{1}{2} R_S H_0^2 \left(\frac{a}{l}\right)^2 \left(\frac{a}{l}\right)(b). \tag{10.139}$$

An expression identical to (10.139) for the power loss through the back Wall $P_{\rm L}$ (back wall) is obtained in a similar way as:

$$P_{\rm L}(\text{back wall}) = \frac{1}{2} R_S H_0^2 \left(\frac{a}{l}\right)^2 \left(\frac{a}{l}\right)(b). \tag{10.140}$$

Finally, adding (10.134), (10.135), (10.137)-(10.140), we estimate the losses contributed by all the six walls to obtain the following expression for the wall loss of the resonator:

$$P_{\rm L} = R_S H_0^2 \left\{ (b)\left(\frac{l}{2}\right) + \left(\frac{a}{l}\right)\left(\frac{l}{2}\right)\left[\left(\frac{a}{l}\right)^2 + 1\right] + \left(\frac{a}{l}\right)^2\left(\frac{a}{l}\right)(b)\right\},$$

which can be alternatively expressed rearranging the terms as

$$P_{\rm L} = R_S H_0^2 \left(\frac{a}{2}\right)\left[\frac{a^2}{l}\left(\frac{b}{l} + \frac{1}{2}\right) + l\left(\frac{b}{a} + \frac{1}{2}\right)\right]. \quad (\text{TE}_{101}\text{ mode}) \tag{10.141}$$

The expression (10.141) for the power loss, together with the expression (10.124) for the time-averaged energy stored in a resonator will be used in the section to follow (Section 10.5) to find an important property of a resonator, namely, the quality factor.

10.5 Quality Factor of a Resonator

The quality factor of a resonator is a measure of the energy stored in the resonator relative to the energy lost in the resonator. The quantitative definition of the quality factor is

$$Q = 2\pi \frac{W}{W_L|_{T=\omega/2\pi}} \tag{10.142}$$

which is a dimensionless quantity, where W is the time-averaged energy stored in the resonator and $W_L|_{T=2\pi/\omega}$ is the energy lost in the wave time period $T = 2\pi/\omega$. Since $P_{\rm L}$ is the power lost or the energy lost in unit time, we can find $W_L|_{T=2\pi/\omega}$ by multiplying $P_{\rm L}$ by the time period T as

$$W_L|_{T=2\pi/\omega} = P_L T = P_L \frac{2\pi}{\omega}. \tag{10.143}$$

Substituting (10.143) in (10.142), we then have an alternative definition of the quality factor of the resonator as

$$Q\left(=2\pi\frac{W}{P_L\frac{2\pi}{\omega}}\right)=\omega\frac{W}{P_L}. \tag{10.144}$$

For a rectangular waveguide resonator excited in the TE_{101} mode, we can substitute W from (10.65) and P_L from (10.141) in (10.144) to obtain the following expression for the quality factor:

$$Q\left(=\omega\frac{W}{P_L}\right)=\omega\frac{\frac{\mu_0}{8}abl\left(1+\frac{a^2}{l^2}\right)}{R_S\left(\frac{a}{2}\right)\left[\frac{a^2}{l}\left(\frac{b}{l}+\frac{1}{2}\right)+l\left(\frac{b}{a}+\frac{1}{2}\right)\right]}.\quad (\text{TE}_{101}\text{ mode}) \tag{10.145}$$

We can interpret ω occurring in (10.145) as the resonant frequency $\omega=\omega_r$, which can be expressed with the help of (10.121) as

$$\omega_r=2\pi f_r=\frac{\pi}{(\mu_0\varepsilon_0)^{1/2}}\left(\frac{1}{a^2}+\frac{1}{l^2}\right)^{1/2}.\quad (\text{TE}_{101}\text{ mode}) \tag{10.146}$$

Now, substituting (10.146) in (10.145) we obtain

$$Q=\frac{\pi}{(\mu_0\varepsilon_0)^{1/2}}\left(\frac{1}{a^2}+\frac{1}{l^2}\right)^{1/2}\frac{\frac{\mu_0}{8}abl\left(1+\frac{a^2}{l^2}\right)}{R_S\left(\frac{a}{2}\right)\left[\frac{a^2}{l}\left(\frac{b}{l}+\frac{1}{2}\right)+l\left(\frac{b}{a}+\frac{1}{2}\right)\right]}.$$

$$(\omega=\omega_r)\quad (\text{TE}_{101}\text{ mode}) \tag{10.147}$$

With a little algebra, expression (10.147) simplifies to

$$Q=\frac{\pi\eta_0}{4R_S}\frac{2b(a^2+l^2)^{3/2}}{al(a^2+l^2)+2b(a^3+l^3)}\quad(\omega=\omega_r)\quad (\text{TE}_{101}\text{ mode}) \tag{10.148}$$

where $\eta_0=(\mu_0/\varepsilon_0)^{1/2}$. As can be seen from (6.98) and (6.99), for a good conductor at relatively high frequencies, the surface resistance R_S of the conducting material of the resonator occurring in (10.148) is given by

$$R_S=\frac{1}{\sigma\delta}=\left(\frac{\pi f\mu_0}{\sigma}\right)^{1/2}$$

where $f=f_r=\omega_r/2\pi$ is obtainable from (10.146) in terms of the cavity dimensions, σ and δ being the conductivity and the skin-depth of the cavity material. Therefore, while examining the expression (10.148) for the dependence of the quality factor Q on cavity dimensions, which is explicitly clear from the expression, one must also take into account the dependence of the quantity R_S on the structure dimensions through the term f, which at resonant frequency of the cavity ($f = f_r$) is given by (10.146).

In an actual application, some part of energy stored in a cavity is coupled out from the cavity to an external load and the cavity thereby becomes loaded. For such a system, we can write

$$P_{L,\,\text{loaded}} = P_{L,\,\text{ohmic}} + P_{L,\,\text{ext}} \tag{10.149}$$

where $P_{L,\,\text{ohmic}}$ is the ohmic power loss due to the finite conductivity of the cavity material for the unloaded cavity, $P_{L,\,\text{ext}}$ is part of stored energy that is lost per unit time in the form of power coupled out of the cavity to the external load. $P_{L,\,\text{loaded}}$ With the help of (10.144) therefore we can write

$$\left.\begin{aligned} P_{L\,\text{ohmic}} &= \frac{\omega W}{Q_{\text{unloaded}}} \\ P_{L\,\text{ext}} &= \frac{\omega W}{Q_{\text{ext}}} \\ P_{L\,\text{loaded}} &= \frac{\omega W}{Q_{\text{loaded}}} \end{aligned}\right\} \tag{10.150}$$

where Q_{unloaded}, Q_{ext}, and Q_{loaded} are the unloaded quality factor, the external quality factor, and the loaded quality factor of the cavity respectively. Therefore, substituting (10.150) in (10.149) we obtain

$$\frac{\omega W}{Q_{\text{loaded}}} = \frac{\omega W}{Q_{\text{unloaded}}} + \frac{\omega W}{Q_{\text{ext}}}$$

which simplifies to

$$\frac{1}{Q_{\text{loaded}}} = \frac{1}{Q_{\text{unloaded}}} + \frac{1}{Q_{\text{ext}}}. \tag{10.151}$$

The waveguide resonator can be represented by a resonant circuit comprising an inductor of inductance L having reactance $j\omega\, L$, a capacitor of capacitance C having reactance $1/j\omega C$, and a resistor of resistance R, all in parallel. Energy is transferred back and forth between the inductor and the capacitor during each cycle, with the total energy W remaining the same and which therefore can be taken as the energy stored in the capacitor given by [see (8.26)]

$$W = \frac{1}{2}CV^2 \tag{10.152}$$

where V is the voltage across the parallel LCR circuit. The loss in the resonant circuit is estimated by the ohmic loss P_L in the resistor given by

$$P_L = \frac{V^2}{2R}. \tag{10.153}$$

It would be of interest to interpret the quality factor Q of the resonant circuit at the resonant frequency $\omega = \omega_r$ by substituting (10.152) and (10.153) in (10.144) to obtain

$$Q = \omega_r C R \tag{10.154}$$

where the resonant frequency of the parallel LC circuit is well known as

$$\omega_r = \frac{1}{(LC)^{1/2}}. \tag{10.155}$$

Substituting

$$C = \frac{1}{L\omega_r^2}$$

obtainable from (10.155) in (10.154), we obtain an expression alternative to (10.154) for the quality factor Q as

$$Q = \frac{R}{\omega_r L}. \tag{10.156}$$

We are next going to find an expression for the impedance Z_{eq} of the parallel L,C,R circuit, which satisfies the following law of parallel impedances:

$$\frac{1}{Z_{\text{eq}}} = \frac{1}{j\omega\, L} + \frac{1}{\dfrac{1}{j\omega\, C}} + \frac{1}{R} \tag{10.157}$$

where the first through the last terms of its right-hand side represent the reciprocals of the reactance of L, the reactance of C, and the resistance of R respectively. We can then obtain by simple algebra from (10.157)

$$Z_{\text{eq}} = \frac{R}{1 - j\dfrac{R}{\omega\, L}(1 - \omega^2 L\, C)}$$

which, with the help of (10.155), simplifies to

$$Z_{\text{eq}} = \frac{R}{1 - j\frac{R}{\omega\, L}\left(1 - \frac{\omega^2}{\omega_r^2}\right)}. \tag{10.158}$$

Substituting $R/L = Q\omega_r$ from (10.158), we can also express (10.158) in terms of Q as

$$Z_{\text{eq}} = \frac{R}{1 + j\dfrac{Q\omega_r}{\omega}\left(\dfrac{\omega^2}{\omega_r^2} - 1\right)}. \tag{10.159}$$

At resonance, $\omega = \omega_r$, we obtain from (10.159)

$$Z_{\text{eq}} = R. \; (\omega = \omega_r). \tag{10.160}$$

Further, we can normalise (10.159) as

$$\frac{Z_{\text{eq}}}{R} = \frac{1}{1 + j\dfrac{Q}{\omega}\dfrac{(\omega + \omega_r)(\omega - \omega_r)}{\omega_r}},$$

which gives the normalised impedance magnitude and its square as

$$\left.\begin{aligned} \frac{|Z_{\text{eq}}|}{R} &= \frac{1}{\left[1 + \left(\dfrac{Q}{\omega}\dfrac{(\omega + \omega_r)(\omega - \omega_r)}{\omega_r}\right)^2\right]^{1/2}} \\ \left(\frac{|Z_{\text{eq}}|}{R}\right)^2 &= \frac{1}{1 + \left(\dfrac{Q}{\omega}\dfrac{(\omega + \omega_r)(\omega - \omega_r)}{\omega_r}\right)^2} \end{aligned}\right\}. \tag{10.161}$$

The impedance magnitude $|Z_{\text{eq}}|$ is maximum at resonance $\omega = \omega_r$ in the parallel L,C,R circuit but decreases both below and above resonance. Correspondingly, the voltage across the circuit is also at a peak at resonance since voltage is proportional to the impedance. The bandwidth of the circuit is measured between the half power points above and below the frequency of resonance. Since the power is proportional to the square of impedance, because of the power being proportional to the square of voltage and the voltage proportional to impedance, we can find the half-power points with the help of (10.161) by putting therein $|Z_{\text{eq}}|/R = 1/\sqrt{2}$ as follows:

$$\frac{1}{1 + \left(\dfrac{Q}{\omega}\dfrac{(\omega + \omega_r)(\omega - \omega_r)}{\omega_r}\right)^2} = \frac{1}{2}. \tag{10.162}$$

We expect two solutions of (10.162) around $\omega = \omega_r$ close to ω_r as $\omega_1 = \omega_r + \Delta\omega$ and $\omega_2 = \omega_r - \Delta\omega$, where $\Delta\omega << \omega$ is taken as positive. Therefore, for the solution $\omega = \omega_1 = \omega_r + \Delta\omega$, we can put $\omega_1 \approx \omega_r$, $\omega_1 + \omega_r \approx 2\omega_r$ and $\omega_1 - \omega_r \approx \Delta\omega$ in (10.162) to obtain the relation

$$\frac{4Q^2}{\omega_r^2}(\Delta\omega)^2 = 1,$$

which can be rearranged as

$$Q = \frac{\omega_r}{2\Delta\omega}. \tag{10.163}$$

Similarly, for the solution $\omega = \omega_2 = \omega_r - \Delta\omega$, we obtain $\omega_2 \approx \omega_r$, $\omega_2 + \omega_r \approx 2\omega_r$ and $\omega_2 - \omega_r \approx -\Delta\omega$, which can be substituted in (10.162) to obtain one and the same expression as (10.163) for the quality factor. Interpreting $2\Delta\omega$ as the bandwidth of the frequency response of half the value of $(|Z_{\text{eq}}|/R)^2$ of the resonant circuit, we can interpret (10.163) as

$$Q = \frac{\omega_r}{\text{angular frequency bandwidth}}$$

which can be put as

$$Q = \frac{f_r}{\text{frequency bandwidth}}. \tag{10.164}$$

EXAMPLE E10-8

Find the dimension of a cubical cavity made of copper ($\sigma = 5.8 \times 10^7$ mho/m) excited in the TE_{101} mode so that it resonates at 15 GHz. Calculate also its quality factor and bandwidth.

SOLUTION

Taking $a = b = l$ (cubical dimensions) and $m = 1$, $p = 1$, $n = 0$ (TE_{101} mode), we can calculate the resonant frequency $f_r (= \omega_r / 2\pi)$ from (10.92) as

$$f_r = \frac{1}{\sqrt{2}\pi}\frac{\pi}{a}c,$$

whence you can find $a (= b = l)$ as

$$a = \frac{1}{\sqrt{2}}\frac{1}{f_r}c. \tag{E10-8.1}$$

Putting $f_r = 15 \times 10^9$ Hz (= 15 GHz) and $c = 3 \times 10^8$m/s in (E10-8.1), we can calculate the cavity dimension as

$$a(= b = l) = \frac{1}{\sqrt{2}}\frac{1}{f_r}c = \frac{1}{\sqrt{2}}\frac{3 \times 10^8}{15 \times 10^9} = 1.414 \times 10^{-2} \text{ m}.$$

Similarly, putting $a = b = l$ in (10.148) you can write the expression for the quality factor for the TE_{101} mode as

$$Q = \frac{\pi\eta_0}{4R_S}\frac{2a(a^2+a^2)^{3/2}}{aa(a^2+a^2)+2a(a^3+a^3)},$$

which simplifies to

$$Q = \frac{\pi\eta_0}{4R_S}\frac{2a(a^2+a^2)^{3/2}}{aa(a^2+a^2)+2a(a^3+a^3)} = \frac{\pi\eta_0}{3\sqrt{2}R_S}. \tag{E10-8.2}$$

You can calculate R_S occurring in (2), putting $f = f_r = 15 \times 10^9$ Hz (= 15 GHz) and $\sigma = 5.8 \times 10^7$ mho/m (given) in the expression for R_S as follows

$$R_s = \sqrt{\frac{\pi f \mu_0}{\sigma}} = \sqrt{\frac{3.143 \times 15 \times 10^9 \times 4\pi \times 10^{-7}}{5.8 \times 10^7}} = 3.19 \times 10^{-2}\ \text{ohm}. \tag{E10-8.3}$$

Putting (E10-8.3) in (E10-8.2), we can calculate

$$Q = \frac{\pi\eta_0}{3\sqrt{2}R_S} = \frac{3.143 \times 377}{3 \times 1.414 \times 3.19 \times 10^{-2}} = 8756. \tag{E10-8.4}$$

The bandwidth of the cavity can be found by putting the value $Q = 8756$ from (E10-8.4) in (10.164) as

$$\text{frequency bandwidth} = \frac{f_r}{Q} = \frac{15 \times 10^9}{8756} = 1.713 \times 10^6\ \text{Hz} = 1.713\ \text{MHz}.$$

Appendix A10.1

Impedance maxima and minima on a transmission line

Dividing (10.48) by (10.55), we have the expression for the impedance $Z_{\max}$ at the voltage maxima as

$$Z_{\max} = \frac{|V_{\max}|}{|I_{\min}|} = Z_0\frac{1+|\Gamma_l|}{1-|\Gamma_l|}$$

which, in view of (10.157), becomes

$$Z_{\max} = (Z_0)(\text{VSWR}). \quad (\theta - 2\beta d = 0, 2\pi, 4\pi, ...) \tag{A10.1}$$

Similarly, dividing (10.49) by (10.54), we obtain

$$Z_{\min} = \frac{|V_{\min}|}{|I_{\max}|} = Z_0\frac{1}{\frac{1+|\Gamma_l|}{1-|\Gamma_l|}}$$

which in view of (10.57) becomes

$$Z_{\text{min}} = \frac{Z_0}{\text{VSWR}}. \quad (\theta - 2\beta\, d = \pi, 3\pi, 5\pi, ...) \tag{A10.2}$$

Appendix A10.2

Load impedance from the shift in the standing-wave pattern due to a short replacing the load

Let us rewrite the reflection coefficient Γ_l of a transmission line terminated in a complex load Z_l, as defined preceding (10.45) as:

$$\Gamma_l = |\Gamma_l| \exp(j\theta), \tag{A10.3}$$

where Γ_l is related to Z_l [see (10.32)] as

$$\Gamma_l = \frac{Z_l - Z_0}{Z_l + Z_0}. \quad \text{(for a complex load)} \tag{A10.4}$$

Next, recall the condition for the voltage minimum at a point on the line that is distant d from the terminating load end towards the source end (see Section 10.1.7):

$$\theta - 2\beta\, d = \pi, 3\pi, 5\pi, ... \quad \text{(for a complex load)} \tag{A10.5}$$

Further, consider a situation when the complex load is replaced by a short, for which we take $Z_l = 0$ in (A10.4) to write

$$\Gamma_l = \frac{0 - Z_0}{0 + Z_0} = -1. \quad \text{(for a short replacing a complex load)} \tag{A10.6}$$

Comparing (A10.6) with (A10.3), we can then write

$$\left. \begin{aligned} |\Gamma_l| &= 1 \\ \theta &= \pi \end{aligned} \right\} \quad \text{(for a short replacing a complex load)} \tag{A10.7}$$

noting that $\exp(j\theta) = -1$ for $\theta = \pi$. Therefore, putting $\theta = \pi$ in view of (A10.7) for a short replacing a complex load, we can modify the condition (A10.5) for the voltage minima at a point on the line to put it as

$$\pi - 2\beta\, d' = \pi, 3\pi, 5\pi, ... \text{(for a short replacing a complex load)} \tag{A10.8}$$

where d' is the new distance towards the source end of the voltage minimum from the terminating short replacing the complex load. Subtracting (A10.5) from (A10.8) you then obtain

$$\theta = \pi + 2\beta(d - d') = \pi + 2\beta\Delta d \tag{A10.9}$$

where $d - d' = \Delta d$ is the shift in the voltage minima pattern when the complex impedance at the load end is replaced by a short. Since the distances d and d' are reckoned from the load end, $d' < d$ and consequently Δd becomes positive when the shift is towards the load end when the short replaces the complex load at the terminating end. This would be the situation when the complex load is such that if the line is terminated in the load, the first extremum from the load end is a minimum in the voltage standing-wave pattern. Similarly, you have to take Δd as negative if the shift in the minima is towards the source end instead of the load end if the terminating complex load is such that the first extremum from the load end is a maximum instead of a minimum in the voltage standing-wave pattern.

Further, let us put (10.57) as

$$\frac{\text{VSWR}}{1} = \frac{1 + |\Gamma_l|}{1 - |\Gamma_l|}$$

and then algebraically manipulate it as

$$\frac{\text{VSWR} - 1}{\text{VSWR} + 1} = \frac{(1 + |\Gamma_l|) - (1 - |\Gamma_l|)}{(1 + |\Gamma_l|) + (1 - |\Gamma_l|)} = \frac{2|\Gamma_l|}{2} = |\Gamma_l| \,. \tag{A10.10}$$

Substituting (A10.10) in (A10.3) we then have

$$\Gamma_l = \frac{\text{VSWR} - 1}{\text{VSWR} + 1} \exp(j\theta)$$

which can be expressed in view of the relation

$$\exp(j\theta) = \cos\theta + j\sin\theta$$

as

$$\Gamma_l = \left(\frac{\text{VSWR} - 1}{\text{VSWR} + 1}\right)(\cos\theta + j\sin\theta). \tag{A10.11}$$

Let us substitute (A10.9) in (A10.11) to write

$$\Gamma_l = \left(\frac{\text{VSWR} - 1}{\text{VSWR} + 1}\right)[\cos(\pi + 2\beta\Delta d) + j\sin(\pi + 2\beta\Delta d)]$$

which simplifies to

$$\Gamma_l = -\left(\frac{\text{VSWR} - 1}{\text{VSWR} + 1}\right)(\cos 2\beta\Delta d + j\sin 2\beta\Delta d). \tag{A10.12}$$

In order to obtain an expression for Z_l in terms of Γ_l given by (A10.12) we can next put (A10.4) as

$$\frac{\Gamma_l}{1} = \frac{Z_l - Z_0}{Z_l + Z_0}$$

and manipulate it as

$$\frac{\Gamma_l + 1}{\Gamma_l - 1} = \frac{(Z_l - Z_0) + (Z_l + Z_0)}{(Z_l - Z_0) - (Z_l + Z_0)} = \frac{2Z_l}{-2Z_0} = -\frac{Z_l}{Z_0},$$

whence we obtain

$$Z_l = \frac{1 + \Gamma_l}{1 - \Gamma_l} Z_0. \tag{A10.13}$$

In summary, let us put together (A10.12) and (A10.13) as

$$\left.\begin{aligned} Z_l &= \frac{1 + \Gamma_l}{1 - \Gamma_l} Z_0 \\ \Gamma_l &= -\left(\frac{\text{VSWR} - 1}{\text{VSWR} + 1}\right)(\cos 2\beta\Delta d + j \sin 2\beta\Delta d) \end{aligned}\right\}. \tag{A10.14}$$

We have used (A10.14) derived here in Section 10.1.7.A.

Appendix A10.3

Algebraic steps leading to the derivation of equations for generating constant-r and constant-x circles of the Smith chart

Let us, for the sake of convenience recall (10.68):

$$r = \frac{1 - u^2 - v^2}{(1 - u)^2 + v^2}$$

which, by cross multiplication becomes

$$r[(1 - u)^2 + v^2] = 1 - u^2 - v^2. \tag{A10.15}$$

By simple algebra, we can express (A10.15) as

$$ru^2 - 2ur + r + u^2 - 1 + v^2(1 + r) = 0. \tag{A10.16}$$

At this stage, let us tacitly add the quantity $1/(1 + r)$ to both sides of (A10.16) to put it as

$$ru^2 - 2ur + r + u^2 - 1 + v^2(1 + r) + \frac{1}{1 + r} = \frac{1}{1 + r}. \tag{A10.17}$$

We can further express (A10.17) as

$$ru^2 - 2ur + \left(r - 1 + \frac{1}{1+r}\right) + u^2 + v^2(1+r) = \frac{1}{1+r}$$

which simplifies to

$$ru^2 - 2ur + \frac{r^2}{1+r} + u^2 + v^2(1+r) = \frac{1}{1+r}. \tag{A10.18}$$

We can then express (A10.18) as

$$u^2(1+r) - 2ur + \frac{r^2}{1+r} + v^2(1+r) = \frac{1}{1+r}$$

which next can also be put as

$$(1+r)\left(u^2 - 2u\frac{r}{1+r} + \frac{r^2}{(1+r)^2} + v^2\right) = \frac{1}{1+r}. \tag{A10.19}$$

Let us next write (A10.19) as

$$u^2 - 2u\frac{r}{1+r} + \frac{r^2}{(1+r)^2} + v^2 = \left(\frac{1}{1+r}\right)^2$$

which by simply combining the first three terms of its left-hand side becomes:

$$\left(u - \frac{r}{r+1}\right)^2 + v^2 = \left(\frac{1}{1+r}\right)^2. \tag{A10.20}$$

The expression (A10.20) is the same as (10.70) in the text.

Similarly, next let us recall (10.69):

$$x = \frac{2v}{(1-u)^2 + v^2}$$

which, by cross multiplication and rearrangement of terms becomes

$$x[(1-u)^2 + v^2] - 2v = 0. \tag{A10.21}$$

Let us multiply (A10.21) by the quantity x to write it as

$$x^2[(1-u)^2 + v^2] - 2vx = 0. \tag{A10.22}$$

Further, by adding the quantity 1 to both sides of (A10.22), we can write it after rearrangement of terms as

$$x^2[(u-1)^2 + v^2] - 2vx + 1 = 1. \tag{A10.23}$$

We can further arrange the terms of (A10.23) to put it as

$$x^2(u-1)^2 + x^2v^2 - 2vx + 1 = 1,$$

which can also be expressed as

$$x^2(u-1)^2 + (xv-1)^2 = 1. \tag{A10.24}$$

Now, if we divide (A10.24) by the quantity x^2, we obtain

$$(u-1)^2 + (v - \frac{1}{x})^2 = \frac{1}{x^2}. \tag{A10.25}$$

The expression (A10.25) is the same as (10.71) in the text.

Summarising Notes

✓ Waveguide resonator at a defined resonant frequency can be made out of an appropriately chosen length of a waveguide with its both ends either closed by a conductor or kept open.

✓ Transmission line theory is an easy approach of treating a waveguide resonator.

✓ Resonator length has been found by transmission line theory as an integral multiple of half the guide wavelength for both closed-ended and open-ended resonators.

✓ Basic concepts of transmission line theory have been developed such as

- distributed transmission line parameters;
- telegrapher's equation;
- condition for distortionless transmission;
- input impedance of the line terminated in a load impedance;
- characteristic impedance of the line;
- voltage standing-wave ratio (VSWR) of the line;
- impedance matching such as in radome for the protection of an antenna and radar duplexer of a radar system; and
- Smith chart: theory and application to transmission line problems to make them simpler.

✓ Resonant frequency of the waveguide resonator can be found with the help of transmission line theory using the dispersion relation of the waveguide.

✓ Field theory can be applied to treat a waveguide resonator as an alternative to transmission line theory.

✓ Field solutions and electromagnetic boundary conditions, typically for a rectangular waveguide resonator—closed-ended and excited in the dominant TE_{10} mode—has yielded (i) the same resonator length as predicted by the transmission line theory and (ii) an additional mode number p of the waveguide resonator, to be read with reference to TE_{10}-mode excitation of the waveguide as TE_{10p} mode of the resonator (which may be generalised as TE_{mnp} mode of the resonator with reference to TE_{mn}-mode excitation of the waveguide).

✓ Field solution and relevant electromagnetic boundary conditions can be used to obtain (i) the expression for the time-averaged energy stored in electric and magnetic fields and (ii) the expression for the power loss in resonator walls.

✓ Expression for the quality factor of a resonator in the TE_{10p} mode at the resonant frequency, in terms of the resonator dimensions and the surface resistance of the conducting material making the resonator, has been derived with the help of the expressions for the time-averaged energy stored and power loss in the resonator.

✓ Relation between the unloaded quality factor, external quality factor and loaded quality factor of a cavity has been obtained keeping in view some part of energy stored in a cavity being coupled out from it to an external load in practice.

✓ Quality factor of a cavity resonator can also be expressed in terms of the resonant frequency and the bandwidth of the frequency response of half the value of the square of the ratio of the magnitude of the equivalent impedance of the resonant circuit comprising an inductor, capacitor and resistor in parallel.

Review Questions

Narrative type

10.1 How can you represent a transmission line by a set of primary line parameters R, L, C and G uniformly distributed over the length of the line? How can you reduce the error of regarding the line parameters to be constant with distance and time over an element of length of the line?

10.2 Derive, using the basic circuit laws, the telegrapher's equations of a transmission line.

10.3 Starting from the telegrapher's equations, obtain the expressions for the circuit potential and current on a transmission line terminated in a load impedance.

10.4 How can the phase propagation constant and the attenuation constant of a transmission line be determined in terms of the line parameters?

10.5 Show that the attenuation constant α and the phase velocity v_p of a transmission line are given by $\alpha = \sqrt{RG}$ and $v_p = 1/\sqrt{LC}$ respectively if the line parameters satisfy the condition: $RC = LG$.

10.6 What do you understand by distortionless transmission through a transmission line? Show that the line parameters should satisfy the condition $RC = LG$ for distortionless transmission.

10,7 Define the reflection coefficient of a transmission line in terms of the terminating load impedance of the line.

10.8 Derive an expression for the input impedance of a lossy transmission line in terms of the terminating load impedance of the line.

10.9 Write down the expression for the input impedance of a lossless transmission line in terms of the terminating load impedance of the line and interpret the corresponding expression for a lossy line.

10.10 How do you define the characteristic impedance of a transmission line of a finite length?

10.11 What is the input impedance of an infinitely long transmission line?

10.12 Construct a vector diagram as the sum of two vectors, one of which is fixed and the other rotating in a circular path with its base fixed on the tip of the fixed vector to represent a quantity proportional to the voltage on a transmission line.

Hint: See the vector diagram depicted in Fig. 10.3.

10.13 What do you understand by the reflection coefficient and the voltage standing wave ratio VSWR of a transmission line? Deduce a relation between them.

10.14 Develop a theory that leads to a method of finding the value of unknown complex load impedance from the study of the shift of the voltage standing wave pattern of a transmission line as the complex load terminating the line is replaced by a short.

Hint: See Appendix A10.2.

10.15 In the method of finding the value of unknown complex load impedance mentioned in the question 10-14 above, will (a) the first voltage amplitude extremum from the load end of the standing-wave pattern of the transmission line, being a maximum or minimum, depend on the value of the complex impedance? (b) Will the shift in the voltage standing-wave pattern, when the terminating short replaces the complex load—being towards the load or the source end of the line—depend on the value of the complex impedance? Give reasons for your answer.

10.16 In the method of finding the value of unknown complex load impedance mentioned in 10-14, would you give preference to estimating the shift in the voltage standing-wave pattern over finding the location of the first maximum or minimum from the load end? Give reasons for your answer.

10.17 Show that on the standing-wave pattern of a lossless transmission line of characteristic impedance (resistance) Z_0 terminated in a load impedance, the impedances $Z_{\max}$ and $Z_{\min}$ at the voltage maxima and minima are related to the standing-wave ratio VSWR as: $Z_{\max} = (Z_0)(\text{VSWR})$ and $Z_{\min} = (Z_0)/(\text{VSWR})$ respectively.

Hint: See Appendix A10.1.

10.18 Taking a lossless transmission line, prove the relation $Z_0 = \sqrt{Z_{SC} Z_{OC}}$, where Z_0 is the characteristic impedance (resistance) of the line and Z_{SC} and Z_{OC} are the input impedances of the line when the line is terminated in short and open circuits respectively.

10.19 How are the systems of constant-$|\Gamma_l|$ circles and constant-θ radial lines generated on the reflection coefficient θ plane of the Smith chart representing the behaviour of a lossless transmission line terminated in a complex load impedance, where $|\Gamma_l|$ and θ are the magnitude and the phase angle of the reflection coefficient $\Gamma_l (= |\Gamma_l| \exp j\theta)$. Hence, what does a point on this plane represent?

10.20 For a lossless transmission line terminated in a complex load, derive the expressions: $[u - r/(r+1)]^2 + v^2 = 1(r+1)^2$ and $(u-1)^2 + (v - 1/x)^2 = 1/x^2$ representing the constant-r and constant-x circles respectively, of the Smith chart, where μ and ν are the real and imaginary parts of the reflection coefficient, and r and x are the real and imaginary parts of the load impedance of the line normalised with respect to the characteristic resistance.

Hint: See Appendix A10.3 for the details of mathematical steps for the required derivation.

10.21 Examining the expressions representing the constant-r and constant-x circles, find the radius as well as the intercepts on u and v axes on the reflection coefficient Λ_l plane of a constant-r circle of the Smith chart.

10.22 Examining the constant-r circle, find the value of r which the largest constant-r circle of the Smith chart refers to?

10.23 Identify the point on the v versus u plot of the Smith chart where the centre of the constant $r = 0$ circle is located vis-à-vis the intersection of the constant $r = 1$ circle with the real u axis of the chart.

10.24 Find the common point of meeting of all the constant-r circles on the real u axis of the Smith chart.

10.25 Show that the constant $r = 0$ and $r = \infty$ circles pass through the extreme left and right points of the real u axis respectively of the Smith chart.

10.26 Show that the constant $r = \infty$ circle of the Smith chart is a circle of zero radius that shrinks to a point.

10.27 Explain how the radius of the constant-r circle of the Smith chart changes with the change in the value of r.

10.28 Explain how the radius of the constant-x circle of the Smith chart changes with the change in the value of x.

10.29 Find the common point of the meeting of all the constant-x circles on the real u axis of the Smith chart.

10.30 Identify the axis of the Smith chart across which two constant-x circles of the Smith chart, which are assigned the same values of x but of opposite signs, may be considered as the mirror images of each other.

10.31 How do you read the value of VSWR of a lossless transmission line on the Smith chart if you know the value of the reflection coefficient $|\Lambda_l|$ of the line?

10.32 How do you locate the point on the Smith chart that represents a value of the impedance at a point on a lossless transmission line?

10.33 Find the appropriate values of r of the constant-r circles of the Smith chart that would give the values of the impedance maxima and minima on a lossless transmission line.

10.34 (i) Which constant-$|\Gamma_l|$ circle or which constant-r circle and (ii) which points on it represent the short circuit and open circuits at the load end on the Smith chart?

10.35 Discuss how you can represent purely inductive, purely capacitive, and matched loads on the Smith chart.

10.36 Describe the method of using the Smith chart to find the input impedance of a lossless transmission line terminated in a complex load for a given operating frequency or wavelength.

10.37 How do you locate the point on the Smith chart that would represent the value of the admittance at a point of a lossless transmission line?

10.38 Discuss how the method of using the Smith chart to find the input impedance from the load impedance of a lossless line of a given length at a known operating frequency or wavelength can be extended to find a method for estimating the value of both the attenuation constant α and the phase constant β of a line with losses.

10.39 Discuss with a relevant theory how to choose the value of the characteristic resistance of a lossless quarter-wave ($\lambda/4$) impedance transmission-line transformer to be connected to a resistive load to be matched with a lossless transmission line.

10.40 Describe the role of quarter-wave impedance transformer in a branch-type radar duplexer that uses TR and ATR switches.

10.41 Describe the circuit arrangement of single-stub matching and the method of implementing it using the Smith chart with the relevant theory.

10.42 Describe the circuit arrangement of double-stub matching and the method of implementing it using the Smith chart with the relevant theory.

10.43 Treating a rectangular waveguide resonator as a transmission line, derive the expressions for the resonant length and resonant frequency of the resonator in terms of its dimensions and modes. Take two situations: one in which the waveguide is closed by the conducting walls at both of its ends and the other in which it is open.

10.44 Starting from first principles, obtain the expressions for fields E_r, H_z and H_z of a close-ended rectangular waveguide resonator excited in the TE_{101} mode.

10.45 Derive an expression for the time-averaged energy stored in the electric and magnetic fields at resonance of a close-ended rectangular waveguide resonator excited in the TE_{101} mode in terms of the waveguide dimensions.

10.46 Derive an expression for the power lost in the walls of a close-ended rectangular waveguide resonator excited in the TE_{101} mode in terms of the resonator dimensions and the surface resistance of the material of the resonator walls.

10.47 How is the quality factor of a resonator defined in terms of (i) the time-averaged energy stored in the resonator and (ii) the energy lost in the resonator walls in a single wave time period or, alternatively, the power lost in the resonator walls and wave angular frequency?

10.48 Derive an expression for the quality factor at resonance of a close-ended rectangular waveguide resonator excited in the TE_{101} mode in terms of the resonator dimensions and the surface resistance of the material of the resonator walls.

10.49 Obtain a relationship between the unloaded, external and loaded quality factors of a cavity resonator when the cavity becomes loaded; t some part of energy stored in the cavity is coupled out from it to an external load.

10.50 Representing a waveguide resonator as a resonant circuit comprising an inductor, a capacitor and a resistor in parallel, show that the quality factor of the resonator becomes equal to the ratio of the resonant frequency to the angular frequency bandwidth of the frequency response of half the value of $(|Z_{eq}/R)^2$ of the resonant circuit, where Z_{eq} is the equivalent impedance of the parallel resonant circuit and R is the resistance of the resistor in the circuit, which is also the value of Z_{eq} at resonance.

Review questions with solutions

10.51 An open-ended cavity is partially reflecting at both of its ends, the energy diffracting out of cavity mostly from one of these ends. Define the diffractive quality factor of the cavity as $Q_D = \omega\tau_{\text{rad}}$, where τ_{rad} is the radiative energy loss time constant, which, in turn, is defined as the ratio of the energy stored in the cavity W_s to radiative energy loss per second or radiative power loss $P_{l,\text{rad}}$. Hence, find the diffractive quality factor of the cavity Q_D in terms of the reflection coefficients R_1 and R_2 at both the ends of the cavity, the cavity length l and the wavelength λ in free space.

Answer: For the sake of convenience, the two definitions to start with may be put as

$$Q_D = \omega\tau_{\text{rad}} \tag{1}$$

and

$$\tau_{\text{rad}} = \frac{W_s}{P_{l,\text{rad}}} \tag{2}$$

where τ_{rad} is the radiative energy loss time constant, W_s is the energy stored in the cavity and $P_{l,\text{rad}}$ is the radiative power loss from the cavity due to diffraction from the cavity ends.

A part of the energy stored in the cavity is lost due to radiation, the remaining part being present in the cavity due to reflection. The phenomenon can be represented as

$$W_s = W_{l,\text{rad}} + W_{\text{reflected}} \tag{3}$$

where W_s, $W_{l,\text{rad}}$ and $W_{\text{reflected}}$ are the energy stored in the cavity, the energy lost due to radiation from the cavity and the energy reflected back into the cavity from the two ends of the cavity respectively.

Since the energy moves with the group velocity v_g, the power lost due to radiation $P_{l,\text{rad}}$ is obtained by dividing the energy lost due to radiation $W_{l,\text{rad}}$ by the time of travel l/v_g through the cavity of length l as follows:

$$P_{l,\text{rad}} = \frac{W_{l,\text{rad}}}{l/v_g} = \frac{W_{l,\text{rad}}v_g}{l}. \tag{4}$$

With the help of (1), (2) and (4), we can show that

$$Q_D = \omega\frac{W_s}{\dfrac{W_{l,\text{rad}}v_g}{l}} = \omega\frac{W_s l}{W_{l,\text{rad}v_g}}. \tag{5}$$

Next, putting $W_{l,\text{rad}} = W_S - W_{\text{reflected}}$, the latter obtained from (3), in (5), we can write

$$Q_D = \omega\frac{W_s}{W_s - W_{\text{reflected}}}\frac{l}{v_g},$$

which can be expressed as

$$Q_D = \omega = \frac{1}{1 - \dfrac{W_{\text{reflected}}}{W_s}} \frac{1}{v_g}. \tag{6}$$

Further, $W_{\text{reflected}}$ is related to W_s through the reflection coefficients R_1 and R_2 of the two ends respectively of the cavity as

$$\frac{W_{\text{reflected}}}{W_s} = R_1 R_2 \tag{7}$$

where $R_1 R_2$ is the joint reflection coefficient being the product of the individual reflection coefficients.

In view of (7), we can then read (6) as

$$Q_D = \omega \frac{1}{1 - R_1 R_2} \frac{l}{v_g} \quad . \tag{8}$$

Now, for the fundamental-mode cavity (,$p = 1$) we know from (10.105) and the discussion following it that the length of the cavity l should satisfy the relation:

$$l = \frac{\pi}{\beta}. \tag{9}$$

We can then express v_g occurring in (8) with the help of the relation $v_{ph} v_g = c^2$, given by (9.66), and the relation $v_{ph} = \omega/\beta$ and the relation (9), as follows:

$$v_g = \frac{c^2}{v_{ph}} \frac{c^2}{\omega/\beta} = \frac{c^2 \beta}{\omega} = \frac{c^2 \pi}{\omega l} \tag{10}$$

Plugging (10) in (8) gives

$$Q_D = \omega = \frac{1}{1 - R_1 R_2} \frac{1}{V_g} = \omega \frac{1}{1 - R_1 R_2} \frac{l}{\frac{c^2 \pi}{\omega l}} = \frac{1}{1 - R_1 R_2} \frac{\omega^2 l^2}{c^2 \pi}. \tag{11}$$

Putting $\omega = 2\pi f = 2\pi c/\lambda$ in (11), we then obtain the relation

$$Q_D = \frac{1}{1 - R_1 R_2} \frac{4\pi l^2}{\lambda^2}$$

Numerical type

10.52 A transmission line of characteristic resistance 75 Ω is connected to a voltage source of zero internal resistance at the input end and terminated in a resistive load of 100 Ω. Calculate the voltage reflection coefficients at the load and source ends respectively.

Answer:

$(100 - 75)/(100 + 75) = 1/7$ at the load end and $(0 - 75)/(0 + 75) = -1$ at the source end.

10.53 A lossless transmission line of length 3 cm and characteristic resistance 100 Ω is fed by a voltage source at a frequency of 10 GHz and is terminated in a load impedance of $10 - j10\ \Omega$. Using the expression for the input impedance of a lossless line in terms of the terminating load impedance, verify that the input impedance of the line is the same as the given load impedance $10 - j10\ \Omega$. How will the value of the input impedance of the line modify if you changed the line characteristic resistance from 100 to 50 Ω?

Answer: Let us recall the expression (10.38) and write it as

$$Z_{\text{in}} = Z_0 \frac{Z_l + jZ_0 \tan \beta l}{Z_0 + jZ_i \tan_l \tan \beta l} \tag{1}$$

where the symbols have been defined in the text of the chapter.

In the present problem, the given operating frequency is $f = 10$ GHz $10 \times 10^9 = 10^{10}$ Hz, and the length of the line is $l = 3$ cm. That gives the wavelength and the quantity βl as

$$\left.\begin{aligned} \lambda &= c/f = (3 \times 10^{10})/(10^{10}) = 3 \text{ cm} \\ \beta l &= (2\pi/\lambda)l = (2\pi/3) \times (3) = 2\pi \end{aligned}\right\} \tag{2}$$

Substituting βl from (2) in (1), we then obtain

$$Z_{\text{in}} = Z_0 \frac{Z_l + jZ_0 \tan 2\pi}{Z_0 + jZ_l \tan 2\pi} Z_0 \frac{Z_l + jZ_0 \times 0}{Z_0 + jZ_l \times 0} = z_l. \tag{3}$$

Substituting $Z_l = 10 - j10\ \Omega$ in (3), we then obtain the value of the input impedance $Z_l = 10 - j10\ \Omega$. Further, in the present case it can be seen from (3) that Z_{in} is independent of the characteristic impedance Z_0. Therefore, the value of the input impedance $Z_{\text{in}} = 10 - j10\ \Omega$ obtained would not have changed had we changed Z_0 from 100 to 50 Ω.

10.54 Calculate the VSWR of a lossless transmission line of characteristic resistance 50 Ω if it is terminated in load impedance $Z_l = 75 + j40\ \Omega$.

Answer: Substituting $Z_l = 75 + j40\ \Omega$ (given) and $Z_0 = 50\ \Omega$ in (10.32), we can find the reflection coefficient of the line Γ_l as

$$\begin{aligned} \Gamma_l &= \left(\frac{Z_l - Z_0}{Z_l - Z_0}\right) = \frac{25 + j40}{125 + j40} = \frac{25 + j40}{125 + j40} \times \frac{25 - j40}{125 - j40} \\ &= \frac{(25 + j40)(125 - j40)}{(125)^2 + (140)^2} \frac{(25 + j40)(125 - j40)}{1562 + 1600} \qquad (1) \\ &= \frac{4725 + j4000}{17225} = 0.2743 + j0.2322 \end{aligned}$$

The magnitude of the reflection coefficient $|\Gamma_l|$ can be calculated from (1) as

$$|\Gamma_l| = \sqrt{(0.2743)^2 + (0.2322)^2} = \sqrt{0.1291} = 0.359 \tag{2}$$

Putting the value of $|\Gamma_l|$ from (2) in (10.57), we can find

$$\text{VSWR} = \frac{1 + |\Gamma_l|}{1 - |\Gamma_l|} = \frac{1.359}{0.641} = 2.12$$

10.55 A lossless transmission line of characteristic resistance 50 Ω is connected to a load impedance of $40 - j50\ \Omega$. Find the values of maximum and minimum resistances in the line standing-wave.

Answer: In the problem, the load impedance given is $Z_l = 40 + j50\,\Omega$. The reflection coefficient Γ_l can be found, in view of the given line characteristic impedance (resistance) $Z_0 = 50\ \Omega$, as follows:

$$\Gamma_l = \frac{Z_l - Z_0}{Z_l + Z_0} = \frac{40 - j50 - 50}{40 - j50 + 50} = \frac{-10 - j50}{90 - j50}. \tag{1}$$

Rationalising the numerator and denominator of (1), we get the reflection coefficient as

$$\Gamma_l = \frac{-10 - j50}{90 - j50} \times \frac{90 + j50}{90 + j50}$$

$$= \frac{-900 - j500 - j4500 + 2500}{(90)^2 + (50)^2} = \frac{1600 - j5000}{8100 + 2500} = 0.1509 - j0.471. \tag{2}$$

From (2) the magnitude of the reflection coefficient can be found as

$$|\Gamma_l| = \sqrt{(0.1509)^2 + (0.4717)^2} = \sqrt{0.02277 + 0.2250} = \sqrt{0.24527} = 0.495. \tag{3}$$

Using (3), we can find the voltage standing-wave ratio of the line as

$$\text{VSWR} = \frac{1 + 0.495}{1 - 0.495} = \frac{1.495}{0.505} = 2.9604. \tag{4}$$

As explained in the text of the chapter [see Section 10.1.7 and the discussion following (10.57)], multiplying $Z_0 (= 50)\ \Omega$ (given in the problem) by VSWR $(= 2.9604)$ given by (4), we can find the value of the maximum resistance on the line as (i) $50 \times 2.9604 = 148.02\ \Omega$. Similarly, dividing $Z_0 (= 50)\ \Omega$ (given in the problem) by VSMR $(=2.9604)$ given by (4), we can find the value of the minimum resistance on the line as (ii) $50/2.9604 = 16.89\ \Omega$.

10.56 Apply the Smith chart for the solution to the problem in Example E10-4.

Answer: The VSWR = 1.5 of the line is given. Draw the constant VSWR = 1.5 circle, which is essentially the constant-$|\Gamma_l|$ circle passing through the point on the real u axis through which the constant $r = 1.5$ circle passes (see Section 10.1.8.D).

The extreme left end point of the constant $VSWR = 1.5$ circle will then represent the point of a voltage minimum on the line (Section 10.1.8.G), here, to be taken as the first voltage minimum. Next, move around this constant VSWR = 1.5 circle through the distance 0.21 wavelengths (given) 'anticlockwise towards the load' to reach the point P representing the complex load. The scale to measure this movement is provided at the periphery of the outermost constant-r circle (constant $r = 0$ circle). With reference to the present problem you will see that the constant $r = 1.4$ and constant $x = -0.28$ circles will pass through the point P representing the load. Therefore, the load impedance normalised with respect to the characteristic impedance comes out to be $z_i (= z_l/z_0) = 1.4 - j0.28$, which, in turn gives, remembering $z_0 = 50\ \Omega$, the value of the load impedance as $Z_l = (1.4 - j0.28) \times 50 = 70 - j14\ \Omega$, a value that comes close to that found analytically in Example E10-4.

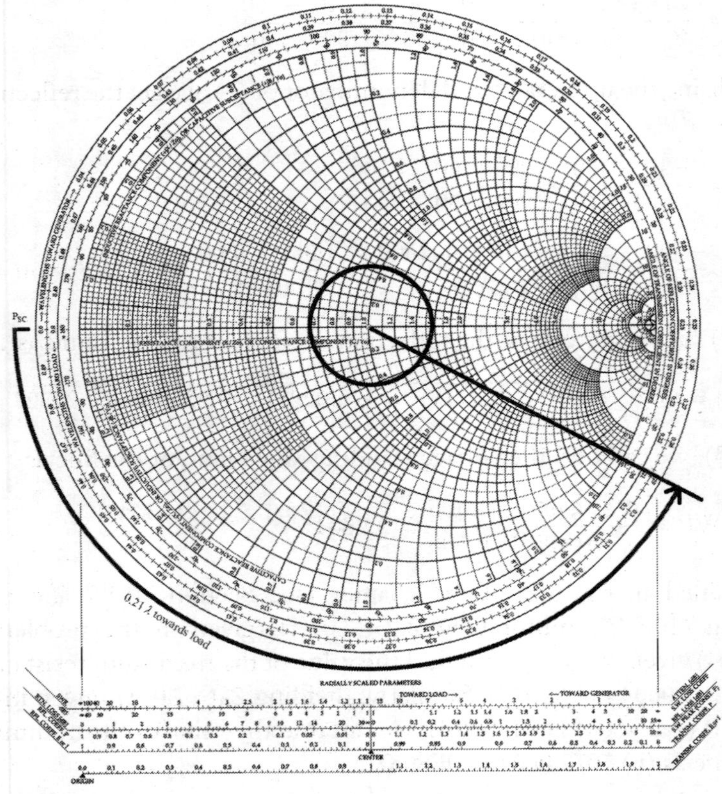

10.57 A rectangular cavity has the dimensions 3 cm (broad), 2 cm (narrow) and 5 cm (linear) and is excited in the mode TE_{101}. Find its frequency of resonance. What will be its value if the broad dimension is changed from 3 to 4 cm?

Answer: With the help of (10.92), we can find the resonant frequency f_r for the TE_{101} mode as

$$f_r = \frac{1}{2}\left[\left(\frac{1}{2}\right)^2 + \left(\frac{1}{2}\right)^2\right]^{1/2} c. \quad (1)$$

Substituting the given data in (1), we get

$$f_r = \frac{1}{2}\left[\left(\frac{100}{3}\right)^2 + \left(\frac{100}{5}\right)^2\right]^{1/2} \times 3 \times 10^8 \text{ Hz}. \quad (2)$$

Solving the above, we obtain the resonant frequency as

$$f_r = 58.3 \times 10^8 \text{ Hz} = 5.83 \text{ GHz}$$

The value of the resonant frequency is independent of the broad dimension b for the mode TE_{101} [as can be seen from (1)] and, therefore, this value will not change if the broad dimension are changed from 3 to 4 cm.

10.58 Show that the cavity resonator (width 33 cm, depth 33.8 cm, height 26.8 cm) excited in the mode TE_{333} of a typical domestic microwave oven supports waves close to the ISM frequency band (2.45 GHz). State the number of regions of concentration of high electric field in the cavity.

Hint: Calculate the resonant frequency with the help of (10.92) taking $m = n = p = 3$ and the dimensions $a = 26.8 \times 10^{-2}$ m, $b = 33.0 \times 10^{-2}$ m and $l = 33.8 \times 10^{-2}$ m and hence verify if this frequency value is close to the ISM (Industrial, Scientific and Medical) frequency band. Appreciate, with the help of the interpretation of mode numbers for the TE_{333} mode (see Sections 9.3 and 10.2), that there are 27 = (3 × 3 × 3) regions of maximum field concentration.

Multiple-choice questions

10.59 The attenuation constant α and the phase constant β of a transmission line characterised by the distributed parameters is given by $\alpha = 0$ and $\beta = \omega(LC)^{1/2}$. With reference to this line, one of the following statements is incorrect.

(a) $R = 0$ (b) $G = 0$

(c) The line is distortion-free (d) The phase velocity varies with frequency.

10.60 An infinitely long lossless transmission line of characteristic resistance 75 Ω is terminated in a load resistance of Ω. The input impedance of the line is

(a) 0 Ω (b) 75 Ω (c) 150 Ω (d) ∞

10.61 The input impedance of a half-wave lossless transmission line of characteristic resistance of 50 Ω is terminated in a load resistance of 75 Ω is

(a) 75 Ω (b) 50 Ω (c) 33.33 Ω (d) 11.25 Ω

10.62 A lossless transmission quarter-wave ($\lambda/4$) line of characteristic resistance 75 Ω is open-ended (that is terminated in an infinite impedance). The input impedance of the line is

(a) 0 (b) 75 Ω (c) $75 \tan \pi/8\,\Omega$ (d) ∞

10.63 A lossless transmission quarter-wave ($\lambda/4$) line of characteristic resistance 75 Ω is short-ended, that is terminated in zero impedance. The input impedance of the line is

(a) 0 (b) 75 Ω (c) $75 \tan \pi/8\,\Omega$ (d) ∞

10.64 A short-circuited load of a lossless transmission line is located on the Smith chart at

(a) extreme left point on the real u axis of the constant $r = \infty$ circle
(b) extreme right point on the real u axis of the constant $r = \infty$ circle
(c) extreme left point on the real u axis of the constant $r = 0$ circle
(d) extreme right point on the real u axis of the constant $r = 0$ circle

10.65 An open-circuited load of a lossless transmission line is located on the Smith chart at

(a) extreme left point on the real u axis of the constant $r = 0$ circle
(b) extreme right point on the real u axis of the constant $r = 0$ circle
(c) extreme left point on the real u axis of the constant $r = \infty$ circle
(d) extreme right point on the real u axis of the constant $r = \infty$ circle

10.66 The constant-$x = 1$ circle of the Smith chart with reference to a lossless transmission line has the radius, on the reflection coefficient $\Gamma_l = (u + jv)$ plane, as follows.

(a) ∞ (b) 0.5 (c) 1 (d) 0

10.67 A short-ended rectangular cavity is constructed out of a length 200 mm of rectangular waveguide of guide wavelength 400 mm by shorting both its ends along the length by conducting planes. The component can work as an open-ended rectangular cavity by removing the shorting conducting planes with

(a) no modification of its length
(b) modification of its length to 150 mm
(c) modification of its length to 100 mm
(d) modification of its length to 500 mm

10.68 The resonant wavelength of the mode TE_{101} of a rectangular cavity

(a) depends on its both broad and narrow dimensions a and b respectively
(b) depends on its both broad and length dimensions a and l respectively
(c) depends only on its broad dimension a
(d) depends only on its length dimension l

Answers

10.59 (d) 10-60 (b) 10-61 (a) 10-62 (a) 10-63 (d) 10.64 (c) 10-65 (b)
10-66 (c) 10-67 (a) 10-68 (b)

11

Summary

WHat this book is about and what the author's objective is in writing this book have been explained in the introductory chapter. We have seen the relevance and importance of electromagnetics—the subject of this book—in electrical, electronics and communication engineering. We have also appreciated how mathematics does not make the subject of electromagnetics difficult but simpler. The subject has been developed from first principles, providing adequate mathematical steps. We have also seen the scope of the book. Care has been taken to select topics of the book such that readers are able to grasp the essence of engineering electromagnetics within a reasonable period of time without getting lost in lengthy exercises on each of these topics (Chapter 1).

We can choose rectangular, cylindrical or spherical-polar coordinate system depending on whether the electromagnetic problem dealt with enjoys rectangular, cylindrical or spherical symmetry (Chapter 2). While formulating and solving the problem, we often need to use relevant vector calculus expressions such as the gradient of a scalar quantity, like the electric potential, and the divergence and curl of a vector quantity, like the electric field, in the desired coordinate system. The ability to write these vector calculus expressions any time in the desired coordinate system (Chapter 2) gives one the confidence to solve problems. We have deduced the vector calculus expressions from first principles in the generalised curvilinear coordinate system (Chapter 2). We have seen that each of these expressions in the curvilinear system is not only easy to deduce, but are also easy to remember in the sense that if one remembers one of the terms of the expression, the remaining terms can be written simply by permutation. Further, the expression for the Laplacian of a scalar or a vector quantity is obtained by combining the gradient and the divergence expressions (Chapter 2).

We have presented some of the basic laws and concepts of static electric fields on which a number of concepts of electromagnetic theory are based (Chapter 3). We have reviewed Coulomb's and Gauss's laws. We have seen how Gauss's law can be used to find the electric field in problems that enjoy geometrical symmetry. We have given a method of relating the electric field with the electric potential in vector form (Chapter 3).

We have also deduced Poisson's and Laplace's equations. The vector calculus expressions can be used to read these equations in the desired coordinate system. A few illustrative examples have been worked out; for example, we have seen how these basic laws and equations can be applied to find the electric fields due to a line charge, a sheet of charge and a charged conductor (Chapter 3).

Further, we have shown how, for a space-charge limited diode, the anode current is related to the anode potential as well as to the distance between the cathode and the anode by what is known as Child-Langmuir's law (Chapter 3). We have also seen how we can find the electric field inside a beam of electrons and, in the limit, at the beam edge. Such concepts have useful applications, for instance in studying the formation and confinement of an electron beam in an electron beam device and understanding its interaction with electromagnetic waves in the device (Chapter 3).

We have also dealt with some basic concepts of static magnetic fields (Chapter 4). We have introduced the magnetostatic quantities analogous to the corresponding electrostatic quantities. We have developed Coulomb's and Gauss's laws as well as Poisson's and Laplace's equations of magnetostatics on the same line as electrostatics (Chapter 4).

Further, we have brought out concepts concerning the steady magnetic field due to a steady current flowing through a conductor, namely, Biot-Savart's law and Ampere's circuital law, as well as the concept of the Lorentz force on a moving charge in a magnetic field and that of the force on a current-carrying conductor placed in a magnetic field. We have also provided examples to illustrate these concepts. For instance, we have demonstrated how the use of Ampere's circuital law makes the problems of static magnetic fields rather easy to treat, as does Gauss's law for the problems of electrostatics (Chapter 4).

Furthermore, we have appreciated that Gauss's law of magnetostatics and Poisson's equation of magnetostatics are manifestation of the absence of free magnetic charges (poles), or that of the continuity of magnetic flux lines (Chapter 4).

The development of the basic concepts of time-varying electric and magnetic fields begins with the continuity equation (Chapter 5). We started from the conservation of charge in a region of charge flow and obtained the continuity equation at a point by relating the divergence of the current density with the time rate of variation of volume charge density at the point. The continuity equation has been subsequently used to find the relaxation time of a medium, which measures how long a charge injected into a medium stays in the bulk of the medium. The relaxation time depends on the conductivity and permittivity of the medium. That a conductor can be charged only at its surface and that a dielectric can be charged throughout its volume can be understood from the fact that the relaxation time of a conductor is very short, while that of a dielectric is very long (Chapter 5).

We have also formulated Faraday's law for time-varying magnetic fields and expressed the law in both its integral and differential forms. We have seen how the electric field in electromagnetic induction is associated with the time-varying magnetic field and thus how electric and magnetic fields are coupled in the time-varying phenomena. The concept of time-varying electric fields has further helped us understand how a medium, even if it is non-conducting, for instance a dielectric, allows a current in the form of the displacement current to pass through it (Chapter 5).

The concept of the displacement current has been developed with the help of the continuity equation, the Poisson's equation and the differential form of Ampere's circuital law. That a capacitor allows time-varying current to pass through can also be understood from the concept of displacement current. Further, we have also found an expression for the loss tangent of a lossy dielectric in terms of the frequency of a time-periodic electric field and dielectric conductivity and permittivity (Chapter 5).

Importantly, we have formulated Maxwell's equations in both their differential and integral forms, which comprise the backbone of the subject of electromagnetics. The relation between the electric field and the electric potential for the electrostatic case has been modified for a time-varying situation by introducing magnetic vector potential and defining it such that its curl represents magnetic flux density. For this purpose, we have used Maxwell's equation that relates the electric field with the time rate of variation of magnetic flux density, and the relation between the magnetic flux density and the vector potential. Moreover, we have used the relation between the scalar and the vector potentials according to Lorentz condition to form two wave equations in scalar and vector potentials. We have obtained the solutions of these wave equations (Chapter 5) that have application in the analysis of conduction current antennas; this has been taken up later in Chapter 8.

Two of Maxwell's equations have in them electric and magnetic fields coupled. These equations in differential form can be combined, thereby decoupling these fields to obtain the wave equation in electric or magnetic field (Chapter 6). We have obtained the solutions of these equations for uniform, plane electromagnetic waves, considering first the wave propagation through an unbounded free-space medium. Then we considered the corresponding solutions for wave propagation through a semi-infinite conducting medium that makes an interface with a free-space medium, the waves being incident from the free-space medium on the surface of the conducting medium which extends to infinity. Further, the study of the propagation of electromagnetic waves through a conducing medium of finite resistively has led to the concepts of the skin depth, the surface resistance and the ac or RF resistance of the medium (Chapter 6).

Furthermore, with reference to wave propagation through an unbounded free-space medium, we have used the differential form of Maxwell's equations to show that the component of the electric and that of the magnetic field, each in the direction of propagation, is zero. We have also shown that the directions of the electric field, the magnetic field and the wave propagation are mutually perpendicular to each other. In other words, the unbounded free-space medium supports transverse electromagnetic

(TEM) mode of propagation. We have also shown how the intrinsic impedance and the wave phase velocity of the medium are each related to the permeability and the permittivity of free-space (Chapter 6).

Based on the analysis of propagation through a semi-infinite conducting medium, we have shown how the skin depth as well as the surface resistance of a conducting medium depends on conductivity and wave frequency. In an illustrative example, we have also shown how to find the ac or RF resistance of a conducting wire of circular cross section, considering the wire to be of high conductivity and/or for high wave frequencies, when the skin depth of the conductor becomes small enough to be able to make a planar conductor approximation for the conducting wire (Chapter 6).

In addition, we have studied the propagation through unbounded sea-water considering the medium of propagation as a dielectric of finite conductivity, and hence shown that for sea-water communication, lower frequencies, say, ~10 kHz is preferred to higher frequencies, say, ~10 GHz in view of the lower attenuation of the waves at such lower frequencies (Chapter 6).

We have next deduced, starting from first principles, the 'general' electromagnetic boundary conditions at a point on the interface between two media without regard to whether the media are made of a dielectric or a conductor, and irrespective of whether the situations are time-independent or time-dependent (Chapter 7). We have deduced them in vector form involving a unit vector directed from one medium to another. We have also subsequently seen how to read these general electromagnetic boundary conditions at the dielectric-dielectric interface and also at the conductor-dielectric interface for both time-independent and time-dependent situations. In electromagnetic problems, we often come across such boundaries where the field quantities get modified in view of the abrupt change in the medium properties at the interface between the two media. We have also seen how a steady current from one conducting medium to another of a dissimilar conductivity gets modified at the interface between the media (Chapter 7).

While deducing the general electromagnetic boundary conditions at the interface between two media, the definition of surface charge density and that of surface current density emerged. Thus, the definition of surface charge density has been found to be the product of the volume charge density, which is a scalar quantity, and the infinitesimal thickness over which the charge is spread at the interface, in the limit of the thickness tending to zero. Therefore, the charge density is a scalar quantity and its unit is C/m^2. Similarly, the definition of the current charge density has been found as the product of current density, which is a vector quantity, and the infinitesimal thickness at the interface over which the current density is significant, in the limit of the thickness tending to zero. Therefore, the surface current density is a vector quantity and its unit is A/m (Chapter 7).

While interpreting the general boundary conditions for the dielectric-dielectric and the conductor-dielectric interfaces, we have used the following findings from our analysis. The relaxation time being large, the electric charge can stay longer inside the bulk of a dielectric. Thus, the bulk of a dielectric can be electrically charged and a 'finite' volume charge density can be established inside the bulk of a dielectric. The finite volume charge density, in turn, when put in the definition of the surface charge density, results in zero surface charge density at the dielectric surface in the limit of the charge thickness tending to zero. On the other hand, the relaxation time of a good conductor is very small and therefore the charge inside the bulk of a good conductor decays very fast to appear with a very large volume charge density at the conductor surface and gets concentrated over an infinitesimally thin layer on the surface of the conductor. This, in turn, in the limit of the volume charge density tending to infinity and the charge thickness tending to zero, manifest as a finite surface charge density on the conductor surface. This happens for both time-independent and time-dependent situations. Therefore, the bulk of a conductor cannot be electrically charged, and consequently, no electric field or electric displacement is present inside the conductor for both time-independent and time-dependent situations. Further, we can establish a finite magnetic field or magnetic flux density inside a dielectric independently of electric field, for both time-independent and time-dependent situations. The magnetic field and the magnetic flux density each become zero inside a conductor for time-dependent situations since for such situations the magnetic field is coupled to the electric field which is absent in a conductor. However, for time-independent situations, a finite magnetic field or magnetic flux density can be established independently of an electric field in a conductor. Further, at the dielectric surface, the surface current density is zero since the dielectric, supposedly ideal, does not conduct current. For time-independent situations, a finite current can be made to flow through the volume of a conductor which in turn makes the surface current density at the conductor surface zero, while, for time-dependent situations, a large current density can be concentrated over a thin layer on the conductor surface, resulting in a finite surface current density (Chapter 7).

For the sake of convenience in recalling, the general electromagnetic boundary conditions as well as the boundary conditions at the dielectric-dielectric interface and those at the conductor-dielectric interface for both time-independent and time-dependent situations have been listed in tabular form in Chapter 7. These boundary conditions have also been illustrated by a number of examples. Further, the boundary conditions at the interface between two conducting media have given us at the interface the law of refraction of current density as well as an expression for the surface charge density developed. The law of refraction of current density relates the angles of incidence and refraction of current at the interface with the conductivities of the media. Similarly, the expression for the surface charge density obtained involves the relaxation times of the media (Chapter 7).

It was also interesting to see that the circuit law of parallel resistances can be easily derived from one of the electromagnetic boundary conditions that states that the tangential component of electric field is continuous at the interface between two media (Chapter 7).

We have demonstrated, using relevant electromagnetic boundary condition, that standing waves are formed in free-space when a uniform plane electromagnetic wave from the free-space region is incident on the surface of a conducting medium. In such standing waves, the minima of the amplitudes of electric field coincide with the maxima of the amplitudes of magnetic field, and vice versa. Further, it has been appreciated that there would be no power flow associated with such standing waves (Chapter 7).

We have also studied the phenomenon of reflection and refraction of electromagnetic waves at a dielectric-dielectric interface with the help of the relevant electromagnetic boundary conditions with reference to both parallel and perpendicular polarisations. We have analysed the problem to understand that for parallel polarisation there would be no reflection at Brewster's angle of incidence, while there would be total internal reflection for angle of incidence greater than the critical angle, both angles depending on the permittivities of the dielectrics comprising the interface, for both parallel and perpendicular polarisations (Chapter 7).

Further, we developed from first principles the concept of energy stored in electric and magnetic fields and used these concepts to derive an energy balance theorem called the Poynting theorem (Chapter 8). We also derived the complex form of the Poynting theorem that can be used to study time-averaged power flow through a bounded or an unbounded medium and looked at the associated power loss due to the presence of a lossy conducting medium (Chapter 8).

The expression for the energy density in an electric field has been obtained in terms of the electric field and the electric displacement (or the electric flux density), while that for the energy density in magnetic field has been obtained in terms of the magnetic field and the magnetic flux density. In an illustrative example, we used the expression for the energy density in electric field to derive an expression for the energy stored in a capacitor in terms of the capacitance of the capacitor and the charge of the capacitor or, alternatively, the voltage across it. Further, in the typical example of a parallel-plate capacitor that is characterised by its capacitance, the expression for the energy density stored in electric field in terms of the electric field and the electric displacement has been found to be valid. Similarly, with typical reference to the example of a solenoid that is characterised by its inductance, the expression for the energy density stored in the magnetic field in terms of the magnetic field and the magnetic flux density has been found to be valid (Chapter 8).

The Poynting theorem stems from the basic energy balance principle and it encapsulates the phenomenon of the storage, loss and flow of electromagnetic energy. It is stated in terms of the flux of the power density vector or the Poynting vector in or out of a closed region. We have established the Poynting theorem involving instantaneous Poynting vector. As an illustrative example, we used the theorem to appreciate Joule's circuit law for the power loss in a current-carrying wire of finite resistance. Further, by applying the theorem to the problem of a parallel-plate capacitor of circular cross section, we deduced the expression for the energy density in an electric field. We have also similarly obtained the expression for the energy in a magnetic field by applying the theorem to the problem of an inductor in the form of a solenoid of circular cross section (Chapter 8).

More importantly, we have also derived the complex Poynting theorem and thus developed the concept of the time averaged electromagnetic power flow. We have obtained the complex Poynting vector as half the cross product of the electric field vector and the complex conjugate of the magnetic field vector, while the time averaged complex Poynting vector is identified as the real part of the complex Poynting vector. One of the most extensively used concepts of electromagnetic engineering developed is that one can find the average power going out of a volume enclosure as the outward flux of the time averaged complex Poynting vector through the enclosure. For instance, this result is used to obtain power transmitted through a waveguide (Chapter 9) and radiation characteristics of an antenna. We also developed the concept of reactive power flowing into a volume enclosure and its relevance to the average energies stored in the electric and magnetic fields in the volume. These concepts with their applications have been further generalised by studying the Poynting theorem in the presence of an external source (Chapter 8). Further, the concept of power flow has been amply exemplified in its application to conduction current antennas, which have been introduced in depth to develop the concepts of the gain and effective aperture area of an antenna. The study has characterised not only the fundamental Hertzian infinitesimal dipole, but also finite-length dipole, antenna arrays, including Yagi-Uda, and loop-current antennas (Chapter 8). Further, Friis transmission equation that relates the radiated power from a transmitting antenna to the power delivered to the load connected to a receiving antenna has been derived (Chapter 8) that finds extensive application in communication engineering.

Furthermore, we came across problems in which electromagnetic waves propagating in the vicinity of a lossy conductor get absorbed in the conductor due to the existence of a component of power density vector normal to the surface of the conductor. Considering uniform plane uniform waves in such a situation, we found the average power which enters the conducting medium normally and which causes a resistive loss in the conductor. We thus obtained an expression for the power loss per unit area in the conductor in terms of the surface resistance and surface current density of the conductor, the latter, in turn, being obtainable from the magnetic field distribution at the conducting surface (Chapter 8). We have subsequently used this expression (Chapter 8) to find the wave attenuation constant of a hollow-pipe waveguide (Chapter 9).

A hollow-pipe waveguide made of a conducting material is extensively used for the transmission of power in the microwave frequency range. Such a waveguide can support the TE mode characterised by non-zero axial magnetic field and zero axial electric field, as well as the TM mode, characterised by non-zero axial electric field and zero axial magnetic field. However, the hollow-pipe waveguide made of a single conductor as its boundary wall cannot support the TEM mode, for which the axial electric field and the axial magnetic field are each zero. The analysis of such a waveguide is based on the solution of the wave equation in a bounded medium (Chapter 9). For the analysis of a rectangular waveguide which enjoys rectangular geometrical symmetry, the rectangular coordinate system is used. Similarly, the cylindrical coordinate system is used for the analysis of a cylindrical waveguide that enjoys cylindrical geometrical symmetry. While for the TE mode of excitation, we start from the wave equation in the axial magnetic field, for the TM mode we start from the wave equation in the axial electric field. We have used the solution for the wave equation as well as Maxwell's equations to obtain the TE-mode and the TM mode field solutions for both the rectangular and the cylindrical waveguides. The background of Bessel equation and its solution involving ordinary Bessel function, necessary for finding the solution of the wave equation in the cylindrical coordinate system, has been provided in Chapter 9 for the analysis of a cylindrical waveguide.

The characteristic equation or the dispersion relation of the waveguide is found with the help of the field solutions and the electromagnetic boundary condition that the tangential component of the electric field is zero at the conducting surface of the waveguide wall. We found that the dispersion relation between the wave angular frequency and the phase propagation constant for the TE and the TM modes involving their respective cutoff frequencies (Chapter 9) to be one and the same.

From the dispersion relation of the waveguide, we can plot its angular frequency versus phase propagation constant or the phase velocity versus frequency (dispersion) characteristics. The cutoff frequency of the waveguide is found as the frequency corresponding to zero phase propagation constant of the wave (Chapter 9). The cutoff frequency depends on the waveguide size and they both depend on the waveguide mode chosen. Further, the waveguide modes TE and TM are each assigned two mode numbers, m and n, and are represented as the TE_{mn} and the TM_{mn} modes respectively. We have correlated these mode numbers of the TE_{mn} and the TM_{mn} modes with their respective field patterns across the waveguide cross section for both rectangular and cylindrical waveguides. For a rectangular waveguide, the mode number m indicates the number of maxima (of any field component) along the broad dimension of the waveguide, while the mode number n indicates the number of maxima (of any field component) along the narrow dimension of the waveguide. Alternatively, we can interpret m and n as the numbers of half-wave field patterns across the broad and the narrow dimensions of the waveguide respectively. For a cylindrical waveguide, we find the meaning of the mode number m as the number of half-wave field pattern around the half circumference, and n as the number of positive or negative maxima across the waveguide radius (Chapter 9).

We have presented the waveguide sizes and the cutoff frequency values of commercial rectangular waveguides. The dominant mode of a waveguide is the mode that has the lowest cutoff frequency; we have identified the dominant modes of the rectangular and the cylindrical waveguides. The characteristic parameters of a waveguide are the guide wavelength, phase propagation constant, phase velocity, group velocity, and wave impedance, each of them depending on the operating frequency relative to the cutoff frequency of the waveguide (Chapter 9).

We have seen that the waveguide supports propagating and evanescent modes according to whether the waveguide operates above or below its cutoff frequency, and thus, the waveguide behaves as a high-pass filter. The evanescent mode below the waveguide cutoff frequency of a waveguide has been found to be associated with no component of the average Poynting vector in the direction of wave propagation. Therefore, there is no power flow in the evanescent mode of a waveguide. For propagation above the cutoff frequency, the analysis of a rectangular waveguide excited in the dominant mode has yielded an expression for the power flow through the waveguide, whence an estimate of the power handling capability of the waveguide has been made in terms of the breakdown voltage of the medium filling the waveguide. Further, there is power lost in the wave propagating through a waveguide due to the finite resistivity of the conducting material of the four waveguide walls. In Chapter 8, using the expression deduced for the power lost per unit area of a conducting surface and noting the area of each waveguide wall covering unit waveguide length in the direction of propagation, an expression for the power loss per unit length of each of the four waveguide walls has been found for a waveguide excited in the dominant mode. Hence, the expression for the attenuation constant of such a dominant-mode-excited rectangular waveguide has been derived using the expression for the power transmitted through the waveguide and the expression for the power loss per unit length of the waveguide. The attenuation constant depends on the operating frequency and the waveguide size, and accordingly, for low waveguide attenuation constant we have to appropriately choose the waveguide mode and frequency (Chapter 9).

A waveguide is called a resonator if both its ends are either closed by a conductor or are open and, at the same time, its length is so selected that it resonates at a defined frequency called the resonant frequency (Chapter 10). One of the approaches to find the length of a waveguide for a closed-ended or an open-ended waveguide to behave as a resonator is to treat the problem by the transmission line theory. We have therefore provided the necessary basics of the transmission line theory including the telegrapher's equation and its application to find the input impedance of the transmission line terminated in a load impedance. Thus, the length of a transmission line that is short-circuited at its load end such that its input impedance is zero becomes equal to the length of a closed-ended waveguide resonator. Similarly, the length of open-ended waveguide resonator is found as the length of a transmission line that is open-circuited at its load end. Thus, we have found by this transmission line analogy the length of both closed-ended and an open-ended resonators as an integral multiple of half the guide wavelength (Chapter 10). Further, an occasion to present the theory and application of a transmission line has made clear many relevant concepts of transmission line such as the characteristic and input impedances and VSWR of the line, distortionless transmission, and impedance matching in practical applications such as the radome and duplexer of a radar system (Chapter 10). The theory of Smith chart from first principles and the application of the chart have made it simpler to solve transmission line problems (Chapter 10).

We have applied the field theory to the problem of the rectangular waveguide resonator as well, taking it, typically, closed-ended and excited in the TE mode. We have obtained the field solution subject to the electromagnetic boundary condition that the tangential component of electric field is zero at each of the conducting resonator ends. Interestingly, the field theory has also yielded the same length of the resonator as predicted by the transmission line theory. In addition, the waveguide-mode dispersion relation (Chapter 9) has been used to find the resonant frequency of the resonator (Chapter 10).

We have also used the field solution and the relevant electromagnetic boundary conditions to obtain the expression for the time-averaged energy stored in electric and magnetic fields and the expression for the power loss in all the resonator walls put together in a rectangular cavity resonator excited in the dominant TE mode in terms of resonator dimensions. We have subsequently used these expressions to find an important parameter, namely the quality factor of the resonator at the resonant frequency in terms of the resonator dimensions and the surface resistance of the conducting material of the resonator, the latter, in turn, depending on the material conductivity and the frequency. Further, since in actual practice some part of the energy stored in a cavity is coupled out from it to an external load, the cavity thereby becoming loaded, we defined the unloaded quality factor, the external quality factor and the loaded quality factor of the cavity and the relationship between them. We also represented the waveguide resonator by a resonant circuit comprising an inductor, a capacitor and a resistor, all in parallel. In this representation, the quality factor of the cavity resonator is expressed in terms of the resonant frequency and the half-power bandwidth of the frequency response of the resonator (Chapter 10).

You will feel immensely confident if after reading this book you are ready to write the basic expressions/equations/relations for the following (and of course, have the necessary knowledge of how they have been obtained or derived and subsequently applied to practical problems), among others:

- Expressions for the gradient of a scalar quantity, the divergence of a vector quantity, the curl of a vector quantity and the Laplacian of a scalar or a vector quantity in generalised curvilinear coordinate system and their interpretation in rectangular, cylindrical and spherical polar coordinate systems that subsequently enables one to write the various basic expressions/equations required to understand the various aspects of electromagnetic engineering;
- current density expression in terms of the volume charge density and electron velocity;
- continuity equation in the form of the divergence of the current density;
- Maxwell's equations in integral and differential forms;
- expression for the voltage/current induced in a circuit;
- general electromagnetic boundary conditions at the interface between any two media and their interpretation for the electromagnetic boundary condition at the dielectric-dielectric and conductor-dielectric interfaces for both time-independent (dc) and time-dependent (time-varying) situations;
- wave equation in electric and magnetic fields;
- surface resistance in terms of the skin depth of a conductor;
- power loss per unit area in a conductor in terms of the surface resistance and surface current density;
- average Poynting vector contributed by the real part of the complex Poynting vector, the integration of which over an area leads to the power flow through the area;
- relation between the magnetic field and the vector potential, that between the electric field and the electric scalar and magnetic vector potentials, and that between the electric field and the magnetic vector potential alone subject to Lorentz condition, all for time-varying situations.

A number of appendixes provided at the end of relevant chapters as well as worked-out illustrative examples in the text of chapters must have helped in consolidating the concepts developed while reading the book. The review questions of the narrative, numerical and objective (multiple-choice) types with answers/solutions/hints provided at the end of chapters would have helped in realising what is expected of you, and the process of answering these questions would have indeed summarised the knowledge gained from these chapters.

The present chapter has summarised the contents and approach of the book. We have selectively chosen topics that are necessary to project the essentials of electromagnetic theory for their engineering applications. An attempt has been made to restrict the length of the book so that the reader is not lost in going through lengthy exercises before appreciating the core of electromagnetics necessary for understanding its basics as well as its practical applications. For example, in microwave engineering, the applications covered encompass antennas, wave propagation, microwave circuits, vacuum and solid-state electron devices, typically encountered in practical engineering situations. It is hoped that the timeline of the progress in electromagnetics presented at the outset of the book, with a special accolade to Sir J. C. Bose at the end, will motivate young practitioners of electromagnetics to take up challenging problems of electromagnetic engineering of practical relevance.

Constants, Properties and Relations

Physical Constants

Permittivity of free space (vacuum) ε_0	$8.854 \times 10^{-12} \approx (1/36\pi) \times 10^{-9}$ F/m
Permeability of free space (vacuum) μ_0	$4\pi \times 10^{-7}$ H/m
Intrinsic impedance of free space (vacuum) η_0	$376.7 \approx 120\pi$ ohm
Velocity of light c	2.998×10^{8} m/s
Charge of electron e	-1.602×10^{-19} C
Mass of electron m	9.107×10^{-31} kg
Charge to mass ratio of electron e/m	1.76×10^{11} C/kg

Relative Permittivity ε_r and Loss Tangent $\tan\delta (= \omega/\mu\varepsilon)$ of Dielectric Materials

Material	ε_r	$10^4 \times \tan\delta$
Alumina 96%	9	6
Beryllia	6.6	1
Epoxy resin	3.5	200
Ferrite (typical)	15	2
Glass (pyrex)	4	6
Mica	5.4	6
Nylon	3.5	200
Paper	3	80
Perspex	2.6	70
Polythene	2.54	1.6-2.5
Porcelain (dry processed)	6	140
Quartz (fused)	3.8	1
Silica (fused)	3.8	7.5
Teflon	2.1	3.7
Wood (dry)	1.5-4.0	100

Conductivity of Materials σ (mho/m)

Aluminum	3.54×10^7
Brass	1.1×10^7
Copper	5.7×10^7
Ferrite (typical)	~ 100
Gold	4.1×10^7
Iron	$\sim 0.97 \times 10^7$
Quartz (fused)	$< 2 \times 10^{-17}$
Silver	6.17×10^7
Silver (german)	0.3×10^7
Solder	0.7×10^7
Steel	0.57×10^7
Steel (stainless)	0.11×10^7
Water (distilled)	2×10^{-4}
Water (sea water) (typical)	4

Relative Permeability μ_r of Materials

Aluminum	1.000
Bismuth	0.999
Cobalt	60
Ferrite (typical)	1000
Iron (cast)	60
Iron (pure)	4000
Iron (powdered)	100
Mumetal	20000
Nickel	50
Paraffin	0.999
Silicon iron	3500
Supermalloy	100000
Wood	0.999

Conversion Factors

$1\ \text{C} = 3 \times 10^9$ statcoulomb (esu)

$1\ \text{V} = 1/300$ statvolt (esu)

$1\ \text{A} =$ statampere (esu)

$1\ \text{J} = 10^7$ erg

$1\ \text{W} = 10^7$ erg/s

$1\ \text{eV} = 1.6 \times 10^{-19}\ \text{J}$

$1\ \text{T} = 10^4$ gauss

1 Wb (weber) $= 10^8$ maxwell

$1\ \text{N} = 10^5$ dyne

$1\ \text{Np} = 8.68\ \text{dB}$

$\ln n\ (= \log_e n) = 2.3\ \log n\ (= 2.3\ \log_{10} n)$

Vector Identities

$$\vec{A} \cdot B \times \vec{C} = \vec{B} \cdot \vec{C} \times \vec{A} = \vec{C} \cdot \vec{A} \times \vec{B}$$

$$\vec{A} \times (\vec{B} \times C) = \vec{B}(\vec{A} \cdot \vec{C}) - \vec{C}(\vec{A}.\vec{B})$$

$$\nabla(\psi V) = \psi \nabla(V) + V \nabla \psi$$

$$\nabla \cdot \psi \vec{A} = \psi \nabla \cdot \vec{A} + \vec{A} \cdot \nabla \psi$$

$$\nabla \times \psi \vec{A} = \psi \nabla \times \vec{A} + \nabla \psi \times \vec{A}$$

$$\nabla \cdot \nabla V = \nabla^2 V$$

$$\nabla \times \nabla \times \vec{A} = \nabla \nabla \cdot \vec{A} - \nabla^2 \vec{A}$$

$$\nabla \times \nabla V = 0$$

$$\nabla \cdot (\nabla \times \vec{A}) = 0$$

$$\int_\tau (\nabla \cdot \vec{A}) d\tau = \oint_s \vec{A} \cdot d\vec{S}$$

$$\int_s (\nabla \times \vec{A}) \cdot d\vec{S} = \oint_l \vec{A} \cdot d\vec{l}$$

Vector Calculus Expressions

Gradient of Scalar V

Curvilinear $\nabla V = \frac{1}{h_1}\frac{\partial V}{\partial u_1}\vec{a}_1 + \frac{1}{h_2}\frac{\partial V}{\partial u_2}\vec{a}_2 + \frac{1}{h_3}\frac{\partial V}{\partial u_3}\vec{a}_3$

Rectangular (x, y, z) $h_1 = 1,\ h_2 = 1,\ h_3 = 1$	$\nabla V = \frac{\partial V}{\partial x}\vec{a}_x + \frac{\partial V}{\partial y}\vec{a}_y + \frac{\partial V}{\partial z}\vec{a}_z$
Cylindrical (r, θ, z) $h_1 = 1,\ h_2 = r,\ h_3 = 1$	$\nabla V = \frac{\partial V}{\partial r}\vec{a}_r + \frac{1}{r}\frac{\partial V}{\partial \theta}\vec{a}_\theta + \frac{\partial V}{\partial z}\vec{a}_z$
Spherical-polar (r, θ, ϕ) $h_1 = 1,\ h_2 = r,\ h_3 = r\sin\theta$	$\nabla V = \frac{\partial V}{\partial r}\vec{a}_r + \frac{1}{r}\frac{\partial V}{\partial \theta}\vec{a}_\theta + \frac{1}{r\sin\theta}\frac{\partial V}{\partial \phi}\vec{a}_\phi$

Divergence of a Vector $\vec{A}$

Curvilinear $\nabla \cdot \vec{A} = \frac{1}{h_1 h_2 h_3}\left[\frac{\partial(h_2 h_3 A_1)}{\partial u_1} + \frac{\partial(h_3 h_1 A_2)}{\partial u_2} + \frac{\partial(h_1 h_2 A_3)}{\partial u_3}\right]$

Rectangular (x, y, z) $h_1 = 1,\ h_2 = 1,\ h_3 = 1$	$\nabla \cdot \vec{A} = \frac{\partial A_x}{\partial x} + \frac{\partial A_y}{\partial y} + \frac{\partial A_z}{\partial z}$
Cylindrical (r, θ, z) $h_1 = 1,\ h_2 = r,\ h_3 = 1$	$\nabla \cdot \vec{A} = \frac{1}{r}\frac{\partial(rA_r)}{\partial r} + \frac{1}{r}\frac{\partial A_\theta}{\partial \theta} + \frac{\partial A_z}{\partial z}$
Spherical-polar (r, θ, ϕ) $h_1 = 1,\ h_2 = r,\ h_3 = r\sin\theta$	$\nabla \cdot \vec{A} = \frac{1}{r^2}\frac{\partial(r^2 A_r)}{\partial r} + \frac{1}{r\sin\theta}\frac{\partial(\sin\theta\, A_\theta)}{\partial \theta} + \frac{1}{r\sin\theta}\frac{\partial A_\phi}{\partial \phi}$

Curl of a Vector $\vec{A}$

Curvilinear

$$\nabla \times \vec{A} = \frac{1}{h_2 h_3}\left(\frac{\partial(h_3 A_3)}{\partial u_2} - \frac{\partial(h_2 A_2)}{\partial u_3}\right)\vec{a}_1 + \frac{1}{h_3 h_1}\left(\frac{\partial(h_1 A_1)}{\partial u_3} - \frac{\partial(h_3 A_3)}{\partial u_1}\right)\vec{a}_2 + \frac{1}{h_1 h_2}\left(\frac{\partial(h_2 A_2)}{\partial u_1} - \frac{\partial(h_1 A_1)}{\partial u_2}\right)\vec{a}_3$$

Rectangular (x, y, z) $h_1 = 1,\ h_2 = 1,\ h_3 = 1$	$\nabla \times \vec{A} = \left(\frac{\partial A_z}{\partial y} - \frac{\partial A_y}{\partial z}\right)\vec{a}_x + \left(\frac{\partial A_x}{\partial z} - \frac{\partial A_z}{\partial x}\right)\vec{a}_y + \left(\frac{\partial A_y}{\partial x} - \frac{\partial A_x}{\partial y}\right)\vec{a}_z$
Cylindrical (r, θ, z) $h_1 = 1,\ h_2 = r,\ h_3 = 1$	$\nabla \times \vec{A} = \left(\frac{1}{r}\frac{\partial A_z}{\partial \theta} - \frac{\partial A_\theta}{\partial z}\right)\vec{a}_r + \left(\frac{\partial A_r}{\partial z} - \frac{\partial A_z}{\partial r}\right)\vec{a}_\theta + \frac{1}{r}\left(\frac{\partial(rA_\theta)}{\partial r} - \frac{\partial A_r}{\partial \theta}\right)\vec{a}_z$
Spherical-polar (r, θ, ϕ) $h_1 = 1,\ h_2 = r,\ h_3 = r\sin\theta$	$\nabla \times \vec{A} = \frac{1}{r\sin\theta}\left(\frac{\partial(\sin\theta A_\phi)}{\partial \theta} - \frac{\partial A_\theta}{\partial \phi}\right)\vec{a}_r + \frac{1}{r}\left(\frac{1}{\sin\theta}\frac{\partial A_r}{\partial \phi} - \frac{\partial(rA_\phi)}{\partial r}\right)\vec{a}_\theta + \frac{1}{r}\left(\frac{\partial(rA_\theta)}{\partial r} - \frac{\partial A_r}{\partial \theta}\right)\vec{a}_\phi$

Laplacian of a Scalar V

Curvilinear

$$\nabla^2 V = \frac{1}{h_1 h_2 h_3}\left[\frac{\partial}{\partial u_1}\left(\frac{h_2 h_3}{h_1}\frac{\partial V}{\partial u_1}\right) + \frac{\partial}{\partial u_2}\left(\frac{h_3 h_1}{h_2}\frac{\partial V}{\partial u_2}\right) + \frac{\partial}{\partial u_3}\left(\frac{h_1 h_2}{h_3}\frac{\partial V}{\partial u_3}\right)\right]$$

Rectangular (x, y, z) $h_1 = 1,\ h_2 = 1,\ h_3 = 1$	$\nabla^2 V = \frac{\partial^2 V}{\partial x^2} + \frac{\partial^2 V}{\partial y^2} + \frac{\partial^2 V}{\partial z^2}$
Cylindrical (r, θ, z) $h_1 = 1,\ h_2 = r,\ h_3 = 1$	$\nabla^2 V = \frac{1}{r}\frac{\partial}{\partial r}\left(r\frac{\partial V}{\partial r}\right) + \frac{1}{r^2}\frac{\partial^2 V}{\partial \theta^2} + \frac{\partial^2 V}{\partial z^2}$
Spherical-polar (r, θ, ϕ) $h_1 = 1,\ h_2 = r,\ h_3 = r\sin\theta$	$\nabla^2 V = \frac{1}{r^2}\frac{\partial}{\partial r}\left(r^2\frac{\partial V}{\partial r}\right) + \frac{1}{r^2\sin\theta}\frac{\partial}{\partial \theta}\left(\sin\theta\frac{\partial V}{\partial \theta}\right) + \frac{1}{r^2\sin^2\theta}\frac{\partial^2 V}{\partial \phi^2}$

Bibliography

Balanis C. A., Antenna Theory: *Analysis and Design* (Harper and Row, New York, 1982).

Basu B. N., *Electromagnetic Theory and Application in Beam-Wave Electronics* (World Scientific, Singapore, 1996).

Carter R. G., *Electromagnetic Waves: Microwave Components and Devices* (Chapman and Hill, London, 1990).

Cheng David K., *Field and Wave Electromagnetics* (Pearson Education, Delhi, 1999).

Chodorow M., and Susskind C., *Fundamentals of Microwave Electronics* (McGraw-Hill, New York, 1964).

Collin R. E., *Foundations for Microwave Engineering* (McGraw-Hill Kogakusha, Tokyo, 1966).

Coulomb's Law Committee, "The teaching of electricity and magnetism at the college level," *American J. Phys.*, vol. 18, pp. 1-25, Feb. 1950.

Croissettee D. Le, *Transistors* (Prentice-Hall of India, New Delhi, 1958).

Das A., and Das S. K., *Microwave Engineering* (Tata Mc-Graw Hill, New Delhi, 2000).

Emerson D. T., "The work of Jagadis Chunder Bose: 100 years of mm-wave research," *IEEE Trans. Microwave Theory Tech.*, vol. 45, No. 12, pp. 2267-2273, Dec. 1997.

Gandhi O. P., *Microwave Engineering and Applications* (Pergamon Press, New York, 1981).

Halliday D., and Resnick R., *Physics, Part II* (John Wiley, New York, 1962).

Hayt W. H., *Engineering Electromagnetics* (Mc Graw-Hill, Kogakusha, 1974).

Jordan E. C., and Balmain K. G., *Electromagnetic Waves and Radiating Systems* (Prentice Hall, Englewood Cliffs, 1968).

Krauss J. D., *Antennas* (Mc Graw-Hill, New York, 1988).

Maxwell J. C., *A Treatise on Electricity and Magnetism* (Clarendon, Oxford, 1873).

Nasar Syed A., *Electromagnetics* (McGraw-Hill, New York, 1992).

Pipes L. A., and Harvill L. R., *Applied Mathematics for Engineers and Physicists* (McGraw-Hill Kogakusha, Tokyo, 1970).

Plonsey R., and Collin R. E., *Principles and Applications of Electromagnetic Fields* (Tata McGraw-Hill, New-York, 1973).

Plonus M. A., *Applied Electromagnetics* (McGraw-Hill, New York, 1978).

Poynting J. H., "On the transfer of energy in the electromagnetic field," *Phil. Trans. Roy. Soc*. vol. 175, pp. 343-361, Jan. 1884.

Ramo S., "Space-charge and field waves in an electron beam," *Phys. Rev.*, vol. 56, pp. 276-283, Aug. 1939.

Ramo S., Whinnery J. R., and Van Duzer T., *Fields and Waves in Communication Electronics* (John Wiley, New York, 1962).

Shevgaonkar R. K., *Electromagnetics Waves* (Tata McGraw-Hill, New Delhi, 2006).

Solymar L., *Lectures on Electromagnetic Theory: A Short Course for Engineers* (Oxford University Press, London, 1976).

Soohoo R. F., *Microwave Electronics* (Addison-Wesley, Reading, Boston, MA , 1971).

Stratton J. A., *Electromagnetic Theory* (McGraw-Hill, New York, 1941)

Weber E., *Electromagnetic Fields* (John Wiley, New York, 1950).

Whinnery J. R., "The teaching of electromagnetics," *IEEE Trans. Education*, vol. 33, No. 1, pp. 3-7, Feb. 1990.

http://archive.worldhistoria.com/compass-in-india_topic17077.html

http://en.wikipedia.org/wiki/History_of_electromagnetic_theory

http://www.hackingtheuniverse.com/science/history-of-science-and-technology/genarticles/timeline-of-electromagnetism

http://inventors.about.com/od/timelines/a/Electromagnets.htm

http://en.wikipedia.org/wiki/Timeline_of_electromagnetic_theory

http://www.physicsoftheuniverse.com/dates.html

http://www.britannica.com/EBchecked/topic/407705/Alexander-Neckam

http://en.wikipedia.org/wiki/Daniel_Gralath

http://en.wikipedia.org/wiki/History_of_electromagnetic_theory

http://www.hackingtheuniverse.com/science/history-of-science-and-technology/genarticles/timeline-of-electromagnetism

http://inventors.about.com/od/timelines/a/Electromagnets.htm

http://en.wikipedia.org/wiki/Timeline_of_electromagnetic_theory

http://www.physicsoftheuniverse.com/dates.html

http://www.britannica.com/EBchecked/topic/407705/Alexander-Neckam

http://www.biography.com/people

Index

About the Author

B. N. Basu, who superannuated as Professor and Head, Electronics Engineering Department, Banaras Hindu University (BHU), Varanasi, is presently Professor Emeritus at the Sir J. C. Bose School of Engineering, Mankundu, West Bengal. Professor Basu, who was Distinguished Visiting Scientist at Central Electronics Engineering Research Institute (CEERI), Pilani, is presently Visiting Professor at CEERI under the Academy of Scientific and Innovative Research of CSIR.

He received his BTech, MTech and PhD degrees from the Institute of Radiophysics and Electronics, Calcutta University, in 1965, 1966 and 1976 respectively. As a research scholar, he received tutelage from Professor N. B. Chakarabarty of IIT Kharagpur. He has served Defence Electronics Research Laboratory, Hyderabad; the Regional Institute of Technology, Jamshedpur; CEERI Pilani; the College of Engineering and Technology, Moradabad; and BHU, Varanasi, in various capacities. At BHU, besides heading the Department of Electronics Engineering, he has been Coordinator, Centre of Research in Microwave Tubes; Professor-in-Charge, University Science and Instrumentation Centre; and Emeritus Fellow of the All India Council of Technical Education.

He is Consultant at Microwave Tube Research and Development Centre (MTRDC), Bengaluru (DRDO), and is associated with a number of national laboratories (CSIR and DRDO) as an advisor and reviewer of projects. He is a member of the DST Steering Committee of the multi-institutional project, Gyrotron, as well as the Research Council of DRDO, Ministry of Defence at MTRDC.

Professor Basu has authored or co-authored more than 115 research papers in journals of international repute and six monograph chapters in the area of microwave tubes. He has also authored two books: *Electromagnetic Theory and Applications in Beam-Wave Electronics* (World Scientific, Singapore/New Jersey/London/Hong Kong, 1996) and *Technical Writing* (Prentice-Hall of India, New Delhi, 2007). He has served the technical committee on Vacuum Electron Devices of the IEEE Electron Devices Society and is a member of the editorial board of the *Journal of Electromagnetic Waves and Applications* (USA). He is mentioned in the Golden List of reviewers of IEEE Transactions.

He is Fellow of the Institution of Electronics and Telecommunication Engineers (IETE), India, and past President, Vacuum Electron Devices and Applications Society, India (2099-2011). Following his superannuation from BHU, Professor Basu was felicitated at a special function of CSIR-CEERI Pilani for his immense contribution to the area of microwave tubes. He is a recipient of the S. V. C. Aiya Memorial Award of the IETE for outstanding contributions to motivating research in the area of electronics and telecommunication engineering.